VOLUME SIXTY SEVEN

Advances in
BOTANICAL RESEARCH

Metabolomics Coming of Age with its Technological Diversity

ADVANCES IN BOTANICAL RESEARCH

Series Editors

Jean-Pierre Jacquot
Professeur, Membre de L'Institut Universitaire de France, Unité Mixte de Recherche INRA, UHP 1136 "Interaction Arbres Microorganismes", Université de Lorraine, Faculté des Sciences, Vandoeuvre, France

Pierre Gadal
Professor honoraire, Université Paris-Sud XI, Institut Biologie des Plantes, Orsay, France

VOLUME SIXTY SEVEN

ADVANCES IN BOTANICAL RESEARCH

Metabolomics Coming of Age with its Technological Diversity

Volume editor

DOMINIQUE ROLIN

Bordeaux University
France

Academic Press is an imprint of Elsevier
32 Jamestown Road, London NW1 7BY, UK
Radarweg 29, PO Box 211, 1000 AE Amsterdam, The Netherlands
The Boulevard, Langford Lane, Kidlington, Oxford, OX5 1GB, UK
225 Wyman Street, Waltham, MA 02451, USA
525 B Street, Suite 1800, San Diego, CA 92101-4495, USA

First edition 2013

ISBN: 978-0-12-397922-3
ISSN: 0065-2296

For information on all Academic Press publications
visit our website at store.elsevier.com

Printed and bound in UK

13 14 15 16 17 10 9 8 7 6 5 4 3 2 1

CONTENTS

CONTRIBUTORS

Serge Akoka
EBSI Team, Chimie et Interdisciplinarité: Synthèse, Analyse, Modélisation (CEISAM), CNRS, UMR 6230, LUNAM Université de Nantes, France

Sandra Alves
Equipe de spectrométrie de masse, Institut Parisien de Chimie Moléculaire, UMR 7201, Université Pierre et Marie Curie, Paris, France

Patricia Ballias
INRA, UMR 1332 Fruit Biology and Pathology and Metabolome Facility of Bordeaux, Functional Genomics Centre, INRA Centre of Bordeaux, Villenave d'Ornon, France

Bertrand Beauvoit
Bordeaux University, UMR1332 Fruit Biology and Pathology, INRA Centre of Bordeaux, Villenave d'Ornon

Benoit Biais
INRA, UMR 1332 Fruit Biology and Pathology, INRA Centre of Bordeaux, Villenave d'Ornon, France

Samia Boudah
Laboratoire d'Etude du Métabolisme des Médicaments, CEA- Centre d'étude de Saclay DSV/iBiTec-S/SPI/LEMM, Gif-sur-Yvette, France

Cécile Cabasson
University of Bordeaux, UMR 1332 Fruit Biology and Pathology, and Metabolome Facility of Bordeaux, Functional Genomics Centre, INRA Centre of Bordeaux, Villenave d'Ornon, France

Cécile Canlet
INRA, UMR 1331 TOXALIM, Research Centre in Food Toxicology, and University of Toulouse, INP, UMR 1331, TOXALIM, Toulouse, France

Emma Cantos-Villar
Instituto de Investigación y Formación Agraria y Pesquera (IFAPA), Rancho de la Merced, Junta de Andalucía, Jerez de la Frontera, Cadiz, Spain

Benoit Colsch
Laboratoire d'Etude du Métabolisme des Médicaments, CEA- Centre d'étude de Saclay DSV/iBiTec-S/SPI/LEMM Gif-sur-Yvette, France

Catherine Deborde
INRA, UMR 1332 Fruit Biology and Pathology and Metabolome Facility of Bordeaux, Functional Genomics Centre, INRA Centre of Bordeaux, Villenave d'Ornon, France

Marianne Defernez
Analytical Science Unit, Institute of Food Research, Norwich Research Park, Colney, Norwich, United Kingdom

Florence Fauvelle
IRBA-CRSSA/EBR, Laboratoire de RMN, La Tronche, France

Laetitia Fouillen
University of Bordeaux, CNRS UMR 5200, and Metabolome Facility of Bordeaux, Functional Genomics Centre, INRA Centre of Bordeaux, Villenave d'Ornon, France

Jaleh Ghashghaie
Laboratoire d'Ecologie, Systématique et Evolution (ESE), CNRS UMR8079, Université de Paris-Sud 11, Orsay, France

Yves Gibon
INRA, UMR 1332 Fruit Biology and Pathology and Metabolome Facility of Bordeaux, Functional Genomics Centre, INRA Centre of Bordeaux, Villenave d'Ornon, France

Patrick Giraudeau
EBSI Team, Chimie et Interdisciplinarité: Synthèse, Analyse, Modélisation (CEISAM), CNRS, UMR 6230, LUNAM Université de Nantes, France

Daniel Jacob
INRA, UMR 1332 Fruit Biology and Pathology and Metabolome Facility of Bordeaux, Functional Genomics Centre, INRA Centre of Bordeaux, Villenave d'Ornon, France

Fabien Jourdan
INRA, UMR 1331 TOXALIM (Research Center in Food Toxicology), Metabolism of Xenobiotics (MeX), Toulouse, France

Cyril Jousse
UBP-UFR Sciences, ICCF-UMR6296, Plateforme d'Exploration du Métabolisme, Aubière, France

Christophe Junot
Laboratoire d'Etude du Métabolisme des Médicaments, CEA- Centre d'étude de Saclay DSV/iBiTec-S/SPI/LEMM, Gif-sur-Yvette, France

Daniel Just
INRA, UMR 1332 Fruit Biology and Pathology, INRA Centre of Bordeaux, Villenave d'Ornon, France

Gwénaëlle Le Gall
Analytical Science Unit, Institute of Food Research, Norwich Research Park, Colney, Norwich, United Kingdom

René Lessire
University of Bordeaux, CNRS UMR 5200, and Metabolome Facility of Bordeaux, Functional Genomics Centre, INRA Centre of Bordeaux, Villenave d'Ornon, France

Mickael Maucourt
University of Bordeaux, UMR 1332 Fruit Biology and Pathology, and Metabolome Facility of Bordeaux, Functional Genomics Centre, INRA Centre of Bordeaux, Villenave d'Ornon, France

Jean-Pierre Mazat
Université de Bordeaux, IBGC-CNRS UMR 5095, Bordeaux, France

Guillaume Ménard
INRA, UMR 1332 Fruit Biology and Pathology, INRA Centre of Bordeaux, Villenave d'Ornon, France

Annick Moing
INRA, UMR 1332 Fruit Biology and Pathology, and Metabolome Facility of Bordeaux, Functional Genomics Centre, INRA Centre of Bordeaux, Villenave d'Ornon, France

Jean-Pierre Monti
Université de Bordeaux, ISVV, Groupe d'Etude des Substances Végétales à Activité Biologique, EA 3675, Villenave d'Ornon, France

Alain Paris
INRA, UR1204, Méthodologies d'Analyse du Risque Alimentaire, Paris, France

Johann Petit
INRA, UMR 1332 Fruit Biology and Pathology, INRA Centre of Bordeaux, Villenave d'Ornon, France

Duyen Prodhomme
INRA, UMR 1332 Fruit Biology and Pathology, INRA Centre of Bordeaux, Villenave d'Ornon, France

Estelle Pujos-Guillot
INRA, UMR 1019, Plateforme d'Exploration du Métabolisme, Nutrition Humaine, Saint Genès Champanelle, and Clermont Université, UFR Médecine, UMR 1019 Nutrition Humaine, Clermont-Ferrand, France

Estelle Rathahao-Paris
INRA, and Agroparistech, UMR 1145 Ingénerie Procédés Aliments, Paris, France

Tristan Richard
Université de Bordeaux, ISVV, Groupe d'Etude des Substances Végétales à Activité Biologique, EA 3675, Villenave d'Ornon, France

Dominique Rolin
University of Bordeaux, UMR 1332 Fruit Biology and Pathology, and Metabolome Facility of Bordeaux, Functional Genomics Centre, INRA Centre of Bordeaux, Villenave d'Ornon, France

Christophe Rothan
INRA, UMR 1332 Fruit Biology and Pathology, INRA Centre of Bordeaux, Villenave d'Ornon, France

Nabil Semmar
UMR INRA 1260/INSERM 1025/Plateau BioMet/Marseille, France, Université Tunis El Manar. Institut Supérieur des Sciences Biologiques Appliquées de Tunis (ISSBAT), Tunis, Tunisia

Jean Claude Tabet
Equipe de spectrométrie de masse, Institut Parisien de Chimie Moléculaire, UMR 7201, Université Pierre et Marie Curie, Paris, France

Guillaume Tcherkez
Institut de Biologie des Plantes (IBP), CNRS UMR 8618, Université de Paris-Sud 11, Orsay, and Institut Universitaire de France, Paris, France

Hamza Temsamani
Université de Bordeaux, ISVV, Groupe d'Etude des Substances Végétales à Activité Biologique, EA 3675, Villenave d'Ornon, France

CHAPTER ONE

High-Resolution ^{1}H-NMR Spectroscopy and Beyond to Explore Plant Metabolome

Dominique Rolin[*,†,1], Catherine Deborde[†,‡], Mickael Maucourt[*,†], Cécile Cabasson[*,†], Florence Fauvelle[§], Daniel Jacob[†,‡], Cécile Canlet[¶,||], Annick Moing[†,‡]

[*]University of Bordeaux, UMR 1332 Fruit Biology and Pathology, INRA Centre of Bordeaux, Villenave d'Ornon, France
[†]Metabolome Facility of Bordeaux, Functional Genomics Centre, INRA Centre of Bordeaux, Villenave d'Ornon, France
[‡]INRA, UMR 1332 Fruit Biology and Pathology, INRA Centre of Bordeaux, Villenave d'Ornon, France
[§]IRBA-CRSSA/EBR, Laboratoire de RMN, La Tronche, France
[¶]INRA, UMR 1331 TOXALIM, Research Centre in Food Toxicology, Toulouse, France
[||]University of Toulouse, INP, UMR 1331, TOXALIM, Toulouse, France
[1]Corresponding author: e-mail address: rolin@bordeaux.inra.fr

Contents

Advances in Botanical Research, Volume 67
ISSN 0065-2296
http://dx.doi.org/10.1016/B978-0-12-397922-3.00001-0

Abstract

During the past decade, metabolomics has become a crucial functional genomic tool. Tremendous advances in instrumentation, automation, data handling capabilities and statistical analyses have boosted this discipline. Today, most investigations of the plant metabolome tend to be based on either nuclear magnetic resonance (NMR) spectrometry or mass spectrometry (MS), with or without hyphenation with chromatography or capillary electrophoresis. Although less sensitive than MS, NMR provides a powerful complementary technique for the identification and quantification of metabolites in plant extracts. High-resolution one-dimensional (1D) proton NMR spectroscopy (^{1}H-NMR) is a robust technique, highly reproducible, non-selective and non-destructive, able to produce structural information and quantitative data. In this review, we highlight important technical prerequisites for sample preparation, acquisition parameters, as well as recent developments in NMR and resources available for spectra annotation, which are necessary to pursue plant ^{1}H-NMR metabolomics studies. ^{1}H-NMR is the ideal nucleus for most metabolomics studies. On the one hand, multivariate analysis of unassigned ^{1}H-NMR spectra (concept of metabolite fingerprinting) is used to compare the overall metabolic composition of plant extracts. On the other hand, identification and quantification of 20–60 metabolites (concept of metabolite profiling) in unfractionated extracts are used to spot the major metabolites in plant extracts. Both strategies generate key information concerning the plant metabolome, but in an incomplete way. Several applications of NMR fingerprinting and profiling approaches are presented from the characterization of food or medicinal plants to plant functional genomics, showing the key role that ^{1}H-NMR plays in data provision for food and plant sciences. The future challenges for ^{1}H-NMR metabolomics are linked with the developments in NMR spectroscopy technology that provide improvement in signal detection and quantitation and also the facility to use shared databases.

1. INTRODUCTION

Metabolomics is the systematic and comprehensive analysis of large numbers of low-molecular-weight compounds (of less than about 1500–3000 Da) from a biological sample, which are collectively called the metabolome. Initial metabolomics studies were pioneered by Nicholson and co-workers (Nicholson, Connelly, Lindon, & Holmes, 2002; Nicholson, Foxall, Spraul, Farrant, & Lindon, 1995; Nicholson, Lindon, & Holmes,

1999), Oliver and co-workers (Oliver, Winson, Kell, & Baganz, 1998; Raamsdonk et al., 2001) and others (Fiehn, 2002; Hall et al., 2002). Nicholson coined the word metabonomics to define the dynamic metabolic response of living systems to biological stimuli (Nicholson et al., 1999), while Oliver favoured metabolomics as the metabolite analysis of complex biological samples (Oliver et al., 1998). Today, metabolomics is preferentially used in the literature.

By identifying and quantifying the metabolome of a biological sample, metabolomics defines the steady-state levels of the intermediates of metabolic networks that constitute the sample, that is, the metabolic phenotype. Plant metabolomics can reveal the metabolic consequence of an environmental or genetic perturbation of a metabolic regulatory network and provides insights into the structure and regulation of that network. For these reasons, metabolomics has been predicted to play a significant and important role in bridging the phenotype–genotype gap. From the beginning, plant metabolomics by providing biochemical composition of plants and plant-derived products has been viewed as a tool that can be used in basic science as well as applied biology. To have a large review of plant metabolomics applications, see the recent book edited by Hall (2011a) and also these two reviews (Hall, 2006; Saito & Matsuda, 2010).

But in contrast to other omics approaches such as genomics or transcriptomics, which are dealing with the same chemical entities, it is currently difficult to capture the complete metabolome of a plant or tissue because plants produce an astonishing wealth of metabolites from both primary and secondary metabolic pathways. It is estimated that metabolome size for an individual species varies in the 5000–25,000 range, while the total number is estimated to exceed over 500,000. Hundred thousand metabolites have been identified in the plant kingdom (Hadacek, 2002; Verpoorte, van der Heijden, & Memelink, 2000). To assess this large and complex chemical diversity, and because no single technique can achieve the complete separation of all types of molecules and no unique analytical instrument can characterize all types of molecules from the same preparation, metabolomics has evolved into many different directions. Initially, concepts of untargeted and targeted analyses were used to measure, respectively, the concentration of all assayable metabolites present in a biological sample and specific compounds with nuclear magnetic resonance (NMR) spectroscopy and mass spectrometry (MS), the two most common spectroscopic analytical techniques employed in metabolomics (Biais et al., 2009; Griffiths et al., 2010; Moco & Vervoort, 2012; Moing et al., 2011).

In metabolomics studies, two distinct strategies have emerged in the quest for a more detailed understanding of metabolic phenotypes. The first most common, "metabolic profiling" requires the identification and quantification of as broad a class of metabolites as possible, while the second "metabolic fingerprinting" refers to the use of machine output as potentially recognizable chemical pattern, specific of an individual sample. Metabolic profiling, which relies on sophisticated analytical techniques to identify and evaluate the concentrations of large number of known metabolites, can be partial and limited to a short list of metabolites. Indeed, the identification of metabolites is often time consuming and difficult to carry out because pure standards of many compounds are not available and also the cost of setting up large collections of standard is prohibitive. Generally, metabolic fingerprinting is used to compare up to several hundred extracts in order to quickly assess the degree of variation and, often, select the most divergent samples or genotypes for more detailed study (Kusano et al., 2011; Lopez-Gresa et al., 2012, 2010; Sanchez, Schwabe, Erban, Udvardi, & Kopka, 2012; Sánchez-Perez et al., 2011; Steinfath et al., 2010). Metabolic profiling and fingerprinting approaches can be combined in the same biological experiment (Bernillon et al., 2012).

During the past decade, tremendous advances in the instrumentation, automation, data handling capabilities and statistical analyses have boosted metabolomics (Aharoni & Brandizzi, 2012). A continual state of modifications and improvement in analytical tools has been achieved, and today, most metabolomics investigations stand on a combination of separation and detection technologies that can also be used in serial or even in parallel combinations. (See the list of the most common technological abbreviations likely to be encountered in metabolomics literature in Table 1.2 of Hall, 2011b.) Today, most investigations of the plant metabolome tend to be based on either NMR or MS with or without hyphenation of chromatography or capillary electrophoresis such as LC-NMR, LC-SPE-NMR, LC-MS, GC-MS, GC-SPE-MS, CE-MS, and Fourier Transform (FT)-MS. MS has established itself as the method of choice (Koek, Jellema, van der Greef, Tas, & Hankemeier, 2011; Scalbert et al., 2009; Vogeser & Seger, 2008). MS provides information on the molecular masses and the fragment patterns of compounds and can detect and resolve a broad range of metabolites with speed, sensitivity and accuracy (Lei, Huhman, & Sumner, 2011; Weckwerth, 2010). A large variety of equipment is available depending on separation modes, ionization techniques and mass detection systems (see Chapters IV and V in this book). Although less sensitive than MS, NMR provides a powerful complementary technique for the identification and quantification of plant metabolites either in plant extracts or *in vivo* (Ratcliffe & Shachar-Hill, 2005). The most sensitive commonly occurring magnetic isotope is proton. For metabolomics study, high-resolution one-dimensional (1D) proton NMR (^{1}H-NMR) spectra present many advantages. It is quantitative, highly reproducible, non-selective and non-destructive. Proton is the ideal nucleus for most metabolite fingerprinting and profiling applications of large-scale plant collections requiring the participation of multiple laboratories (Ward, Baker, et al., 2010). For metabolite profiling, 20–60 metabolites can be identified from ^{1}H-NMR spectrum in an unfractionated plant extract (Ben Akal-Ben Fatma et al., 2012; Le Gall, Colquhoun, Davis, Collins, & Verhoeyen, 2003; Le Gall, Colquhoun, & Defernez, 2004; Mounet et al., 2007; Weljie, Newton, Mercier, Carlson, & Slupsky, 2006; Zulak, Weljie, Vogel, & Facchini, 2008). Indeed, NMR spectra contain a wealth of information (chemical shift, intensity, magnetic relaxation properties) about the identity of the molecules in the sample. And it is on this basis that NMR can be used to identify and quantify metabolites in plant extracts. Generally, in metabolite fingerprinting by ^{1}H-NMR, multivariate analysis of unassigned ^{1}H-NMR spectra is used to compare metabolite composition of wild type, mutants and transgenic plants and to assess the impact of environmental conditions on the plant metabolome (Lopez-Gresa et al., 2012; Steinfath et al., 2010; Ward, Baker, Llewellyn, Hawkins, & Beale, 2011). ^{1}H-NMR metabolite fingerprinting is fast, convenient and very effective to discriminate between groups through variation of the spectral pattern. Over the past decade, numerous statistical and data processing methods have been explored to handle the large amount of data generated by metabolomics studies (Kohl et al., 2012; Mercier, Lewis, Chang, Baker, & Wishart, 2011). Spectral alignment techniques (Staab, O'Connell, & Gomez, 2010; Veselkov et al., 2009) followed by multivariable statistical approaches (Robinette et al., 2009; see Chapter XI in this book) are generally used to quickly identify spectral regions or spectral patterns of interest. In this approach, the compounds of interest are only identified and possibly quantified after the statistical analyses revealing discriminant spectral regions.

This chapter provides a critical introduction to the application of high-resolution ^{1}H-NMR spectroscopy to explore plant metabolome, on the one hand providing background NMR information on major parameters for NMR acquisition data, and on the other hand discussing the strengths and weaknesses of the ^{1}H-NMR for metabolic profiling and metabolic fingerprinting illustrated with examples of use in food and plant science. Comprehensive accounts of the application of high-resolution ^{1}H-NMR spectroscopy to explore plant metabolome can be found below, as well as elsewhere (Krishnan, Kruger, & Ratcliffe, 2005; Ratcliffe & Shachar-Hill, 2005; Ward, Baker, & Beale, 2007).

2. NMR SIGNALS: MEASURABLE PROPERTIES

2.1. What is nuclear magnetic resonance spectroscopy?

NMR has its origin in the magnetic properties of the atomic nucleus. In 1952, physicists Bloch (Bloch, Hansen, & Packard, 1946) and Purcell (Purcell, Torrey, & Pound, 1946) shared a Nobel Prize for putting NMR to practical use by constructing the first crude NMR spectrometer. During the following years, NMR has completely revolutionized the study of chemistry and biochemistry and significantly impacted a host of other areas such as biology, physiology, imaging and, today, metabolomics. NMR has arguably become the most widely used technique for elucidation of molecular structures in chemistry.

Certain atomic nuclei, such as the hydrogen nucleus ^{1}H, possess a property known as spin. This can be visualized as a spinning motion of the nucleus around its own axis. Associated with the spin is a magnetic property so that the nucleus can be regarded as a tiny bar magnet with its axis along the axis of rotation. If a magnetic field (B0) is applied to a sample containing such nuclei, for example, by introducing this biological sample into a cryo-magnet, the nuclear magnet aligns along the field just as a compass needle aligns along a magnetic field. Many isotopes have such properties. In the basic NMR experiment, the sample is placed in a magnetic field, and by irradiating the sample with a suitable oriented radio frequency (RF) field, the NMR signal is generated. The magnetization created by the RF pulse generates a time-dependent signal associated to voltage decay to zero. This signal is recorded by the spectrometer in the form of a digitized free induction decay (FID). FID is routinely converted by computers into the conventional frequency domain spectrum by FT. To have a deeper introduction to the theory and practice of NMR, see the following books or articles (Bothwell & Griffin, 2011; Claridge, 1999; Keeler, 2010; Macomber, 1998; Pochapsky & Pochapsky, 2007). Finally, NMR can be considered as the measurement of the interaction of an oscillating RF electromagnetic field generated by a collection of nuclei immersed in a strong external magnetic field. The interaction between the nuclear magnetic moment and the applied magnetic field is experimentally observable. The result is a family of NMR analytical techniques with an unusually wide range of applications, and some of these methods can be applied directly to living systems (*in vivo* or *ex vivo* NMR) or to biological extracts (*in vitro* NMR).

NMR signals have a great number of measureable properties. The intensity, frequency, line shape, line width and relaxation time of the NMR signals are the main features of interest. NMR signal intensities are important analytically because they can be related to the sample content of the molecules that produce the signals. However, there is no simple relation between the molecule content and the signal intensity because the strength of an NMR signal also depends on the pulse sequence that was used to generate it. Quantitative information can only be obtained by correcting the distorting effects of the pulse sequence by using the signal from a standard solution or by choosing a pulse sequence that allows quantitative acquisition (see Section 2.6). The frequency of an NMR signal depends at the same time on the identity of the magnetic isotope, the strength of the applied magnetic field and the chemical environment of the nucleus. The effect of the chemical environment on the resonance frequency is determinant to detect similar nuclei from different molecules. The resonance frequency also called chemical shift is measured in parts per million (ppm) on a scale that is independent from the field strength of the magnet, but the spectral width in Hz will change depending upon the spectrometer used. This chemical shift effect is fundamental to identify metabolites on the basis of the characteristic NMR signal patterns. NMR line shapes and line widths have to be optimized by properly shimming the sample (i.e. to get the most homogenous field across the sample) to limit overlapping with neighbouring signals. Because line broadening can be generated by diverse factors such as the presence of paramagnetic ions in the sample or reduced molecular mobility, some precautions have to be applied during extraction procedures (see Section 2.4). Two more measurable properties of an NMR signal, the spin-lattice or longitudinal relaxation time (T_1) and the spin–spin or transverse relaxation time (T_2) need to be briefly mentioned. The RF irradiation used to generate the NMR signals perturbs the nuclear spin system and the subsequent return to equilibrium is often slow. The relaxation time needed to return to equilibrium viewed as T_1 and T_2 are important for two reasons. On the one hand, relaxation times can influence the intensity of an NMR signal and this factor needs to be considered when optimizing the conditions to record a spectrum. On the other hand, the relaxation times determined by the physical and chemical environment of the magnetic nucleus can give insights into the nature of the intracellular medium. These pieces of information can be very relevant for *in vivo* NMR where there is only one solvent matrix.

NMR has a long history in analytical and natural products research. Along with MS and single crystal X-ray diffraction, it has become one of the main analytical tools for structure analyses in chemistry. Although most pioneer work in NMR has been carried out by chemists with pure molecules or a simple mixture of synthesized compounds, NMR can be easily applied to mixtures and complex extracts prepared from plant tissues or cells. 1D ^{1}H-NMR spectroscopy is widely used in metabolomics studies especially in food and plant science. This is linked to the following advantages. Spectral acquisition is relatively rapid, reproducible and robust. Peak area provides a direct measure of metabolite concentration allowing all detectable metabolites to be quantified using a single or a few internal standards. Furthermore, the analytical versatility of NMR, for example, two-dimensional (2D) NMR has been useful to improve compound identification and quantification in complex matrices generated by plant extraction.

2.2. Choosing the isotope of biological interest detected by NMR

Any molecule containing one or more atoms with a non-zero magnetic moment is potentially detectable by NMR. Because isotopes with non-zero magnetic moments include ^{1}H, ^{2}H, ^{13}C, ^{14}N, ^{15}N and ^{31}P, theoretically all biological molecules should give NMR signals. But ultimately what can and cannot be achieved by NMR techniques depends on the magnetic properties of the nucleus that is the source of the NMR signal, the molecular weight of the molecules and the quantity of nucleus that can be detected in the spectrometer. Indeed, to be detected, biological molecules must be of low molecular weight, freely mobile and present in sufficient quantity to exceed the detection threshold for the magnetic isotope of interest. This latter point is related to the fact that the natural abundance of some biologically relevant magnetic isotopes (^{2}H, ^{13}C and ^{15}N) is low. ^{1}H- and ^{13}C-NMR are the two most used techniques for structural determination of biological molecules, reflecting the fact that most metabolites in plant cells contain carbon and hydrogen. In contrast, ^{31}P-NMR has been used in *in vivo* NMR to monitor phosphorylated metabolites (glucose-6-P, fructose-6-P, fructose-1-6-P, ATP, ADP, UDPG, PolyP) and inorganic phosphate in the two major compartments of the plant cell (vacuolar and cytosol), also providing an excellent tool to measure intracellular pH changes.

Although less sensitive than ^{1}H-NMR, ^{13}C-NMR offers two main advantages for identification and quantification of metabolites. Firstly, the chemical shift scale of ^{13}C nucleus (240 ppm) is more than an order of magnitude greater compared to ^{1}H (12 ppm), reducing the overlap in the spectra and thus facilitating detection and assignation of signals from particular metabolites. Secondly, the low natural abundance of ^{13}C of biological molecules (1.1%) opens the possibility of labelling tissues and monitoring the operation of specific metabolic pathways. Finally, the widespread occurrence of carbon atoms in biomolecules has convinced scientists to use this method to detect metabolites in living plant tissue by *in vivo* NMR as well as in plant extracts, despite the fact that ^{13}C-NMR shows a relatively low sensitivity. Compared to ^{31}P- and ^{13}C-NMR, the high sensitivity of ^{1}H-NMR has made it an attractive choice for the detection of metabolites especially in metabolomics. Protons are widely distributed in biological molecules, but the large number of observable metabolites and the small chemical shift dispersion (10 ppm) make the problem of assignment difficult. For ^{1}H-NMR, the concentration threshold for routine detection of metabolites in an extract using a modern high-field cryo-spectrometer is around 10 μM (Krishnan et al., 2005). In reality, the achievable sensitivity is strongly dependent on the field strength of the magnet and on the design of the probe head that allows the signals to be detected.

2.3. Choosing the field and probe technology

The major problem with NMR technology as applied to metabolomics is its low sensibility compared to MS technology. The two main approaches aiming at increasing sensitivity of NMR involve the use of higher field magnets and the development of more sensitive NMR probes (Bothwell & Griffin, 2011). Since theory predicted that energy differences between nuclear spin states varied with external magnetic field strength, ^{1}H-NMR spectra of biological samples containing hundreds of molecules require high magnetic field strengths to get a good spectral dispersion with limited overlapping signals (Bothwell & Griffin, 2011; Lindon, Nicholson, Holmes, & Everett, 2000). Today, a large choice of commercial cryo-magnets is available, generating fields up to 23.4 Tesla (T) corresponding to instruments with an observation frequency for ^{1}H of 1 GHz (with higher frequencies on the way). For small-molecule studies as in metabolomics, the increased sensitivity and dispersion that result are of great value but the cost of such a system is prohibitive for most laboratories, as magnet costs increase rapidly with field strength. For most metabolomics studies, the routine fields used have a ^{1}H frequency of 400–600 MHz corresponding, respectively, to 9.4–14.1 T.

The probe head is the sensor positioned in the centre of the magnet containing a coil that is used both to send RF pulses to the sample and to detect the NMR signals returning from the excited atomic nuclei of the sample (Ross, Schlotterbeck, Dieterle, & Senn, 2007). Today, a large choice of NMR probe heads is available for any possible application (large choice of frequencies for solid or liquid NMR in spectroscopy or imaging mode) with a large choice of sample volumes. Over the past few years, there has been a significant growth in demand for miniaturization in all areas of biological research in order to analyze very small amount of samples. A 1-mm microprobe can operate with disposable 1-mm capillary sample tubes and the sample volume of 5 μl enables use of the lowest amounts of sample to run all high-resolution NMR experiments with outstanding sensitivity. On the other extreme, vertical wide bore magnets have been used for imaging studies allowing an inner space up to 72 mm. For metabolomics studies where the endogenous molecules in plant extracts can be diluted, the probe should be optimized for ^{1}H detection and the NMR observed volume of the probe coil has to be filled completely with the sample to allow for highest sensitivity and proper shimming. The standard NMR spectrometer is usually equipped with an automatic shimming and tuning 5 mm inverse probe head requiring 350–600 μl sample volume. Indeed, shimming and tuning procedures used to be a time-consuming manual process that has been recently automated on modern machines to speed up acquisition procedures and restrain operator proceedings (see Section 2.7). For large-scale studies, influx probes can be chosen to link a liquid handler to the flow probe allowing direct transfer of the sample into the NMR probe and contribute to high-throughput metabolomics.

NMR has been presented as a method of choice for metabolomics, but its application has so far been limited because of its currently much poorer sensitivity compared to MS-based method. While NMR analysis can easily detect major metabolites in plant extracts such as major sugars, organic acids or amino acids, many other important compounds will typically be missed because they occur in the plant extract at levels below the usual NMR detection threshold. An immense number of secondary compounds produced by plants are concerned. To overcome this obstacle, solid-phase extraction (SPE) sample preparation procedure is applied to purify and concentrate the secondary compounds of interest before NMR acquisition (see

Chapter II in this book). The signal-to-noise ratio can be increased by electronic noise reduction using cryogenic probes. This approach relies on cooling the NMR RF detector and pre-amplifier to 20 K or less. Probe cooling reduces the contribution of electronic and thermal noise and provides a theoretical up-to-fourfold increase in signal-to-noise ratio. The increased sensitivity of cryogenic probes has the added benefit of allowing the use of nuclei with lower sensitivity (i.e. lower gyromagnetic ratio) than ^{1}H, which would normally be prohibited due to the time required to acquire a sufficient signal (Keun et al., 2002).

One of the most striking elements about NMR is the ability to acquire information non-invasively and non-destructively from an exceptionally broad range of samples including intact tissues. But, intact tissues are inhomogeneous material, having different physical phases characterized by distinct magnetic susceptibilities. Therefore, anisotropic effects can broaden the NMR signals with the loss of spectroscopic information. The values of anisotropic NMR parameters depend on the angle at which the sample is placed with respect to the B0 field. If the solid sample is spun rapidly at the so-called magic angle, 54.7°, the angular term becomes zero reducing the line width of resonances. This approach has been applied to solid-state NMR and also to a range of biological tissues, and line widths comparable to solution-state NMR spectroscopy can be obtained.

Recently, high-resolution magic angle spinning–nuclear magnetic resonance (HRMAS-NMR), which represents a method bridging the liquid-state and solid-state NMR spectroscopy has been proposed as a reliable system for plant metabolomics study (Pérez, Iglesias, Ortiz, Pérez, & Galeraet, 2010) and foodstuff characterization (Ritota, Casciani, Failla, & Valentini, 2012; Valentini et al., 2011) in the future (see Section 6.1).

2.4. Adapting the extraction method to the biological objective

The purpose of plant metabolomics is to capture instant pictures of metabolism changes in plants according to various culture conditions or genetic variations. Therefore, the generation of high-quality metabolomics data by ^{1}H-NMR spectroscopy will depend strongly on the experiment being correctly designed (see checklist for experimental design; Gibon & Rolin, 2012), and plant tissue collection correctly sampled and stored before extraction (Biais et al., 2012). The ideal procedure to extract water-soluble metabolites from plant cells needs to be easy to set up with a limited number of steps, reproducible, robust and easy to automatize. The main objective of the procedure is to halt enzyme activity as quickly as possible in order to avoid any metabolic changes. The problem of halting enzyme activity was addressed a long time ago (Bieleski, 1964; Stitt & Ap Rees, 1978), and extraction methods have long been used in biochemistry and revisited, optimized and compared in all fields since the emergence of metabolomics, for plants (Gromova & Roby, 2010; Gullberg, Jonsson, Nordstrom, Sjostrom, & Moritz, 2004; Kaiser, Barding, & Larive, 2009; Kim, Choi, & Verpoorte, 2010; Kim & Verpoorte, 2010; Kim, Wilson, Choi, & Verpoorte, 2010; Moing et al., 2004; Sekiyama, Chikayama, & Kikuchi, 2011), animals (Brown, McKelvie, Simpson, & Simpson, 2010; Lin et al., 2009), yeast (Villas-Bôas, Rasmussen, & Lane, 2005) and bacteria (Bolten, Kiefer, Letisse, Portais, & Wittmann, 2007; Maharjan & Ferenci, 2003). In the following text, we describe the essential steps of extraction procedures used in plant metabolomics. These include a list of technical recommendations of extraction procedures that are often critical to the ultimate success of metabolomics by ^{1}H-NMR spectroscopy.

2.4.1 Is the extraction step necessary?

NMR spectroscopy makes it possible to get recordings from living cells, tissues and organs avoiding extraction step, to minimize metabolism disturbances occurring during the extraction. Thereby, the measurement of levels of certain labile metabolites and information on metabolite subcellular compartmentalization can be obtained by *in vivo* NMR (Ratcliffe & Shachar-Hill, 2001, 2005). Nevertheless, despite this enormous potential for *in vivo* NMR, the bulk of NMR spectroscopy is carried out on samples in solution. In medicine, some *ex vivo* samples such as blood, serum, urine or cerebrospinal fluid are already in solution and may be observed almost directly. In plants, it is also possible to record NMR spectra directly from crude extracts, like sap or juice (Belton et al., 1998; Biais et al., 2009; Sobolev et al., 2010; Spraul et al., 2009). In these cases, sample preparation is rapid, but nevertheless room temperature or hot deuterated methanol is still necessary to deactivate enzymes in order to quench the metabolism.

2.4.2 Extraction challenges for high-resolution ^{1}H-NMR spectroscopy

Compared to nucleotides or proteins, metabolites have a wide variety of physico-chemical properties, ranging from ionic inorganic species to hydrophilic carbohydrates, hydrophobic lipids and complex natural products (Fernie, 2007; Saito & Matsuda, 2010; Sumner, Mendes, & Dixon,

2003). With NMR spectroscopy, chemically different molecules can be detected at once in a crude extract if the molecules are NMR-visible and above the NMR detection threshold. A limited number of molecules can be chelated by paramagnetic ions (Mn^{2+}) and become invisible. As mentioned above, for ^{1}H-NMR, the concentration threshold for routine detection of a metabolite in an extract using a modern high-field spectrometer is probably 10 μM (Krishnan et al., 2005) and the achievable sensitivity is strongly dependent on the field strength of the magnet and design of the probe head (see Section 2.3).

To analyze plant metabolomes from tissues or cells, the extraction method has a crucial bearing on the proceeding detection of metabolites (Fernie, 2007). Indeed, plant tissue is inherently heterogeneous because it consists of various cell types, each having a dedicated metabolism. In order to guarantee high-quality final data set, it is necessary to minimize any variation of metabolites caused by enzymes released during tissue extraction (Kopka, Fernie, Weckwerth, Gibon, & Stitt, 2004). Considering dynamic range variation of metabolites in plants (i.e. relative abundance up to 6 orders of magnitude concentration difference) and saturation threshold during extraction, no single extraction methodology will provide an extract truly quantitative and representative of the plant sample. Therefore, without a robust, efficient and accurate extraction procedure, identification and quantification of plant metabolites in tissue can be challenging (Fernie, 2007).

The usual procedure is normally limited to a two-step procedure that involves, initially, restricting enzyme activity by a quick cooling of the material to extremely low temperature (most of the time it is achieved by transferring the samples in liquid nitrogen) and subsequently disrupting the tissue and inactivating enzymes by denaturation with acid or organic solvents. For small samples of plant material, dropping them directly into liquid nitrogen is the easiest way; for larger samples of bulky tissue, freeze-clamping them is more adapted. The favoured method for inactivating enzymes and extracting metabolites is to grind the frozen tissue in organic solvents. The choice of the extraction solvent may be critically linked to the fact that the solubility of relatively abundant compounds can be different according to used solvents.

2.4.3 Solvent choice: A critical step for extraction

The most common extraction procedures generally involve a multi-step extraction in water or D_2O, in organic solvents such as methanol, ethanol, chloroform and acetonitrile, in acids such as perchloric acid or in different types of mixtures such as chloroform:methanol, chloroform:methanol:water and ethanol:water (Allwood et al., 2011). Our experience suggests that the optimum preparation protocol should be developed on a case-by-case basis depending on the targeted metabolites, the number of metabolites examined and their respective quantities.

Extraction in D_2O buffer solution is suitable for high-throughput analysis; for example, after extraction and centrifugation, the supernatant is directly transferred into NMR tube for spectra acquisition. But many drawbacks remain, among them the difficulty to control pH (in fact pD) and ionic force of the samples within a series. It is also unsatisfactory because this extraction procedure does not quench all enzyme activity. Our experience with perchloric acid procedure suggests that it is a very good procedure to extract polar metabolite efficiently (sugar, sugar phosphate, etc.) and to effectively denature and precipitate proteins (Fiehn, 2002). This procedure has been used for years for ^{31}P-NMR spectroscopy of plant extracts (Roberts & Jardetsky, 1981) and recently proposed for metabolomics studies (Kruger, Troncoso-Ponce, & Ratcliffe, 2008). However, this procedure is not appropriate for plant metabolomics because it facilitates acid-catalyzed cleavage of sucrose complicating biological interpretation of the resulting metabolic profile and provides unstable pH extract complicating the ^{1}H-NMR spectrum interpretation (Kaiser et al., 2009). Indeed, sample pH is an important parameter in metabolic profiling experiments because the extraction efficiency of some metabolites can be affected by pH, but also because the chemical shifts of any metabolites depend on pH. Methanol: D_2O:chloroform allows the greatest degree of analytical variability. However, a two-phase solvent system yielding a non-polar and a polar extract requires an additional separating step, the evaporation of solvents and redissolving of the residue in appropriate NMR-deuterated solvent. Carbohydrates including sucrose, fructose, glucose, arabinose, etc. are metabolites that may be found in the water-soluble fraction of plant extracts as well as the main amino acids, related amines and organic acids, while plant lipids, triglycerides and waxes, highly soluble in chloroform, may be found in the organic soluble fraction. This procedure can be useful for some studies, but internal standards have to be used in order to separate biological variability from extraction procedure variability (Kaiser et al., 2009).

The most popular alternative extraction procedures involve extraction with aqueous methanol or ethanol mixtures in one-step (Le Gall et al., 2003, 2004; Moing et al., 2004; Ward, Harris, Lewis, & Beale, 2003) or two-step procedure (Ben Akal-Ben Fatma et al., 2012). These procedures have the advantages to be simple, quick, stable, reproducible and easily adapted for automated processing, when handling a large number of samples. Recently, a technical focus on four commonly used extraction protocols (perchloric acid, boiling ethanol, methanol and methanol:chloroform) to establish the hydrophilic metabolome by quantitative ^{1}H-NMR has been published with heterotrophic *Arabidopsis thaliana* cells grown in suspension (Gromova & Roby, 2010). These quantitative studies demonstrated that the four extraction protocols commonly used lead to quite similar molecular composition as analyzed by ^{1}H-NMR. The impacts of extraction protocols (solvent composition, pH, one- vs. two-step processes, lyophilization) on the stability of extracts and on the quality of metabolomics profiles of *A. thaliana* mature leaves were described by Kaiser et al. (2009).

2.4.4 Other determining factors for high-quality ^{1}H-NMR spectrum

In order to optimize extraction, homogeneous sampling of ground material is required. Considering the detection threshold of 5 nmol for ^{1}H-NMR, the amount of sample material required for NMR analysis has to weigh approximately 20–50 mg dry weight corresponding to 200–500 mg fresh weight which is higher than the sample material necessary for MS spectroscopy. As water is the main component of plant tissue, it is important to remove it by lyophilization or cryodessication or to use appropriate NMR methods for solvent signal suppression. To be more efficient, this step should be performed after cryogrinding, as fine plant powder facilitates the freeze-drying process as well as that of extraction (Kim & Verpoorte, 2010). The abundance of macromolecules in some plant species—polysaccharides or proteins—may cause NMR peak broadening and compromise further analysis. The extraction step should eliminate these unwanted molecules. Because impurities are present in solvent and plastic ware and absorption of metabolites on glass wall can occur (Gottlieb, Kotlyar, & Nudelman, 1997), solvent blank and extraction blank (all the process without the biological sample) need to be carried out for each series of samples. The presence of paramagnetic ions often has a profound effect on the NMR spectrum as a result of the changes in relaxation properties and chemical shifts that can occur when a magnetic nucleus interacts with a paramagnetic ion. In chlorophyll-containing tissues such as leaves, paramagnetic ions, mainly manganese, need to be sequestered by addition of EDTA or CDTA during the extraction procedure. The choice of the extraction temperature is a compromise between the extraction efficiency and the putative denaturation of some metabolites. The pH of the extracts should be controlled to avoid shifts of NMR signals. Several buffer solutions are available for NMR measurements; the most commonly used for ^{1}H-NMR is phosphate buffer solution. Buffer solution salinity may have a further impact on shift of metabolite NMR signals and also needs to be controlled. Adequate numbers of replicates should be defined to control the technical variability at extraction step, and in relation to the aim of the experiment. Replication of biological samples is indispensable and should be favoured to extraction and analytical replication, if all of those are not feasible due to analytical costs. It is recommended to always analyze immediately after extraction in order to avoid metabolite degradation through time. It would be helpful to propose a few compounds as indicators of storage condition quality and duration (Llewellyn et al., 2012).

In conclusion, the extraction step is probably the most difficult step to reach consensus. Several extraction methods have been applied in NMR metabolomics analyses. Each method has its advantages, unique applications and also limitations. If further data mining is planned with previously obtained data of the same plant species, obviously, the same extraction procedure has to be applied with spectra measured in the same solvent, at the same pH and temperature, to avoid shifts of metabolite resonances.

2.5. Choosing the appropriate method for solvent signal suppression

2.5.1 Using selective pulse sequences

Plant cells are characterized by high water content. If after extraction, water remains in the sample extract, the water peak in the ^{1}H-NMR spectrum can be so huge as to obscure other molecular information. Indeed, all NMR experiments, which rely on the detection of proton in biological samples, face the problem of molecules having to be detected in an aqueous sample at micromolar or lower concentrations against a background of 110 M water protons (Bothwell & Griffin, 2011). The large NMR signal arising from water protons in water extract can be easily eliminated by using appropriate standard NMR solvent suppression methods that have been reviewed in the NMR literature (Antoine, Coron, & Dereppe, 2000; Aranibar, Ott, Roongta, & Mueller, 2006; Hwang & Shaka, 1995; Liu et al., 1998; Simpson & Brown, 2005; Smallcombe, Patt, & Keifer, 1995; Zheng & Price, 2010). In NMR metabolomics, excellent water suppression in a single reference sample is not sufficient. Metabolomics requires high reproducibility among a wide range of sample compositions and without the need for exhaustive manual adjustment of acquisition or processing parameters to detect subtle variations that are due to treatment effects. Solvent suppression techniques and related pulse sequences have to be applied with care. The technique used for solvent suppression is of high importance for the interpretation of NMR data. NMR spectra obtained under different solvent suppression conditions can be erroneously misinterpreted if water suppression is not carefully applied (Potts et al., 2001).

The most widespread solvent suppression technique in use is the so-called standard presaturation technique known as noesy–presat NMR pulse sequence (Nicholson et al., 1995; Ross et al., 2007). With this technique, the water resonance is saturated by frequency-selective irradiation during the inter-scan delay and the mixing time of the sequence, thus leading to a reduction of the equilibrium population difference of the spin species resonating at the frequency of the weak (<50 Hz) B1 field applied. The big advantage of this method is given by chemical shift selectivity superior to any other technique allowing reliable quantification of signals resonating close to water such as anomeric proton of glucose if the frequency of the water signal is determined with accuracy. But the main disadvantage of this presaturation technique comes from the fact that hydrogen atoms from different molecules in chemical exchange with water will also be saturated. Although water presaturation technique is considered a simple, robust and highly reproducible technique, one should keep in mind that absolute value quantification is possible only with limited accuracy when using this technique.

WATERGATE (water suppression by gradient-tailored excitation) and WET (water suppression enhanced through T_1 effects) are two other pulse sequences that can help suppress water signals (Araníbar et al., 2006; Liu et al., 1998; Smallcombe et al., 1995). WATERGATE sequence is close to a spin-echo sequence with the refocusing pulse flanked by two symmetrical gradient pulses, while WET sequence suppresses the solvent signal by the application of a series of weak frequency-selective B1 pulses, interleaved by gradient pulses Gz. The flip angles (durations) of the selective pulses and the strengths of the gradient pulses are optimized to achieve high-frequency selectivity, robustness and a flat baseline. Other alternative pulse sequences such as excitation sculpting (ES) and associated sequences (ESCW and ESWGL) have been suggested in the literature (Araníbar et al., 2006; Hwang & Shaka, 1995; Simpson & Brown, 2005; Smallcombe et al., 1995).

For metabolomics use, Aranibar and colleagues have assessed robustness, repeatability, sensitivity, selectivity and practicality of commonly used solvent peak suppression methods in NMR analysis. ES and noesy–presat sequences show very similar performances when used in pattern recognition analysis. Noesy–presat sequence is a well-established method, but it obtains reduced sensitivity when used in statistical tests and requires more manual reprocessing of data. The two ES sequences named ESCW and ESWGL have proven to be excellent complementary techniques (Araníbar et al., 2006).

When a sample contains more than one type of solvent or a solvent has multiple resonances, multiple-resonance suppression methods must be used. Some of the solvent suppression methods mentioned above can be used to suppress multiple resonances (Smallcombe et al., 1995; Zheng & Price, 2010). Politi and colleagues analyzed cannabis extracts without evaporation and purification steps using 1D noesy–presat pulse sequence with multiple offset presaturation to suppress the ethanol and water signals (Politi et al., 2008).

2.5.2 Using D_2O exchange

As mentioned in Section 2.4, lyophilization and reconstitution in deuterated solvent are frequently practical sample preparation methods for biological samples. This practice has many advantages. Firstly, freeze-dried samples reconstituted in solvents in which protons (^1H) have been replaced by deuterons (^2H) show spectra with reducing signals from water. Solvent suppression pulse sequences with these samples are not necessary anymore. The remaining water signal due to non-exchangeable solvation water around molecules is small and compatible with other signals. Contrary to all suppression water sequences, using D_2O exchange prior to NMR spectra acquisition in aqueous solutions allows absolute value quantification with high accuracy. Secondly, deuterium will be used to operate NMR spectrometers in a deuterium-locked mode where the known frequency of the deuterium signal is used to calibrate the other frequencies observed during the NMR experiment. However, loss of volatile or unstable components from biological samples may occur during the lyophilization process, for example, ethanol, acetate or formate. Selective deuteration of acidic protons on reconstitution with D_2O may further complicate spectral interpretation.

2.5.3 Relaxation and diffusion-editing techniques

Proton NMR metabolomics can be applied to biological samples coming from a broad range of sources. In plants, the bulk of NMR spectroscopy is carried out on sample extracts, while in medicine a part of it is carried out on *ex vivo* samples (biofluids) such as blood, urine and serum that are already in solution. The presence of macromolecules such as lipoproteins and lipids in biological samples can cause overlapping resonances and broad peaks in ^1H-NMR spectra and may mask the detection of low-level small-molecule metabolites. To improve the detection of small metabolites, a number of NMR spectral editing methods can be employed utilizing the variation of diffusion properties and spin-relaxation (Bothwell & Griffin, 2011; Lindon et al., 2000; Ross et al., 2007). Generally, molecules with greater molecular weight diffuse slowly but have relatively short spin-relaxation times (T_1), whereas small molecules diffuse quickly but have longer relatively spin-relaxation times (unless they are bound to macromolecules). Therefore, the complex NMR spectra of biological tissues can be edited by taking advantage of the differences in diffusivity and spin-relaxation times.

For spin-relaxation editing, a "relaxation filter" can be inserted prior to detection to attenuate or eliminate the resonances having relatively short relaxation times. The magnetization can be prepared using a spin-echo sequence, typically a CPMG sequence (*C*arr *P*urcell *M*eiboom *G*ill sequence), $(90°_x - (\tau - 180°_y - \tau)_n)$, by choosing τ and n such that signals can be separated according to their T_2. Generally, the broad NMR signals from macromolecules or bound small molecules are attenuated or even eliminated, leaving the sharper resonances from the mobile small molecules unaffected (Lucas, Larive, Wilkinson, & Huhn, 2005). For diffusion-edited NMR spectroscopy, a "diffusion filter" has been inserted to eliminate resonances from the fast-diffusing molecules prior to the detection. This technique is referred to as PGSE (*P*ulsed *G*radient *S*pin *E*cho), whereby bipolar gradients up to 100 G/cm are used. An extension to 2D by incrementation of the gradient strength is known by the acronym DOSY (*D*iffusion *O*rdered *S*pectroscopy). This technique allows "the sorting" of the compounds in a mixture according to their translation diffusion.

Leon and colleagues reported an alternative analytical DOSY NMR experiment that was used to carry out a virtual separation of some components in the acetone extract of the rhizomes of *L. porteri*, which aimed to determine the presence of major dimeric phthalides (Leon, Chavez, & Delgado, 2011). This technique is based on different translation diffusion coefficients, *D*, which depend on the effective molecular weight, size and shape of each compound.

2.6. Quantitative ^{1}H-NMR: Precautions during extraction, acquisition and processing

Several recent publications review and detail the advantages and pitfalls of quantitative NMR, and the technical and experimental advice to follow (Allwood et al., 2011; Barding, Fukao, Beni, Bailey-Serres, & Larive,

2012; Pauli, Godecke, Jaki, & Lankin, 2012). To sum up, NMR acquisition parameters should establish quantitative conditions so that proton signal with the longest longitudinal relaxation rate T_1 (i.e. the slowest relaxing nuclei) is not saturated. The repetition delay should be at least five times this T_1 value to avoid FID truncation. All the resonances should be properly relaxed. The probe should be well tuned and matched. The signal-to-noise ratio (S/N) of the spectrum should also be at least higher than 100:1 (Barding et al., 2012) to achieve accurate absolute quantitation. Special care to the digital resolution should be taken; well-resolved resonances of interest should be defined at least by 5 datapoints. The digital resolution depends on the number of datapoints acquired during the acquisition time and the spectral width chosen.

Two standards are required for quantitative ^{1}H-NMR: a chemical shift reference and a calibration standard. Both can be internal or external standards. In some cases, the chemical shift reference is also used as calibration standard. The chemical shift reference is used to adjust the chemical shift of the resonances. TSP also called TMSP (3-trimethylsilylpropionic-2,2,3,3-d_4 acid sodium salt) and DSS (3-trimethylsilyl-1-propanesulfonic-1,1,2,2,3,3-d_6 acid sodium salt) are the most used internal chemical shift reference in ^{1}H-NMR plant metabolomics.

The choice of internal standards is a difficult task. Pauli and collaborators reviewed all the considerations to be taken into account such as chemical stability, solubility, pH independency, no signal overlap and all the compounds available for ^{1}H-NMR (Pauli et al., 2012; Pauli, Jaki, & Lankin, 2005).

Two alternative approaches to the use of internal chemical calibration standards are electronic standard or artificial signal. Historically, Electronic REference To access *In vivo* Concentration (ERETIC) was the first electronic external calibration standard method for quantification by NMR (Akoka, Barantin, & Trierweiler, 1999; Barantin, Le Pape, & Akoka, 1997). ERETIC involves the synthesis of an electronic reference signal by an extra electronic device and changes in the spectrometer hardware, its transmission by an additional antenna and its acquisition in the spectrum simultaneously with the signals from the metabolites. The intensity and localization of the ERETIC signal can be freely changed by the spectroscopist. The nominal concentration of this signal is determined with calibration curves (samples with known concentrations). This method has been used in plant metabolomics for quantification of primary metabolites of tomato (Deborde et al., 2009; Moing et al., 2004; Mounet et al., 2007;

Neily et al., 2011), grapes (Fernandez et al., 2006), strawberry (Lerceteau-Kohler et al., 2006; Moing et al., 2004), maize (Cossegal et al., 2008), *Arabidopsis* (Allen et al., 2010) and melon (Bernillon et al., 2012; Biais et al., 2009). Nowadays, "inverse ERETIC" (also called ERETIC 2) and QUANTAS for QUANTification by Artificial Signal (Farrant et al., 2010) are preferred since no extra electronic device is needed. These software tools can be easily implemented in modern spectrometer. Both tools are based on PULCON for PULse Length-based CONcentration measurement method (Dreier & Wider, 2006), which correlates the absolute intensities of two different spectra. The concentration measurements use the principle of reciprocity which indicates that the lengths of a 90° or 360° pulse are inversely proportional to the NMR signal intensity (Hoult & Richards, 1976). The 90° pulse of all the samples should be well calibrated, and the probe should be tuned and matched for each sample. In this case, if the concentration of one of the samples is known precisely, the unknown concentrations of the others samples can be obtained with an accuracy and a precision of less than 2% (Dreier & Wider, 2006).

In conclusion, absolute quantification of metabolites by ^{1}H-NMR can be carried out if some rules are respected (Moing et al., 2004). Firstly, elimination of water in samples during extractions is necessary through lyophilization and reconstitution in deuterated solvent. Contrary to all suppression water sequences, using D_2O exchange prior to acquisition of NMR spectra in aqueous solutions allows absolute value quantification with high accuracy. Secondly, the choice of an unsaturating recycle time is absolutely necessary to avoid peak distortion. T_1 values for each extract have to be measured by the inversion-recovery method (Keeler, 2010). To obtain undistorted spectra, a full 90° pulse and a 5T1 repetition time have to be applied during acquisition. Thirdly, for absolute quantification of ^{1}H-NMR spectra, calibration curves have to be recorded using external reference(s) (chemical or electronic references). Improved quantification from ^{1}H-NMR spectra using reduced repetition times can also be applied (Bharti et al., 2008). Today, ^{1}H-NMR coupled with an electronic reference signal has become a routine method that allows absolute quantification of individual major metabolites in complex plant tissues. One of the main advantages of such an approach is that structural and quantitative information with the same quality as HPLC can be contained on numerous chemical species with a wide range of concentrations in a single NMR experiment (Deborde et al., 2009; Moing et al., 2004; Mounet et al., 2007).

2.7. Automation of ^{1}H-NMR analysis for high-throughput metabolomics

2.7.1 Automatic sample preparation and transfer to the spectrometer

Metabolomics studies involve a large sample number, which have to be prepared and measured as much as possible with the higher reproducibility. A way to achieve this requirement is to use lab automation when available. In urine metabolomics analysis preparation, a robot is commercially available; this device dispenses urine, buffer solution and deuterated water into the NMR tube. These tubes with or without barcode are placed in autosampler or robotic sample changer (with or without controlled temperature, with or without UV protection). Use of such preparation robot for aqueous-based plant extracts is limited since pH measurement and titration are required. Devices combining automated sample preparation and sample titration are under development.

2.7.2 Automatic probe tuning and magnetic field adjustment

For the same reasons as those mentioned above and to reduce the dependence of spectral quality on the operator, automatically tuneable probe heads are highly recommended. In addition, use of nitrogen gas flush (coming from a nitrogen separator which enriches N_2 to about 98% of the compressed air) and probe head temperature control unit are highly recommended. The chemical shift of water is temperature dependent; that is why some care should be taken with sample temperature equilibration before NMR measurement. A delay of 2–3 min is required for this equilibration to take place, especially with samples containing salts and buffers. Inverse or indirect detection probe head measures optimal signal-to-noise ratio for ^{1}H detection. The diameter of such standard probe heads is usually 5 mm and requires a sample volume of 500–600 μl (Ross et al., 2007). Shimming operation has to be conducted on each tube since this operation will generate homogeneous magnetic field along the sample volume to obtain good line shapes of all resonances in the NMR spectrum. Automatic field adjustment by gradient autoshimming of on-axis shims is routinely used in metabolomics. A non-finely tuned system (poorly optimized shims and poorly optimized probe head tuning) will impair the water suppression.

2.7.3 Automatic data acquisition and processing

Special care should also be taken with spectrometer adjustments for each sample regarding temperature equilibrium, shimming and determination of the 90° pulse angle. With recent NMR spectrometers, the determination and optimization of these parameters are routinely achieved by automation. Recommended spectral acquisition and processing parameters for quantitative 1D ^{1}H-NMR spectrum and for semi-quantitative 1D ^{1}H-NMR with presaturation (noesy–presat) of polar metabolites using a 500-MHz NMR spectrometer with an automatically tuneable inverse probe are summarized in Allwood et al. (2011) and Pauli et al. (2005). Automation of phasing and baseline correction can be used, but only with limited reliability. Besides, human visual inspection of processed spectra is compulsory to detect artefacts of automated processing.

3. METABOLOMICS NMR FINGERPRINTING FOR DISCRIMINATING PLANT METABOLOMES

3.1. Challenging tasks of metabolomics fingerprinting

Fingerprinting studies are usually conducted at organ level, but specific tissues can be characterized. As for all metabolomics studies, preliminary studies may help to optimize the experimental design and choose the wet-lab and dry-lab-operating procedures. Crude extracts of plant materials have complex ^{1}H-NMR spectra, as mentioned above, with many overlapping signals. This complexity is dealt with using the fingerprinting approach where the ^{1}H-NMR signature is studied as a whole in a non-targeted way and often before spectra annotation. These signatures of samples from a given population are compared using multivariate analyses after data reduction in order to see if groups of samples can be distinguished and, if so, to highlight the spectral regions that discriminate the sample groups.

Several factors affecting the robustness of metabolite fingerprinting using ^{1}H-NMR spectra have been detailed previously (Defernez & Colquhoun, 2003), including pH control of the extracts and post-acquisition peak alignment. Increase of throughput and robustness are key points for fingerprinting and are often addressed using automation of each step of the experiment pipeline from weighing, extraction and pH control to spectra pre-processing.

As for the analysis of all high-dimensional omics data sets, several other types of dangers related to the statistical analysis of NMR fingerprints have to be taken into account to avoid the detection of spurious biomarkers. These dangers have been summarized previously, for example, (Broadhurst & Kell, 2006) and will not be detailed here. They include bias, inadequate sample size, excessive false discovery rate due to multiple hypothesis testing, inappropriate choice of particular numerical methods, and overfitting.

Fingerprinting first provides anonymous biomarker(s) (i.e. one or several given spectral regions), but generally the discriminant spectral regions are later submitted to a detailed annotation work using NMR alone (see Section 4) or a combination of NMR and MS analytical approaches.

3.2. The basis of metabolic fingerprinting

As for any other fingerprinting experiment, for NMR fingerprinting the design of the experiment is crucial in order to extract pertinent information from the data (Gibon & Rolin, 2012). Depending on the objectives, analyses can use parallel extraction methods that can each provide several fractions (e.g. methanol and chloroform fractions) from the same samples. Numerous studies use polar extracts only, for which pH control, with buffer and sometimes manual or automatic titration, are crucial to avoid uncontrolled chemical shift variations. NMR acquisition is usually carried out with short acquisition times to increase sample throughput or sample number analyzed within a given time. As NMR acquisition is highly reproducible, increasing the number of biological replicates is usually preferable to the addition of technological replicates. NMR spectra have to be acquired in a randomized order and in the same analytical series in order to avoid any bias. ^{1}H-NMR fingerprint pre-processing and processing consist in FT with line broadening, phase correction, alignment using a reference signal, baseline correction, data reduction and normalization prior to multivariate statistical analyses. Spectra pre-processing and processing before chemometrics analysis have to be achieved exactly in the same way for the entire experiment. Multivariate analyses can be diverse and more or less sophisticated, that is, requiring the expertise of a chemometrician or easily performed by a biologist (Kopka, Walther, Allwood, & Goodacre, 2011), but the "classical" principal component analysis (PCA) should not be forgotten. Indeed, if the factor of interest contributes to the majority of the variance within the total explained variance, PCA appears to be the method of choice to identify spectral regions, which contribute to that variance. Web or open source software tools for ^{1}H-NMR spectra processing and analysis have recently been suggested (Izquierdo-Garcia et al., 2011; Wang et al., 2009). Minimum reporting standards for data analysis have been recommended by the MSI for each step of a metabolomics study, from experimental design to model validation and testing (Goodacre et al., 2007). Some key points for NMR fingerprinting are detailed below.

3.2.1 Design of fingerprinting experiments

Formalizing the experimental design is mandatory for the biological study as well as the analytical design (Goodacre et al., 2007). Moreover, the metadata describing these two parts of the experimental design have to be carefully recorded in a machine-readable way, and important metadata have to be released at the time of results publication using manuscript supplemental information, Web sites or databases. The reader is referred to the checklist proposed by Gibon and Rolin (2012) to identify the critical steps of a plant NMR fingerprinting experimental design and use Xeml Lab, a software suite for a standardized description of the growth environment of plants (Hannemann et al., 2009).

3.2.2 Acquisition and post-treatment of NMR spectra

Depending on the nature of the plant extracts, routine and large-scale metabolomics, fingerprinting studies use either noesy–presaturation method for aqueous samples (Biais et al., 2009; Lee, Hwang, Lee, & Hong, 2009; Lee, Shaykhutdinov, et al., 2009; Mattoo et al., 2006) or one-pulse sequence method for samples depleted of water (Pereira, Gaudillere, Pieri, et al., 2006; Pereira, Gaudillere, van Leeuwen, et al., 2006). In both cases, the determination of the 90° pulse angle is required for each sample, but in one-pulse sequence an Ernst angle of 30° or 45° can be used (Pauli et al., 2005). Recommended spectral acquisitions are summarized in Allwood et al. (2011). In metabolomics fingerprinting experiments, the main goal is to classify groups of samples. In some cases, identification and relative quantification of the biomarkers responsible for the classification could be of great interest. The relative quantification can be achieved when the S/N ratio is higher than 10:1 (Barding et al., 2012).

The spectra pre-processing steps are essential to take as much as possible advantage of all information contained in the NMR spectra. Because pre-processing methods usually attempt to reduce variance and any other possible source of bias such as phase correction, peak shifting or misalignment and baseline correction, NMR spectra require pre-processing steps before applying statistical pattern recognition methods in order to detect subtle variations. Many publications have discussed this point (Craig, Cloarec, Holmes, Nicholson, & Lindon, 2006; De Meyer et al., 2010; Ebbels, Lindon, & Coen, 2011; Zhang et al., 2009), and in particular, a recent review gives a good picture of the topic (Smolinska, Blanchet, Buydens, & Wijmenga, 2012). Therefore, we will only introduce the most frequently used methods in this section.

Before any treatment, apodization and zero filling of the FID is performed, followed by an FT. The phase is then corrected to obtain absorption line shape. The FT step usually produces a complex vector with a real and imaginary part. Subsequently, the next steps of data pre-processing only concern the real part. Generally, pre-processing steps include baseline correction, noise elimination and alignment. Noise reduction can be performed with the help of a Savitzky–Golay filter (Savitzky & Golay, 1964), which gives good results. For baseline correction, an approach relying on a smoothed spectrum can be used for baseline recognition and modelling. To complete the model, an interpolation technique is employed over the signal area; then the model is subtracted from the spectrum giving a flat baseline (Bao et al., 2012; Golotvin & Williams, 2000). The alignment step is undoubtedly one of the most tedious parts to solve. The misalignments are the results of changes in chemical shifts of NMR peaks largely due to differences in pH and other physico-chemical interactions. To solve this prickly problem, some methods have been developed such as ones based on correlation optimized warping and dynamic time warping (Veselkov et al., 2009), on cross-correlation between segments (Savorani, Tomasi, & Engelsen, 2010) or even those based on the Bayesian statistics (Bum Kim, Wang, & Hiremath, 2010). Another possibility is to align interval by interval interactively, each interval being chosen by the user and the alignment based on a least-squares algorithm. Although this method is more tedious, it can be easily implemented. After baseline correction, spectral regions, which do not contain relevant information, have to be removed, especially those belonging to solvent (Smolinska et al., 2012).

NMR spectrum may contain several thousand points, and therefore variables. In order to reduce the data dimensionality, binning is commonly used. In binning, the spectra are divided into bins (also called buckets), and the total area within each bin is calculated to represent the original spectrum. The more simple approach consists in sub-dividing all the spectra with an uniform area width (typically 0.04 ppm; Lindon, Holmes, & Nicholson, 2001). Due to the arbitrary division of peaks, one bin may contain pieces from two or more peaks, which may affect the data analysis. Many other approaches attempt to split the spectra so that each area region common to all spectra contains the same resonance, that is, belongs to the same metabolite (Anderson et al., 2011; Davis et al., 2007; De Meyer et al., 2008). In such methods, the width of each spectra region is then determined by the maximum difference of chemical shift among all spectra. A good review is given by Anderson and colleagues (Anderson et al., 2011). The advantage of these methods is to give a better significance for each variable thus generated, and moreover, most of them can avoid the alignment step if the misalignments are small enough. Whatever the chosen method, the merging of adjacent buckets may be necessary if some significant chemical shifts remain misaligned.

To make all spectra comparable to one another, the variations of the overall concentrations of samples should be taken into account. In NMR plant metabolomics, the total intensity normalization is often used so that all spectra correspond to the same overall concentration. It simply consists in normalizing the total intensity of each individual spectrum to a same value (i.e. 100; Craig et al., 2006).

Metabolites can range in concentration over several orders of magnitude. Moreover, the variation in metabolites level is often linked to concentration in such a way that higher concentration metabolites have higher variations (Ebbels et al., 2011). Therefore, such metabolites have the highest influence on the data analysis. Therefore, it is important to scale metabolite intensities before analysis to avoid selection of the most abundant metabolites without biological signification. A number of scaling methods are commonly used, namely, mean-centering, autoscaling, Pareto scaling, range scaling, vast scaling and level scaling (van den Berg, Hoefsloot, Westerhuis, Smilde, & van der Werf, 2006). A good practice is to apply a thresholding of spectra, that is, zeros all buckets below a threshold value. Thus, thanks to noisy bucket elimination, several scaling methods can be applied, enabling the detection of low-intensity compounds (Halouska & Powers, 2006).

3.2.3 Highlighting biomarkers from NMR data with chemometrics

Depending on the data reduction method used, ^{1}H-NMR fingerprinting yields from about 100 to 1000 variables. Therefore, multivariate analyses are mandatory to get an overview of the data and highlight discriminant spectral regions. PCA is usually used in a first step and can be complemented with partial least square discriminant analysis (PLS-DA) for modelling (Lindon et al., 2001). PCA can be performed on all variables or on significant variables selected after variance analysis or non-parametric univariate analyses. PCA score plots allow visualizing the clustering of the different samples without any knowledge of their group membership. Therefore, PCA is classified as an unsupervised method. The comparison of PCA score plots and loading plots allows pointing to spectral regions that contribute to discriminate sample groups. PLS-DA is classified as a supervised method. It allows calculating synthetic variables, called latent variables, along which the scores are maximized to discriminate the groups of samples according to the studied discriminating factor. Recently, PCA and PLS-DA of vanilla pod extracts showed that the different cultivated accessions could be differentiated according to their glucovanillin and glucosides A and B contents (Palama et al., 2011). It must be kept in mind that PLS-DA models have to be carefully validated. Moreover, the tendencies highlighted with multivariate analyses such as PCA have to be verified using univariate analyses. Of course, in a following step, the discriminant spectral regions have to be annotated (see Section 5) to provide relevant biological information instead of "anonymous" biomarkers.

Methods derived from PCA such as multi-block PCA (Biais et al., 2009) may be of interest for certain experimental designs and especially for the fusion of NMR fingerprinting data with other types of fingerprinting data. In a study of fruit juice authentication with ^{1}H-NMR (Cuny et al., 2008), independent component analysis (ICA) consistently gave better results than PCA. The authors suggested that since ICA is based on the idea of demixing the spectra into a sum of "pure" signals, ICA may prove better than PCA which only determines vectors indicating the direction of the greatest dispersion of the population of samples in the multidimensional space constituted by the fingerprint variables. Methods derived from PLS-DA such as supervised orthogonal projection to latent structure discriminant analysis (OPLS-DA) are more and more used (Consonni, Cagliani, Stocchero, & Porretta, 2009; Lee et al., 2011).

Among non-linear computational methods, Artificial Neural Networks (ANNs) are a possibility for NMR fingerprinting and have already been used for food analysis (Marini, 2009). ANNs have been used to classify the biochemical mode of action for herbicides after crude extracts from the treated plants were subjected to ^{1}H-NMR spectroscopy (Di Anibal, Callao, & RuisÃ¡nchez, 2011; Ott, Aranibar, Singh, & Stockton, 2003). In a case study on spice adulteration with Sudan dyes, genetic algorithm (GA) technique was tested for variable selection and gave satisfactory classification and prediction results (Di Anibal et al., 2011). As GA methods are conceptually more complex than classical analyses such as PCA, a chemometrician is required. It has been reminded in a review based on GC-MS analysis of wine that it is quite impossible to propose a generalized standard operating procedure for the statistical data mining of metabolome experiments, but that case-by-case choices and sequences of statistical tools have to be applied (Kopka et al., 2011). The same applies to NMR fingerprinting data. Moreover, specific tools are needed to explore time-resolved, dynamic or temporal metabolomics data (Smilde et al., 2010).

3.3. A wide range of applications as evidence of success

Applications of NMR fingerprinting range from the characterization of food or medicinal plants to functional genomics. As references are numerous, we will focus on pioneering works and recent articles. For a more exhaustive list of applications, see Hall (2011a), Kim, Choi, and Verpoorte (2011) and Krishnan et al. (2005).

3.3.1 Food origin and authentication

NMR fingerprinting has been used in food analysis for origin discrimination in a range of plant species using polar or non-polar extracts. A pioneering work on orange juice (Vogels et al., 1996) proposed the application of proton NMR spectroscopy as a screening tool for the determination of the authenticity of orange juices. A few years ago, an NMR-based, fully automated system has been proposed as a screening method for the quality control of fruit juices (Spraul et al., 2009). Recently, an NMR-based approach allowed studying the authenticity of saffron (Yilmaz, Nyberg, Mølgaard, Asili, & Jaroszewski, 2010) or the geographical origin of durum wheat (Lamanna, Cattivelli, Miglietta, & Troccoli, 2011). Similar approaches are used to characterize plant-derived products, for example, characterization of wine or beer (Hong, 2011; Rodrigues & Gil, 2011), detection of adulteration of olive oil (Mannina et al., 2009), characterization of soy sauces of different origins and ages (Ko, Ahn, van den Berg, Lee, & Hong, 2009) or identification of specific markers for the botanical origin of a plant- and insect-derived product, honey (Schievano, Stocchero, Morelato, Facchin, & Mammi, 2012). Most of the time, liquid NMR has been used, but HRMAS-NMR is another possibility for food fingerprinting (Perez et al., 2011; Ritota, Marini, Sequi, & Valentini, 2010).

3.3.2 Medicinal plant characterization

The use of NMR fingerprinting for the characterization of medicinal plants has been exploding over the past few years. Only a few representative examples are given below. ^{1}H-NMR has been used to discriminate between three *Echinacea* (Asteraceae) species (Frederich et al., 2010). The combination of NMR and MS metabolite profiling and fingerprinting has been used to characterize resins and extracts from 13 hop cultivars (Farag, Porzel, Schmidt, & Wessjohann, 2012). The combination of ^{1}H-NMR with targeted analysis allowed to discriminate thyme samples on a chemotype basis (Pieri, Sturm, Seger, Franz, & Stuppner, 2012).

3.3.3 Ecotoxicological and environment effect

^{1}H-NMR fingerprinting has been used to characterize the effect of environment on plant composition. For instance in biotic environments, it has been used over the past few years to identify candidate compounds for host plant resistance in several plant species (Browne & Brindle, 2007; Leiss, Choi, Verpoorte, & Klinkhamer, 2011; Lima et al., 2010), study rice *Magnaporthe grisea* interaction (Jones et al., 2011), study metabolic changes during disease following infection of *A. thaliana* by *Pseudomonas syringae* (Ward, Forcat, et al., 2010) and characterize rice plant hopper infestation (Liu et al., 2010). In abiotic environments, ^{1}H-NMR fingerprinting has been used to characterize the terroir or microclimate effect on grape berries (Pereira, Gaudillere, Pieri, et al., 2006; Pereira, Gaudillere, van Leeuwen, et al., 2006), characterize salt stress effect in maize (Gavaghan et al., 2011), study drought stress in pea (Charlton et al., 2008), study osmotic stress in *Thellungiella* (Lugan et al., 2010) or cadmium exposure effect in *Silene* (Bailey, Oven, Holmes, Nicholson, & Zenk, 2003). ^{1}H-NMR fingerprinting has been used on the model organism *Lemna minor* to characterize the effect of phytotoxic substances in the context of ecotoxicological studies (Aliferis, Materzok, Paziotou, & Chrysayi-Tokousbalides, 2009; Broyart et al., 2010).

3.3.4 Functional genomics and genetics

In the context of functional genomics, ^{1}H-NMR fingerprinting has been used for the characterization of mutants (Broyart et al., 2010). ^{1}H-NMR fingerprinting contributed to substantial equivalence studies of GMOs of different species, including *A. thaliana*, tomato, potato, lettuce, maize and grapes (Beale & Sussman, 2011; Defernez et al., 2004; Le Gall et al., 2003; Noteborn, Lommen, van der Jagt, & Weseman, 2000; Picone et al., 2011; Sobolev et al., 2010). In the context of functional genomics of opium poppy, ^{1}H-NMR fingerprinting associated with targeted profiling highlighted the interplay between primary and secondary metabolism in cultured cells treated with a fungal elicitor (Zulak et al., 2008). ^{1}H-NMR profiling of rice seeds has contributed to the exploration of the correlation between metabolic and genomic diversity in rice (Calingacion et al., 2012; Mochida, Furuta, Ebana, Shinozaki, & Kikuchi, 2009) in order to understand the genetic bases of phenotypic variation among natural populations. Mochida and colleagues selected 18 accessions from the world rice collection and genotyped 128 restriction fragment length polymorphism markers to calculate the genetic distance among the accessions (Mochida et al., 2009). In parallel, the accessions were phenotyped using metabolic fingerprinting of seed extracts. The data presented the first correlative relationship for the chromosomal distribution of each NMR metabolic profile in rice. In three commercially relevant waxy rice cultivars, complementary metabolomics and mineral platforms were combined with genome-wide genotyping at 1536 single nucleotide polymorphism loci to determine the extent of biochemical and genetic diversity (Calingacion et al., 2012). For each metabolomics platform, including ^{1}H-NMR, metabolomics diversity was highly associated with genetic distance between the varieties, demonstrating that multiple complementary metabolomics platforms have potential as phenotyping tools to assist rice breeders to combine key yield and quality characteristics. For modelling in a systems biology approach, absolute quantification is needed; therefore, fingerprinting is not the golden standard, but metabolomics profiling is recommended.

4. METABOLOMICS NMR PROFILING AND SPECTRA ANNOTATION

In contrast to fingerprinting metabolomics, where only chemical patterns of the analytical machine output are used, profiling metabolomics relies on the simultaneous chemical identification and characterization of numerous metabolites from biological samples. The detailed metabolite profiling of thousands of plant samples has been a recurrent goal for a whole community of scientists for the past 15 years. However, complete and comprehensive analysis of metabolome at high throughput has been difficult to achieve simultaneously due to the wide diversity of metabolites in plants and the difficulty to handle thousands of plant samples with sophisticated technologies.

Analysis of metabolites has long been a key component in life and food science research, regardless of the organism investigated. Conventional approaches to metabolite analysis were laborious, time consuming, requiring elaborate sample preparation and multiple procedures for a limited number of identified compounds. Equally important, only predetermined metabolites could be analyzed through these conventional methods, while unexpected or unknown metabolites were easily overlooked. In recent years, many significant advances have been made in the area of analytical technology development. The real breakthrough came with the advent of high-performance gas and liquid chromatography coupled to high-resolution MS to analyze crude plant extracts after double derivatization. Our capacity to simultaneously analyze a chemically diverse range of metabolites in a complex mixture has never been greater. For NMR spectroscopy, there have been continuous and spectacular improvements in super-conducting materials, stable and robust electronic equipment allowing NMR technology to become a robust and reproducible analytical technique for metabolomics. Analysis of metabolites in complex mixtures by 1D- or 2D NMR analysis has been explored long before the metabolomics concept had been invented (Fan, 1996; Fan, Higashi, Lane, & Jardetzky, 1986; Shachar-Hill, Pfeffer, & Germann, 1996).

4.1. 1D spectra annotation by using spectra libraries of pure compounds

Although NMR has been used as an analytical tool to identify metabolites in biological samples for many years, multivariate analysis of ^{1}H-NMR spectra is a relatively recent development. Extracting meaningful information from complex spectroscopic data of metabolite mixtures is an area of active research in the field of metabolomics. NMR signals are characterized by their frequency (chemical shift), intensity and magnetic relaxation properties indicating the precise information on molecular environment of the detected nucleus. This information can be used to identify molecules in the sample. Assigning the molecular origins of the observed resonances (chemical shift) is the key to interpreting the spectra. In NMR, the exact chemical shift value of resonances gives a reasonable indication of the molecule identity. However, depending on the nucleus, the molecular assignment ranges from trivial, for example, in ^{23}Na and ^{39}K-NMR where the resonances can come only from the corresponding cations, to very difficult, for example, in ^{1}H-NMR where there are many potential contributors to the spectra. In addition, many resonances at a given chemical shift can be split into two, three or more sub-peaks, called, respectively, doublets, triplets and multiplets in ^{1}H-NMR. This splitting generates a characteristic spectral pattern for each molecule. In the past, final assignment was made only with knowledge of the composition of the tissue based on an analysis of tissue extract by conventional analytical techniques such as GC or LC-MS (Fan et al., 1986). Today, a number of NMR spectral commercial and open source databases have been set up to help this metabolite assignation (see below for resources and data standards for NMR metabolomics). The open source databases are Web libraries of standard reference spectra of pure compounds. Because each molecule has a characteristic spectral pattern, a simple comparison of unidentified peaks with standard reference spectra often allows identification of the metabolite in a sample. A usual laboratory practice is to record ^{1}H-NMR spectrum of an extract before and after the addition of a small quantity of the considered molecules, if commercially available, in order to determine whether the molecule is present in the extract. Because the differences in chemical shift are relatively small for ^{1}H-containing biological molecules, and ^{1}H-NMR spectra are often cluttered with many overlapping resonances, this procedure can be of great help but is not an absolute demonstration of the presence of the molecules in the sample. To resolve the problem of identification of overlapped resonances in complex extract, 2D NMR methods have been applied in many ways, that is, 2D-J-Resolved (Ludwig & Viant, 2010).

1D ^{1}H-NMR spectroscopy remains a leading analytical technology in metabolomics, including for food and plant sciences. This approach is indeed relatively fast and robust providing a direct measure of metabolite concentration. However, the full potential of ^{1}H-NMR metabolomics cannot be achieved without accurate identification and quantification of metabolites in complex biological samples.

4.2. Spectra annotation using different multidimensional NMR pulse sequences

The versatility of NMR as analytical technique is greatly increased by many options that exist for manipulating the detected signals. 2D NMR spectroscopy retains many benefits of 1D NMR and offers better signal resolution. Indeed, the main advantage comes from the fact that 2D NMR spreads the overlapping resonances into a second dimension, reducing congestion and increasing metabolite specificity (Kim et al., 2011). However, the long acquisition times for most 2D spectra and their multi-impulsional character have been a limitation for the implementation of 2D NMR spectroscopy in high-throughput metabolomics. For example, a typical 2D TOCSY spectrum (*TO*tal *C*orrelation *S*pectroscop*Y* is a homo-nuclear experiment generally used for protons) of a biological sample requires around 18 h to provide the same sensitivity as compared to 1D spectrum obtained in only few minutes (Sobolev, Segre, & Lamanna, 2003; Tang, Wang, Nicholson, & Lindon, 2004).

4.3. Resources and data standards for NMR metabolomics

The use of NMR technology for metabolite identification and quantification depends on NMR libraries of pure metabolite spectra stored in databases accessible from the Web (Banimustafa & Hardy, 2012). These databases include the Spectral Database for Organic Compounds (SDBS Web site at http://riodb01.ibase.aist.go.jp/sdbs/ from the National Institute of Advanced Industrial

Science and Technology) and the Platform for Riken Metabolomics (http://prime.psc.riken.jp; PRIMe) in Japan, the Birmingham Metabolite Library Nuclear Magnetic Resonance database (BML-NMR, a publicly accessible database of 1-D ^{1}H and 2-D ^{1}H J-resolved NMR spectra of authentic metabolite standards) in the United Kingdom (Ludwig et al., 2012), the Human Metabolome Database (HMDB) in Canada (Wishart et al., 2009, 2007), the Madison Metabolomics Consortium Database, the Carbon TOCSY NMR Metabolomics Database (TOCCATA, http://spinportal.magnet.fsu.edu/toccata/webquery.html), and the BioMagResBank (BMRB) for sequence-specific protein NMR data (MMCD) in the United States (Bingol, Zhang, Bruschweiler-Li, & Brüschweiler, 2012; Cui et al., 2008; Seavey, Farr, Westler, & Markley, 1991), the NMR metabolomics database of Linkoping in Sweden (Lundberg et al., 2005), and the Metabolomic Repository Bordeaux (Mery-B), a database specialized for plants in France (Ferry-Dumazet et al., 2011). They contain extensive resources in the form of experimental 1D ^{1}H-NMR, 1D^{13}C-NMR, 2D ^{1}H, ^{1}H TOCSY, 2D JRES, 2D ^{1}H, ^{13}C HSQC, etc. of pure compounds or complex extracts from biological samples. Not all of these online databases provide graphical support for peak filtering, processing, comparative display or annotation.

Today, the tendency is to develop software tools or Web-based tools that are able to use these databases for automatic or semi-automatic identification of metabolites in complex biological extracts from NMR spectra. Mery-B, a platform devoted to plants manages all the data generated by NMR-based plant metabolomics experiments, from description of the biological source to identification of the metabolites and determinations of their concentrations. It is the first database allowing the display and overlay of NMR metabolomics profiles selected through queries on data or metadata for plants. This database contains lists of plant metabolites, mostly primary metabolites and unknown compounds, with information about experimental conditions, the factors studied and metabolite concentrations for 16 different plant species (Arabidopsis, Broccoli, Daphne, Grape, Maize, Barrel Clover, Melon, Ostreococcus tauri, Palm date, Palm tree, Peach, Rice, Strawberry, Sugar beet, Tomato, Vanilla), compiled from more than 2000 annotated NMR profiles for various organs or tissues deposited by 26 different private or public contributors on March 2012 (Deborde & Jacob, 2013; Ferry-Dumazet et al., 2011). MetaboMiner (Xia, Bjorndahl, Tang, & Wishart, 2008) and MetaboLab (Ludwig & Günther, 2011) are two good examples of softwares, which are able to handle^{1}H,^{1}H TOCSY and^{1}H,^{13}C HSQC data to identify compounds by comparing 2D spectra patterns in the

NMR spectrum of a biofluid mixture with specially constructed libraries containing reference spectra of pure molecules. These programmes greatly facilitate the process of metabolite identification in complex 2D NMR spectra of complex mixture such as biofluids (Xia et al., 2008). MetaboAnalyst 2.0 is a new comprehensive Web server for metabolomics data analysis. It provides a user-friendly, Web-based analytical pipeline for high-throughput metabolomics studies with a variety of commonly used procedures for metabolomics data processing, normalization, multivariate statistical analysis, as well as data annotation (Xia, Mandal, Sinelnikov, Broadhurst, & Wishart, 2012). Commercial databases and specialized NMR-based metabolomics tools from Bruker, Bio-Rad and Chenomx are available and facilitate automatic peak processing, rapid compound identification and spectrum annotation. Chenomx NMR suite (V 4.6), a software package capable of identifying and quantifying individual compounds based on their respective pattern spectra, has been used in a metabolic study in elicitor-treated opium poppy cell cultures (Zulak et al., 2008). In this study, 148 analytes were detected and 42 metabolites were quantitatively monitored by ^{1}H-NMR over a 100-h time course experiment. A high-resolution map of primary and secondary metabolism was built describing the response of the opium poppy cell cultures to elicitor treatment. Other developments in metabolomics field aim at providing Web portals with detailed protocols and tutorials on conducting plant metabolomics experiments to promote metabolomics in the community and newsletter to keep the communities informed. PlantMetabolomics.org is a good example of Web portal and database for exploring, visualizing and downloading MS-based plant metabolomics data (Bais et al., 2010). The Metabolomics Innovation Centre (TMIC) produces newsletters, which intend to keep worldwide metabolomics researchers and other professionals informed about new technologies, software, databases, events, job postings, conferences, training opportunities, interviews, publications, awards and other newsworthy items concerning metabolomics (http://www.metabonews.ca/).

These repository databases enormously increase the amount of data that can be accessed to extract new information and decrease the number of additional experiments to perform. Data mining to extract relevant information from complex data set is becoming one of the most challenging field in biology (de Almeida Vieira Lima, Nunes, da Silva Dias, & Marques, 2012). Standards are critical to facilitate the reporting and archiving of experimental results, including efficient querying, accurate data exchange and subsequent re-use in the analysis of data from multiple experiments. Without standardized repository databases, there will be no way to benefit from all this valuable information. The importance of data reporting standards in metabolomics has been mentioned very early by metabolomics pioneers and has led to the formation of the Metabolomics Standards Initiative (MSI; Bino et al., 2004; Fernie et al., 2011; Fiehn et al., 2007; Jenkins et al., 2004, 2005). Under the MSI sponsoring, minimum reporting standards for chemical analysis by a range of analytical techniques have been proposed by Sumner et al. (2007) and Goodacre et al. (2007) and detailed specifically for NMR-based metabolomics experiments by Rubtsov et al. (2007). An NMR-ML standard data format is under development in order to facilitate NMR spectra exchange (see http://cosmos-fp7.eu), as done for MS spectra with the mzML format (Martens et al., 2011).

Standardization is not only needed for analytical data; it is also important to have standardized descriptions of the genetic material, the growth conditions through environmental conditions and the experimental design (Gibon & Rolin, 2012). The latter is referred to as "metadata" (data about data). The generation of high-quality molecular and physiological data depends on the experiment being correctly designed, and interpretation of the data depends on the experiment being adequately described. Because plant phenotypes typically result from complex interactions between the genotype and a highly dynamic environment, it is crucial to access machine-readable metadata that describe all environmental conditions during the plant growth as well as sampling times and analytical acquisition data. All plant researchers agree on the usefulness of this metadata. But there are few tools that allow linking the metadata in a machine-readable form with analytical data. Xeml Lab is a tool that supports in a user-friendly interface mode the design of experiments with a graphical interface and generates computer-readable metadata files, which capture information about genotypes, growth conditions, environmental perturbations and sampling strategy (Hannemann et al., 2009). This information technology tool could provide an efficient solution to handle the complex metadata describing plant growth conditions and to link with "omics" data such as metabolomics.

5. APPLICATION OF ^{1}H-NMR PROFILING

5.1. A wide range of applications as evidence of success

In a few years, the ability of high-resolution ^{1}H-NMR spectroscopy to discriminate between numerous compounds in complex solutions has found many new applications in plant science from fundamental research to applied biotechnology. At the beginning of the XXI century, metabolomics appears as a potential functional genomics tool allowing the link between genotypes and phenotypes (Fiehn, 2002; Hall et al., 2002) as well as a new tool to discover biomarkers and to analyze gene function (Hall, 2011a; Hall, Vos, Verhoeven, & Bino, 2005; Steinfath et al., 2010). Today, all fields of biological science have used metabolomics tools and concepts to explore their domain ranging from metabolism exploration (Kusano et al., 2011; Sanchez et al., 2012) to bioengineering strategy (Liu, Zhu, & Jiang, 2009), from functional genomics (Bino et al., 2004; Carrari et al., 2006; Mounet et al., 2009) to systems biology (Saito & Matsuda, 2010; Wang, Wu, Chen, Chen, & Zhao, 2006), from food authenticity (Cuny et al., 2008; Schievano et al., 2012; Vogels et al., 1996) to quality control in food science and technology and nutritional sciences (Davies, 2010; Scalbert et al., 2009), from plant population screening (Schauer et al., 2008; Steinfath et al., 2010) to chemical biodiversity exploration (Politi et al., 2008; Safer et al., 2011), from genetically modified organisms characterization (Ben Akal-Ben Fatma et al., 2012; Kusano et al., 2011; Le Gall et al., 2003; Ren, Wang, Peng, Xia, & Qu, 2009; Schauer et al., 2006) to plant developmental studies (Mounet et al., 2007; Urano et al., 2009), from ecotoxicology (Simpson & McKelvie, 2009; Yuk, Simpson, & Simpson, 2012) to environmental applications (Viant et al., 2009), from plant–pathogen interactions (Choi et al., 2006; Leiss et al., 2011; Lopez-Gresa et al., 2012) to plant–herbivore interactions (Jansen et al., 2009). In most of these studies, ^{1}H-NMR metabolomics profiling has been used to characterize specific networks, including those associated with biotic and abiotic stresses (Gavaghan et al., 2011; Sanchez et al., 2011), primary and secondary metabolism (Kirk, Cheng, Choi, Vrieling, & Klinkhamer, 2012) such as flavonoids (Schutz, Quitschau, Hamburger, & Potterat, 2011; Timmers & Urban, 2012), phenylpropanoids (Wolfram et al., 2010) and alkaloid biosynthesis (Gathungu et al., 2012; Hagel, Weljie, Vogel, & Facchini, 2008), and plant products of medical relevance (Gad, El-Ahmady, Abou-Shoer, & Al-Azizi, 2013; Ioset et al., 2011; Sharma, Cardoso-Taketa, Choi, Verpoorte, & Villarreal, 2012; Yang, Cheng, Yang, & Guo, 2012).

5.1.1 *Application of ^{1}H-NMR profiling in functional genomics and plant breeding*

The strategy of globally comparing outcomes of gene expression at different molecular levels (i.e. transcriptomics and proteomics) has been at the heart of functional genomics (Bino et al., 2004; Fernie & Keurentjes, 2011; Fernie & Stitt, 2012; Redestig et al., 2011; Sumner et al., 2003). Because there is an increased reliance on genetically modified organisms as a functional genomic tool to elucidate the role of genes and their protein products in the phenotype, mutant or transgenic plants have been used to understand how the transcriptional or proteomic network may conspire to alter the expected phenotype (Griffin, 2004). ^{1}H-NMR metabolomics profiling has provided a useful technique to identify qualitative changes in major metabolites according to genotype alteration, and many studies have been conducted with transgenic plants. Metabolism of genetically modified lettuce (*Lactuca sativa* L.) leaves was investigated by comparing NMR metabolic profiles of wild type (WT) and overexpressing *E. coli* asparagine synthetase A lines (Sobolev et al., 2010). MS and NMR metabolomics techniques have also been applied to differentiate WT versus transgenic organisms in tobacco (Choi et al., 2004), potatoes (Defernez et al., 2004), tomato (Le Gall et al., 2003; Mattoo et al., 2006), peas (Charlton et al., 2004), maize (Broyart et al., 2010; Castro, Motto, Rossi, & Manetti, 2008; Piccioni, Capitani, Zolla, & Mannina, 2009) and albino *A. thaliana* mutants (Tian et al., 2007).

Numerous plant breeders have long wished for identified metabolic biomarkers for the prediction of crop product quality. Today, discovering plant metabolic biomarkers for phenotype prediction has become a reality (Fernie & Schauer, 2009). The procedure consists in established techniques such as metabolomics profiling to screen untargeted for a large amount of metabolites in combination with machine learning methods (Steinfath et al., 2010). This procedure has been successfully applied to potato tubers, but it should be applicable to any crop. In a wheat breeding programme for resistance to fusarium, five commercial wheat cultivars and 116 wheat genotypes have been explored by ^{1}H-NMR spectroscopy. Biochemical correlation of partial resistance to fungal species implicated in the fusarium head blight disease complex was identified by NMR fingerprinting, while metabolites related to increased disease resistance were characterized by ^{1}H-NMR profiling (Browne & Brindle, 2007).

Hybridization between plant species can have a number of biological consequences, and inter-specific hybridization can be tied to speciation events, biological invasions and diversification at the level of genes, metabolites and phenotypes. The changes of primary and secondary metabolisms in hybrids with transgressive segregation between *Jacobaea vulgaris* and *J. aquaticus* (Kirk et al., 2012) and *Senecio aquaticus* and *S. jacobaea* (Kirk, Choi, Kim, Verpoorte, & van der Meijden, 2005) have been explored by

^{1}H-NMR profiling. For *Senecio* hybridization study, NMR profiling did not detect any novel compounds among the tested hybrid compared to parent plants, while for *Jacobaea* hybridization study, a number of F2 hybrid genotypes exhibited metabolomics profiles that were outside the range encompassed by parental species. The authors concluded that metabolomics profiling is a useful technique for identifying qualitative changes in major metabolites according to plant species and/or genotype, but is less useful for identifying small differences between plant groups, or differences in compounds expressed in low concentrations (Kirk et al., 2012, 2005).

5.1.2 Following metabolic adaption in plant submitted to stress conditions

Profiling metabolomics gives a biochemical snapshot of an organism at a given time, showing which metabolites are present and in what quantities. Considering metabolites as the result of the interaction of the systems genome with its environment and as a part of the regulatory system, profiling metabolomics appears as an integrative tool to identify metabolite patterns or markers that are characteristic for a physiological status impacted by biotic or abiotic stress such as drought and salinity stress, etc. As references are numerous in the literature, we will focus on recent articles where ^{1}H-NMR profiling metabolomics have been used to characterize plant adaption to biotic or abiotic stress conditions.

High salinity and drought stress, caused by either natural (e.g. climatic changes) or anthropic factors (e.g. cultural practices in agriculture), are widespread environmental stressors that can affect development and growth of agronomic plants. The application of NMR profiling metabolomics to the investigation of saline-induced or drought stress provides a powerful tool to identify specific adaptations of primary and secondary metabolisms (Ruan & Teixeira da Silva, 2011; Sanchez, Siahpoosh, Roessner, Udvardi, & Kopka, 2008; Sanchez et al., 2011). Marked changes in several primary metabolites were observed using ^{1}H-NMR profiling in maize (Gavaghan et al., 2011), in kiwi fruits (Capitani et al., 2010) and in drought tolerant and sensitive soybean genotypes (Silvente, Sobolev, & Lara, 2012). The metabolomes of *Oryza sativa* ssp. *japonica* cv. M202 and cv. M202 (Sub1) were profiled using ^{1}H-NMR to compare the metabolic effect of submergence stress and recovery on rice in the presence or absence of SUB1A, a gene that helps to stabilize crop production in submergence-prone fields (Barding et al., 2012). Significant changes were observed in the NMR resonances of compounds in pathways important for carbohydrate metabolism. The presence of SUB1A in M202 (Sub1) was correlated with suppression of carbohydrate metabolism in shoot tissue, consistent with the role of SUB1A in limiting starch catabolism to fuel elongation growth. The identification and characterization of an unusual di-amino acid alanylglycine is also reported in this rice genotype (Barding et al., 2012).

5.1.3 Application of ^{1}H-NMR profiling in production of grapevines and wines

The numerous uses of the grapevine fruit, especially for wine and beverages, have made it one of the most important plants worldwide. The phytochemistry of grapevine and wine is rich in a wide range of compounds (Hong, 2011). Many of them are renowned for their numerous medicinal uses (van Dorsten et al., 2010). The production of grapevine metabolites is highly conditioned by many factors like environment or pathogen attack. Grapevine is a good example to show how ^{1}H-NMR spectroscopy (fingerprinting and profiling) have been used to follow and to characterize plant development or plant adaption to biotic or abiotic stress conditions (Fortes et al., 2011). Environmental vineyard conditions can affect the chemical composition or metabolites of grapes and their wines. ^{1}H-NMR fingerprinting with pattern recognition methods was used to investigate the metabolic differences in pulp, skin and seed of grapes grown in different regions (Pereira, Gaudillere, Pieri, et al., 2006; Pereira et al., 2005; Pereira, Gaudillere, van Leeuwen, et al., 2006; Son et al., 2009). Wine contains a number of metabolites that are produced during alcoholic and malolactic fermentations (MLFs) or ageing, which are important compounds for determining wine quality. Metabolomic characterization of MLF and fermentative behaviours of wine yeasts in grape wine can be monitored by ^{1}H-NMR (Howell, Cozzolino, Bartowsky, Fleet, & Henschke, 2006). ^{1}H-NMR fingerprinting allowed to differentiate clearly between non- and induced-malolactic fermented wines by wine lactic acid bacteria and between wines fermented with various wine yeast strains. While ^{1}H-NMR profiling allowed identifying metabolites responsible for the differentiation of induced MLF wines from non-MLF wines (Son et al., 2009), chemical profiles of wines produced from different grape varieties and production areas were obtained by ^{1}H-NMR and conventional physico-chemical analyses (Caruso et al., 2012; Lee, Hwang, et al., 2009; Lee, Shaykhutdinov, et al., 2009; Son et al. 2008). Fungi infection such as *Botrytis cinerea* of grape berries leads to changes in the chemical composition of grape and the corresponding wine and, thus, affects wine quality. The metabolic effect of *Botrytis* infection in grape and Champagne wine was investigated through a combination of fingerprinting and profiling ^{1}H-NMR metabolomic approach (Hong et al., 2011). Metabolic differentiations between healthy and botrytized wines were identified by ^{1}H-NMR fingerprinting by using multivariate statistical analysis such as PCA, while ^{1}H-NMR profiling listed the main discriminant metabolites induced by *Botrytis* infection and responsible for the fermentative retardation during alcoholic fermentation.

5.1.4 Application of ^{1}H-NMR profiling for characterization of medicinal plants

Herbal medicines and their preparations have been widely used for many years all over the world. However, they have not been officially recognized due to a lack of adequate or accepted research methodology for composition certification. Today, chemometrics and analytical tools have tried to solve the discrimination of many herbs from closely related species and from adulterants, based on the principal bioactive components and phytochemical diversity (Gad et al., 2013; Sharma et al., 2012; Yang et al., 2012).

In most studies, ^{1}H-NMR- and MS-based metabolite profiling are used to characterize medical plant extracts that contain bioactive molecules with many pharmacological properties such as anti-inflammatory, anti-conceptive, anti-oxidant and anti-cancer activity (Kim, Choi, et al., 2010; Kim, Wilson, et al., 2010). Metabolomics is seen as a tool that can improve the efficiency of lead-finding from natural products and thus reinstate this prolific source of potential new drugs for medicine. Elicited *Eschscholzia californica* (California poppy) cell cultures were used as alternative to whole plants for obtaining natural products of medical relevance. A rapid assay for profiling cell culture using a micro-scale LC-UV-MS-NMR platform was set up to follow alkaloid production. Eight benzophenanthridine alkaloids were identified at the sub-microgram level. This chapter demonstrates the utility of the nanosplitter LC-MS/microdroplet NMR platform when establishing cell culture expression systems (Gathungu et al., 2012). Another example is *Rhodiola rosea*, which is a broadly used medicinal plant with largely unexplored natural variability in secondary metabolite levels. ^{1}H-NMR fingerprinting of rhizome extracts was applied for pattern recognition analysis and ^{1}H-NMR profiling for identification of secondary metabolites responsible for differences in sample composition from three different geographic areas (Ioset et al., 2011). *Cimicifugae Rhizoma*, a well-known botanical dietary supplement, has been the subject of intense interest due to its potential application for alleviating menopausal symptom. The main triterpenoid of the rhizome is cimicidol-3-O-β-d-xyloside (CX), its toxicological effects have been studied in rats by ^{1}H-NMR (He et al., 2012).

5.1.5 Application of ^{1}H-NMR in food authenticity and quality control

The requirements of reliability, expeditiousness, accuracy, consistency and simplicity for quality and safety assessment of food products encouraged the development of non-destructive technologies to meet the demands of consumers to obtain superior food qualities (Elmasry, Kamruzzaman, Sun, & Allen, 2012; Koehn, 2008). ^{1}H-NMR spectroscopy is one of the techniques currently investigated for quality evaluation purposes in numerous sorts of applications. ^{1}H-NMR can be used in food analysis for origin discrimination and biomarker discovery. The botanical origin of honey has been investigated by ^{1}H-NMR. 353 honey samples of seven floral origins have been used. Specific markers were identified for each monofloral origin (Schievano et al., 2012). A same type of study has been carried out to investigate the metabolite content of roasted *Coffea arabica* samples from the three main production areas, America, Africa and Asia (Consonni, Cagliani, & Cogliati, 2012). Quality assessment of tea has been the focus of many research teams (Le Gall et al., 2004; van der Hooft, Akermi, et al., 2012; van der Hooft, de Vos, et al., 2012). Advanced analytical approaches consisting of both LC-LTQ-Orbitrap FT-MS and LC-time-of-flight-MS coupled to SPE-NMR were used to obtain more insight into the complex phenolic composition of tea. 177 phenolic compounds were annotated from 17 commercially available black, green and white tea (van der Hooft, Akermi, et al., 2012a).

In agriculture production, quality assessment has been the focus of an important increase of attention. Quantitative ^{1}H-NMR metabolic profiling of strawberry grape has been applied to determine molecular composition contents of the methanolic extracts of peel, pulp, seed, leaf and stalk components of the plant and evaluate anti-oxidant and anti-proliferative properties in human cell lines (Pacifico et al., 2011). In order to establish quality control code protocol based on biochemical phenotype of ginger, metabolite profiling and fingerprint analysis by ^{1}H-NMR spectroscopy were used to identify potential biomarkers capable of distinguishing different ginseng species, varieties and commercial products. Glutamine, arginine, sucrose, malate and myo-inositol are the major metabolites in ginseng samples, but the combination of metabolite fingerprinting and profiling suggested that several compounds including glucose, fumarate and various amino acids could serve as biomarkers for quality assurance in ginseng (Lee, Hwang, et al., 2009; Lee, Shaykhutdinov, et al., 2009). ^{1}H-NMR profiling have been applied also to fleshy fruit for quality assessment during fruit development or after harvest: tomato (Deborde et al., 2009; Mounet et al., 2007), melon (Biais et al., 2009), watermelon (Tarachiwin, Masako, & Fukusaki, 2008), citrus (Slisz, Breksa, Mishchuk, McCollum, & Slupsky, 2012), mango (Gil et al., 2000; Koda, Furihata, Wei, Miyakawa, & Tanokura, 2012), cherry (Clausen, Pedersen, Bertram, & Kidmose, 2011), peach (Nakatani et al., 2000) and apple (Cuthbertson, Andrews, Reganold, Davies, & Lange, 2012; Karaman, Tutem, Baskan, & Apak, 2013).

6. NEW CHALLENGES FOR ^{1}H-NMR METABOLOMICS

Metabolomics is a demanding field in terms of equipment and competences. Technological advances in large-scale analytical tools (e.g. NMR-, and MS, based technologies), appropriate computer systems and mathematical tools have pushed the development of metabolomics during the past 10 years forward. The future use of ^{1}H-NMR metabolomics is linked to the developments in NMR spectroscopy that provide improvement in detection, higher sensitivity and reliability, quantitative and accurate data. The future use of ^{1}H-NMR metabolomics will also depend on the capability to integrate ^{1}H-NMR data to other metabolomics information provided by other analytical tools.

6.1. Technological challenges

6.1.1 Cryogenically cooled probes and microprobes

NMR probes with coil and pre-amplifier cryogenically cooled to 25 K with helium have provided a new hope for metabolomics with their increase in S/N ratio by a factor of 4–5 compared to conventional RT probes (Kovacs, Moskau, & Spraul, 2005). This technology has been widely applied to metabolomics (Griffin, 2003). For structural determination of purified metabolite in small or limited amounts, NMR data acquisition times have been shortened by a factor of 25 with such cryoprobes (Lindon & Nicholson, 2008). However, these probes also have some disadvantages such as an expensive cost of maintenance and the fact that such probes require cool-down and warm-up 0.5-day process; therefore this equipment could not be installed and de-installed as rapidly as RT probes and often needs to be used with a dedicated spectrometer. Moreover, a reduction in the S/N ratio has been observed in the case of aqueous solutions containing salt, such as plant-buffered extracts, making the use of cryogenic probes less appealing for plant metabolomics routine analysis (Barding et al., 2012; Mannina, Sobolev, & Viel, 2012). The versatility of the NMR platform is then limited. But recently, Bruker proposed a new type of cryoprobe with nitrogen-cooled RF coils and pre-amplifiers, which delivers a very high-performance boost (sensitivity enhancement of a factor of 2.5 corresponding to a significant "300 MHz-equivalent" boost in sensitivity) with minimum operating and maintenance costs. This innovation should be rapidly spread to metabolomics platforms and increase the attractiveness of NMR approach for metabolomics.

The other avenue of probe sensitivity improvement in the future involves the development of micro-coil technology (Grimes & O'Connell, 2011; Olson et al., 2004) and is generally used. Within these probes, the sample volumes are down in the micro litre range. This small size enables the use of the more sensitive transverse solenoid coil design, which yields a significant sensitivity increase. Sensitivity gains are also made possible in these probes by having the receiver coil physically closer to the sample. This coil design increases the amount of sample that is within or very near the detection coil, called the observed factor. It has been determined that the mass sensitivity (sensitivity of the probe to receiving signals from a defined number of nuclei) is approximately 10-fold greater than with a conventional NMR probe (Olson et al., 2004). Microcryoprobe combines the advantages of cryoprobes and microprobes and can be very useful to increase the number of metabolites that can be identified by ^{1}H-NMR in plant extracts (Molinski, 2010). For a given amount of material, a 14-fold gain over a conventional 5-mm probe is obtained.

6.1.2 HRMAS-NMR for intact plant sample studies

HRMAS (High-Resolution Magic Angle Spinning) NMR derives from the well-known solid-state MAS technology. However, HRMAS-NMR is a technique dedicated to heterogeneous samples with relative molecular mobility. HRMAS-NMR is well suited for pre-clinical and clinical applications, where biopsies are usually already sampled for histopathology. This is the reason why most of HRMAS studies concern biomedical applications (Lindon, Beckonert, Holmes, & Nicholson, 2009), mainly in the field of cancer (Moestue, Sitter, Bathen, Tessem, & Gribbestad, 2011) and neuroscience research.

Classical NMR spectra of heterogeneous samples are composed of broad resonances, mainly due to anisotropy of magnetic susceptibility. This effect can be averaged by spinning at the so-called magic angle of 54.7° relative to the static magnetic field B0. Low spin rates are sufficient to obtain highly resolved spectra (Kiefer, Baltusis, Rice, Timiak, & Shoolery, 1996; Lippens et al., 1999), but higher spinning rates are needed to displace spinning side bands outside the spectral range of interest, that is, 3 KHz at 400 MHz proton frequency.

HRMAS-NMR allows obtaining highly resolved NMR metabolic profiles of intact rough samples, of small size (approximately 3–20 mg), avoiding the extraction step. Most of the proton acquisitions for metabolomics studies are performed using CPMG sequences in order to minimize lipid and macromolecule contributions, and using presaturation for water suppression (Beckonert et al., 2010). Conversely, diffusion-weighed sequences can be used for visualization of mobile lipids only, giving important additional information, for example, in tumoral tissues (Zietkowski et al., 2010) or in plant residues (Kelleher, Simpson, & Simpson, 2006).

The sample volume is limited by the size of zirconium rotors, usually 50 μl for 4-mm Bruker probe head, or approximately 35 μl for disposable inserts. The sampling step is then crucial, and may become a limiting step in the case of heterogeneous organs or tissues. Homogenization has been proposed to counteract this drawback. Another crucial point is tissue degradation during acquisition: HRMAS-NMR needs molecular mobility that excludes frozen samples, and a temperature of 4 °C is generally used. For rapid ^{1}H acquisition, this degradation can be neglected, whereas it is not the case for time-consuming experiments like ^{13}C experiments. The fixation of tissue with micro-waves could constitute a possible issue (Detour et al., 2011).

Finally, HRMAS-NMR is used for metabolomics studies as an analytical tool of intact samples instead of extracts, but for the same purpose. In addition, it enables the possibility to explore small sub-structures that can be difficult to extract, and to give access simultaneously to lipid and metabolic profiles in a given sample (Shintu, Ziarelli, & Caldarelli, 2004), an advantage that has not been not fully exploited so far. In a close future, HRMAS-NMR should be used with intact plant samples, mimicking *in vivo* conditions and providing key information on metabolites in plant tissues.

Indeed, less numerous but increasing applications of HRMAS-NMR concern food characterization, either for control quality and traceability evaluation, or to improve nutritional and gustative quality of food (Valentini et al., 2011). The changes during storage of smoked salmon (Castejon et al., 2010) have been investigated. The origin of parmesan or Emmental cheese (Shintu & Caldarelli, 2005, 2006), of meat (Ritota et al., 2012; Shintu, Caldarelli, & Franke, 2007) and of Italian garlic (Ritota et al., 2012) could be distinguished. In the plant kingdom, fruit material has been essentially studied: mango (Gil et al., 2000), strawberry (Otero & Préstamo, 2009) and apple pulp (Vermathen, Marzorati, Baumgartner, Good, & Vermathen, 2011). Interestingly, in this latter study, the intra-apple variability appeared as significantly lower than inter-apple variability. Different cultivars could be distinguished, even in absence of data about the ripening stage and storage conditions. To limit the sample heterogeneity, peel puree instead of intact fruit tomato has been used (Pérez et al. 2010; Sanchez Pérez et al. 2011). Structural information on biopolymer structure and organization, like cutin from tomato (Deshmukh, Simpson, & Hatcher, 2003), agave (Deshmukh, Simpson, Hadad, & Hatcher, 2005) or key lime (Fang et al., 2001) were gained using HRMAS-NMR of the polymer swollen in DMSO. HRMAS-NMR has also been carried out on intact cereal grains. The impact of storage conditions on lipid composition has been evaluated in a half barley grain (Seefeldt, Larsen, Viereck, Petersen, & Engelsen, 2011) and the impact of abiotic stress in a single kernel (Winning et al., 2009).

HRMAS-NMR application to environmental research is in constant development. In ecotoxicology, small sentinel organism, like the shellfish Daphnia magna, has been studied *in vivo* in order to evaluate the impact of xenobiotic stressors on its metabolism (Bunescu et al., 2010). However, proton HRMAS-NMR spectra of shellfish or of intact insects like fly (Righi et al., 2010) or carabe (Desmoulin, Bon, Martino, & Malet-Martino, 2008) are mainly composed of lipid resonances, which limit the potential of the method. The impact of an insecticide accumulation in water on heart clam metabolism has been studied by ^{31}P-NMR and ^{1}H-HRMAS-NMR (Hanana, Simon, Kervarec, Mohammadou, & Cerantola, 2012).

Magnetic resonance imaging and HRMAS-NMR were associated to investigate alterations of the oilseed shrub *Jatropha curcas* L. after viral infection (Sidhu et al., 2010). Stem and leaf tissue of infected plant exhibited lower concentrations of carbohydrates while amino acids and citrate accumulated. In order to investigate the process involved in biomass decomposition and plant accumulation in the environment, Kelleher (Kelleher et al., 2006) used ^{13}C and ^{15}N-enriched pine and wheat grass. Multinuclear HRMAS-NMR (^{1}H, ^{13}C and ^{15}N) in DMSO provided important results unattainable by other methods, for example, the rapid disappearance of tannins after 6 months of incubation. Identically, both fresh *alga brauuni* homogenized in DMSO inside the HRMAS rotor and the alga residue after

201 days of oxic biodegradation were analyzed by multinuclear HRMAS-NMR (Ruhl, Salmon, & Hatcher, 2011).

In conclusion, HRMAS-NMR is a very powerful method that should find increasing applications in the future in plant science, food and environmental research. However, it should be kept in mind that it requires specific attention to the sampling step and to the tissue degradation during acquisition.

6.1.3 Hyphenated NMR techniques for identification

A procedure where a separation technique is coupled with an online spectroscopic detection technology is known as hyphenated technique, for example, GC-MS, LC-PDA, LC-MS, LC-FTIR, etc. Recent advances in hyphenated analytical techniques using LC-NMR, LC-SPE-NMR or LC-MS-NMR have remarkably widened their applications to the analysis of complex biomaterials, especially natural products. This technique should also play a key role in the future in compound identification and this will lead to more annotation of the 1D NMR fingerprinting. Chapter 2 of this book reviews the recent contributions brought to the field of LC-NMR-MS to identify secondary metabolites in plants and provides a survey of recent studies of metabolic samples using this methodology.

6.1.4 Spectra annotation using fast 2D NMR pulse sequences

The long acquisition times for most 2D spectra and their multi-impulsional character have been a limitation for the implementation of 2D NMR spectroscopy in metabolomics. To solve this problem, numerous methods have been published that describe how to record 2D NMR spectra much faster (Ludwig & Viant, 2010; Ludwig et al., 2009). Ultrafast 2D NMR makes it possible to record multidimensional spectra in a single scan. This methodological breakthrough opens new ways to metabolomics and clearly represents a new key methodology within the metabolomics toolbox. Chapter 4 of this book reviews the recent contributions brought to the field of quantitative 2D NMR, and provides a survey of recent studies of metabolic samples using this methodology.

6.2. Integrating ^{1}H-NMR with other metabolomics and omics technologies

The ultimate goal of plant biology is the integration of data acquired from living plants at gene, protein and metabolite levels. With the possibility to combine data from transcriptomics, proteomics and metabolomics, it is hoped that improved understanding of plant functioning will result. But we still have a long way to go before reaching this goal. As mentioned by Redestig et al. (2011), one of the characteristic features of metabolomics is that data integration and subsequent network analysis are the major issues in the research. The challenge for metabolomics is, on the one hand, how we will integrate metabolome data from different analytical platforms (concept of intra-metabolomics integration) and, on the other hand, how we will integrate transcriptome, proteome and metabolome data (concept of intra-omics integration). Since no single technology currently available (or likely in the near future) is able to detect all compounds found in plants, it becomes critical to provide a combination of multiple analytical techniques in metabolomics platforms. Multidisciplinary and multiparallel approaches are required and these must be performed in a cohesive and coordinated manner in the future.

The challenge for ^{1}H-NMR metabolomics is to continue to be a key method for high-throughput comparative analysis of plant extracts by providing accurate and reproducible data sets in a standardized way in order to be exchanged between platforms and laboratories, and be reanalyzed in meta-analyses. It is clear that the now standard ^{1}H-NMR fingerprinting or profiling strategies will continue to provide large data sets. But we can expect that hyphenated NMR techniques and fast 2D NMR techniques will boost compound identification and will lead to more annotation of 1D NMR spectra. Considering the easiness to produce absolute quantitative data with ^{1}H-NMR, the intra-metabolomics integration will be easy to set up.

Plants present a very large biochemical diversity and plant metabolomes are complex and difficult to explore. Metabolites are involved in nutrient storage as well as in structural, defensive or signalling roles. In the future, analytical challenges will be concerned by the spatial and temporal variation in metabolite measurements. It is crucial that scientists develop methods to go beyond taking snapshots of metabolite levels of mixtures provided by different plant cells or subcellular compartments. For this type of analysis, an increasing use of MS-based analysis is expected, especially in targeted studies aimed at specific groups of metabolites.

7. CONCLUDING REMARKS

Although compromised to some extent by it sensitivity, ^{1}H-NMR spectroscopy has become a key method in plant metabolomics. Initially considered as a method of choice for natural product structure determination, now it has been recognized as a robust and reproducible technique in plant metabolomics. NMR spectra contain a wealth of information about the identity of the molecules and can be used to quantify the molecules in a targeted or untargeted way. Today, ^{1}H-NMR is used as a fast, convenient and effective tool to discriminate between groups of related plants or food samples when used in the fingerprinting mode, while between 30–50 metabolites can be identified in an unfractionated extract in the profiling mode. The challenge of this technique will be to increase its sensitivity in order to remain competitive compared to MS-based technologies.

REFERENCES

Aharoni, A., & Brandizzi, F. (2012). High-resolution measurements in plant biology. *The Plant Journal*, *70*(1), 1–4.

Akoka, S., Barantin, L., & Trierweiler, M. (1999). Concentration measurement by proton NMR using the ERETIC method. *Analytical Chemistry*, *71*(13), 2554–2557.

Aliferis, K. A., Materzok, S., Paziotou, G. N., & Chrysayi-Tokousbalides, M. (2009). *Lemna minor* L. as a model organism for ecotoxicological studies performing H-1 NMR fingerprinting. *Chemosphere*, *76*(7), 967–973.

Allen, E., Moing, A., Ebbels, T. M., Maucourt, M., Tomos, A. D., Rolin, D., et al. (2010). Correlation Network Analysis reveals a sequential reorganization of metabolic and transcriptional states during germination and gene-metabolite relationships in developing seedlings of Arabidopsis. *BMC Systems Biology*, *4*, 62.

Allwood, J. W., De Vos, R. C., Moing, A., Deborde, C., Erban, A., Kopka, J., et al. (2011). Plant metabolomics and its potential for systems biology research background concepts, technology, and methodology. *Methods in Enzymology*, *500*, 299–336.

Anderson, P. E., Mahle, D. A., Doom, T. E., Reo, N. V., DelRaso, N. J., & Raymer, M. L. (2011). Dynamic adaptive binning: An improved quantification technique for NMR spectroscopic data. *Metabolomics*, 7(2), 179–190.

Antoine, J. P., Coron, A., & Dereppe, J. M. (2000). Water peak suppression: Time-frequency vs time-scale approach. *Journal of Magnetic Resonance*, *144*(2), 189–194.

Araníbar, N., Ott, K. H., Roongta, V., & Mueller, L. (2006). Metabolomic analysis using optimized NMR and statistical methods. *Analytical Biochemistry*, *355*(1), 62–70.

Bailey, N. J. C., Oven, M., Holmes, E., Nicholson, J. K., & Zenk, M. H. (2003). Metabolomic analysis of the consequences of cadmium exposure in *Silene cucubalus* cell cultures via ^{1}H-NMR spectroscopy and chemometrics. *Phytochemistry*, *62*, 851–858.

Bais, P., Moon, S. M., He, K., Leitao, R., Dreher, K., Walk, T., et al. (2010). PlantMetabolomics.org: A web portal for plant metabolomics experiments. *Plant Physiology*, *152*(4), 1807–1816.

Banimustafa, A. H., & Hardy, N. W. (2012). A strategy for selecting data mining techniques in metabolomics. *Methods in Molecular Biology*, *860*, 317–333.

Bao, Q., Feng, J., Chen, F., Mao, W., Liu, Z., Liu, K., et al. (2012). A new automatic baseline correction method based on iterative method. *Journal of Magnetic Resonance*, *218*, 35–43.

Barantin, L., Le Pape, A., & Akoka, S. (1997). A new method for absolute quantitation of MRS metabolites. *Magnetic Resonance in Medicine*, *38*(2), 179–182.

Barding, G. A., Jr., Fukao, T., Beni, S., Bailey-Serres, J., & Larive, C. K. (2012). Differential metabolic regulation governed by the rice SUB1A gene during submergence stress and identification of alanylglycine by ^{1}H-NMR spectroscopy. *Journal of Proteome Research*, *11*(1), 320–330.

Beale, M. H., & Sussman, M. R. (2011). Metabolomics of *Arabidopsis thaliana*. In R. D. Hall (Ed.), *Biology of plant metabolomics. Annual plant reviews*, Vol. 43, (pp. 157–180). Chichester, West Sussex, UK: Wiley-Blackwell.

Beckonert, O., Coen, M., Keun, H. C., Wang, Y., Ebbels, T. M., Holmes, E., et al. (2010). High-resolution magic-angle-spinning NMR spectroscopy for metabolic profiling of intact tissues. *Nature Protocols*, *5*(6), 1019–1032.

Belton, P. S., Colquhoun, I. J., Kemsley, E. K., Delgadillo, I., Roma, P., Dennis, M. J., et al. (1998). Application of chemometrics to the ^{1}H-NMR spectra of apple juices: Discrimination between apple varieties. *Food Chemistry*, *61*, 207–213.

Ben Akal-Ben Fatma, Y., Pianelli, K., Dieuaide-Noubhani, M., Le Menn, A., Deborde, C., Maucourt, M., et al. (2012). ^{1}H-NMR metabolomics: Profiling method for a rapid and efficient screening of transgenic plants. *African Journal of Biotechnology*, *11*(52), 11386–11399.

Bernillon, S., Biais, B., Deborde, C., Maucourt, M., Cabasson, C., Gibon, Y., et al. (2012). Metabolomic and elemental profiling of melon fruit quality as affected by genotype and environment. *Metabolomics*, *9*(1), 57–77.

Bharti, S., Sinha, N., Joshi, B., Mandal, S., Roy, R., & Khetrapal, C. (2008). Improved quantification from ^{1}H-NMR spectra using reduced repetition times. *Metabolomics*, *4*(4), 367–376.

Biais, B., Allwood, J. W., Deborde, C., Xu, Y., Maucourt, M., Beauvoit, B., et al. (2009). ^{1}H NMR, GC-EI-TOFMS, and data set correlation for fruit metabolomics: Application to spatial metabolite analysis in melon. *Analytical Chemistry*, *81*(8), 2884–2894.

Biais, B., Bernillon, S., Deborde, C., Cabasson, C., Rolin, D., Tadmor, Y., et al. (2012). Precautions for harvest, sampling, storage, and transport of crop plant metabolomics samples. *Methods in Molecular Biology*, *860*, 51–63.

Bieleski, R. L. (1964). The problem of halting enzyme action when extracting plant tissues. *Analytical Biochemistry*, *9*, 431–442.

Bingol, K., Zhang, F., Bruschweiler-Li, L., & Brüschweiler, R. (2012). TOCCATA: A customized carbon total correlation spectroscopy NMR metabolomics database. *Analytical Chemistry*, *84*, 9395–9401.

Bino, R. J., Hall, R. D., Fiehn, O., Kopka, J., Saito, K., Draper, J., et al. (2004). Potential of metabolomics as a functional genomics tool. *Trends in Plant Science*, *9*, 418–425.

Bloch, F., Hansen, W. W., & Packard, M. (1946). Nuclear introduction. *Physical Review*, *69*, 127.

Bolten, C. J., Kiefer, P., Letisse, F., Portais, J. C., & Wittmann, C. (2007). Sampling for metabolome analysis of microorganisms. *Analytical Chemistry*, *79*(10), 3843–3849.

Bothwell, J. H., & Griffin, J. L. (2011). An introduction to biological nuclear magnetic resonance spectroscopy. *Biological Reviews of the Cambridge Philosophical Society*, *86*(2), 493–510.

Broadhurst, D. I., & Kell, D. B. (2006). Statistical strategies for avoiding false discoveries in metabolomics and related experiments. *Metabolomics*, *2*(4), 191–196.

Brown, S. A., McKelvie, J. R., Simpson, A. J., & Simpson, M. J. (2010). ^{1}H-NMR metabolomics of earthworm exposure to sub-lethal concentrations of phenanthrene in soil. *Environmental Pollution*, *158*(6), 2117–2123.

Browne, R. A., & Brindle, K. M. (2007). ^{1}H-NMR-based metabolite profiling as a potential selection tool for breeding passive resistance against Fusarium head blight (FHB) in wheat. *Molecular Plant Pathology*, *8*(4), 401–410.

Broyart, C., Fontaine, J. X., Molinie, R., Cailleu, D., Terce-Laforgue, T., Dubois, F., et al. (2010). Metabolic profiling of maize mutants deficient for two glutamine synthetase isoenzymes using ^{1}H-NMR-based metabolomics. *Phytochemical Analysis*, *21*(1), 102–109.

Bum Kim, S., Wang, Z., & Hiremath, B. (2010). A Bayesian approach for the alignment of high-resolution NMR spectra. *Annals of Operations Research*, *174*, 19–32.

Bunescu, A., Garric, J., Vollat, B., Canet-Soulas, E., Graveron-Demilly, D., & Fauvelle, F. (2010). In vivo proton HR-MAS NMR metabolic profile of the freshwater cladoceran Daphnia magna. *Molecular Biosystems*, *6*(1), 121–125.
Calingacion, M. N., Boualaphanh, C., Daygon, V. D., Anacleto, R., Sackville Hamilton, R., Biais, B., et al. (2012). A genomics and multi-platform metabolomics approach to identify new traits of rice quality in traditional and improved varieties. *Metabolomics*, *8*, 771–783.
Capitani, D., Mannina, L., Proietti, N., Sobolev, A. P., Tomassini, A., Miccheli, A., et al. (2010). Monitoring of metabolic profiling and water status of Hayward kiwifruits by nuclear magnetic resonance. *Talanta*, *82*(5), 1826–1838.
Carrari, F., Baxter, C., Usadel, B., Urbanczyk-Wochniak, E., Zanor, M. I., Nunes-Nesi, A., et al. (2006). Integrated analysis of metabolite and transcript levels reveals the metabolic shifts that underlie tomato fruit development and highlight regulatory aspects of metabolic network behavior. *Plant Physiology*, *142*(4), 1380–1396.
Caruso, M., Galgano, F., Castiglione Morelli, M. A., Viggiani, L., Lencioni, L., Giussani, B., et al. (2012). Chemical profile of white wines produced from 'Greco bianco' grape variety in different Italian areas by nuclear magnetic resonance (NMR) and conventional physicochemical analyses. *Journal of Agricultural and Food Chemistry*, *60*(1), 7–15.
Castejon, D., Villa, P., Calvo, M. M., Santa-Maria, G., Herraiz, M., & Herrera, A. (2010). ^{1}H-HRMAS NMR study of smoked Atlantic salmon (*Salmo salar*). *Magnetic Resonance in Chemistry*, *48*(9), 693–703.
Castro, C., Motto, M., Rossi, V., & Manetti, C. (2008). Variation of metabolic profiles in developing maize kernels up- and down-regulated for the hda101 gene. *Journal of Experimental Botany*, *59*(14), 3913–3924.
Charlton, A. J., Donarski, J. A., Harrison, M., Jones, S. A., Godward, J., Oehlschlager, S., et al. (2008). Responses of the pea (*Pisum sativum* L.) leaf metabolome to drought stress assessed by nuclear magnetic resonance spectroscopy. *Metabolomics*, *4*(4), 312–327.
Charlton, A., Allnutt, T., Holmes, S., Chisholm, J., Bean, S., Ellis, N., Oehlschlager, S., et al. (2004). NMR profiling of transgenic peas. *Plant Biotechnology Journal*, *2*(1), 27–35. http://dx.doi.org/10.1046/j.1467-7652.2003.00045.x.
Choi, H. K., Choi, Y. H., Verberne, M., Lefeber, A. W., Erkelens, C., & Verpoorte, R. (2004). Metabolic fingerprinting of wild type and transgenic tobacco plants by ^{1}H NMR and multivariate analysis technique. *Phytochemistry*, *65*(7), 857–864.
Choi, Y. H., Kim, H. K., Linthorst, H. J., Hollander, J. G., Lefeber, A. W., Erkelens, C., et al. (2006). NMR metabolomics to revisit the tobacco mosaic virus infection in *Nicotiana tabacum* leaves. *Journal of Natural Products*, *69*(5), 742–748.
Claridge, T. D. W. (1999). *High-resolution NMR techniques in organic chemistry*. San Diego: Elsevier.
Clausen, M. R., Pedersen, B. H., Bertram, H. C., & Kidmose, U. (2011). Quality of sour cherry juice of different clones and cultivars (*Prunus cerasus* L.) determined by a combined sensory and NMR spectroscopic approach. *Journal of Agricultural and Food Chemistry*, *59*(22), 12124–12130.
Consonni, R., Cagliani, L. R., & Cogliati, C. (2012). NMR based geographical characterization of roasted coffee. *Talanta*, *88*, 420–426.
Consonni, R., Cagliani, L. R., Stocchero, M., & Porretta, S. (2009). Triple concentrated tomato paste: Discrimination between Italian and Chinese products. *Journal of Agricultural and Food Chemistry*, *57*(11), 4506–4513.
Cossegal, M., Chambrier, P., Mbelo, S., Balzergue, S., Martin-Magniette, M. L., Moing, A., et al. (2008). Transcriptional and metabolic adjustments in ADP-glucose pyrophosphorylase-deficient bt2 maize kernels. *Plant Physiology*, *146*(4), 1553–1570.
Craig, A., Cloarec, O., Holmes, E., Nicholson, J. K., & Lindon, J. C. (2006). Scaling and normalization effects in NMR spectroscopic metabonomic data sets. *Analytical Chemistry*, *78*(7), 2262–2267.

Cui, Q., Lewis, I. A., Hegeman, A. D., Anderson, M. E., Li, J., Schulte, C. F., et al. (2008). Metabolite identification via the Madison Metabolomics Consortium Database. *Nature Biotechnology*, *26*(2), 162–164.

Cuny, M., Vigneau, E., Le Gall, G., Colquhoun, I., Lees, M., & Rutledge, D. (2008). Fruit juice authentication by ^{1}H-NMR spectroscopy in combination with different chemometrics tools. *Analytical and Bioanalytical Chemistry*, *390*(1), 419–427.

Cuthbertson, D. J., Andrews, P. K., Reganold, J. P., Davies, N. M., & Lange, B. M. (2012). Utility of metabolomics toward assessing the metabolic basis of quality traits in apple fruit with an emphasis on antioxidants. *Journal of Agricultural and Food Chemistry*, *60*(35), 8552–8560.

Davies, H. (2010). A role for "omics" technologies in food safety assessment. *Food Control*, *21*(12), 1601–1610.

Davis, R. A., Charlton, A. J., Godward, J., Jones, S. A., Harrison, M., & Wilson, J. C. (2007). Adaptive binning: An improved binning method for metabolomics data using the undecimated wavelet transform. *Chemometrics and Intelligent Laboratory Systems*, *85*(1), 144–154.

de Almeida Vieira Lima, L. M., Nunes, N. G., da Silva Dias, P. G., & Marques, F. J. (2012). Implemented data mining and signal management systems on spontaneous reporting systems' databases and their availability to the scientific community—A systematic review. *Current Drug Safety*, 7(2), 170–175.

Deborde, C., & Jacob, D. (2013). MeRy-B, a metabolomic database and knowledge base for exploring plant primary metabolism. In G. Sriram (Ed.), *Plant Metabolism: Methods and Protocols*, (In Press). Humana Press.

Deborde, C., Maucourt, M., Baldet, P., Bernillon, S., Biais, B., Talon, G., et al. (2009). Proton NMR quantitative profiling for quality assessment of greenhouse-grown tomato fruit. *Metabolomics*, *5*(2), 183–198.

Defernez, M., & Colquhoun, I. J. (2003). Factors affecting the robustness of metabolite fingerprinting using ^{1}H NMR spectra. *Phytochemistry*, *62*(6), 1009–1017.

Defernez, M., Gunning, Y. M., Parr, A. J., Shepherd, L. W. T., Davies, H. V., & Colquhoun, I. J. (2004). NMR and HPLC-UV profiling of potatoes with genetic modifications to metabolic pathways. *Journal of Agricultural and Food Chemistry*, *52*, 6075–6085.

De Meyer, T., Sinnaeve, D., Van Gasse, B., Rietzschel, E. R., De Buyzere, M. L., Langlois, M. R., et al. (2010). Evaluation of standard and advanced preprocessing methods for the univariate analysis of blood serum 1H-NMR spectra. *Analytical and Bioanalytical Chemistry*, *398*(4), 1781–1790.

De Meyer, T., Sinnaeve, D., Van Gasse, B., Tsiporkova, E., Rietzschel, E. R., De Buyzere, M. L., et al. (2008). NMR-based characterization of metabolic alterations in hypertension using an adaptive, intelligent binning algorithm. *Analytical Chemistry*, *80*(10), 3783–3790.

Deshmukh, A. P., Simpson, A. J., Hadad, C. M., & Hatcher, P. G. (2005). Insights into the structure of cutin and cutan from Agave americana leaf cuticle using HRMAS NMR spectroscopy. *Organic Geochemistry*, *36*(7), 1072–1085.

Deshmukh, A. P., Simpson, A. J., & Hatcher, P. G. (2003). Evidence for cross-linking in tomato cutin using HR-MAS NMR spectroscopy. *Phytochemistry*, *64*(6), 1163–1170.

Desmoulin, F., Bon, D., Martino, R., & Malet-Martino, M. (2008). Étude critique de l'utilisation de la RMN HR-MAS pour l'analyse des tissus biologiques. *Comptes Rendus de Chimie*, *11*(4–5), 423–433.

Detour, J., Elbayed, K., Piotto, M., Moussallieh, F. M., Nehlig, A., & Namer, I. J. (2011). Ultra fast in vivo microwave irradiation for enhanced metabolic stability of brain biopsy samples during HRMAS NMR analysis. *Journal of Neuroscience Methods*, *201*(1), 89–97.

Di Anibal, C. V., Callao, M. P., & RuisÃ¡nchez, I. (2011). ¹H NMR variable selection approaches for classification. A case study: The determination of adulterated foodstuffs. *Talanta, 86*, 316–323.

Dreier, L., & Wider, G. (2006). Concentration measurements by PULCON using X-filtered or 2D NMR spectra. *Magnetic Resonance in Chemistry, 44 Spec No*, S206–S212.

Ebbels, T. M., Lindon, J. C., & Coen, M. (2011). Processing and modeling of nuclear magnetic resonance (NMR) metabolic profiles. *Methods in Molecular Biology, 708*, 365–388.

Elmasry, G., Kamruzzaman, M., Sun, D. W., & Allen, P. (2012). Principles and applications of hyperspectral imaging in quality evaluation of agro-food products: a review. *Critical Reviews in Food Science and Nutrition, 52*(11), 999–1023. http://dx.doi.org/10.1080/10408398.2010.543495.

Fan, T. W.-M. (1996). Metabolite profiling by one- and two-dimensional NMR analysis of complex mixtures. *Progress in Nuclear Magnetic Resonance Spectroscopy, 28*(2), 161–219.

Fan, T. W., Higashi, R. M., Lane, A. N., & Jardetzky, O. (1986). Combined use of ¹H-NMR and GC-MS for metabolite monitoring and in vivo ¹H-NMR assignments. *Biochimica et Biophysica Acta, 882*(2), 154–167.

Fang, X., Qiu, F., Yan, B., Wang, H., Mort, A. J., & Stark, R. E. (2001). NMR studies of molecular structure in fruit cuticle polyesters. *Phytochemistry, 57*(6), 1035–1042.

Farag, M. A., Porzel, A., Schmidt, J., & Wessjohann, L. A. (2012). Metabolite profiling and fingerprinting of commercial cultivars of *Humulus lupulus* L. (hop): A comparison of MS and NMR methods in metabolomics. *Metabolomics, 8*(3), 492–507.

Farrant, R. D., Hollerton, J. C., Lynn, S. M., Provera, S., Sidebottom, P. J., & Upton, R. J. (2010). NMR quantification using an artificial signal. *Magnetic Resonance in Chemistry, 48*(10), 753–762.

Fernandez, L., Romieu, C., Moing, A., Bouquet, A., Maucourt, M., Thomas, M. R., et al. (2006). The grapevine fleshless berry mutation. A unique genotype to investigate differences between fleshy and nonfleshy fruit. *Plant Physiology, 140*(2), 537–547.

Fernie, A. R. (2007). The future of metabolic phytochemistry: Larger numbers of metabolites, higher resolution, greater understanding. *Phytochemistry, 68*(22–24), 2861–2880.

Fernie, A. R., & Schauer, N. (2009). Metabolomics-assisted breeding: a viable option for crop improvement?. *Trends Genet, 25*(1), 39–48 http://dx.doi.org/S0168-9525(08)00301-6 [pii].

Fernie, A. R., Aharoni, A., Willmitzer, L., Stitt, M., Tohge, T., Kopka, J., et al. (2011). Recommendations for reporting metabolite data. *The Plant Cell, 23*(7), 2477–2482.

Fernie, A. R., & Keurentjes, J. J. B. (2011). Genetics, genomics and metabolomics. In R. Hall (Ed.), *Biology of plant metabolomics. Annual plant reviews*, Vol. 43, (pp. 219–246). Chichester, West Sussex, UK: Wiley-Blackwell.

Fernie, A. R., & Stitt, M. (2012). On the discordance of metabolomics with proteomics and transcriptomics: Coping with increasing complexity in logic, chemistry, and network interactions scientific correspondence. *Plant Physiology, 158*(3), 1139–1145.

Ferry-Dumazet, H., Gil, L., Deborde, C., Moing, A., Bernillon, S., Rolin, D., et al. (2011). MeRy-B: A web knowledgebase for the storage, visualization, analysis and annotation of plant NMR metabolomic profiles. *BMC Plant Biology, 11*, 104–116.

Fiehn, O. (2002). Metabolomics—The link between genotypes and phenotypes. *Plant Molecular Biology, 48*(1–2), 155–171.

Fiehn, O., Sumner, L., Rhee, S., Ward, J., Dickerson, J., Lange, B., et al. (2007). Minimum reporting standards for plant biology context information in metabolomic studies. *Metabolomics, 3*(3), 195–201.

Fortes, A. M., Agudelo-Romero, P., Silva, M. S., Ali, K., Sousa, L., Maltese, F., et al. (2011). Transcript and metabolite analysis in Trincadeira cultivar reveals novel information regarding the dynamics of grape ripening. *BMC Plant Biology, 11*, 149–183.

Frederich, M., Jansen, C., de Tullio, P., Tits, M., Demoulin, V., & Angenot, L. (2010). Metabolomic analysis of Echinacea spp. by ^{1}H nuclear magnetic resonance spectrometry and multivariate data analysis technique. *Phytochemical Analysis*, *21*(1), 61–65.

Gad, H. A., El-Ahmady, S. H., Abou-Shoer, M. I., & Al-Azizi, M. M. (2013). Application of chemometrics in authentication of herbal medicines: A review. *Phytochemical Analysis*, *24*(1), 1–24.

Gathungu, R. M., Oldham, J. T., Bird, S. S., Lee-Parsons, C. W., Vouros, P., & Kautz, R. (2012). Application of an integrated LC-UV-MS-NMR platform to the identification of secondary metabolites from cell cultures: Benzophenanthridine alkaloids from elicited *Eschscholzia californica* (california poppy) cell cultures. *Analytical Methods*, *4*(5), 1315–1325.

Gavaghan, C. L., Li, J. V., Hadfield, S. T., Hole, S., Nicholson, J. K., Wilson, I. D., et al. (2011). Application of NMR-based metabolomics to the investigation of salt stress in maize (Zea mays). *Phytochemical Analysis*, *22*(3), 214–224.

Gibon, Y., & Rolin, D. (2012). Aspects of experimental design for plant metabolomics experiments and guidelines for growth of plant material. *Methods in Molecular Biology*, *860*, 13–30.

Gil, A. M., Duarte, I. F., Delgadillo, I., Colquhoun, I. J., Casuscelli, F., Humpfer, E., et al. (2000). Study of the compositional changes of mango during ripening by use of nuclear magnetic resonance spectroscopy. *Journal of Agricultural and Food Chemistry*, *48*(5), 1524–1536.

Golotvin, S., & Williams, A. (2000). Improved baseline recognition and modeling of FT NMR spectra. *Journal of Magnetic Resonance*, *146*(1), 122–125.

Goodacre, R., Broadhurst, D., Smilde, A., Kristal, B., Baker, J., Beger, R., et al. (2007). Proposed minimum reporting standards for data analysis in metabolomics. *Metabolomics*, *3*(3), 231–241.

Gottlieb, H. E., Kotlyar, V., & Nudelman, A. (1997). NMR chemical shifts of common laboratory solvents as trace impurities. *Journal of Organic Chemistry*, *62*(21), 7512–7515.

Griffin, J. L. (2003). Metabonomics: NMR spectroscopy and pattern recognition analysis of body fluids and tissues for characterisation of xenobiotic toxicity and disease diagnosis. *Current Opinion in Chemical Biology*, 7(5), 648–654.

Griffin, J. L. (2004). Metabolic profiles to define the genome: Can we hear the phenotypes? *Philosophical Transactions of the Royal Society B: Biological Sciences*, *359*(1446), 857–871.

Griffiths, W. J., Koal, T., Wang, Y., Kohl, M., Enot, D. P., & Deigner, H. P. (2010). Targeted metabolomics for biomarker discovery. *Angewandte Chemie International Edition*, *49*(32), 5426–5445.

Grimes, J. H., & O'Connell, T. M. (2011). The application of micro-coil NMR probe technology to metabolomics of urine and serum. *Journal of Biomolecular NMR*, *49*(3–4), 297–305.

Gromova, M., & Roby, C. (2010). Toward *Arabidopsis thaliana* hydrophilic metabolome: Assessment of extraction methods and quantitative ^{1}H-NMR. *Physiology Plantarum*, *140*(2), 111–127.

Gullberg, J., Jonsson, P., Nordstrom, A., Sjostrom, M., & Moritz, T. (2004). Design of experiments: An efficient strategy to identify factors influencing extraction and derivatization of *Arabidopsis thaliana* samples in metabolomic studies with gas chromatography/mass spectrometry. *Analytical Biochemistry*, *331*(2), 283–295.

Hadacek, F. (2002). Secondary metabolites as plant traits: Current assessment and future perspectives. *Critical Reviews in Plant Sciences*, *21*(4), 273–322.

Hagel, J. M., Weljie, A. M., Vogel, H. J., & Facchini, P. J. (2008). Quantitative ^{1}H nuclear magnetic resonance metabolite profiling as a functional genomics platform to investigate alkaloid biosynthesis in opium poppy. *Plant Physiology*, *147*(4), 1805–1821.

Hall, R. D. (2006). Plant metabolomics: From holistic hope, to hype, to hot topic. *New Phytologist*, *169*(3), 453–468.

Hall, R. (2011a). H. Robert (Ed.), *Biology of plant metabolomics* (p. 420). *Annual plant reviews*, Vol. 43, (p. 420). Chichester, West Sussex, UK: Wiley-Blackwell.

Hall, R. D. (2011b). Plant metabolomics in a nutshell: Potential and future challenges. In *Biology of plant metabolomics. Annual plant reviews*, Vol. 43, (pp. 1–24). Chichester, West Sussex, UK: Wiley-Blackwell.

Hall, R., Beale, M., Fiehn, O., Hardy, N., Sumner, L., & Bino, R. (2002). Plant metabolomics: The missing link in functional genomics strategies. *The Plant Cell, 14*(7), 1437–1440.

Hall, R. D., Vos, C. H. R., Verhoeven, H. A., & Bino, R. J. (2005). Metabolomics for the assessment of functional diversity and quality traits in plants. In S. Vaidyanathan, G. G. Harrigan, & R. Goodacre (Eds.), *Metabolome analyses: Strategies for systems biology*. New York: Springer.

Halouska, S., & Powers, R. (2006). Negative impact of noise on the principal component analysis of NMR data. *Journal of Magnetic Resonance, 178*(1), 88–95.

Hanana, H., Simon, G., Kervarec, N., Mohammadou, B. A., & Cerantola, S. (2012). HRMAS NMR as a tool to study metabolic responses in heart clam Ruditapes decussatus exposed to Roundup(R). *Talanta, 97*, 425–431.

Hannemann, J., Poorter, H., Usadel, B., Blasing, O. E., Finck, A., Tardieu, F., et al. (2009). Xeml Lab: A tool that supports the design of experiments at a graphical interface and generates computer-readable metadata files, which capture information about genotypes, growth conditions, environmental perturbations and sampling strategy. *Plant, Cell & Environment, 32*(9), 1185–1200.

He, C. C., Dai, Y. Q., Hui, R. R., Hua, J., Chen, H. J., Luo, Q. Y., et al. (2012). NMR-based metabonomic approach on the toxicological effects of a Cimicifuga triterpenoid. *Journal of Applied Toxicology, 32*(2), 88–97.

Hong, Y. S. (2011). NMR-based metabolomics in wine science. *Magnetic Resonance in Chemistry, 49*(Suppl. 1), S13–S21.

Hong, Y. S., Cilindre, C., Liger-Belair, G., Jeandet, P., Hertkorn, N., & Schmitt-Kopplin, P. (2011). Metabolic influence of *Botrytis cinerea* infection in champagne base wine. *Journal of Agricultural and Food Chemistry, 59*(13), 7237–7245.

Hoult, D. I., & Richards, R. E. (1976). The signal-to-noise ratio of the nuclear magnetic resonance experiment. *Journal of Magnetic Resonance, 24*, 71–85.

Howell, K. S., Cozzolino, D., Bartowsky, E. J., Fleet, G. H., & Henschke, P. A. (2006). Metabolic profiling as a tool for revealing *Saccharomyces* interactions during wine fermentation. *FEMS Yeast Research, 6*, 91–101.

Hwang, T. L., & Shaka, A. J. (1995). Water suppression that works. Excitation sculpting using arbitrary wave-forms and pulsed-field gradients. *Journal of Magnetic Resonance Series A, 112*(2), 275–279.

Ioset, K. N., Nyberg, N. T., Van Diermen, D., Malnoe, P., Hostettmann, K., Shikov, A. N., et al. (2011). Metabolic profiling of *Rhodiola rosea* rhizomes by ^{1}H-NMR spectroscopy. *Phytochemical Analysis, 22*(2), 158–165.

Izquierdo-Garcia, J. L., Villa, P., Kyriazis, A., del Puerto-Nevado, L., Perez-Rial, S., Rodriguez, I., et al. (2011). Descriptive review of current NMR-based metabolomic data analysis packages. *Progress in Nuclear Magnetic Resonance Spectroscopy, 59*(3), 263–270.

Jansen, J. J., Allwood, W. J., Marsden-Edwards, E., van de Putten, W. H., Goodacre, R., & van Dam, N. M. (2009). Metabolomic analysis of the interaction between plants and herbivores. *Metabolomics, 5*, 150–161.

Jenkins, H., Hardy, N., Beckmann, M., Draper, J., Smith, A. R., Taylor, J., et al. (2004). A proposed framework for the description of plant metabolomics experiments and their results. *Nature Biotechnology, 22*, 1601–1606.

Jenkins, H., Johnson, H., Kular, B., Wang, T., & Hardy, N. (2005). Toward supportive data collection tools for plant metabolomics. *Plant Physiology, 138*(1), 67–77.

Jones, O. A. H., Maguire, M. L., Griffin, J. L., Jung, Y. H., Shibato, J., Rakwal, R., et al. (2011). Using metabolic profiling to assess plant-pathogen interactions: An example using rice (*Oryza sativa*) and the blast pathogen *Magnaporthe grisea*. *European Journal of Plant Pathology, 129*(4), 539–554.

Kaiser, K. A., Barding, G. A., Jr., & Larive, C. K. (2009). A comparison of metabolite extraction strategies for ^{1}H-NMR-based metabolic profiling using mature leaf tissue from the model plant *Arabidopsis thaliana*. *Magnetic Resonance in Chemistry, 47*(Suppl. 1), S147–S156.

Karaman, S., Tutem, E., Baskan, K. S., & Apak, R. (2013). Comparison of antioxidant capacity and phenolic composition of peel and flesh of some apple varieties. *Journal of the Science of Food and Agriculture, 93*(4), 867–875.

Keeler, J. (2010). In J. Keeler (Ed.), *Understanding NMR spectroscopy*. Chichester, West Sussex, UK: John Wiley & Sons Inc.

Kelleher, B. P., Simpson, M. J., & Simpson, A. J. (2006). Assessing the fate and transformation of plant residues in the terrestrial environment using HR-MAS NMR spectroscopy. *Geochimica et Cosmochimica Acta, 70*, 4080–4094.

Keun, H. C., Beckonert, O., Griffin, J. L., Richter, C., Moskau, D., Lindon, J. C., et al. (2002). Cryogenic probe ^{13}C NMR spectroscopy of urine for metabonomic studies. *Analytical Chemistry, 74*(17), 4588–4593.

Kiefer, P. A., Baltusis, L., Rice, D. M., Timiak, A. A., & Shoolery, J. N. (1996). A comparison of NMR spectra obtained for solid-phase-synthesis resins using conventional high-resolution, magic-angle-spinning, and high-resolution magic-angle-spinning probe. *Journal of Magnetic Resonance, 119*(1), 65–75.

Kim, H. K., Choi, Y. H., & Verpoorte, R. (2010). NMR-based metabolomic analysis of plants. *Nature Protocols, 5*(3), 536–549.

Kim, H. K., Choi, Y. H., & Verpoorte, R. (2011). NMR-based plant metabolomics: Where do we stand, where do we go? *Trends in Biotechnology, 29*(6), 267–275.

Kim, H. K., & Verpoorte, R. (2010). Sample preparation for plant metabolomics. *Phytochemical Analysis, 21*(1), 4–13.

Kim, H. K., Wilson, E. G., Choi, Y. H., & Verpoorte, R. (2010). Metabolomics: A tool for anticancer lead-finding from natural products. *Planta Medica, 76*(11), 1094–1102.

Kirk, H., Cheng, D., Choi, Y. H., Vrieling, K., & Klinkhamer, P. G. (2012). Transgressive segregation of primary and secondary metabolites in F(2) hybrids between *Jacobaea aquatica* and *J. vulgaris*. *Metabolomics, 8*(2), 211–219.

Kirk, H., Choi, Y. H., Kim, H. K., Verpoorte, R., & van der Meijden, E. (2005). Comparing metabolomes: The chemical consequences of hybridization in plants. *New Phytologist, 167*(2), 613–622.

Ko, B.-K., Ahn, H.-J., van den Berg, F., Lee, C.-H., & Hong, Y.-S. (2009). Metabolomic insight into soy sauce through ^{1}H NMR spectroscopy. *Journal of Agricultural and Food Chemistry, 57*(15), 6862–6870.

Koda, M., Furihata, K., Wei, F., Miyakawa, T., & Tanokura, M. (2012). Metabolic discrimination of mango juice from various cultivars by band-selective NMR spectroscopy. *Journal of Agricultural and Food Chemistry, 60*(5), 1158–1166.

Koehn, F. E. (2008). High impact technologies for natural products screening. *Progress in Drug Research, 65*(175), 177–210.

Koek, M. M., Jellema, R. H., van der Greef, J., Tas, A. C., & Hankemeier, T. (2011). Quantitative metabolomics based on gas chromatography mass spectrometry: Status and perspectives. *Metabolomics*, 7(3), 307–328.

Kohl, S. M., Klein, M. S., Hochrein, J., Oefner, P. J., Spang, R., & Gronwald, W. (2012). State-of-the art data normalization methods improve NMR-based metabolomic analysis. *Metabolomics, 8*(Suppl. 1), 146–160.

Kopka, J., Fernie, A., Weckwerth, W., Gibon, Y., & Stitt, M. (2004). Metabolite profiling in plant biology: Platforms and destinations. *Genome Biology, 5*(6), 109.

Kopka, J., Walther, D., Allwood, J. W., & Goodacre, R. (2011). Progress in chemometrics and biostatistics for plant applications, or a good red wine is a bad white wine. In R. D. Hall (Ed.), *Biology of plant metabolomics: Vol. 43*. Oxford, UK: Wiley-Blackwell.

Kovacs, H., Moskau, D., & Spraul, M. (2005). Cryogenically cooled probes: A leap in NMR technology. *Progress in Nuclear Magnetic Resonance Spectroscopy, 46*, 131–155.

Krishnan, P., Kruger, N. J., & Ratcliffe, R. G. (2005). Metabolite fingerprinting and profiling in plants using NMR. *Journal of Experimental Botany, 56*(410), 255–265.

Kruger, N. J., Troncoso-Ponce, M. A., & Ratcliffe, R. G. (2008). 1H NMR metabolite fingerprinting and metabolomic analysis of perchloric acid extracts from plant tissues. *Nature Protocols, 3*(6), 1001–1012.

Kusano, M., Tohge, T., Fukushima, A., Kobayashi, M., Hayashi, N., Otsuki, H., et al. (2011). Metabolomics reveals comprehensive reprogramming involving two independent metabolic responses of *Arabidopsis* to UV-B light. *The Plant Journal, 67*(2), 354–369.

Lamanna, R., Cattivelli, L., Miglietta, M. L., & Troccoli, A. (2011). Geographical origin of durum wheat studied by H-1-NMR profiling. *Magnetic Resonance in Chemistry, 49*(1), 1–5.

Lee, J. E., Hwang, G. S., Lee, C. H., & Hong, Y. S. (2009). Metabolomics reveals alterations in both primary and secondary metabolites by wine bacteria. *Journal of Agricultural and Food Chemistry, 57*(22), 10772–10783.

Lee, J. E., Lee, B. J., Hwang, J. A., Ko, K. S., Chung, J. O., Kim, E. H., et al. (2011). Metabolic dependence of green tea on plucking positions revisited: A metabolomic study. *Journal of Agricultural and Food Chemistry, 59*(19), 10579–10585.

Lee, E. J., Shaykhutdinov, R., Weljie, A. M., Vogel, H. J., Facchini, P. J., Park, S. U., et al. (2009). Quality assessment of ginseng by ^{1}H-NMR metabolite fingerprinting and profiling analysis. *Journal of Agricultural and Food Chemistry, 57*(16), 7513–7522.

Le Gall, G., Colquhoun, I. J., Davis, A. L., Collins, G. J., & Verhoeyen, M. E. (2003). Metabolite profiling of tomato (*Lycopersicon esculentum*) using ^{1}H NMR spectroscopy as a tool to detect potential unintended effects following a genetic modification. *Journal of Agricultural and Food Chemistry, 51*(9), 2447–2456.

Le Gall, G., Colquhoun, I. J., & Defernez, M. (2004). Metabolite profiling using ^{1}H-NMR spectroscopy for quality assessment of green tea, *Camellia sinensis* (L.). *Journal of Agricultural and Food Chemistry, 52*(4), 692–700.

Lei, Z., Huhman, D. V., & Sumner, L. W. (2011). Mass spectrometry strategies in metabolomics. *Journal of Biological Chemistry, 286*(29), 25435–25442.

Leiss, K. A., Choi, Y. H., Verpoorte, R., & Klinkhamer, P. G. (2011). An overview of NMR-based metabolomics to identify secondary plant compounds involved in host plant resistance. *Phytochemistry Reviews, 10*(2), 205–216.

Leon, A., Chavez, M. I., & Delgado, G. (2011). ^{1}H and DOSY NMR spectroscopy analysis of *Ligusticum porteri* rhizome extracts. *Magnetic Resonance in Chemistry, 49*(8), 469–478.

Lerceteau-Kohler, E., Moing, A., Guerin, G., Renaud, C., Maucourt, M., Rolin, D., et al. (2006). QTL analysis for sugars and organic acids in strawberry fruits. In: G. Waite (Ed.), *Proceedings of the Vth international strawberry symposium* (pp. 573–577).

Lima, M. R., Felgueiras, M. L., Graça, G., Rodrigues, J. E., Barros, A., Gil, A. M., et al. (2010). NMR metabolomics of esca disease-affected *Vitis vinifera* cv. Alvarinho leaves. *Journal of Experimental Botany, 61*(14), 4033–4042.

Lin, C. Y., Anderson, B. S., Phillips, B. M., Peng, A. C., Clark, S., Voorhees, J., et al. (2009). Characterization of the metabolic actions of crude versus dispersed oil in salmon smolts via NMR-based metabolomics. *Aquatic Toxicology, 95*(3), 230–238.

Lindon, J. C., Beckonert, O. P., Holmes, E., & Nicholson, J. K. (2009). High-resolution magic angle spinning NMR spectroscopy: Application to biomedical studies. *Progress in Nuclear Magnetic Resonance Spectroscopy, 55*(2), 79–100.

Lindon, J. C., Holmes, E., & Nicholson, J. K. (2001). Pattern recognition methods and applications in biomedical magnetic resonance. *Progress in Nuclear Magnetic Resonance Spectroscopy, 39*, 1–40.

Lindon, J. C., & Nicholson, J. K. (2008). Spectroscopic and statistical techniques for information recovery in metabonomics and metabolomics. *Annual Review of Analytical Chemistry, 1*, 45–69.

Lindon, J. C., Nicholson, J. K., Holmes, E., & Everett, J. R. (2000). Metabonomics: Metabolic processes studied by NMR spectroscopy of biofluids. *Concepts in Magnetic Resonance, 12*(5), 289–320.

Lippens, G., Bourdonneau, M., Dhalluin, C., Warrass, R., Richert, T., Seetharaman, C., et al. (1999). Study of compound attached to solid supports using high resolution magic angle spinning NMR. *Current Organic Chemistry, 3*, 147–169.

Liu, C. X., Hao, F. H., Hu, J., Zhang, W. L., Wan, L. L., Zhu, L. L., et al. (2010). Revealing different systems responses to brown planthopper infestation for pest susceptible and resistant rice plants with the combined metabonomic and gene-expression analysis. *Journal of Proteome Research, 9*(12), 6774–6785.

Liu, M., Mao, X.-A., Ye, C., Huang, H., Nicholson, J. K., & Lindon, J. C. (1998). Improved WATERGATE pulse sequences for solvent suppression in NMR spectroscopy. *Journal of Magnetic Resonance, 132*, 125–129.

Liu, G. N., Zhu, Y. H., & Jiang, J. G. (2009). The metabolomics of carotenoids in engineered cell factory. *Applied Microbiology and Biotechnology, 83*(6), 989–999.

Llewellyn, A. M., Lewis, J., Miller, S. J., Corol, D. I., Beale, M. H., & Ward, J. L. (2012). Tissue preparation using *Arabidopsis*. *Methods in Molecular Biology, 860*, 65–81.

Lopez-Gresa, M. P., Lison, P., Kim, H. K., Choi, Y. H., Verpoorte, R., Rodrigo, I., et al. (2012). Metabolic fingerprinting of tomato mosaic virus infected *Solanum lycopersicum*. *Journal of Plant Physiology, 169*(16), 1586–1596.

Lopez-Gresa, M. P., Maltese, F., Belles, J. M., Conejero, V., Kim, H. K., Choi, Y. H., et al. (2010). Metabolic response of tomato leaves upon different plant-pathogen interactions. *Phytochemical Analysis, 21*(1), 89–94.

Lucas, L. H., Larive, C. K., Wilkinson, P. S., & Huhn, S. (2005). Progress toward automated metabolic profiling of human serum: Comparison of CPMG and gradient-filtered NMR analytical methods. *Journal of Pharmaceutical and Biomedical Analysis, 39*(1–2), 156–163.

Ludwig, C., Easton, J. M., Lodi, A., Tiziani, S., Manzoor, S. E., Southam, A. D., et al. (2012). Birmingham Metabolite Library: A publicly accessible database of 1-D ^{1}H and 2-D ^{1}H J-resolved NMR spectra of authentic metabolite standards (BML-NMR). *Metabolomics, 8*(1), 8–18.

Ludwig, C., & Günther, U. L. (2011). MetaboLab-advanced NMR data processing and analysis for metabolomics. *BMC Bioinformatics, 12*, 366.

Ludwig, C., & Viant, M. R. (2010). Two-dimensional J-resolved NMR spectroscopy: Review of a key methodology in the metabolomics toolbox. *Phytochemical Analysis, 21*(1), 22–32.

Ludwig, C., Ward, D. G., Martin, A., Viant, M. R., Ismail, T., Johnson, P. J., et al. (2009). Fast targeted multidimensional NMR metabolomics of colorectal cancer. *Magnetic Resonance in Chemistry, 47*(Suppl. 1), S68–S73.

Lugan, R., Niogret, M.-F., Leport, L., Guégan, J.-P., Larher, F. R., Savouré, A., et al. (2010). Metabolome and water homeostasis analysis of *Thellungiella salsuginea* suggests that dehydration tolerance is a key response to osmotic stress in this halophyte. *The Plant Journal, 64*(2), 215–229.

Lundberg, P., Vogel, T., Malusek, A., Lundquist, P.-O., Cohen, L., & Dahlqvist, O. (2005). MDL—The magnetic resonance metabolomics database. Available online http://mdl.imv.liu.se.

Macomber, R. S. (1998). In R. S. Macomber (Ed.), *A complete introduction to modern NMR spectroscopy*. Chichester, West Sussex, UK: John Wiley & Sons Inc.

Maharjan, R. P., & Ferenci, T. (2003). Global metabolite analysis: The influence of extraction methodology on metabolome profiles of *Escherichia coli*. *Analytical Biochemistry*, *313*(1), 145–154.

Mannina, L., D'Imperio, M., Capitani, D., Rezzi, S., Guillou, C., Mavromoustakos, T., et al. (2009). ^{1}H-NMR-based protocol for the detection of adulterations of refined olive oil with refined hazelnut oil. *Journal of Agricultural and Food Chemistry*, *57*(24), 11550–11556.

Mannina, L., Sobolev, A. P., & Viel, S. (2012). Liquid state ^{1}H high field NMR in food analysis. *Progress in Nuclear Magnetic Resonance Spectroscopy*, *66*, 1–39.

Marini, F. (2009). Artificial neural networks in foodstuff analyses: Trends and perspectives A review. *Analytica Chimica Acta*, *635*(2), 121–131.

Martens, L., Chambers, M., Sturm, M., Kessner, D., Levander, F., Shofstahl, J., et al. (2011). mzML–a community standard for mass spectrometry data. *Molecular Cell Proteomics*, *10*(1), R110.000133. http://dx.doi.org/10.1074 mcp.R110.000133.

Mattoo, A. K., Sobolev, A. P., Neelam, A., Goyal, R. K., Handa, A. K., & Segre, A. L. (2006). Nuclear magnetic resonance spectroscopy-based metabolite profiling of transgenic tomato fruit engineered to accumulate spermidine and spermine reveals enhanced anabolic and nitrogen-carbon interactions. *Plant Physiology*, *142*(4), 1759–1770.

Mercier, P., Lewis, M. J., Chang, D., Baker, D., & Wishart, D. S. (2011). Towards automatic metabolomic profiling of high-resolution one-dimensional proton NMR spectra. *Journal of Biomolecular NMR*, *49*(3–4), 307–323.

Mochida, K., Furuta, T., Ebana, K., Shinozaki, K., & Kikuchi, J. (2009). Correlation exploration of metabolic and genomic diversity in rice. *BMC Genomics*, *10*, 10.

Moco, S., & Vervoort, J. (2012). Chemical identification strategies using liquid chromatography-photodiode array-solid-phase extraction-nuclear magnetic resonance/mass spectrometry. *Methods in Molecular Biology*, *860*, 287–316.

Moestue, S., Sitter, B., Bathen, T. F., Tessem, M. B., & Gribbestad, I. S. (2011). HRMAS NMR spectroscopy in metabolic characterization of cancer. *Current Topics in Medicinal Chemistry*, *11*(1), 2–26.

Moing, A., Aharoni, A., Biais, B., Rogachev, I., Meir, S., Brodsky, L., et al. (2011). Extensive metabolic cross-talk in melon fruit revealed by spatial and developmental combinatorial metabolomics. *New Phytologist*, *190*(3), 683–696.

Moing, A., Maucourt, M., Renaud, C., Gaudillere, M., Brouquisse, R., Lebouteiller, B., et al. (2004). Quantitative metabolic profiling by 1-dimensional ^{1}H-NMR analyses: Application to plant genetics and functional genomics. *Functional Plant Biology*, *31*(9), 889–902.

Molinski, T. F. (2010). Microscale methodology for structure elucidation of natural products. *Current Opinion in Biotechnology*, *21*(6), 819–826.

Mounet, F., Lemaire-Chamley, M., Maucourt, M., Cabasson, C., Giraudel, J. L., Deborde, C., et al. (2007). Quantitative metabolic profiles of tomato flesh and seeds during fruit development: Complementary analysis with ANN and PCA. *Metabolomics*, *3*(3), 273–288.

Mounet, F., Moing, A., Garcia, V., Petit, J., Maucourt, M., Deborde, C., et al. (2009). Gene and metabolite regulatory network analysis of early developing fruit tissues highlights new candidate genes for the control of tomato fruit composition and development. *Plant Physiology*, *149*(3), 1505–1528.

Nakatani, N., Kayano, S., Kikuzaki, H., Sumino, K., Katagiri, K., & Mitani, T. (2000). Identification, quantitative determination, and antioxidative activities of chlorogenic acid isomers in prune (*Prunus domestica* L.). *Journal of Agricultural and Food Chemistry*, *48*(11), 5512–5516.

Neily, M. H., Matsukura, C., Maucourt, M., Bernillon, S., Deborde, C., Moing, A., et al. (2011). Enhanced polyamine accumulation alters carotenoid metabolism at the

transcriptional level in tomato fruit over-expressing spermidine synthase. *Journal of Plant Physiology*, *168*(3), 242–252.

Nicholson, J. K., Connelly, J., Lindon, J. C., & Holmes, E. (2002). Metabonomics: A platform for studying drug toxicity and gene function. *Nature Reviews. Drug Discovery*, *1*(2), 153–161.

Nicholson, J. K., Foxall, P. J., Spraul, M., Farrant, R. D., & Lindon, J. C. (1995). 750 MHz ^{1}H and ^{1}H-^{13}C NMR spectroscopy of human blood plasma. *Analytical Chemistry*, *67*(5), 793–811.

Nicholson, J. K., Lindon, J. C., & Holmes, E. (1999). Metabonomics: Understanding the metabolic responses of living systems to pathophysiological stimuli via multivariate statistical analysis of biological NMR spectroscopic data. *Xenobiotica*, *29*(11), 1181–1189.

Noteborn, H., Lommen, A., van der Jagt, R. C., & Weseman, J. M. (2000). Chemical fingerprinting for the evaluation of unintended secondary metabolic changes in transgenic food crops. *Journal of Biotechnology*, 77(1), 103–114.

Oliver, S. G., Winson, M. K., Kell, D. B., & Baganz, F. (1998). Systematic functional analysis of the yeast genome. *Trends in Biotechnology*, *16*, 373–378.

Olson, D. L., Norcross, J. A., O'Neil-Johnson, M., Molitor, P. F., Detlefsen, D. J., Wilson, A. G., et al. (2004). Microflow NMR: Concepts and capabilities. *Analytical Chemistry*, *76*(10), 2966–2974.

Otero, L., & Préstamo, G. (2009). Effects of pressure processing on strawberry studied by nuclear magnetic resonance. *Innovative Food Science & Emerging Technologies*, *10*, 434–440.

Ott, K. H., Aranibar, N., Singh, B. J., & Stockton, G. W. (2003). Metabonomics classifies pathways affected by bioactive compounds. Artificial neural network classification of NMR spectra of plant extracts. *Phytochemistry*, *62*, 971–985.

Pacifico, S., D'Abrosca, B., Scognamiglio, M., Gallicchio, M., Potenza, N., Piccolella, S., et al. (2011). Metabolic profiling of strawberry grape (*Vitis* x *labruscana* cv. 'Isabella') components by nuclear magnetic resonance (NMR) and evaluation of their antioxidant and antiproliferative properties. *Journal of Agricultural and Food Chemistry*, *59*(14), 7679–7687.

Palama, T. L., Khatib, A., Choi, Y. H., Come, B., Fock, I., Verpoorte, R., et al. (2011). Metabolic characterization of green pods from *Vanilla planifolia* accessions grown in La Reunion. *Environmental and Experimental Botany*, 72(2), 258–265.

Pauli, G. F., Godecke, T., Jaki, B. U., & Lankin, D. C. (2012). Quantitative ^{1}H NMR. Development and potential of an analytical method: An update. *Journal of Natural Products*, *75*(4), 834–851.

Pauli, G. F., Jaki, B. U., & Lankin, D. C. (2005). Quantitative 1H NMR: Development and potential of a method for natural products analysis. *Journal of Natural Products*, *68*(1), 133–149. http://dx.doi.org/10.1021/np0497301.

Pereira, G. E., Gaudillere, J. P., Pieri, P., Hilbert, G., Maucourt, M., Deborde, C., et al. (2006). Microclimate influence on mineral and metabolic profiles of grape berries. *Journal of Agricultural and Food Chemistry*, *54*(18), 6765–6775.

Pereira, G. E., Gaudillere, J. P., Van Leeuwen, C., Hilbert, G., Lavialle, O., Maucourt, M., et al. (2005). ^{1}H-NMR and chemometrics to characterize mature grape berries in four wine-growing areas in Bordeaux, France. *Journal of Agricultural and Food Chemistry*, *53*(16), 6382–6389.

Pereira, G. E., Gaudillere, J. P., van Leeuwen, C., Hilbert, G., Maucourt, M., Deborde, C., et al. (2006). ^{1}H-NMR metabolite fingerprints of grape berry: Comparison of vintage and soil effects in Bordeaux grapevine growing areas. *Analytica Chimica Acta*, *563*(1–2), 346–352.

Pérez, E. M. S., Iglesias, M. J., Ortiz, F. L., Pérez, I. S., & Galera, M. M. (2010). Study of the suitability of HRMAS NMR for metabolic profiling of tomatoes: Application to tissue differentiation and fruit ripening. *Food Chemistry*, *122*(3), 877–887.

Perez, E. M. S., Lopez, J. G., Iglesias, M. J., Ortiz, F. L., Toresano, F., & Camacho, F. (2011). HRMAS-nuclear magnetic resonance spectroscopy characterization of tomato "flavor varieties" from Almeria (Spain). *Food Research International*, *44*(10), 3212–3221.

Piccioni, F., Capitani, D., Zolla, L., & Mannina, L. (2009). NMR metabolic profiling of transgenic maize with the Cry1Ab gene. *Journal of Agricultural and Food Chemistry*, *57*(14), 6041–6049.

Picone, G., Mezzetti, B., Babini, E., Capocasa, F., Placucci, G., & Capozzi, F. (2011). Unsupervised principal component analysis of NMR metabolic profiles for the assessment of substantial equivalence of transgenic grapes (*Vitis vinifera*). *Journal of Agricultural and Food Chemistry*, *59*(17), 9271–9279.

Pieri, V., Sturm, S., Seger, C., Franz, C., & Stuppner, H. (2012). ^{1}H-NMR-based metabolic profiling and target analysis: A combined approach for the quality control of *Thymus vulgaris*. *Metabolomics*, *8*(2), 335–346.

Pochapsky, S., & Pochapsky, T. (2007). *NMR for physical and biological scientists*. New York: Taylor & Francis.

Politi, M., Peschel, W., Wilson, N., Zloh, M., Prieto, J. M., & Heinrich, M. (2008). Direct NMR analysis of cannabis water extracts and tinctures and semi-quantitative data on delta9-THC and delta9-THC-acid. *Phytochemistry*, *69*(2), 562–570.

Potts, B. C., Deese, A. J., Stevens, G. J., Reily, M. D., Robertson, D. G., & Theiss, J. (2001). NMR of biofluids and pattern recognition: Assessing the impact of NMR parameters on the principal component analysis of urine from rat and mouse. *Journal of Pharmaceutical and Biomedical Analysis*, *26*(3), 463–476.

Purcell, E. M., Torrey, H. C., & Pound, R. V. (1946). Resonance absorption by nuclear magnetic moments in a solid. *Physical Review*, *69*, 37–38.

Raamsdonk, L. M., Teusink, B., Broadhurst, D., Zhang, N., Hayes, A., Walsh, M. C., et al. (2001). A functional genomics strategy that uses metabolome data to reveal the phenotype of silent mutations. *Nature Biotechnology*, *19*(1), 45–50.

Ratcliffe, R. G., & Shachar-Hill, Y. (2001). Probing plant metabolism with NMR. *Annual Review of Plant Physiology and Plant Molecular Biology*, *52*, 499–526.

Ratcliffe, R. G., & Shachar-Hill, Y. (2005). Revealing metabolic phenotypes in plants: Inputs from NMR analysis. *Biological Reviews of the Cambridge Philosophical Society*, *80*(1), 27–43.

Redestig, H., Kusano, M., Ebana, K., Kobayashi, M., Oikawa, A., Okazaki, Y., et al. (2011). Exploring molecular backgrounds of quality traits in rice by predictive models based on high-coverage metabolomics. *BMC Systems Biology*, *5*, 176.

Ren, Y., Wang, T., Peng, Y., Xia, B., & Qu, L. J. (2009). Distinguishing transgenic from non-transgenic *Arabidopsis* plants by ^{1}H-NMR-based metabolic fingerprinting. *Journal of Genetics and Genomics*, *36*(10), 621–628.

Righi, V., Apidianakis, Y., Mintzopoulos, D., Astrakas, L., Rahme, L. G., & Tzika, A. A. (2010). *In vivo* high-resolution magic angle spinning magnetic resonance spectroscopy of *Drosophila melanogaster* at 14.1 T shows trauma in aging and in innate immune-deficiency is linked to reduced insulin signaling. *International Journal of Molecular Medicine*, *26*(2), 175–184.

Ritota, M., Casciani, L., Failla, S., & Valentini, M. (2012). HRMAS-NMR spectroscopy and multivariate analysis meat characterisation. *Meat Science*, *92*(4), 754–761.

Ritota, M., Marini, F., Sequi, P., & Valentini, M. (2010). Metabolomic characterization of Italian sweet pepper (*Capsicum annum* L.) by means of HRMAS-NMR spectroscopy and multivariate analysis. *Journal of Agricultural and Food Chemistry*, *58*(17), 9675–9684.

Roberts, J. K., & Jardetzky, O. (1981). Monitoring of cellular metabolism by NMR. *Biochimica et Biophysica Acta*, *639*(1), 53–76.

Robinette, S. L., Veselkov, K. A., Bohus, E., Coen, M., Keun, H. C., Ebbels, T. M., et al. (2009). Cluster analysis statistical spectroscopy using nuclear magnetic resonance generated metabolic data sets from perturbed biological systems. *Analytical Chemistry*, *81*(16), 6581–6589.

Rodrigues, J. E., & Gil, A. M. (2011). NMR methods for beer characterization and quality control. *Magnetic Resonance in Chemistry*, *49*, S37–S45.

Ross, A., Schlotterbeck, G., Dieterle, F., & Senn, H. (2007). NMR spectroscopy techniques for application to metabonomics. In J. C. Lindon, J. K. Nicholson, & E. Holmes (Eds.), *The handbook of metabonomics and metabolomics* (pp. 55–112): Kidlington, UK: Elsevier Press.

Ruan, C. J., & Teixeira da Silva, J. A. (2011). Metabolomics: Creating new potentials for unraveling the mechanisms in response to salt and drought stress and for the biotechnological improvement of xero-halophytes. *Critical Reviews in Biotechnology*, *31*(2), 153–169.

Rubtsov, D., Jenkins, H., Ludwig, C., Easton, J., Viant, M., Günther, U., et al. (2007). Proposed reporting requirements for the description of NMR-based metabolomics experiments. *Metabolomics*, *3*(3), 223–229.

Ruhl, I. D., Salmon, E., & Hatcher, P. G. (2011). Early diagenesis of *Botryococcus braunii* race A as determined by high resolution magic angle spinning (HRMAS) NMR. *Organic Geochemistry*, *42*, 1–14.

Safer, S., Cicek, S. S., Pieri, V., Schwaiger, S., Schneider, P., Wissemann, V., et al. (2011). Metabolic fingerprinting of *Leontopodium* species (*Asteraceae*) by means of ^{1}H-NMR and HPLC-ESI-MS. *Phytochemistry*, *72*(11–12), 1379–1389.

Saito, K., & Matsuda, F. (2010). Metabolomics for functional genomics, systems biology, and biotechnology. *Annual Review of Plant Biology*, *61*, 463–489.

Sanchez, D. H., Pieckenstain, F. L., Escaray, F., Erban, A., Kraemer, U., Udvardi, M. K., et al. (2011). Comparative ionomics and metabolomics in extremophile and glycophytic Lotus species under salt stress challenge the metabolic pre-adaptation hypothesis. *Plant, Cell & Environment*, *34*(4), 605–617.

Sanchez, D. H., Schwabe, F., Erban, A., Udvardi, M. K., & Kopka, J. (2012). Comparative metabolomics of drought acclimation in model and forage legumes. *Plant, Cell & Environment*, *35*(1), 136–149.

Sanchez, D. H., Siahpoosh, M. R., Roessner, U., Udvardi, M., & Kopka, J. (2008). Plant metabolomics reveals conserved and divergent metabolic responses to salinity. *Physiology Plantarum*, *132*(2), 209–219.

Sánchez Pérez, E. M., García López, J., Iglesias, M. J., López Ortiz, F., Toresano, F., & Camacho, F. (2011). HRMAS-nuclear magnetic resonance spectroscopy characterization of tomato "flavor varieties" from Almería (Spain). *Food Research International*, *44*(10), 3212–3221.

Savitzky, A., & Golay, M. J. E. (1964). Smoothing and differentiation of data by simplified least-squares procedures. *Analytical Chemistry*, *36*(8), 1627–1639.

Savorani, F., Tomasi, G., & Engelsen, S. B. (2010). Icoshift: A versatile tool for the rapid alignment of 1D NMR spectra. *Journal of Magnetic Resonance*, *202*(2), 190–202.

Scalbert, A., Brennan, L., Fiehn, O., Hankemeier, T., Kristal, B. S., van Ommen, B., et al. (2009). Mass-spectrometry-based metabolomics: Limitations and recommendations for future progress with particular focus on nutrition research. *Metabolomics*, *5*(4), 435–458.

Schauer, N., Semel, Y., Balbo, I., Steinfath, M., Repsilber, D., Selbig, J., et al. (2008). Mode of inheritance of primary metabolic traits in tomato. *The Plant Cell*, *20*(3), 509–523.

Schauer, N., Semel, Y., Roessner, U., Gur, A., Balbo, I., Carrari, F., et al. (2006). Comprehensive metabolic profiling and phenotyping of interspecific introgression lines for tomato improvement. *Nature Biotechnology*, *24*(4), 447–454.

Schievano, E., Stocchero, M., Morelato, E., Facchin, C., & Mammi, S. (2012). An NMR-based metabolomic approach to identify the botanical origin of honey. *Metabolomics*, *8*(4), 679–690.

Schutz, C., Quitschau, M., Hamburger, M., & Potterat, O. (2011). Profiling of isoflavonoids in *Iris germanica* rhizome extracts by microprobe NMR and HPLC-PDA-MS analysis. *Fitoterapia, 82*(7), 1021–1026.

Seavey, B. R., Farr, E. A., Westler, W. M., & Markley, J. L. (1991). A relational database for sequence-specific protein NMR data. *Journal of Biomolecular NMR, 1*(3), 217–236.

Seefeldt, H. F., Larsen, F. H., Viereck, N., Petersen, M. A., & Engelsen, S. B. (2011). Lipid composition and deposition during grain filling in intact barley (*Hordeum vulgare*) mutant grains as studied by ^{1}H HR MAS NMR. *Journal of Cereal Science, 54*(3), 442–449.

Sekiyama, Y., Chikayama, E., & Kikuchi, J. (2011). Evaluation of a semipolar solvent system as a step toward heteronuclear multidimensional NMR-based metabolomics for ^{13}C-labeled bacteria, plants, and animals. *Analytical Chemistry, 83*(3), 719–726.

Shachar-Hill, Y., Pfeffer, P. E., & Germann, M. W. (1996). Following plant metabolism in vivo and in extracts with heteronuclear two-dimensional nuclear magnetic resonance spectroscopy. *Analytical Biochemistry, 243*(1), 110–118.

Sharma, A., Cardoso-Taketa, A., Choi, Y. H., Verpoorte, R., & Villarreal, M. L. (2012). A comparison on the metabolic profiling of the Mexican anxiolytic and sedative plant *Galphimia glauca* four years later. *Journal of Ethnopharmacology, 141*(3), 964–974.

Shintu, L., & Caldarelli, S. (2005). High-resolution MAS NMR and chemometrics: Characterization of the ripening of Parmigiano Reggiano cheese. *Journal of Agricultural and Food Chemistry, 53*(10), 4026–4031.

Shintu, L., & Caldarelli, S. (2006). Toward the determination of the geographical origin of Emmental(er) cheese via high resolution MAS NMR: A preliminary investigation. *Journal of Agricultural and Food Chemistry, 54*(12), 4148–4154.

Shintu, L., Caldarelli, S., & Franke, B. M. (2007). Pre-selection of potential molecular markers for the geographic origin of dried beef by HR-MAS NMR spectroscopy. *Meat Science, 76*(4), 700–707.

Shintu, L., Ziarelli, F., & Caldarelli, S. (2004). Is high-resolution magic angle spinning NMR a practical speciation tool for cheese samples? Parmigiano Reggiano as a case study. *Magnetic Resonance in Chemistry, 42*(4), 396–401.

Sidhu, O. P., Annarao, S., Pathre, U., Snehi, S. K., Raj, S. K., Roy, R., et al. (2010). Metabolic and histopathological alterations of Jatropha mosaic begomovirus-infected *Jatropha curcas* L. by HR-MAS NMR spectroscopy and magnetic resonance imaging. *Planta, 232*(1), 85–93.

Silvente, S., Sobolev, A. P., & Lara, M. (2012). Metabolite adjustments in drought tolerant and sensitive soybean genotypes in response to water stress. *PLoS One*, 7(6), e38554.

Simpson, A. J., & Brown, S. A. (2005). Purge NMR: Effective and easy solvent suppression. *Journal of Magnetic Resonance, 175*(2), 340–346.

Simpson, M. J., & McKelvie, J. R. (2009). Environmental metabolomics: New insights into earthworm ecotoxicity and contaminant bioavailability in soil. *Analytical and Bioanalytical Chemistry, 394*(1), 137–149.

Slisz, A. M., Breksa, A. P., Mishchuk, D. O., McCollum, G., & Slupsky, C. M. (2012). Metabolomic analysis of citrus infection by '*candidatus liberibacter*' reveals insight into pathogenicity. *Journal of Proteome Research, 11*(8), 4223–4230.

Smallcombe, S. H., Patt, S. L., & Keifer, P. A. (1995). WET solvent suppression and its applications to LC NMR and high-resolution NMR spectroscopy. *Journal of Magnetic Resonance Series A, 117*(2), 295–303.

Smilde, A., Westerhuis, J., Hoefsloot, H., Bijlsma, S., Rubingh, C., Vis, D., et al. (2010). Dynamic metabolomic data analysis: A tutorial review. *Metabolomics, 6*(1), 3–17.

Smolinska, A., Blanchet, L., Buydens, L. M., & Wijmenga, S. S. (2012). NMR and pattern recognition methods in metabolomics: From data acquisition to biomarker discovery: A review. *Analytical Chimica Acta*, *750*, 82–97.

Sobolev, A. P., Capitani, D., Giannino, D., Nicolodi, C., Testone, G., Santoro, F., et al. (2010). NMR-metabolic methodology in the study of GM foods. *Nutrients*, *2*(1), 1–15.

Sobolev, A. P., Segre, A., & Lamanna, R. (2003). Proton high-field NMR study of tomato juice. *Magnetic Resonance in Chemistry*, *41*(4), 237–245.

Son, H. S., Hwang, G. S., Kim, K. M., Ahn, H. J., Park, W. M., Van Den Berg, F., et al. (2009). Metabolomic studies on geographical grapes and their wines using ^{1}H NMR analysis coupled with multivariate statistics. *Journal of Agricultural and Food Chemistry*, *57*(4), 1481–1490.

Son, H. S., Kim, K. M., van den Berg, F., Hwang, G. S., Park, W. M., Lee, C. H., et al. (2008). 1H nuclear magnetic resonance-based metabolomic characterization of wines by grape varieties and production areas. *Journal of Agricultural and Food Chemistry*, *56*(17), 8007–8016. http://dx.doi.org/10.1021/jf801424u.

Spraul, M., Schutz, B., Humpfer, E., Mortter, M., Schafer, H., Koswig, S., et al. (2009). Mixture analysis by NMR as applied to fruit juice quality control. *Magnetic Resonance in Chemistry*, *47*(Suppl. 1), S130–S137. http://dx.doi.org/10.1002/mrc.2528.

Staab, J. M., O'Connell, T. M., & Gomez, S. M. (2010). Enhancing metabolomic data analysis with Progressive Consensus Alignment of NMR Spectra (PCANS). *BMC Bioinformatics*, *11*, 123.

Steinfath, M., Strehmel, N., Peters, R., Schauer, N., Groth, D., Hummel, J., et al. (2010). Discovering plant metabolic biomarkers for phenotype prediction using an untargeted approach. *Plant Biotechnology Journal*, *8*(8), 900–911.

Stitt, M., & Ap Rees, T. (1978). Pathways of carbohydrate oxidation in leaves of *Pisum-Sativum* and *Triticum-Aestivum*. *Phytochemistry*, *17*, 1251–1256.

Sumner, L., Amberg, A., Barrett, D., Beale, M., Beger, R., Daykin, C., et al. (2007). Proposed minimum reporting standards for chemical analysis. *Metabolomics*, *3*(3), 211–221.

Sumner, L. W., Mendes, P., & Dixon, R. A. (2003). Plant metabolomics: Large-scale phytochemistry in the functional genomics era. *Phytochemistry*, *62*(6), 817–836.

Tang, H., Wang, Y., Nicholson, J. K., & Lindon, J. C. (2004). Use of relaxation-edited one-dimensional and two dimensional nuclear magnetic resonance spectroscopy to improve detection of small metabolites in blood plasma. *Analytical Biochemistry*, *325*(2), 260–272.

Tarachiwin, L., Masako, O., & Fukusaki, E. (2008). Quality evaluation and prediction of *Citrullus lanatus* by ^{1}H NMR-based metabolomics and multivariate analysis. *Journal of Agricultural and Food Chemistry*, *56*(14), 5827–5835.

Tian, C., Chikayama, E., Tsuboi, Y., Kuromori, T., Shinozaki, K., Kikuchi, J., et al. (2007). Top-down phenomics of *Arabidopsis thaliana*: Metabolic profiling by one- and two-dimensional nuclear magnetic resonance spectroscopy and transcriptome analysis of albino mutants. *Journal of Biological Chemistry*, *282*(25), 18532–18541.

Timmers, M., & Urban, S. (2012). On-line (HPLC-NMR) and off-line phytochemical profiling of the Australian plant, *Lasiopetalum macrophyllum*. *Natural Product Communications*, 7(5), 551–560.

Urano, K., Maruyama, K., Ogata, Y., Morishita, Y., Takeda, M., Sakurai, N., et al. (2009). Characterization of the ABA-regulated global responses to dehydration in *Arabidopsis* by metabolomics. *The Plant Journal*, *57*(6), 1065–1078.

Valentini, M., Ritota, M., Cafiero, C., Cozzolino, S., Leita, L., & Sequi, P. (2011). The HRMAS-NMR tool in foodstuff characterisation. *Magnetic Resonance in Chemistry*, *49*(Suppl. 1), S121–S125.

van den Berg, R. A., Hoefsloot, H. C., Westerhuis, J. A., Smilde, A. K., & van der Werf, M. J. (2006). Centering, scaling, and transformations: Improving the biological information content of metabolomics data. *BMC Genomics*, 7, 142.

van der Hooft, J. J., Akermi, M., Unlu, F. Y., Mihaleva, V., Roldan, V. G., Bino, R. J., et al. (2012). Structural annotation and elucidation of conjugated phenolic compounds in black, green, and white tea extracts. *Journal of Agricultural and Food Chemistry, 60*(36), 8841–8850.

van der Hooft, J. J., de Vos, R. C., Mihaleva, V., Bino, R. J., Ridder, L., de Roo, N., et al. (2012). Structural elucidation and quantification of phenolic conjugates present in human urine after tea intake. *Analytical Chemistry, 84*(16), 7263–7271.

van Dorsten, F. A., Grün, C. H., van Velzen, E. J., Jacobs, D. M., Draijer, R., & van Duynhoven, J. P. (2010). The metabolic fate of red wine and grape juice polyphenols in humans assessed by metabolomics. *Molecular Nutrition & Food Research, 54*(7), 897–908.

Vermathen, M., Marzorati, M., Baumgartner, D., Good, C., & Vermathen, P. (2011). Investigation of different apple cultivars by high resolution magic angle spinning NMR. A feasibility study. *Journal of Agricultural and Food Chemistry, 59*(24), 12784–12793.

Verpoorte, R., van der Heijden, R., & Memelink, J. (2000). Engineering the plant cell factory for secondary metabolite production. *Transgenic Research, 9*(4–5), 323–343.

Veselkov, K. A., Lindon, J. C., Ebbels, T. M., Crockford, D., Volynkin, V. V., Holmes, E., et al. (2009). Recursive segment-wise peak alignment of biological ^{1}H-NMR spectra for improved metabolic biomarker recovery. *Analytical Chemistry, 81*(1), 56–66.

Viant, M. R., Bearden, D. W., Bundy, J. G., Burton, I. W., Collette, T. W., Ekman, D. R., et al. (2009). International NMR-based environmental metabolomics intercomparison exercise. *Environmental Science and Technology, 43*(1), 219–225.

Villas-Bôas, S. G., Rasmussen, S., & Lane, G. A. (2005). Metabolomics or metabolite profiles? *Trends in Biotechnology, 23*(8), 385–386.

Vogels, J. T. W. E., Terwel, L., Tas, A. C., van den Berg, F., Dukel, F., & van der Greef, J. (1996). Detection of adulteration in orange juices by a new screening method using proton NMR spectroscopy in combination with pattern recognition techniques. *Journal of Agricultural and Food Chemistry, 44*(1), 175–180.

Vogeser, M., & Seger, C. (2008). A decade of HPLC-MS/MS in the routine clinical laboratory—Goals for further developments. *Clinical Biochemistry, 41*(9), 649–662.

Wang, T., Shao, K., Chu, Q. Y., Ren, Y. F., Mu, Y. M., Qu, L. J., et al. (2009). Automics: An integrated platform for NMR-based metabonomics spectral processing and data analysis. *BMC Bioinformatics, 10*, 15.

Wang, Q. Z., Wu, C. Y., Chen, T., Chen, X., & Zhao, X. M. (2006). Integrating metabolomics into a systems biology framework to exploit metabolic complexity: Strategies and applications in microorganisms. *Applied Microbiology and Biotechnology, 70*(2), 151–161.

Ward, J. L., Baker, J. M., & Beale, M. H. (2007). Recent applications of NMR spectroscopy in plant metabolomics. *FEBS Journal, 274*(5), 1126–1131.

Ward, J. L., Baker, J. M., Llewellyn, A. M., Hawkins, N. D., & Beale, M. H. (2011). Metabolomic analysis of *Arabidopsis* reveals hemiterpenoid glycosides as products of a nitrate ion-regulated, carbon flux overflow. *Proceedings of the National Academy of Sciences USA, 108*(26), 10762–10767.

Ward, J. L., Baker, J. M., Miller, S. J., Deborde, C., Maucourt, M., Biais, B., et al. (2010). An interlaboratory comparison demonstrates that ^{1}H-NMR metabolite fingerprinting is a robust technique for collaborative plant metabolomic data collection. *Metabolomics, 6*(2), 263–273.

Ward, J. L., Forcat, S., Beckmann, M., Bennett, M., Miller, S. J., Baker, J. M., et al. (2010). The metabolic transition during disease following infection of *Arabidopsis thaliana* by *Pseudomonas syringae* pv. tomato. *The Plant Journal, 63*(3), 443–457.

Ward, J. L., Harris, C., Lewis, J., & Beale, M. H. (2003). Assessment of ^{1}H-NMR spectroscopy and multivariate analysis as a technique for metabolite fingerprinting of *Arabidopsis thaliana*. *Phytochemistry, 62*, 949–957.

Weckwerth, W. (2010). Metabolomics: An integral technique in systems biology. *Bioanalysis, 2*(4), 829–836.

Weljie, A. M., Newton, J., Mercier, P., Carlson, E., & Slupsky, C. M. (2006). Targeted profiling: Quantitative analysis of ^{1}H NMR metabolomics data. *Analytical Chemistry*, *78*(13), 4430–4442.

Winning, H., Viereck, N., Wollenweber, B., Larsen, F. H., Jacobsen, S., Sondergaard, I., et al. (2009). Exploring abiotic stress on asynchronous protein metabolism in single kernels of wheat studied by NMR spectroscopy and chemometrics. *Journal of Experimental Botany*, *60*(1), 291–300.

Wishart, D. S., Knox, C., Guo, A. C., Eisner, R., Young, N., Gautam, B., et al. (2009). HMDB: A knowledgebase for the human metabolome. *Nucleic Acids Research*, *37*(Database issue), D603–D610.

Wishart, D. S., Tzur, D., Knox, C., Eisner, R., Guo, A. C., Young, N., et al. (2007). HMDB: The human metabolome database. *Nucleic Acids Research*, *35*(Database issue), D521–D526.

Wolfram, K., Schmidt, J., Wray, V., Milkowski, C., Schliemann, W., & Strack, D. (2010). Profiling of phenylpropanoids in transgenic low-sinapine oilseed rape (*Brassica napus*). *Phytochemistry*, *71*(10), 1076–1084.

Xia, J., Bjorndahl, T. C., Tang, P., & Wishart, D. S. (2008). MetaboMiner—Semi-automated identification of metabolites from 2D NMR spectra of complex biofluids. *BMC Bioinformatics*, *9*, 507.

Xia, J., Mandal, R., Sinelnikov, I. V., Broadhurst, D., & Wishart, D. S. (2012). MetaboAnalyst 2.0—A comprehensive server for metabolomic data analysis. *Nucleic Acids Research*, *40*(Web Server issue), W127–W133.

Yang, M., Cheng, C., Yang, J., & Guo, D. A. (2012). Metabolite profiling and characterization for medicinal herbal remedies. *Current Drug Metabolism*, *13*(5), 535–557.

Yilmaz, A., Nyberg, N., Mølgaard, P., Asili, J., & Jaroszewski, J. (2010). ^{1}H NMR metabolic fingerprinting of saffron extracts. *Metabolomics*, *6*(4), 511–517.

Yuk, J., Simpson, M. J., & Simpson, A. J. (2012). Coelomic fluid: A complimentary biological medium to assess sub-lethal endosulfan exposure using ^{1}H-NMR-based earthworm metabolomics. *Ecotoxicology*, *21*(5), 1301–1313.

Zhang, S., Zheng, C., Lanza, I. R., Nair, K. S., Raftery, D., & Vitek, O. (2009). Interdependence of signal processing and analysis of urine ^{1}H NMR spectra for metabolic profiling. *Analytical Chemistry*, *81*(15), 6080–6088.

Zheng, G., & Price, W. S. (2010). Solvent signal suppression in NMR. *Progress in Nuclear Magnetic Resonance Spectroscopy*, *56*(3), 267–288.

Zietkowski, D., Davidson, R. L., Eykyn, T. R., De Silva, S. S., Desouza, N. M., & Payne, G. S. (2010). Detection of cancer in cervical tissue biopsies using mobile lipid resonances measured with diffusion-weighted ^{1}H magnetic resonance spectroscopy. *NMR in Biomedicine*, *23*(4), 382–390.

Zulak, K. G., Weljie, A. M., Vogel, H. J., & Facchini, P. J. (2008). Quantitative ^{1}H NMR metabolomics reveals extensive metabolic reprogramming of primary and secondary metabolism in elicitor-treated opium poppy cell cultures. *BMC Plant Biology*, *8*, 5.

CHAPTER TWO

Application of LC–MS and LC–NMR Techniques for Secondary Metabolite Identification

Tristan Richard[*,1], **Hamza Temsamani**[*], **Emma Cantos-Villar**[†], **Jean-Pierre Monti**[*]

[*]Université de Bordeaux, ISVV, Groupe d'Etude des Substances Végétales à Activité Biologique, EA 3675, Villenave d'Ornon, France

[†]Instituto de Investigación y Formación Agraria y Pesquera (IFAPA), Rancho de la Merced, Junta de Andalucía, Jerez de la Frontera, Cadiz, Spain

[1]Corresponding author: e-mail address: tristan.richard@u-bordeaux2.fr

Contents

Abstract

An overview of advancements and applications of liquid chromatography coupled to nuclear magnetic resonance (LC–NMR) is given and discussed for secondary metabolites in the plant kingdom: terpenoids, alkaloids and phenolic compounds. The different LC–NMR operating modes are presented in order to structure compound elucidation. On-flow and stop-flow modes are described, together with technical improvements such as LC–SPE–NMR and capLC–NMR. Synergetic combination of hyphenated techniques coupled to mass spectrometry and NMR is illustrated and discussed with an example.

Advances in Botanical Research, Volume 67
ISSN 0065-2296
http://dx.doi.org/10.1016/B978-0-12-397922-3.00002-2

1. INTRODUCTION

In the last two decades, rapid and efficient characterization of secondary metabolites has become a great challenge for drug research and natural product investigations. A large number of known natural products have been identified by classical techniques. Compounds were generally isolated from plant extracts by semi-preparative liquid chromatography and identified using classical spectrometric techniques such as ultraviolet–visible (UV–Vis), infrared, mass spectrometry (MS) or nuclear magnetic resonance (NMR). Nevertheless, this process is time-consuming. In order to eliminate the isolation and purification of compounds from the matrix, hyphenated analytical techniques such as high-performance liquid chromatography (HPLC) coupled to mass spectrometry (LC–MS) or NMR (LC–NMR) have been developed for natural product investigations.

LC–MS is widely used in phytochemical studies for compound identification in complex plant matrices as it presents both high sensitivity and efficiency (Garcia-Beneytez, Cabello, & Revilla, 2003; Guerrero et al., 2009). LC–MS can also be used for compounds characterization by using multiple stage MS experiments (Hossain, Rai, Brunton, Martin-Diana, & Barry-Ryan, 2010; Tsimogiannis, Samiotaki, Panayotou, & Oreopoulou, 2007). Nevertheless, LC–MS alone cannot successfully identify all unknown complex compounds in plant extracts. For example, MS is unable to distinguish between geometrical or optical stereoisomers. In those cases, NMR allows unambiguous structure determination for the individually isolated compounds.

NMR is widely used for studying secondary metabolites from natural plant extracts. It is the most powerful technique for structural identification of unknown complex compounds. Nevertheless, classical NMR studies involve isolation and purification of compounds before NMR spectra acquisition. To reduce these time-consuming steps, physical coupling between liquid chromatography and NMR has been developed over the last 20 years. In practical, LC-NMR routine applications have been successfully applied only in the last decade (Emerson, Kurt, & Jean-Luc, 2007; Exarchou et al., 2005; Silva Elipe, 2003; Sturm & Seger, 2012; Wolfender, Ndjoko, & Hostettmann, 2001).

The present chapter gives an overview of the LC–NMR techniques for identifying natural compounds from the three major classes of secondary metabolites: terpenoids, alkaloids and phenolic compounds. The different LC–NMR techniques are presented from the more basic one, on-flow

NMR measurements, to more complex ones such as Liquid chromatography solid-phase extraction nuclear magnetic resonance (LC-SPE-NMR) or capLC–NMR techniques.

Currently, synergistic combination of LC–MS and LC–NMR techniques yields an efficient way for identifying natural products in complex plant matrices (Acevedo De la Cruz et al., 2012; Cogne et al., 2003). LC–MS provides a rapid preliminary screening of plant crude extract. LC–NMR analysis allows the complete structure determination of unknown natural products. To illustrate this process, an example of LC–MS and LC–NMR synergic combination is presented in order to identify secondary metabolites in plant extract.

2. PRINCIPAL SECONDARY METABOLITES IN PLANTS

Plant compounds are divided into primary and secondary metabolites. Primary metabolites, such as organic acids, amino acids, lipids and other compounds, are directly involved in growth and development or other primary functions. In contrast, secondary metabolites are not directly involved in primary functions. They have specific roles such as protection against infections and pollinator attractants. These organic compounds are widely distributed in the plant kingdom. Moreover, many of these compounds have biological effects on human health and are suggested as medicines. Thus, they are usually called as 'plant natural products' (Croteau, Kutchan, & Lewis, 2000; Kaufman, Brielmann, Cseke, Setzer, & Kirakosyan, 2006).

Secondary metabolites can be distinguished on the basis of their chemical structure, precursor molecules or synthetic pathway. Currently, many thousands have been successfully isolated and characterized. These phytochemical studies have induced the progress of separation techniques and spectrometric approaches such as LC–MS and LC–NMR.

Secondary metabolites can be classified into three major classes: terpenoids (55%), alkaloids (27%) and phenolic compounds (18%) (Fig. 2.1). Secondary metabolites are mainly synthesized by two pathways: mevalonic acid and shikimic acid or aromatic amino acid (Fig. 2.2).

2.1. Terpenoids

Currently, more than 30,000 terpenoids have been reported from natural products and among them are 1000 monoterpenes, 7000 sesquiterpenes and 3000 diterpenes (Bohlmann, Meyer-Gauen, & Croteau, et al., 1998).

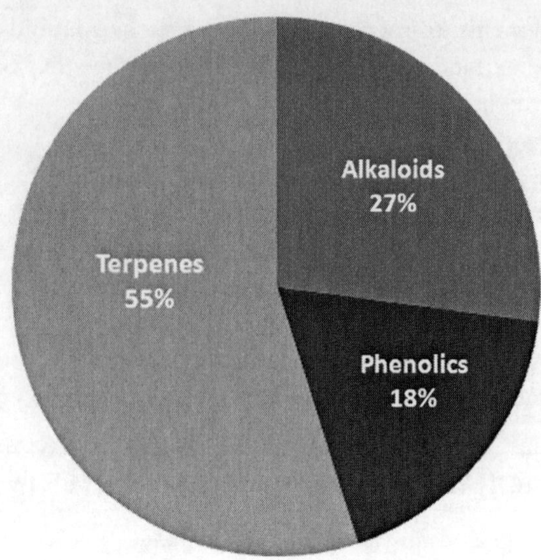

Figure 2.1 Pie chart representing the major groups of plant secondary metabolites according to Croteau et al. (2000). Based on their numbers and diversity, terpenes offer much potential in an array of industrial and medicinal applications.

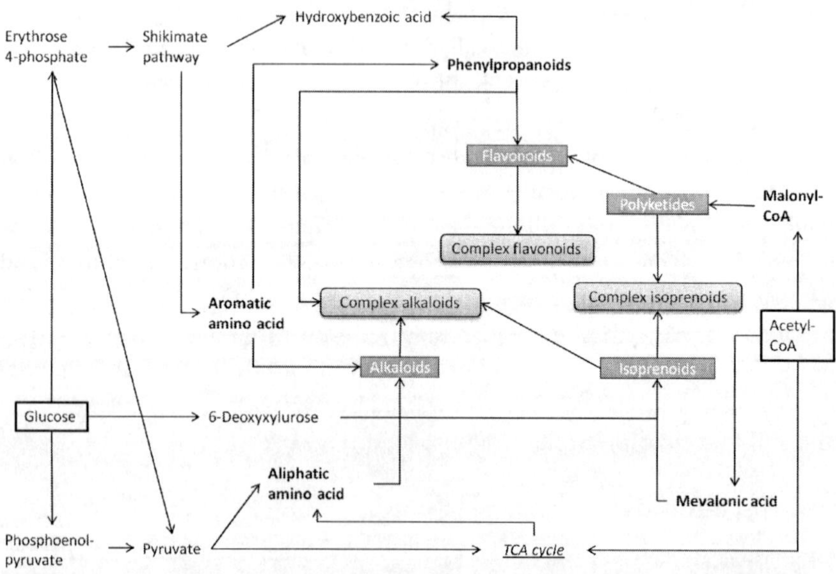

Figure 2.2 Secondary metabolites synthetic pathway. (For colour version of this figure, the reader is referred to the online version of this chapter.)

The main biosynthetic route is known as the mevalonic acid pathway (Fig. 2.2). The second biosynthesis route to terpenes is referred to as either the 1-deoxy-D-xylulose or methylerythriol-4-phosphate pathway (Dubey, Bhalla, & Luthra, 2003; Eisenreich, Bacher, Arigoni, & Rohdich, 2004). These pathways lead to the synthesis of the isopenthyl diphosphate, which is the terpenoid fundamental precursor. They are present in the entire plant kingdom, since they are involved in various vital functions such as defence against many predators, pathogens and competitors (Gershenzon & Dudareva, 2007).

Terpenes, also called isoterpenoids, are defined as a unique group of hydrocarbon-based natural products that possess a structure that may be hypothetically derived from isoprene, giving rise to structures that may be divided into isopentane (2-methylbutane) units.

This is a structurally varied class of secondary metabolites from hemiterpenes (half-terpene) to polyterpenes (Fig. 2.3). Classification of terpenoids has been established according to the number of isoprene units (Table 2.1). Monoterpenes, C10 compounds, are the main constituents of both the volatile essences of flowers and of the essential oils. Examples of acyclic monoterpenes include citral, linalool and geraniol. Well-known cyclic monoterpenes include menthol, limonene and pinene (Fig. 2.3 and Table 2.1). Sesquiterpenes, C15 or compounds having 3-isoprene units, exist in aliphatic bicyclic and tricyclic frameworks. A member of this series, farnesol, is a key intermediate in terpenoid biosynthesis. The diterpenes, C20 compounds, are not considered as essential oils. They are composed of four isoprene units. Taxol and gibberellic acid, a plant growth regulator, are diterpenes. Triterpenes, C30 or compounds having 6-isoprene units, are biosynthetically derived from squalene (Ramawat, Dass, & Mathur 2009).

Sesquiterpenes and diterpenes act as phytoalexins, antimicrobial compounds biosynthesized by plants (Stoessl, Stothers, & Ward, 1976). Some others terpenes have roles in growth and development. Plant hormones such as gibberellins, brassinosteroids and abscisic acid belong to terpenoid family. Finally, sterols and carotenoids, involved in structural function and pigmentation, are triterpenes and tetraterpenes, respectively (Ramawat et al., 2009).

2.2. Alkaloids

Alkaloids are secondary metabolites originally defined as pharmacologically active compounds, primarily composed of nitrogen (Croteau et al., 2000;

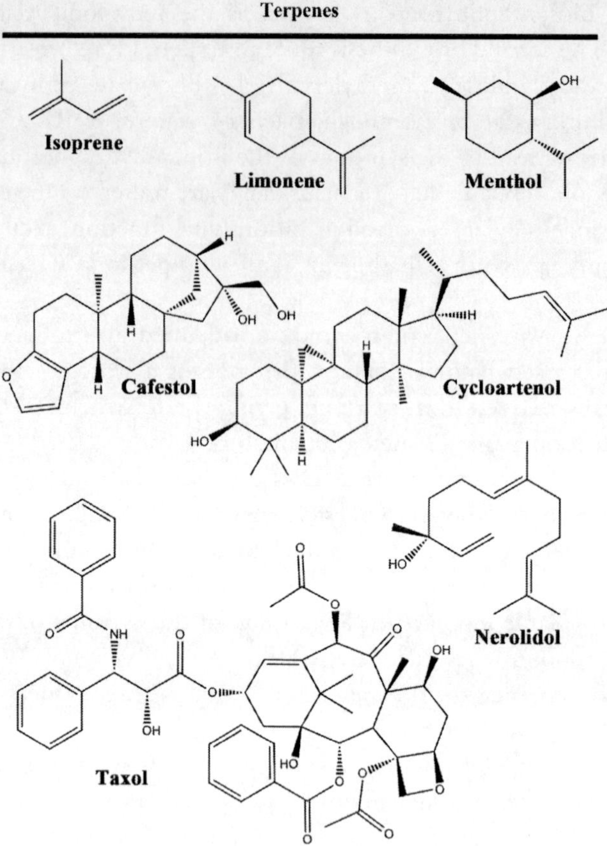

Figure 2.3 Structures of representative terpenoid compounds.

Ziegler & Facchini, 2008). They are synthesized from one of the few common amino acids: lysine, tyrosine and tryptophan. More than 12,000 alkaloids, including more than 150 families, have been identified in plants; and around 20% of the 'species of flowering plants' contain alkaloids. In plants, alkaloids generally exist as salts of organic acids like acetic, malic, lactic, citric, oxalic, tartaric, tannic and other acids. Some weak basic alkaloids (such as nicotine) occur freely in nature (Fig. 2.4). A few alkaloids also occur as glycosides of sugar such as glucose, rhamnose and galactose, for example, alkaloids of the solanum group (solanine), as amides (piperine) and as esters (atropine, cocaine) of organic acids (Ramawat et al., 2009).

The biological properties of various alkaloids have also led to their use as pharmaceuticals, narcotics, stimulants and poisons (Table 2.2). Since the morphine (a well-known alkaloid from *Papaver somniferum* poppy)

Table 2.1 Example of isoprenoids in plants

Class	Compounds	Source	Effects and uses
Monoterpene (C_{10})	Menthol	*Mentha piperita*	Antiseptic, anaesthetic
	Linalool	Lamiaceae (mentha) Lauraceae (laurel) Rutaceae (citrics)	Olfactory pollinator, attractant
	Eugenol	Clove oil	Perfumes, insect attractant
	Citral	Lemon grass oil	Lemon-like odour
Sesquiterpene (C_{15})	Acorone	Sweet flag oil	Insect repellent
	Artemisinin	*Artemisia annua* sweet wormwood	Anti-malaria
Diterpene (C_{20})	*p*-Camphorene	*Kaempferia rotunda* and *angustifolia*	Fragrances, food additive for flavouring
	Taxol	Pacific yew tree *Taxus brevifolia*	Anticancer
Triterpene (C_{30})	Azadirachtin	Neem seeds *Azadirachta indica*	Biological activities among pest insects
	Glycyrrhizin	Liquorice root	Antihypertensive, antiedema
	Ginsenoside rb2	Plant genus *Panax* (ginseng)	Protect memory impairment

discovery, alkaloids are widely used as drugs in modern medicine such as morphine, quinine and codeine. Thus, alkaloids at lower doses are pharmacologically useful, but are poisoning agents at high doses (Table 2.2). In fact, alkaloids can also be highly effective toxins, such as strychnine, because of their wide range of physiological effects. For that reason, this class of secondary metabolites is involved in plant chemical defences mechanisms, especially against mammals.

2.3. Phenolic compounds

Phenolic compounds are found in nearly all the plant kingdom and placed in nearly all parts of the plant. Phenolic compounds are generally synthesized via the shikimate pathway (Fig. 2.2), but the polyketide pathway can also provide some phenolics, such as orcinols and quinones.

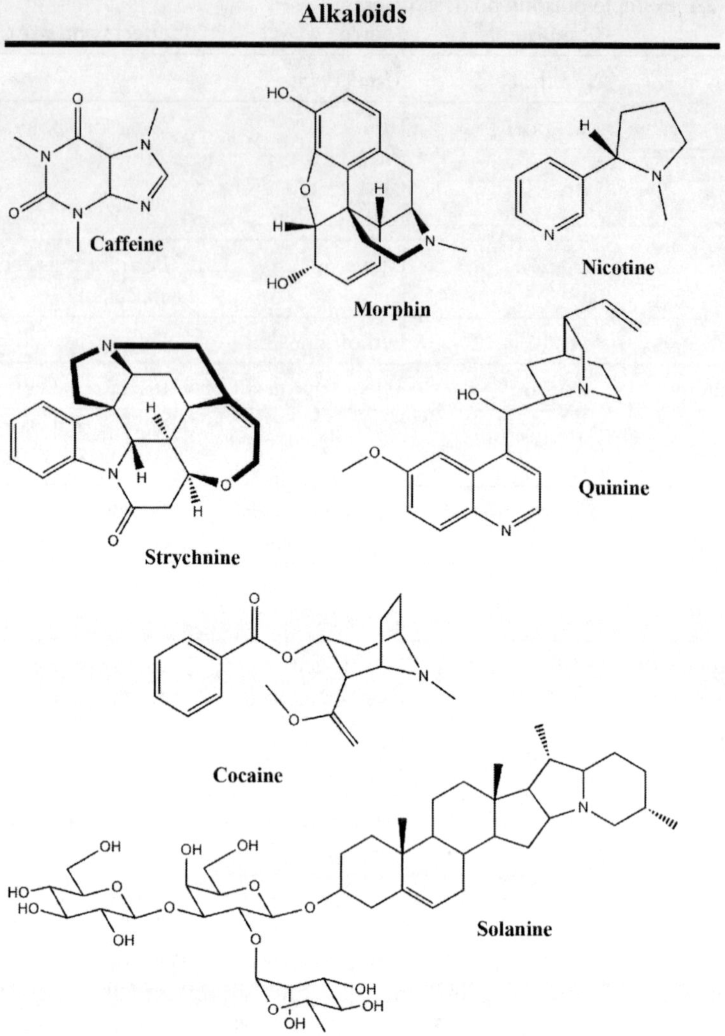

Figure 2.4 Example of alkaloids.

There is a large variability in their structure and occurrence. Phenolic compounds are aromatic metabolites that possess at least one hydroxyl group attached to an aromatic ring. More than 10,000 phenolic compounds have been described. Polyphenols can basically be divided into two groups: flavanoids and non-flavanoids (Fig. 2.5). Flavonoids have a $C_6C_3C_6$ carbon skeleton, whereas non-flavonoids show more simple structures. Flavonoids are brightly coloured compounds generally present in plants as their

Table 2.2 Example of some alkaloids in plants

Compounds	**Source**	**Effects and uses**
Morphine	*Papaver somniferum*	Analgesic
Camptothecin	*Camptotheca acuminata*	Anticancer
Atropine	*Hyoscyamus niger*	Prevention of intestinal spams, antidote to other poisons
Vinblastine	*Catharanthus roseus*	Anticancer
Codeine	*Papaver somniferum*	Analgesic, antitusive
Caffeine	*Coffea arabica*	Stimulant, natural pesticides
Nicotine	*Nicotiana tabacum*	Stimulant, tranquillizer
Cocaine	*Erythroxylon coca*	Stimulant of the central nervous system local anesthetic

glycoside derivatives. Different classes within this group differ by additional oxygen-containing heterocyclic rings and hydroxyl groups and include numerous compounds such as flavonols, flavones, isoflavones, flavanones and anthocyanins. Anthocyanins impart red and blue pigment to flowers and fruits and can make up as much as 30% of the dry weight of some flowers. Flavonols, flavanones and anthocyanins are present as their glycoside derivatives (Ramawat et al., 2009).

Moreover, phenolic compounds are known to have a wide range of biochemical properties and health benefits (Quideau, Deffieux, Douat-Casassus, & Pouységu, 2011). Additionally, they are really important in food since they are responsible for food quality. Finally, they impact on plant growth and defence mechanisms (Table 2.3).

3. HYPHENATED CHROMATOGRAPHIC TECHNIQUES IN METABOLITE IDENTIFICATION

High-performance liquid chromatography coupled to mass detection (HPLC–MS) is the most frequently used technique for analyzing secondary metabolites. HPLC allows powerful and rapid separations of metabolites in a mixture, and MS immediate and highly sensitive detection. HPLC–MS has become an important tool for secondary metabolites characterization, providing mass spectrum of intact molecular and fragment ions. This technique is fast and the short exposure of the compounds to light and air limits their

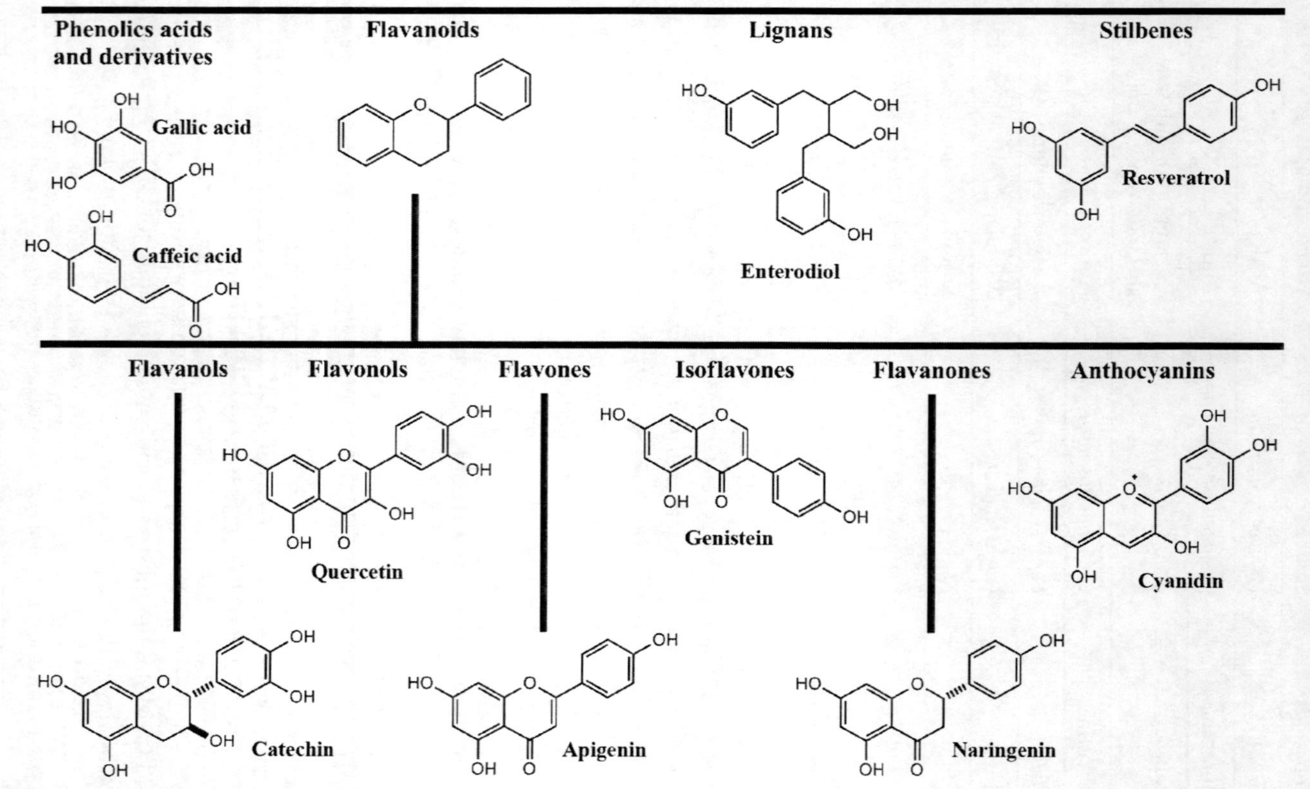

Figure 2.5 Chemical structures of polyphenols.

Table 2.3 Example of some phenolic in food

Class	**Compounds**	**Source**	**Effects and uses**
Anthocyanins	Cyanidin	Red fruits	Attract animals (pigments)
Flavonols	Quercetin	Tea, apple, onion	Protect from UV light damage
Flavanols	Catechin	Cocoa, tea	Antibiotic, herbicide
Flavanones	Naringenin	Grape, orange, tomato	Bitter flavour
Flavones	Apigenin	Persil, celery, green pepper	Protect from UV light damage
Isoflavones	Genistein	Soya	Antimicrobial activity
Lignanes	Enterodiol	Cereals, soybean, strawberry	Structural function
Stilbenes	Resveratrol	Grape, peanut	Defence (phytoalexin)
Coumarins	Umbeliferone	Carrot	Defence (phytoalexin)
Simple phenylpropanoid	Caffeic acid	Strawberry, grape	Precursor of many other phenolics
Benzoic acid derivatives	Salicylic acid	Lettuce, apple, plumb, artichoke	Resistance to plant pathogen

degradation. However, MS data do not give detailed and conclusive structural information, especially when isomeric compounds are studied like *cis* and *trans* conformers. In such cases, the combination of both MS and NMR spectrometries leads to unequivocal identification of the compound individual structures.

NMR spectrometry is a powerful technique allowing the complete structural elucidation and identification of natural compounds. Generally, in addition to 1D-NMR experiments (such as ^{1}H- and ^{13}C NMR spectra), 2D-NMR measurements are needed to identify organic compounds. Neighbourhood relationships within the proton scaffold are identified by correlated spectroscopy (COSY) and/or total correlation spectroscopy (TOCSY) measurements. Connections between proton and carbon scaffolds were investigated by ^{1}H–^{13}C heteronuclear NMR measurements: heteronuclear single quantum correlation (HSQC) and heteronuclear multi-quantum correlation (HMBC) experiences. Finally, through-space correlations between protons close to each other are studied by nuclear

Overhauser effect spectroscopy (NOESY) or rotational Overhauser effect spectroscopy (ROESY). These techniques are the centrepiece of structure determination using NMR spectrometry.

However, fractionation of the extract prior to its isolation is necessary for the NMR analysis. Liquid chromatography coupled with NMR detection (LC-NMR) has become an important technique for food and natural products analysis and may allow the complete assignment and structure determination of individual metabolites without the need of their isolation.

Ideally, the use of hyphenated chromatographic techniques coupled to mass and NMR spectrometry should allow the complete spectrometry characterization of different compounds in a mixture in a single analysis (Fig. 2.6). In practice, these techniques offer a rapid way for characterizing new unknown metabolites or for dereplicating known analytes. However, some structure details will remain undetermined.

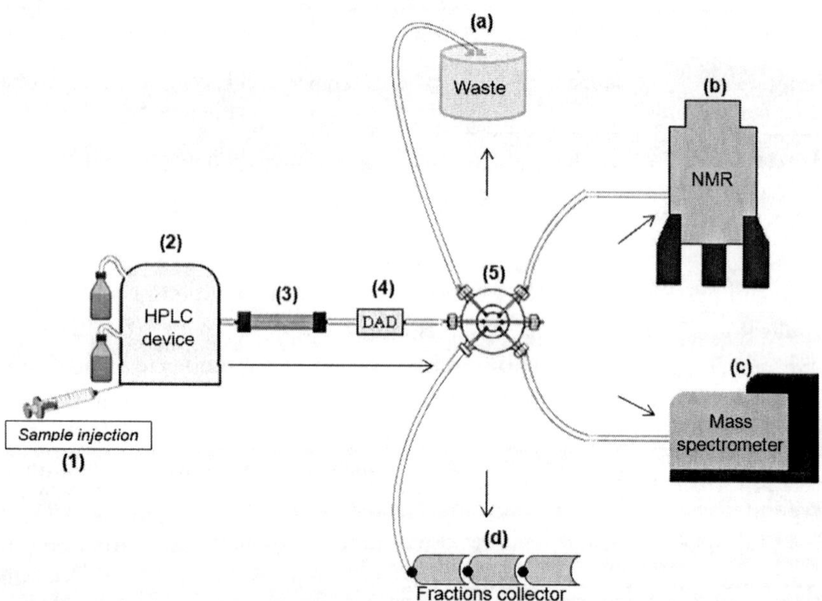

Figure 2.6 Layout of hypenated acquisitions. The sample is injected (1) by the pump inside the HPLC device (2). It is thus micro-fractionated by elution trough the inverse phase column (3). Exiting peaks are then detected by the DAD (4) and directed towards a splitting valve (5). The user can then choose whether he wants to direct the peak towards the waste container (a) or towards de fractions collector (d). If the peak has to be analysed, it can be injected into the NMR flow-cell for structure elucidation (b) or towards the mass spectrometer for further ionization and mass analysis (c). (For colour version of this figure, the reader is referred to the online version of this chapter.)

In general terms, LC–MS is used as a first dereplication step for the chemical profiling of crude plant extracts. Compounds are tentatively identified based on the comparison of their molecular weight and fragment information with those recorded into natural product libraries. LC–NMR is mainly used in a second step for a more detailed structural investigation of compounds presenting original structural features or displaying interesting activities according to previously performed bioassays (Fig. 2.7).

4. LC–NMR TECHNIQUES

4.1. LC–NMR spectrometry in natural product identification

NMR is a technique of choice to quickly determine the complete structure of natural compounds. The coupling of this technique with the HPLC high separative abilities allows an online and complete identification of compounds present in a vegetal extract (Emerson et al., 2007; Exarchou et al., 2005; Silva Elipe, 2003; Sturm & Seger, 2012; Wolfender et al., 2001). LC–NMR provides that information without needing a particular handling to treat the eluted peak for an NMR analysis. This is why LC–NMR is a technique under a constant development since its first uses at the end of the 1970s (Watanabe & Niki, 1979).

Coupling HPLC with NMR requires some modifications on the NMR detector's hardware. It is indeed necessary to use a particular probe connected to the HPLC capillary. This probe is called an LC–NMR probe and commonly has an internal volume ranged from 60 to 220 μL. Acquisitions in LC–NMR can be realized through several acquisition modes (Fig. 2.8).

4.2. Principal modes of acquisition for LC–NMR

4.2.1 On-flow and stop-flow acquisition

4.2.1.1 On-flow mode

The mode of acquisition based on on-flow LC–NMR experiments is the most basic one (Spring et al., 1995; Wolfender, Rodriguez, Hostettmann, & Hiller, 1997). It consists of setting online an HPLC apparatus and an NMR detector. The sample eluted with a deuterated solvent is transferred directly and uninterruptedly towards the NMR detection flow-cell (Fig. 2.8). This continuous elution implies that the exposure of an elution peak within the detection cell has a very short duration, approximately 3.6 s, with a flow of 1 mL/min using a 4.6-mm column (Silva Elipe, 2003).

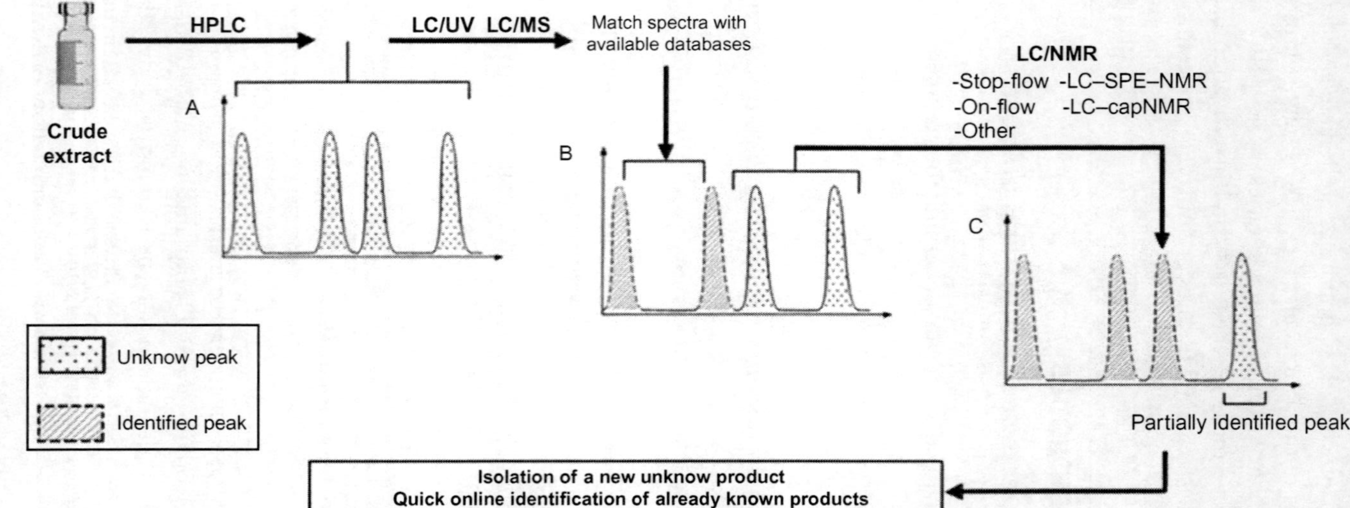

Figure 2.7 Sample peaks dereplication and NMR use optimization. The crude extract is micro-fractionated after a run through HPLC. Several peaks need then to be identified (A). Using LC/UV or LC/MS provides data on the peak's compounds. Those data can be compared to the available databases in order to identify the peaks corresponding to already known compounds (B). Only the peaks that are still unidentified need to be analysed by NMR. NMR spectroscopy can provide additional data, which will help in identifying those peaks (C) through matching with the databases. Nevertheless, some peaks will not be identified by this way, indicating that they potentially contain a new unknown compound. This step-by-step analysis allows a quick on-line identification of known products and isolation of an unknown one, which can be then studied deeper.

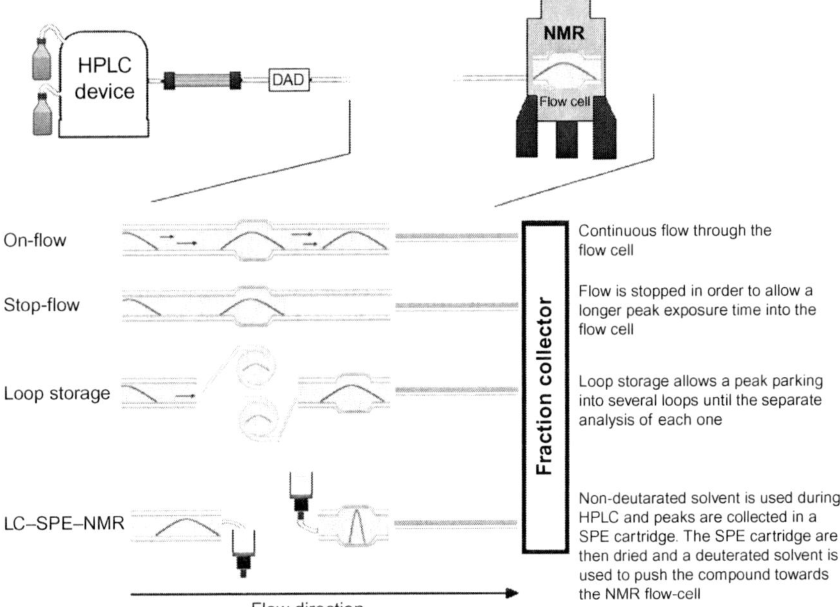

Figure 2.8 LC–NMR and acquisition techniques.

On-flow LC–NMR measurements were recorded successfully for secondary metabolites characterization in various natural product extracts. Spring et al. report the identification of sesquiterpene lactones and flavanones by the combination of LC–MS and on-flow LC–RMN mode techniques in plant extracts (Spring et al., 1995, Spring, Heil, & Vogler, 1997). For example, in *Zaluzania grayana*, application of on-flow LC–NMR allowed the identification of three major compounds in the microgram range by the observation of ^{1}H NMR spectra. Concerning alkaloids, in *Erythroxylum vacciniifolium*, on-flow LC-^{1}H NMR data allowed the identification of five main alkaloids of the crude extract (Zanolari et al., 2003). With regard to phenolic compounds, on-flow LC–NMR analysis was successfully used. In order to rapidly identify the active compounds from *Blumea gariepina* extracts on-flow LC–^{1}H NMR spectra were recorded (Queiroz et al., 2005). Eight compounds were identified including flavonoid derivatives. Online couplings of LC–NMR and LC–MS were used to identify secondary metabolite products of *Streptomyces violaceoruber* (Pham, Vater, Rotard, & Mügge, 2005). Eight main compounds were identified by the combination of on-flow LC–^{1}H NMR experiments and offline 2D-NMR spectrometry. In particular, on-flow

LC–^{1}H NMR experiments allow the direct identification of some flavonoids such as daidzein and genistein.

Nevertheless, NMR is a poorly sensitive method of detection: (i) acquisitions in LC-NMR are thus limited to the only ^{1}H or ^{19}F spectra (Ghosh & Konishi, 2007) and (ii) only the major peaks of the mixture are correctly detected. Moreover, the detection sensitivity lowers as the width of the elution peak increases. Indeed, only a fraction of the elution peak is observed by the NMR detector if the peak width is larger than the flow-cell capacity, implying that only part of the compound's eluted quantity is analyzed. This limited exploitation of the peak is a cause of sensitivity loss. Another disadvantage of this acquisition mode is the need for carrying out the removal of solvent signal, causing then the loss of analyte signals if this last is masked by the solvent. A solution used is the acquisition of NMR data in the presence of two different solvents during two independent experiments. The two solvent signals parasitize the spectra at different chemical shifts, but superposing the two spectra allows the deduction of a hypothetical third spectrum that has the entirety of the analyte signals (Ndjoko, Wolfender, Röder, & Hostettmann, 1999). The acquisitions results are presented in the shape of 2D graphic. One of the two dimensions of the graph is represented by a scale of time to which are associated elution times of each peak, and each elution time is matched with its 1D-NMR spectra.

One must notice that the on-flow mode is dedicated to analyze high abundance constituents of mixture by acquiring ^{1}H or ^{19}F NMR data. Currently, this mode is not used for secondary metabolites identification because of its limitations. Indeed, the acquisition methods development in LC–NMR made it possible to decrease the peaks width and to transfer more important concentrations of analyte within the flow-cell. These improvements increased LC–NMR sensitivity and also allowed to record more complete NMR experiments by carrying out, for example, 2D correlation spectra on the peaks of interest.

4.2.1.2 Stop-flow mode

By using stop-flow mode, it is possible to considerably increase the exposure time of the elution peak within the flow-cell. Realizing acquisitions in stop-flow mode requires the connection of a UV–Vis or diode array detector (DAD), between the HPLC and the NMR devices (Fig. 2.8). When a peak leaves the HPLC column, it is detected by the DAD. The duration necessary to the passage of this peak from the column exit until the arrival within the NMR probe's flow-cell is recorded. By knowing the value of this delay, it is possible to stop the elution after the peak detection by the DAD. The peak can then be parked into the LC–NMR probe cell for up to few days. This long duration exposure to NMR detection allows enhancing the number of acquisitions, thus increasing considerably the signal-to-noise ratio. It also allows carrying out 2D-NMR experiments (Bringmann, Wohlfarth, Rischer, Schlauer, & Brun, 2002; Schefer et al., 2006), which can lead to obtain additional details about the structure of the analyte presented in the elution peak.

Stop-flow mode is generally used when metabolite quantity is reasonable or for unstable compounds. To illustrate this mode, the analysis of major metabolites of vine stalk extract is presented (Fig. 2.9). The extract was eluted with a gradient of deuterated water (A) and acetonitrile (B) according to the following gradient program (v/v): 0 min 85% A 15% B, 15 min 100% B. The column was a Prontosil C_{18} (5 μm, 250 mm × 4.6 mm), and flow rate was typically 0.6 mL/min. ^{1}H NMR spectra were acquired in stopped-flow mode. Solvent suppression was achieved by a one-dimensional version of the NOESY pulse sequence with presaturation during the relaxation delay and mixing time at two frequencies simultaneously. ^{1}H NMR spectra (Fig. 2.10) on the LC–NMR system (Bruker Avance III 600 MHz equipped with equipped with an ^{1}H–^{13}C inverse-detection flow probe with an active volume of 120 μL) revealed the presence of two phenolic compounds belonging to the stilbene family: resveratrol (**1**) and ε-viniferin (**2**).

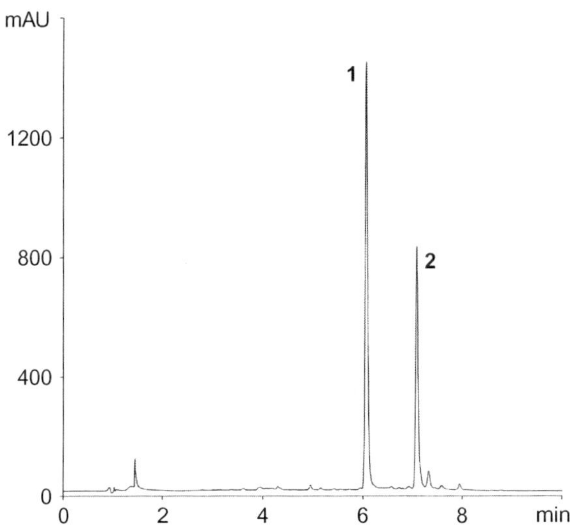

Figure 2.9 HPLC chromatograms in vine stalk extract recorded at a signal of 360 nm.

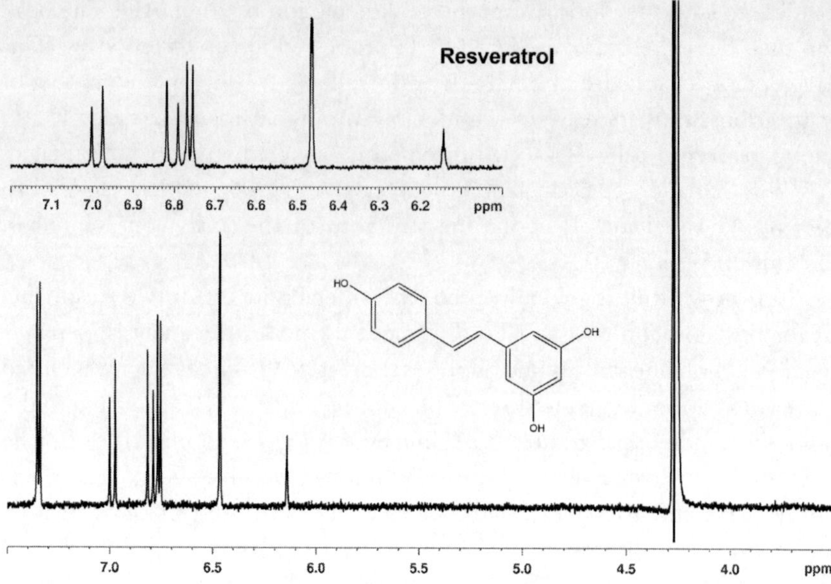

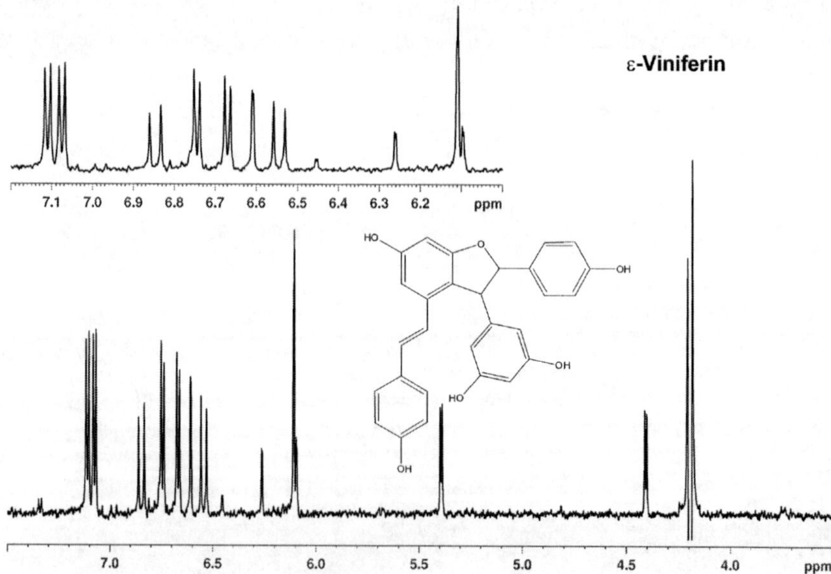

Figure 2.10 LC–^{1}H NMR spectra of resveratrol (**1**) and ε-viniferin (**2**) in stop-flow mode.

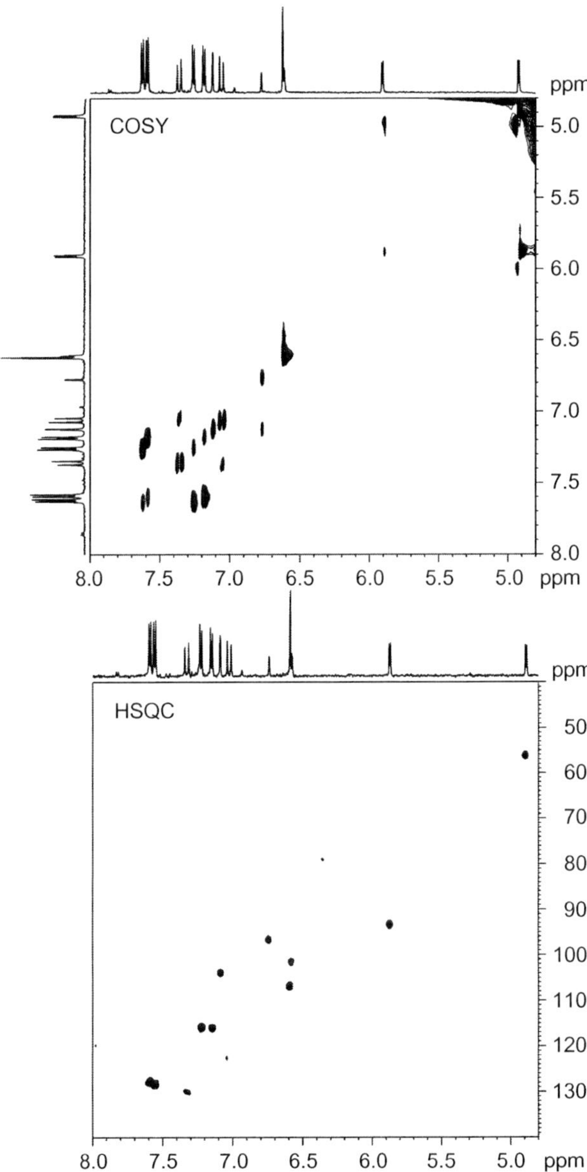

Figure 2.11 COSY and HSQC spectra of ε-viniferin (**2**) in stop-flow mode.

To confirm identification of ε-viniferin, 2D-NMR experiments were realized (including COSY, ROESY and HSQC) on the LC–NMR system over night. COSY and HSQC experiments were presented in Fig. 2.11. Solvent suppression was achieved by using shaped pulse for off-resonance presaturation. This illustration emphasizes the relevance of stop-flow experiments for structural analyses of secondary metabolites. This mode is widely used for natural products and pharmaceutical investigations (Elipe, Huskey, & Zhu, 2003; Iwasa et al., 2003; Stintzing, Conrad, Klaiber, Beifuss, & Carle, 2004).

Stop-flow LC–NMR experiments were developed for characterizing alkaloids in various plant extracts. Combination of stop-flow LC–^{1}H NMR and LC–MS analysis was applied for identifying new unstable iridoid glycosides and for monitoring their transformations (Cogne et al., 2003, 2005). Stop-flow LC–^{1}H NMR and LC–2D-NMR measurements (COSY experiences) have been recorded for the analysis of carotenoids in food products such as palm oil, tomato, mandarin or orange juices (Tode, Maoka, & Sugiura, 2009). In this study, various kinds of carotenoids were identified by comparing NMR spectra with those of standard. In *Ancistrocladus griffithii* extracts, 2D-NMR experiments, such as TOCSY and ROESY, were recorded to characterize new monomeric naphthylisoquinoline alkaloids. Stop-flow LC–NMR has been successfully applied in the phytochemical investigation of *Laurencia filiformis* a marine organism (Dias, White, & Urban, 2009). Several sesquiterpenes were identified and their online chemical conversions monitored. In combination with 2D-NMR experiments, stop-flow LC–^{1}H NMR measurements allowed the identification of new chamigenes and sesquiterpenes in Australian marine alga *Laurencia elata* (Dias & Urban, 2011). Concerning phenolic compounds, several classes have been successfully characterized by stop-flow LC–NMR such as phenolic acids, flavonoids, stilbenes, or lignans (de Rijke, de Kanter, Ariese, Brinkman, & Gooijer, 2004; Qu et al., 2008; Timmers & Urban, 2012; Zhao et al., 1999). In *Torreya jackii* extract, seven lignans were analyzed by stop-flow LC–^{1}H NMR analysis (Zhao et al., 1999). The structures were confirmed thanks to conventional NMR measurements. Stop-flow LC–NMR analysis was performed for identifying isoflavone glucoside malonates in *Trifolium pratense* (de Rijke et al., 2004). Complete compound determination was achieved by offline COSY and NOESY experiments to determine the positions of the glucose moiety.

The analysis of each peak implies stopping the elution at each time. Many stops and long detection duration can give place to a diffusion phenomenon within the column and the NMR flow-cell (Keifer, 2010). This passive diffusion can cause a widening of the elution peaks, having then negative effects on the recorded spectrum quality. However, this resolution loss seems to have weak incidences (Wolfender et al., 2001).

An often used technique in order to optimize this mode of acquisition is the loop-storage-assisted acquisition (Tseng, Braumann, Godejohann, Lee, & Albrecht, 2000). Each peak is collected and stored in a capillary loop for a later analysis by NMR (Fig. 2.8). Peaks collection can be thus created and each one will be then isolated, preventing from any dilution or mixing phenomenon. That also avoids undergoing the diffusion phenomenon in the HPLC column and in the capillary. This technique is adapted for the simultaneous analysis of several components from the same extract.

These two traditional methods of stop-flow and loop-collection suffer from several limitations that impact the spectra quality. However, it is possible to largely improve the acquisitions sensitivity while refining the peaks width and decreasing the quantity of biological material. This last parameter is particularly important during the secondary metabolites analysis.

4.2.2 LC-SPE-NMR

LC-SPE-NMR is a mode of acquisition first described by Exarchou, Godejohann, van Beek, Gerothanassis, and Vervoort (2003). When analyte concentration is too low, the sensitivity of the LC–NMR precludes its structure determination. A solution to this problem consists in inserting a solid-phase extraction (SPE) apparatus between the HPLC system and the NMR spectrometer in order to trap and accumulate the compounds onto SPE cartridges (Fig. 2.8). The matrix of those cartridges is usually hydrophobic when it is about phytochemical compounds studies. Depending on compound classes, different cartridge can be used such as GP (general phase, a polystyrene-type polymer) or C18 phase (Lambert, Stærk, Hansen, & Jaroszewski, 2005). Each compound can be eluted with a specific deuterated solvent before being trapped into the NMR probe.

Basically, the detected peak exiting from the column is diluted by water, in order to induce the analyte adsorption on the matrix, and then transferred towards the SPE. The cartridge is then dried by nitrogen in order to eliminate chromatographic solvents. Lastly, a deuterated elution solvent is used to inject the peak towards the flow-cell for NMR analysis.

This method thus makes it possible to pre-concentrate the analyte present in the peak inside a low volume of deuterated solvent; this volume can be about the capacity of the NMR flow-cell. The pre-concentration provide peaks sharper than those obtained at the HPLC column exit (Jaroszewski, 2005). The entirety of the peak will be thus analyzed by the LC–NMR flow-cell, which results in better detection results. LC-SPE-NMR thus also prevents from using deuterated solvents in great and expansive quantity during the liquid chromatography. Exchanging solvents volume and/or nature at the column exit has other advantages; it becomes possible to use a more suitable solvent for the NMR detection. Moreover, the injection of a lower solvent volume in the LC–NMR probe considerably reduces the issues than one can meet during the removal of the solvent signal.

A major asset of acquisitions through LC-SPE-NMR is the possibility to increase the detectable analyte mass through numerous trappings of the same peak on the same SPE cartridge. Successive injections are carried out and the analyte peak is each time trapped on the same cartridge. The spectrum quality can be then improved thanks to the availability of more analyte quantity inside the LC–NMR probe.

The secondary metabolites are often minor metabolites within the vegetal extract. Thus it is often necessary to use great extract volume and quantity in order to be able to work with sufficient metabolites quantities for NMR detection. This can involve an overload of the HPLC column. LC-SPE-NMR allows splitting the extract volume while making it possible to recover an identical quantity, if not higher, of secondary metabolite at the end of the successive injections of the volume fractions.

LC-SPE-NMR offers several advantages compared to the other more conventional techniques. Nevertheless, an essential parameter to take into account during the use of this method is the choice the elution solvent and the nature of the matrix (Clarkson, Madikane, Hansen, Smith, & Jaroszewski, 2007). Those have to be appropriated to the studied analyte. Indeed it can happen that the analyte is retained too little or too much within the matrix, involving consequent elution issues.

To date LC-SPE-NMR is widely used in natural products investigations. Numerous studies have been successfully conducted by using this technique. Various terpenoids (including diterpenes, iridoids and sesquiterpenes) have been successfully investigated by LC-SPE-NMR (Clarkson, Hansen, Smith, & Jaroszewski, 2006a, 2006b; Clarkson et al., 2007; Sorensen et al., 2006). LC-SPE-NMR measurements allowed the identification of new tricyclic diterpenes in *Harpagophytum procumbens* crude extracts (Clarkson et al., 2006a). Structure elucidations were achieved by using ^{1}H NMR, NOESY and HMBC spectra. NOESY and HMBC spectra were recorded in the LC-SPE-NMR mode by multi-trapping (ten and eight trappings, respectively). NOESY and HMBC measurement allowed the structural identification of these secondary metabolites by unequivocal assignment of all signals. Numerous alkaloids, such as isomeric tropane alkaloids, have been identified by LC-SPE-NMR experiments (Staerk et al., 2009; Sturm, Seger, Godejohann, Spraul, & Stuppner, 2007). Concerning phenolic constituents of plant extracts, in Greek *Hypericum perforatum*, Tatsis et al. have reported the identification of several complex compounds including naphtodianthrones, phloroglucinols, flavonoids and phenolic acids (Tatsis et al., 2007). Direct analysis by LC-SPE-NMR analysis of the polar part of olive oil permitted structure elucidation of simple phenols, lignans and flavonoids (Christophoridou, Dais, Tseng, & Spraul, 2005). In greek oregano, multiple flavonoids have been successfully identified (Exarchou et al., 2003). 2D-NMR experiments were acquired in this study including homonuclear TOCSY and heteronuclear HMQC and HMBC experiments, thanks to the multiple trapping on the same SPE cartridge.

4.2.3 Other acquisition techniques

4.2.3.1 CapLC-NMR

NMR probes are characterized by their mass sensitivity, which is a parameter directly related to the signal-to-noise ratio. Mass sensitivity is inversely proportional to the detection coil volume. This link can be explained by the fact that the transverse magnetic field intensity per volume unit is more important within a detection coil of small size (Ghosh & Konishi, 2007). The capLC–NMR is a method in which the traditional LC–NMR probe is replaced by a capillary presenting a zone of detection of 1.5 μL surrounded by a small detection coil (Wolfender, Queiroz, & Hostettmann, 2005). The eluted peaks are collected in 96-well plates before being dried by evaporation. The analyte remaining in the wells are then diluted with an appropriated deuterated solvent and finally injected towards the capLC–NMR probe. A push volume of solvent is used to maintain the peak in the centre of the microcoil.

Numbers of studies have established that high throughput analysis based on capLC–NMR leads to a way better spectra quality when compared to those obtained through other acquisition modes (Exarchou et al., 2005). Another advantage is that a reduced amount of analyte is needed for this kind of NMR detection, approximately 10 μg (Xu & Alexander, 2005). A discussable disadvantage of capLC–NMR is the transfer efficiency between the HPLC apparatus and the capLC–NMR probe. Nevertheless fully automated capLC–capNMR experiences are possible (Jansma et al., 2005), where the transfer seems to be sufficiently efficient.

Improving the NMR instrumentation with a home-made capLC–NMR instrumentation allowed to separate and identify four terpenoids with only 25 nmol of each one using ^{1}H NMR. A home-made capNMR probe was then made by wrapping a detection capillary with a flat copper wire, which lead to an NMR observable volume of 1.1 μL. The liquid chromatography capillaries were then transferred to this improved microcoil (Lacey, Tan, Webb, & Sweedler, 2001).

Concerning polyphenols, LC–capNMR was able to provide great quality spectrum with using only thirty micrograms of each compound (Xiao, Krucker, Putzbach, & Albert, 2005). Passing this quantity through a 1.5-μL active-volume capillary within the microprobe allowed identifying four isoflavones in *Radix astragali* by using ^{1}H capNMR-LC experiments.

Alkanoids were also studied through LC–capNMR. A library of taxol-derivatives molecules have been fully dereplicated by analyzing 5–10 μg of compounds by COSY-based experiments. A 1.5-μL capillary flow-probe was then used. For additional experiment in order to determine unknown compounds structure, 50 μg were used for HMQC and HMBC experiments (Eldridge et al., 2002).

4.2.3.2 Column-trapping

Pre-concentrating the peak of elution is a method that increases the quantity of material seen within the volume of the coil, increasing significantly the NMR signal quality. Column-trapping is a parallel method to the SPE that also provides pre-concentrated analyte. The column trapping was described by Djukovic, Liu, Henry, Tobias, and Raftery (2006); it is based on the connection of a second chromatography column into the LC–RMN system. Peaks eluted out of the first column, with a non-deuterated solvent, are stored in a loop. A syringe pushes then a peak out of the loop towards a second column using 90% deuterated water and 10% deuterated acetonitrile solvents. This mixture aims to decrease the peak organic solvents concentration. Once the peak has entered into the column, a syringe injects a 10%-water and 90%-deuterated acetonitrile solvent. This high concentration in acetonitrile allows rushing the analyte into the second column, eluting it into a more concentrated sharper peak. This peak is then transferred towards the NMR detector. The column trapping improves the signal by a factor 15 (Djukovic et al., 2006), compared to factor ranging between 8 and 30 for SPE (Xu & Alexander, 2005). This method also allows avoiding the use of deuterated solvents during the liquid chromatography.

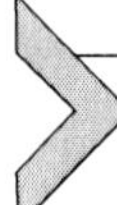

5. COMBINATION OF LC–MS AND LC–NMR IN METABOLITE PROFILING

In many cases, the LC–NMR analyses are performed independently from the LC–MS runs. LC–MS is used as a first dereplication step for the chemical profiling of plant extracts and compounds are tentatively identified based on their molecular weight and fragment information after a manual search in the secondary metabolite libraries. LC–NMR is mainly used in a second step for a more detailed structural investigation of compounds presenting original structural features.

To illustrate this methodology, the obtained results concerning composition of anthocyanins in various *Vitis* are presented. Based on LC–MS and LC–NMR data, it was possible to identify a total of 33 compounds in the different extracts (Acevedo De la Cruz et al., 2012). Anthocyanins have simple structures derivating from five anthocyanidins: cyanidin, delphinidin, petunidin, peonidin and malvidin (Fig. 2.12). Nevertheless, the glycosylation pattern of grape anthocyanins could be complex. In wild *Vitis*, glycosylation could occur at both the 3- and 5-positions giving complex structures.

LC–MS experiments give crucial structural information about each structure. MS/MS mode provides information on the aglycone and its corresponding sugar due to the observed *m*/*z* characteristic fragmentation values (Fig. 2.13). Moreover, observation of MS data indicates the presence of some substituent groups, such as *p*-coumaroyl or acetyl units. LC–MS is a rapid and cost-effective technique for the identification of secondary metabolites. Nevertheless, mass data alone do not allow the complete structure elucidations. Isomers, such as *cis*- and *trans*-coumaroyl-glucosides derivatives, cannot be differentiated from each other only based on mass measurements.

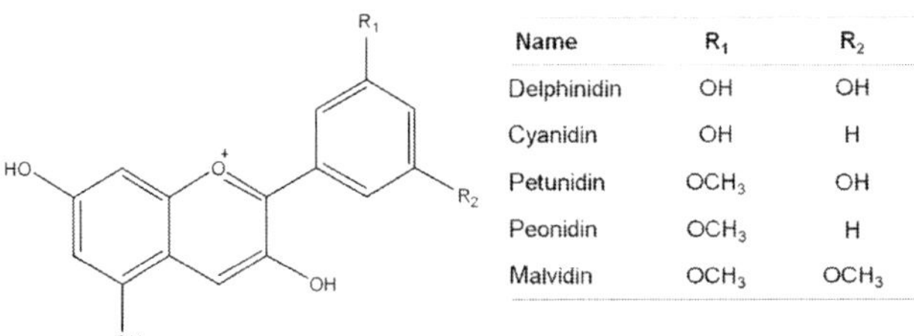

Name	R_1	R_2
Delphinidin	OH	OH
Cyanidin	OH	H
Petunidin	OCH_3	OH
Peonidin	OCH_3	H
Malvidin	OCH_3	OCH_3

Figure 2.12 Chemical structures of anthocyanin aglycones in vine.

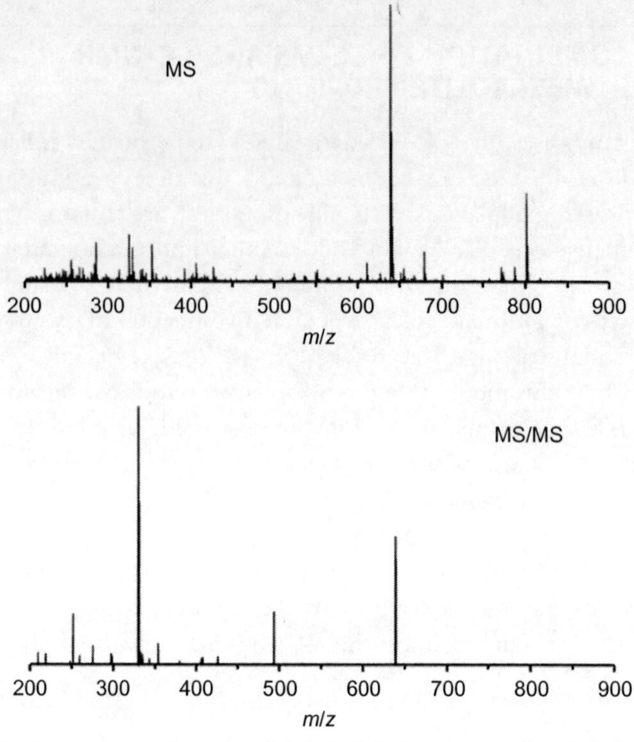

Figure 2.13 Mass spectra of *trans*-malvidin-3-*O*-(6-*O*-*p*-coumaryl)-5-*O*-diglucoside.

To complete structure elucidation, NMR experiments were performed by means of stop-flow LC–NMR and 2D-NMR in accordance with the results of LC–MS. Prior ^{1}H NMR spectra were recorded by using stop-flow mode (Fig. 2.14). As previously mentioned, NMR experiments were performed on a Bruker 600-MHz spectrometer equipped with an inverse-detection flow probe. The flow rate was typically 0.6 mL/min. Fractions were eluted with a gradient of deuterated water (A), acetonitrile–0.25% TFA (B) and water–0.25% TFA (C) according to the following gradient program (v/v): 0 min 30% A 10% B, 60% C, 55 min 30% A 55% B, 15% C, 60 min 100% B linear for 10 min, followed by 10 min for re-equilibration. Deuterated water flow was maintained constant to facilitate shimming process. By comparison of ^{1}H NMR data with reference compounds and with literature data, the complete identification of the major part of the anthocyanins was achieved.

However, when the concentration of the eluted compound was not in sufficient quantity, the sensitivity of LC–NMR prevented further

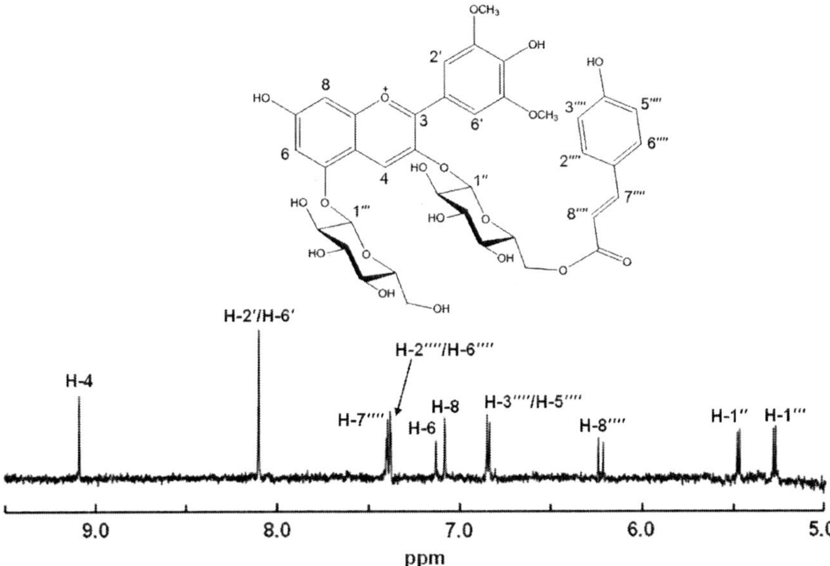

Figure 2.14 ^{1}H NMR spectrum of *trans*-malvidin-3-*O*-(6-*O*-*p*-coumaryl)-5-*O*-diglucoside.

2D-NMR experiments such as COSY or ROESY experiments. LC-SPE-NMR method was applied in order to obtain minor compounds spectra or to complete structure identification by mean of 2D-NMR experiments. Each unknown compound was trapped onto SPE cartridges, concentrated and eluted into NMR probe with deuterated solvent. Fractions were eluted with the same gradient condition as for stop-flow mode without deuterated water.

Nevertheless, the trapping conditions constituted a critical factor, in particular with polar compounds like anthocyanins. Moreover, despite the multiple trapping, it was not possible to accumulate enough amount of each anthocyanin to successfully perform more sophisticated 2D-NMR experiments for unambiguous spectral and structural assignment. To solve this problem, ambiguous or new compounds were directly collected after on-flow ^{1}H LC–NMR analysis onto an Isco Foxy fraction collector, lyophilized and subsequently analyzed with a classical NMR probe. This workflow allowed multiple collections according to ^{1}H LC–NMR spectra; deuterated solvent of choice was able to use and 2D-NMR was able to perform.

Anthocyanin identification by NMR was based on protocol of Andersen and Fossen (2003). To illustrate how the structure of a relatively complex anthocyanin is elucidated by ^{1}H NMR and 2D-NMR spectra, the detailed procedure applied on the *trans*-malvidin-3-*O*-(6-*O*-*p*-coumaryl)-5-*O*-diglucoside is described below. The ^{1}H NMR spectrum of *trans*-malvidin-3-*O*-(6-*O*-*p*-coumaryl)-5-*O*-diglucoside shows six resonances in accordance with the aglycone malvidin (Fig. 2.14). The singlet at δ 9.10 ppm is typical of the H-4 of most aglycones. The two signals at δ 7.14 and 7.11 ppm assigned to H-6 and H-8, respectively, have a *meta*-coupling to each other ($^4J_{HH}$=1.8 Hz). The 2H singlet at 8.10 ppm assigned to H-2′ and H-6′, indicate an anthocyanin with a symmetrically substituted ring. The 6H singlet at δ 4.12 ppm of the two *O*-methyl groups confirmed the identity of the aglycone to be malvidin. With regard to the sugar moiety, the anomeric protons are considerably downfield of the other sugar resonances, and thus the two doublets at δ 5.48 (H-1″, $^4J_{HH}$=7.9 Hz) and 5.28 ppm (H-1‴, $^4J_{HH}$=7.9 Hz) together with integration data, indicate two monosaccharides. The chemical shifts together with ^{1}H–^{1}H coupling constants are in agreement with two β-glucopyranosyl units. COSY and TOCSY spectra were recorded to assign the protons of sugar units. Finally, concerning the coumaroyl moiety, the two signals at δ 7.40 and 6.24 ppm assigned to H-7⁗ and H-8⁗, respectively, have large coupling constants ($^3J_{HH}$=16.1 Hz), indicating *p*-coumaric acid to have the *trans* configuration. NOESY experiments were recorded to confirm the connection sites between the glucose units, the aglycone moiety and between the *p*-coumaric acid.

6. CONCLUSION

Online LC–MS and LC–NMR synergic combination techniques have been shown in several studies to be a successful tool for the characterization of complex plant extracts. These techniques provide efficient structure identification without preliminary isolation and purification steps for the three major classes of secondary metabolites: terpenoids, alkaloids and phenolic compounds.

Development of LC–NMR techniques yields to structural identification by reducing to a minimum the restrictions of NMR such as solvent cost and sensitivity. For identification of major analytes in the extract, stop-flow mode provides appropriate NMR data. For complete identification of minor compounds needing 2D-NMR measurements, LC-SPE-NMR or capLC–NMR techniques are necessary. Each technique has advantages and limitations; the aims of the analysis lead the choice of the appropriated LC–NMR mode to solve the structural issues.

REFERENCES

Acevedo De la Cruz, A., Hilbert, G., Rivière, C., Mengin, V., Ollat, N., Bordenave, L., et al. (2012). Anthocyanin identification and composition of wild Vitis spp. accessions by using LC-MS and LC-NMR. *Analytica Chimica Acta*, *732*, 145–152.

Andersen, Ø. M., & Fossen, T. (2003). Characterization of anthocyanins by NMR. In *Current Protocols in Food Analytical Chemistry*. New York: John Wiley & Sons Inc.

Bohlmann, J., Meyer-Gauen, G., & Croteau, R. (1998). Plant terpenoid synthases: Molecular biology and phylogenetic analysis. *Proceedings of the National Academy of Sciences of the United States of America*, *95*(8), 4126–4133.

Bringmann, G., Wohlfarth, M., Rischer, H., Schlauer, J., & Brun, R. (2002). Extract screening by HPLC coupled to MS-MS, NMR, and CD: A dimeric and three monomeric naphthylisoquinoline alkaloids from Ancistrocladus griffithii. *Phytochemistry*, *61*(2), 195–204.

Christophoridou, S., Dais, P., Tseng, L.-H. , & Spraul, M. (2005). Separation and identification of phenolic compounds in olive oil by coupling high-performance liquid chromatography with postcolumn solid-phase extraction to nuclear magnetic resonance spectroscopy (LC-SPE-NMR). *Journal of Agricultural and Food Chemistry*, *53*(12), 4667–4679.

Clarkson, C., Hansen, S. H., Smith, P. J., & Jaroszewski, J. W. (2006a). Discovering new natural products directly from crude extracts by HPLC-SPE-NMR: Chinane diterpenes in Harpagophytum procumbens. *Journal of Natural Products*, *69*(4), 527–530.

Clarkson, C., Hansen, S. H., Smith, P. J., & Jaroszewski, J. W. (2006b). Identification of major and minor constituents of Harpagophytum procumbens (Devil's Claw) using HPLC-SPE-NMR and HPLC-ESIMS/APCIMS. *Journal of Natural Products*, *69*(9), 1280–1288.

Clarkson, C., Madikane, E. V., Hansen, S. H., Smith, P. J., & Jaroszewski, J. W. (2007). HPLC-SPE-NMR characterization of sesquiterpenes in an antimycobacterial fraction from Warburgia salutaris. *Planta Medica*, *73*(6), 578–584.

Cogne, A. L., Queiroz, E. F., Marston, A., Wolfender, J. L., Mavi, S., & Hostettmann, K. (2005). On-line identification of unstable iridoids from Jamesbrittenia fodina by HPLC–MS and HPLC–NMR. *Phytochemical Analysis*, *16*(6), 429–439.

Cogne, A. L., Queiroz, E. F., Wolfender, J. L., Marston, A., Mavi, S., & Hostettmann, K. (2003). On-line identification of unstable catalpol derivatives from Jamesbrittenia fodina by LC-MS and LC-NMR. *Phytochemical Analysis*, *14*(2), 67–73.

Croteau, R., Kutchan, T. M., & Lewis, N. G. (2000). Natural products (secondary metabolites). In B. Buchanan, W. Gruissem, & R. Jones (Eds.), *Biochemistry & molecular biology of plants* (pp. 1250–1318). Rockville, Maryland: American Society of Plant Physiologists.

de Rijke, E., de Kanter, F., Ariese, F., Brinkman, U. A., & Gooijer, C. (2004). Liquid chromatography coupled to nuclear magnetic resonance spectroscopy for the identification of isoflavone glucoside malonates in T. pratense L. leaves. *Journal of Separation Science*, *27*(13), 1061–1070.

Dias, D. A., & Urban, S. (2011). Phytochemical studies of the southern Australian marine alga, Laurencia elata. *Phytochemistry*, *72*(16), 2081–2089.

Dias, D. A., White, J. M., & Urban, S. (2009). Laurencia filiformis: Phytochemical profiling by conventional and HPLC-NMR approaches. *Natural Product Communications*, *4*(2), 157–172.

Djukovic, D., Liu, S., Henry, I., Tobias, B., & Raftery, D. (2006). Signal enhancement in HPLC/microcoil NMR using automated column trapping. *Analytical Chemistry*, *78*(20), 7154–7160.

Dubey, V., Bhalla, R., & Luthra, R. (2003). An overview of the non-mevalonate pathway for terpenoid biosynthesis in plants. *Journal of Biosciences*, *28*(5), 637–646.

Eisenreich, W., Bacher, A., Arigoni, D., & Rohdich, F. (2004). Biosynthesis of isoprenoids via the non-mevalonate pathway. *Cellular and Molecular Life Sciences: CMLS, 61*(12), 1401–1426.

Eldridge, G. R., Vervoort, H. C., Lee, C. M., Cremin, P. A., Williams, C. T., Hart, S. M., et al. (2002). High-throughput method for the production and analysis of large natural product libraries for drug discovery. *Analytical Chemistry, 74*(16), 3963–3971.

Elipe, M. V. S., Huskey, S.-E. W. , & Zhu, B. (2003). Application of LC-NMR for the study of the volatile metabolite of MK-0869, a substance P receptor antagonist. *Journal of Pharmaceutical and Biomedical Analysis, 30*(5), 1431–1440.

Emerson, Q., Kurt, H., & Jean-Luc, W. (2007). Development and application of LC-NMR techniques to the identification of bioactive natural products. In *Bioactive Natural Products* (pp. 143–190). FL USA: Taylor & Francis group.

Exarchou, V., Godejohann, M., van Beek, T. A., Gerothanassis, I. P., & Vervoort, J. (2003). LC-UV-solid-phase extraction-NMR-MS combined with a cryogenic flow probe and its application to the identification of compounds present in Greek Oregano. *Analytical Chemistry, 75*(22), 6288–6294.

Exarchou, V., Krucker, M., van Beek, T. A., Vervoort, J., Gerothanassis, I. P., & Albert, K. (2005). LC–NMR coupling technology: Recent advancements and applications in natural products analysis. *Magnetic Resonance in Chemistry, 43*(9), 681–687.

Garcia-Beneytez, E., Cabello, F., & Revilla, E. (2003). Analysis of grape and wine anthocyanins by HPLC-MS. *Journal of Agricultural and Food Chemistry, 51*(19), 5622–5629.

Gershenzon, J., & Dudareva, N. (2007). The function of terpene natural products in the natural world. *Nature Chemical Biology, 3*(7), 408–414.

Ghosh, D., & Konishi, T. (2007). Anthocyanins and anthocyanin-rich extracts: Role in diabetes and eye function. *Asia Pacific Journal of Clinical Nutrition, 16*(2), 200–208.

Guerrero, R. F., Liazid, A., Palma, M., Puertas, B., González-Barrio, R., Gil-Izquierdo, Á., et al. (2009). Phenolic characterisation of red grapes autochthonous to Andalusia. *Food Chemistry, 112*(4), 949–955.

Hossain, M. B., Rai, D. K., Brunton, N. P., Martin-Diana, A. B., & Barry-Ryan, C. (2010). Characterization of phenolic composition in Lamiaceae spices by LC-ESI-MS/MS. *Journal of Agricultural and Food Chemistry, 58*(19), 10576–10581.

Iwasa, K., Kuribayashi, A., Sugiura, M., Moriyasu, M., Lee, D.-U. , & Wiegrebe, W. (2003). LC-NMR and LC-MS analysis of 2,3,10,11-oxygenated protoberberine metabolites in Corydalis cell cultures. *Phytochemistry, 64*(7), 1229–1238.

Jansma, A., Chuan, T., Albrecht, R. W., Olson, D. L., Peck, T. L., & Geierstanger, B. H. (2005). Automated microflow NMR: Routine analysis of five-microliter samples. *Analytical Chemistry*, 77(19), 6509–6515.

Jaroszewski, J. W. (2005). Hyphenated NMR methods in natural products research, part 2: HPLC-SPE-NMR and other new trends in NMR hyphenation. *Planta Medica, 71*(9), 795–802.

Kaufman, P., Brielmann, H., Cseke, L., Setzer, W., & Kirakosyan, A. (2006). Phytochemicals. In L. Cseke, A. Kirakosyan, P. Kaufman, S. Warber, J. Duke, & H. Brielmann (Eds.), *Natural Products from Plants* (2nd ed., pp. 1–49). FL USA: Taylor & Francis group.

Keifer, P. A. (2010). Chemical-shift referencing and resolution stability in methanol:water gradient LC-NMR. *Journal of Magnetic Resonance, 205*(1), 130–140.

Lacey, M. E., Tan, Z. J., Webb, A. G., & Sweedler, J. V. (2001). Union of capillary high-performance liquid chromatography and microcoil nuclear magnetic resonance spectroscopy applied to the separation and identification of terpenoids. *Journal of Chromatography. A, 922*(1–2), 139–149.

Lambert, M., Stærk, D., Hansen, S. H., & Jaroszewski, J. W. (2005). HPLC–SPE–NMR hyphenation in natural products research: Optimization of analysis of Croton membranaceus extract. *Magnetic Resonance in Chemistry*, *43*(9), 771–775.

Ndjoko, K., Wolfender, J.-L. , Röder, E., & Hostettmann, K. (1999). Determination of pyrrolizidine alkaloids in senecio species by liquid chromatography/thermospray-mass spectrometry and liquid chromatography/nuclear magnetic resonance spectroscopy. *Planta Medica*, *65*(06), 562–566.

Pham, L. H., Vater, J., Rotard, W., & Mügge, C. (2005). Identification of secondary metabolites from Streptomyces violaceoruber TÜ22 by means of on-flow LC–NMR and LC–DAD–MS. *Magnetic Resonance in Chemistry*, *43*(9), 710–723.

Qu, J., Hu, Y. C., Li, J. B., Wang, Y. H., Zhang, J. L., Abliz, Z., et al. (2008). Structural characterization of constituents with molecular diversity in fractions from Lysidice brevicalyx by liquid chromatography/diode-array detection/electrospray ionization tandem mass spectrometry and liquid chromatography/nuclear magnetic resonance. *Rapid Communications in Mass Spectrometry: RCM*, *22*(6), 755–765.

Queiroz, E. F., Ioset, J. R., Ndjoko, K., Guntern, A., Foggin, C. M., & Hostettmann, K. (2005). On-line identification of the bioactive compounds from Blumea gariepina by HPLC-UV-MS and HPLC-UV-NMR, combined with HPLC-micro-fractionation. *Phytochemical Analysis*, *16*(3), 166–174.

Quideau, S., Deffieux, D., Douat-Casassus, C., & Pouységu, L. (2011). Plant polyphenols: Chemical properties, biological activities, and synthesis. *Angewandte Chemie, International Edition*, *50*(3), 586–621.

Ramawat, K. G., Dass, S., & Mathur, M. (2009). The chemical diversity of bioactive molecules and therapeutic potential of medicinal plants. In *Herbal Drugs: Ethnomedicine to Modern Medicine* (pp. 7–32). Berlin Heidelberg: Springer.

Schefer, A. B., Braumann, U., Tseng, L.-H. , Spraul, M., Soares, M. G., Fernandes, J. B., et al. (2006). Application of high-performance liquid chromatography-nuclear magnetic resonance coupling to the identification of limonoids from mahogany tree (Switenia macrophylla, Meliaceae) by stopped-flow 1D and 2D NMR spectroscopy. *Journal of Chromatography. A*, *1128*(1–2), 152–163.

Silva Elipe, M. V. (2003). Advantages and disadvantages of nuclear magnetic resonance spectroscopy as a hyphenated technique. *Analytica Chimica Acta*, *497*(1–2), 1–25.

Sorensen, D., Raditsis, A., Trimble, L. A., Blackwell, B. A., Sumarah, M. W., & Miller, J. D. (2006). Isolation and structure elucidation by LC-MS-SPE/NMR: PR toxin- and cuspidatol-related eremophilane sesquiterpenes from Penicillium roqueforti. *Journal of Natural Products*, *70*(1), 121–123.

Spring, O., Buschmann, H., Vogler, B., Schilling, E. E., Spraul, M., & Hoffmann, M. (1995). Sesquiterpene lactone chemistry of Zaluzania grayana from on-line LC-NMR measurements. *Phytochemistry*, *39*(3), 609–612.

Spring, O., Heil, N., & Vogler, B. (1997). Sesquiterpene lactones and flavanones in Scalesia species. *Phytochemistry*, *46*(8), 1369–1373.

Staerk, D., Kesting, J. R., Sairafianpour, M., Witt, M., Asili, J., Emami, S. A., et al. (2009). Accelerated dereplication of crude extracts using HPLC-PDA-MS-SPE-NMR: Quinolinone alkaloids of Haplophyllum acutifolium. *Phytochemistry*, *70*(8), 1055–1061.

Stintzing, F. C., Conrad, J., Klaiber, I., Beifuss, U., & Carle, R. (2004). Structural investigations on betacyanin pigments by LC NMR and 2D NMR spectroscopy. *Phytochemistry*, *65*(4), 415–422.

Stoessl, A., Stothers, J. B., & Ward, E. W. B. (1976). Sesquiterpenoid stress compounds of the solanaceae. *Phytochemistry*, *15*(6), 855–872.

Sturm, S., & Seger, C. (2012). Liquid chromatography-nuclear magnetic resonance coupling as alternative to liquid chromatographyâ-mass spectrometry hyphenations: Curious option or powerful and complementary routine tool? *Journal of Chromatography. A, 1259*, 50–61.

Sturm, S., Seger, C., Godejohann, M., Spraul, M., & Stuppner, H. (2007). Conventional sample enrichment strategies combined with high-performance liquid chromatography-solid phase extraction-nuclear magnetic resonance analysis allows analyte identification from a single minuscule Corydalis solida plant tuber. *Journal of Chromatography. A, 1163*(1–2), 138–144.

Tatsis, E. C., Boeren, S., Exarchou, V., Troganis, A. N., Vervoort, J., & Gerothanassis, I. P. (2007). Identification of the major constituents of Hypericum perforatum by LC/SPE/NMR and/or LC/MS. *Phytochemistry, 68*(3), 383–393.

Timmers, M., & Urban, S. (2012). On-line (HPLC-NMR) and off-line phytochemical profiling of the Australian plant, Lasiopetalum macrophyllum. *Natural Product Communications*, 7(5), 551–560.

Tode, C., Maoka, T., & Sugiura, M. (2009). Application of LC-NMR to analysis of carotenoids in foods. *Journal of Separation Science, 32*(21), 3659–3663.

Tseng, L., Braumann, U., Godejohann, M., Lee, S., & Albrecht, K. (2000). Structure identification of aporphine alkaloids by on-line coupling of HPLC-NMR with Loop-storage. *Journal of the Chinese Chemical Society, 47*(6), 1231–1236.

Tsimogiannis, D., Samiotaki, M., Panayotou, G., & Oreopoulou, V. (2007). Characterization of flavonoid subgroups and hydroxy substitution by HPLC-MS/MS. *Molecules, 12*(3), 593–606.

Watanabe, B. N., & Niki, E. (1979). An experiment on direct combination of high performance liquid chromatography with FT-NMR (LC-NMR). *JEOL News, 15*(A), 194–199.

Wolfender, J.-L. , Ndjoko, K., & Hostettmann, K. (2001). The potential of LC-NMR in phytochemical analysis. *Phytochemical Analysis, 12*(1), 2–22.

Wolfender, J.-L. , Queiroz, E. F., & Hostettmann, K. (2005). Phytochemistry in the microgram domain—A LC–NMR perspective. *Magnetic Resonance in Chemistry, 43*(9), 697–709.

Wolfender, J.-L. , Rodriguez, S., Hostettmann, K., & Hiller, W. (1997). Liquid chromatography/ultra violet/mass spectrometric and liquid chromatography/nuclear magnetic resonance spectroscopic analysis of crude extracts of Gentianaceae species. *Phytochemical Analysis, 8*(3), 97–104.

Xiao, H. B., Krucker, M., Putzbach, K., & Albert, K. (2005). Capillary liquid chromatography-microcoil ^{1}H nuclear magnetic resonance spectroscopy and liquid chromatography-ion trap mass spectrometry for on-line structure elucidation of isoflavones in Radix astragali. *Journal of Chromatography. A, 1067*(1–2), 135–143.

Xu, F., & Alexander, A. J. (2005). The design of an on-line semi-preparative LC–SPE–NMR system for trace analysis. *Magnetic Resonance in Chemistry, 43*(9), 776–782.

Zanolari, B., Wolfender, J. L., Guilet, D., Marston, A., Queiroz, E. F., Paulo, M. Q., et al. (2003). On-line identification of tropane alkaloids from Erythroxylum vacciniifolium by liquid chromatography-UV detection-multiple mass spectrometry and liquid chromatography-nuclear magnetic resonance spectrometry. *Journal of Chromatography. A, 1020*(1), 75–89.

Zhao, Y., Nookandeh, A., Schneider, B., Sun, X., Schmitt, B., & Stöckigt, J. (1999). Lignans from Torreya jackii identified by stopped-flow high-performance liquid chromatography-nuclear magnetic resonance spectroscopy. *Journal of Chromatography. A, 837*(1–2), 83–91.

Ziegler, J., & Facchini, P. J. (2008). Alkaloid biosynthesis: Metabolism and trafficking. *Annual Review of Plant Biology, 59*(1), 735–769.

CHAPTER THREE

Fast and Ultrafast Quantitative 2D NMR: Vital Tools for Efficient Metabolomic Approaches

Patrick Giraudeau[1], Serge Akoka

EBSI Team, Chimie et Interdisciplinarité: Synthèse, Analyse, Modélisation (CEISAM), CNRS, UMR 6230, LUNAM Université de Nantes, France

[1]Corresponding author: e-mail address: patrick.giraudeau@univ-nantes.fr

Contents

Abstract

Nuclear magnetic resonance (NMR) offers great potential for quantitative analysis of metabolic samples, but accurate and precise quantitative analysis is often made difficult by a high degree of spectral overlap. Multi-dimensional NMR offers an interesting alternative, as it provides a much better discrimination of resonances. However, its use for quantitative analysis is limited by its long acquisition duration and its multi-impulsional character. Nevertheless, numerous improvements in the quantitative performance of 2D NMR have been made during the past decade, facilitating a wide range of applications.

This review assesses the recent contributions brought to the field of quantitative 2D NMR, and provides a survey of recent studies of metabolic samples using this

Advances in Botanical Research, Volume 67
ISSN 0065-2296
http://dx.doi.org/10.1016/B978-0-12-397922-3.00003-4

methodology. The ongoing need for developing and optimizing powerful tools is first examined. It highlights the necessity of reducing the experimental duration, resulting in a fast 2D NMR approach. This methodology is then described, relying on recent applications to metabolic samples. The last section of the chapter describes a more recent approach, ultrafast 2D NMR, which makes it possible to record multi-dimensional spectra in a single scan. The principles and the analytical performances of this novel methodology are described, as well as its perspectives in terms of quantitative analysis.

1. INTRODUCTION

Nuclear magnetic resonance (NMR) is a highly versatile spectroscopic technique with applications in a large range of disciplines, from physics and chemistry to biology and medicine. It is widely used in all chemistry laboratories for its potential to provide information for the elucidation of complex molecular structures with a high degree of confidence, but NMR is also well recognized in biochemistry for its ability to determine the spatial structure of large macromolecules. While a great deal of attention has been paid to the capabilities of NMR for structural elucidation, its ability to determine the concentration of analytes in mixtures is generally less well known. However, because NMR is a quantitative technique, and has the recognized advantage of being both non-destructive and non-specific (i.e. all molecular species present can be detected simultaneously), its potential for quantitative analysis is considerable. In addition, it is also the technique of choice to obtain site-specific information.

The application of quantitative NMR has been reported in fields such as pharmaceutical analysis (Holzgrabe, 2010; Kwakye, 1985), metabolic studies (Lindon, Nicholson, Holmes, & Everett, 2000; Wishart, 2008; Zhang et al., 2008) or the authentication of natural products (Le Grand, George, & Akoka, 2005; Tenailleau, Lancelin, Robins, & Akoka, 2004). In the vast majority of studies reported so far, the so-called qNMR (Pauli, 2001) methodology is used. It often exploits a very standardized 1D ^{1}H NMR protocol, often associated with statistical analysis in the field of metabolomics (Kruger, Troncoso-Ponce, & Ratcliffe, 2008; Le Gall, Colquhoun, Davis, Collins, & Verhoeyen, 2003), to compare samples from different sets. The most widely used strategy consists of recording the metabolic fingerprint of a high number of individuals, followed by multivariate statistical analysis in order to discriminate between individuals and to identify the spectral regions responsible for this discrimination. This untargeted approach has been widely applied in fields including plant metabolic studies (Kim, Choi, & Verpoorte, 2010; Krishnan, Kruger, & Ratcliffe, 2005; Kruger et al., 2008). While this method provides an efficient discrimination between individuals, it does not allow for the quantification of metabolites responsible for this discrimination. A more targeted strategy, metabolic profiling, consists in identifying and quantifying a few tens of metabolites in unfractionated extracts (Le Gall et al., 2003; Weljie, Newton, Mercier, Carlson, & Slupsky, 2006; Wishart, 2008). However, a true quantitative analysis of metabolic samples (i.e. the precise and accurate determination of the analyte concentrations) is made very difficult by the high degree of signal overlap characterizing complex samples.

Multi-dimensional NMR, proposed by Jeener in 1971, and applied as a routine tool for the elucidation of small organic molecules or macromolecular structures, has the high advantage of offering, in addition to indispensible correlation information, a better discrimination of resonances than 1D NMR (Aue, Bartholdi, & Ernst, 1976; Jeener, 1971). Consequently, 2D NMR offers promising potential for quantitative analysis of complex mixtures. That its use for quantitative analysis seems to have been expanding only over the past decade (Giraudeau, Guignard, Hillion, Baguet, & Akoka, 2007; Koskela, Kilpeläinen, & Heikkinen, 2005; Koskela & Väänänen, 2002; Lewis et al., 2010, 2007; Martineau, Giraudeau, Tea, & Akoka, 2011; Martineau, Tea, Akoka, & Giraudeau, 2012; Massou, Nicolas, Letisse, & Portais, 2007a, 2007b; Zhang & Gellerstedt, 2007) is probably due to the main intrinsic characteristics of 2D NMR experiments, namely, their long acquisition duration due to their incremental nature, and their multi-impulsional character. These factors make the use of 2D NMR far from trivial for quantitative purposes. Nevertheless, numerous methodological and analytical improvements in the quantitative performance of 2D NMR have been made during the past decade, facilitating a wide range of applications. In particular, recent years have witnessed efforts geared at decreasing the duration of 2D NMR experiments, while improving their precision and accuracy for quantitative measurements.

This review assesses the contribution that NMR spectroscopists are making to the field of quantitative 2D NMR, and provides a survey of recent studies of metabolic samples using this methodology. In the first section, the ongoing need for developing and optimizing powerful methods for quantitative 2D NMR is examined. It highlights the necessity of reducing the experimental duration, resulting in a fast 2D NMR approach which has been developed concomitantly by several research groups.

This methodology is described in the second section, insisting on the optimization logic of this kind of experiment, and relying on recent applications to metabolic samples. The last section of the chapter describes a much more recent approach, ultrafast 2D NMR, which makes it possible to record multi-dimensional spectra in a single scan. The principles and the analytical performances of this novel methodology are described, as well as its perspectives in terms of quantitative analysis.

2. WHY QUANTITATIVE 2D NMR?

Before analyzing the reasons which have led to the development of quantitative approaches by 2D NMR, it is necessary to describe the current limitations of 1D NMR for quantitative studies. The notion of quantitative nature has been associated with NMR spectroscopy since its very early days, initially in the structural elucidation of organic compounds for the measurement of integrals for determining the number of protons on each site of a molecule (Jungnickel & Forbes, 1963; Rizzo & Pinciroli, 2005). In the field of analytical chemistry, the first quantitative analysis of a mixture by ^{1}H NMR was reported by Hollis (1963). Since then, ^{1}H NMR has continuously been used for quantitative measurements in a wide range of domains, including but not limited to pharmaceutical analysis (Kwakye, 1985), natural products (Pauli, Jaki, & Lankin, 2005), *in vivo* spectroscopy (Podo et al., 1998) and metabolomics (Wishart, 2008). While the overwhelming majority of studies have focussed on ^{1}H NMR, quantitative measurement is, at least in principle, compatible with the whole range of nuclei observable by NMR, provided that parameters such as relaxation and the Overhauser effect are taken into account. In particular, quantitative ^{13}C NMR has been widely described (Cookson & Smith, 1984; Mareci & Scott, 1977; Shoolery, 1977) and applied in a variety of fields, from food science (Mavromoustakos et al., 1997; Vlahov, 1999) or metabolic studies (Aursand, Jorgensen, & Grasdalen, 1995) to the isotopic analysis of natural or synthetic molecules (Bussy et al., 2011; Caytan, Remaud, Tenailleau, & Akoka, 2007; Tenailleau et al., 2004). Quantitative studies relying on other nuclei have also been reported, such as ^{15}N (Levy, Pehk, & Srinivasan, 1980), ^{31}P (Dais & Spyros, 2007) or ^{17}O (Lonnon & Hook, 2003). Hetero-nuclei present high potential for quantitative analysis, as their NMR spectra are characterized by a much larger spectral width than ^{1}H, thus leading to a much greater degree of discrimination. However, they are generally characterized by a low NMR sensitivity, due to their low natural abundance and/or to their small gyromagnetic ratio. As a consequence, relying on nuclei other than ^1H for quantitative measurements requires either highly concentrated samples or very long measurement times. The latter can be shortened by the use of paramagnetic agents (Caytan et al., 2007) or by employing time-shortening acquisition strategies (Giraudeau, Wang, & Baguet, 2006), while sensitivity can be enhanced by polarization-transfer strategies (Thibaudeau, Remaud, Silvestre, & Akoka, 2010). But when considering low-concentrated samples, as is very likely to be the case in metabolomic studies, quantitative analysis is still limited to ^1H spectroscopy due to its high sensitivity, with other nuclei providing additional help for the assignment or structural elucidation of metabolites.

2.1. Limitations of 1D NMR for quantitative metabolomics

Thanks to its high sensitivity, ^1H NMR is widely used in the growing field of metabolomics. In the vast majority of studies reported so far, a very standardized 1D ^1H NMR protocol is employed for data acquisition, associated with multivariate statistical methods to perform an efficient discrimination between samples of various origins (Ludwig et al., 2009; Zhang et al., 2008). In addition to simple discrimination, however, measuring metabolite concentrations is often required to fully understand metabolic complexities and to determine characteristic biomarkers. However, precise and accurate quantification is often made difficult by the high degree of overlap characterizing ^1H NMR spectra of biological samples. In addition to the limited range of ^1H chemical shifts, two main reasons explain this overlap. First, biological samples often contain closely related compounds, for which the ^1H NMR spectra are very close. This example is illustrated in Fig. 3.1 on a model mixture of tropine and nortropine, two tropane alkaloids that are members of a group of compounds with important pharmaceutical activity. The 1D spectra of these two compounds present a large degree of overlap, while the resonances arising from tropine and nortropine are clearly separated on a homo-nuclear 2D spectrum (Fig. 3.1). The second reason explaining the complexity of ^1H NMR spectra is the high number of compounds showing peaks in the same spectral region. This is particularly the case for mixtures of metabolites, as illustrated in Fig. 3.2. Even in a simple model mixture of 11 metabolites, a number of resonances overlap in the 1D spectrum, such as those arising from myo-inositol, taurine and arginine, whereas the 2D TOCSY spectrum provides a better discrimination between these resonances, thus avoiding overlap. These examples point out the

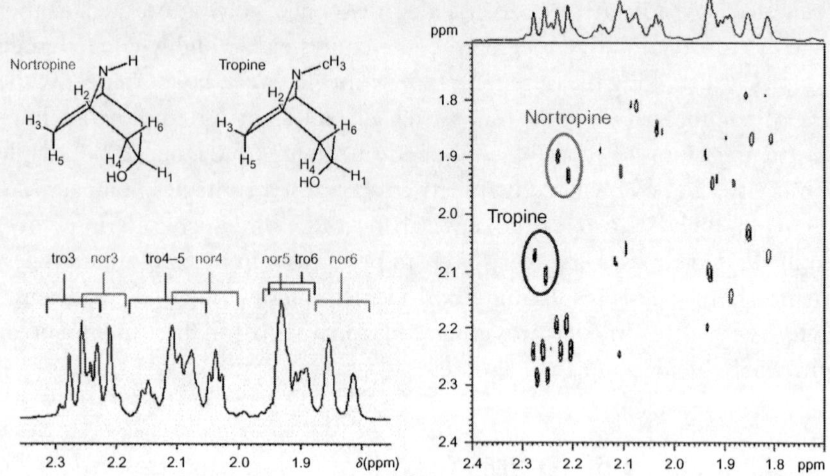

Figure 3.1 1D (left) and 2D DQF–COSY (right, positive signals only) spectra of an equimolar mixture of tropine and nortropine. ^{1}H sites are numbered in decreasing chemical shift. Spectra were acquired at 298 K on a 400-MHz spectrometer (Giraudeau et al., 2007). (For colour version of this figure, the reader is referred to the online version of this chapter.)

limitations of 1D NMR, even in the case of simple model mixtures, thus prefiguring what happens in real biological samples. Figures 3.1 and 3.2 also highlight the potential of 2D NMR for quantitative analysis of such mixtures, as is discussed below. An alternative means to obtain better information from 1D spectra is the 1D targeted approach, designed to quantify individual metabolites by fitting the 1D spectra of complex samples from the sum of individual metabolite 1D spectra (Weljie et al., 2006). However, recent studies have shown that precision is still limited by the complexity of spectral patterns when addressing complex biological samples (Slupsky et al., 2007; Tredwell, Behrends, Geier, Liebeke, & Bundy, 2011).

2.2. Advantages and constraints of 2D NMR for quantitative analysis

The concept of 2D NMR, proposed by Jeener (1971), revolutionized the field of magnetic resonance. His original idea was to reconstruct a 2D temporal signal from a series of 1D experiments recorded with the following sequence of events: preparation—evolution (t_1)—mixing—detection (t_2), as illustrated in Fig. 3.3. The repetition of this sequence with incremented values of t_1 leads to a 2D FID $S(t_1, t_2)$ whose double Fourier transform gives rise to a 2D spectrum $S(\Omega_1, \Omega_2)$. Depending on the spin evolution during

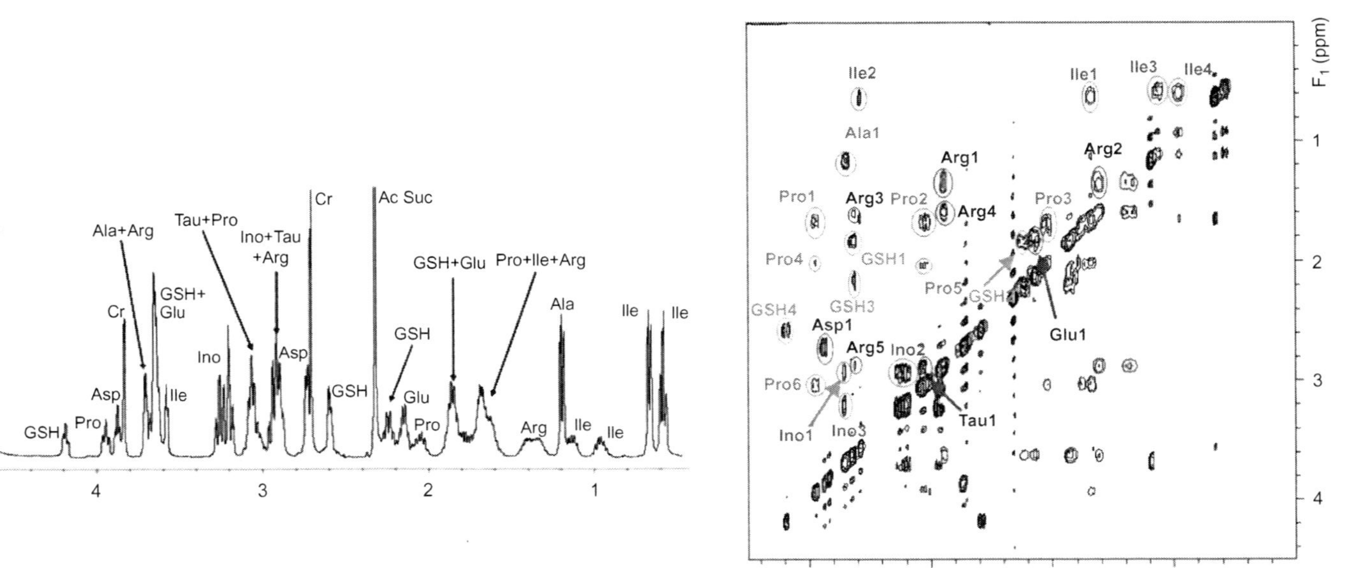

Figure 3.2 1D (left) and 2D TOCSY (right) spectra of a model mixture of metabolites, recorded at 298 K on a 400-MHz spectrometer. *PhD thesis of Estelle Martineau, Université de Nantes, 2011*. (For colour version of this figure, the reader is referred to the online version of this chapter.)

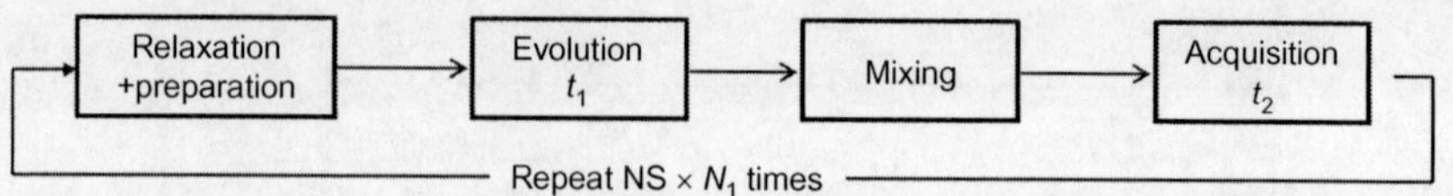

Figure 3.3 Acquisition strategy initially proposed by Jeener (1971) to record 2D NMR spectra.

each period and on the preparation and mixing periods, an *a priori* infinite number of experiments can be designed to correlate the parameters of two identical or different nuclei, giving rise to homo- or hetero-nuclear 2D correlations, respectively. The idea of 2D NMR was further developed by Ernst and co-workers (Aue, Bachmann, Wokaun, & Ernst, 1978; Aue, Bartholdi, et al., 1976; Aue, Karhan, & Ernst, 1976; Ernst, Bodenhausen, & Wokaun, 1987), and this success story was rapidly and widely implemented. This popularity reflected the capacity of 2D NMR to better separate signals in complex molecules and to correlate information (chemical shifts, couplings, etc.) between different nuclei, thus providing incredibly helpful tools to organic, inorganic and analytical chemists for helping with the elucidation of molecular structures.

From the point of view of quantitative analysis, 2D NMR has the great advantage of allowing a much better separation of resonances than 1D NMR, partially or even fully resolving the overlap present on 1D spectra. The capacity to separate signals varies depending on the pulse sequence considered. Homo-nuclear 2D experiments reduce the 1D overlap limitation by splitting the 1D ^{1}H spectrum in a 2D plane where the two dimensions are generally equivalent (i.e. both contain the ^{1}H chemical shifts). This is the case of the well-known correlation spectroscopy (COSY) (Bax & Freeman, 1981) and total correlation spectroscopy (TOCSY) (Braunschweiler & Ernst, 1983) experiments. Also noteworthy is the *J*-resolved pulse sequence (Aue, Karhan, et al., 1976), where resonances are split according to *J*-coupling patterns in the indirect dimension. The potential of homo-nuclear 2D NMR for analyzing complex mixtures of metabolites is illustrated in Fig. 3.4 for the case of a double-quantum-filtered COSY (DQF–COSY) experiment (Shaka & Freeman, 1983). Hetero-nuclear experiments offer an even greater capacity to discriminate signals, as the indirect dimension generally contains information on hetero-nuclei, such as ^{13}C or ^{15}N, for which the resonances are spread over a much greater chemical shift range. Popular experiments include hetero-nuclear single

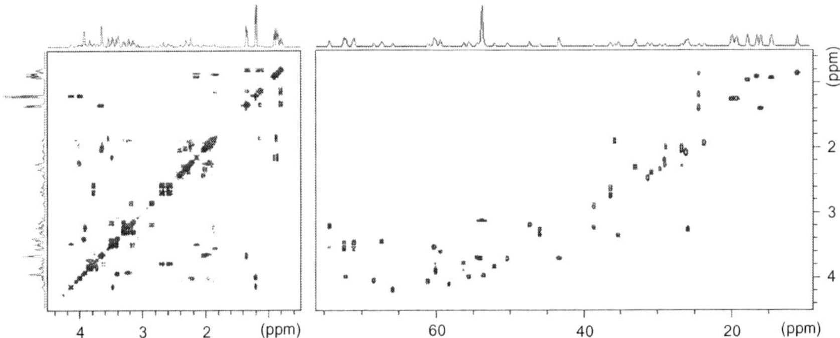

Figure 3.4 500 MHz 2D spectra of a 50-mM model mixture of metabolites (alanine, glutathione, myo-inositol, lactic acid, proline and taurine). On the left, homo-nuclear correlation (COSY–DQF) and on the right, hetero-nuclear correlation (HSQC). (For colour version of this figure, the reader is referred to the online version of this chapter.)

quantum correlation (HSQC) (Bodenhausen & Ruben, 1980), hetero-nuclear multiple quantum correlation (HMQC) (Bax, Griffey, & Hawkins, 1983) and hetero-nuclear multiple bond correlation (HMBC). The HSQC spectrum presented in Fig. 3.4 highlights the resolving power of hetero-nuclear spectroscopy. The spectrum exhibits more resonances than its homo-nuclear counterpart, and these resonances are better separated. The evaluation of various homo- and hetero-nuclear experiments is discussed in Section 3.

In spite of its high potential, the use of 2D NMR for quantitative purposes is quite recent. Even though early works reported the application of DQF–COSY (Alonso, Arus, Westler, & Markley, 1989) or TOCSY (Morvan, Demidem, Papon, & Madelmont, 2003) for measuring relative concentrations in complex samples, the number of papers on quantitative 2D NMR has really only taken off in the past 7 years, thanks to the works of Koskela et al. (Koskela, 2009; Koskela et al., 2005; Koskela & Väänänen, 2002), Zhang et al. (Zhang & Gellerstedt, 2007), Massou et al. (Massou et al., 2007a, 2007b), Markley et al. (Lewis et al., 2010, 2007), ourselves (Giraudeau et al., 2007; Martineau et al., 2011, 2012) and others. This somewhat late success is due to several constraints, which are specific to multi-dimensional experiments and highly restrain their ease of application for quantitative analysis.

The most evident limitation of 2D NMR is the long experimental duration, due to the numerous t_1 increments (typically several hundreds) needed to sample the indirect dimension with sufficient resolution. This is further increased by the possible need to accumulate several transients for each t_1 increment for sensitivity and/or coherence selection purposes. The impact of the incrementation procedure is particularly serious in the case of quantitative measurements, where the recovery delay generally has to be set to a long value (typically five times the highest longitudinal relaxation time T_1) in order to allow for the full relaxation of all longitudinal magnetizations, except when working under partial saturation conditions. Long experimental durations have several consequences, the first being the overload of spectrometer schedules, which translate into non-negligible costs. A more fundamental consequence is the impossibility of studying samples in which composition evolves within the timescale of the *n*D experiment. This is not only the case for samples undergoing chemical reactions or dynamic processes but also for biological samples with limited lifetimes. From the quantitative point of view, a last, but not least, repercussion of the experimental duration is that long experiments are more likely to be affected by spectrometer instabilities over time (Mehlkopf, Korbee, Tiggelman, & Freeman, 1984; Morris, 1992). These include electronic variations (variations in pulse angle, phase, receiver gain, etc.), lock instabilities, as well as magnetic field variations inside or outside the magnet. They generate additional noise in the indirect dimension, due to the long time interval (several seconds) between the acquisition of two successive points of the pseudo-FID. As a consequence of this "t_1 noise", the signal-to-noise ratio (SNR) is always lower in the indirect F_1 dimension, leading to noise ridges parallel to the F_1 axis, as illustrated in Fig. 3.5. While t_1 noise can be removed by subtraction or symmetrization operations in routine experiments (Baumann, Wider, Ernst, & Wütrich, 1981), the application of such mathematical treatments in quantitative experiments may affect quantification. Therefore, t_1 noise, which highly depends on the spectrometer and hardware configuration, potentially affects the precision of quantitative 2D experiments and also decreases the limit of concentration that can be measured precisely.

These arguments tend to suggest a close relationship between the experimental duration and the quantitative nature of multi-dimensional experiments. The optimization of the experimental time is therefore a key feature for generalizing the use of 2D NMR for quantitative analysis, as described in Sections 3 and 4.

Beyond the long experimental duration, another characteristic feature of multi-dimensional experiments is their multi-impulsional character, which has two main consequences on the quantitative nature of *n*D experiments.

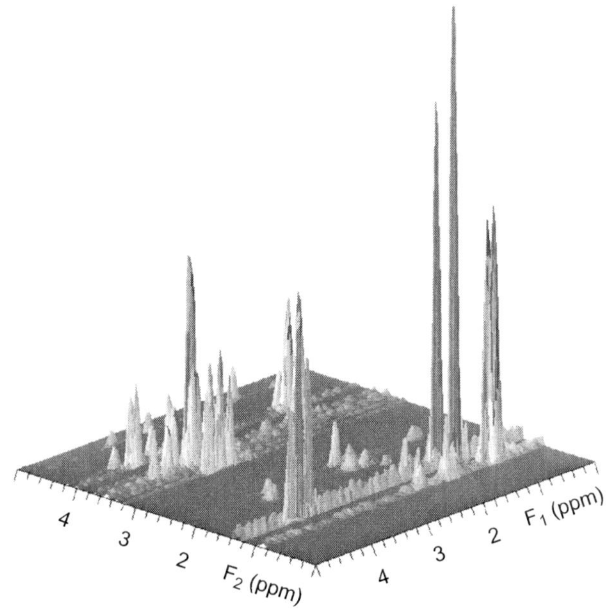

Figure 3.5 Stacked plot representation of a 2D COSY spectrum recorded on a model mixture of metabolites, on a 500-MHz spectrometer with a cryoprobe. t_1 noise ridges parallel to the indirect F_1 dimension are clearly visible, whereas the signal-to-noise ratio is much greater in the direct F_2 dimension (M. Pathan, S. Akoka, & P. Giraudeau, unpublished data). (For colour version of this figure, the reader is referred to the online version of this chapter.)

The first is that quantitative 2D NMR experiments are sensitive to pulse imperfections which affect their precision. This effect can be reduced by employing pulses that compensate for their own imperfections such as composite pulses (Freeman, Kempsell, & Levitt, 1980; Fujirawa & Nagayama, 1988; Levitt & Freeman, 1979, 1981) or adiabatic pulses (Cano, Smith, & Shaka, 2002; Hwang, Van Zijl, & Garwood, 1997; Kupce & Freeman, 1995a, 1995b). Several studies in 1D and 2D NMR have shown the benefits of introducing such pulses in multi-impulsional experiments in order to increase the repeatability for peak area or volume measurements (Boyer, Johnson, & Krishnamurthy, 2003; Heikkinen, Toikka, Karhunen, & Kilpeläinen, 2003; Koskela et al., 2005; Tenailleau & Akoka, 2007; Thibaudeau et al., 2010). On the other hand, the consequence of the multi-impulsional nature of 2D experiments is that peak volumes depend on several factors, such as homo- or hetero-nuclear *J*-couplings and transverse relaxation times T_2. This dependence has been fully described by

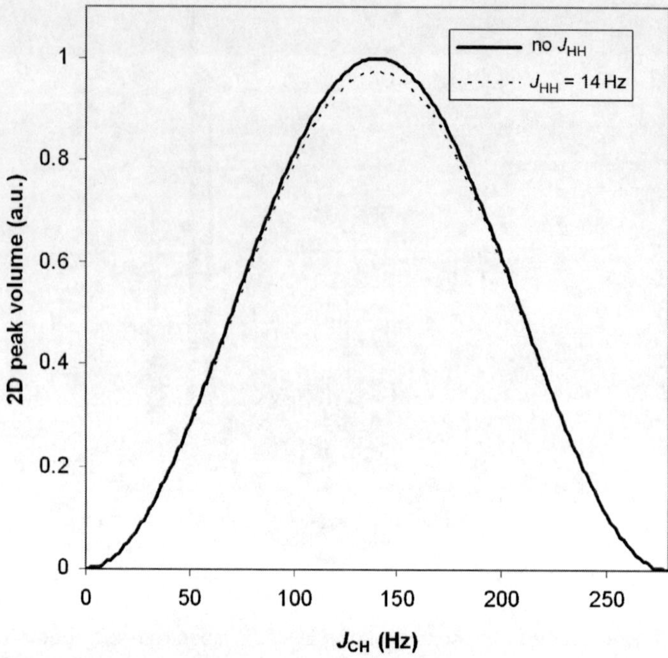

Figure 3.6 Influence of a mismatch of the J_{CH} coupling constant on the 2D peak volume in an HSQC experiment where the INEPT delay was adapted to $J_{CH}^{opt} = 140$ Hz. The much smaller influence of a homo-nuclear J_{HH} coupling is also represented.

Koskela and co-workers in recent studies (Koskela, 2009; Koskela et al., 2005). As an example, Fig. 3.6, derived from Eq. (6) in Koskela (2009) describes the impact of *J*-couplings in the case of an HSQC experiment, highlighting how a mismatch between the polarization transfer delays and the *J*-coupling values affects the 2D peak volumes.

An important consequence of the numerous factors affecting peak volumes in 2D NMR experiments is that the latter are not directly proportional to concentration and number of spins as is the case in 1D NMR. Therefore, the peak volumes for two compounds in identical concentrations in a mixture are, except in particular cases, different. Indeed, even for the same molecule showing several correlations on the 2D spectrum, peak volumes will generally differ from one site to another. Quantitative experimental protocols need to deal with this aspect. Two kinds of approaches have been described in the literature. The first consists in measuring peak volumes and correcting them by a factor accounting for their dependence on *J*-couplings, relaxation times, etc. (Rai, Tripathi, & Sinha, 2009). This approach, however, requires that all coupling constants and relaxation times are previously determined with a good precision, thus leading in a long and tedious optimization procedure for each different sample to be analyzed. A more realistic approach relies on a calibration procedure, consisting of recording spectra of model mixtures at different concentrations and plotting the peak volumes versus the concentrations for each peak of interest (Giraudeau et al., 2007; Gronwald et al., 2008; Hu, Furihata, Kato, & Tanokura, 2007; Lewis et al., 2007; Martineau et al., 2011). Provided that the method is linear, concentrations can be measured for unknown mixtures relying on such calibration curves. The efficiency of this procedure for studying metabolic samples has been demonstrated (Gronwald et al., 2008; Lewis et al., 2007). Unfortunately, its duration is considerably lengthened by the long experimental time characteristic of 2D NMR experiments, particularly when parameters are not optimized. This highlights again the need for optimizing the experimental duration, as described above. Sections 3 and 4 aim at describing several approaches which have been proposed to reduce the time needed to perform quantitative 2D NMR measurements while optimizing their analytical performance.

3. FAST CONVENTIONAL 2D NMR APPROACHES FOR QUANTITATIVE ANALYSIS OF COMPLEX MIXTURES

As previously mentioned, a major drawback of 2D NMR is that it is often time consuming. Numerous t_1 increments and the collection of several independent transients are required to obtain a spectrum with a good resolution, which leads to very long acquisition times, up to several hours when following the standard protocol given by the manufacturer. Furthermore, since the factors contributing to the peak volume are numerous (multiplicity, coupling constants, relaxation, etc.), a calibration procedure is indispensable for each compound of interest in order to obtain precise and accurate concentration measurements (Giraudeau et al., 2007; Lewis et al., 2007). Reduction of the experimental duration is therefore paramount.

Moreover, reducing the experimental duration is also essential to obtain a low standard deviation. The SNR plays a crucial role in the precision of NMR spectra, thus in the precision of measurement. From a statistical point of view, the maximum error σ on the signal intensity is inversely proportional to the SNR (Ernst et al., 1987). The acquisition of N transients is necessary to obtain a 2D FID with $N = N_1 \cdot \mathrm{NS}$. Here, N_1 is the number of t_1 values in the F_1 dimension and NS is the number of transients accumulated in order to increase the SNR of each 1D FID and to perform the phase cycling. This makes 2D NMR experiments particularly sensitive to spectrometer instabilities, leading to the formation of t_1 noise as described in Section 2. Long experiments are also subject to sample variations caused by biochemical and biological instability.

Therefore, pulse sequences need to be chosen and optimized to ensure a good repeatability on 2D NMR peak volumes, while acquisition and processing parameters should be carefully set to ensure the optimum quantitative performance in the shortest possible time.

3.1. Optimization strategies

As the acquisition of 2D NMR data requires N transients with $N = N_1 \cdot \mathrm{NS}$, while two consecutive transients are spaced by TR, optimizing the acquisition time requires finding experimental conditions where TR, N_1 and NS are as low as possible. The total sampling time of the 2D FID in the F_1 dimension (AQ_1) is proportional to the number of t_1 increments (N_1) and inversely proportional to the frequency range observed in the F_1 dimension (SW_1). Thus, larger ranges induce larger N_1 for a given value of AQ_1. If AQ_1 is too low, this leads to truncation artefacts in the F_1 dimension (Ernst et al., 1987). Therefore, a wide range in the F_1 dimension often leads to a long experimental duration since it is proportional to N_1.

On the other hand, sensitivity considerations will govern NS; so, these have to be chosen to obtain the minimal value giving the target SNR. Moreover, for a given NS, $\mathrm{SNR} = k \cdot \gamma_{\mathrm{exc}} \cdot \gamma_{\mathrm{obs}}^{1.5} \cdot \left(1 - \mathrm{e}^{-\mathrm{TR}/T_1^{\mathrm{exc}}}\right)$ (Ernst et al., 1987), where γ_{exc} and γ_{obs} are the gyromagnetic ratios of the excited and observed nuclei, respectively, and T_1^{exc} is the longitudinal relation time of the excited nucleus.

SW_1, γ_{exc}, γ_{obs} and T_1^{exc} highly depend on the 2D method used. We will therefore discuss in the following part some general aspects of the different sequences which can potentially be used in the field of metabolic studies.

3.1.1 Choice of the type of correlation

In view of the arguments discussed above, the choice of the 2D NMR method for quantitative analysis is not always straightforward, as many factors govern the duration and sensitivity of such experiments. The most influent factors are summarized in Table 3.1 for the most common 2D correlation strategies.

Table 3.1 Relative sensitivity (S_{rel}) and parameters governing the F_1 dimension (F_1 par.) for the different kinds of 2D NMR correlations with potential usefulness in metabolomic studies

	Homo-nuclear ^1H		Hetero-nuclear ^1H–^{13}C	
			Carbon detected	Proton detected
	J-resolved	(COSY, TOCSY, INADEQUATE-^1H)	2D INEPT	(HSQC, HMQC, HMBC)
S_{rel}	++	++	−	+
F_1 par.	J_{HH}	$\delta\ ^1H$	$\delta\ ^1H$	$\delta\ ^{13}C$

In homo-nuclear *J*-resolved spectroscopy, the evolution period is a spin echo. Thus, during the time t_1, only the ^1H–^1H coupling is expressed as the effect of the chemical shift is refocussed. During the detection period, the ^1H NMR signal is detected ($\gamma_{exc}=\gamma_{obs}=\gamma_{1H}$ and T_1^{exc} is the proton longitudinal relaxation time). The ^1H chemical shift and ^1H–^1H coupling are therefore expressed during t_2. The positions of the lines in the F_2 dimension are determined by the ^1H chemical shifts and ^1H–^1H couplings. The positions of the lines in the F_1 dimension are, on the other hand, determined by ^1H–^1H couplings only. The range which must be properly sampled in the F_1 dimension is therefore very low (approximately 40 Hz) which leads to low values for N_1.

When the resolution in the F_1 dimension is not sufficient, as with the *J*-resolved method, it can be improved by using homo-nuclear COSY (Aue, Bartholdi, et al., 1976; Bax & Freeman, 1981; Bax, Freeman, & Morris, 1981; Hurd, 1990; von Kienlin, Moonen, van der Toorn, & van Zilj, 1991). In this sequence, the evolution period is a free evolution delay t_1. During this period, the ^1H chemical shifts and ^1H–^1H couplings are freely expressed. During the detection period, the ^1H NMR signal is detected. The ^1H chemical shifts and ^1H–^1H couplings are therefore expressed in t_2 ($\gamma_{exc}=\gamma_{obs}=\gamma_{1H}$ and $T_1{}^{exc}$ is the ^1H longitudinal relaxation time). The positions of the lines in both dimensions are therefore determined by the chemical shift of ^1H and ^1H–^1H coupling. Because the frequency dispersion induced by chemical shifts is greater than that induced by ^1H–^1H couplings, the frequency range which must be properly sampled in the F_1 dimension generally leads to greater values for N_1 compared to in *J*-resolved spectroscopy. The properties mentioned here are common to other 2D homo-nuclear methods, such as DQF–COSY (Piantini, Sorensen, & Ernst,

1982; Rance et al., 1983), TOCSY (Braunschweiler & Ernst, 1983; Davis & Bax, 1985) or 2D ^{1}H INADEQUATE (Debart, Rayner, Imbach, Chang, & Lown, 1986; Turner, 1982).

Compared with ^{1}H, ^{13}C has a much larger chemical shift range. Therefore, 2D ^{1}H–^{13}C NMR spectra offer a better data scattering (Fig. 3.4). Such spectra correlate ^{1}H chemical shifts to ^{13}C chemical shifts and can be obtained by two kinds of sequences: the carbon-detected experiments and the proton-detected experiments.

The former are experiments based on the INEPT sequence (Bax & Sarkar, 1984; Bendall, Pegg, & Doddrell, 1981; Ding, 1999). In these sequences, ^{1}H magnetization is excited. The evolution period is a spin echo in which only the ^{1}H chemical shifts are expressed. The ^{1}H magnetization is then transferred to carbons directly linked to protons. During the detection period, the ^{13}C NMR signal is detected in the presence of ^{1}H decoupling and therefore only the ^{13}C chemical shifts are expressed during t_2 ($\gamma_{\text{exc}} = \gamma_{1\text{H}}$, $\gamma_{\text{obs}} = \gamma_{13\text{C}}$ and T_1^{exc} is the ^{1}H longitudinal relaxation time). The positions of the lines in the F_2 dimension are determined by the chemical shifts of the carbon considered, and the position of the lines in the F_1 dimension are determined by the chemical shifts of protons directly linked to this carbon. The frequency range which must be properly sampled in the F_1 dimension is, therefore, governed by ^{1}H chemical shifts.

The most popular proton-detected 2D ^{1}H–^{13}C NMR method is HSQC (Bodenhausen & Ruben, 1980; Ernst et al., 1987). In this sequence, ^{1}H magnetization is excited and transferred to carbon. The evolution period is a spin echo in which only the carbon chemical shifts are expressed. Then, magnetization is flipped back to ^{1}H and, during the detection period, the ^{1}H NMR signal is detected in the presence of ^{13}C decoupling. Therefore, only the ^{1}H chemical shifts and ^{1}H–^{1}H couplings are expressed in t_2 ($\gamma_{\text{exc}} = \gamma_{\text{obs}} = \gamma_{1\text{H}}$.and T_1^{exc} is also the ^{1}H longitudinal relaxation time; however, only ^{1}H linked to ^{13}C is observed). The positions of the lines in the F_2 dimension are determined by the chemical shifts of the protons considered, and the positions of the lines in the F_1 dimension are determined by the chemical shifts of carbons directly linked to these protons. The frequency range which must be properly sampled in the F_1 dimension is, therefore, governed by ^{13}C chemical shifts. Because the frequency dispersion induced by carbon chemical shifts is greater than the one induced by proton chemical shifts, this leads generally to greater values for N_1 compared to the carbon-detected sequences. The properties mentioned here are common to other proton-detected 2D ^{1}H–^{13}C NMR methods, such as HMQC (Bax &

Subramanian, 1986; Hurd & John, 1991; Müller, 1979; Ruiz-Cabello et al., 1992) or HMBC (Bax & Summers, 1986; Willker, Leibfritz, Kerssebaum, & Bermel, 1993).

As summarized in Table 3.1, the choice of the pulse sequence depends not only on the target sensitivity, and therefore on the sample concentration, but also on its spectral complexity. In respect to sensitivity, 2D ^{1}H homonuclear correlations are the best choice since the high gyromagnetic ratio of ^{1}H governs excitation and detection. This is still true for the proton-detected 2D ^{1}H–^{13}C NMR methods; however, only protons linked to ^{13}C are detected and the natural abundance of ^{13}C is close to 1.1%. On the other hand, these methods allow a better data scattering because of the large ^{13}C chemical shift range. Carbon-detected 2D ^{1}H–^{13}C NMR methods offer the same advantage, and the frequency range which must be correctly sampled in the F_1 dimension is smaller. However, their sensitivity is eight-fold lower since it is the ^{13}C gyromagnetic ratio which governs detection. These methods should therefore be used when the quantity of analyte is not an issue (Koskela, 2009).

3.1.2 Optimization of the number of t_1 increments (N_1)

As the number of points of the pseudo-FID in the F_1 dimension (N_1) is of the utmost importance for the total duration of the experiment, it should be as short as possible, since it has a proportional influence on the experimental time. In order to study the effect of decreasing N_1 in quantitative 2D *J*-resolved acquisitions, we cut off a 128-data-point FID at various lengths to simulate different N_1 values (Giraudeau et al., 2007). Data were acquired on an equimolar mixture of nortropine and tropine, and the standard deviation for the nortropine:tropine peak volume ratio was determined from five consecutive spectra. Figure 3.7 shows the evolution of this standard deviation with N_1. It is apparent that $N_1 = 7$ is sufficient to keep a good standard deviation value, as this remains constant at about 5% when N_1 is greater than 7. $N_1 = 12$ was finally chosen to preserve a minimum resolution. New experiments performed with this value made it possible to decrease the standard deviation to 2.2%. Hence, the experimental duration was divided by 10, thus minimizing the influence of spectrometer instabilities.

In the same article, we also evaluated the optimum value for N_1 in the case of DQF–COSY spectra. For the DQF–COSY cross-peaks, the FID envelope in both dimensions has a sinusoidal modulation in *J*. In the F_1 dimension, the signal reaches a maximum value for $t_1 = 1/(2J)$. In contrast to *J*-resolved spectroscopy, the relevant information is not located at the

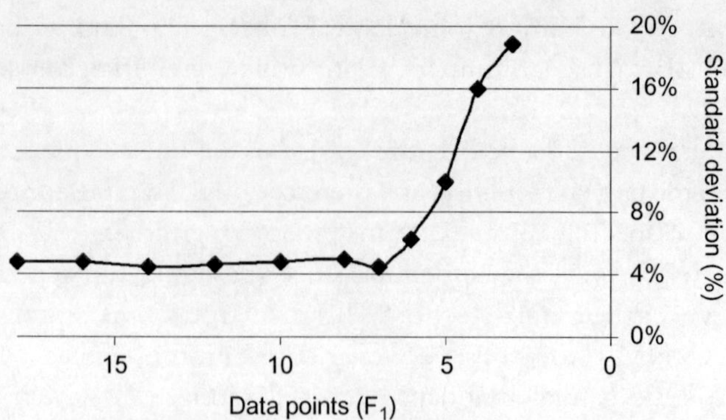

Figure 3.7 Influence of the number of increments (N_1) in the F_1 dimension on the standard deviation, for five *J*-resolved spectra (Giraudeau et al., 2007).

beginning of the FID. As a consequence, it is not possible to reduce the number of points in the F_1 dimension as much as for *J*-resolved spectroscopy. An N_1 value of 64 increments was found to be the minimum value to preserve essential information.

In the same way, we used 64 t_1 increments in the indirect dimension in a 2D ^{1}H INADEQUATE protocol for fast determination of metabolite concentrations in complex mixtures (Martineau et al., 2011). Greater N_1 values did not significantly improve SNR and resolution, whereas for $N_1 = 32$, a significant alteration of the resolution along the F_1 dimension was observed. Thanks to this optimization, the total experimental time was reduced to 7 min, a duration that makes it possible to obtain a complete calibration curve in a reasonable time.

In the case of proton-detected 2D ^{1}H–^{13}C NMR methods, the frequency range is greater in the F_1 dimension than for ^{1}H homo-nuclear correlations. Consequently, the best value for N_1 is higher. Still, Lewis et al. showed that concentrations can be accurately determined in 12 min using HSQC and 128 t_1 increments (Lewis et al., 2007).

3.1.3 Optimization of the NS

When N_1 has been optimized, the signal envelope does not decrease significantly over AQ_1 and the SNR is therefore proportional to $(N_1 \cdot \mathrm{NS})^{1/2}$. NS must therefore be adjusted to reach the target SNR which depends on the expected precision, since the maximum error σ on the signal intensity is inversely proportional to the SNR.

When the amount of material available is not an issue, NS = 1 could be sufficient from this point of view, however, the value of NS must be sufficient to allow for a complete phase cycling. Field gradient pulses can be used to select coherence transfer pathways (Price, 1996) but do not always compensate for RF pulses imperfections. The minimum value for NS is then determined by the basic phase cycling (Kessler, Gehrke, & Griesinger, 1988).

In order to evaluate the impact of NS on the precision, Martineau et al. have measured, in the case of DQF–COSY, the relative standard deviation (RSD) on 2D peak volume ratios as a function of the number of scans (Martineau et al., 2011). Figure 3.8 shows the results obtained for the lactic acid/alanine peak volume ratio on six consecutive spectra obtained on a mixture of metabolites (similar results were obtained for other 2D peak ratios). As can be seen in Fig. 3.2, a tendency for precision and experimental duration is to be inversely correlated. As already mentioned, long experiments are more sensitive to spectrometer instabilities over time, leading to a degradation of precision. Contrariwise, short experiments are less susceptible to such instabilities and show a better repeatability, as long as the SNR and the resolution are sufficient to quantify relevant peaks with the target precision. However, the RSD notably increases for NS = 1 which illustrates that in this case the phase cycling needs at least two transients.

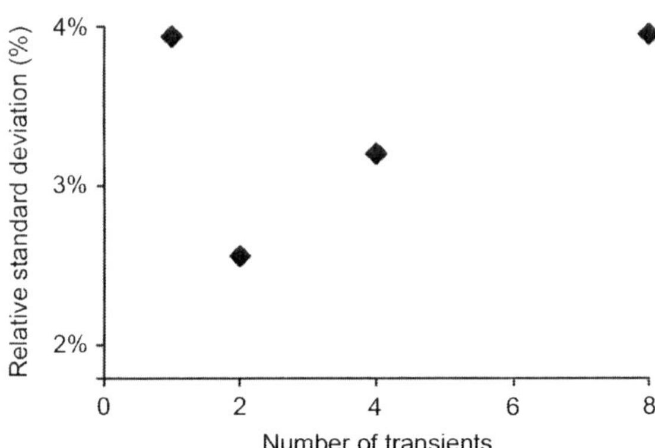

Figure 3.8 Influence of the number of transients on the relative standard deviation (lactic acid/alanine peak volume ratio) for DQF–COSY spectra recorded with 128 data points in the F_1 dimension, for a 50-mM metabolite mixture (Martineau et al., 2011).

3.1.4 Optimization of the processing before integration

Before Fourier transformation, the FID is generally multiplied by a mathematical function called the weighting function. Various weighting functions can be used according to the expected effect (Delikatny, Hull, & Mountford, 1991). First, weighting the time-domain signal reduces the effect of truncation (Freeman, 1998). Indeed, when the acquisition duration is too short (which is often the case in 2D NMR), signal does not decrease to zero at the end of the sampling period. This truncation gives rise to "wiggles" on each side of the peak. By weighting the FID by applying a function, the amplitude of which decreases up to zero at the end of the acquisition period, the FID is forced to return to the baseline. Thus this unwanted effect is avoided. Also, the decreasing functions (such as the exponential function) for which the amplitude is 1 for $t=0$ give an improvement of the SNR. This is because they do not affect the beginning of the FID containing the main part of the signal, but minimize the influence of the last points of the FID which essentially contain noise. The sensitivity is thus improved, but weighting with this kind of function induces a broadening of the lines and thus a decrease of resolution. When a higher resolution is needed, it is better to use a function presenting a maximum for $t>0$ (but returning to 0 at the end of the FID to limit the effect of a possible truncation (Tiziani, Lodi, Ludwig, Parsons, & Viant, 2008). However, such improvement of the resolution is detrimental to the SNR. A good compromise between SNR and resolution can be obtained by using a Lorentz-Gauss transformation in which the maximum can easily be adjusted (Ferrige & Lindon, 1978). In quantitative NMR, the preservation of the area or the volume of the signals, determined by the maximal amplitude of the FID, is indispensable (Canet, Boubel, & Canet Soulas, 2002). The weighting function must therefore be optimized at the same time as the number of increments in the F_1 dimension by respecting two conditions: its amplitude must decrease down to zero at the end of the sampling period and its maximum has to coincide with the maximum of the NMR signal which is generally at the beginning of the FID.

Measuring the peak volumes for quantitative analysis is also very sensitive to baseline distortions. In 2D NMR, this effect is important, since the baseline distortions occur in both dimensions. The sources of baseline distortion are numerous, some being of instrumental origin and others resulting directly from the sample analyzed (Freeman, 1998). The distortions of instrumental origin can arise from various instabilities or imperfections of the spectrometer (Marion & Bax, 1988; Tang, 1994). For example, an error in the first point of the FID creates an offset, while if several points are erroneous at the beginning of the FID, distortions dependent on the frequency can be introduced (rolling). On the other hand, as Otting, Widmer, Wagner, and Wüthrich (1986) demonstrated, the baseline offset can also have its source in the properties of the discrete Fourier transformation. The contribution of the sample to baseline distortion is particularly problematic in those cases where large peaks are present in the spectrum. A notable example is distortion that can be caused by the water peak during experiments in aqueous solution.

While baseline correction can be made directly on the FID (Braun, Kalinowski, & Berger, 1998), mostly it is applied in the frequency domain. Numerous algorithms have been proposed for this operation (Braun et al., 1998; Cobas, Bernstein, Martín-Pastor, & Tahoces, 2006; Dietrich, Rüdel, & Neumann, 1991) using a modelling of the baseline by a polynomial or a spline function. The baseline correction can itself have a major influence on the precision of the measurements. Again, a particular difficulty arises for mixtures in aqueous solution, since correcting for the major distortions due to the residual water signal is not facile. Furthermore, this distortion differs between two successive experiments and thus reduces the repeatability of the measurements. We recently showed, in the case of the *J*-resolved sequence, (Giraudeau et al., 2007), that a better precision is obtained when a polynomial correction of order 2 is applied. Furthermore, the correction is generally more efficient when (i) it is restricted to the zone of interest and (ii) the polynomial degree is lower than 3. Indeed, when a polynomial of too high a degree is used, the baseline correction function may consider certain peaks as being a part of the baseline, inducing a distortion of peaks detrimental to the quantitative analysis.

Finally, various symmetrization procedures have been proposed in order to improve the quality of certain homo-nuclear correlation spectra (COSY, TOCSY, NOESY, *J*-resolved; Baumann, Kumar, Ernst, & Wütrich, 1981; Baumann, Wider, et al., 1981). These procedures are generally particularly efficient to suppress t_1 noise. However, they could create artificial cross-peaks, and recovering small signals buried in noise may be difficult (Ernst et al., 1987). Furthermore, this operation leads to a modification of the peak volumes which can greatly affect the precision of measurements. For instance, in the case of measurements performed on DQF–COSY spectra, the RSD shifts from 0.4% to 5.2% when symmetrization is applied (P. Giraudeau and S. Akoka, unpublished results). In the same manner, in the case of the *J*-resolved sequence, the "tilt" operation leads to a partial overlap of the integration zones, thus altering the determination of the peak volumes. Furthermore, the alignment of multiplets in the F_1 dimension is not really necessary, as the peak volumes are directly determined on the 2D massifs and not on the projections. Therefore, like symmetrization, the tilt is not an efficient processing in the case of a quantitative analysis by 2D NMR.

3.1.5 Peak integration

Methods of data processing and analysis also play a vital role in implementing a reproducible and high throughput strategy for quantitative analysis. Different methods of peak integration are currently used.

The simplest is the direct summation of the spectral data points. It does not need assumption on the peak shape, and integration tools are integrated in all spectrometer software. However, this method is very sensitive to baseline distortions, t_1 noise and tailing of adjacent peaks. It is particularly difficult to apply in cases of peak overlap, although Romano, Paris, Acernese, Barone, and Motta (2008) have proposed an approach to circumvent this problem.

A variant of the previous method is to make projections of the lines containing the peak to quantify. For each peak, positive and negative intensities are projected separately, forming two 1D spectra for each 2D signal. The areas are then determined using the 1D integration tool on each of the projections and their values are summed (Massou et al., 2007b).

However, for performing reproducible analysis of 2D NMR spectra, a parametric model-fitting approach to spectral deconvolution is often the best choice (de Graaf, Chowdhury, & Behar, 2011; Miller & Greene, 1989), as it accounts for spectral overlap when estimating intensities and also makes an effective use of *a priori* information, for example, the assumption of approximately uniform chemical shifts and line widths for corresponding signals within related spectra acquired on the same sample.

Recent papers reported the use of maximum likelihood for analyzing 2D NMR spectra. The first applications of this method in multiple dimensions focussed on pure time-domain modelling of the FID for purposes of accurate NOESY cross-peak modelling (Jeong, Borer, Wang, & Levy, 1993). The work of Chylla, Volkman, and Markley (1998) in a protein biomolecular context demonstrated the practicality of using hybrid time-domain, frequency-domain maximum likelihood fitting methods in a series of 2D ^{1}H–^{15}N-HSQC relaxation experiments. Recently, the same team has developed an algorithm called fast maximum likelihood reconstruction (FMLR) that performs spectral deconvolution of 2D NMR spectra for

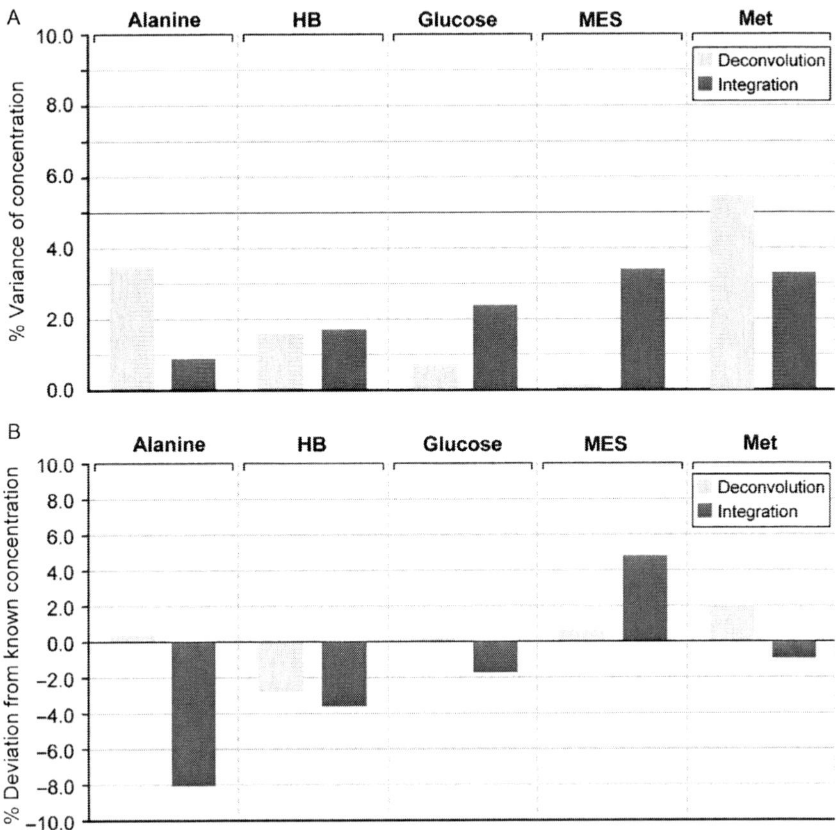

Figure 3.9 (A) Precision (% variance) for the relative molar concentration and (B) accuracy (% deviation from known concentration) for the molar concentration for deconvolution (light bars) and manual peak integration (dark bars) (Chylla et al., 2011). (For colour version of this figure, the reader is referred to the online version of this chapter.)

the purpose of accurate signal quantification (Chylla, Hu, Ellinger, & Markley, 2011). FMLR demonstrates greater accuracy (0.5–5.0% error) than peak height analysis and peak integral analysis with greatly reduced operator intervention (Fig. 3.9).

3.2. Applications for metabolic studies

Although the use of 2D NMR for studying complex metabolic mixtures has been generally limited to metabolite identification, recent publications have reported a quantitative analysis of metabolite mixtures by 2D NMR. One of these chapters (Rai et al., 2009) suggested bypassing the calibration procedure by measuring all the relaxation times and *J*-couplings for a given sample in order to obtain, after mathematical correction, 2D peak volumes reflecting the exact metabolite concentration. However, this appears to be a very long and tedious procedure, which is probably difficult to generalize to a large number of metabolic samples. Therefore, the approach relying on a calibration curve seems more appropriate for robust and precise quantitative analysis.

Several studies relying on homo-nuclear 2D NMR have been reported recently. Ludwig et al. have studied extracts from fish gonads by using the *J*-resolved sequence (Ludwig & Viant, 2010). For the quantification, they established the correlation between the concentrations calculated from the 2D spectra and those calculated from the 1D spectra and showed that this sequence could be envisaged for the quantification of metabolites. A similar approach has also been used by other groups for the analysis of liver tissues (Wang et al., 2003) and for the study of plant metabolism (Hendrawati et al., 2006; Sanchez-Sampedro, Kim, Choi, Verpoorte, & Corchete, 2007; Widarto et al., 2006). Similarly, Alonso et al., 1989 calculated the metabolite concentrations in frog muscles from COSY spectra, while Nouaille, Matulova, Delort, and Forano (2004) identified and quantified maltodextrin-1-phosphate in a strain of *Fibrobacter succinogenes* (rumen bacterium) using the DQF–COSY sequence. Massou et al. working with ^{13}C-enriched glucose in *Escherichia coli* analyzed quantitatively the metabolic flux: for this, they used the TOCSY sequence optimized by a zero quantum filter (Massou et al., 2007a, 2007b). Gowda *et al.* also used enriched compounds by marking plasma with ^{15}N-éthanolamine and ^{13}C-formic acid in order to increase the sensitivity and the resolution of TOCSY and HSQC experiments (^{1}H–^{15}N and ^{1}H–^{13}C; Gowda et al., 2010). In certain samples, they added a model mixture then measured the metabolite concentrations compared with an internal reference: the comparison between the concentrations of the samples with and without addition of the model mixture showed the good adequacy of the results and thus the feasibility of the quantification.

The potential of hetero-nuclear 2D NMR, and particularly of the HSQC pulse sequence, for the quantification of metabolites in complex mediums has also been developed in a number of publications. In order to understand the cellular metabolism in breast cancer, Richardson, Yang, Osterman, & Smith, 2008 quantified the metabolites after extraction. Hu et al. (2007) worked with ^{1}H–^{13}C HSQC on milk quality control, after optimization on a model mixture of metabolites to verify the proportionality between the measured signal and the concentration. Application to milk components gave successful results. By using a model mixture, Xi, de Ropp, Viant, Woodruff, and Yu (2008) were able to develop a program for peak recognition and calculation of the intensity. After comparison with results obtained in 1D for quantification, they successfully applied their method to different biological matrices (abalone muscles, fish eggs and trout liver). A similar approach was taken to quantify the metabolites in urine by Gronwald et al. (2008), who built calibration curves and established the limits of detection and quantification of different metabolites.

Other research groups tried to take into account the parameters that may influence the quantification. For example, Heikkinen et al. quantified the metabolites of wood lignin using the Q-HSQC sequence (with removal of the J dependence of polarization transfer). They used several values of the polarization transfer time to obtain a uniform response on all the J_{CH} values (Heikkinen et al., 2003). Tests on a reference mixture were used to verify the quantitative nature of the Q-HSQC sequence before applying it to the lignin samples. Recently, Koskela et al. published an improved version of this sequence (Q-OCCAHSQC), which is less sensitive to off-resonance effects, by the introduction of optimized RF pulses (Koskela, Heikkilä, Kilpeläinen, & Heikkinen, 2010). In this chapter, a human blood plasma sample of a healthy volunteer was measured with Q-OCCAHSQC, and the spectrum was compared with ^{1}H NMR and 1D CPMG spectra.

Different peaks of α-glucose, β-glucose, alanine, lactate, and valine were integrated, and the integral of anomeric proton H_1 of α-glucose was set as the reference (Fig. 3.10). The Q-OCCAHSQC integration results were in good agreement with the 1D CPMG results but were slightly higher. However, this small difference can be explained by signal attenuation occurring during the T_2-filtering period of the 1D CPMG sequence.

In all the studies summarized above, the optimization of the experimental duration was not considered, although it is one of the major drawbacks of 2D NMR, as previously explained. However, since 2007, a few studies have proposed strategies to reduce the experimental duration of 2D NMR experiments while preserving their quantitative character. In 2007, we demonstrated that J-resolved spectroscopy and DQF–COSY are efficient techniques for precise and accurate quantitative analysis of complex mixtures in a short time, when acquisition and post-processing parameters are carefully optimized (Giraudeau et al., 2007). In this study, J-resolved and DQF–COSY quantitative spectra were obtained on an equimolar mixture of tropine and nortropine in 2.7 min and 12 min, respectively (Fig. 3.11).

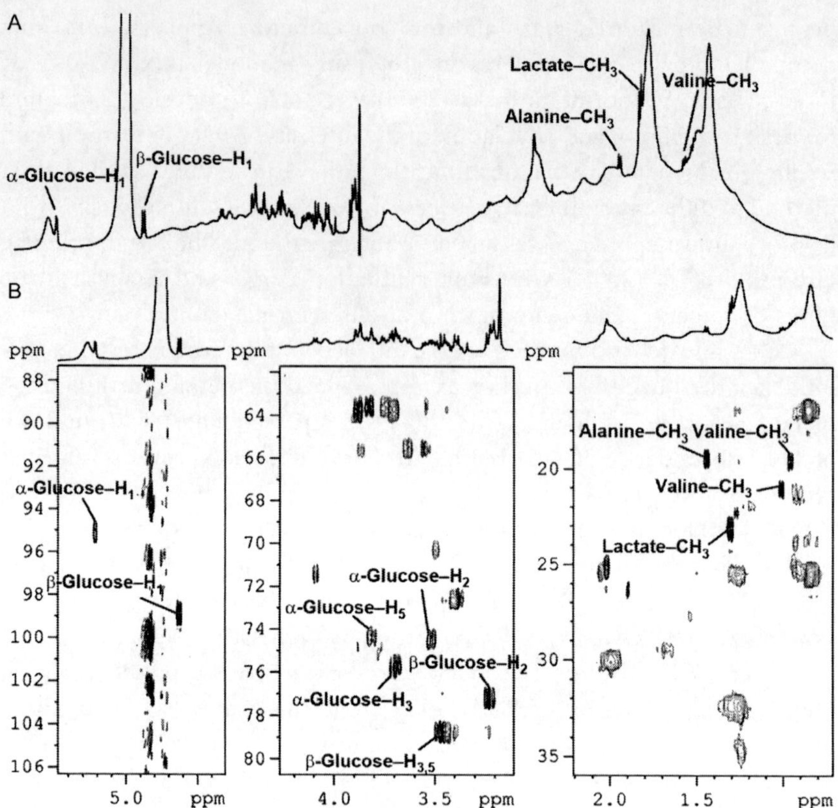

Figure 3.10 NMR spectra from a human blood plasma sample: (A) ^{1}H NMR spectrum, (B) expansions of Q-OCCAHSQC spectrum. The ^{1}H NMR spectrum was measured at 300 K with water signal presaturation, using 90° excitation pulse, and the repetition time was set to five times the longest T_1^H of the sample (7.2 s). The Q-OCCAHSQC spectrum was measured with water signal presaturation and using 32 scans per increment. The spectral widths in the proton and carbon dimensions were 10 and 200 ppm, respectively. The repetition time was set to 7.2 s. Total measurement time was 26 h 41 min. *Adapted from Koskela et al. (2010).*

In a recent article, we introduced the ^{1}H-INADEQUATE sequence for quantitative applications (Martineau et al., 2011). In this work, quantitative ^{1}H INADEQUATE 2D spectra of metabolite mixtures were obtained in 7 min with a repeatability better than 2% for metabolite concentrations as low as 100 μM, and with an excellent linearity. A high analytical performance was demonstrated for the homo-nuclear INADEQUATE pulse sequence (Fig. 3.12A) compared to more common experiments such as COSY or TOCSY. The high degree of precision reached by this technique

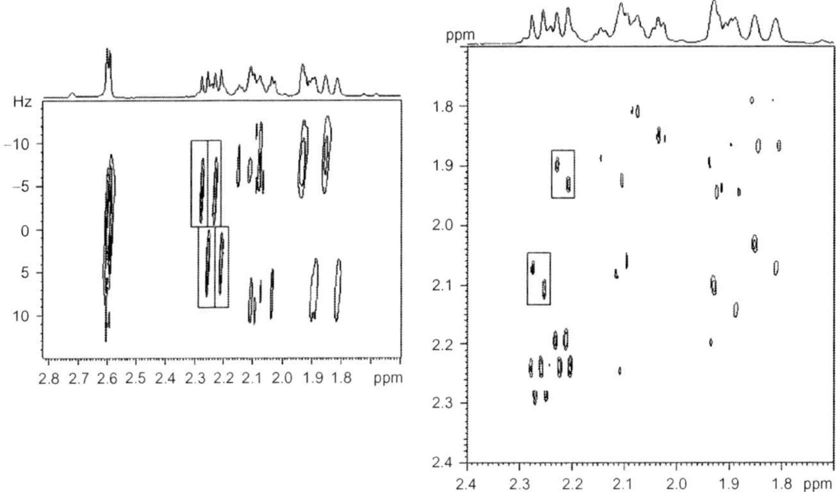

Figure 3.11 400 MHz 2D *J*-resolved spectrum (left, 2.7 min) and COSY–DQF spectrum (right, 12 min) of an equimolar mixture of tropine and nortropine, with water signal presaturation. Framed areas correspond to integration zones (Giraudeau et al., 2007).

was essentially attributed to a much cleaner diagonal in the case of INADEQUATE.

This fast protocol was then applied to compare and contrast the metabolite content in three different breast cancer cell lines expressing different hormonal and tyrosine kinase receptors (SKBR3, MCF-7 and MDA-MB-468; Martineau et al., 2012). Thanks to the cleanness of the diagonal, it was possible to integrate peak volumes for metabolites close to the diagonal, such as taurine, myo-inositol and serine (Fig. 3.12B), which would have not been possible with a more classical 2D pulse sequence such as COSY or TOCSY.

In order to establish the calibration curves (necessary to determine absolute metabolite concentrations), we chose to rely on a standard addition procedure (Bader, 1980). A model solution containing the 15 target metabolites previously identified was used and three additions were performed on the initial cell extract.

For the initial cell extracts, and after each addition, the peak volume was determined, and for each peak, this volume was plotted as a function of the added concentration, resulting in 15 graphs for each initial cell extract. As an example, the graph obtained from the threonine peak of a SKBR3 cell extract is shown in Fig. 3.13. The absolute concentrations of 15 metabolites were then determined for the different breast cancer cell lines, revealing significant differences between them (Fig. 3.14).

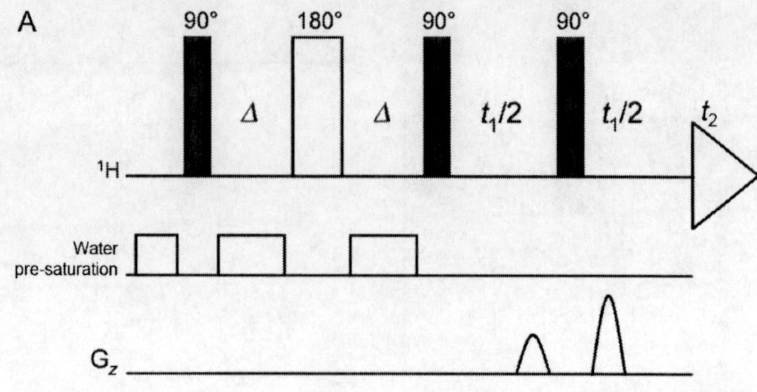

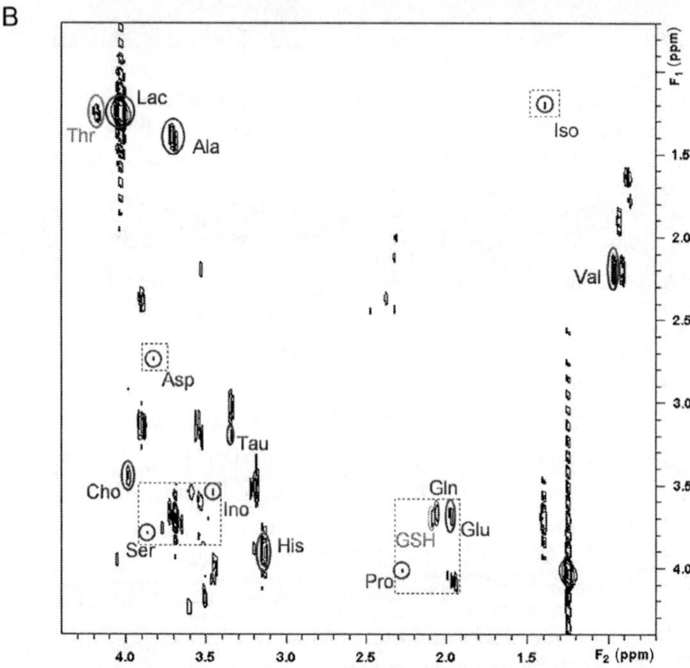

Figure 3.12 Pulse sequence for the acquisition of quantitative 2D ^{1}H INADEQUATE spectra (A) and 2D spectrum obtained on a cellular extract of MCF-7 breast cancer cells (B). The 2D spectrum was acquired in 13 min (1 transient, 128 t_1 increments), at 298 K, with water signal pre-saturation, on an 11.7 T spectrometer with a cryogenic (^{1}H/^{13}C) probe. The peaks chosen for integration are indicated for each metabolite. Ala, alanine; Lac, lactate; Thr, threonine; GSH, glutathione; Ino, myo-inositol; Tau, taurine; Gln, glutamine; Val, valine; Iso, isoleucine; Pro, proline; Asp, aspartate; His, histidine; Ser, serine; Glu, glutamate; Cho, choline (Martineau et al., 2012). (For colour version of this figure, the reader is referred to the online version of this chapter.)

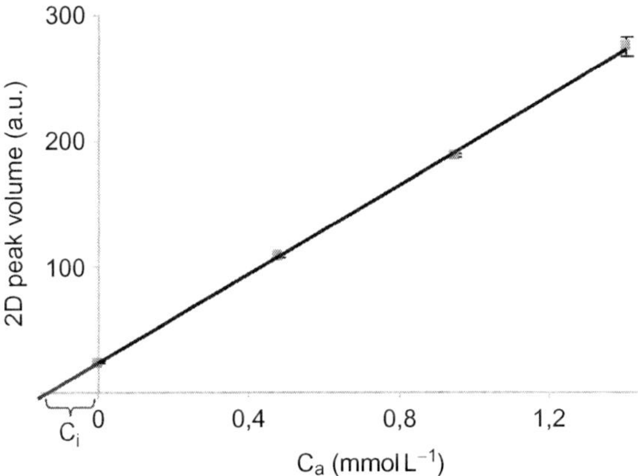

Figure 3.13 Standard addition curve obtained for the threonine signal from a ^1H INADEQUATE 2D spectrum of a SKBR3 cell extract. The 2D peak volume is plotted as a function of the added concentration C_a. C_i represents the initial metabolite concentration. Similar curves were obtained for other metabolites and cell lines. The high correlation coefficient ($r^2 = 0.9996$) and the small standard deviations demonstrate that 2D INADEQUATE NMR is reliable for precise and accurate quantitative analysis (Martineau et al., 2012). (For colour version of this figure, the reader is referred to the online version of this chapter.)

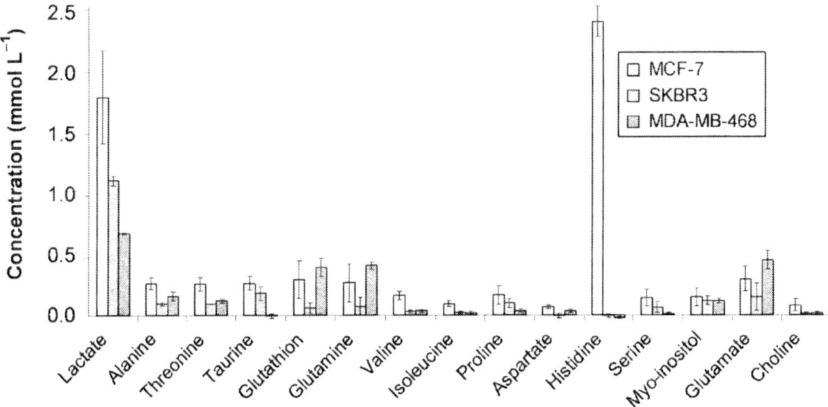

Figure 3.14 Metabolite concentrations of intracellular extracts obtained from three cell lines: SKBR3, MCF-7 and MDA-MB-468 by a quantitative 2D ^1H INADEQUATE protocol. Characteristic biomarkers of cell lines can be identified: histidine, threonine, valine, isoleucine, glutathione, proline, alanine and lactate. Standard deviations represent the biological variability (calculated on three growth times per cell line for each metabolite; Martineau et al., 2012).

The metabolite concentrations measured were in good agreement with previous studies regarding metabolite profile changes in breast cancer. While providing a high degree of discrimination, this methodology offers a powerful tool for the determination of relevant biomarkers.

Employing a similar approach, Lewis et al. (2010) introduced a fast protocol for measuring metabolite concentrations in complex solutions using 2D ^{1}H–^{13}C NMR. This protocol, called FMQ (for fast metabolite quantification) uses the HSQC pulse sequence. It was validated for a model mixture of 26 metabolites and a spectrum was obtained in 12 min (Fig. 3.15). Calibration curves plotting absolute peak intensity against concentration were established for each metabolite. The average error on calculated concentrations was 2.7% with a maximum of 10.3% while equivalent values were 16.2% and 44.5%, respectively for 1D ^{1}H analysis. The FMQ protocol was then applied to biological extracts (*Arabidopsis thaliana*, *Saccharomyces*

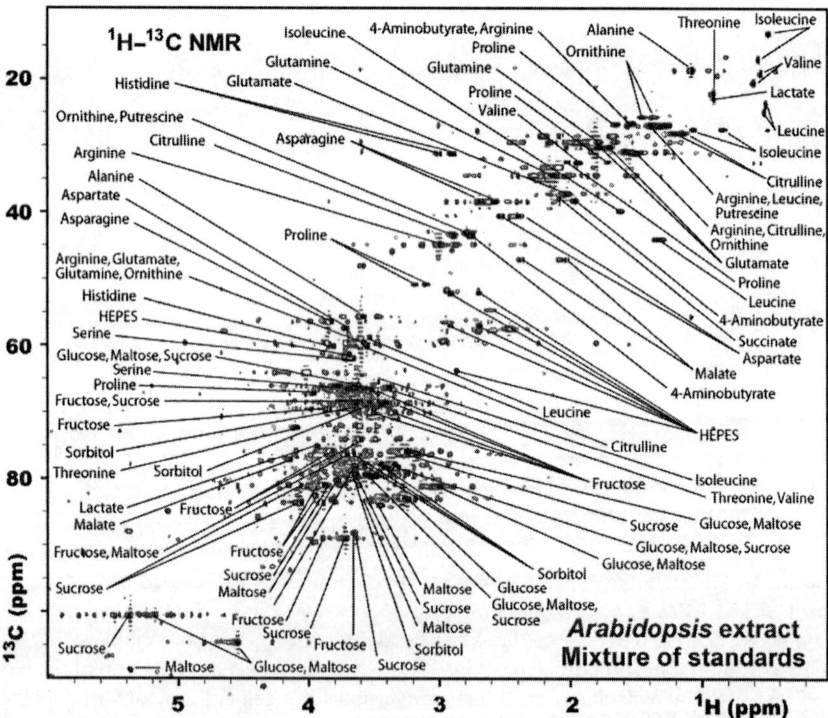

Figure 3.15 2D ^{1}H–^{13}C HSQC NMR spectrum of a synthetic mixture (equimolar mixture of 26 metabolites, red) overlaid onto a spectrum of aqueous whole-plant extract from *A. thaliana* (blue) (Lewis et al., 2010). (See Colour Insert.)

cerevasiae, *Medicago sativa*) to measure about 40 metabolite concentrations (depending of the extract) ranging from 40 μM to 230 mM. This method seems therefore to be an alternative to 1D ^{1}H NMR to measure metabolite concentrations in cases when at least 50 mg (extract dry weight) sample can be obtained.

4. GOING FASTER

The results presented in Section 3 considered the optimization of conventional 2D NMR to obtain precise and accurate quantitative data in the shortest possible time. In spite of their high analytical quality, these experiments are still limited in terms of experimental time due to the nature of 2D NMR experiments themselves. Even when optimized, their duration remains above a few minutes, due to the need to record a sufficient number of t_1 increments to reach the target resolution. While this duration may be sufficient for many studies, other cases may require going below this time limit, as, for example, in samples undergoing chemical or biological modifications in the course of the experiment. Moreover, reducing even further the experimental duration could potentially help to improve the precision of quantitative measurements. In this section, we describe recent alternatives to the conventional 2D NMR methods, which offer promising new tools towards fast 2D quantitative analysis of complex mixtures.

4.1. Strategies for reducing the experimental duration

During the past two decades, NMR spectroscopists have proposed numerous strategies to reduce the duration of 2D NMR experiments. A first family of approaches consists of optimizing conventional pulse sequences in order to reduce the recovery delay separating two successive t_1 increments. Ross and co-workers proposed to use Ernst-angle pulses in the HMQC pulse sequence to study ligand–protein interactions (Ross, Salzmann, & Senn, 1997). Schanda and Brutscher carefully optimized the delays and pulse angles in the HMQC pulse sequence, and managed to obtain hetero-nuclear 2D spectra of proteins in a few seconds (Schanda & Brutscher, 2005). More recently, Vitorge et al. implemented a clever gradient-based coherence selection scheme allowing a dramatic reduction of the recovery delay in homo-nuclear 2D NMR experiments (Vitorge, Bodenhausen, & Pelupessy, 2010). In another approach, Jeannerat proposed a procedure based on optimized aliasing in order to reduce the number of increments required to sample the indirect dimension of hetero-nuclear experiments (Jeannerat, 2003).

Another possibility to reduce the experimental duration needed to obtain a given resolution in the indirect dimension is to record a very limited number of t_1 increments and to apply, after acquiring the data, a specific processing to reach the resolution which would have been obtained with a larger number of increments. This is the case of the linear prediction (LP) (Barkhuijsen, De Beer, Bovée, & Van Ormondt, 1985; Stern, Li, & Hoch, 2002) and maximum entropy reconstruction (Hoch, 1985; Stern et al., 2002) approaches, where algorithms are applied to reconstruct a complete FID in the F_1 dimension from a limited number of points. Another alternative is the covariance NMR methodology (Brüschweiler & Zhang, 2004), where the covariance matrix of the experimental data is calculated, thus resulting in a high resolution improvement. A regularization procedure was proposed to improve the quantitative nature of this procedure (Chen et al., 2007).

The approaches described above aim at reducing the number of t_1 increments in the indirect dimension, but all of them are still based on the conventional 2D NMR approach. An alternative consists in sampling in a sparse way the (t_1, t_2) space in order to reduce the number of increments. Several strategies have been proposed, including exponential (Barna, Laue, Mayger, Skilling, & Worrall, 1987), radial (Kupce & Freeman, 2005) or random sampling (Kamzimierczuk, Zawadzka, Kozminski, & Zhukov, 2007). These approaches are often combined with one of the non-conventional processing methods described above (Lafon, Hu, Amoureux, & Lesot, 2011). Also noteworthy is the Hadamard 2D NMR spectroscopy, proposed by Kupče and Freeman, relying on an excitation in the frequency domain, thus departing from the Fourier-based methodology (Kupce & Freeman, 2003). However, this approach requires an *a priori* knowledge of the resonances to be excited: therefore, its application to complex mixtures is not straightforward.

While the strategies outlined above limit the sampling of the indirect dimension, they all require the repetition of several experiments. The experimental duration is therefore often dictated by the relaxation delay, and several tens of seconds are still necessary to obtain the resulting 2D spectrum. This is already much faster than the time required to record a conventional 2D spectrum, and these approaches form promising tools in various application fields of NMR spectroscopy. However, none of them has been evaluated for quantitative analysis or applied for quantitative purposes.

In 2002, a completely different strategy, inspired from imaging techniques, was proposed, which has revolutionized the way of acquiring multi-dimensional spectra. The "ultrafast NMR" methodology (Frydman, Lupulescu, & Scherf, 2003; Frydman, Scherf, & Lupulescu, 2002), proposed by Frydman and co-workers, makes it possible to record any kind of 2D NMR spectra in a single scan, and therefore in a fraction of a second. Not only is this strategy much faster than the others described above but also it is the only one which has been evaluated for and applied to quantitative analysis so far (Giraudeau, Remaud, and Akoka (2009); Giraudeau, Massou, et al., 2011; Pathan, Akoka, Tea, Charrier, & Giraudeau, 2011). In the following paragraphs, we describe the main features of this important breakthrough, as well as its performances for quantitative analysis.

4.2. The fastest way of acquiring 2D NMR spectra: Ultrafast 2D NMR

The ultrafast methodology initially proposed by Frydman relies on spatial encoding (Frydman et al., 2003). Its main idea is that instead of repeating N successive experiments on a sample, this sample is divided into N virtual slices where the spins undergo different evolution periods, but simultaneously for all slices. The basic scheme of ultrafast experiments is presented in Fig. 3.16. After a common preparation period (typically consisting of a 90° excitation pulse, possibly followed by a polarization transfer), different slices undergo different evolution times by means of spatial encoding. This

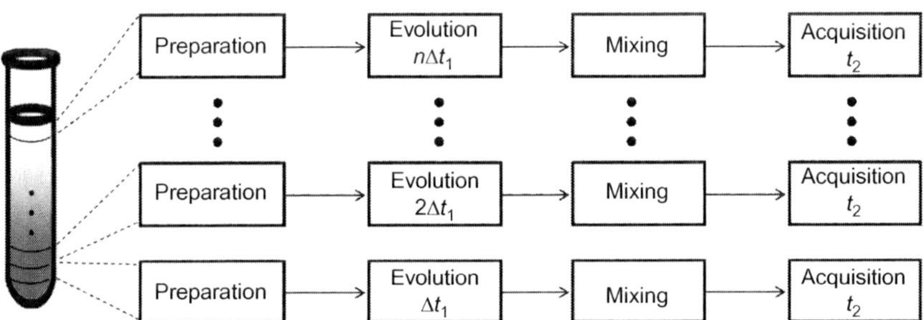

Figure 3.16 General principle of ultrafast 2D NMR experiments. Instead of repeating N transients on the same sample (as in conventional 2D NMR), the sample is virtually divided into N infinitesimal slices, which undergo different evolution periods within the same experiment by means of spatial encoding followed by an echo planar imaging-based detection. (For colour version of this figure, the reader is referred to the online version of this chapter.)

step is followed by a mixing period, generally identical to the one employed in the corresponding conventional experiment. Finally, the signal is detected thanks to an echo-planar-imaging (EPI) scheme inspired from imaging experiments (Mansfield, 1977). An appropriate processing procedure leads to the desired 2D spectrum. Depending on how the different periods are designed, this methodology makes it possible to obtain any kind of homo- or hetero-nuclear 2D spectrum in a single scan. The principles of ultrafast 2D NMR have been extensively described in the recent literature (Gal & Frydman, 2009; Mishkovsky & Frydman, 2009; Tal & Frydman, 2010); consequently, only its main features are described below.

The key feature of the ultrafast methodology is its capacity to obtain an evolution period depending on the spatial location of the spins in the NMR tube. This was originally performed by applying a series of frequency-selective pulses together with pairs of bipolar gradients along the z-axis (Frydman et al., 2003, 2002). This procedure was affected by several drawbacks (ghost peaks, hardware requirements, off-resonance effects, etc.) (Shrot & Frydman, 2003) and was almost immediately replaced by continuous spatial encoding schemes, such as the one proposed by Pelupessy (Pelupessy, 2003), of which the principle is described in Fig. 3.17. A magnetic field gradient G_e is applied along the z-axis, under the effect of which spins situated at different positions undergo different resonance frequencies. A chirp pulse (Basus, Ellis, Hill, & Waugh, 1979; Böhlen & Bodenhausen, 1992) with a linear frequency sweep is applied simultaneously, resulting in spins being excited at a different time $t_{(z)}$ according to their position z.

The pulse/gradient combination described in Fig. 3.17A is not sufficient to perform the spatial encoding step required by ultrafast experiments, as it creates a quadratic z^2 dephasing which cannot be further refocussed by linear gradients. Therefore, the first pulse is generally followed by a second identical one, applied during an opposite gradient. In the spatial-encoding scheme proposed by Pelupessy (Fig. 3.17B), two 180° chirp pulses are applied following a non-selective 90° excitation, thus resulting in a constant-time spatial encoding. This scheme leads to a linear dephasing that is proportional to the position along the z-axis, as illustrated in Fig. 3.17B. It is important to note that this dephasing also depends on the resonance frequency Ω_1. Several other schemes have been proposed to reach such a linear dephasing (Andersen & Köckenberger, 2005; Pelupessy, 2003; Shrot & Frydman, 2008b; Shrot, Shapira, & Frydman, 2004; Tal, Shapira, & Frydman, 2005). However, that proposed by Pelupessy leads to the best

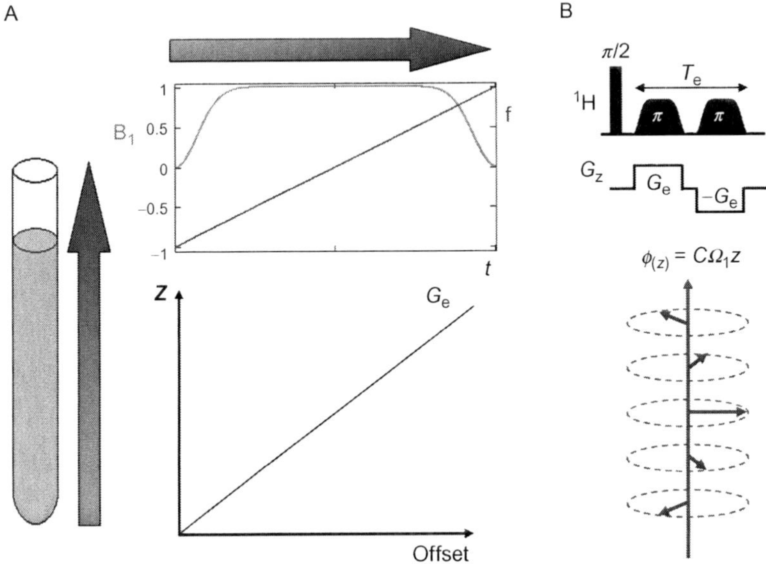

Figure 3.17 (A) Principle of continuous spatial encoding in ultrafast 2D NMR, as proposed by Pelupessy (2003). A chirp pulse with a linear frequency sweep is applied together with a magnetic field gradient G_e. As a result, spins located at different positions z are excited at different times $t_{(z)}$. (B) When two identical 180° pulses are applied together with opposite gradients after a 90° excitation, it results in a linear dephasing $\phi_{(z)}$ which also depends on the resonance frequency Ω_1. (For colour version of this figure, the reader is referred to the online version of this chapter.)

compromise between resolution and sensitivity (Giraudeau & Akoka, 2008a), and is the preferred choice for ultrafast pulse sequences.

After reaching this linear dephasing, a mixing period is generally applied, allowing a transfer of information between scalarly or dipolarly coupled spins, like in conventional 2D NMR. This period is therefore identical to its conventional counterpart while preserving the linear dephasing obtained after spatial encoding. In order to refocus this dephasing and obtain an observable signal, a gradient G_a is applied while the receiver is open, leading to a series of echoes at positions which are proportional to the resonance frequencies (Fig. 3.18A). In other words, a signal equivalent to a 1D spectrum is observed during this gradient (which typically lasts a few hundreds of microseconds), while no Fourier transform is needed. This signal forms the first dimension of the 2D spectrum, and is generally called "ultrafast dimension", as it results from the spatial-encoding process characterizing this method.

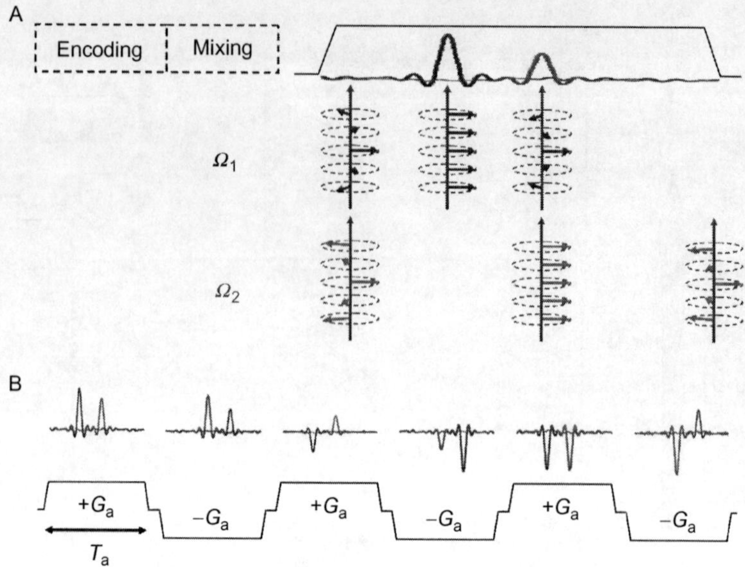

Figure 3.18 Signal detection in ultrafast 2D NMR. (A) Acquisition of the ultrafast dimension. A magnetic field gradient G_a is applied while the receiver is open, which refocuses the dephasing induced in the course of spatial encoding. It leads to the formation of echo peaks whose positions are proportional to resonance frequencies. (B) Acquisition of the conventional dimension. A series of sub-spectra are detected during a train of bipolar gradient pulses, while the system evolves under the influence of conventional parameters (relaxation, resonance frequency, couplings, etc.). Data rearrangement and Fourier transform along the conventional dimension are necessary to obtain the final 2D spectrum. (See Colour Insert.)

In order to obtain the second dimension of the 2D spectrum, a technique similar to EPI (Mansfield, 1977) is employed. It consists in applying a series of bipolar gradient pairs (Fig. 3.18B) while detecting the signal. The alternated gradients provoke a series of refocussings and defocussings, leading to the formation of mirror-image echoes, that is, to a series of symmetrical 1D spectra. During this gradient train, which typically lasts a hundred milliseconds, the system continues evolving under the influence of conventional parameters: T_2 relaxation, resonance frequency, J-couplings, etc. Therefore, an appropriate data rearrangement followed by a conventional Fourier transform is required to separate the different resonance frequencies, resulting in the second dimension of the 2D NMR spectrum. As this dimension arises from evolution under the effect of conventional interactions, it is generally referred to as "conventional dimension".

As a consequence of this particular data acquisition procedure, a specific processing is necessary to obtain the resulting 2D spectrum (Frydman et al., 2003). It includes separating the mirror-image data acquired during positive and negative acquisition gradients, correcting for possible shearing effects, Fourier transforming the resulting data and adding, after inversion of one of them, the two resulting 2D spectra. While such a processing is not commercially available, the use of several home-made processing programs has been reported (Frydman et al., 2003; Giraudeau & Akoka, 2007; Pelupessy, 2003; Wu, Zhao, Cai, Lin, & Chen, 2010). We have designed a complete processing program in Python language, freely available on demand, and working under Bruker Topspin® commercial software. It processes a 2D ultrafast spectrum in a few seconds, including conventional processing such as apodization, zero-filling, LP and baseline correction. It has already been successfully implemented in several research laboratories.

The ultrafast methodology described above makes it possible to obtain in a single scan any kind of 2D NMR correlation, provided that the different periods of the pulse sequence are designed to encode the same interactions as their conventional counterpart. Figure 3.19 presents a few examples of ultrafast spectra, recently obtained using commercial spectrometers, together with the corresponding pulse sequences. These examples highlight the potential of ultrafast 2D NMR, and also point out some of its characteristic features, such as resolution issues, or the need to accumulate several transients (Fig. 3.19C) for sensitivity issues. The analytical performance of ultrafast 2D NMR is further discussed in the next paragraphs.

4.3. State-of-the-art in improving the quality of ultrafast experiments

In spite of its impressive efficiency in reducing the duration of 2D experiments, ultrafast 2D NMR suffers from several limitations which impact on the design and the performance of the corresponding NMR pulse sequences. In the past 5 years, these limitations have been carefully analyzed and solutions have been proposed to optimize the efficiency of ultrafast 2D NMR experiments.

The first issue which has been addressed in the literature is the question of sensitivity versus resolution (Giraudeau & Akoka, 2008a, 2008c; Pelupessy, Duma, & Bodenhausen, 2008). In the conventional dimension, the resolution is dictated by the number of acquisition gradients that can be applied, depending on the performance of the hardware gradient system. Typically, 256 gradient pairs are applied, resulting in a resolution equivalent to the

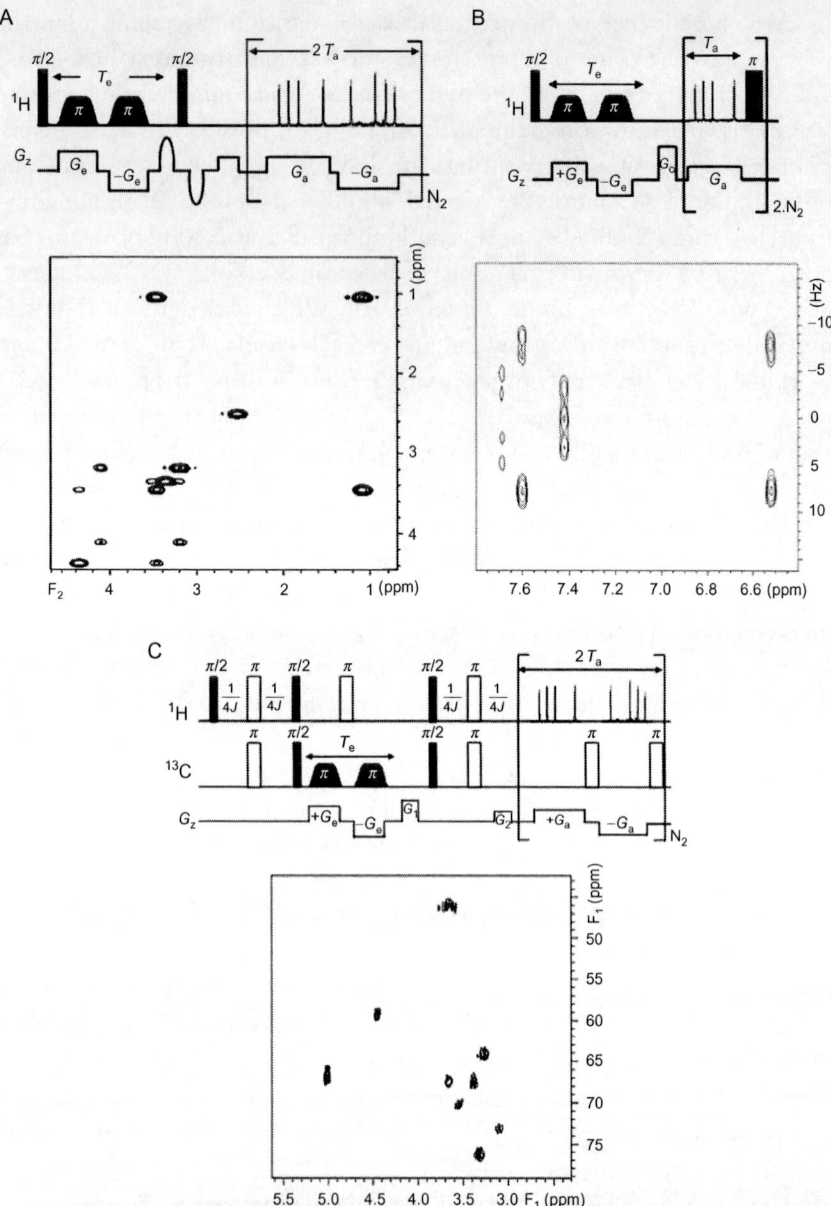

Figure 3.19 Examples of 2D spectra recorded in a single scan, illustrated together with the corresponding NMR pulse sequence, showing the potential and the versatility of ultrafast 2D NMR. (A) Ultrafast COSY spectrum recorded in 100 ms on a 100-mM methanol/ethanol mixture in DMSO-d6. (B) Ultrafast *J*-resolved spectrum recorded in 500 ms on a 100-mM cinnamic acid sample in DMSO-d_6 (Giraudeau & Akoka, 2007). (C) Ultrafast

indirect dimension of a 2D spectrum obtained with 256 t_1 increments. In contrast, the resolution appears much lower in the ultrafast dimension, as evidenced by the COSY spectrum presented in Fig. 3.19A, where the ultrafast dimension is represented horizontally. The factors impacting on the resolution in this dimension have been carefully analyzed and we have shown that it is only dictated by the duration of spatial encoding (Giraudeau & Akoka, 2008c). Independently of the spatial-encoding scheme, the resolution is inversely proportional to the duration of spatial encoding T_e. As an example, $T_e = 30$ ms leads to a theoretical peak width of ca. 20 Hz when the spatial-encoding scheme proposed by Pelupessy is employed. Consequently, the resolution should be easily increased by applying a greater spatial-encoding duration. However, this leads to large sensitivity losses, mainly due to the effects of molecular diffusion in the presence of gradients (Giraudeau & Akoka, 2008b; Shrot & Frydman, 2008a), and to a lesser extent to transverse relaxation. These effects also affect line shape, in a way that depends on the spatial encoding scheme. As a consequence, a compromise between resolution and sensitivity has to be made, as shown in Fig. 3.20.

From the practical point of view, a spatial encoding duration of 30 ms is generally chosen, as it offers a reasonably good resolution while limiting sensitivity losses. Moreover, increasing T_e further increases the demand on acquisition gradients to refocus the dephasing induced by the spatial encoding. However, a solution has been proposed to decrease the susceptibility of ultrafast experiment to diffusion losses that is particularly useful when a long T_e is needed. It consists in replacing the two 180° chirp pulses by a succession of shorter 180° pulse pairs applied during alternated gradients (Giraudeau & Akoka, 2008a). For the same value of T_e, the resolution is identical, but the sensitivity is greater and line shape distortions are reduced. The efficiency of this "multi-echo" excitation is illustrated in Fig. 3.21.

In addition to the inevitable compromise between sensitivity and resolution, the question of the spectral width accessible by spatially encoded experiments has also been extensively studied. Shrot and Frydman have demonstrated that the spectral widths SW^{UF} and SW^{Conv} in the ultrafast

HSQC spectrum recorded in 1 min 22 s (16 transients) on a 2 M D-glucose sample in D_2O (Giraudeau, Lemeunier, et al., 2011). Spectra A and B were recorded at 500 MHz with an inverse probe equipped with *z*-axis gradients. Spectrum C was recorded at 298 K on a 400-MHz spectrometer equipped with a $^1H/^{13}C$ dual+ probe including gradients along the *z*-axis.

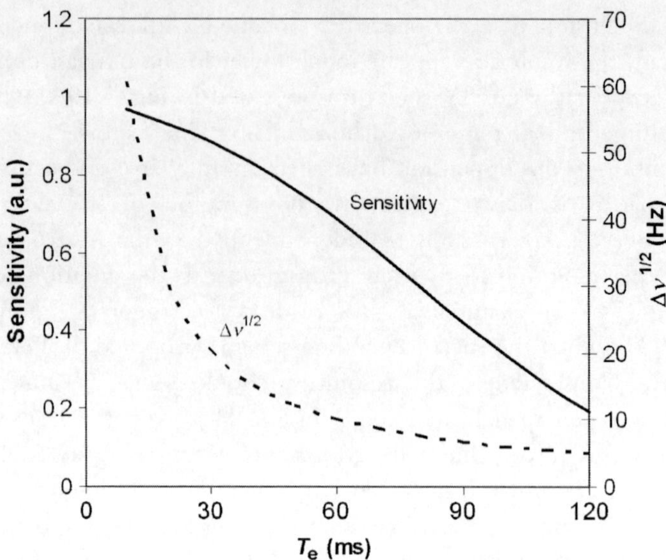

Figure 3.20 Simulated sensitivity and resolution of homo-nuclear ultrafast 2D experiments based on Pelupessy's spatial encoding scheme (Pelupessy, 2003), as a function of the spatial encoding duration T_e. The simulation was carried out with $T_2 = 300$ ms and a diffusion coefficient $D = 2.3 \times 10^{-9}$ m^2/s. The graph shows the necessary compromise between the sensitivity and the peak width $\Delta\nu^{1/2}$ in the spatially encoded dimension. Similar results are obtained for other spatial-encoding schemes.

and conventional dimensions, respectively, are related to the resolution in the ultrafast dimension by the following equation (Shrot & Frydman, 2009):

$$\gamma_a G_a L = \frac{2 \cdot \mathrm{SW}^{\mathrm{UF}} \cdot \mathrm{SW}^{\mathrm{Conv}}}{\Delta\nu}$$

where $\Delta\nu$ is the peak width in the ultrafast dimension, γ_a the gyromagnetic ratio of the detected nuclei, L the height of the detection coil and G_a the amplitude of the acquisition gradients. This relation shows that increasing the spectral width in any of the two dimensions without altering the resolution is limited by the maximum gradient amplitude available (ca. 53 G/cm on our spectrometers). Moreover, even if higher G_a values could be reached, the subsequent larger frequency dispersion induced by the acquisition gradients would make it necessary to increase the receiver bandwidth, thus resulting in sensitivity loss.

As a consequence of this additional compromise between spectral width and resolution, the maximum spectral width that can be sampled while preserving a reasonable resolution is limited. When $T_e = 30$ ms, the maximum

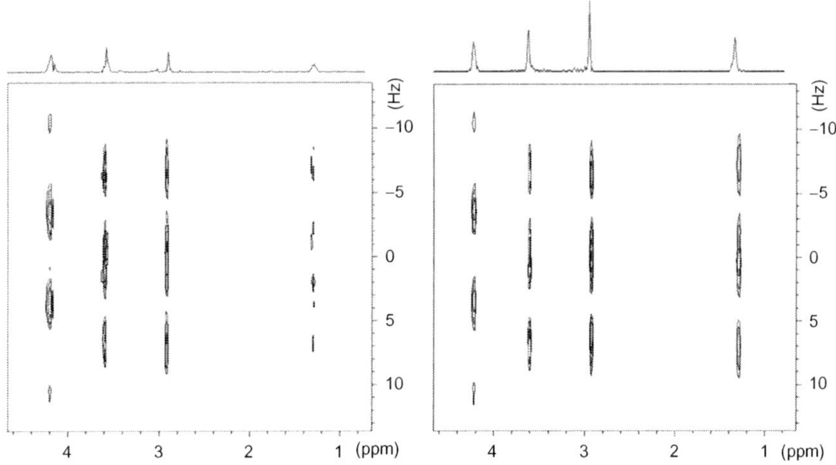

Figure 3.21 500 MHz ultrafast *J*-resolved spectra obtained on a 100-mM 3-ethyl bromopropionate sample in $CDCl_3$ at 298 K. Both spectra were obtained in 500 ms, with the same average excitation time $T_e = 120$ ms. (Left) Spectrum obtained with Pelupessy's encoding scheme consisting of two 60-ms π pulses. (Right) Spectrum obtained with multi-echo excitation scheme formed by six 20-ms π pulses. Sensitivity losses due to molecular diffusion are reduced in the latter case. *Adapted from Giraudeau and Akoka (2008a).*

spectral width at 400 MHz is approximately 4 ppm for ^{1}H and 40 ppm for ^{13}C. These values can be sufficient for certain samples, such as mixtures of metabolites where the relevant resonances are generally in the 0–4 ppm range for ^{1}H. However, in most cases, greater spectral width needs to be accessed in order to fully characterize a sample. Several methods have been proposed to increase the spectral width accessible in ultrafast 2D NMR. Concerning the conventional dimension, peaks lying outside of the observed spectral width can easily be detected by means of spectral aliasing, which automatically occurs when the Nyquist condition is violated. Provided that this aliasing procedure is optimized, the whole spectral range can be observed within a limited width, an approach which has been proposed by Jeannerat in conventional 2D NMR (Jeannerat, 2007) and applies equally in the conventional dimension of ultrafast experiments. In the ultrafast dimension, the problem is much more complex, as regular aliasing is intrinsically impossible due to the non-FT character of this dimension. In 2009, we proposed a solution relying on the spatial/spectral encoding of interactions (Giraudeau, Shrot, & Frydman, 2009). This methodology consists in replacing the 90° excitation by a series of spectrally selective pulses followed by gradients imparting site-specific spatial windings (Shrot & Frydman, 2009). If suitably chosen, these gradients will shift the ultrafast-domain resonances into arbitrary positions within the ultrafast 2D spectrum; these spatial/spectral manipulations are done so as to place all peaks in a reduced final window, compatible with the minimal G_a acquisition gradients desired. However, this approach requires *a priori* knowledge of the precise position of the resonances to be encoded, a prerequisite which is not compatible with unknown, and *a fortiori* complex, samples. In 2010, we described a slightly less efficient, but much more general approach for folding resonances in the ultrafast dimension (Giraudeau & Akoka, 2010). It relies on suitably chosen gradients placed on each side of the mixing period and can be applied to almost all ultrafast pulse sequences. Figure 3.22 highlights the potential of this approach in the case of DQF–COSY.

While the previously described methods do not give access to the real chemical shift of folded peaks (which could be, in principle, calculated from gradient parameters), this is not problematic when the folded peaks belong to molecules that are already known and well characterized, which is often the case when metabolic samples are studied.

These approaches to increased resolution, sensitivity or spectral width all rely on modifications of the acquisition procedure. The quality of ultrafast spectra can be further improved by appropriate processing protocols, such as that which we have recently proposed for enhancing the line shape

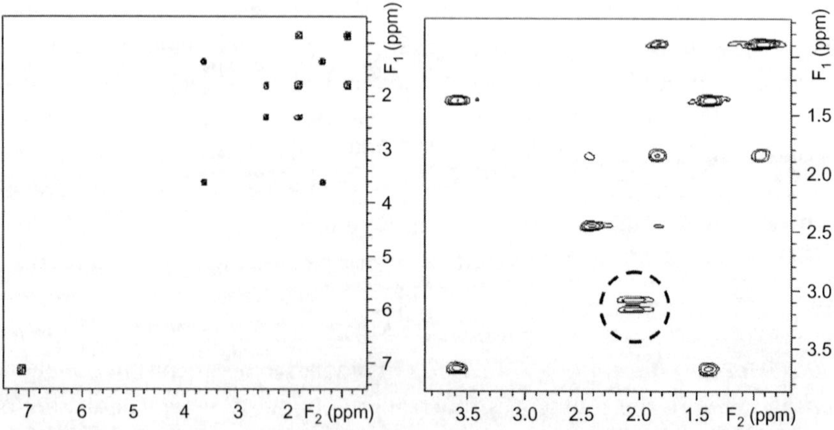

Figure 3.22 (Left) 400 MHz conventional DQF–COSY spectrum of a 100-mM ibuprofen sample in DMSO-d_6 recorded in 13 min, and (right) 400-MHz ultrafast DQF–COSY spectrum of the same sample obtained in 0.2 s with folding of the aromatic proton signal (dotted circle). *Adapted from Giraudeau and Akoka (2010).*

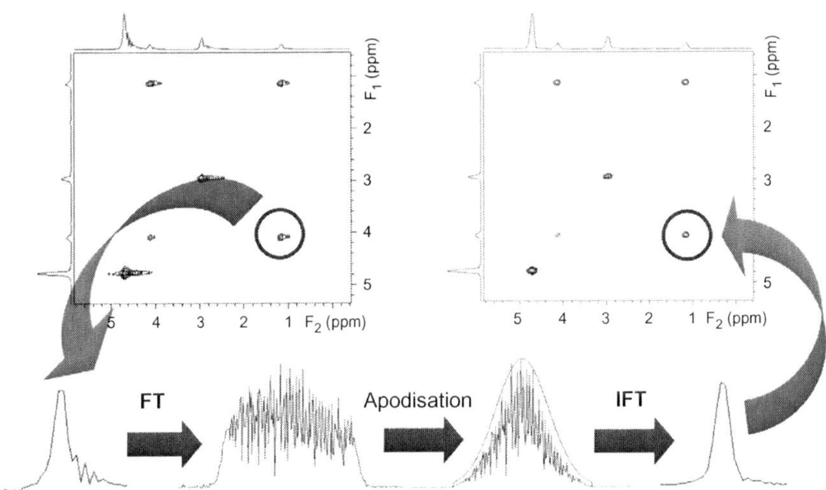

Figure 3.23 Apodization procedure designed to improve lineshape in the spatially encoded dimension of ultrafast spectra, illustrated on the ultrafast 2D COSY spectrum of a 100-mM ethanol sample in D_2O/DMSO-d_6 (1:4). Each line of the 2D spectrum is Fourier transformed, multiplied by an optimized mathematical function minimizing the non-uniformities on the edge of the sample, and finally inverse-Fourier transformed, giving rise to lineshape almost exempt from asymmetric distortions. *Adapted from Giraudeau and Akoka (2011).* (For colour version of this figure, the reader is referred to the online version of this chapter.)

and sensitivity in the ultrafast dimension (Giraudeau & Akoka, 2011). We observed that with certain hardware configurations, the peak shapes were characterized by asymmetric sinc wiggles in the ultrafast dimension. We showed, by means of numerical simulations, that the origin of these wiggles is phase distortions due to non-idealities of excitation or gradient coils on the edge of the sample. We therefore proposed a processing approach to improve line shape while also increasing sensitivity. The method, described in Fig. 3.23, first consists in Fourier transforming each line of the 2D spectrum, resulting in a smoothed square excitation profile modulated by a phase dispersion due to the different resonance frequencies present in the sample. This profile is then multiplied by an optimized mathematical function designed to minimize the contribution from the edges of the sample. Finally, an inverse Fourier transform regenerates the original peak, but the asymmetry is almost completely removed. Not only does this procedure improve the line shape in the ultrafast dimension but also it increases the sensitivity by a factor of 2 at no cost in terms of resolution. This processing approach is easily implemented in routine and greatly helps in improving the quality of ultrafast spectra.

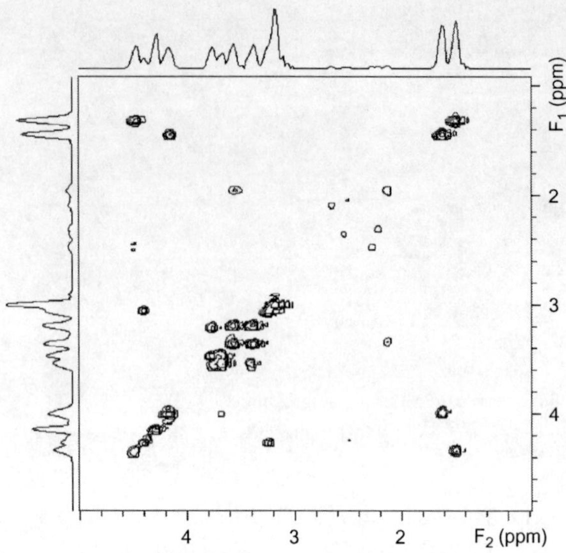

Figure 3.24 Ultrafast 2D COSY spectrum of a 50-mM mixture of six metabolites in D_2O, recorded in 100 ms on a Bruker Avance III 500-Hz spectrometer equipped with a cryogenic probe, highlighting the spectral quality that can be obtained with currently available ultrafast techniques. *Adapted from Giraudeau and Akoka (2011).*

Thanks to the numerous improvements made to the acquisition and processing of ultrafast 2D NMR spectra, it is now possible to record single-scan spectra for which the quality is not significantly different from that obtained with their conventional counterparts. While the first ultrafast experiments were generally limited to pure model molecules, it is now possible to record, in a single scan, 2D spectra of complex diluted mixtures. This performance is illustrated in Fig. 3.24, on the COSY spectrum of a model mixture of metabolites.

4.4. Quantitative ultrafast 2D NMR: Analytical performances and applications

Ultrafast 2D NMR offers promising perspectives for quantitative analysis. First, it potentially allows a significant reduction in the experimental duration, which is not negligible when considering the calibration procedure required for quantitative 2D NMR, and the frequent need to analyze a large number of samples. Moreover, ultrafast experiments are potentially less sensitive to hardware temporal instabilities which generate the t_1 noise described in Section 1. In 2009, we performed the first analytical evaluation of ultrafast 2D NMR on model mixtures (Giraudeau, Remaud, et al., 2009). Two homo-nuclear ultrafast techniques, *J*-resolved spectroscopy and TOCSY, were evaluated on model mixtures in terms of repeatability and linearity. Repeatability better than 1% for ultrafast *J*-resolved spectra and better than 7% for TOCSY spectra was obtained. Moreover, both methods were characterized by an excellent linearity, thus opening promising perspectives for this new quantitative methodology called ufo-qNMR (ultrafast optimized quantitative NMR).

The ultrafast quantitative approach is potentially much faster than its conventional counterpart provided that the sensitivity is sufficient. However, the question of the relative sensitivity per unit of time of ultrafast versus conventional 2D NMR needs to be discussed. It is fairly intuitive that a one-scan ultrafast experiment is less sensitive than a conventional experiment where signal is accumulated during several minutes or hours. But a fair and relevant comparison has to consider the SNR per unit of time, in order to answer the following question: for a given experiment time (e.g. 30 min), should we run a conventional 2D experiment or is it preferable to accumulate ultrafast acquisitions for 30 min, thus resulting in a "multi-scan single shot" approach? This is of particular pertinence for metabolic samples which are generally characterized by low concentrations, and where averaging several ultrafast experiments will obviously be indispensible. The two acquisition strategies are compared in Fig. 3.25 in the case of constant-time COSY (Girvin, 1994), a more sensitive version of the well-known COSY experiment.

In order to compare these two approaches, we recently carried out a systematic analytical study on model mixtures of metabolites (Pathan et al., 2011). Surprisingly, it turned out that for the same experimental duration, the multi-scan single shot approach was more sensitive than conventional 2D NMR. This was attributed to the absence of t_1 noise in ultrafast experiments, where all the points of the FID in the indirect dimension are acquired in a very short period of time, thus leaving almost no time for the expression of hardware instabilities. Figure 3.26 shows the conventional and ultrafast COSY spectra of a metabolite mixture, both acquired in the same duration. t_1 Noise is clearly visible on the conventional spectrum but is notably absent from the ultrafast spectrum. In this particular example, the sensitivity was multiplied by a factor of 3 in the case of the ultrafast approach (Pathan et al., 2011). This is particularly interesting from the quantitative point of view, as a greater precision can thus be expected from ultrafast experiments.

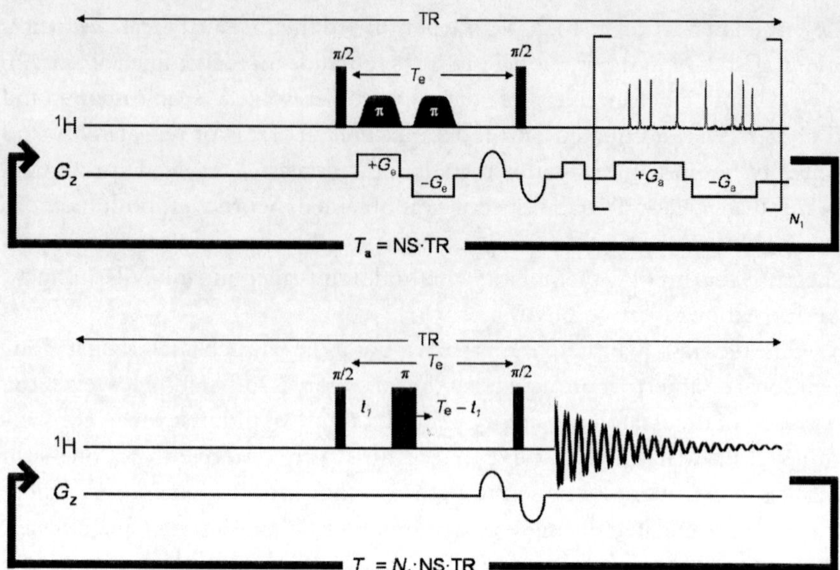

Figure 3.25 Strategies for recording quantitative 2D NMR constant-time COSY spectra, (top) by accumulating NS ultrafast experiments or (bottom) by accumulating NS scans for each N_1 increment of a conventional 2D pulse sequence. *Adapted from Pathan et al. (2011).*

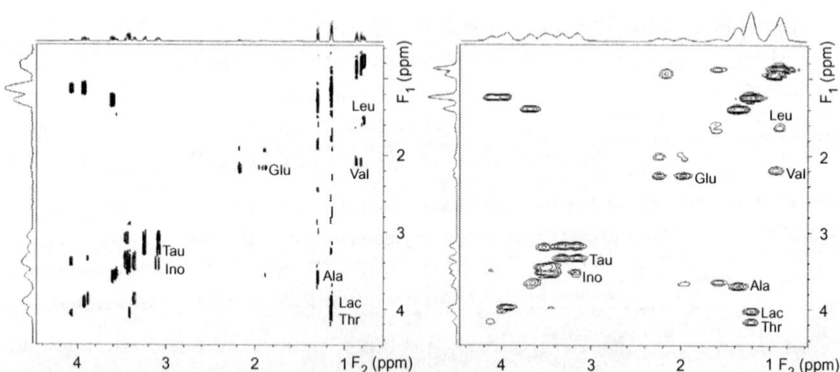

Figure 3.26 Conventional (left) and ultrafast (right) constant-time COSY spectra of a 5-mM metabolite mixture, acquired in 34 min on a 500-MHz spectrometer with a cryoprobe. The spectra were acquired with the pulse sequences and accumulation strategies described in Fig. 3.25. The metabolite peaks are indicated using the standard international three-letter code. *Adapted from Pathan et al. (2011).*

These results highlight the potential of ultrafast 2D NMR for quantitative analysis. A number of applications to the quantitative study of metabolic samples are currently being investigated. Recently, we have presented the first application of ultrafast 2D NMR to real metabolic samples, in the context of fluxomics. As recently demonstrated by Massou and co-workers, 2D NMR is a promising tool for studying metabolic fluxes by measuring ^{13}C enrichments in complex mixtures of ^{13}C-labelled metabolites (Massou et al., 2007a, 2007b). However, the methods reported so far were hampered by long experimental durations which limited their use as a quantitative tool for fluxomics. In order to take advantage of the time saving offered by ultrafast 2D NMR, we recently proposed several acquisition strategies to measure ^{13}C enrichments from single-scan 2D spectra (Giraudeau, Massou, et al., 2011; Pathan, Akoka, & Giraudeau, 2012). In collaboration with Massou and Portais, we have designed two homo-nuclear experiments, ultrafast COSY and zTOCSY, where the site-specific ^{13}C enrichments can be measured by reading lines of the 2D spectrum (Giraudeau, Massou, et al., 2011). As shown in Fig. 3.27, the ultrafast spectra recorded on a model alanine mixture give the same isotopic enrichments as those measured by conventional NMR.

The two approaches were compared in terms of analytical performance, showing accuracy and precision of a few percent and an excellent linearity. The most sensitive approach is the COSY experiment, which is less affected by molecular diffusion effects. It was applied to the measurement of ^{13}C enrichments on a biomass hydrolysate (Fig. 3.28), showing the first known application of ultrafast 2D NMR to a real biological extract. The experimental duration was divided by 200 compared to the conventional approach, while yielding the same information on site-specific isotopic enrichments.

4.5. Perspectives for metabolomic studies

The recent analytical results presented above emphasize the high potential of ultrafast NMR for quantitative analysis of metabolic mixtures. A crucial question is the limit of concentration that can be detected and quantified by these methods, as metabolic samples often contain low concentrations of metabolites whose amounts are important probes for understanding metabolism. There is no unique solution that can be proposed, as the sensitivity of these experiments depends, like any NMR experiment, on parameters such as the magnetic field strength, the probe, etc. and also on the

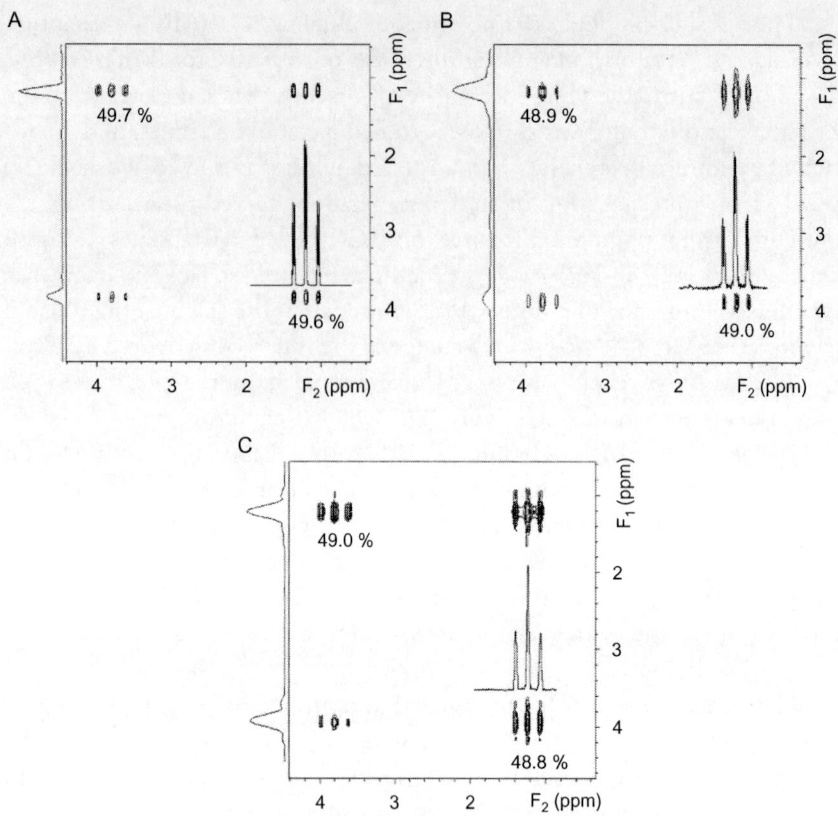

Figure 3.27 Conventional zTOCSY (A), ultrafast zTOCSY (B) and ultrafast COSY (C) spectra of a model mixture of labelled and unlabelled alanine. The ultrafast spectra were acquired in a single scan of 3 s duration (including water signal presaturation), whereas the conventional spectrum was recorded in 2.5 h. The spatially encoded dimension of the ultrafast spectra is represented vertically. Examples of 1D cross-sections are indicated in the conventional dimension where the quantification was performed. ^{13}C-enrichments measured for CH (3.8 ppm) and CH_3 (1.2 ppm) are indicated close to the relevant signals. *Adapted from Giraudeau, Massou, et al. (2011).*

performances of the gradient system. As an example, the limit concentration that can be detected in a single scan on our 500 MHz spectrometer equipped with a cryoprobe is 10 mM for a COSY experiment, but this value is decreased to 1.5 mM when signal is accumulated for 3 min (64 scans with a 3-s recovery delay). Working at higher field strength should help in increasing this sensitivity, but it must be kept in mind that this will also reduce the accessible spectral width. Alternative acquisition strategies will therefore have to be designed, such as the possibility of interleaving several

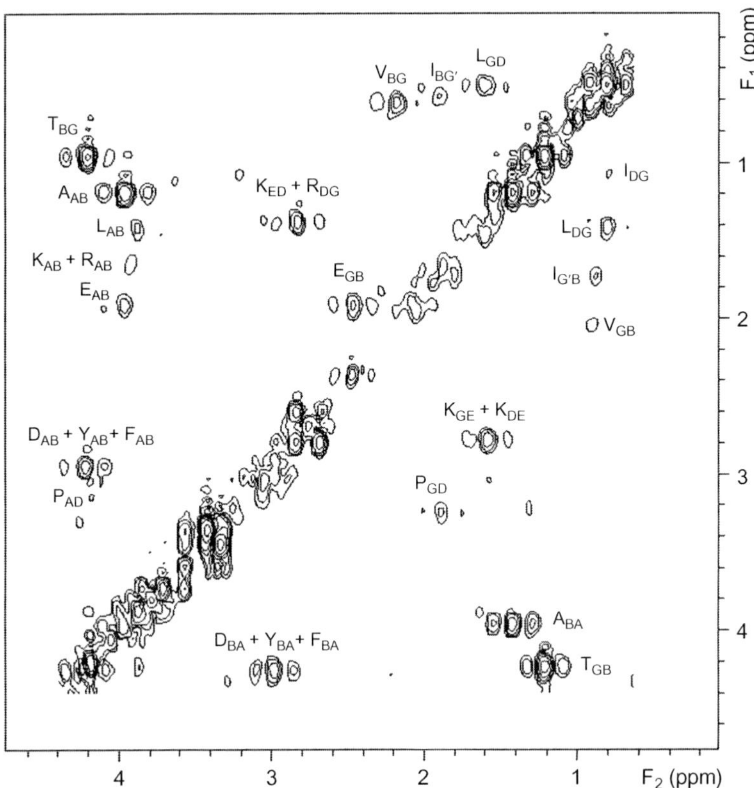

Figure 3.28 Ultrafast COSY spectrum of a biomass hydrolysate from *E. coli* cells grown on a 1:1 [U–^{13}C]-glucose:glucose mixture (12 g/L). The sample contains mainly amino acids released from the hydrolysis of cellular proteins. The spectrum was recorded in 40 s (eight scans) on a 500-MHz spectrometer equipped with a cryoprobe. Cross-peaks are annotated according to the corresponding carbon in the amino acids, using the international one-letter code for amino acids and using standard letters for the position of the protonated carbon in the molecular backbone, that is, A, B, G, D and E for α, β, γ, δ and ε, respectively. *Adapted from Giraudeau, Massou, et al. (2011).*

ultrafast acquisitions with different evolution delays to increase the accessible spectral width (Frydman et al., 2003), or the use of alternative sampling schemes (Shrot & Frydman, 2011).

In order to make ultrafast 2D NMR applicable as a routine technique for users who are not familiar with spatial-encoding methods, efforts are currently directed towards automating the acquisition and processing of ultrafast data. While processing routines are already quite efficient, more user-friendly programs will be designed to offer interfaces similar to those available for conventional experiments, including automatic settings of specific ultrafast parameters from common parameters such as spectral widths. Research towards new pulse sequences is also in constant progress, including the design of fast 3D NMR experiments taking advantage of the spatial-encoding strategies. These developments aim at being able to obtain routine 2D NMR tools for quantitative analysis of metabolite mixtures, applicable by the majority of members of the researcher community.

5. CONCLUSION

The fast and ultrafast quantitative strategies described in this chapter underline the interest of using 2D NMR for quantitative analysis, as well as the diversity and versatility of tools available for analytical chemists or biochemists who need to determine precisely the concentrations of various components in a mixture. These include, as described above, conventional 2D NMR approaches optimized for quantitative analysis, more recent ultrafast 2D NMR strategies, and intermediate approaches such as multi-scan single shot experiments. While the examples given in this chapter are not exhaustive, we hope that they provide a representative overview of the advances in quantitative 2D NMR studies as performed over the past 10 years. Facing the number of approaches available, a key question is the choice of which would be the optimal technique. As often, the answer is not unique and depends on many factors such as the initial biological or metabolic question, the characteristics of the sample, and the available hardware. Future studies and applications will further highlight the advantages and drawbacks of the different approaches, making this choice easier.

Several questions will need to be addressed in the following years. The first concerns the apparent non-quantitative nature of 2D NMR experiments, which often requires a calibration procedure to obtain quantitative results. While many studies have proved the efficiency of this approach, it is still quite a weighty procedure which increases the experimental time and introduces potential sources of error in the quantitative results. Alternative approaches are currently being investigated, such as the introduction of the electronic reference ERETIC™ (Akoka, Barantin, & Trierweiler, 1999), which was already successfully implemented in 2D NMR pulse sequences (Michel & Akoka, 2004).

All the examples described in this review point out the main limitation of quantitative 2D NMR (and of NMR in general): its low sensitivity compared with other analytical techniques such as mass spectrometry. It is particularly problematic in multi-dimensional NMR where the experimental duration is a central question. Fortunately, after providing solutions to the time issue, the community of NMR spectroscopists is now concentrating on the development of hyperpolarisation approaches where the bulk nuclear polarisations depart from their usual (ca. 10^{-5}) Boltzmann distributions and approach unity values, which increases the sensitivity of NMR experiments by several orders of magnitude. They include the use of parahydrogen (Eisenschmid et al., 2002), the optical pumping of noble gases (Albert et al., 1994; Navon et al., 1996) and microwave-driven dynamic nuclear polarization (DNP) from unpaired electrons (Hausser & Stehlik, 1968; Wolber et al., 2004). The last option has the appeal of providing high sensitivity enhancements (ca. 10^4), while making few requirements in terms of chemical substrate or sample preparation (Ardenkjaer-Larsen et al., 2003). DNP is currently revolutionizing the NMR community (Griffin & Prisner, 2010), as did 2D spectroscopy in the 1970s. While methodological developments are still necessary to make it applicable in routine, it has already been successfully coupled to ultrafast 2D NMR (Frydman & Blazina, 2007; Giraudeau, Shrot, et al., 2009), thus combining speed and sensitivity to detect analytes present at low concentrations in a very short time. Moreover, spatial-encoding strategies coupled with hyperpolarisation open promising perspectives for the real-time study of metabolic reactions (Harris, Giraudeau, & Frydman, 2011; Sarkar et al., 2009). We are confident that these new approaches, after being optimized, will dramatically enhance the sensitivity of NMR-based metabolomic studies, thus providing great help for applications which are currently limited by their sensitivity, such as metabolomics of low-sensitive nuclei or *in vivo* metabolomics.

ACKNOWLEDGEMENTS

The authors are grateful to their co-authors for the research work on quantitative 2D NMR published in this review: Evelyne Baguet, Edern Cahoreau, Nadia Guignard, Emilie Hillion, Estelle Martineau, Stéphane Massou, Meerakhan Pathan, Jean-Charles Portais, Gérald Remaud, Yoann Robin, Illa Tea. We also thank Richard Robins for useful comments and corrections. Research was supported in part by funding from the "Agence Nationale de la Recherche" for young researchers (ANR Grant 2010-JCJC-0804-01).

REFERENCES

Akoka, S., Barantin, L., & Trierweiler, M. (1999). Concentration measurement by proton NMR using the ERETIC method. *Analytical Chemistry*, *71*, 2554–2557.

Albert, M. S., Cates, G. D., Driehuys, B., Happer, W., Saam, B., Springer, C. S., et al. (1994). Biological magnetic resonance imaging using laser-polarized ^{129}Xe. *Nature*, *370*, 199–201.

Alonso, J., Arus, C., Westler, W. M., & Markley, J. L. (1989). Two-dimensional correlated spectroscopy (COSY) of intact frog muscle: Spectral pattern characterization and lactate quantitation. *Magnetic Resonance in Medicine, 11*, 316–330.

Andersen, N. S., & Köckenberger, W. (2005). A simple approach for phase-modulated single-scan 2D NMR spectroscopy. *Magnetic Resonance in Chemistry, 43*, 791–794.

Ardenkjaer-Larsen, J. H., Fridlund, B., Gram, A., Hansson, G., Hansson, L., Lerche, M. H., et al. (2003). Increase in signal-to-noise ratio of > 10,000 times in liquid-state NMR. *Proceedings of the National Academy of Sciences of the United States of America, 100*, 10158–10163.

Aue, W. P., Bachmann, P., Wokaun, A., & Ernst, R. R. (1978). Sensitivity of two-dimensional NMR spectroscopy. *Journal of Magnetic Resonance, 29*, 523–533.

Aue, W. P., Bartholdi, E., & Ernst, R. R. (1976). Two-dimensional spectroscopy. Application to nuclear magnetic resonance. *Journal of Chemical Physics, 64*, 2229–2246.

Aue, W. P., Karhan, J., & Ernst, R. R. (1976). Homonuclear broad band decoupling and two-dimensional J-resolved NMR spectroscopy. *Journal of Chemical Physics, 64*, 4226–4227.

Aursand, M., Jorgensen, L., & Grasdalen, H. (1995). Quantitative high-resolution ^{13}C nuclear magnetic resonance of anserine and lactate in white muscle of Atlantic salmon (salmo salar). *Comparative Biochemistry and Physiology, 112B*, 315–321.

Bader, M. (1980). A systematic approach to standard addition methods in instrumental analysis. *Journal of Chemical Education, 57*, 703–706.

Barkhuijsen, H., De Beer, R., Bovée, W. M. M. J., & Van Ormondt, D. (1985). Retrieval of frequencies, amplitudes, damping factors, and phases from time-domain signals using a linear least-squares procedure. *Journal of Magnetic Resonance, 61*, 465–481.

Barna, J. C. J., Laue, E. D., Mayger, M. R., Skilling, J., & Worrall, S. J. P. (1987). Exponential sampling, an alternative method for sampling in two-dimensional NMR experiments. *Journal of Magnetic Resonance, 73*, 69–77.

Basus, V. J., Ellis, P. D., Hill, H. D. W., & Waugh, J. S. (1979). Utilization of chirp frequency modulation for heteronuclear spin decoupling. *Journal of Magnetic Resonance, 35*, 19–37.

Baumann, R., Kumar, A., Ernst, R. R., & Wütrich, K. (1981). Improvement of 2D NOE and 2D correlated spectra by triangular multiplication. *Journal of Magnetic Resonance, 44*, 76–83.

Baumann, R., Wider, G., Ernst, R. R., & Wütrich, K. (1981). Improvement of 2D NOE and 2D correlated spectra by symmetrization. *Journal of Magnetic Resonance, 44*, 402–406.

Bax, A., & Freeman, R. (1981). Investigation of complex networks of spin–spin coupling by two-dimensional NMR. *Journal of Magnetic Resonance, 44*, 542–561.

Bax, A., Freeman, R., & Morris, G. A. (1981). Correlation of proton chemical shifts by two-dimensional Fourier transform NMR. *Journal of Magnetic Resonance, 42*, 164–168.

Bax, A., Griffey, R. H., & Hawkins, B. L. (1983). Correlation of proton and nitrogen-15 chemical shifts by multiple quantum NMR. *Journal of Magnetic Resonance, 55*, 301–315.

Bax, A., & Sarkar, S. K. (1984). Elimination of refocusing pulses in NMR experiments. *Journal of Magnetic Resonance, 60*, 170–176.

Bax, A., & Subramanian, S. (1986). Sensitivity-enhanced two-dimensional heteronuclear shift correlation NMR spectroscopy. *Journal of Magnetic Resonance, 67*, 565–569.

Bax, A., & Summers, M. F. (1986). Proton and carbon-13 assignments from sensitivity-enhanced detection of heteronuclear multiple-bond connectivity by 2D multiple quantum NMR. *Journal of the American Chemical Society, 108*, 2093–2094.

Bendall, M. R., Pegg, D. T., & Doddrell, D. M. (1981). Polarization transfer pulse sequences for two-dimensional NMR by Heisenberg vector analysis. *Journal of Magnetic Resonance, 45*, 8–29.

Bodenhausen, G., & Ruben, D. J. (1980). Natural abundance nitrogen-15 NMR by enhanced heteronuclear spectroscopy. *Chemical Physics Letters, 69*, 185–189.

Böhlen, J. M., & Bodenhausen, G. (1992). Experimental aspects of chirp NMR spectroscopy. *Journal of Magnetic Resonance Series A*, *102*, 293–301.

Boyer, R. D., Johnson, R., & Krishnamurthy, K. (2003). Compensation of refocusing inefficiency with synchronized inversion sweep (CRISIS) in multiplicity-edited HSQC. *Journal of Magnetic Resonance*, *165*, 253–259.

Braun, S., Kalinowski, H.-O., & Berger, S. (1998). *150 and more basic NMR experiments: A practical course*. Weinheim: Wiley-VCH.

Braunschweiler, L., & Ernst, R. R. (1983). Coherence transfer by isotropic mixing: Application to proton correlation spectroscopy. *Journal of Magnetic Resonance*, *53*, 521–528.

Brüschweiler, R., & Zhang, F. (2004). Covariance nuclear magnetic resonance spectroscopy. *Journal of Chemical Physics*, *120*, 5253–5260.

Bussy, U., Thibaudeau, C., Thomas, F., Desmurs, J.-R., Jamin, E., Remaud, G., et al. (2011). Isotopic finger-printing of active pharmaceutical ingredients by ^{13}C NMR and polarization transfer techniques as a tool to fight against counterfeiting. *Talanta*, *85*, 1909–1914.

Canet, D., Boubel, J.-C., & Canet Soulas, E. (2002). *La RMN, concepts, méthodes et applications*. Paris: Dunod.

Cano, E. C., Smith, M. A., & Shaka, A. J. (2002). Adjustable, broadband, selective excitation with uniform phase. *Journal of Magnetic Resonance*, *155*, 131–139.

Caytan, E., Remaud, G. S., Tenailleau, E., & Akoka, S. (2007). Precise and accurate quantitative ^{13}C NMR with reduced experimental time. *Talanta*, *71*, 1016–1021.

Chen, Y., Zhang, F., Snyder, D., Gan, Z., Brüschweiler-Li, L., & Brüschweiler, R. (2007). Quantitative covariance NMR by regularization. *Journal of Biomolecular NMR*, *38*, 73–77.

Chylla, R. A., Hu, K., Ellinger, J. J., & Markley, J. L. (2011). Deconvolution of two-dimensional NMR spectra by fast maximum likelihood reconstruction: Application to quantitative metabolomics. *Analytical Chemistry*, *83*, 4871–4880.

Chylla, R. A., Volkman, B. F., & Markley, J. L. (1998). Practical model fitting approaches to the direct extraction of NMR parameters simultaneously from all dimensions of multi-dimensional NMR spectra. *Journal of Biomolecular NMR*, *12*, 277–297.

Cobas, J. C., Bernstein, M. A., Martín-Pastor, M., & Tahoces, P. G. (2006). A new general-purpose fully automatic baseline-correction procedure for 1D and 2D NMR data. *Journal of Magnetic Resonance*, *183*, 145–151.

Cookson, D. J., & Smith, B. E. (1984). Optimal conditions for obtaining quantitative ^{13}C NMR data. *Journal of Magnetic Resonance*, *57*, 355–368.

Dais, P., & Spyros, A. (2007). ^{31}P NMR spectroscopy in the quality control and authentication of extra-virgin olive oil: A review of recent progress. *Magnetic Resonance in Chemistry*, *45*, 367–377.

Davis, D. G., & Bax, A. (1985). Assignment of complex ^{1}H NMR spectra via two-dimensional homonuclear Hartmann-Hahn spectroscopy. *Journal of the American Chemical Society*, *107*, 2820–2821.

Debart, F., Rayner, B., Imbach, J. L., Chang, D. K., & Lown, J. W. (1986). Structure and conformation of the duplex consensus acceptor exon:intron junction d[(CpTpApCpApGpGpT). (ApCpCpTpGpTpApG)] deduced from high-field ^{1}H-NMR of non-exchangeable and imino protons. *Journal of Biomolecular Structure & Dynamics*, *4*, 343–363.

de Graaf, R. A., Chowdhury, G. M. I., & Behar, K. L. (2011). Quantification of high-resolution ^{1}H NMR spectra from rat brain extracts. *Analytical Chemistry*, *83*, 216–224.

Delikatny, E. J., Hull, W. E., & Mountford, C. E. (1991). The effect of altering time domains and window functions in two-dimensional proton COSY spectra of biological specimens. *Journal of Magnetic Resonance*, *94*, 563–573.

Dietrich, W., Rüdel, C. H., & Neumann, M. (1991). Fast and precise automatic baseline correction of one and two-dimensional spectra. *Journal of Magnetic Resonance*, *91*, 1–11.
Ding, K. (1999). Accurate measurements of multiple-bond ^{13}C–^{1}H coupling constants from phase-sensitive 2D INEPT spectra. *Journal of Magnetic Resonance*, *140*, 495–498.
Eisenschmid, T. C., Kirss, R. U., Deutsch, P. P., Hommeltoft, S. I., Eisenberg, R., Bargon, J., et al. (2002). Para hydrogen induced polarization in hydrogenation reactions. *Journal of the American Chemical Society*, *109*, 8089–8091.
Ernst, R. R., Bodenhausen, G., & Wokaun, A. (1987). *Principles of nuclear magnetic resonance in one and two dimensions*. Oxford: Oxford Science Publications.
Ferrige, A. G., & Lindon, J. C. (1978). Resolution enhancement in FT NMR through the use of a double exponential function. *Journal of Magnetic Resonance*, *31*, 337–340.
Freeman, R. (1998). *A handbook of nuclear magnetic resonance*. Harlow: Longman.
Freeman, R., Kempsell, S. P., & Levitt, M. H. (1980). Radiofrequency pulse sequences which compensate their own imperfections. *Journal of Magnetic Resonance*, *38*, 453–479.
Frydman, L., & Blazina, D. (2007). Ultrafast two-dimensional nuclear magnetic resonance spectroscopy of hyperpolarized solutions. *Nature Physics*, *3*, 415–419.
Frydman, L., Lupulescu, A., & Scherf, T. (2003). Principles and features of single-scan two-dimensional NMR spectroscopy. *Journal of the American Chemical Society*, *125*, 9204–9217.
Frydman, L., Scherf, T., & Lupulescu, A. (2002). The acquisition of multidimensional NMR spectra within a single scan. *Proceedings of the National Academy of Sciences of the United States of America*, *99*, 15858–15862.
Fujirawa, T., & Nagayama, K. (1988). Composite inversion pulses with frequency switching and their application to broadband decoupling. *Journal of Magnetic Resonance*, 77, 55–63.
Gal, M., & Frydman, L. (2009). Ultrafast multidimensional NMR: Principles and practice of single-scan methods. In G. A. Morris & J. W. Emsley (Eds.), *Multidimensional NMR methods for the solution state* (pp. 43–60). Chichester: John Wiley & Sons Ltd.
Giraudeau, P., & Akoka, S. (2007). A new detection scheme for ultrafast J-resolved spectroscopy. *Journal of Magnetic Resonance*, *186*, 352–357.
Giraudeau, P., & Akoka, S. (2008a). Resolution and sensitivity aspects of ultrafast J-resolved 2D NMR spectra. *Journal of Magnetic Resonance*, *190*, 339–345.
Giraudeau, P., & Akoka, S. (2008b). Sensitivity losses and line shape modifications due to molecular diffusion in continuous encoding ultrafast 2D NMR experiments. *Journal of Magnetic Resonance*, *195*, 9–16.
Giraudeau, P., & Akoka, S. (2008c). Sources of sensitivity losses in ultrafast 2D NMR. *Journal of Magnetic Resonance*, *192*, 151–158.
Giraudeau, P., & Akoka, S. (2010). A new gradient-controlled method for improving the spectral width of ultrafast 2D NMR experiments. *Journal of Magnetic Resonance*, *205*, 171–176.
Giraudeau, P., & Akoka, S. (2011). Sensitivity and lineshape improvement in ultrafast 2D NMR by optimized apodization in the spatially encoded dimension. *Magnetic Resonance in Chemistry*, *49*, 307–313.
Giraudeau, P., Guignard, N., Hillion, H., Baguet, E., & Akoka, S. (2007). Optimization of homonuclear 2D NMR for fast quantitative analysis: Application to tropine–nortropine mixtures. *Journal of Pharmaceutical and Biomedical Analysis*, *43*, 1243–1248.
Giraudeau, P., Lemeunier, P., Coutand, M., Doux, J.-M., Gilbert, A., Remaud, G. S., et al. (2011). Ultrafast 2D NMR applied to the kinetic study of D-glucose mutarotation in solution. *Journal of Spectroscopy and Dynamics*, *1*, 2.
Giraudeau, P., Massou, S., Robin, Y., Cahoreau, E., Portais, J.-C., & Akoka, S. (2011). Ultrafast quantitative 2D NMR: An efficient tool for the measurement of specific isotopic enrichments in complex biological mixtures. *Analytical Chemistry*, *83*, 3112–3119.

Giraudeau, P., Remaud, G. S., & Akoka, S. (2009). Evaluation of ultrafast 2D NMR for quantitative analysis. *Analytical Chemistry*, *81*, 479–484.

Giraudeau, P., Shrot, Y., & Frydman, L. (2009). Multiple ultrafast, broadband 2D NMR spectra of hyperpolarized natural products. *Journal of the American Chemical Society*, *131*, 13902–13903.

Giraudeau, P., Wang, J. L., & Baguet, E. (2006). Improvement of the inverse-gated decoupling sequence for a faster quantitative analysis of various samples by ^{13}C NMR spectroscopy. *Journal of Magnetic Resonance*, *180*, 110–117.

Girvin, M. E. (1994). Increased sensitivity of COSY spectra by use of constant time t_1 periods. *Journal of Magnetic Resonance Series A*, *108*, 99–102.

Gowda, G. A. N., Tayyari, F., Ye, T., Suryani, Y., Wei, S., Shanaiah, N., et al. (2010). Quantitative analysis of blood plasma metabolites using isotope enhanced NMR methods. *Analytical Chemistry*, *82*, 8983–8990.

Griffin, R. G., & Prisner, T. (2010). High field dynamic nuclear polarization—The renaissance. *Physical Chemistry Chemical Physics*, *12*, 5737–5740.

Gronwald, W., Klein, M. S., Kaspar, H., Fagerer, S. R., Nurnberger, N., Dettmer, K., et al. (2008). Urinary metabolite quantification employing 2D NMR spectroscopy. *Analytical Chemistry*, *80*, 9288–9297.

Harris, T., Giraudeau, P., & Frydman, L. (2011). Metabolic kinetics From indirectly-detected hyperpolarized NMR using spatially-selective coherence transfers. *Chemistry—A European Journal*, *17*, 697–703.

Hausser, K. H., & Stehlik, D. (1968). Dynamic nuclear polarization in liquids. *Advances in Magnetic Resonance*, *3*, 79.

Heikkinen, S., Toikka, M. M., Karhunen, P. T., & Kilpeläinen, A. (2003). Quantitative 2D HSQC (Q-HSQC) via suppression of J-dependence of polarization transfer in NMR spectroscopy: Application to wood lignin. *Journal of the American Chemical Society*, *125*, 4362–4367.

Hendrawati, O., Yao, Q., Kim, H. K., Linthorst, H. J. M., Erkelens, C., Lefeber, A. W. M., et al. (2006). Metabolic differentiation of Arabidopsis treated with methyl jasmonate using nuclear magnetic resonance spectroscopy. *Plant Science*, *170*, 1118–1124.

Hoch, J. C. (1985). Maximum entropy signal processing of two-dimensional NMR data. *Journal of Magnetic Resonance*, *64*, 436–440.

Hollis, D. P. (1963). Quantitative analysis of aspirin, phenacetin, and caffeine mixtures by nuclear magnetic resonance spectrometry. *Analytical Chemistry*, *35*, 1682–1684.

Holzgrabe, U. (2010). Quantitative NMR spectroscopy in pharmaceutical applications. *Progress in Nuclear Magnetic Resonance Spectroscopy*, *57*, 229–240.

Hu, F., Furihata, K., Kato, Y., & Tanokura, M. (2007). Nondestructive quantification of organic compounds in whole milk without pretreatment by two-dimensional NMR spectroscopy. *Journal of Agricultural and Food Chemistry*, *55*, 4307–4311.

Hurd, R. E. (1990). Gradient-enhanced spectroscopy. *Journal of Magnetic Resonance*, *87*, 422–428.

Hurd, R. E., & John, B. K. (1991). Gradient-enhanced proton-detected heteronuclear multiple-quantum coherence spectroscopy. *Journal of Magnetic Resonance*, *91*, 648–653.

Hwang, T.-L., Van Zijl, P. C. M., & Garwood, M. (1997). Broadband adiabatic refocusing without phase distorsion. *Journal of Magnetic Resonance*, *124*, 250–254.

Jeannerat, D. (2003). High resolution in heteronuclear ^{1}H–^{13}C NMR experiments by optimizing spectral aliasing with one-dimensional carbon data. *Magnetic Resonance in Chemistry*, *41*, 3–17.

Jeannerat, D. (2007). Computer optimized spectral aliasing in the indirect dimension of ^{1}H–^{13}C heteronuclear 2D NMR experiments. A new algorithm and examples of applications to small molecules. *Journal of Magnetic Resonance*, *186*, 112–122.

Jeener, J. (1971). Lecture presented at Ampere International Summer School II, Basko Polje, Yugoslavia.

Jeong, G. W., Borer, P. N., Wang, S. S., & Levy, G. C. (1993). Maximum-likelihood-constrained deconvolution of two-dimensional NMR spectra. Accuracy of spectral quantification. *Journal of Magnetic Resonance Series A*, *103*, 123–134.

Jungnickel, J. L., & Forbes, J. W. (1963). Quantitative measurement of hydrogen types by integrated nuclear magnetic resonance intensities. *Analytical Chemistry*, *35*, 938–942.

Kamzimierczuk, K., Zawadzka, A., Kozminski, V., & Zhukov, I. (2007). Lineshapes and artifacts in multidimensional Fourier transform of arbitrary sampled NMR data sets. *Journal of Magnetic Resonance*, *188*, 344–356.

Kessler, H., Gehrke, M., & Griesinger, C. (1988). Two-dimensional NMR spectroscopy—Background and overview of the experiments. *Angewandte Chemie International Edition*, *27*, 490–536.

Kim, H. K., Choi, Y. H., & Verpoorte, R. (2010). NMR-based metabolomic analysis of plants. *Nature Protocols*, *5*, 536–549.

Koskela, H. (2009). Quantitative 2D NMR studies. *Annual Reports on NMR Spectroscopy*, *66*, 1–31.

Koskela, H., Heikkilä, O., Kilpeläinen, I., & Heikkinen, S. (2010). Quantitative two-dimensional HSQC experiment for high magnetic field NMR spectrometers. *Journal of Magnetic Resonance*, *202*, 24–33.

Koskela, H., Kilpeläinen, I., & Heikkinen, S. (2005). Some aspects of quantitative 2D NMR. *Journal of Magnetic Resonance*, *174*, 237–244.

Koskela, H., & Väänänen, T. (2002). Quantitative determination of aliphatic hydrocarbon compounds by 2D NMR. *Magnetic Resonance in Chemistry*, *40*, 705–715.

Krishnan, P., Kruger, N. J., & Ratcliffe, R. G. (2005). Metabolite fingerprinting and profiling in plants using NMR. *Journal of Experimental Botany*, *56*, 255–265.

Kruger, N. J., Troncoso-Ponce, M. A., & Ratcliffe, R. G. (2008). ^{1}H NMR metabolite fingerprinting and metabolomic analysis of perchloric acid extracts from plant tissues. *Nature Protocols*, *3*, 1001–1012.

Kupce, E., & Freeman, R. (1995a). Adiabatic pulses for wideband inversion and broadband decoupling. *Journal of Magnetic Resonance Series A*, *115*, 273–276.

Kupce, E., & Freeman, R. (1995b). Stretched adiabatic pulses for broadband decoupling spin inversion. *Journal of Magnetic Resonance Series A*, *117*, 246–256.

Kupce, E., & Freeman, R. (2003). Two-dimensional Hadamard spectroscopy. *Journal of Magnetic Resonance*, *162*, 300–310.

Kupce, E., & Freeman, R. (2005). Fast multidimensional NMR: Radial sampling of evolution space. *Journal of Magnetic Resonance*, *173*, 317–321.

Kwakye, J. K. (1985). Use of NMR for quantitative analysis of pharmaceuticals. *Talanta*, *32*, 1069–1071.

Lafon, O., Hu, B., Amoureux, J.-P., & Lesot, P. (2011). Fast and high-resolution stereochemical analysis by nonuniform sampling and covariance processing of anisotropic natural abundance 2D ^{2}H NMR datasets. *Chemistry—A European Journal*, *17*, 6716–6724.

Le Gall, G., Colquhoun, I. J., Davis, A. L., Collins, G. J., & Verhoeyen, M. E. (2003). Metabolite profiling of tomato (*Lycopersicon esculentum*) using ^{1}H NMR spectroscopy as a tool to detect potential unintended effects following a genetic modification. *Journal of Agricultural and Food Chemistry*, *51*, 2447–2456.

Le Grand, F., George, G., & Akoka, S. (2005). Natural abundance ^{2}H-ERETIC-NMR authentication of the origin of methyl salicylate. *Journal of Agricultural and Food Chemistry*, *53*, 5125–5129.

Levitt, M. H., & Freeman, R. (1979). NMR population inversion using a composite pulse. *Journal of Magnetic Resonance*, *33*, 473–476.

Levitt, M. H., & Freeman, R. (1981). Compensation for pulse imperfections in NMR spin-echo experiments. *Journal of Magnetic Resonance*, *43*, 65–80.
Levy, G. C., Pehk, T., & Srinivasan, P. R. (1980). Quantitative ^{15}N NMR spectroscopy. *Organic Magnetic Resonance*, *14*, 129–132.
Lewis, I. A., Karsten, R. H., Norton, M. E., Tonelli, M., Westler, W. M., & Markley, J. L. (2010). NMR method for measuring carbon-13 isotopic enrichment of metabolites in complex solutions. *Analytical Chemistry*, *82*, 4558–4563.
Lewis, I. A., Schommer, S. C., Hodis, B., Robb, K. A., Tonelli, M., Westler, W., et al. (2007). Method for determining molar concentrations of metabolites in complex solutions from two-dimensional ^{1}H–^{13}C NMR spectra. *Analytical Chemistry*, *79*, 9385–9390.
Lindon, J. C., Nicholson, J. K., Holmes, E., & Everett, J. R. (2000). Metabonomics: Metabolic processes studied by NMR spectroscopy of biofluids. *Concepts in Magnetic Resonance*, *12*, 289–320.
Lonnon, D. G., & Hook, J. M. (2003). ^{17}O quantitative nuclear magnetic resonance spectroscopy of gasoline and oxygenated additives. *Analytical Chemistry*, *75*, 4659–4666.
Ludwig, C., & Viant, M. R. (2010). Two-dimensional J-resolved NMR spectroscopy: Review of a key methodology in the metabolomics toolbox. *Phytochemical Analysis*, *21*, 22–32.
Ludwig, C., Ward, D. G., Martin, A., Viant, M. R., Ismail, T., Johnson, P. J., et al. (2009). Fast targeted multidimensional NMR metabolomics of colorectal cancer. *Magnetic Resonance in Chemistry*, *47*, S68–S73.
Mansfield, P. (1977). Multi-planar image formation using NMR spin echoes. *Journal of Physics C: Solid State Physics*, *10*, 55–58.
Mareci, T. H., & Scott, K. N. (1977). Quantitative analysis of mixtures by carbon-13 nuclear magnetic resonance spectroscopy. *Analytical Chemistry*, *49*, 2130–2136.
Marion, D., & Bax, A. (1988). Baseline distortion in real-Fourier-transform NMR spectra. *Journal of Magnetic Resonance*, *79*, 352–356.
Martineau, E., Giraudeau, P., Tea, I., & Akoka, S. (2011). Fast and precise quantitative analysis of metabolic mixtures by 2D ^{1}H INADEQUATE NMR. *Journal of Pharmaceutical and Biomedical Analysis*, *54*, 252–257.
Martineau, E., Tea, I., Akoka, S., & Giraudeau, P. (2012). Absolute quantification of metabolites in breast cancer cell extracts by quantitative 2D ^{1}H INADEQUATE NMR. *NMR in Biomedicine*, *25*, 985–992.
Massou, S., Nicolas, C., Letisse, F., & Portais, J.-C. (2007a). Application of 2D-TOCSY NMR to the measurement of specific ^{13}C-enrichments in complex mixtures of ^{13}C-labeled metabolites. *Metabolic Engineering*, *9*, 252–257.
Massou, S., Nicolas, C., Letisse, F., & Portais, J.-C. (2007b). NMR-based fluxomics: Quantitative 2D NMR methods for isotopomers analysis. *Phytochemistry*, *68*, 2330–2340.
Mavromoustakos, T., Zervou, M., Theodoropoulou, E., Panagiotopoulos, D., Bonas, G., Day, M., et al. (1997). ^{13}C NMR analysis of the triacylglycerol composition of Greek virgin olive oils. *Magnetic Resonance in Chemistry*, *35*, S3–S7.
Mehlkopf, A. F., Korbee, D., Tiggelman, T. A., & Freeman, R. (1984). Sources of t_1 noise in two-dimensional NMR. *Journal of Magnetic Resonance*, *58*, 315–323.
Michel, N., & Akoka, S. (2004). The application of the ERETIC method to 2D-NMR. *Journal of Magnetic Resonance*, *168*, 118–123.
Miller, M. I., & Greene, A. S. (1989). Maximum-likelihood estimation for nuclear magnetic resonance spectroscopy. *Journal of Magnetic Resonance*, *83*, 525–548.
Mishkovsky, M., & Frydman, L. (2009). Principles and progress in ultrafast multidimensional nuclear magnetic resonance. *Annual Review of Physical Chemistry*, *60*, 429–448.
Morris, G. A. (1992). Systematic sources of signal irreproducibility and t_1 noise in high field NMR spectrometers. *Journal of Magnetic Resonance*, *100*, 316–328.

Morvan, D., Demidem, A., Papon, J., & Madelmont, J.-C. (2003). Quantitative HRMAS proton total correlation spectroscopy applied to cultured melanoma cells treated by chloroethyl nitrosourea: Demonstration of phospholipid metabolism alterations. *Magnetic Resonance in Medicine*, *49*, 241–248.

Müller, L. (1979). Sensitivity enhanced detection of weak nuclei using heteronuclear multiple quantum coherence. *Journal of the American Chemical Society*, *101*, 4481–4484.

Navon, G., Song, Y.-Q., Rõõm, T., Appelt, S., Taylor, R. E., & Pines, A. (1996). Enhancement of solution NMR and MRI with laser-polarized xenon. *Science*, *271*, 1848–1851.

Nouaille, R., Matulova, M., Delort, A.-M., & Forano, E. (2004). Production of maltodextrin 1-phosphate by Fibrobacter succinogenes S85. *FEBS Letters*, *576*, 226–230.

Otting, G., Widmer, H., Wagner, G., & Wüthrich, K. (1986). Origin of t_1 and t_2 ridges in 2D NMR spectra and procedures for suppression. *Journal of Magnetic Resonance*, *66*, 187–193.

Pathan, M., Akoka, S., & Giraudeau, P. (2012). Ultrafast hetero-nuclear 2D J-resolved spectroscopy. *Journal of Magnetic Resonance*, *214*, 335–339.

Pathan, M., Akoka, S., Tea, I., Charrier, B., & Giraudeau, P. (2011). "Multi-scan single shot" quantitative 2D NMR: A valuable alternative to fast conventional quantitative 2D NMR. *Analyst*, *136*, 3157–3163.

Pauli, G. F. (2001). qNMR—A versatile concept for the validation of natural product reference compounds. *Phytochemical Analysis*, *12*, 28–42.

Pauli, G. F., Jaki, B. U., & Lankin, D. C. (2005). Quantitative ^{1}H NMR: Developpement and potential of a method for natural products analysis. *Journal of Natural Products*, *68*, 133–149.

Pelupessy, P. (2003). Adiabatic single scan two-dimensional NMR spectroscopy. *Journal of the American Chemical Society*, *125*, 12345–12350.

Pelupessy, P., Duma, L., & Bodenhausen, G. (2008). Improving resolution in single-scan 2D spectroscopy. *Journal of Magnetic Resonance*, *194*, 169–174.

Piantini, U., Sorensen, O. W., & Ernst, R. R. (1982). Multiple quantum filters for elucidating NMR coupling networks. *Journal of the American Chemical Society*, *104*, 6800–6801.

Podo, F., Henriksen, O., Bovée, W. M. M. J., Leacj, M. O., Leibfritz, D., & De Certaines, J. D. (1998). Absolute metabolite quantification by in vivo NMR spectroscopy: I. Introduction, objectives and activities of a concerted action in biomedical research. *Magnetic Resonance Imaging*, *16*, 1085–1092.

Price, W. S. (1996). Gradient NMR. *Annual Reports on NMR Spectroscopy*, *32*, 51–142.

Rai, R. K., Tripathi, P., & Sinha, N. (2009). Quantification of metabolites from two-dimensional nuclear magnetic resonance spectroscopy: Application to human urine samples. *Analytical Chemistry*, *81*, 10232–10238.

Rance, M., Sorensen, O. W., Bodenhausen, G., Wagner, G., Ernst, R. R., & Wüthrich, K. (1983). Improved spectral resolution in COSY ^{1}H NMR spectra of proteins via double quantum filtering. *Biochemical and Biophysical Research Communications*, *117*, 479–485.

Richardson, A., Yang, C., Osterman, A., & Smith, J. (2008). Central carbon metabolism in the progression of mammary carcinoma. *Breast Cancer Research and Treatment*, *110*, 297–307.

Rizzo, V., & Pinciroli, V. (2005). Quantitative NMR in synthetic and combinatorial chemistry. *Journal of Pharmaceutical and Biomedical Analysis*, *38*, 851–857.

Romano, R., Paris, D., Acernese, F., Barone, F., & Motta, A. (2008). Fractional volume integration in two-dimensional NMR spectra: CAKE, a Monte Carlo approach. *Journal of Magnetic Resonance*, *192*, 294–301.

Ross, A., Salzmann, M., & Senn, H. (1997). Fast-HMQC using Ernst angle pulses: An efficient tool for screening of ligand binding to target proteins. *Journal of Biomolecular NMR*, *10*, 389–396.

Ruiz-Cabello, J., Wuister, G. W., Moonen, C. T. W., van Gelderen, P., Cohen, J. S., & van Zilj, P. C. M. (1992). Gradient-enhanced heteronuclear correlation spectroscopy. Theory and experimental aspects. *Journal of Magnetic Resonance, 100*, 282–302.

Sanchez-Sampedro, A., Kim, H. K., Choi, Y. H., Verpoorte, R., & Corchete, P. (2007). Metabolomic alterations in elicitor treated *Silybum marianum* suspension cultures monitored by nuclear magnetic resonance spectroscopy. *Journal of Biotechnology, 130*, 133–142.

Sarkar, R., Comment, A., Vasos, P. R., Jannin, S., Gruetter, R., Bodenhausen, G., et al. (2009). Proton NMR of ^{15}N-choline metabolites enhanced by dynamic nuclear polarization. *Journal of the American Chemical Society, 131*, 16014–16015.

Schanda, P., & Brutscher, B. (2005). Very fast two-dimensional NMR spectroscopy for real-time investigation of dynamic events in proteins on the time scale of seconds. *Journal of the American Chemical Society, 127*, 8014–8015.

Shaka, A. J., & Freeman, R. (1983). Simplification of NMR spectra by filtration through multiple-quantum coherence. *Journal of Magnetic Resonance, 51*, 169–173.

Shoolery, J. N. (1977). Some quantitative applications of ^{13}C NMR spectroscopy. *Progress in Nuclear Magnetic Resonance Spectroscopy, 11*, 79–93.

Shrot, Y., & Frydman, L. (2003). Ghost-peak suppression in ultrafast two-dimensional NMR spectroscopy. *Journal of Magnetic Resonance, 164*, 351–357.

Shrot, Y., & Frydman, L. (2008a). The effects of molecular diffusion in ultrafast two-dimensional nuclear magnetic resonance. *Journal of Chemical Physics, 128*, 164513.

Shrot, Y., & Frydman, L. (2008b). Spatial encoding strategies for ultrafast multidimensional nuclear magnetic resonance. *Journal of Chemical Physics, 128*, 052209.

Shrot, Y., & Frydman, L. (2009). Spatial/spectral encoding of the spin interactions in ultrafast multidimensional NMR. *Journal of Chemical Physics, 131*, 224516.

Shrot, Y., & Frydman, L. (2011). Compressed sensing and the reconstruction of ultrafast 2D NMR data: Principles and biomolecular applications. *Journal of Magnetic Resonance, 209*, 352–358.

Shrot, Y., Shapira, B., & Frydman, L. (2004). Ultrafast 2D NMR spectroscopy using a continuous spatial encoding of the spin interactions. *Journal of Magnetic Resonance, 171*, 163–170.

Slupsky, C. M., Rankin, K. N., Wagner, J., Fu, H., Chang, D., Weljie, A. M., et al. (2007). Investigations of the effects of gender, diurnal variation and age in human urinary metabolomic profiles. *Analytical Chemistry, 79*, 6995–7004.

Stern, A. S., Li, K.-B., & Hoch, J. C. (2002). Modern spectrum analysis in multidimensional NMR spectroscopy: Comparison of linear-prediction extrapolation and maximum-entropy reconstruction. *Journal of the American Chemical Society, 124*, 1982–1993.

Tal, A., & Frydman, L. (2010). Single-scan multidimensional magnetic resonance. *Progress in Nuclear Magnetic Resonance Spectroscopy, 57*, 241–292.

Tal, A., Shapira, B., & Frydman, L. (2005). A continuous phase-modulated approach to spatial encoding in ultrafast 2D NMR spectroscopy. *Journal of Magnetic Resonance, 176*, 107–114.

Tang, C. (1994). An analysis of baseline distortion and offset in NMR spectra. *Journal of Magnetic Resonance Series A, 109*, 232–240.

Tenailleau, E., & Akoka, S. (2007). Adiabatic ^{1}H decoupling scheme for very accurate intensity measurements in ^{13}C NMR. *Journal of Magnetic Resonance, 185*, 50–58.

Tenailleau, E., Lancelin, P., Robins, R. J., & Akoka, S. (2004). Authentication of the origin of vanillin using quantitative natural abundance ^{13}C NMR. *Journal of Agricultural and Food Chemistry, 52*, 7782–7787.

Thibaudeau, C., Remaud, G., Silvestre, V., & Akoka, S. (2010). Performance evaluation of quantitative adiabatic ^{13}C NMR pulse sequences for site-specific isotopic measurements. *Analytical Chemistry, 2010*, 5582–5590.

Tiziani, S., Lodi, A., Ludwig, C., Parsons, H. M., & Viant, M. R. (2008). Effects of the application of different window functions and projection methods on processing of ^{1}H J-resolved nuclear magnetic resonance spectra for metabolomics. *Analytica Chimica Acta, 610*, 80–88.

Tredwell, G. D., Behrends, V., Geier, F. M., Liebeke, M., & Bundy, J. G. (2011). Between-person comparison of metabolite fitting for NMR-based quantitative metabolomics. *Analytical Chemistry, 83*, 8683–8687.

Turner, D. L. (1982). Carbon-13 autocorrelation using double-quantum coherence. *Journal of Magnetic Resonance, 49*, 175–178.

Vitorge, B., Bodenhausen, G., & Pelupessy, P. (2010). Speeding up nuclear magnetic resonance spectroscopy by the use of SMAll Recovery Times—SMART NMR. *Journal of Magnetic Resonance, 207*, 149–152.

Vlahov, G. (1999). Application of NMR to the study of olive oils. *Progress in Nuclear Magnetic Resonance Spectroscopy, 35*, 341–357.

von Kienlin, M., Moonen, C. T. W., van der Toorn, A., & van Zilj, P. C. M. (1991). Rapid recording of solvent-suppressed 2D COSY spectra with inherent quadrature detection using pulsed field gradients. *Journal of Magnetic Resonance, 93*, 423–429.

Wang, Y., Bollard, M. E., Keun, H., Antti, H., Beckonert, O., Ebbels, T. M., et al. (2003). Spectral editing and pattern recognition methods applied to high-resolution magic-angle spinning ^{1}H nuclear magnetic resonance spectroscopy of liver tissues. *Analytical Biochemistry, 323*, 26–32.

Weljie, A. M., Newton, J., Mercier, P., Carlson, E., & Slupsky, C. M. (2006). Targeted profiling: Quantitative analysis of ^{1}H NMR metabolomics data. *Analytical Chemistry, 78*, 4430–4442.

Widarto, H., Van Der Meijden, E., Lefeber, A., Erkelens, C., Kim, H., Choi, Y., et al. (2006). Metabolomic differentiation of Brassica rapa; following herbivory by different insect instars using two-dimensional nuclear magnetic resonance spectroscopy. *Journal of Chemical Ecology, 32*, 2417–2428.

Willker, W., Leibfritz, D., Kerssebaum, R., & Bermel, W. (1993). Gradient selection in inverse heteronuclear correlation spectroscopy. *Magnetic Resonance in Chemistry, 31*, 287–292.

Wishart, D. S. (2008). Quantitative metabolomics using NMR. *Trends in Analytical Chemistry, 27*, 228–237.

Wolber, J., Ellner, F., Fridlund, B., Gram, A., Johannesson, H., Hansson, G., et al. (2004). Generating highly polarized nuclear spins in solution using dynamic nuclear polarization. *Nuclear Instruments & Methods in Physics Research A, Section A: Accelerators, Spectrometers, Detectors and Associated Equipment, 526*, 173–181.

Wu, C., Zhao, M., Cai, S., Lin, Y., & Chen, Z. (2010). Ultrafast 2D COSY with constant-time phase-modulated spatial encoding. *Journal of Magnetic Resonance, 204*, 82–90.

Xi, Y., de Ropp, J. S., Viant, M. R., Woodruff, D., & Yu, P. (2008). Improved identification of metabolites in complex mixtures using HSQC NMR spectroscopy. *Analytica Chimica Acta, 614*, 127–133.

Zhang, L., & Gellerstedt, G. (2007). Quantitative 2D HSQC NMR determination of polymer structures by selecting suitable internal standard references. *Magnetic Resonance in Chemistry, 45*, 37–45.

Zhang, S., Nagana Gowda, G. A., Asiago, V., Shanaiah, N., Barbas, C., & Raftery, D. (2008). Correlative and quantitative ^{1}H NMR-based metabolomics reveals specific metabolic pathway disturbances in diabetic rats. *Analytical Biochemistry, 383*, 76–84.

CHAPTER FOUR

Liquid Chromatography Coupled to Mass Spectrometry-Based Metabolomics and the Concept of Biomarker

Samia Boudah*, Alain Paris†, Christophe Junot*,1

*Laboratoire d'Etude du Métabolisme des Médicaments, CEA- Centre d'étude de Saclay DSV/iBiTec-S/SPI/LEMM, Gif-sur-Yvette, France
†INRA, UR1204, Méthodologies d'Analyse du Risque Alimentaire, Paris, France
[1]Corresponding author: e-mail address: christophe.junot@cea.fr

Contents

Abstract

The metabolome is the set of small molecular weight compounds found in biological fluids and metabolomics/metabonomics is known as the large-scale, qualitative, and quantitative study of all metabolites in a given biological system. It is a data-driven approach combining analytical chemistry, biostatistics, informatics, and biochemistry. It complements tools already available to biologists for the characterization of gene functions, that is, transcriptomics and proteomics. After having been mostly used by analytical chemists and chemometricians, metabolomics is now part of the tools available to biologists working in the fields of agronomy, environmental sciences, and medicine. The aim of this review is to address liquid chromatography coupled to mass spectrometry (LC/MS)-based metabolomics and its expected input to discovery and validation of biomarkers. It starts with an introduction of the concept of biomarker for medicine and drug discovery. LC/MS-based metabolomics is then detailed with a particular emphasis on analytical chemistry and statistics. Bioinformatics issues, which are mandatory to enable the large-scale exploitation of metabolomics data by biologists and clinicians, are at last addressed.

Advances in Botanical Research, Volume 67
ISSN 0065-2296
http://dx.doi.org/10.1016/B978-0-12-397922-3.00004-6

1. INTRODUCTION

1.1. The concept of biomarker

1.1.1 What is a biomarker?

Generally speaking, a biomarker can be viewed as an indicator of a given biological state. Although the whole biological field may be concerned, biomarkers have been and are extensively used in the field of clinical chemistry applied to medicine and toxicology. Biological markers or biomarkers are measurable internal indicators of molecular and/or cellular alterations that may appear in an organism during or after the progression of a disease or a possible exposure to a toxicant (Bennett & Waters, 2000; Paustenbach, 2001). This definition is used in environmental and occupational toxicology and is larger than that of the National Institute of Health (NIH) that focuses on drug development and defines a biomarker as "*a characteristic that is objectively measured and evaluated as an indicator of normal biologic processes, pathogenic processes, or pharmacological processes to a therapeutic intervention*" (Biomarkers Definitions Working Group, 2001).

The concept of biomarker originates in ancient and medieval medicine, as emphasized by the recognition of the importance of urine as a diagnostic tool by several of the earliest civilizations, with analysis of taste, smell, and sediments in relation to diseases such as diabetes (e.g. Armstrong, 2007; Haber, 1988). It is based on observations that living beings are able to maintain their own stability regarding their external environment. The French physiologist Claude Bernard introduced the notion of "milieu intérieur" (i.e. internal environment or internal milieu) in 1878. He wrote "*it is the fixity of the 'milieu intérieur' which is the condition of free and independent life*" and also "*all the vital mechanisms have only one object, that of preserving constant the conditions of life in the internal environment*" (Bernard, 1878). In the following years, Walter Bradford-Cannon was the first to introduce the word "homeostasis", which refers to as the coordinated physiological reactions that maintain the steady states in the body of living organisms (Cannon, 1932). He stated that many diseases are a result of homeostasis disturbance.

1.1.2 The different kinds of biomarkers

Biomarkers could be divided into several categories depending on their nature. First of all, a biomarker can be any physical, chemical or biological parameter. Originally, biomarkers were simple and single physiological or laboratory parameters such as body temperature, any metabolite or protein concentration measured in a given biological fluid and associated to a given biological state. Concerning small molecules or metabolites, increased glucose and cholesterol concentrations in blood have long been associated with diabetes and risks of cardiovascular diseases, respectively. In the same manner, for proteins, prostate-specific antigen (PSA) and C-reactive protein are linked to prostate cancer and inflammatory processes, respectively. All these analytes can be routinely quantified in clinical chemistry laboratories.

Many biomarkers have also been developed using imaging technologies such as nuclear magnetic resonance (NMR) imaging or positron emission tomography (PET), which are non-invasive and help clinicians for disease diagnostic. By injecting [^{18}F]-fluorodeoxyglucose, a radiolabelled tracer enabling PET experiments, to patients, it is possible to track glucose consumption, which is a marker of increased metabolic activity that can be induced by tumour growth.

Since the 1990s, technological improvements in the fields of informatics, physics, analytical chemistry, and molecular biology led to the obtention of multiparametric imaging, genomic, and proteomic biomarkers (Baker, 2005), the two latter being referred to as molecular biomarkers. As a matter of fact, by analysing primary breast tumors of 117 young patients using DNA microarray coupled to supervised multivariate statistical analyses, the Netherland Cancer Institute obtained a 70-gene expression signature highly predictive of cancer prognosis. Such a gene expression profile provides a useful tool to adapt the therapeutic strategy, to reduce adverse side effects of anticancer chemotherapies, and also to limit health care expenditure (van't Veer et al., 2002).

1.2. The main application of biomarkers

Most of biomarker applications are in the field of health. They address toxicology, medicine, and drug discovery and development.

1.2.1 Biomarkers for toxicology

Biomarkers enable to measure the exposure to a toxic agent and the extent of the toxic response, and also to predict the likely response. They are compounds or a set of compounds that must be quantitatively, sensitively, specifically, and easily measurable on non-invasively collected biological media (Timbrell, 1998). They could be divided into several categories that include biomarkers of exposure, biomarkers of effect, and biomarkers of susceptibility. A biomarker of exposure is an indication of the occurrence and extent of exposure. It depends on the chemical fate of the exposed toxicant in the body. The biomonitoring of exposure has been used for a long time in occupational settings, for example, for the determination of lead (Kehoe, Thamann, & Cholak, 1933) or benzene metabolites (Pearce, Schrenk, & Yant, 1936) in blood or urine.

Biomarkers of (biochemical) effect(s) indicate that exposure has resulted in an interaction between the toxicant and a biological target. Mutagenic and carcinogenic substances that possess electrophilic function(s) bind to macromolecules such as proteins, DNA or lipids. Haemoglobin is often used in biomonitoring because of its long life span and ease of access. Oxidative stress perturbs the homeostasis of cell and leads to the production of specific substances such as 8-hydroxy-2′-deoxyguanosine or to an imbalance of glutathione pathway (Angerer, Ewers, & Wilhelm, 2007).

Biomarkers of susceptibility describe inter-individual differences in response to toxicants from genetic causes or from non-genetics factors (age, liver disease, kidney disease, diet, dietary supplementation, etc.). Polymorphisms of activating/detoxificating enzymes have been identified as key factors in the relationship between external (e.g. ambient air) and internal exposure (e.g. urinary excretion). Haufroid et al. (2002) demonstrated the relationship between the urinary excretion of phenylhydroxyethylmercapturic acid (a mercapturic acid metabolite of styrene) and the genetic polymorphism of glutathione *S*-transferase M1.

1.2.2 Biomarkers for medicine

Biomarkers have been used a long time ago by clinicians for disease diagnosis and follow-up. They are biomolecules such as metabolites and proteins that arised from basic and applied research in the field of experimental medicine, genetics, and biochemistry, and also from technological improvements achieved from the second part of the twentieth century in analytical chemistry, with the development of immunoassays (Yalow & Berson, 1960a, 1960b) and physicochemical analysis tools such as chromatography (Vassella, Hellstrom, & Wengle, 1962) and mass spectrometry (MS) (Bader, Lehnert, & Angerer, 1994). One of the oldest biomarkers is glucose whose measurement in blood samples enables diagnosis and follow-up of diabetes. However, many other biomolecules can now be quantified in clinical laboratories, including small molecules such as steroid hormones, cholesterol, neurotransmitters, and also peptides or proteins such as PSA, a glycoprotein secreted by epithelial cells of the prostate gland that is used for risk-stratification of prostate cancer, cardiac troponin that is released into blood after a myocardial infarction or many proteins involved in the control of hormonal secretions such as thyroid-stimulating hormone, follicle-stimulating hormone, luteinizing hormone, or human chorionic gonadodotropin.

Nowadays, gas and liquid chromatography coupled to mass spectrometry are available in laboratories of clinical chemistry found in many hospitals. They are reference methods for the quantification of small molecules, such as organic acids, amino acids, acylcarnitins, steroid hormones, and drugs (for therapeutic drug monitoring). This is especially the case in the United States of America, where MS is used for the screening of inborn errors of metabolism in neonates (Chace, Lim, Hansen, Adam, & Hannon, 2009; Chace, Lim, Hansen, De Jesus, et al., 2009; Chace, Singleton, Diperna, Aiello, & Foley, 2009). Concerning proteins, immunoassay is still the reference method approved by Food and Drug Administration (FDA) for sensitive detection and quantification of proteins in biofluids (Surinova et al., 2011). However, development of a given immunoassay is expensive and time consuming, and stable isotope dilution (SID) coupled to LC/MS should also become a reference method, although it is still less sensitive than immunoassays by at least a factor of 10 (Carr & Anderson, 2008).

In the late 1990s, technological improvements in molecular biology, analytical chemistry, and also in informatics led to the advent of the so-called omics-based approaches which give an overview of the genotype (genomics) and also the phenotype at the different levels of biological organization (i.e. transcriptome, proteome, and metabolome, for the measurement of the whole mRNA, proteins, and metabolites contained in a biological medium, respectively). These new technologies offer a new paradigm for disease biomarker discovery in that they are not driven by any prior biological hypothesis but rather by data, thus enabling to detect events that could not have been anticipated from any biological reasoning. Such an approach is of relevance to "stratify" patients suffering from complex and multifactorial diseases or syndroms. This is exemplified by a study aimed at identifying new metabolic abnormalities in patients with complex neurodegenerative disorders of unknown aetiology. By analysing cerebrospinal fluid (CSF) samples of 149 adult patients with "unexplained encephalopathies" using NMR spectroscopy, a new form of cerebellar ataxia characterized by abnormal elevation of free sialic acid in CSF was identified (Mochel et al., 2009).

1.2.3 Biomarker as a tool for drug research and development

Drug research and development begin with a discovery step, which aims at selecting drug candidates for further pre-clinical and clinical studies, before approval by registration authorities. The drug discovery and development process suffers from a high attrition rate mainly due to an unfavorable efficacy to toxicity ratio. Thanks to the development of omics-based approaches, biomarkers are expected to maximize the drug discovery. They are thus present at each step of the drug discovery and development process (Baker, 2005).

Drug discovery programs start with the implementation of a high-throughput screening test, whose objective is to select hit compounds that are active towards a target (i.e. often a receptor or an enzyme) in an *in vitro* or a cell model. Among these hits, lead compounds, which are active on *in vivo* models, are further selected for optimization of their pharmacological (efficacy and selectivity against the target), toxicological, and pharmacokinetic properties; this represents the so-called lead optimization phase. Thereafter, few selected drug candidate undergoes pre-clinical development including formulation and safety studies and also clinical development. Different kinds of biomarkers can be implemented throughout this process: target biomarkers documenting the interaction of lead compounds with their target; disease biomarkers that indicate the presence of a particular disease in animal models; efficacy biomarkers that correlate with the expected or the desired drug effect and toxicity biomarkers indicating potentially harmful effects of the drug candidate (see Fig. 4.1) (Baker, 2005).

However, the use of biomarker is not restricted to the sole drug discovery process. Predictive biomarkers, also called "companion diagnostics", are used to limit the toxicity of treatment in a personalized or stratified medicine perspective to target a specific sub-population of patients who are likely to benefit from the drug therapy on the basis of genetic or clinical characteristics (Trusheim et al., 2011). This is especially the case in the field of oncology which has to face dramatic toxicity issues and important treatment costs. Since 1998, some successful companion biomarkers have been released, some of them being approved by regulatory agencies and included in a drug label (Alymani, Smith, Williams, & Petty, 2010).

This is, for example, the case with the Herceptin/Herceptest couple, developed by Roche and GenenTech/Dako, and the Erbitux (Merck/Imclone)/KRAS test couple. In the first case, the protein HER2 is the target of trastuzumab (brand-named Herceptin), a therapeutic monoclonal antibody used against breast cancer. The overexpression of this protein, occurring in 25% of breast cancer patient, is used as predictive biomarker for patient selection (Alymani et al., 2010; Trusheim et al., 2011). In the second example, Cetuximab (brand-named Erbitux) is a monoclonal antibody

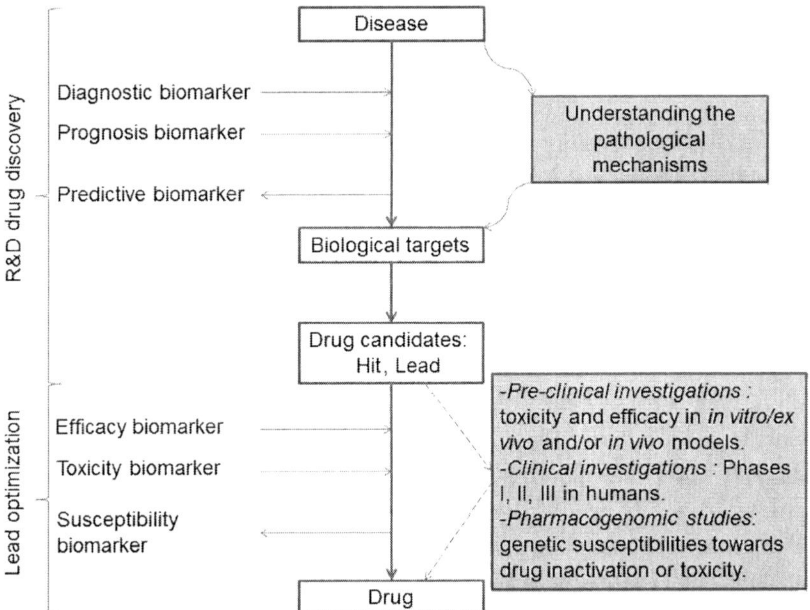

Figure 4.1 Biomarkers throughout the *drug discovery process*. (For colour version of this figure, the reader is referred to the online version of this chapter.)

active as an inhibitor of an epidermal growth factor receptor (EGFR) and used in the treatment of colorectal cancer. KRAS is a small G-protein involved in signal transduction of EGFR. The detection of its mutations, which are significantly associated with the absence of treatment response, can serve as a predictor of resistance to Cetuximab-based therapies (Lievre et al., 2006).

1.3. Biomarker discovery and development

Most of the biomarkers available are proteins, genes or expression profiles of genes. Despite the technological improvements achieved in genomics, transcriptomics, and proteomics, only few biomarkers have been released to the marked and have a widespread clinical use. In a recent paper, Anderson reported that among the 109 FDA cleared or approved protein assayed in plasma available in 2009, 80% of them were measured using immunoassay and 87% of them were introduced before 1993. The average rate of FDA approval of new protein-based test between 1993 and 2009 was approximately 1 per year (Anderson, 2010). As a matter of fact, many studies present some lists of candidate biomarkers rather than truly validated biomarkers.

Reasons of these failures have been extensively analysed over the last 10 years, especially for protein biomarkers (Anderson, 2010; Carr & Anderson, 2008; Rifai, Gillette, & Carr, 2006; Surinova et al., 2011; Whiteaker et al., 2011). The main factors usually pointed out are (i) technological issues, that is, the lack of effective platforms other than immunoassays for candidate validation in large sample sets, (ii) the lack of availability of large cohorts for candidate validation, and (iii) the lack of integrated and formalized pipeline for biomarker discovery and development.

Pepe et al. were among the first authors to propose that biomarker development for early detection of cancer could occur in five consecutive steps, with phase 1 being related to pre-clinical exploratory studies, using immunohistochemistry, western blot or gene expression profiles; phase 2 dealing with clinical assay validation; phase 3 corresponding to retrospective longitudinal studies in which a positive screening rule is defined; phase 4 corresponding to prospective screening studies in which stage or nature of cancer is taken into account and the false referral rate sources are identified; and at last, phase 5 which is related to cancer control studies and whose aim is to address whether or not the developed biomarker allows reducing the burden of cancer on the population (Pepe et al., 2001).

More recently, the same kind of pipeline has been proposed for protein biomarkers (Addona et al., 2011; Rifai et al., 2006; Surinova et al., 2011; Whiteaker et al., 2011) (see Fig. 4.2). The *discovery phase* is aimed at identifying protein leads as biomarker candidates, mainly by using protein depletion or fractionation coupled with shotgun proteomics (i.e. analysis of a protein extract using liquid chromatography coupled to high-resolution MS). During the *qualification phase*, analytical methods and biological materials are made suitable for further verification, using targeted proteomics tools such as SID and liquid chromatography coupled to a triple quadrupole mass spectrometer operated in the multiple reaction monitoring (MRM) scanning mode. The goal of *verification studies* is to assess the clinical performance (such as specificity or sensitivity) of biomarker candidates. Then, in the course of a *research assay optimization phase*, an assay of protein signature, usually an immunoassay, which complies with the requirement of an *in vitro* diagnostic test, has to be developed before the *biomarker validation step and commercialization* (Rifai et al., 2006; Surinova et al., 2011).

Such an approach is illustrated by a published study aimed at identifying protein biomarkers of myocardial injury from samples of peripheral plasma (Addona et al., 2011). The discovery phase was performed using blood

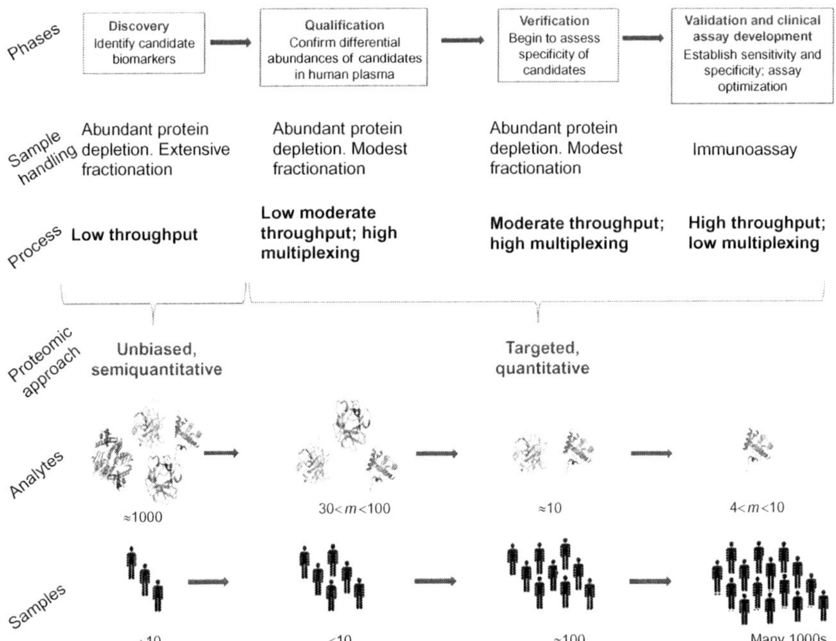

Figure 4.2 Protein biomarker discovery pipeline. The *discovery phase* aims at identifying protein biomarker candidates. This process is global, unbiased, and semiquantitative. Target-driven and quantitative approaches are then applied in the course of *qualification and verification phases*. Qualification confirms the differential expression of the biomarker using an alternative targeted method then the *verification studies* assess specificity and sensitivity. Then, in the course of a *research assay optimization phase*, an assay of protein signature, usually an immunoassay, which complies with the requirement of an *in vitro* diagnostic test, has to be developed before *biomarker validation and commercialization. Adapted from Rifai et al. (2006).* (For colour version of this figure, the reader is referred to the online version of this chapter.)

collected from coronary sinuses from three patients undergoing a therapeutic and planned myocardial infarction for the treatment of hypertrophic cardiomyopathy. Such coronary blood samples were obtained at baseline, 10 and 60 min following therapeutic myocardial infarction, and they were analysed using AIMS (i.e., Accurate Inclusion Mass Screening) using an LTQ-Orbitrap mass spectrometer. Briefly, consecutively to proteolysis of the protein extract, masses related to proteotypic peptides of interest are included on a software inclusion list and monitored in each scan on the Orbitrap MS system. MS/MS spectra for sequence confirmation are acquired only when a peptide from the list is detected with both the correct accurate mass and charge state (Jaffe et al., 2008). By these means, 1105 unique proteins were detected, 70% of them being observed in all three patients, and the levels of 40 of them were increased more than fivefold as compared to baseline. Furthermore, 81 proteins of interest were also detected in 2 out of the 3 patients. These lists contained already known markers such as myoglobine, myeloperoxidase or creatine kinase-myocardial isoforms B. A qualification phase was then implemented using peripheral blood samples. Among the 121 proteins of interest initially observed in coronary blood samples, 83 of them were detected in peripheral blood samples of the three patients, using AIMS. At last, 53 candidate proteins were selected on the basis of a Pearson correlation test to check correlations in temporal abundance. The verification phase was then performed, still with peripheral blood samples of the same three patients, but now using SID–MRM–LC–MS/MS. This led to the selection of 12 protein candidate (i.e. 8 already known to be relevant and 7 being novel potential candidates) for a proof of concept study. Protein changes were verified by western blot analyses and immunoassay, when available. Finally, the clinical validation of these biomarker candidates could be assessed on a cohort of 52 individuals (26 controls and 26 with inducible ischemia).

1.4. Metabolomics and biomarker discovery

The metabolome is the set of small molecular mass compounds found in biological media. It includes all organic substances naturally occurring from the metabolism of the studied living organism, except biological polymers, but also xenobiotics (i.e. chemicals from food, beverages, drugs, and more generally speaking from the environment, referring to as "xenometabolome") and their biotransformation products. Polymerized structures such as proteins and nucleic acids are excluded from the metabolome, but small peptides such as the tripeptide glutathione are included. Molecules that constitute the metabolome are called metabolites. Metabolomics/metabonomics is the analysis of metabolome and its variations under a given condition. It is a data-driven approach combining analytical chemistry for the acquisition of metabolic fingerprints and biostatistics, informatics, and biochemistry for the interpretation of such a rich information data. After having been mostly used by analytical chemists and chemometricians, metabolomics is now part of the tools available to biologists working in the fields of agronomy, environmental sciences, and medicine, and it opens new perspectives in the fields of nutrition, biomarker discovery in pathology, and toxicity or in the development of "system biology".

Actually, despite breakthroughs achieved in genetics, molecular bases of many diseases remain poorly understood, and it appears unlikely that they will be obtained from the genetic information alone. The contribution of genetic and environmental factors to the development and evolution of many diseases is now well accepted, and one of the most important challenges of today's medicine is to understand and highlight interactions between genetic variations and environmental factors such as lifestyle, nutrition, infections, or exposure to pollutants, for biomarker discovery. Metabolomics offers this possibility by detecting information related to both genetic and environmental contributions in biofluids or cell extracts of individuals, by using mainly MS- and NMR-based technologies. For these reasons, metabolomics will complement other omics tools for personalized medicine (i.e. matching the right drug to the right patient at the right time for safer drugs and improved healthcare).

Contrary to what is observed with genomics, transcriptomics, and now with proteomics, the contributions of metabolomics to biomarkers available for medicine, clinical chemistry, or drug discovery remain at the exploratory level. In pilot studies, metabolomics has been already proved to be a promising phenotyping technique for biomarker discovery in the field of clinical chemistry. Thus, different medical areas such as cardiology (Lewis et al., 2008; Sabatine et al., 2005; Turer et al., 2009), transplantation (Li et al., 2008), human reproduction (Brison et al., 2004; Marhuenda-Egea et al., 2009), diabetes (Oresic et al., 2008; Wang et al., 2011), central nervous system diseases (Ibanez et al., 2012; Mochel et al., 2009; Quinones & Kaddurah-Daouk, 2009), or oncology (Dang et al., 2009; Ritchie et al., 2010; Sreekumar et al., 2009; Vander Heiden et al., 2010) have been addressed. However, despite these proof of concept studies, metabolomics is still emerging and has to face different pitfalls that are inherent not only to the technique itself but also to the reduced size of cohorts on which it has been applied. First, some progresses of the technique are awaited before being routinely used in clinical investigation. These include analytical issues such as the necessity to cover a wide range of structurally different metabolites, the lack of reproducibility, problems related to metabolite identification, the defect of extensive relevant chemical standard structure databases, and also a lack of multiplexed assays providing absolute quantification for biomarker validation studies. These methodological problems concern also statistical issues such as the lack of appropriate modelling tools and the false discovery rate estimation. Second, as biological issues are the crucial ones that claim an extensive validation of these new techniques before a routine use in clinical investigations, it is necessary to generate and validate the different sets of metabolic biomarkers on large cohorts. At the present time, many published studies involve only reduced cohorts. This is indeed rather detrimental to get very robust biomarkers usable in clinical medicine and nutritional prevention. Yet, one of the most promising applications of metabolomics is to extensively document host–environment interactions through an in-depth metabolic phenotyping of the general population, or some targeted population like children, pregnant women, elderly, or disease patients in the frame of nutritional and clinical epidemiological studies.

In the following paragraphs of this chapter, we focus on LC/MS-based technologies, which are expected to provide the most extensive coverage of metabolome. We review the current technological issues regarding analytical chemistry and statistics, and we report and discuss on technological advances and future developments in a perspective of biomarker discovery and development.

2. LC/MS-BASED METABOLOMICS FOR BIOMARKER DISCOVERY

Because of the chemical diversity of metabolites, which involves important differences in physicochemical properties such as molecular size, hydrophobicity, pK_a, etc., across a huge range of concentrations (from nanomolar to millimolar), no single chemical analysis is able to measure the entire metabolome simultaneously (Allwood & Goodacre, 2010; Dunn, Bailey, & Johnson, 2005; Fukushima, Kusano, Redestig, Arita, & Saito, 2009) (see Fig. 4.3). The size of the metabolome is estimated to be in the range of 3000 metabolites in human metabolome (Duarte et al., 2007) and more than 200,000 in plants (Fiehn, 2001). This results in the necessity of combining various analytical approaches in order to detect the largest possible number of metabolites, or on the contrary, to focus on a subset of them. In the following paragraphs, we discuss about this compromise applied at each stage of the experiment: sample preparation in the course of targeted or global approaches, fingerprints acquisition using different LC/MS platforms, and data mining.

2.1. The LC/MS-based metabolomics process

Metabolomics is an interdisciplinary approach, which combines analytical chemistry, mathematics, statistics, and bioinformatics for data interpretation in a systems biology perspective.

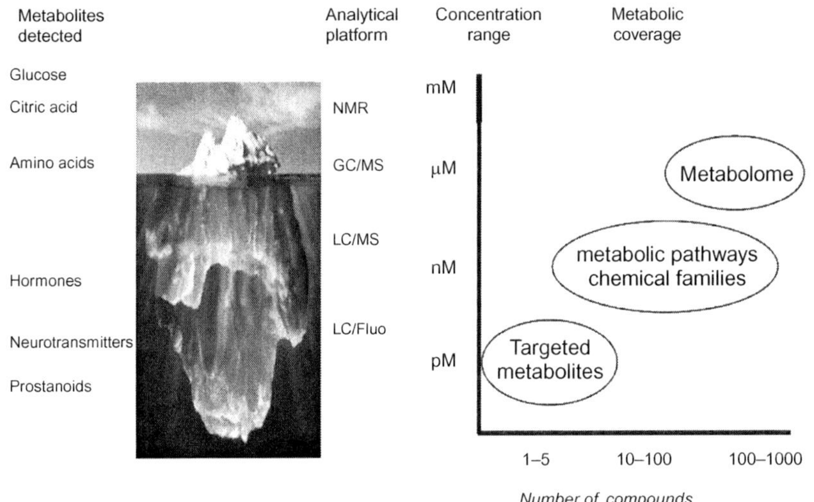

Figure 4.3 Metabolome technologies. Metabolites encompass different chemical classes and occur at a wide concentration range; their measurement needs thus several analytical platforms. The selection of the most suitable technology is generally a compromise between speed, selectivity, and sensitivity which give rise to several approaches. *Adapted from Sumner, Mendes, and Dixon (2003).* (For colour version of this figure, the reader is referred to the online version of this chapter.)

Depending on the biological medium, which will be investigated, samples are processed (extracted, cleaned, concentrated, etc.) before injection into the appropriate LC/MS system. Different MS instruments are commercially available differing on ionization sources and mass analysers. The choice of the MS technology is challenging since it depends on the kind of experiments to perform: global or targeted metabolic profiling, metabolite identification, or metabolite quantification. This will be discussed in Section 2.1.3. After LC/MS process, two dimension fingerprints (i.e. retention time vs. m/z ratio) are obtained, then submitted to a preprocessing step through automatic peak detection software to give a data matrix suitable for statistical analyses. Once a discriminating signal is highlighted, it undergoes an identification step involving MS, MS/MS spectra analyses, and databases queries (see Fig. 4.4).

2.1.1 Design of experiment

Metabolomic experiments are driven by a biological investigation. First of all, an adequate design of experiment has to be implemented to prevent, or at least to manage, the occurrence of confounding factors, which are

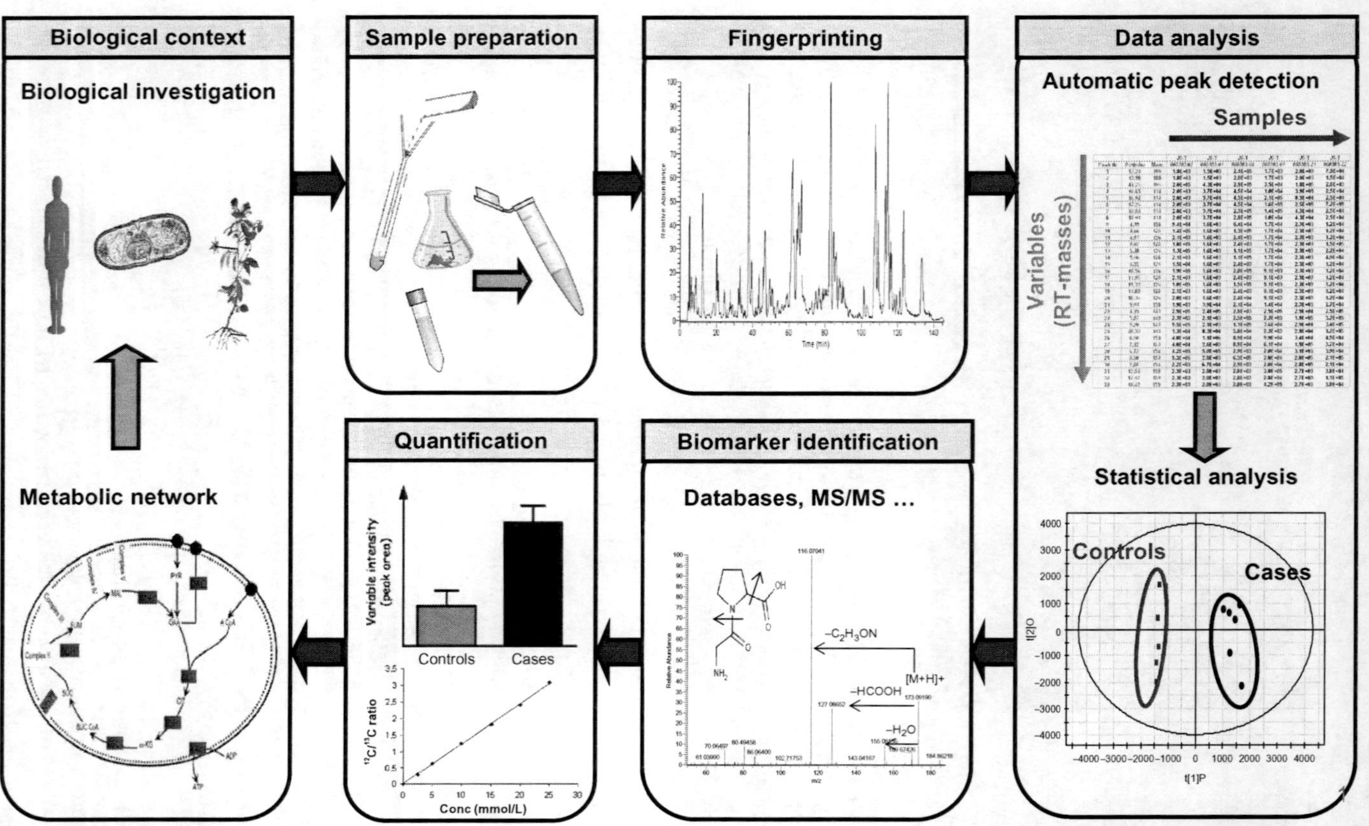

Figure 4.4 LC/MS metabolomic analysis pipeline. *Adapted from Roux, Lison, Junot, and Heilier (2011).* (For colour version of this figure, the reader is referred to the online version of this chapter.)

hidden dependent variables that artefactually correlate with investigated biological factors, leading to erratic biological interpretations. They can be of analytical (i.e. analysis batch effect or bias induced by the source clogging at the end of an analytical batch) or biological origin, as it is the case with physiological parameters such as gender, age, or body mass index. Whereas the impact of confounding factors of analytical origin can be minimized by randomizing the analysis of biological samples within an analytical batch, or by implementing algorithms to correct from batch to batch effects (Dunn et al., 2011), and the impact of physiological confounding factors can be balanced within the different groups of the cohorts, the issue of unanticipated confounding factors arising in the course of biological interpretation is the most delicate to solve.

2.1.2 Sample preparation

The success of a metabolomic experiment not only depends on LC/MS conditions but also involves upstream steps dealing with sample handling (i.e. the choice of the adequate medium, sample collection and transport, metabolite extraction, and stability). This was emphasized by Canelas and colleagues in their study whose objective was to compare two growth conditions of yeast using five different extraction methods. They observed that the detected metabolites and their concentrations were tightly depending on extraction protocols. This could lead to erratic biological interpretations as some metabolites could be falsely found as up- or down-regulated depending on the extraction protocol (Canelas et al., 2009).

Some common sample preparation method incorporates different steps: sample collection, metabolism quenching, and metabolite extraction. Despite their impact on the quality of the results, the first two steps are often neglected, whereas the latter is well documented in the literature. Nevertheless, no clear standardized extraction protocol has been established for biofluids or cells.

2.1.2.1 Sample collection

Several mammalian biofluids can be investigated in metabolomics, including plasma, serum, urine, or CSF, depending on the biological field of investigation. For example, CSF is relevant regarding neurological disorders because of its proximity with neuronal and glial cells (Mochel et al., 2009; Oreskovic & Klarica, 2010; Sinclair et al., 2010), and erythrocytes are of interest to investigate metabolic alterations in patients suffering from genetic diseases of the red blood cells (Darghouth, Koehl, Heilier, et al., 2011;

Darghouth, Koehl, Madalinski, et al., 2011). Nevertheless, urine and blood are still the most widely used biological fluids.

Urine contains by-products generated by the cellular metabolism which have to be cleared from the body by the kidneys. Many physiological and pathophysiological conditions are reflected by metabolite concentration changes in urine, indicating breakdown of homeostasis (Chace, 2001; Sun et al., 2008; Wu et al., 2004). The collection of urine is non-invasive. It may be achieved by either punctual or 12 or 24 h period collections. In the latter case, the use of preservative is recommended in order to avoid bacterial contamination (Want et al., 2010).

Blood is an attractive source for biomarkers which reflects the whole organism metabolic status at a given sampling time. Blood samples are centrifuged to remove blood cells so as to obtain either plasma (by inactivating the clotting cascade using anticoagulants) or serum (after blood clotting). These two biofluids can be used interchangeably as both seem to generate similar results in clinical and biological studies (Yu et al., 2011). Indeed, when comparing LC/MS profiles by simple visual inspection and further using Mass Profiler Professional software, Denery, Nunes, and Dickerson (2011) only found subtle differences between the various blood preparation methods.

However, regarding serum, the clot contact time impacts on energy metabolism, due to the anaerobic metabolism of red blood cells. These changes are delayed on ice, as demonstrated by a principal component analysis (PCA) of NMR spectra (Teahan et al., 2006). Furthermore, concerning plasma samples, the choice of the anticoagulant may impact metabolite detection. There is still no clear consensus about the kind of anticoagulant to be used for metabolomics studies. For example, it is known that heparin, which contains sodium bisulfite as preservative, can impair the quantification of cysteine and sulfocysteine due to the formation of sulfocysteine and homosulfocysteine (Dettmer et al., 2010; Parvy, Bardet, Rabier, Bonnefont, & Kamoun, 1989). The use of citrate is not recommended because this compound is a metabolite involved in the Krebs cycle. By chelating metals (Ca^{2+}) which are necessary for enzyme activation in the coagulation cascade, EDTA impacts on the MS protein profile without affecting the metabolome. Bando et al. (2010) recommended its use as anticoagulant because peaks derived from heparin might overlap with endogenous metabolites in GC–TOF analysis. However, EDTA may interfere with metabolite detection since it is present in MS and NMR spectra. At last, when selected, the anticoagulant should be kept the same within a study to avoid artefactual sample clustering in the course of statistical analyses.

For the moment, plasma is the most used for metabolomics studies (Kaddurah-Daouk et al., 2012; Lawton et al., 2008; Zhang et al., 2011), whereas serum is mentioned in two consortia: the Human Metabolome Project which is an american partnership whose objective is to develop centralized metabolite reference resources for certain important biofluids including serum (Psychogios et al., 2011) and the HUSERMET consortium which is a collaboration between Manchester's University and 2 industrial partners designed to characterize the human serum metabolome and identify biomarkers for the onset, progression, and treatment of the Alzheimer disease and ovarian cancer (Cottingham, 2008).

Otherwise, sample preparation for cells or tissue metabolomic studies consists of separating cells of interest from their biological environment (i.e. culture medium, neighbouring cells, etc.) while preserving their integrity. According to the sample type, centrifugation or filtration on appropriate membranes should achieve this goal. Regarding microbiology samples, sterile conditions have to be maintained. At last, washing the collected cells is recommended to remove medium or extracellular compounds (Meyer, Liebeke, & Lalk, 2010).

2.1.2.2 Metabolism quenching

A metabolism-quenching step may have to be incorporated into the process. It stops the ongoing enzymatic reactions occurring after sample collection in order to measure the true metabolome composition at sampling time. Quenching is especially relevant for studies dealing with compounds exhibiting a high metabolic turnover, as it is the case with energy metabolism. The use of heat, freeze (using liquid nitrogen), the addition of acids, or the transferring of cells suspensions into a cold organic solvent can stop the residual enzymatic activity, but these techniques have to cope with the timescale of the metabolic process of interest without affecting the stability of the metabolites (Vuckovic, 2012). Procedures based on cold organic solvents are the most common, but they have the disadvantage to release intracellular metabolites due to membrane lysis. Last, the issue of metabolic quenching is routinely addressed in bacterial or plant metabolomics (Bolten, Kiefer, Letisse, Portais, & Wittmann, 2007), but it is frequently omitted for global approaches performed in biological fluid such as blood or urine.

Other kinds of precaution are taken into account for the collection of mammalian biofluids such as the addition of glycolysis inhibitors such as fluoride that inhibits the enzyme enolase and thereby prevents glucose degradation and lactate production (Mikesh & Bruns, 2008).

2.1.2.3 Metabolites extraction

Traditionally, in the course of targeted analyses, sample extraction is intended to recover the sole metabolites of interest rather than large sets of molecules. Nowadays, with the increasing interest towards large-scale analysis of metabolites in biological media, the objective of such a global analysis is to detect as many metabolites as possible. However, because of the huge number and the large chemical diversity of the metabolites involved, the analytical chemist has to define a compromise between a sufficient purity of the extract, which drives optimal sensitivity in MS detection of ions, and the extent of the information collected.

Although selective metabolite extraction protocols are mandatory for the detection of some metabolite classes exhibiting specific physicochemical properties, as it is, for example, the case with lipids, or for the detection of metabolites occurring at trace level, sample treatment for global approaches has to be minimal in order to avoid external source of variability and loss of information. Anyway, it is always necessary to remove macromolecules such as proteins and lipoproteins that may induce source clogging, and thus decrease the sensitivity of detection, and that could also alter the chromatographic columns and thus degrade the quality of separation. To this end, different categories of sample processing such as liquid–liquid, solid-phase extraction (SPE), protein precipitation using organic solvents, or ultra-filtration are considered. Furthermore, disruption of cell membranes is required for cellular metabolome investigations, to release intracellular metabolites. This could be done either mechanically using either glass beads or ultrasonification (Meyer et al., 2010), freeze and thaw cycles, or transferring the cells into an hypotonic medium such as boiling water (Darghouth, Koehl, Heilier, et al., 2011; Darghouth, Koehl, Madalinski, et al., 2011).

a. *Liquid–liquid extraction*, also known as solvent partitioning, is based on difference solubilities of compounds between two non-miscible liquids (i.e. aqueous and organic phases). Usually, the biological medium is considered as the aqueous phase and an organic solvent is added. This technique has long been used for the determination of drug such as benzodiazepine in whole blood (Simonsen, Hermansson, Steentoft, & Linnet, 2010; Simonsen, Steentoft, Buck, Hansen, & Linnet, 2010), doping agent in urine (Deventer, Pozo, Van, & Delbeke, 2009), or pesticides analysis (Gilbert-Lopez, Garcia-Reyes, Fernandez-Alba, & Molina-Diaz, 2010). The use of salting out reagents is a potential concern for further atmospheric pressure ionization (API)–MS analyses, and Wu, Zhang, Norem, and El-Shourbagy (2008) have demonstrated that acetate ammonium is compatible with API–MS detection. Otherwise, liquid–liquid extraction using chloroform and methanol (Bligh & Dyer, 1959; Folch, Ascoli, Meath, & LeBaron, 1951) is still popular for global lipidomic approaches. It leads to a biphasic system in which the lower phase contains the lipids of interest. At last, Eikel and colleagues have successfully combined micro-liquid extraction from a solid surface with nano-electrospray MS. The so-called liquid extraction surface analysis-mass spectrometry was found to be more informative than matrix-assisted laser desorption-ionization (MALDI)-imaging MS study for drug distribution and biotransformation studies on mice tissue sections (Eikel et al., 2011). Such an *in vivo* sampling is attractive because it reduces analytical variations inherent to the multi-step sample handling and leaves the sample in its biological environment.

b. SPE and *solid-phase microextraction* (SPME) are known to be valuable tools for sample clean-up and analyte pre-concentration in a context of targeted analyses. They appear to be less appropriate to untargeted metabolomics because metabolic coverage is limited by the selectivity brought by the sorbent material (Daykin, Foxall, Connor, Lindon, & Nicholson, 2002). Nevertheless, SPE is of value to analyse metabolites occurring at trace levels in biological media. As a matter of fact, Vuckovic et al. have recently evaluated the extraction performances in human plasma of 42 SPME coatings for the analysis of 36 metabolites from different chemical classes and varying polarities. They found that ionization suppression in MS analysis is reduced by SPME, and that some coatings can improve plasma metabolic coverage (Vuckovic & Pawliszyn, 2011).

A new trend in SPE extraction is the "turbulent-flow-chromatography" which is an automatic sample treatment processing allowing high-throughput metabolite extraction. It enables direct on line injection of crude biological matrices. TFC columns packed with large particles retain small molecules at high flow rate of aqueous mobile phase and wash macromolecules and phospholipids away. Molecules that are retained on the TF column are then directly eluted onto the analytical column where enhanced chromatographic separation occurs. On-line extracting systems using TFC have already been used in targeted LC–MS/MS toxicological screening in urine (Mueller, Duretz, Espourteille, & Rentsch, 2011). It has also been successfully used for a rapid global LC/MS metabolomics analysis of serum and plasma samples (Michopoulos, Edge, Theodoridis, & Wilson, 2010).

c. The treatment of biological samples with *organic solvents*, *high salt concentrations*, or *strong acids* or *base* induces alterations of the tridimensional structure of proteins and thus their precipitation due to a lack of solubility. Want et al. have evaluated acid precipitation using perchloric and sulfosalicylic acid, heat in a water bath at 90 °C for 30 min and different organic solvents, including methanol, ethanol, acetonitrile, and acetone. The method involving acetonitrile was the less reproducible one, whereas sample treatment with strong acids and heat gave the lowest number of extracted metabolites features. With regard to the reproducibility and the number of features detected, methanol protein precipitation was found to be the most effective approach (Want et al., 2006). Bruce et al. assessed the amount of remaining proteins after solvent precipitation by using gel electrophoresis. They concluded that the combination of MeOH/acetone/ACN in equal volumes was the most appropriate to remove proteins and extends column lifetime (Bruce et al., 2009). Of note, the issue of neutral lipids that are not eluted from the chromatographic columns was not efficiently addressed by these studies. Want and colleagues took into account the hydrophobicity of features when comparing extraction methods with regard to lipids. They displayed a density distribution plot which represents feature distribution along chromatographic run time in reversed-phase (RP) column (Want et al., 2006). However, the most polar lipids such as lysophospholipids and gangliosides could be extracted by organic solvents without any adapted extraction.

Otherwise, the use of dried blood spot is an emerging technique in the field of metabolomics. It is well suited for small sample volumes and has long been used in the USA for the screening of inborn errors of metabolism in neonates (Chace, Lim, Hansen, Adam, et al., 2009; Chace, Lim, Hansen, De Jesus, et al., 2009; Chace, Singleton, Diperna, Aiello, et al., 2009). Briefly, a drop of whole blood/plasma or urine is adsorbed on filter paper then the dried spot is extracted with solvent. Using such an approach, Michopoulos, Theodoridis, Smith, and Wilson (2010) obtained similar sample classification patterns in LC/MS using PCA as with conventional solvent precipitation, except for some acidic compounds in urine. Issues about the background noise effect of the paper used and the stability of samples have to be addressed in the course of method development (Michopoulos, Theodoridis, Smith, & Wilson, 2011).

d. *Ultrafiltration*: Using ^{1}H NMR, Daykin et al. (2002) observed that proteins and lipoproteins were completely removed from blood using ultrafiltration. Although this technique ensures a greater sensitivity for some metabolite classes, some hydrophobic species may be adsorbed on the filtration membrane. To address this issue, Tiziani et al. (2008) propose to add a filter wash step and chloroform extraction after ultrafiltration in order to desorb the metabolites remaining on the filter.

Even if there is no consensus concerning metabolite extraction regarding biofluids, some trends are clearly pinpointed. In most studies, urine samples are processed according to the "dilute-and-shoot strategy", by simply diluting the crude sample in water before injection into the LC/MS system (Plumb et al., 2003; Williams, Major, Lock, Lenz, & Wilson, 2005). Regarding plasma and serum, solvent precipitation-based methods are the most widely used. Acetonitrile performs the better protein removal, but it has been reported to exhibit poor reproducibility. Furthermore, pure methanol or methanol-containing mixtures ensure a better metabolite coverage (Bruce et al., 2009; Jiye et al., 2005; Want et al., 2006). At last, for cell metabolomics, protocols are generally optimized for specific studies and no standardized methods exist (Merlo, Jankevics, Takano, & Breitling, 2011).

2.1.2.4 Sample storage and stability considerations

Following protein removal, cell or biofluid extracts are often pre-aliquoted (to avoid freeze–thaw cycles), dried, and stored at −80 °C in order to minimize degradation. They are reconstituted in an appropriate solvent, which ensures solubilization of extracted metabolites and compatibility with the analytical system prior to analysis. It is crucial to assess storage conditions and stability of both initial and extracted sample prior to MS analysis. However, only few studies are available and there is a lack of long-term stability studies in metabolomics. Using visual inspection of LC–ESI/MS data and PCA score plots, Gika et al. proved that urines are stable without preservative after 1 and even after 6 months of storage at −20 or −80 °C, but are not after 6 days at 4 °c (Gika, Theodoridis, & Wilson, 2008; Gika, Theodoridis, Wingate, & Wilson, 2007). However, PCA only provides users with a summarized representation of the structure of the dataset; some metabolites, for example, occurring at low concentration, may exhibit concentration variations with limited impact on PCA score plots. It is thus necessary to investigate stability issues at the levels of annotated or identified metabolites.

2.1.2.5 How to assess sample preparation?

Until now, no exhaustive analysis covering all the metabolite classes has been reported. In most publications, the evaluation of extraction efficiency is performed at the chemometric level, that is, by considering the number of features detected (which are potential metabolites) (Bruce et al., 2009; Want et al., 2006), and/or by including small lists of known metabolites (Daykin et al., 2002; Jiye et al., 2005; Vuckovic & Pawliszyn, 2011) and/or labelled internal standards (ISs) added to samples (Jiye et al., 2005; Want et al., 2006). These parameters could not assess the quality of data at the metabolomic level but should be considered as the first-level assessment of data quality. There is a need for quality control (QC) procedures to monitor sample preparation performance. As proposed by Vuckovic, extraction methods could be first of all evaluated with regard to ionization suppression using post-column infusion experiments or serially diluted samples as an indication of matrix component removal (Vuckovic, 2012). The recovery of several ISs could also be assessed (Bruce et al., 2009), or biological reference material could be used, like SRM 1950 for human plasma. For studies involving cell extracts and dealing with energy metabolism, another strategy could also rely on evaluating the levels of metabolites linked to the cellular energy status, such as energy-charge ratio ([ATP] + 0.5[ADP])/([ATP] + [ADP] + [AMP]) or catabolic reduction charge ratio [NADH]/([NAD+] + [NADH]) which should be within reference ranges (Vuckovic, 2012).

2.1.3 A wide range of MS instruments for the acquisition of metabolic fingerprints, metabolite identification, and quantification

The choice of the analytical platform used in metabolomics is challenging. Indeed, it should enable not only the detection of the highest number of metabolites in biological media, but also their identification and quantification. Since metabolomics has emerged, many analytical methods were used as fingerprinting platform: infrared spectroscopy "IR" (Oliver, Winson, Kell, & Baganz, 1998), "NMR" (Roberts, 2000), MS (Smedsgaard & Frisvad, 1996), "LC–MS" (Idborg-Bjorkman, Edlund, Kvalheim, Schuppe-Koistinen, & Jacobsson, 2003), gas chromatography coupled to mass spectrometry "GC/MS" (Horning & Horning, 1971), and capillary electrophoresis coupled to mass spectrometry "CE/MS" (Soga et al., 2002). All these techniques allow chemometrical approaches that help to classify the samples, but only NMR- and MS-based approaches enable metabolite identification. ^{1}H NMR spectroscopy is rapid (i.e. it requires minimal sample preparation), non-destructive, and reproducible. It is suitable for high-throughput analyses and allows metabolite detection, identification, and quantification. Although initially developed in the field of medicine and toxicology (Nicholson, Lindon, & Holmes, 1999), it has proven to be an appropriate tool for plant metabolomics, especially when focusing upon abundant bulk metabolite species; for example, sugars in fruits (Biais et al., 2009). However, ^{1}H NMR spectroscopy exhibits limits of detection in the micromolar range, which is not sensitive enough to obtain information on metabolites occurring at trace levels. Due to their greater sensitivity, MS-based technologies enable to analyse a larger number of metabolites than ^{1}H NMR, especially those whose physiological concentrations are in the nanomolar range.

2.1.3.1 MS profiling platforms

MS is an analytical technique that allows the measurement of molecules after their conversion into ions. Once a sample has been introduced into the mass spectrometer either directly or after sample preparation or separation, molecules are ionized and the resulting ions are separated according to their mass-to-charge (m/z) ratio.

A mass spectrometer consists of five modules: a system for introducing the analytes, an ion source that ionizes the analytes, one or several mass analysers operating at high vacuum to separate ions according to their m/z values, a detector which counts ions, and finally a data processing system for recording and viewing mass spectra (i.e. plots of ion abundance as a function of m/z values) and to control instrument settings. The ion sources used in MS convert gas-phase sample molecules into ions, or move ions produced in solution into the gas phase. The ionization can be carried out in the gas phase under vacuum using electron impact (EI) sources (Horning, Horning, Carroll, Dzidic, & Stillwell, 1973) or at atmospheric pressure, as it is the case with atmospheric pressure chemical ionization and atmospheric pressure photo-ionization sources (Horning et al., 1973; Marchi, Rudaz, & Veuthey, 2009; Robb, Covey, & Bruins, 2000). Ions can also be produced by desorption–ionization of analytes under vacuum on solid supports as it is the case with MALDI sources for molecules of mass greater than 1000 Da or at atmospheric pressure on nebulisate, using electrospray sources (ESI) (Fenn, Mann, Meng, Wong, & Whitehouse, 1989).

Among these *sources*, EI has long been the most used one in metabolic profiling. Indeed, the energy transmitted to the analytes in EI sources is very important and leads to a highly reproducible and informative fragmentation that helps for metabolite identification since universal spectral libraries are available, like those of the National Institute of Standards and Technology (NIST) (Pasikanti, Ho, & Chan, 2008). Furthermore, EI–MS is associated with gas chromatography which efficiently separates the analytes with a high peak resolution. Volatile compounds can be analysed directly using GC/MS, whereas a chemical derivatization is required for compounds of molecular weight up to 400 Da. However, it is often difficult to detect the molecular species that provide information about the molecular mass of the compound of interest, which makes it difficult to characterize molecules that are not present in spectral libraries. In addition, GC/MS analysis is not compatible with thermolabile and non-volatile compounds, which is quite restrictive in the biochemical field.

The introduction of sources designed to operate at atmospheric pressure such as ESI has extended the range of detected molecules to peptides, proteins, and lipids without any chemical modification requirement, and it has also enabled the coupling of MS with liquid chromatography. Atmospheric pressure ionization sources ensure soft ionization. This enables the detection of protonated (i.e. $[M+H]^+$) or deprotonated (i.e. $[M-H]^-$) ions which give access to molecular masses of intact molecules. However, ions produced in API sources provide only limited structural information. It is therefore necessary to perform either MS/MS or sequential MS^n experiments to go deeper into structural characterization of compounds.

Moreover, different *analyser technologies* can be implemented with ESI–MS: "in time" instruments based on ion storage and enabling MS^n sequential fragmentation for multistage structural elucidation and "in space" analysers operating with and without scanning devices (Werner, Heilier, et al., 2008). The different combinations available between sources and analysers enable a large panel of targeted and global analyses. Indeed, analysers, with regard to the selected scan mode, can either ensure global profiling approach by transmitting the full spectrum range or focus on the detection of only some ions of interest using selected ion monitoring in MS mode (by transmitting a limited mass range) or selected reaction monitoring in the MS/MS mode. Thanks to this versatility, MS became the method of choice in the field of metabolomics.

2.1.3.1.1 The evolution of the profiling approaches using MS platform The efficacy of MS-based approaches has long been limited by sensitivity issues and the difficulty for identifying quickly and unambiguously the metabolites of biological interest. In the last decade, global approaches were performed using *low-resolution* instruments such as ion trap and triple quadrupole mass spectrometers. Ion trap analysers were the instruments of choice for metabolite identification thanks to their MS^n sequential fragmentation capabilities that facilitate the interpretation of collision-induced dissociation (CID) spectra. Conversely, triple quadripole instruments are used for *targeted approaches*. Indeed, when operated in the precursor ion and/or the constant neutral loss scanning modes, they enable the selective detection of metabolites belonging to particular chemical families such as phospholipids (Yoon, Kim, & Cho, 2003), sulfoconjugates (Lafaye et al., 2004), or acylcarnitines (McClellan, Quarmby, & Yost, 2002; see Fig. 4.5). Otherwise, the SRM mode is the most sensitive as no scan occurs and the MS analysis time is only focused on the selected recording of events consisted of parent-ion-to-fragment-ion transitions (Kitteringham, Jenkins, Lane, Elliott, & Park, 2009). It thus enables sensitive detection and quantification of tens to

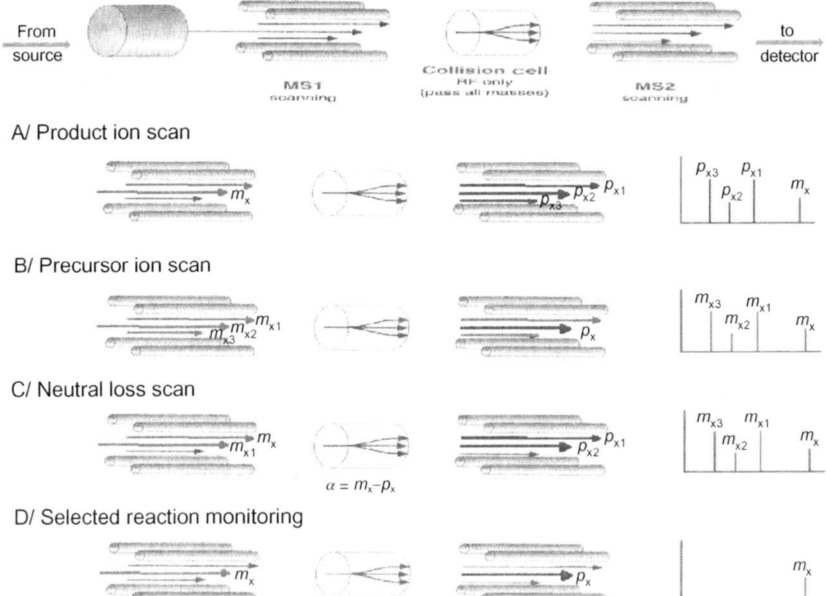

Figure 4.5 Multi-dimensional mass spectrometry-based analyses. Tandem mass spectrometric modes available on triple quadrupole analyser: The principle of MS/MS is to isolate a characteristic ion (precursor ion) in the first quadrupole, and then to fragment it in the second thanks to collision with atoms of inert gas. The resulted ions are isolated in the third quadrupole. There are several ways of acquiring MS/MS: (i) Product ion scan: the precursor ion selected in the first *Q* will fragment and generate product ions which would be separated and analysed in the third quadrupole, (ii) precursor ion scan: it enables to detect ions that give the same product ion in the course of their fragmentation process. (iii) Neutral loss scan: the first and the third quadrupoles are scanning in parallel with a constant difference in *m/z*. It enables to detect ions that produce a constant and characteristic neutral mass (α) in the course of the fragmentation process. (iv) Selected reaction monitoring is a specific case of these three scan modes as the precursor, the product ions and the neutral loss are selected at the same time. *Adapted from Han, Yangm, and Gross (2012).* (For colour version of this figure, the reader is referred to the online version of this chapter.)

hundreds of metabolites, as exemplified by the work of Bajad et al., who used liquid chromatography coupled to a triple quadripole mass spectrometer to detect 141 metabolites and quantify 69 of them from an *Escherichia coli* cell extract, respectively (Bajad et al., 2006).

The advent of *high-resolution* analysers (i.e. time of flight, Fourier transform ion cyclotron resonance, and Orbitrap instruments), enabling accurate mass measurements with sub-ppm errors, improved the sensitivity and information content of global approaches. By using these instruments, it is possible to detect metabolites having the same nominal mass (referred to as isobaric metabolites) in complex biological media and also to obtain their elemental composition, which facilitates database inquiries for identification purposes (Aharoni et al., 2002). This prompted the use of MS-based untargeted metabolomic experiments.

Two kinds of approaches can be implemented with high-resolution mass spectrometers: (i) direct injection MS (DIMS), referred to as shotgun experiments in the field of lipidomics, and (ii) MS hyphenated with separative methods such as chromatography or capillary electrophoresis (Wolfender, Glauser, Boccard, & Rudaz, 2009).

2.1.3.1.2 Direct injection MS DIMS enables a rapid and direct analysis of crude extracts and is suitable with high-throughput screening. The sample is introduced to the spectrometer by a syringe pump (infusion MS) or using LC pump (flow injection analysis). DIMS with low-resolution instruments was applied in the 1990s for profiling amino acids, organic acids, and carnitine derivatives for the diagnosis of inborn errors of metabolism in neonates (Chace, 2001). More recently, it was used in the field of plant metabolomics (Castrillo, Hayes, Mohammed, Gaskell, & Oliver, 2003). However, as many isobaric metabolites occur in biological media, DIMS requires the use of high-resolution mass spectrometers for an optimal use (Aharoni et al., 2002). Nevertheless, it suffers from several drawback: (i) its sensitivity is limited by ion suppression induced by matrix effects, which occur in biological media, (ii) it is difficult to identify ions as coming from an intact metabolites or an in-source fragment ion, and at last, (iii) it does not resolve isomers. This justifies the need for an additional separation dimension brought by LC/MS or GC/MS. Note that CE/MS represents an interesting alternative for the analysis of very polar metabolites unsuccessfully supported in RP chromatography (Allwood, Ellis, & Goodacre, 2008). However, it is up till now less robust and less sensitive.

2.1.3.1.3 Liquid chromatography coupled to mass spectrometry

Different kinds of chromatographic retention mechanisms are available: exclusion chromatography, for which analytes are separated according to their size, affinity chromatography, ion exchange chromatography, and also partition chromatography, including normal and RP chromatography, the latter being widely used for LC/MS.

Normal phase chromatography, which is well suited for the detection of polar compounds, involves stationary phases consisting of unmodified silica or alumina and non-aqueous mobile phases which are unfortunately not suitable for API–MS detection. Concerning RP chromatography, silica-based stationary phases are functionalized to make them hydrophobic and thus compatible with the aqueous and organic solvents, which are suitable with API–MS. Methods relying on RP chromatography are highly reproducible; separation mechanisms have long been investigated and are well understood (Kirkland, 2004; Stella et al., 2007; Vervoort et al., 2002). Columns containing alkyl hydrocarbon chains covalently bonded to silica are the most popular. However, in order to achieve a good separation in RP chromatography, it is necessary for the analytes to exhibit hydrophobic properties so that they may interact with the stationary phase. Whereas RP chromatography performs well with metabolites of low-to-intermediate polarity, highly polar metabolites are poorly retained on these columns and are thus prone to matrix effect, which limits the sensitivity of detection of analytes not retained by the column and thus eluted in the void volume.

Alternative retention mechanisms such as ion pair chromatography, aqueous normal phase (ANP) chromatography or hydrophilic interaction chromatography (HILIC) are developed to address this key issue. Concerning ion pair chromatography, a volatile counter ion is introduced in the mobile phase to form an ion pair with the analytes. The resulting ion pairs are neutral, hydrophobic, and well retained on RP columns. The use of anionic and cationic ion pairing agents for the detection and quantification of basic and acid metabolites has already been extensively published (Dai & Carr, 2005; Dai, Mendonsa, Bowser, Lucy, & Carr, 2005; Shibue, Mant, & Hodges, 2005a, 2005b, 2005c). However, the two main drawbacks of this approach are the limited volatility of the ion pair and/or the too strong affinity between the ion pairing reagent and the analyte. This prevents the dissociation of the ion pair in the electrospray source and induces signal suppression effects (Apffel, Fischer, Goldberg, Goodley, & Kuhlmann, 1995; Gustavsson, Samskog, Markides, & Langstrom, 2001; Lu, Bennett, & Rabinowitz, 2008; Sysi-Aho, Katajamaa, Yetukuri, & Oresic, 2007).

Another alternative is the use of other chromatographic supports than the RP, such as cyano-, diol-, amino-bonded silica or perfluorophases (such as pentafluorophenylpropyl) which may exhibit complementary selectivity profiles. For example, pentafluorophenylpropyl, which enhances pi–pi and steric interactions due to the phenyl ring and exhibits polar interaction due to the presence of fluorine, is well suited for the retention of aromatic compounds and also nitrogen-containing metabolites (Narainsamy, Cassier-Chauvat, Junot, & Chauvat, 2011). The orthogonality of the perfluorophases compared to the conventional RP was established for the basic analytes which were considerably more retained on perflurophases compared to the non-perfluorophases (Euerby & Petersson, 2003).

Recently, a promising new technology referred to as "Aqueous Normal Phase Chromatography" has emerged. While most silica materials used for chromatography have a surface composed primarily of silanols (—Si—OH), the terminal groups in "hydride surface" are —Si—H. Mobile phases for ANPC, which are based on an organic solvent mixtures (such as methanol or acetonitrile) containing a small amount of water, are close to that used in normal phase chromatography and compatible with API–MS detection. Thus, polar solutes (such as aliphatic acids and amines) are efficiently retained on ANPC columns, with retention times decreasing as the amount of water in the mobile phase increases. Pesek et al. demonstrated that ANP mode coupled with MS detection ensures retention and separation of hydrophilic compounds such as sugar, aminoacids, and carboxylic acids in both synthetic (a mixture of endogenous metabolites standards) and human urine samples (Pesek, Matyska, & Fischer, 2011; Pesek, Matyska, Fischer, & Sana, 2008).

Another alternative rely on HILIC, which has become more popular as columns retain polar analytes using MS-friendly solvents. In fact, molecules retention occurs thanks to a thin water layer established on the stationary phase so that hydrogen bonding and dipole–dipole interactions are possible. Different applications have been published using this particular retention mode since the early 2000s: analysis of oligosaccharides, amino acids, and sugar nucleotides from plant leaves (Tolstikov & Fiehn, 2002); metabolic fingerprinting of rat urines (Idborg, Zamani, Edlund, Schuppe-Koistinen, & Jacobsson, 2005); the quantification of metabolites involved in the glutathione biosynthesis pathway in yeast (Lafaye, Labarre, Tabet, Ezan, & Junot, 2005); and the detection and quantification of polar metabolite from *E. coli* cell extracts (Bajad et al., 2006). New HILIC supports such as HILIC-amide and the zwitterionic phase ZIC-pHILIC (Chirita, West, Zubrzycki, Finaru, & Elfakir, 2011) have been developed to improve the retention of polar metabolites exhibiting a wide range of chemical structures. By using the HILIC-amide stationary phase which enhances hydrogen bonding as the O and the N of the carbamoyl moiety can act as hydrogen bonding acceptors, more than 90 polar metabolites including amino acids, organic acids, carbohydrates, and polyols could be quantified in grape extracts (Gika, Theodoridis, Vrhovsek, & Mattivi, 2012).

ANP and HILIC techniques were recently evaluated by Zhang et al. using urine samples and compared to the reference RP chromatographic platform. With regard to peak shape, the kinds of detected metabolites, and the separation of isomers, the ZIC–pHILIC column combined with basic mobile phases ($pH > 10$) ensures the detection of the widest range of polar metabolites in human urines compared to the other chromatographic systems (Zhang, Creek, Barrett, Blackburn, & Watson, 2012).

Otherwise, the introduction of ultra-high-pressure liquid chromatography (UHPLC) copes with high pressure (>600 bars), thanks to column packed with sub-2-μm particles which enables faster separations. Chromatographic peaks are thinner and higher than those observed using conventional LC systems. This increases the signal/noise ratio (Wilson et al., 2005), peak capacity, and the efficiency of the column (Leandro, Hancock, Fussell, & Keely, 2006). UHPLC allows covering 20% more molecules than HPLC in serum extracts (Nordstrom, O'Maille, Qin, & Siuzdak, 2006). However, the sharpness of the chromatographic peaks has to be compatible with the acquisition time of the analysers. This is the case with Q-TOF instruments, but the situation could be more problematic with LTQ-Orbitrap mass spectrometer operated at a higher mass resolution power and with a relatively long cycle time (about 1 s), and more particularly with FTICR instruments (Metz et al., 2007).

2.1.3.2 Identification of signals of biological interest

Metabolite identification is tedious. It often requires several complementary analytical tools such as NMR spectroscopy, IR, UV, and MS. However, with regard to the difficulty to purify the metabolites of interest, their low abundance in media, and the inability of some of them to be detected by other methods, the implementation of a multidimensional approach is not always possible and, as a consequence, structural elucidation is achieved using the analytical method used for the acquisition of metabolic profiles.

2.1.3.2.1 From dataset annotation to metabolite identification In LC/API–MS approaches, identification of variables of interest starts with an annotation step achieved by querying biochemical databases like KEGG (Kanehisa & Goto, 2000) and HMDB (Wishart et al., 2007), or in-house spectral databases (Roux et al., 2012) with accurate measured masses. Then, annotated variables have to be identified using MS/MS experiments. The information provided by MS/MS spectra can rule out or reinforce an annotation proposal, and it may also help to discriminate between isomers. At last, formal metabolite identification requires authentic compounds. However, many metabolites are not commercially available, and many of them may require tedious and expensive chemical synthesis which often hampers their definitive metabolite identification.

To address these issues, the Chemical Analysis Working Group in the metabolomics Standards Initiatives has proposed four different levels of identification (Sumner et al., 2007):

1. "Identified compounds" correspond to the greatest confidence level of identification. It requires a minimum of two orthogonal criteria relative to an authentic compound analysed under identical experimental conditions: retention time and mass spectrum or retention time and ^{1}H NMR spectra, accurate mass and tandem mass spectra or accurate mass and related isotopic clusters or ^{1}H and/or ^{13}C NMR with 2D NMR spectrum.
2. "Putatively annotated compounds:" this level of identification is for compounds that are not commercially available. The identification of the metabolite is based on the spectral and/or physicochemical properties' similarity with spectral databases.
3. "Putatively characterized compound classes" are compounds for which there is no proposed structure in chemical and spectral databases. However, they exhibit spectral and/or physicochemical similarities with compounds of a given known chemical class.
4. "Unknown compounds" for which no structural hypothesis is issued.

2.1.3.2.2 The input of spectral databases Since metabolite identification process requires comparison of the spectra with those of standards compounds, it underlies the implementation of a centralized or in-house reference spectral repository. API-mass spectral libraries aim at annotating biological datasets, whereas API–MS/MS spectral libraries are useful to confirm peak list annotations and to characterize unknown compounds (Roux et al., 2011, 2012). Spectral libraries for metabolites have initially been built using

NMR and GC/MS instruments which provide users with robust and reproducible data that can be shared and compared within different laboratories. Some of these tools such as those proposed by Brucker (500 compounds, 13,500 NMR spectra), Chenomx (315 compounds, NMR spectra), and the NIST (GC/MS) are commercial, whereas others are free, such as the Golm database (Kopka et al., 2005), HMDB database (Wishart et al., 2007), and Riken Standard spectrum database (Akiyama et al., 2008). The situation is unfortunately different with API-based mass spectrometers as they exhibit poor reproducibility and high inter-instrument variability in fragmentation patterns of metabolites (Oberacher et al., 2009a, 2009b; Palit & Mallard, 2009). Furthermore, and contrary to GC/MS-based instruments, API-mass spectra provide limited structural information, requiring the complementary use of MS/MS experiments. This is the reason why both API–MS and API–MS/MS spectral libraries have to be developed together.

Despite poor reproducibility and high inter-instrument variability in the generation of fragmentation patterns, databases containing API mass spectra combined with MS/MS spectra such as HMDB (Wishart et al., 2007), Metlin (Smith et al., 2005), mass bank from metabolome.jp (Horai, et al., 2010), and lipid maps (Fahy, Sud, Cotter, & Subramaniam, 2007) are beginning to be released. However, the use of such spectra for comparison and identification must be performed carefully and may lead to erroneous results (Werner, Heilier, et al., 2008). Thus, there is an urgent need to standardize experimental conditions of the recording of MS/MS spectra.

2.1.3.3 Metabolite quantification

As previously explained in the field of proteomics, metabolites of biological interest have to undergo a qualification process before moving to the status of biomarker. The validation process should include quantitative analysis of either a subset of the cohort used in the discovery step or another cohort (Koulman, Lane, Harrison, & Volmer, 2009). However, profiling tools using for global approaches are in most cases not suitable for quantification. It is therefore necessary to develop appropriate quantitative approaches which mostly rely on the use of triple quadrupole instruments operated in the SRM or MRM modes.

Nowadays, two different quantification strategies can be implemented: (i) relative quantification that relates to the fold of differences between samples and (ii) absolute quantification that allows measuring the true concentration of metabolites in the samples. The first one, which is less restrictive to implement, enables to correct experimental biases inherent to ESI sources, while the second is more difficult to address but facilitate the use and sharing of metabolomic data with biologists and clinicians.

2.1.3.3.1 Relative quantification API–MS is prone to numerous analytical drifts including degradation of chromatographic performances, "ion suppression" induced by matrix effect, and source clogging over time which alter the ionization efficiency within an experiment (i.e. intra-experiment variability) and/or also across batches (i.e. inter-experiment variability). This complicates the data processing and interpretation, and relative and absolute quantification may address this issue.

Relative quantification refers to as normalization procedures. It involves the use of ISs that are added to the samples to be analysed in order to correct from matrix effect and source clogging. As such analytical biases are depending on the kind of chemical structures, ISs should exhibit a close structural similarity with the compounds of interest.

ISs can be obtained from chemical synthesis of stable isotope-labelled metabolites, but it is expensive and time consuming. ISs can also be obtained from metabolically labelled organisms (Giavalisco, Kohl, Hummel, Seiwert, & Willmitzer, 2009; Kiefer, Portais, & Vorholt, 2008; Lafaye et al., 2005; Mashego et al., 2004), which is restricted to plants and microorganisms (*in vivo* labelling of microbial metabolome). Derivatization strategies using stable isotopic tags have also been developed for metabolites bearing chemical functions such as amine or carboxylic acid functions (Guo, Ji, & Li, 2007; Guo & Li, 2009, 2010; Shortreed et al., 2006). At last, the obtention of an ISs mixture representative of a large number of metabolites remains challenging for high-throughput metabolomic experiments. To address this issue, some authors recommend the use of QC samples consisting of pools of samples that are representative of the sample type under investigation and that are analysed over the whole time course of the study, to provide signal correction (regression for each feature according to the intensities recorded on the QCs across the acquisition sequence) and integration of data from multiple analytical batches (Dunn et al., 2011). Of course, such an approach is relevant as far as metabolites are stable in biological extracts.

2.1.3.3.2 Absolute quantification Providing concentration values as molarity units is an effective way to standardize metabolomics data for their sharing and their use by biologists. To this end, signals (expressed as natural abundance to labelled metabolites intensity ratios) in the samples are compared to those of a signal–concentration curve (i.e. calibration curve). Such an approach is widely used in pharmaceutical research and development to measure drugs or xenobiotics concentrations in biological matrices, and an FDA guidance is available for bioanalytical method validation (Shah et al., 2000). Despite this, absolute quantification of endogenous metabolites remains challenging. Indeed, the large number of metabolites exhibiting a huge chemical diversity requires several mixtures of ISs and calibrators and makes method validation a complex and tedious task; and the endogenous presence of metabolites in biological matrices complicates the construction of standard calibration curves.

The routine use of absolute quantification methods is currently limited by the lack of commercial availability of metabolites labelled with stable isotopes. As proposed above for relative metabolite quantification, an alternative is to use stable isotope-labelled calibrators and ISs obtained from chemical synthesis. Using this approach, Human metabolome Technologies (Japan) has developed the HMT metabolomics solution package-kit based on CE/TOF–MS quantification of ionic compounds and the Biocrates Company (Austria) has developed AbsoluteIDQ™, a kit for absolute quantification of more than 180 metabolites by DIMS, including amino acids, acylcarnitines, hexoses, biogenic amines, and phospholipids. Another alternative is the use of organic extracts obtained from metabolically labelled organisms (Bennett, Yuan, Kimball, & Rabinowitz, 2008; Giavalisco et al., 2009; Godat et al., 2010; Kim, Harada, Bamba, Fukusaki, & Kobayashi, 2005; Lafaye et al., 2005; Mashego et al., 2004), but issues of metabolite stability can occur in biological extracts.

Although triple quadrupole instruments are still considered as a reference for metabolite quantification methods (Bajad et al., 2006; Wei, Li, & Seymour, 2010), HRMS is emerging in this field, especially in proteomics, as suggested by recent publications with FTICR (Collier, Hawkridge, Georgianna, Payne, & Muddiman, 2008), Orbitrap (Savitski et al., 2011), and Q-TOF (Garcia-Villalba et al., 2010) instruments. As a matter of fact, the last improvements achieved with the hydrid Q-Exactive and Q-TOF technologies enable efficient multiplexed quantitative MS and MS/MS experiments, such as the parallel reaction monitoring (Gallien et al., 2012) and the SWATH (Gillet et al., 2012) modes, for the first and the second kind of instrument, respectively. It is clear that these approaches will be used in metabolomics in a near future for the possibility they offer to achieve metabolite detection, identification, and quantification in the same time.

2.1.4 Bioinformatics and data mining

Bioinformatics is a branch of biological science which deals with algorithms, data mining, databases, statistics, and soft computing for the handling and analysis of biological data. Applied to metabolomics, it aims to convert analytical data into biological knowledge. Bioinformatic tools can be divided into two categories: (i) low-level data processing method, which convert raw data into numerical datathat can be used for further statistical and biochemical interpretations and (ii) "mid-level" data processing methods, which generate valuable biological information from these numerical data (Katajamaa & Oresic, 2007). Both categories are addressed with regard to LC/MS metabolomic datasets in the following section.

2.1.4.1 From raw data to processed data

Raw data processing aims at converting raw files into a data format suitable for further statistical and biological analyses. Dedicated algorithms perform (i) automatic peak detection, (ii) alignment of features in the m/z and chromatographic retention time domains, and (iii) results are returned as a peak table containing variable identity (i.e. m/z and retention time) and signal abundances (i.e. peak intensities and/or area of extracted ion chromatographic peaks) in the samples. To achieve this goal, software tools have been released since the middle of the 2000s. Some of these are of commercial origin (i.e. sold by an MS manufacturer or any other bioinformatics company), some are free but with proprietary code, as it is, for example, the case with MetAlign (Lommen, 2009), whereas others are free open source, the most popular ones being MZmine (Katajamaa, Miettinen, & Oresic, 2006) and XCMS (Smith, Want, O'Maille, Abagyan, & Siuzdak, 2006). Of note, data-processing tools have already been reviewed (Katajamaa & Oresic, 2007) and readers who would like further reading regarding codes and their implementation can refer to all the above cited references. However, low-level data processing is no more restricted to peak detection, but nowadays includes algorithms to organize and annotate peak tables.

In the following sections, we would like to address recent advances about software for automatic peak detection and issues related to the evaluation of their performances, and also new algorithms for the annotation of peak tables obtained from LC/MS-based metabolomic datasets.

2.1.4.1.1 Improvements of software tools for automatic peak detection and alignment Since they have been launched, software tools underwent significant changes motivated by the need to support the technological progress of analytical instruments. They now comply with ultra-high-resolution mass spectral data obtained from TOF, Orbitrap, Fourier transform ion cyclotron resonance MS instruments, and fast chromatographic separation brought by UHPLC technologies. Furthermore, initially restricted to peak detection and alignment, many of them now consist of software suites including algorithms developed to address issues such as data normalization or data annotation.

For example, MetaXCMS is an extension of the XCMS software enabling the comparison of several datasets obtained from global metabolomics experiments (Tautenhahn et al., 2011). It gives a Venn diagram for a direct visualization of the number of common features to all datasets, as well as the number of features specific to each dataset. XCMS2 is another extension of XCMS able to handle MS/MS data and to perform automatic matching of experimental MS/MS spectra with those of the METLIN database (Benton, Wong, Trauger, & Siuzdak, 2008). At last, the user-friendly web-based tool "XCMS online" has been introduced to enhance the accessibility of untargeted metabolomics to a large population of non-expert users (Tautenhahn, Patti, Rinehart, & Siuzdak, 2012).

In the same way, MZmine 2 was redesigned in a flexible framework to support modularity. It now processes larger datasets and enables easy incorporation of new algorithms that could handle other tasks (Pluskal, Castillo, Villar-Briones, & Oresic, 2010). MZmine that was initially dedicated to automatic detection and alignment of signals now include modules that manage multiple spectra from the same sample (MS or MS^n), enable database connectivity for metabolite identification, and improve isotope pattern support.

2.1.4.1.2 How to compare automatic peak detection softwares?

To date, no comparative study regarding the performances of available algorithms for LC/MS-based metabolomic data processing has been published.

To address this issue, Tautenhahn et al. proposed a relevant approach to evaluate automatic feature detection from high-resolution LC/MS datasets, through three parameters: (i) "Recall" which takes into account the false negatives and corresponds to the fraction of the relevant features that are extracted within the dataset, (ii) "Precision" which is the percentage of relevant items relative to artefacts and highlights the false positive features and (iii) "run time" estimated from several hours to several days, depending on the size of the dataset (Tautenhahn, Bottcher, & Neumann, 2008).

Pluskal et al. tested the performance of algorithms used in the alignment step of MZmine 2 using this evaluation method. They found that the RANSAC aligner algorithm performs better on a synthetic lipidomics dataset and also on proteomic and metabolomic datasets (Lange, Tautenhahn, Neumann, & Gropl, 2008) than the Join aligner algorithm (Pluskal et al., 2010). Whereas Peters et al. proposed another strategy to evaluate alignment parameters by monitoring spiked compounds in biological samples. Inclusion of non-spiked analogues and pairwise comparison gives the information on the overall quality of data alignment (Peters, van, & Janssen, 2009).

2.1.4.1.3 Software tools for metabolome annotation Thanks to accurate mass measurements provided by high-resolution mass spectrometers, it is possible to efficiently match experimental measured masses contained in peaktables to those referenced in biochemical and metabolomic databases. Nowadays, most automatic peak detection software tools include algorithms enabling automatic peaktable annotations using HMDB (Wishart et al., 2007), KEGG (Kanehisa & Goto, 2000), or Metlin (Smith et al., 2005) databases, and this is considered as the first annotation step, as previously mentioned in Section 2.1.3.2.

However, this first step does not enable confident metabolite identification because (i) many metabolites have the same molecular mass and (ii) atmospheric pressure ionization methods often produce several ions for a given metabolite. These ions may correspond to fragments (in-source CID), adducts, and isotopes. Although it may be helpful for metabolite identification, such a signal redundancy complicates data processing because it may produce erratic annotations (Roux et al., 2011). This highlights the complexity of API–MS-based datasets and a thorough inspection of mass spectra requiring spectral libraries is necessary before biological interpretation, as previously discussed in Section 2.1.3.2.

Otherwise, some tools have been developed to mine structuring of the dataset by grouping features related to the same "parent molecule", like isotope, adduct, and product ions generated by in-source CID. The differential annotation can be carried out by matching mass differences within a spectrum to specific mass differences covering known isotopes, adducts, or neutral losses. The principle of these tools is based on the fact that different signals coming from a single metabolite should exhibit the same shape peak and the same changes in intensities from one experimental condition to another. So the correlation should be calculated between pairs of ion either across (i) several spectra within a sample (spectral correlation) like

CAMERA package (Kuhl, Tautenhahn, Bottcher, Larson, & Neumann, 2012) or across (ii) all the samples of the dataset as done in AStream (Alonso et al., 2011; Matsuda et al., 2009; Tautenhahn, Böttcher, & Neumann, 2007; Werner, Croixmarie, et al., 2008).

Different authors have developed workflows dealing with intensity correlation across samples and/or retention time and/or pairwise *m/z* differences matched to a fixed rule table for annotating isotopes and adduct ions. Chen et al. (2008) have manually calculated correlation between the EIC of interest and coeluted EIC. Werner, Heilier, et al. (2008) applied autocorrelation matrices in order to structure LC/MS datasets. Ipsen, Want, Lindon, & Ebbels, 2010 used χ^2 test to highlight parent-product ion pairs in LC-TOF/MS data.

Nowadays, some statistical software packages are available to address the issue of redundancy. IntelliXtract (ACD/IntelliXtract, Advanced Chemistry Development, Inc. www.acdlabs.com/intellixtract) (Ito, Odake, Katoh, Yamaguchi, & Aoki, 2011) developed by ACD/Labs ensures $[M+H]^+$ or $[M-H]^-$ assignment by considering the expected $^{12}C/^{13}C$ ratios, adducts, multimers, isotopes, neutral losses, and fragment ions to provide reliable spectrum. Kuhl et al. (2012) developed a collection of algorithms for metabolite profile *a*nnotation named CAMERA. It is a Bioconducter-package implemented in R language and used for the postprocess of XCMS peaklists. It first groups peaks in the retention time domain and then calculates the EIC correlation within an Rt width chosen according to the LC. If the correlation is higher than a fixed threshold, peaks will be kept in the same group. Then, accurate compound mass is calculated and isotopes and adducts are annotated. AStream is another R package for annotating LC/MS metabolomic data for which data reduction relies on the intensity correlation across samples. It requires processed data and not raw files (Alonso et al., 2011).

All these tools now begin to be used for metabolome annotation and metabolite identification in biological extracts. As a matter of fact, Roux et al. (2012) proposed a workflow dealing with CAMERA package, auto-correlation matrices, and an in-house ESI–MS database to annotate urine samples. They identified 384 metabolites that represent 659 and 825 annotated ions in the positive and the negative ion mode, respectively. Thanks to these different software tools and workflow, it was possible to exclude the redundant information generated by MS detection so as to focus only on protonated or deprotonated molecules for further metabolite identification.

2.1.4.2 From processed data to biological interpretation

Metabolomics relies on differential analysis of metabolic fingerprints which leads to a semi-quantitative expression of the results (i.e. decreased or increased area or intensity ratios). The challenge is to manage and undertake a reliable comparative analysis of these fingerprints that contain hundreds to thousands of relevant signals. To this end, data-mining approaches, including statistical methods and, more recently, tools to visualize metabolic networks, are developed.

2.1.4.2.1 Statistics for metabolomics Inherent to the production of huge datasets in metabolomics resulting from fingerprinting of different biological matrices by high- or ultra-high-resolution mass spectrometers, the question of how to manage this raw information needs to be studied in details. Clearly, researchers are confronted to data analysis problems dealing with multidimensionality. Classical statistical tools were elaborated before the second World War and were well designed to mine full-rank data matrices M with n statistical individuals far higher than the set of the p independent variables describing them. Algebraically, covariance (or correlation) matrices built from M are invertible and the resulting eigen values are easily used to get few principal components that summarize efficiently the overall variance as in PCA, an unsupervised multivariate method. Now, most of the time, we need to solve similar problems related to the structuring of variance (information), but with the striking difference that it is extracted from matrices with n statistical individuals, a number which is far lower than the p independent variables used to described them. From MS datasets, we are able to get few thousand of different ions in either positive or negative mode detected by high-resolution mass spectrometers. The problem is rather similar with data generated by high field NMR and, hence, data-mining methods developed mainly on the two last decades are more or less transposable from one domain to another.

We present hereafter the different statistical approaches usually required in metabolomics to study the information structuring of datasets, depending on whether there is an explicit factorial design, a possibility to build additional datasets with complementary covariates, a prior simple distribution of individuals in several homogenous subgroups, or to proceed to a longitudinal follow-up of individuals thanks to a coding factor describing a time course dependency (see Fig. 4.6).

2.1.4.2.1.1 The simplest methods Most often, modelling methods used in metabolomics are rather simple and do not lay on highly sophisticated

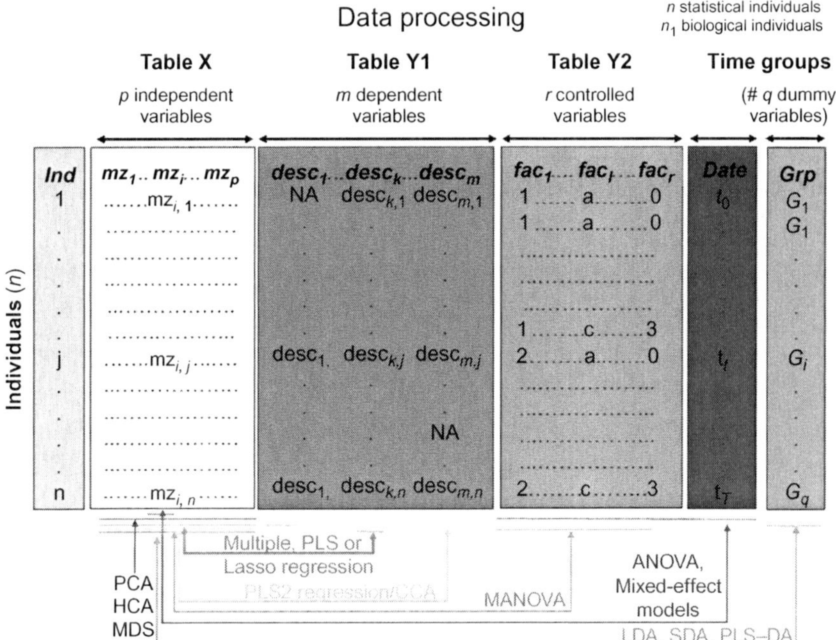

Figure 4.6 Organization of the different datasets used to extract meaningful metabolomic information contained in Table X built from fingerprinting of n statistical individuals coming from n_1 different subjects enrolled in the population sample submitted to a metabolomic study. Different statistical procedures can be achieved to mine metabolomic data depending on the structure of adjacent matrices of co-variates related to Table X, as display at the bottom of the diagram. Abbreviations: PCA, principal component analysis; HCA, hierarchical clustering analysis; MDS, multidimensional scaling; PLS, partial least squares regression; LASSO, least absolute shrinkage and selection operator; CCA, canonical correlation analysis; ANOVA, analysis of variance; MANOVA, multivariate analysis of variance; LDA, linear discriminant analysis; SDA, shrinkage discriminant analysis; PLS-DA, PLS-based discriminant analysis. (For colour version of this figure, the reader is referred to the online version of this chapter.)

models. This concerns mainly the partial least-squares (PLS)-based regression, a robust chemometric method initially developed by Wold (Wold, Martens, & Wold, 1983; Wold, Ruhe, Wold, & Dunn, 1984). This is a method based on a projection onto a small set of latent variables (LV); it is usable even though the data matrix contains correlated variables and displays incomplete data (Trygg, Holmes, & Lundstedt, 2007). PCA can be achieved again easily by using PLS and the different orthogonal components summarize sequentially the variance by decreasing order. Projection of the coordinates of individuals (scores) in this new set of components provides an overview of information structuring, trends or grouping of individuals as well as the possibility to detect outliers which eventually can be discarded from the original dataset. Although it is not really a classification tool, PCA has been used, for example, to detect some species-dependent urinary biomarkers in human, rat, mouse, and rabbit with a classification objective (Bollard, Stanley, Lindon, Nicholson, & Holmes, 2005). Components are easily "explained" by some subsets of initial variables (loadings); they are built from the linear combination of those original variables. Hence, contrasts between some sub-groups of individuals are supported by co-variations of these loadings. Using a distance calculation between either individuals or variables, it is possible to get an information clustering which display also a seemingly homogenous property between individuals or variables found inside a given cluster. This hierarchical clustering analysis (HCA) is a complementary way to detect a hidden structuring inside a dataset without any *a priori* hypothesis requirement. Application of PCA on a matrix of distance (or similarities) between individuals, also called multidimensional scaling (MDS), results again in revealing the hidden structuring of the metabolic information.

However, most of the time, one or several adjacent matrices can be designed from the same subset of metabolomically phenotyped individuals. Two types of matrices may exit. The first one consists in a table of dependent variables (Table Y1, Fig. 4.6), which are used in a complementary way to describe individuals included in the experimental design. Those variables may correspond to anthropometric, biochemical, or clinical descriptors of individuals enrolled into a population sample. A canonical link, which is a significant correlation link between the two first principal components coming from PCA achieved separately on Tables X and Y1, can be established to prove the statistical relationship between a metabolomic description on one side and a more global (integrated) one recorded on an anthropometric, biochemical, or clinical side. No regularization procedure is necessary when we mine full-rank matrices. However, for metabolomic data, most of the time, such a prior regularization (simplication) of the matrix is unavoidable to get by canonical correlation analysis (CCA) the scores of individuals or to detect loadings coming from both datasets. Their close proximity is a decidable rule in order to highlight some correlation between a given Y1 variable and some metabolic variables, which then become candidate biomarkers of this dependent Y1 variable (Le Cao, Martin, Robert-Granie, & Besse, 2009; Rohart et al., 2012). In the same way, but based on different calculation procedures, PLS2 and some related versions integrating a parallel correction of data for discarding unrelated (orthogonal) information towards Tables X or Y1 contained in Tables Y1 and X, respectively, are alternative ways to couple two types of biological information recorded on the same set of individuals (Domange et al., 2008). A second type consists of a table of coding values of some controlled variables (factors) (Table Y2, Fig. 4.6). Combination of those factors leads to the constitution of different homogenous groups (Fig. 4.6) on which a multivariate variance analysis (MANOVA) can be performed from the metabolomic information considered as a "whole" information set contained in Table X. Note that in this case, in term of calculation, the coding variable for the different groups is exploded into a matrix containing the same number of dummy variables taking either 0 or 1 as coding value. However, the MANOVA test also needs to use a full rank matrix X. A geometric procedure corresponding to MANOVA consists of the classical linear discriminant analysis (LDA) when the multi-dimensional variance calculated for every group is constant (homoscedasticity) or in the quadratic discriminant analysis (QDA) in case of heteroscedasticity. The PLS2 version of such a discriminant analysis (DA) is easily used to get a seemingly more "natural" projection of scores. This PLS-DA procedure is also efficient in case of the n number of individuals is far lower than the p number of variables without requiring a prior selection of candidate variables as for the more classical LDA (Dumas, Canlet, Andre, Vercauteren, & Paris, 2002). Recently, an efficient regularization of high-dimension data matrices for DA has been proposed (Le Cao, Boitard, & Besse, 2011).

In all those factorial approaches, building of components such as principal components, discriminant axes, or LV results in the detection of candidate metabolomic biomarkers. The higher the correlation between a metabolic variable and a factorial axis, the higher the probability to be confirmed as a valuable biomarker (in a statistical meaning). In a PLS analyses, complementary tools such as the variable importance in the projection criterion and an S-plot combining the correlation and covariance calculations are helpful to detect such candidate biomarkers (Wiklund et al., 2008). In addition, analysis of correlation between metabolomic variables may also be useful to detect intrinsically related ions corresponding to a unique given analyte as it is possible for redundant signals found in ^{1}H NMR (Blaise et al., 2009). This spectroscopic assessment recoupling a statistical-based selection of biomarkers is really helpful to discard false positive biomarkers (Blaise et al., 2010; Dumas, 2012).

More generally, selection of putative biomarkers resulting from selection of a given statistical model needs to be sustained by a cross-validation analysis, the objective of which is to test the relevance of prediction given by a selected model compared to this resulting from application of the null hypothesis. In case of univariate tests, such an analysis of variance is performed on a large set of independent variables, and the probability to find by chance some significant variables is far from being low. Conservative procedures based on the multiple comparisons such as the Bonferroni test are too stringent to select statistically valuable biomarkers when calculating a probability to get a false positive feature. Yet, some more recent and more sophisticated techniques can control the estimated proportion of incorrectly rejecting null hypotheses (i.e. false discoveries); they are called FDR techniques (Benjamini & Hochberg, 1995; Storey & Tibshirani, 2003; Strimmer, 2008).

In classification procedures based on PLS-DA or LDA, cross-validations in the selection of predictive (non null) models are necessary to confirm that a given feature (variable) can be identified as a candidate biomarker. Cross-validation techniques use iteratively g-1 partitions of the population to build a predictive model, the performances of which being then assessed on the last partition. This way is efficient to measure the performance of prediction models and, hence, to detect putative biomarkers. Yet, even if the size of the population is rather high to get good statistical performances, generalization of the functional character of so-detected biomarkers in the general population may depend on the representativeness of the sampled population. In epidemiological studies, covariates like ethnicity, age, dietary habits, drug exposure, physical activity, environment, etc., which does not represent any *a priori* testing hypothesis, need to be balanced when case and control populations are enrolled to avoid any misinterpretation of biomarkers due to some eventually existing confounding factors. However, as shown in the following example, even though few thousands of subjects were recruited in two independent epidemiological cohorts devoted to the study of the aetiology of diabetes, the search for metabolomic markers of a prediabetic state achieved in these populations has only resulted in an association between metabolism of amino acids and a diabetes risk assessment that should be further confirmed in larger size populations (Wang et al., 2011).

2.1.4.2.1.2 Some more sophisticated methods In parallel to the recent high improvement in spectroscopic data acquisition techniques achieved by high- or ultra-high-resolution spectrometers as well as in calculation performances available now on desk computers, sophisticated methods have arisen in the domain of data mining. This concerns firstly the validation of multivariate models using resampling techniques (Dumas, 2012; Fonville, Richards, et al., 2010). In addition, a better analytical detection of biomarkers can be achieved thanks to the extensive calculations based on non-linear kernel-based models used to get projections of scores to latent structures when considering at the same time a variation in information which is orthogonal to the response (Bylesjo, Rantalainen, Nicholson, Holmes, & Trygg, 2008; Fonville et al., 2011). Application of correlation techniques (Fonville, Maher, et al., 2010; Maher et al., 2012) or Bayesian models (Astle, De Iorio, Richardson, Stephens, & Ebbels, 2011; Hao, Astle, De, & Ebbels, 2012) to NMR spectra is helpful to better characterize and to quantify metabolites. A statistical modelling of the detection of ions achieved by LC–TOF–MS and a definition of a goodness-of-fit test were built to get a better identification of parent-fragment pairs in case of exact coelution of ions, more particularly to distinguish those which are unrelated compounds (Ipsen et al., 2010).

Phenotyping of organisms based on multi-omic techniques needs to cope with this huge information by using adapted multivariate statistics. Most of the techniques are based on correlation, CCA, PLS2-based regressions, multiple factorial analyses, or parallel factor analysis (Richards et al., 2011).

In large-scale molecular epidemiology, use of post-genomic techniques has recently overcome some challenging association problems between disease and molecular markers like genes or metabolites. The metabolome-wide association studies aim to get a significant link between a quantitative variation of a given metabolite and the presence of a quantitative trait locus in genome. More precisely, mapping of metabotype quantitative trait loci corresponds to the detection of an association between a genome marker (microsatellite, single nucleotide polymorphism (SNP), number of copy variants), and a metabolic one (Dumas et al., 2007). Regressions using fixed or mixed (fixed/random) effects modelling are used to detect such metabolic candidates linked to some SNP considering false discovery rates, statistical power, analytical quality of data, and, more satisfactorily, a further confirmation of this association in a comparable but independent study (Nicholson et al., 2011).

The longitudinal metabolomic follow-up of large cohorts is a crucial step that needs to be developed before dedicating some mature techniques of metabolomic phenotyping to the personalized medicine (Nicholson, Everett, & Lindon, 2012). Considering both the sets of metabolomic variables (Table X, Fig. 4.6) and dependent variables (Tables Y1 and Y2,

Fig. 4.6) recorded in a time course perspective (a time-coding factor, Fig. 4.6), it is possible in a first approximation to preselect some interesting variables using a mixed-effects regression model. This model allows assignment of a significant part of the variance to the random factor and is helpful to better adjust the fixed effect variance and, hence, to reduce the residual variance. In this case, repeatedly "observed" individuals in a time course perspective are categorized as a random factor. More interesting is the recent development of a similar statistical framework for the discovery of metabolomic biomarkers displaying a time course variation. This method takes into account every spectrum as a whole and uses functional mixed-effects models by considering within-group variability and between-group differences (Berk, Ebbels, & Montana, 2011).

Although most mid-level data processing tools in the field of metabolomics still deal with statistical analyses, some advanced tools developed to the improve the visualization of metabolite networks and dedicated to LC/MS datasets are released.

2.1.4.2.2 Tools to visualize MS-based metabolic networks Historically, metabolic networks were studied at the levels of metabolic pathways (i.e., subsets of molecules linked by biochemical reactions), or physiological functions involving several metabolic pathways. As described and listed in biochemical databases such as KEGG (Kanehisa & Goto, 2000) or MetaCyc (Karp et al., 2000), metabolic pathways are first considered as separate processes in which selected metabolites involved in the step-by-step modification of the first enzyme substrate to the final product are linked together through a series of biochemical reactions. Although this static approach generates wealth of biochemical information, it gives an erroneous view of the metabolism since most metabolites are in fact connected as they are involved in several pathways in parallel (i.e. intermediate and final products should initiate another anabolic or catabolic pathway). As described below, high-resolution MS-based metabolomics provide biochemists with an alternative way of visualizing metabolic networks.

High-throughput metabolic datasets acquired at high-resolution MS enable *ab initio* inference of biosynthetic relationships between metabolites masses directly from the mass spectrum (Breitling, Pitt, & Barrett, 2006; Jourdan, Breitling, Barrett, & Gilbert, 2008). Indeed, thanks to accurate mass measurements, the detection of mass differences corresponding to common biotransformations between two metabolites (such as oxidation or phosphorylation) may be searched from metabolomics datasets.

More recently, a high-resolution LC-MS analysis in combination with stable-isotope-assisted metabolomics approach has been developed for network-wide metabolic pathway elucidation (Creek et al., 2012). While current stable isotope tracing were limited to targeted approaches, Creek et al. proposed to perform an untargeted detection of isotopic patterns for stable isotope labelled metabolites in a ^{13}C-labelled glucose culture of *Trypanosoma brucei*. Thanks to comparative study of relative isotopomer intensities of each metabolite, unexpected metabolite labellings were observed. This led to the discovery of novel metabolic pathways, such as *de novo* pyrimidine synthesis from glucose-derived aspartate in the procyclic life-cycle stage of this parasite.

3. CONCLUSION AND PERSPECTIVES

Metabolomics is now a powerful tool in the exploration of biological systems with promising applications in various biological fields. Over the last decade, MS has gained a prominent position among metabolomics analytical methods due to its versatility and sensitivity. Nowadays, LC/MS profiles obtained from biological extracts are increasingly informative, but their interpretation remains limited by available knowledge. Indeed, among the thousands of detected features, only a small part of them is annotated, identified, and ultimately used to produce biological knowledge. Similarly, the inability of ensuring complete metabolome coverage in a single analysis technique and to share information from different analytical techniques hampers metabolite identification and makes access to metabolite networks difficult.

However, the versatility of newly released mass spectrometers will enable to perform sensitive metabolite detection, identification, and quantification in the same time in a near future, as it is already the case with proteomics (Gallien et al., 2012; Gillet et al., 2012; Hopfgartner, Tonoli, & Varesio, 2012). Improvements of metabolome coverage will be achieved by efficiently combining data arising from several LC/MS methods, or even from several different analytical platforms. Although it is already the case, it will be facilitated by efficient annotation and identification procedures, and also by the development of dedicated informatics tools enabling the fusion of metabolomic datasets (Vaughan et al., 2012). Furthermore, the development of CE/MS (Ramautar, Mayboroda, Somsen, & de Jong, 2011; Ramautar, Somsen, & de Jong, 2009, 2012) and also ion mobility-MS should help to address the issue of isomers (Shimizu, Ohe, & Chiba, 2012). Although most publications in the latter field address proteomics, some pioneering papers have already been published in metabolomics (Dwivedi, Schultz, & Hill, 2010) and lipidomics (Jackson et al., 2005). Otherwise, the development of imaging MS for metabolites should provide biochemists with tissular localization information, although it is still limited by the difficulty to detect a wide range of small molecules exhibiting different chemical structures at the same time by using MALDI ionization.

Another major analytical obstacle is the lack of reproducibility of the API–MS which greatly complicates analytical validation initiatives and also the sharing of data and their exploitation by biologists. The development of informatics tools for normalization (Dunn et al., 2011) and quantitative metabolomics should address this issue and enable the analysis of large epidemiological cohorts using LC/MS. Furthermore, improvement of current bioinformatic tools such as low-level data-processing software should ensure more efficient use of metabolomic data, and the development of mathematical tools to visualize metabolic networks will assist on the biological interpretation of data.

Finally, the collection and storage of metabolic profiles by the mean of relational databases implementation is now an indispensable source of information for researchers in biology. This is being achieved at the European and American levels by the development of initiatives about data standardization, publication and dissemination workflows, and also by the implementation of data repositories. In this context, the European Bioinformatics Institute has launched MetaboLights, a database for metabolomics experiments and the associated metadata and the NIH in the US is establishing 6–8 metabolomics services cores as well as a national metabolomics repository (Steinbeck et al., 2012).

ACKNOWLEDGEMENTS

The authors would like to thank GlaxoSmithKline and Association Nationale de Recherche et de Technologie to fund the thesis work of Samia BOUDAH and also the Commissariat à l'Energie Atomique-Saclay staff for its expertise and assistance.

REFERENCES

Addona, T. A., Shi, X., Keshishian, H., Mani, D. R., Burgess, M., Gillette, M. A., et al. (2011). A pipeline that integrates the discovery and verification of plasma protein biomarkers reveals candidate markers for cardiovascular disease. *Nature Biotechnology*, *29*(7), 635–643.

Aharoni, A., Ric de Vos, C. H., Verhoeven, H. A., Maliepaard, C. A., Kruppa, G., Bino, R., et al. (2002). Nontargeted metabolome analysis by use of fourier transform ion cyclotron mass spectrometry. *OMICS: A Journal of Integrative Biology*, *6*(3), 217–234.

Akiyama, K., Chikayama, E., Yuasa, H., Shimada, Y., Tohge, T., Shinozaki, K., et al. (2008). PRIMe: A Web site that assembles tools for metabolomics and transcriptomics. *In Silico Biology*, *8*(3–4), 339–345.

Allwood, J. W., Ellis, D. I., & Goodacre, R. (2008). Metabolomic technologies and their application to the study of plants and plant-host interactions. *Physiologia Plantarum*, *132*(2), 117–135.

Allwood, J. W., & Goodacre, R. (2010). An introduction to liquid chromatography-mass spectrometry instrumentation applied in plant metabolomic analyses. *Phytochemical Analysis*, *21*(1), 33–47.

Alonso, A., Julia, A., Beltran, A., Vinaixa, M., Diaz, M., Ibanez, L., et al. (2011). AStream: An R package for annotating LC/MS metabolomic data. *Bioinformatics*, *27*(9), 1339–1340.

Alymani, N. A., Smith, M. D., Williams, D. J., & Petty, R. D. (2010). Predictive biomarkers for personalised anti-cancer drug use: Discovery to clinical implementation. *European Journal of Cancer*, *46*(5), 869–879.

Anderson, N. L. (2010). The clinical plasma proteome: A survey of clinical assays for proteins in plasma and serum. *Clinical Chemistry*, *56*(2), 177–185.

Angerer, J., Ewers, U., & Wilhelm, M. (2007). Human biomonitoring: State of the art. *International Journal of Hygiene and Environmental Health*, *210*(3–4), 201–228.

Apffel, A., Fischer, S., Goldberg, G., Goodley, P. C., & Kuhlmann, F. E. (1995). Enhanced sensitivity for peptide mapping with electrospray liquid chromatography-mass spectrometry in the presence of signal suppression due to trifluoroacetic acid-containing mobile phases. *Journal of Chromatography. A*, *712*(1), 177–190.

Armstrong, J. A. (2007). Urinalysis in Western culture: A brief history. *Kidney International*, *71*(5), 384–387.

Astle, W., De Iorio, M., Richardson, S., Stephens, D., & Ebbels, T. (2012). A Bayesian model of NMR spectra for the deconvolution and quantification of metabolites in complex biological mixtures. *Journal of the American Statistical Association*, *107*, 1259–1271.

Bader, M., Lehnert, G., & Angerer, J. (1994). GC/MS determination of N-phenylvaline, a possible biomarker for benzene exposure in human hemoglobin by the "N-alkyl Edman method" *International Archives of Occupational and Environmental Health*, *65*(6), 411–414.

Bajad, S. U., Lu, W. Y., Kimball, E. H., Yuan, J., Peterson, C., & Rabinowitz, J. D. (2006). Separation and quantitation of water soluble cellular metabolites by hydrophilic interaction chromatography-tandem mass spectrometry. *Journal of Chromatography. A*, *1125*(1), 76–88.

Baker, M. (2005). In biomarkers we trust? *Nature Biotechnology*, *23*(3), 297–304.

Bando, K., Kawahara, R., Kunimatsu, T., Sakai, J., Kimura, J., Funabashi, H., et al. (2010). Influences of biofluid sample collection and handling procedures on GC-MS based metabolomic studies. *Journal of Bioscience and Bioengineering*, *110*(4), 491–499.

Benjamini, Y., & Hochberg, Y. (1995). Controlling the false discovery rate: A practical and powerful approach to multiple testing. *Journal of Royal Statistical Society B*, *57*, 289–300.

Bennett, D. A., & Waters, M. D. (2000). Applying biomarker research. *Environmental Health Perspectives*, *108*(9), 907–910.

Bennett, B. D., Yuan, J., Kimball, E. H., & Rabinowitz, J. D. (2008). Absolute quantitation of intracellular metabolite concentrations by an isotope ratio-based approach. *Nature Protocols*, *3*(8), 1299–1311.

Benton, H. P., Wong, D. M., Trauger, S. A., & Siuzdak, G. (2008). XCMS2: Processing tandem mass spectrometry data for metabolite identification and structural characterization. *Analytical Chemistry*, *80*(16), 6382–6389.

Berk, M., Ebbels, T., & Montana, G. (2011). A statistical framework for biomarker discovery in metabolomic time course data. *Bioinformatics*, *27*(14), 1979–1985.

Bernard, C. (1878). Leçons sur les phénomènes de la vie communs aux animaux et aux végétaux. In Albert Dastre (Ed.). Paris: Baillière J.-B.

Biais, B., Allwood, J. W., Deborde, C., Xu, Y., Maucourt, M., Beauvoit, B., et al. (2009). 1H NMR, GC-EI-TOFMS, and data set correlation for fruit metabolomics: Application to spatial metabolite analysis in melon. *Analytical Chemistry*, *81*(8), 2884–2894.

Biomarkers Definitions Working Group, (2001). Biomarkers and surrogate endpoints: Preferred definitions and conceptual framework. *Clinical Pharmacology and Therapeutics*, *69*(3), 89–95.

Blaise, B. J., Navratil, V., Domange, C., Shintu, L., Dumas, M. E., Elena-Herrmann, B., et al. (2010). Two-dimensional statistical recoupling for the identification of perturbed metabolic networks from NMR spectroscopy. *Journal of Proteome Research*, *9*(9), 4513–4520.

Blaise, B. J., Shintu, L., Elena, B., Emsley, L., Dumas, M. E., & Toulhoat, P. (2009). Statistical recoupling prior to significance testing in nuclear magnetic resonance based metabonomics. *Analytical Chemistry*, *81*(15), 6242–6251.

Bligh, E. G., & Dyer, W. J. (1959). A rapid method of total lipid extraction and purification. *Canadian Journal of Biochemistry and Physiology*, *37*(8), 911–917.

Bollard, M. E., Stanley, E. G., Lindon, J. C., Nicholson, J. K., & Holmes, E. (2005). NMR-based metabonomic approaches for evaluating physiological influences on biofluid composition. *NMR in Biomedicine*, *18*(3), 143–162.

Bolten, C. J., Kiefer, P., Letisse, F., Portais, J. C., & Wittmann, C. (2007). Sampling for metabolome analysis of microorganisms. *Analytical Chemistry*, *79*(10), 3843–3849.

Breitling, R., Pitt, A. R., & Barrett, M. P. (2006). Precision mapping of the metabolome. *Trends in Biotechnology*, *24*(12), 543–548.

Brison, D. R., Houghton, F. D., Falconer, D., Roberts, S. A., Hawkhead, J., Humpherson, P. G., et al. (2004). Identification of viable embryos in IVF by non-invasive measurement of amino acid turnover. *Human Reproduction*, *19*(10), 2319–2324.

Bruce, S. J., Tavazzi, I., Parisod, V., Rezzi, S., Kochhar, S., & Guy, P. A. (2009). Investigation of human blood plasma sample preparation for performing metabolomics using ultrahigh performance liquid chromatography/mass spectrometry. *Analytical Chemistry*, *81*(9), 3285–3296.

Bylesjo, M., Rantalainen, M., Nicholson, J. K., Holmes, E., & Trygg, J. (2008). K-OPLS package: Kernel-based orthogonal projections to latent structures for prediction and interpretation in feature space. *BMC Bioinformatics*, *9*, 106.

Canelas, A. B., ten, P. A., Ras, C., Seifar, R. M., van Dam, J. C., van Gulik, W. M., et al. (2009). Quantitative evaluation of intracellular metabolite extraction techniques for yeast metabolomics. *Analytical Chemistry*, *81*(17), 7379–7389.

Cannon, W. B. (1932). The wisdom of the body. NY: W.W. Norton.

Carr, S. A., & Anderson, L. (2008). Protein quantitation through targeted mass spectrometry: The way out of biomarker purgatory? *Clinical Chemistry*, *54*(11), 1749–1752.

Castrillo, J. I., Hayes, A., Mohammed, S., Gaskell, S. J., & Oliver, S. G. (2003). An optimized protocol for metabolome analysis in yeast using direct infusion electrospray mass spectrometry. *Phytochemistry*, *62*(6), 929–937.

Chace, D. H. (2001). Mass spectrometry in the clinical laboratory. *Chemical Reviews*, *101*(2), 445–477.

Chace, D. H., Lim, T., Hansen, C. R., Adam, B. W., & Hannon, W. H. (2009). Quantification of malonylcarnitine in dried blood spots by use of MS/MS varies by stable isotope internal standard composition. *Clinica Chimica Acta*, *402*(1–2), 14–18.

Chace, D. H., Lim, T., Hansen, C. R., De Jesus, V. R., De, J., & Hannon, W. H. (2009). Improved MS/MS analysis of succinylacetone extracted from dried blood spots when combined with amino acids and acylcarnitine butyl esters. *Clinica Chimica Acta*, *407*(1–2), 6–9.

Chace, D. H., Singleton, S., Diperna, J., Aiello, M., & Foley, T. (2009). Rapid metabolic and newborn screening of thyroxine (T4) from dried blood spots by MS/MS. *Clinica Chimica Acta*, *403*(1–2), 178–183.

Chen, J., Zhao, X., Fritsche, J., Yin, P., Schmitt-Kopplin, P., Wang, W., et al. (2008). Practical approach for the identification and isomer elucidation of biomarkers detected in a metabonomic study for the discovery of individuals at risk for diabetes by integrating the chromatographic and mass spectrometric information. *Analytical Chemistry*, *80*(4), 1280–1289.

Chirita, R. I., West, C., Zubrzycki, S., Finaru, A. L., & Elfakir, C. (2011). Investigations on the chromatographic behaviour of zwitterionic stationary phases used in hydrophilic interaction chromatography. *Journal of Chromatography. A*, *1218*(35), 5939–5963.

Collier, T. S., Hawkridge, A. M., Georgianna, D. R., Payne, G. A., & Muddiman, D. C. (2008). Top-down identification and quantification of stable isotope labeled proteins from Aspergillus flavus using online nano-flow reversed-phase liquid chromatography coupled to a LTQ-FTICR mass spectrometer. *Analytical Chemistry*, *80*(13), 4994–5001.

Cottingham, K. (2008). HUSERMET researchers look to the metabolome for answers. *Journal of Proteome Research*, 7(10), 4213.

Creek, D. J., Chokkathukalam, A., Jankevics, A., Burgess, K. E., Breitling, R., & Barrett, M. P. (2012). Stable isotope-assisted metabolomics for network-wide metabolic pathway elucidation. *Analytical Chemistry*, *84*(20), 8442–8447.

Dai, J., & Carr, P. W. (2005). Role of ion pairing in anionic additive effects on the separation of cationic drugs in reversed-phase liquid chromatography. *Journal of Chromatography. A*, *1072*(2), 169–184.

Dai, J., Mendonsa, S. D., Bowser, M. T., Lucy, C. A., & Carr, P. W. (2005). Effect of anionic additive type on ion pair formation constants of basic pharmaceuticals. *Journal of Chromatography. A*, *1069*(2), 225–234.

Dang, L., White, D. W., Gross, S., Bennett, B. D., Bittinger, M. A., Driggers, E. M., et al. (2009). Cancer-associated IDH1 mutations produce 2-hydroxyglutarate. *Nature*, *462*(7274), 739–744.

Darghouth, D., Koehl, B., Heilier, J. F., Madalinski, G., Bovee, P., Bosman, G., et al. (2011). Alterations of red blood cell metabolome in overhydrated hereditary stomatocytosis. *Haematologica*, *96*(12), 1861–1865.

Darghouth, D., Koehl, B., Madalinski, G., Heilier, J. F., Bovee, P., Xu, Y., et al. (2011). Pathophysiology of sickle cell disease is mirrored by the red blood cell metabolome. *Blood*, *117*(6), e57–e66.

Daykin, C. A., Foxall, P. J., Connor, S. C., Lindon, J. C., & Nicholson, J. K. (2002). The comparison of plasma deproteinization methods for the detection of low-molecular-weight metabolites by (1)H nuclear magnetic resonance spectroscopy. *Analytical Biochemistry*, *304*(2), 220–230.

Denery, J. R., Nunes, A. A., & Dickerson, T. J. (2011). Characterization of differences between blood sample matrices in untargeted metabolomics. *Analytical Chemistry*, *83*(3), 1040–1047.

Dettmer, K., Almstetter, M. F., Appel, I. J., Nurnberger, N., Schlamberger, G., Gronwald, W., et al. (2010). Comparison of serum versus plasma collection in gas chromatography–mass spectrometry-based metabolomics. *Electrophoresis*, *31*(14), 2365–2373.

Deventer, K., Pozo, O. J., Van, E. P., & Delbeke, F. T. (2009). Qualitative detection of diuretics and acidic metabolites of other doping agents in human urine by high-performance liquid chromatography-tandem mass spectrometry: Comparison between liquid-liquid extraction and direct injection. *Journal of Chromatography. A*, *1216*(31), 5819–5827.

Domange, C., Canlet, C., Traore, A., Bielicki, G., Keller, C., Paris, A., et al. (2008). Orthologous metabonomic qualification of a rodent model combined with magnetic resonance imaging for an integrated evaluation of the toxicity of Hypochoeris radicata. *Chemical Research in Toxicology*, *21*(11), 2082–2096.

Duarte, N. C., Becker, S. A., Jamshidi, N., Thiele, I., Mo, M. L., Vo, T. D., et al. (2007). Global reconstruction of the human metabolic network based on genomic and bibliomic data. *Proceedings of the National Academy of Sciences of the United States of America*, *104*(6), 1777–1782.

Dumas, M. E. (2012). Metabolome 2.0: Quantitative genetics and network biology of metabolic phenotypes. *Molecular BioSystems*, *8*(10), 2494–2502.

Dumas, M. E., Canlet, C., Andre, F., Vercauteren, J., & Paris, A. (2002). Metabonomic assessment of physiological disruptions using 1H-13C HMBC-NMR spectroscopy combined with pattern recognition procedures performed on filtered variables. *Analytical Chemistry*, *74*(10), 2261–2273.

Dumas, M. E., Wilder, S. P., Bihoreau, M. T., Barton, R. H., Fearnside, J. F., Argoud, K., et al. (2007). Direct quantitative trait locus mapping of mammalian metabolic phenotypes in diabetic and normoglycemic rat models. *Nature Genetics*, *39*(5), 666–672.

Dunn, W. B., Bailey, N. J., & Johnson, H. E. (2005). Measuring the metabolome: Current analytical technologies. *The Analyst*, *130*(5), 606–625.

Dunn, W. B., Broadhurst, D., Begley, P., Zelena, E., Francis-McIntyre, S., Anderson, N., et al. (2011). Procedures for large-scale metabolic profiling of serum and plasma using gas chromatography and liquid chromatography coupled to mass spectrometry. *Nature Protocols*, *6*(7), 1060–1083.

Dwivedi, P., Schultz, A. J., & Hill, H. H. (2010). Metabolic profiling of human blood by high resolution ion mobility mass spectrometry (IM-MS). *International Journal of Mass Spectrometry*, *298*(1–3), 78–90.

Eikel, D., Vavrek, M., Smith, S., Bason, C., Yeh, S., Korfmacher, W. A., et al. (2011). Liquid extraction surface analysis mass spectrometry (LESA-MS) as a novel profiling tool for drug distribution and metabolism analysis: The terfenadine example. *Rapid Communications in Mass Spectrometry*, *25*(23), 3587–3596.

Euerby, M. R., & Petersson, P. (2003). Chromatographic classification and comparison of commercially available reversed-phase liquid chromatographic columns using principal component analysis. *Journal of Chromatography. A*, *994*(1-2), 13–36.

Fahy, E., Sud, M., Cotter, D., & Subramaniam, S. (2007). LIPID MAPS online tools for lipid research. *Nucleic Acids Research*, *35*(Web Server issue), W606–W612.

Fenn, J. B., Mann, M., Meng, C. K., Wong, S. F., & Whitehouse, C. M. (1989). Electrospray ionization for mass spectrometry of large biomolecules. *Science*, *246*(4926), 64–71.

Fiehn, O. (2001). Combining genomics, metabolome analysis, and biochemical modelling to understand metabolic networks. *Comparative and Functional Genomics*, *2*(3), 155–168.

Folch, J., Ascoli, I., Meath, J. A., & LeBaron, N. (1951). Preparation of lipide extracts from brain tissue. *The Journal of Biological Chemistry*, *191*(2), 833–841.

Fonville, J. M., Bylesjo, M., Coen, M., Nicholson, J. K., Holmes, E., Lindon, J. C., et al. (2011). Non-linear modeling of 1H NMR metabonomic data using kernel-based orthogonal projections to latent structures optimized by simulated annealing. *Analytica Chimica Acta*, *705*(1–2), 72–80.

Fonville, J. M., Maher, A. D., Coen, M., Holmes, E., Lindon, J. C., & Nicholson, J. K. (2010). Evaluation of full-resolution J-resolved 1H NMR projections of biofluids for metabonomics information retrieval and biomarker identification. *Analytical Chemistry*, *82*(5), 1811–1821.

Fonville, J. M., Richards, S. E., Barton, R. H., Boulange, C., Ebbels, T. M. D., Nicholson, J. K., et al. (2010). The evolution of partial least squares models and related chemometric approaches in metabonomics and metabolic phenotyping. *Journal of Chemometrics*, *24*, 636–649.

Fukushima, A., Kusano, M., Redestig, H., Arita, M., & Saito, K. (2009). Integrated omics approaches in plant systems biology. *Current Opinion in Chemical Biology*, *13*(5–6), 532–538.

Gallien, S., Duriez, E., Crone, C., Kellmann, M., Moehring, T., & Domon, B. (2012). Targeted proteomic quantification on quadrupole-orbitrap mass spectrometer. *Molecular & Cellular Proteomics*, *11*, 1709–1723.

Garcia-Villalba, R., Carrasco-Pancorbo, A., Oliveras-Ferraros, C., Vazquez-Martin, A., Menendez, J. A., Segura-Carretero, A., et al. (2010). Characterization and quantification of phenolic compounds of extra-virgin olive oils with anticancer properties by a rapid and resolutive LC-ESI-TOF MS method. *Journal of Pharmaceutical and Biomedical Analysis*, *51*(2), 416–429.

Giavalisco, P., Kohl, K., Hummel, J., Seiwert, B., & Willmitzer, L. (2009). (13)C isotope-labeled metabolomes allowing for improved compound annotation and relative quantification in liquid chromatography-mass spectrometry-based metabolomic research. *Analytical Chemistry*, *81*, 6546–6551.

Gika, H. G., Theodoridis, G. A., Vrhovsek, U., & Mattivi, F. (2012). Quantitative profiling of polar primary metabolites using hydrophilic interaction ultrahigh performance liquid chromatography-tandem mass spectrometry. *Journal of Chromatography. A*, *1259*, 121–127.

Gika, H. G., Theodoridis, G. A., & Wilson, I. D. (2008). Liquid chromatography and ultra-performance liquid chromatography-mass spectrometry fingerprinting of human urine: Sample stability under different handling and storage conditions for metabonomics studies. *Journal of Chromatography. A*, *1189*(1–2), 314–322.

Gika, H. G., Theodoridis, G. A., Wingate, J. E., & Wilson, I. D. (2007). Within-day reproducibility of an HPLC-MS-based method for metabonomic analysis: Application to human urine. *Journal of Proteome Research*, *6*(8), 3291–3303.

Gilbert-Lopez, B., Garcia-Reyes, J. F., Fernandez-Alba, A. R., & Molina-Diaz, A. (2010). Evaluation of two sample treatment methodologies for large-scale pesticide residue analysis in olive oil by fast liquid chromatography-electrospray mass spectrometry. *Journal of Chromatography. A*, *1217*(24), 3736–3747.

Gillet, L. C., Navarro, P., Tate, S., Rost, H., Selevsek, N., Reiter, L., et al. (2012). Targeted data extraction of the MS/MS spectra generated by data-independent acquisition: A new concept for consistent and accurate proteome analysis. *Molecular & Cellular Proteomics*, *11*(6), O111.

Godat, E., Madalinski, G., Muller, L., Heilier, J. F., Labarre, J., & Junot, C. (2010). Mass spectrometry-based methods for the determination of sulfur and related metabolite concentrations in cell extracts. *Methods in Enzymology*, *473*, 41–76.

Guo, K., Ji, C., & Li, L. (2007). Stable-isotope dimethylation labeling combined with LC-ESI MS for quantification of amine-containing metabolites in biological samples. *Analytical Chemistry*, *79*(22), 8631–8638.

Guo, K., & Li, L. (2009). Differential (12)C-/(13)C-isotope dansylation labeling and fast liquid chromatography/mass spectrometry for absolute and relative quantification of the metabolome. *Analytical Chemistry*, *81*, 3919–3932.

Guo, K., & Li, L. (2010). High-performance isotope labeling for profiling carboxylic acid-containing metabolites in biofluids by mass spectrometry. *Analytical Chemistry*, *82*, 8789–8793.

Gustavsson, S. A., Samskog, J., Markides, K. E., & Langstrom, B. (2001). Studies of signal suppression in liquid chromatography-electrospray ionization mass spectrometry using volatile ion-pairing reagents. *Journal of Chromatography. A*, *937*(1–2), 41–47.

Haber, M. H. (1988). Pisse prophecy: A brief history of urinalysis. *Clinics in Laboratory Medicine*, *8*(3), 415–430.

Han, X., Yang, K., & Gross, R. W. (2012). Multi-dimensional mass spectrometry-based shotgun lipidomics and novel strategies for lipidomic analyses. *Mass Spectrometry Reviews*, *31*(1), 134–178.

Hao, J., Astle, W., De, I. M., & Ebbels, T. M. (2012). BATMAN—An R package for the automated quantification of metabolites from nuclear magnetic resonance spectra using a Bayesian model. *Bioinformatics*, *28*(15), 2088–2090.

Haufroid, V., Jakubowski, M., Janasik, B., Ligocka, D., Buchet, J. P., Bergamaschi, E., et al. (2002). Interest of genotyping and phenotyping of drug-metabolizing enzymes for the interpretation of biological monitoring of exposure to styrene. *Pharmacogenetics*, *12*(9), 691–702.

Hopfgartner, G., Tonoli, D., & Varesio, E. (2012). High-resolution mass spectrometry for integrated qualitative and quantitative analysis of pharmaceuticals in biological matrices. *Analytical and Bioanalytical Chemistry*, *402*(8), 2587–2596.

Horai, H., Arita, M., Kanaya, S., Nihei, Y., Ikeda, T., Suwa, K., et al. (2010). MassBank: a public repository for sharing mass spectral data for life sciences. *Journal of Mass Spectrometry*, *45*(7), 703–714.

Horning, E. C., & Horning, M. G. (1971). Metabolic profiles: Gas-phase methods for analysis of metabolites. *Clinical Chemistry*, *17*(8), 802–809.

Horning, E. C., Horning, M. G., Carroll, D. I., Dzidic, I., & Stillwell, R. N. (1973). Chemical ionization mass spectrometry. *Advances in Biochemical Psychopharmacology*, 7, 15–31.

Ibanez, C., Simo, C., Martin-Alvarez, P. J., Kivipelto, M., Winblad, B., Cedazo-Minguez, A., et al. (2012). Toward a predictive model of Alzheimer's disease progression using capillary electrophoresis-mass spectrometry metabolomics. *Analytical Chemistry*, *84*(20), 8532–8540.

Idborg, H., Zamani, L., Edlund, P. O., Schuppe-Koistinen, I., & Jacobsson, S. P. (2005). Metabolic fingerprinting of rat urine by LC/MS Part 1. Analysis by hydrophilic interaction liquid chromatography-electrospray ionization mass spectrometry. *Journal of Chromatography B*, *828*(1–2), 9–13.

Idborg-Bjorkman, H., Edlund, P. O., Kvalheim, O. M., Schuppe-Koistinen, I., & Jacobsson, S. P. (2003). Screening of biomarkers in rat urine using LC/electrospray ionization-MS and two-way data analysis. *Analytical Chemistry*, *75*(18), 4784–4792.

Ipsen, A., Want, E. J., Lindon, J. C., & Ebbels, T. M. (2010). A statistically rigorous test for the identification of parent-fragment pairs in LC-MS datasets. *Analytical Chemistry*, *82*(5), 1766–1778.

Ito, T., Odake, T., Katoh, H., Yamaguchi, Y., & Aoki, M. (2011). High-throughput profiling of microbial extracts. *Journal of Natural Products*, *74*(5), 983–988.

Jackson, S. N., Wang, H. Y., Woods, A. S., Ugarov, M., Egan, T., & Schultz, J. A. (2005). Direct tissue analysis of phospholipids in rat brain using MALDI-TOFMS and MALDI-ion mobility-TOFMS. *Journal of the American Society for Mass Spectrometry*, *16*(2), 133–138.

Jaffe, J. D., Keshishian, H., Chang, B., Addona, T. A., Gillette, M. A., & Carr, S. A. (2008). Accurate inclusion mass screening: A bridge from unbiased discovery to targeted assay development for biomarker verification. *Molecular & Cellular Proteomics*, 7(10), 1952–1962.

Jiye, A., Trygg, J., Gullberg, J., Johansson, A. I., Jonsson, P., Antti, H., et al. (2005). Extraction and GC/MS analysis of the human blood plasma metabolome. *Analytical Chemistry*, 77(24), 8086–8094.

Jourdan, F., Breitling, R., Barrett, M. P., & Gilbert, D. (2008). MetaNetter: Inference and visualization of high-resolution metabolomic networks. *Bioinformatics*, *24*(1), 143–145.

Kaddurah-Daouk, R., McEvoy, J., Baillie, R., Zhu, H., Yao, K., Nimgaonkar, V. L., et al. (2012). Impaired plasmalogens in patients with schizophrenia. *Psychiatry Research*, *198*, 347–352.

Kanehisa, M., & Goto, S. (2000). KEGG: Kyoto encyclopedia of genes and genomes. *Nucleic Acids Research*, *28*(1), 27–30.

Karp, P. D., Riley, M., Saier, M., Paulsen, I. T., Paley, S. M., & Pellegrini-Toole, A. (2000). The EcoCyc and MetaCyc databases. *Nucleic Acids Research*, *28*(1), 56–59.

Katajamaa, M., Miettinen, J., & Oresic, M. (2006). MZmine: Toolbox for processing and visualization of mass spectrometry based molecular profile data. *Bioinformatics*, *22*(5), 634–636.

Katajamaa, M., & Oresic, M. (2007). Data processing for mass spectrometry-based metabolomics. *Journal of Chromatography. A*, *1158*(1–2), 318–328.

Kehoe, R. A., Thamann, F., & Cholak, J. (1933). Lead absorption and excretion in relation to the diagnosis of lead poisoning. *The Journal of Industrial Hygiene and Toxicology*, *15*(5), 320–340.

Kiefer, P., Portais, J. C., & Vorholt, J. A. (2008). Quantitative metabolome analysis using liquid chromatography-high-resolution mass spectrometry. *Analytical Biochemistry*, *382*(2), 94–100.

Kim, J. K., Harada, K., Bamba, T., Fukusaki, E., & Kobayashi, A. (2005). Stable isotope dilution-based accurate comparative quantification of nitrogen-containing metabolites in Arabidopsis thaliana T87 cells using in vivo (15)N-isotope enrichment. *Bioscience, Biotechnology, and Biochemistry*, *69*(7), 1331–1340.

Kirkland, J. J. (2004). Development of some stationary phases for reversed-phase high-performance liquid chromatography. *Journal of Chromatography. A*, *1060*(1–2), 9–21.

Kitteringham, N. R., Jenkins, R. E., Lane, C. S., Elliott, V. L., & Park, B. K. (2009). Multiple reaction monitoring for quantitative biomarker analysis in proteomics and metabolomics. *Journal of Chromatography. B, Analytical Technologies in the Biomedical and Life Sciences*, *877*(13), 1229–1239.

Kopka, J., Schauer, N., Krueger, S., Birkemeyer, C., Usadel, B., Bergmuller, E., et al. (2005). GMD@CSB.DB: The Golm Metabolome Database. *Bioinformatics*, *21*(8), 1635–1638.

Koulman, A., Lane, G. A., Harrison, S. J., & Volmer, D. A. (2009). From differentiating metabolites to biomarkers. *Analytical and Bioanalytical Chemistry*, *394*(3), 663–670.

Kuhl, C., Tautenhahn, R., Bottcher, C., Larson, T. R., & Neumann, S. (2012). CAMERA: An integrated strategy for compound spectra extraction and annotation of liquid chromatography/mass spectrometry data sets. *Analytical Chemistry*, *84*(1), 283–289.

Lafaye, A., Junot, C., Ramounet-Le, G. B., Fritsch, P., Ezan, E., & Tabet, J. C. (2004). Profiling of sulfoconjugates in urine by using precursor ion and neutral loss scans in tandem mass spectrometry. Application to the investigation of heavy metal toxicity in rats. *Journal of Mass Spectrometry*, *39*(6), 655–664.

Lafaye, A., Labarre, J., Tabet, J. C., Ezan, E., & Junot, C. (2005). Liquid chromatography-mass spectrometry and 15N metabolic labeling for quantitative metabolic profiling. *Analytical Chemistry*, *77*(7), 2026–2033.

Lange, E., Tautenhahn, R., Neumann, S., & Gropl, C. (2008). Critical assessment of alignment procedures for LC-MS proteomics and metabolomics measurements. *BMC Bioinformatics*, *9*, 375.

Lawton, K. A., Berger, A., Mitchell, M., Milgram, K. E., Evans, A. M., Guo, L., et al. (2008). Analysis of the adult human plasma metabolome. *Pharmacogenomics*, *9*(4), 383–397.

Leandro, C. C., Hancock, P., Fussell, R. J., & Keely, B. J. (2006). Comparison of ultra-performance liquid chromatography and high-performance liquid chromatography for the determination of priority pesticides in baby foods by tandem quadrupole mass spectrometry. *Journal of Chromatography. A*, *1103*(1), 94–101.

Le Cao, K. A., Boitard, S., & Besse, P. (2011). Sparse PLS discriminant analysis: Biologically relevant feature selection and graphical displays for multiclass problems. *BMC Bioinformatics*, *12*, 253.

Le Cao, K. A., Martin, P. G., Robert-Granie, C., & Besse, P. (2009). Sparse canonical methods for biological data integration: Application to a cross-platform study. *BMC Bioinformatics*, *10*, 34.

Lewis, G. D., Wei, R., Liu, E., Yang, E., Shi, X., Martinovic, M., et al. (2008). Metabolite profiling of blood from individuals undergoing planned myocardial infarction reveals early markers of myocardial injury. *The Journal of Clinical Investigation*, *118*(10), 3503–3512.

Li, Q., Zhang, Q., Xu, G., Yin, P., Li, N., & Li, J. (2008). Metabonomics study of intestinal transplantation using ultrahigh-performance liquid chromatography time-of-flight mass spectrometry. *Digestion*, 77(2), 122–130.

Lievre, A., Bachet, J. B., Le, C. D., Boige, V., Landi, B., Emile, J. F., et al. (2006). KRAS mutation status is predictive of response to cetuximab therapy in colorectal cancer. *Cancer Research*, *66*(8), 3992–3995.

Lommen, A. (2009). MetAlign: Interface-driven, versatile metabolomics tool for hyphenated full-scan mass spectrometry data preprocessing. *Analytical Chemistry*, *81*(8), 3079–3086.

Lu, W., Bennett, B. D., & Rabinowitz, J. D. (2008). Analytical strategies for LC-MS-based targeted metabolomics. *Journal of Chromatography. B, Analytical Technologies in the Biomedical and Life Sciences*, *871*(2), 236–242.

Maher, A. D., Fonville, J. M., Coen, M., Lindon, J. C., Rae, C. D., & Nicholson, J. K. (2012). Statistical total correlation spectroscopy scaling for enhancement of metabolic information recovery in biological NMR spectra. *Analytical Chemistry*, *84*(2), 1083–1091.

Marchi, I., Rudaz, S., & Veuthey, J. L. (2009). Atmospheric pressure photoionization for coupling liquid-chromatography to mass spectrometry: A review. *Talanta*, *78*(1), 1–18.

Marhuenda-Egea, F. C., Martinez-Sabater, E., Gonsalvez-Alvarez, R., Lledo, B., Ten, J., & Bernabeu, R. (2009). A crucial step in assisted reproduction technology: Human embryo selection using metabolomic evaluation. *Fertility and Sterility*, *94*, 772–774.

Mashego, M. R., Wu, L., van Dam, J. C., Ras, C., Vinke, J. L., van Winden, W. A., et al. (2004). MIRACLE: Mass isotopomer ratio analysis of U-13C-labeled extracts. A new method for accurate quantification of changes in concentrations of intracellular metabolites. *Biotechnology and Bioengineering*, *85*(6), 620–628.

Matsuda, F., Yonekura-Sakakibara, K., Niida, R., Kuromori, T., Shinozaki, K., & Saito, K. (2009). MS/MS spectral tag-based annotation of non-targeted profile of plant secondary metabolites. *The Plant Journal*, *57*(3), 555–577.

McClellan, J. E., Quarmby, S. T., & Yost, R. A. (2002). Parent and neutral loss monitoring on a quadrupole ion trap mass spectrometer: Screening of acylcarnitines in complex mixtures. *Analytical Chemistry*, *74*(22), 5799–5806.

Merlo, M. E., Jankevics, A., Takano, E., & Breitling, R. (2011). Exploring the metabolic state of microorganisms using metabolomics. *Bioanalysis*, *3*(21), 2443–2458.

Metz, T. O., Zhang, Q., Page, J. S., Shen, Y., Callister, S. J., Jacobs, J. M., et al. (2007). The future of liquid chromatography-mass spectrometry (LC-MS) in metabolic profiling and metabolomic studies for biomarker discovery. *Biomarkers in Medicine*, *1*(1), 159–185.

Meyer, H., Liebeke, M., & Lalk, M. (2010). A protocol for the investigation of the intracellular Staphylococcus aureus metabolome. *Analytical Biochemistry*, *401*(2), 250–259.

Michopoulos, F., Edge, A. M., Theodoridis, G., & Wilson, I. D. (2010). Application of turbulent flow chromatography to the metabonomic analysis of human plasma: Comparison with protein precipitation. *Journal of Separation Science*, *33*(10), 1472–1479.

Michopoulos, F., Theodoridis, G., Smith, C. J., & Wilson, I. D. (2010). Metabolite profiles from dried biofluid spots for metabonomic studies using UPLC combined with oaToF-MS. *Journal of Proteome Research*, *9*(6), 3328–3334.

Michopoulos, F., Theodoridis, G., Smith, C. J., & Wilson, I. D. (2011). Metabolite profiles from dried blood spots for metabonomic studies using UPLC combined with orthogonal acceleration ToF-MS: Effects of different papers and sample storage stability. *Bioanalysis*, *3*(24), 2757–2767.

Mikesh, L. M., & Bruns, D. E. (2008). Stabilization of glucose in blood specimens: Mechanism of delay in fluoride inhibition of glycolysis. *Clinical Chemistry*, *54*(5), 930–932.

Mochel, F., Sedel, F., Vanderver, A., Engelke, U. F., Barritault, J., Yang, B. Z., et al. (2009). Cerebellar ataxia with elevated cerebrospinal free sialic acid (CAFSA). *Brain*, *132*(Pt 3), 801–809.

Mueller, D. M., Duretz, B., Espourteille, F. A., & Rentsch, K. M. (2011). Development of a fully automated toxicological LC-MS(n) screening system in urine using online extraction with turbulent flow chromatography. *Analytical and Bioanalytical Chemistry*, *400*(1), 89–100.

Narainsamy, K., Cassier-Chauvat, C., Junot, C., & Chauvat, F. (2011). High performance analysis of the cyanobacterial metabolism via liquid chromatography coupled to a LTQ-Orbitrap mass spectrometer: Evidence that glucose reprograms the whole carbon metabolism and triggers oxidative stress. *Metabolomics*, *9*, 21–32.

Nicholson, J. K., Everett, J. R., & Lindon, J. C. (2012). Longitudinal pharmacometabonomics for predicting patient responses to therapy: Drug metabolism, toxicity and efficacy. *Expert Opinion on Drug Metabolism & Toxicology*, *8*(2), 135–139.

Nicholson, J. K., Lindon, J. C., & Holmes, E. (1999). 'Metabonomics': Understanding the metabolic responses of living systems to pathophysiological stimuli via multivariate statistical analysis of biological NMR spectroscopic data. *Xenobiotica*, *29*(11), 1181–1189.

Nicholson, G., Rantalainen, M., Li, J. V., Maher, A. D., Malmodin, D., Ahmadi, K. R., et al. (2011). A genome-wide metabolic QTL analysis in Europeans implicates two loci shaped by recent positive selection. *PLoS Genetics*, 7(9), e1002270.

Nordstrom, A., O'Maille, G., Qin, C., & Siuzdak, G. (2006). Nonlinear data alignment for UPLC-MS and HPLC-MS based metabolomics: Quantitative analysis of endogenous and exogenous metabolites in human serum. *Analytical Chemistry*, *78*(10), 3289–3295.

Oberacher, H., Pavlic, M., Libiseller, K., Schubert, B., Sulyok, M., Schuhmacher, R., et al. (2009a). On the inter-instrument and the inter-laboratory transferability of a tandem mass spectral reference library: 2. Optimization and characterization of the search algorithm. *Journal of Mass Spectrometry*, *44*(4), 494–502.

Oberacher, H., Pavlic, M., Libiseller, K., Schubert, B., Sulyok, M., Schuhmacher, R., et al. (2009b). On the inter-instrument and inter-laboratory transferability of a tandem mass spectral reference library: 1. Results of an Austrian multicenter study. *Journal of Mass Spectrometry*, *44*(4), 485–493.

Oliver, S. G., Winson, M. K., Kell, D. B., & Baganz, F. (1998). Systematic functional analysis of the yeast genome. *Trends in Biotechnology*, *16*(9), 373–378.

Oresic, M., Simell, S., Sysi-Aho, M., Nanto-Salonen, K., Seppanen-Laakso, T., Parikka, V., et al. (2008). Dysregulation of lipid and amino acid metabolism precedes islet autoimmunity in children who later progress to type 1 diabetes. *The Journal of Experimental Medicine*, *205*(13), 2975–2984.

Oreskovic, D., & Klarica, M. (2010). The formation of cerebrospinal fluid: Nearly a hundred years of interpretations and misinterpretations. *Brain Research Reviews*, *64*(2), 241–262.

Palit, M., & Mallard, G. (2009). Fragmentation energy index for universalization of fragmentation energy in ion trap mass spectrometers for the analysis of chemical weapon convention related chemicals by atmospheric pressure ionization-tandem mass spectrometry analysis. *Analytical Chemistry*, *81*(7), 2477–2485.

Parvy, P., Bardet, J., Rabier, D., Bonnefont, J. P., & Kamoun, P. (1989). A new pitfall in plasma amino acid analysis. *Clinical Chemistry*, *35*(1), 178.

Pasikanti, K. K., Ho, P. C., & Chan, E. C. (2008). Gas chromatography/mass spectrometry in metabolic profiling of biological fluids. *Journal of Chromatography. B, Analytical Technologies in the Biomedical and Life Sciences*, *871*(2), 202–211.

Paustenbach, D. J. (2001). The practice of exposure assessment: A state-of-the-art review. *Journal of Toxicology and Environmental Health, Part B. Critical Reviews*, *3*, 179–291.

Pearce, S. J., Schrenk, H. H., & Yant, W. P. (1936). *Microcolorimetric determination of benzene in blood and urine*. United States Bureau of Mines 3302. Chem Abstr 30:5600.

Pepe, M. S., Etzioni, R., Feng, Z., Potter, J. D., Thompson, M. L., Thornquist, M., et al. (2001). Phases of biomarker development for early detection of cancer. *Journal of the National Cancer Institute, 93*(14), 1054–1061.

Pesek, J. J., Matyska, M. T., & Fischer, S. M. (2011). Improvement of peak shape in aqueous normal phase analysis of anionic metabolites. *Journal of Separation Science, 34*(24), 3509–3516.

Pesek, J. J., Matyska, M. T., Fischer, S. M., & Sana, T. R. (2008). Analysis of hydrophilic metabolites by high-performance liquid chromatography-mass spectrometry using a silica hydride-based stationary phase. *Journal of Chromatography. A, 1204*(1), 48–55.

Peters, S., van, V. E., & Janssen, H. G. (2009). Parameter selection for peak alignment in chromatographic sample profiling: Objective quality indicators and use of control samples. *Analytical and Bioanalytical Chemistry, 394*(5), 1273–1281.

Plumb, R. S., Stumpf, C. L., Granger, J. H., Castro-Perez, J., Haselden, J. N., & Dear, G. J. (2003). Use of liquid chromatography/time-of-flight mass spectrometry and multivariate statistical analysis shows promise for the detection of drug metabolites in biological fluids. *Rapid Communications in Mass Spectrometry, 17*(23), 2632–2638.

Pluskal, T., Castillo, S., Villar-Briones, A., & Oresic, M. (2010). MZmine 2: Modular framework for processing, visualizing, and analyzing mass spectrometry-based molecular profile data. *BMC Bioinformatics, 11*, 395.

Psychogios, N., Hau, D. D., Peng, J., Guo, A. C., Mandal, R., Bouatra, S., et al. (2011). The human serum metabolome. *PloS One, 6*, e16957.

Quinones, M. P., & Kaddurah-Daouk, R. (2009). Metabolomics tools for identifying biomarkers for neuropsychiatric diseases. *Neurobiology of Disease, 35*(2), 165–176.

Ramautar, R., Mayboroda, O. A., Somsen, G. W., & de Jong, G. J. (2011). CE-MS for metabolomics: Developments and applications in the period 2008–2010. *Electrophoresis, 32*(1), 52–65.

Ramautar, R., Somsen, G. W., & de Jong, G. J. (2009). CE-MS in metabolomics. *Electrophoresis, 30*(1), 276–291.

Ramautar, R., Somsen, G. W., & de Jong, G. J. (2012). CE-MS for metabolomics: Developments and applications in the period 2010–2012. *Electrophoresis, 34*, 86–98.

Richards, S. E., Dumas, M. E., Fonville, J. M., Ebbels, T. M. D., Holmes, E., & Nicholson, J. K. (2011). Intra- and inter-omic fusion of metabolic profiling data in a systems biology framework. *Chemometrics and Intelligent Laboratory Systems, 104*, S121–S131.

Rifai, N., Gillette, M. A., & Carr, S. A. (2006). Protein biomarker discovery and validation: The long and uncertain path to clinical utility. *Nature Biotechnology, 24*(8), 971–983.

Ritchie, S. A., Ahiahonu, P. W., Jayasinghe, D., Heath, D., Liu, J., Lu, Y., et al. (2010). Reduced levels of hydroxylated, polyunsaturated ultra long-chain fatty acids in the serum of colorectal cancer patients: Implications for early screening and detection. *BMC Medicine, 8*, 13.

Robb, D. B., Covey, T. R., & Bruins, A. P. (2000). Atmospheric pressure photoionization: An ionization method for liquid chromatography-mass spectrometry. *Analytical Chemistry, 72*(15), 3653–3659.

Roberts, J. K. (2000). NMR adventures in the metabolic labyrinth within plants. *Trends in Plant Science, 5*(1), 30–34.

Rohart, F., Paris, A., Laurent, B., Canlet, C., Molina, J., Mercat, M. J., et al. (2012). Phenotypic prediction based on metabolomic data on the growing pig from three main European breeds. *Journal of Animal Science, 90*, 4729–4740.

Roux, A., Lison, D., Junot, C., & Heilier, J. F. (2011). Applications of liquid chromatography coupled to mass spectrometry-based metabolomics in clinical chemistry and toxicology: A review. *Clinical Biochemistry, 44*(1), 119–135.

Roux, A., Xu, Y., Heilier, J. F., Olivier, M. F., Ezan, E., Tabet, J. C., et al. (2012). Annotation of the human adult urinary metabolome and metabolite identification using ultra high performance liquid chromatography coupled to a linear quadrupole ion trap-orbitrap mass spectrometer. *Analytical Chemistry, 84*(15), 6429–6437.

Sabatine, M. S., Liu, E., Morrow, D. A., Heller, E., McCarroll, R., Wiegand, R., et al. (2005). Metabolomic identification of novel biomarkers of myocardial ischemia. *Circulation, 112*(25), 3868–3875.

Savitski, M. M., Sweetman, G., Askenazi, M., Marto, J. A., Lang, M., Zinn, N., et al. (2011). Delayed fragmentation and optimized isolation width settings for improvement of protein identification and accuracy of isobaric mass tag quantification on Orbitrap-type mass spectrometers. *Analytical Chemistry, 83*(23), 8959–8967.

Shah, V. P., Midha, K. K., Findlay, J. W., Hill, H. M., Hulse, J. D., McGilveray, I. J., et al. (2000). Bioanalytical method validation—A revisit with a decade of progress. *Pharmaceutical Research, 17*(12), 1551–1557.

Shibue, M., Mant, C. T., & Hodges, R. S. (2005a). The perchlorate anion is more effective than the trifluoroacetate anion as an ion-pairing reagent for reversed-phase chromatography of peptides. *Journal of Chromatography. A, 1080*(1), 49–57.

Shibue, M., Mant, C. T., & Hodges, R. S. (2005b). Effect of anionic ion-pairing reagent concentration (1–60 mM) on reversed-phase liquid chromatography elution behaviour of peptides. *Journal of Chromatography. A, 1080*(1), 58–67.

Shibue, M., Mant, C. T., & Hodges, R. S. (2005c). Effect of anionic ion-pairing reagent hydrophobicity on selectivity of peptide separations by reversed-phase liquid chromatography. *Journal of Chromatography. A, 1080*(1), 68–75.

Shimizu, A., Ohe, T., & Chiba, M. (2012). A novel method for the determination of the site of glucuronidation by ion mobility spectrometry-mass spectrometry. *Drug Metabolism and Disposition, 40*(8), 1456–1459.

Shortreed, M. R., Lamos, S. M., Frey, B. L., Phillips, M. F., Patel, M., Belshaw, P. J., et al. (2006). Ionizable isotopic labeling reagent for relative quantification of amine metabolites by mass spectrometry. *Analytical Chemistry, 78*(18), 6398–6403.

Simonsen, K. W., Hermansson, S., Steentoft, A., & Linnet, K. (2010). A validated method for simultaneous screening and quantification of twenty-three benzodiazepines and metabolites plus zopiclone and zaleplone in whole blood by liquid-liquid extraction and ultra-performance liquid chromatography-tandem mass spectrometry. *Journal of Analytical Toxicology, 34*(6), 332–341.

Simonsen, K. W., Steentoft, A., Buck, M., Hansen, L., & Linnet, K. (2010). Screening and quantitative determination of twelve acidic and neutral pharmaceuticals in whole blood by liquid-liquid extraction and liquid chromatography-tandem mass spectrometry. *Journal of Analytical Toxicology, 34*(7), 367–373.

Sinclair, A. J., Viant, M. R., Ball, A. K., Burdon, M. A., Walker, E. A., Stewart, P. M., et al. (2010). NMR-based metabolomic analysis of cerebrospinal fluid and serum in neurological diseases—A diagnostic tool? *NMR in Biomedicine, 23*(2), 123–132.

Smedsgaard, J., & Frisvad, J. C. (1996). Using direct electrospray mass spectrometry in taxonomy and secondary metabolite profiling of crude fungal extracts. *Journal of Microbiological Methods, 25*(1), 5–17.

Smith, C. A., O'Maille, G., Want, E. J., Qin, C., Trauger, S. A., Brandon, T. R., et al. (2005). METLIN: A metabolite mass spectral database. *Therapeutic Drug Monitoring, 27*(6), 747–751.

Smith, C. A., Want, E. J., O'Maille, G., Abagyan, R., & Siuzdak, G. (2006). XCMS: Processing mass spectrometry data for metabolite profiling using nonlinear peak alignment, matching, and identification. *Analytical Chemistry, 78*(3), 779–787.

Soga, T., Ueno, Y., Naraoka, H., Ohashi, Y., Tomita, M., & Nishioka, T. (2002). Simultaneous determination of anionic intermediates for Bacillus subtilis metabolic pathways by capillary electrophoresis electrospray ionization mass spectrometry. *Analytical Chemistry*, *74*(10), 2233–2239.

Sreekumar, A., Poisson, L. M., Rajendiran, T. M., Khan, A. P., Cao, Q., Yu, J., et al. (2009). Metabolomic profiles delineate potential role for sarcosine in prostate cancer progression. *Nature*, *457*(7231), 910–914.

Steinbeck, C., Conesa, P., Haug, K., Mahendraker, T., Williams, M., Maguire, E., et al. (2012). MetaboLights: Towards a new COSMOS of metabolomics data management. *Metabolomics*, *8*(5), 757–760.

Stella, C., Rudaz, S., Gauvrit, J. Y., Lanteri, P., Huteau, A., Tchapla, A., et al. (2007). Characterization and comparison of the chromatographic performance of different types of reversed-phase stationary phases. *Journal of Pharmaceutical and Biomedical Analysis*, *43*(1), 89–98.

Storey, J. D., & Tibshirani, R. (2003). Statistical significance for genomewide studies. *Proceedings of the National Academy of Sciences of the United States of America*, *100*(16), 9440–9445.

Strimmer, K. (2008). A unified approach to false discovery rate estimation. *BMC Bioinformatics*, *9*, 303.

Sumner, L. W., Amberg, A., Barrett, D., Beale, M. H., Beger, R., Daykin, C. A., et al. (2007). Proposed minimum reporting standards for chemical analysis. *Metabolomics*, *3*(3), 211–221.

Sumner, L. W., Mendes, P., & Dixon, R. A. (2003). Plant metabolomics: Large-scale phytochemistry in the functional genomics era. *Phytochemistry*, *62*(6), 817–836.

Sun, J., Schnackenberg, L. K., Holland, R. D., Schmitt, T. C., Cantor, G. H., Dragan, Y. P., et al. (2008). Metabonomics evaluation of urine from rats given acute and chronic doses of acetaminophen using NMR and UPLC/MS. *Journal of Chromatography. B, Analytical Technologies in the Biomedical and Life Sciences*, *871*(2), 328–340.

Surinova, S., Schiess, R., Huttenhain, R., Cerciello, F., Wollscheid, B., & Aebersold, R. (2011). On the development of plasma protein biomarkers. *Journal of Proteome Research*, *10*(1), 5–16.

Sysi-Aho, M., Katajamaa, M., Yetukuri, L., & Oresic, M. (2007). Normalization method for metabolomics data using optimal selection of multiple internal standards. *BMC Bioinformatics*, *8*, 93.

Tautenhahn, R., Böttcher, C., & Neumann, S. (2007). Annotation of LC/ESI-MS mass signals. In *BIRD'07 proceedings of the 1st international conference on bioinformatics research and development* (pp. 371–380). Berlin, Heidelberg: Springer-Verlag.

Tautenhahn, R., Bottcher, C., & Neumann, S. (2008). Highly sensitive feature detection for high resolution LC/MS. *BMC Bioinformatics*, *9*, 504.

Tautenhahn, R., Patti, G. J., Kalisiak, E., Miyamoto, T., Schmidt, M., Lo, F. Y., et al. (2011). metaXCMS: Second-order analysis of untargeted metabolomics data. *Analytical Chemistry*, *83*(3), 696–700.

Tautenhahn, R., Patti, G. J., Rinehart, D., & Siuzdak, G. (2012). XCMS Online: A web-based platform to process untargeted metabolomic data. *Analytical Chemistry*, *84*(11), 5035–5039.

Teahan, O., Gamble, S., Holmes, E., Waxman, J., Nicholson, J. K., Bevan, C., et al. (2006). Impact of analytical bias in metabonomic studies of human blood serum and plasma. *Analytical Chemistry*, *78*(13), 4307–4318.

Timbrell, J. A. (1998). Biomarkers in toxicology. *Toxicology*, *129*(1), 1–12.

Tiziani, S., Emwas, A. H., Lodi, A., Ludwig, C., Bunce, C. M., Viant, M. R., et al. (2008). Optimized metabolite extraction from blood serum for 1H nuclear magnetic resonance spectroscopy. *Analytical Biochemistry*, *377*(1), 16–23.

Tolstikov, V. V., & Fiehn, O. (2002). Analysis of highly polar compounds of plant origin: Combination of hydrophilic interaction chromatography and electrospray ion trap mass spectrometry. *Analytical Biochemistry*, *301*(2), 298–307.

Trusheim, M. R., Burgess, B., Hu, S. X., Long, T., Averbuch, S. D., Flynn, A. A., et al. (2011). Quantifying factors for the success of stratified medicine. *Nature Reviews. Drug Discovery*, *10*(11), 817–833.

Trygg, J., Holmes, E., & Lundstedt, T. (2007). Chemometrics in metabonomics. *Journal of Proteome Research*, *6*(2), 469–479.

Turer, A. T., Stevens, R. D., Bain, J. R., Muehlbauer, M. J., van der, W. J., Mathew, J. P., et al. (2009). Metabolomic profiling reveals distinct patterns of myocardial substrate use in humans with coronary artery disease or left ventricular dysfunction during surgical ischemia/reperfusion. *Circulation*, *119*(13), 1736–1746.

Vander Heiden, M. G., Locasale, J. W., Swanson, K. D., Sharfi, H., Heffron, G. J., Amador-Noguez, D., et al. (2010). Evidence for an alternative glycolytic pathway in rapidly proliferating cells. *Science*, *5998*, 1492–1499.

van't Veer, L. J., Dai, H., van de Vijver, M. J., He, Y. D., Hart, A. A., Mao, M., et al. (2002). Gene expression profiling predicts clinical outcome of breast cancer. *Nature*, *415*(6871), 530–536.

Vassella, F., Hellstrom, B., & Wengle, B. (1962). Urinary excretion of tryptophan metabolites in the healthy infant. *Pediatrics*, *30*, 585–591.

Vaughan, A. A., Dunn, W. B., Allwood, J. W., Wedge, D. C., Blackhall, F. H., Whetton, A. D., et al. (2012). Liquid chromatography-mass spectrometry calibration transfer and metabolomics data fusion. *Analytical Chemistry*, *84*(22), 9848–9857.

Vervoort, R. J., Ruyter, E., Debets, A. J., Claessens, H. A., Cramers, C. A., & de Jong, G. J. (2002). Characterisation of reversed-phase stationary phases for the liquid chromatographic analysis of basic pharmaceuticals by thermodynamic data. *Journal of Chromatography. A*, *964*(1–2), 67–76.

Vuckovic, D. (2012). Current trends and challenges in sample preparation for global metabolomics using liquid chromatography-mass spectrometry. *Analytical and Bioanalytical Chemistry*, *403*(6), 1523–1548.

Vuckovic, D., & Pawliszyn, J. (2011). Systematic evaluation of solid-phase microextraction coatings for untargeted metabolomic profiling of biological fluids by liquid chromatography-mass spectrometry. *Analytical Chemistry*, *83*(6), 1944–1954.

Wang, T. J., Larson, M. G., Vasan, R. S., Cheng, S., Rhee, E. P., McCabe, E., et al. (2011). Metabolite profiles and the risk of developing diabetes. *Nature Medicine*, *17*(4), 448–453.

Want, E. J., O'Maille, G., Smith, C. A., Brandon, T. R., Uritboonthai, W., Qin, C., et al. (2006). Solvent-dependent metabolite distribution, clustering, and protein extraction for serum profiling with mass spectrometry. *Analytical Chemistry*, *78*(3), 743–752.

Want, E. J., Wilson, I. D., Gika, H., Theodoridis, G., Plumb, R. S., Shockcor, J., et al. (2010). Global metabolic profiling procedures for urine using UPLC-MS. *Nature Protocols*, *5*(6), 1005–1018.

Wei, R., Li, G., & Seymour, A. B. (2010). High-throughput and multiplexed LC/MS/MRM method for targeted metabolomics. *Analytical Chemistry*, *82*(13), 5527–5533.

Werner, E., Croixmarie, V., Umbdenstock, T., Ezan, E., Chaminade, P., Tabet, J. C., et al. (2008). Mass spectrometry-based metabolomics: Accelerating the characterization of discriminating signals by combining statistical correlations and ultrahigh resolution. *Analytical Chemistry*, *80*(13), 4918–4932.

Werner, E., Heilier, J. F., Ducruix, C., Ezan, E., Junot, C., & Tabet, J. C. (2008). Mass spectrometry for the identification of the discriminating signals from metabolomics: Current status and future trends. *Journal of Chromatography. B, Analytical Technologies in the Biomedical and Life Sciences*, *871*(2), 143–163.

Whiteaker, J. R., Lin, C., Kennedy, J., Hou, L., Trute, M., Sokal, I., et al. (2011). A targeted proteomics-based pipeline for verification of biomarkers in plasma. *Nature Biotechnology*, *29*(7), 625–634.

Wiklund, S., Johansson, E., Sjostrom, L., Mellerowicz, E. J., Edlund, U., Shockcor, J. P., et al. (2008). Visualization of GC/TOF-MS-based metabolomics data for identification of biochemically interesting compounds using OPLS class models. *Analytical Chemistry*, *80*(1), 115–122.

Williams, R. E., Major, H., Lock, E. A., Lenz, E. M., & Wilson, I. D. (2005). D-Serine-induced nephrotoxicity: A HPLC-TOF/MS-based metabonomics approach. *Toxicology*, *207*(2), 179–190.

Wilson, I. D., Plumb, R., Granger, J., Major, H., Williams, R., & Lenz, E. M. (2005). HPLC-MS-based methods for the study of metabonomics. *Journal of Chromatography. B, Analytical Technologies in the Biomedical and Life Sciences*, *817*(1), 67–76.

Wishart, D. S., Tzur, D., Knox, C., Eisner, R., Guo, A. C., Young, N., et al. (2007). HMDB: The Human Metabolome Database. *Nucleic Acids Research*, *35*(Database issue), D521–D526.

Wold, S., Martens, H., & Wold, H. (1983). The multivariate calibration method in chemistry solved by the PLS method. In A. Ruhe, & B. Kagstrom (Eds.), *Lecture Notes in Mathematics. Proceedings of conference on matrix pencils, Piteå, Sweden*. Heidelberg: Springer-Verlag.

Wold, S., Ruhe, A., Wold, H., & Dunn, W. J. (1984). The Collinearity problem in linear regression. The partial least squares approach to generalized inverses. *SIAM Journal on Scientific and Statistical Computing*, *5*, 735–743.

Wolfender, J. L., Glauser, G., Boccard, J., & Rudaz, S. (2009). MS-based plant metabolomic approaches for biomarker discovery. *Natural Product Communications*, *4*(10), 1417–1430.

Wu, J. Y., Kao, H. J., Li, S. C., Stevens, R., Hillman, S., Millington, D., et al. (2004). ENU mutagenesis identifies mice with mitochondrial branched-chain aminotransferase deficiency resembling human maple syrup urine disease. *The Journal of Clinical Investigation*, *113*(3), 434–440.

Wu, H., Zhang, J., Norem, K., & El-Shourbagy, T. A. (2008). Simultaneous determination of a hydrophobic drug candidate and its metabolite in human plasma with salting-out assisted liquid/liquid extraction using a mass spectrometry friendly salt. *Journal of Pharmaceutical and Biomedical Analysis*, *48*(4), 1243–1248.

Yalow, R. S., & Berson, S. A. (1960a). Immunoassay of endogenous plasma insulin in man. *The Journal of Clinical Investigation*, *39*, 1157–1175.

Yalow, R. S., & Berson, S. A. (1960b). Plasma insulin concentrations in nondiabetic and early diabetic subjects. Determinations by a new sensitive immuno-assay technic. *Diabetes*, *9*, 254–260.

Yoon, H. R., Kim, H., & Cho, S. H. (2003). Quantitative analysis of acyl-lysophosphatidic acid in plasma using negative ionization tandem mass spectrometry. *Journal of Chromatography. B, Analytical Technologies in the Biomedical and Life Sciences*, *788*(1), 85–92.

Yu, Z., Kastenmuller, G., He, Y., Belcredi, P., Moller, G., Prehn, C., et al. (2011). Differences between human plasma and serum metabolite profiles. *PloS One*, *6*(7), e21230.

Zhang, T., Creek, D. J., Barrett, M. P., Blackburn, G., & Watson, D. G. (2012). Evaluation of coupling reversed phase, aqueous normal phase, and hydrophilic interaction liquid chromatography with Orbitrap mass spectrometry for metabolomic studies of human urine. *Analytical Chemistry*, *84*(4), 1994–2001.

Zhang, Y., Dai, Y., Wen, J., Zhang, W., Grenz, A., Sun, H., et al. (2011). Detrimental effects of adenosine signaling in sickle cell disease. *Nature Medicine*, *17*(1), 79–86.

CHAPTER FIVE

Potential of Fourier Transform Mass Spectrometry for High-Throughput Metabolomics Analysis

Sandra Alves*, Estelle Rathahao-Paris†,‡, Jean Claude Tabet*,[1]

*Equipe de spectrométrie de masse, Institut Parisien de Chimie Moléculaire, UMR 7201, Université Pierre et Marie Curie, Paris, France

†INRA, UMR 1145 Ingénerie Procédés Aliments, Paris, France

‡Agroparistech, UMR 1145 Ingénerie Procédés Aliments, Paris, France

[1]Corresponding author: e-mail address: jean-claude.tabet@upmc.fr

Contents

Advances in Botanical Research, Volume 67
ISSN 0065-2296
http://dx.doi.org/10.1016/B978-0-12-397922-3.00005-8

Abstract

Metabolomics investigations have driven extensive developments in analytical platforms. Among analytical technologies, mass spectrometry plays a central role into the wide range of metabolomics applications as global metabolite profiling, absolute quantification and also metabolite structural elucidation. This chapter presents an overview of Fourier transform mass spectrometer (FT-MS) advantageous features for the large-scale measurements of metabolites. Principles of FT-MS devices are described with some details on the performance characteristics such as the highest available mass resolving power, mass measurement ability with sub-ppm errors, the most powerful dynamic range and various activation modes. Particularly, we focused onto the usefulness of atmospheric pressure ionization sources coupled to FT analyzers for very high-throughput direct injection mass spectrometry approach. However, different analytical issues should be considered when employing Direct Injection High Resolution Mass Spectrometry approach, including matrix effects and metabolite identification complexity. The collection of large HRMS data sets constitutes only the preliminary step of the metabolomics workflow. Indeed, metabolite identification and ultimately the wished better biological understanding require traditional bioinformatics tools such as the metabolite database query but also new data processing specifically developed on the basis of HRMS data characteristics, that is, accurate mass, mass defect or highly resolved isotopic peaks.

1. INTRODUCTION

Metabolomics involves the characterization of the small molecular mass molecules (<1000 u), which are metabolites found in biological samples at the whole organism level (cells, organ) or in tissue section, in biofluids (urine and blood) or in biopsies (Fiehn, 2002; Goodacre, Vaidyanathan, Dunn, Harrigan, & Kell, 2004). This approach is used for studying profiles of metabolites that reflect at the best biological functions and allow distinction between normal biological events and perturbations. Indeed, metabolic perturbations result in the change in the relative abundance of some metabolites and then, in the modification of metabolic profiles. By providing an overview of the metabolic status, metabolomics represents a powerful tool to reveal normal or pathophysiological states, for example, disease states and their progression.

Metabolomics is an emerging field within the 'omics' science such as genomics and proteomics, developed to give a complete picture of functioning of living organisms. Since the emergence of genomics, which refers to the study of the genome of organisms, several other 'omics' techniques have come into the scene such as proteomics, which involves the global studies of protein expression, or transcriptomics, which is the study of the transcriptome including mRNA, rRNA and tRNA. Metabolomics is the endpoint of the 'omics' cascade as metabolites are the end products of cellular processes and are, in fact, a part of the phenotype description. As all 'omics' studies, the complexity of metabolomics investigations requires interdisciplinary collaboration including life and computational sciences as well as analytical technology. The analytical techniques employed in such studies represent fundamentally important tools for generating informative data. Hence, they should offer large-scale analyses, sensitive and selective detection.

However, the analytical challenge for metabolomics investigations is especially difficult because of the very wide chemical and physical properties and the structural heterogeneity of metabolites that compose the metabolome. Indeed, the metabolome includes both endogenous and exogenous metabolites. Endogenous metabolites are produced by enzymatic processes (e.g. sugars, organic acids, ketones, aldehydes, amines, amino acids, nucleosides, lipids, hormones and alkaloids), whereas exogenous metabolites (or xenometabolites) are generated from the degradative metabolism of xenobiotic compounds coming from food or drug.

The other 'omics' areas focus on macromolecules such as deoxyribonucleic acids (DNA for genomics), ribonucleic acids (RNA for transcriptomics) and proteins (for proteomics). Such large size biomolecules have molecular masses that can reach several millions of mass units. However, their structure is chemically similar, like polymeric assemblages, basically constituted by repetition of limited number of subunits, that is, only five types of nucleotides in DNA strands (and also in RNAs) and 20 natural amino acids in proteins. Therefore, puzzle resolution to reconstitute the primary structure of macromolecules is relatively easier than the structural elucidation of novel or unknown biomarker compounds found in the metabolome. Small-size metabolites display a greater structural variability in the arrangements of atoms and subgroups (only based upon atom valence properties) resulting in one elemental composition (EC) in a large number of possible isomers. Additional complexity in metabolome analysis in terms of structural elucidation is brought by the existence of variable spatial conformations, that is, possible stereoisomers. Therefore, metabolites cannot be sequenced like DNA or proteins.

As other 'omics' approaches, metabolomics implies detection of complex mixture composition in an exhaustive manner. An organism contains several thousands of metabolites (e.g. over 200,000 metabolites including secondary metabolites as estimated in the plant kingdom, Fiehn, 2001). This increases dramatically the possibility to find numerous isobaric compounds. To obtain a maximum coverage of the metabolome, one must take into account that those metabolites are often present in a very wide range of concentrations (i.e. relatively low amount of metabolites <1 pM and large amount metabolites >1 mM, Psychogios et al., 2011), causing possible missing signals (worsen by the ionization discrimination, *vide infra*). So the identification of novel or unknown biomarker metabolite is very challenging.

In addition to the previous limiting features, the metabolome should be considered in a dynamic metabolic perspective, with rapid changes in a short timescale. To compare patterns or fingerprints of metabolites that change due to metabolic perturbations caused by disease, toxin exposure, environmental or genetic alterations, a large number of samples must be analyzed in a short time and in a rigorously repeatable fashion. So, high-throughput analyses of series of samples must be performed to accurately assess the response variation using if possible a single analytical technique.

Global metabolite analysis or untargeted metabolomics focuses on the detection of as many metabolites as possible to generate patterns, which are characteristic of the metabolome. These patterns are called *metabolic fingerprints*. This approach does not directly contribute to biochemical knowledge and understanding of underlying mechanisms of action. However, a supplementary analytical step is needed to identify biomarker metabolites. So, structural elucidation of unknown or novel metabolites is progressively becoming an essential phase, which implies serious analytical difficulties. Alternatively, *metabolite profiling*, which corresponds to targeted analysis focused on a group of metabolites, requires identification and quantification of metabolites of interest. The metabolite quantification represents another challenge for metabolomics investigations because of the limited availability of reference compounds as well as internal standards, which are generally metabolite molecules labelled by stable isotopes.

The mostly used analytical techniques for metabolomics studies are nuclear magnetic resonance (NMR) and MS (Dunn, Bailey, & Johnson, 2005; Lindon & Nicholson, 2008). Those techniques enable simultaneous analysis of a large number of metabolites in a single experiment. Pioneering works in metabolomics were all based on NMR. Although NMR is a rapid and non-destructive method requiring minimal sample preparation, its sensitivity is very low comparatively to that of the MS. In addition, interpretation of NMR spectra is not easy for complex mixtures. At the opposite, because of its intrinsic sensitivity and its specificity, MS allows identification (e.g. using data bank) and quantification of a wide range of molecules present in trace amounts in complex matrices. Today, progresses in metabolomics are more and more based on MS techniques.

MS displays a vast potentiality as analytical tool for this type of investigation. A mass spectrometer, which typically measures the molecular mass of a compound, proceeds through different steps. (i) A sample is introduced into the ionization source of the instrument. (ii) A gas-phase ionization for volatile compounds or desorption/ionization processes for thermolabile and polar (and large size) molecules is achieved, yielding ionic species characterized by their mass-to-charge (*m/z*) ratios. (iii) A charged specie separation according to their *m/z* ratios takes place into the analyzer region. (iv) Separated ions are finally detected and the generated signal is sent to a computer, where the *m/z* ratios are stored together with their relative abundance for a condensed presentation as a *mass spectrum* (de Hoffmann & Stroobant, 2007).

Mass spectrometry-based metabolomics offers quantitative analyses with high selectivity and sensitivity (from the parts per million, ppm to the ppt) and the potentiality for structural elucidation allowing ultimate identification of metabolites (Bedair & Sumner, 2008; Dettmer, Aronov, & Hammock, 2007; Dunn, 2008; Villas-Bôas, Mas, Akesson, Smedsgaard, & Nielsen, 2005). Indeed, arsenal of MS technique is tremendous. It includes the tandem mass spectrometry (MS/MS), which may be the best approach for structural elucidation of metabolites. The ultimate step that could pave the way towards the high-throughput analysis of metabolome is the access to ultrahigh mass resolving power. By this way, it is possible to separate isobaric molecular species and consequently, to determine their molecular weight (MW) with high-mass measurement accuracy, which allows reaching a single EC.

Moreover, easy MS coupling with a separation technique (i.e. gas chromatography (GC, Pasikanti, Ho, & Chan, 2008) and liquid chromatography (LC, Wilson et al., 2005) or capillary electrophoresis (CE, Ramautar, Somsen, & de Jong, 2009) reduces the complexity of MS response thanks to the compound separation in an additional time dimension. It provides possible separation of isobaric and isomeric compounds, and also delivers additional physicochemical information (i.e. retention time). However, ion introduction with separation technique is a delicate step, which must be optimized for each sample. More critically it is time-consuming. Another drawback of coupling methods is the shift in the retention making data processing difficult. Hence, those methods are not well suited to high-throughput metabolomics investigations.

In the hope of finding a single MS-based approach, the art-main has recently examined the ability of direct introduction mass spectrometry (DIMS) for high-throughput metabolomics analysis (Bedair & Sumner,

2008; Dettmer et al., 2007; Dunn, 2008). The direct introduction can be performed by different means depending upon the source properties in terms of vacuum conditions and ionization processes: (i) in vacuum, through a septum introduction as well as direct solid (or liquid) probes whereas (ii) at atmospheric pressure conditions, injection is possible through a syringe pump or using a flow injection analysis (FIA) approach with a chromatographic system. Anyway, the resulting mass spectrum suffers from mass discrimination phenomenon related to limitations occurring in the different parts of instrument, for example, ionization discrimination and matrix effects, saturation effects at the detector, etc.

So, scientists quickly recognized the ultrahigh-resolution mass spectrometry ability to resolve, at least partially, the previous drawbacks of DIMS approach. Time-of-flight (TOF)-based instruments have been primarily tested thanks to its high mass resolving power (of about 10,000), its mass accuracy (few ppm), and above all, its fast scan rate (Dunn, Overy, & Quick, 2005). Besides the accurate mass measurement ability to obtain an EC, Fourier transform mass spectrometry (FT-MS) instrumentation offers ultrahigh peak resolution and consequently facilitates complex mixture analysis by resolving isobaric species in MS and also MS/MS data. Furthermore, the highest available dynamic range of FT-MS, which is the available ratio of the highest to lowest magnitude peaks, impacts directly the metabolic fingerprint and also influences the sensitivity, as well as the mass measurement accuracy for the lowest intensity peaks (Junot, Madalinski, Tabet, & Ezan, 2010).

In this chapter, definitions of some concepts used in the high-resolution mass spectrometry (HRMS) are provided followed by the description of the Fourier transform mass spectrometers, that is, Orbitrap-based and Fourier Transform-Ion Cyclotron Resonance (FT-ICR) instruments. A brief description of ionization techniques that can be applied to metabolomics investigations as well as of their limitations is also given. Afterwards, metabolomics applications based upon FT-MS instrumentations are described. Finally, different feature tools for HRMS-based metabolomics data processing are presented in the last section.

2. DEFINITION OF SOME CONCEPTS USED IN MASS SPECTROMETRY

In this section, will be presented definitions of some terms commonly used in the HRMS (Brenton & Godfrey, 2010). Because metabolites are small size molecules, they generate only singly charged ions in mass spectrometry. Hence, all definitions will be given by considering only singly charged ions ($z=1$). Note that every ion is related to an m/z value, where m is the mass of an ion in unified atomic mass unit (u) or Dalton (Da, non-SI unit) and z is the charge number.

2.1. Mass resolution and mass resolving power

Mass resolution and *mass resolving power* are two terms usually used in mass spectrometry to characterize the performance of a mass spectrometer to separate two mass spectral peaks with closely spaced m/z values.

The IUPAC definition (McNaught & Wilkinson, 1997) for the *mass resolving power* is the capacity of the mass spectrometer to distinguish two ions with small difference in m/z. The mass resolving power (R_p) is defined as the ratio:

$$R_p = \frac{m}{\Delta m}(\text{by considering } z=1) \tag{5.1}$$

when m is the mass at which a given resolution was determined and Δm is defined either as the minimum peak separation or as the peak width at the mass m depending on how it is obtained.

The Δm value can be determined by two different processing methods: the valley definition and the peak width definition (see Fig. 5.1). The valley definition is based on the separation of two mass spectral peaks of equal intensity at $(m/z)_1 = m$ and $(m/z)_2 = (m + \Delta m)$ with a valley between them. The 10% valley definition is used in magnetic sector mass spectrometers. In the peak width definition, the measurement is made on a single mass spectral peak at the mass m at a specified height of the peak. The Δm value is commonly determined at full-width half-height maximum (FWHM) of the peak for quadrupole mass filter, ion trap, time of flight, ion mobility and Fourier transform instruments such as Orbitrap and ICR based-instruments mass spectrometers. High-mass resolving power is related to a R_p value higher than 10,000, excluding quadrupole mass filter and ion trap analyzers.

The *mass resolution* refers to the separation between two closely spaced peaks in a mass spectrum. According to the IUPAC definition (McNaught & Wilkinson, 1997), the mass resolution is the small difference in m/z of two peaks at masses m_1 and m_2, and is defined as: $(m_2 - m_1)$. Those two peaks can be resolved if the $(m_2 - m_1)$ value is higher than Δm, when Δm is the minimum peak separation or peak width as previously defined. To accurately measure the masses m_1 and m_2, the baseline resolution of those two peaks should be reached. This involves a mass resolving power R_p higher than approximately three times the $m_2/(m_2 - m_1)$ ratio.

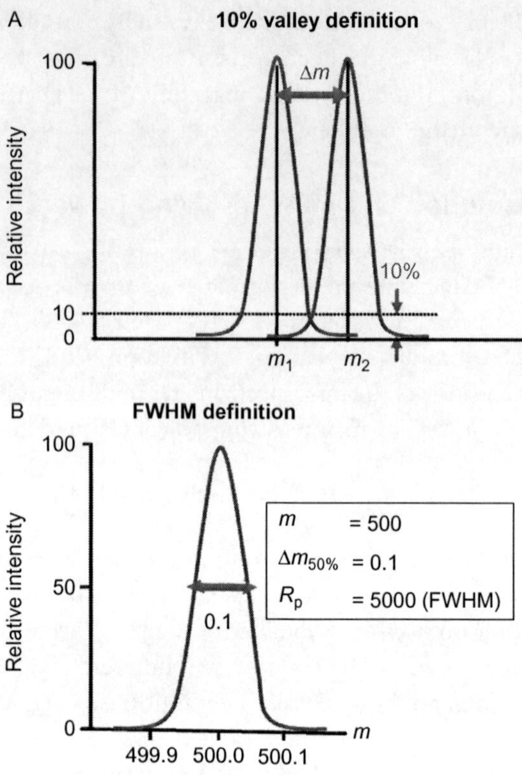

Figure 5.1 Mass resolving power defined at (A) 10% valley definition and (B) FWHM definition. (For colour version of this figure, the reader is referred to the online version of this chapter.)

As shown in Fig. 5.2, the peak width as well as the accurate mass measurement is affected by the R_p value. Hence, increasing R_p leads to the reduction of the peak width. The accurate mass measurement is obtained only at baseline resolution ($R_p = 24{,}000$) as determined by the following calculation:

$$R_p \geq 3 \times \frac{m_2}{(m_2 - m_1)} = 3 \times \frac{287}{(287.2006 - 287.1642)} = 23{,}654$$

$$R_p \approx 24{,}000$$

2.2. Naturally occurring isotopes and mass defects

An isotope is a chemical element having the same number of protons in the nucleus or the same atomic number (Z) but differing by the mass number (A) due to a difference in the number of neutrons: for example, the carbon

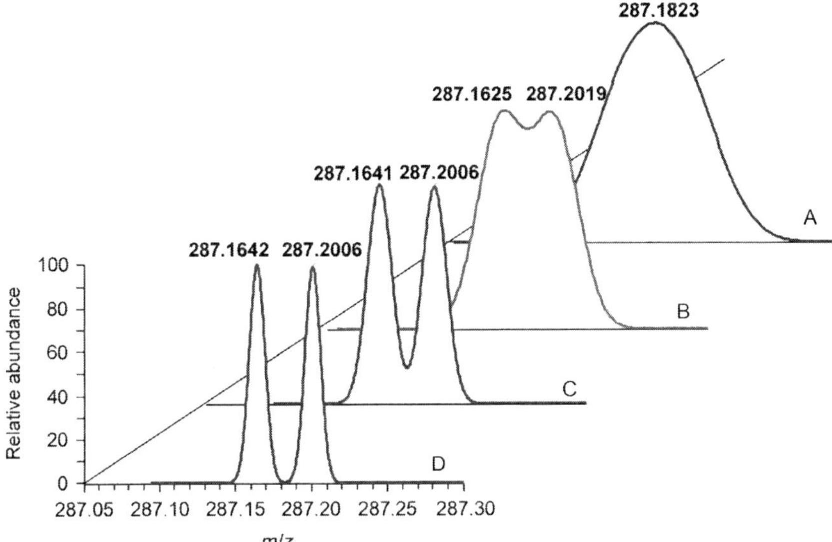

Figure 5.2 Simulated positive ionization mass spectra of methylether estradiol ($[M+H]^+$, $C_{19}H_{27}O_2$, *m/z* 287.2006) and hydroxyestrone ($[M+H]^+$, $C_{18}H_{23}O_3$, *m/z* 287.1642) peaks at mass resolving power of (A) 5000, (B) 8000, (C) 15,000 and (D) 24,000. (For colour version of this figure, the reader is referred to the online version of this chapter.)

($Z=6$) is constituted of two isotopes, ^{12}C ($A=12$) and ^{13}C ($A=13$). The number of isotopes for a given element is variable: for example, the F fluorine atom has only one isotope (i.e. ^{19}F) and the Cl chlorine possesses two isotopes (i.e. ^{35}Cl and ^{37}Cl). Some isotopes are unstable forms of elements; they are radioactive and are called radioisotopes: for example, ^{14}C is a radioisotope of carbon.

For any element, there is one isotope that is the most abundant in nature than the others. According to the abundances of the naturally occurring isotopes, the chemical elements can be classified in three types (Table 5.1):

- *A*: elements having only one isotope with abundance higher than 99.98% (e.g. ^{1}H).
- $A+1$: elements having two isotopes which differ by only one mass unit. ($A+1$) is the mass number of the second isotope when *A* is the mass number of the lightest isotope.
- $A+2$: elements having isotopes that differ by two mass units. The abundance of isotope with ($A+2$) as mass number is non-negligible (>0.2%).

For a given chemical formula, different isotopic compositions of a molecule or an ion can be obtained leading to the ($A+1$), ($A+2$), $\cdots$, ($A+n$) isotopic

Table 5.1 Accurate masses and abundances of naturally occurring isotopes of some elements commonly present in metabolites (no exhaustive list, reported from Audi, Wapstra, & Thibault, 2003; Böhlke et al., 2005)

Type	Element	Symbol	Integer mass	Isotopic mass	Isotopic abundance (%)
A	Hydrogen	^{1}H	1	1.007825	99.9885
	Deuterium	^{2}H	2	2.014102	0.0115
	Fluorine	^{19}F	19	18.998403	100
	Sodium	^{23}Na	23	22.989769	100
	Phosphorus	^{31}P	31	30.973762	100
A+1	Carbon	^{12}C	12	12.000000	98.93
		^{13}C	13	13.003355	1.07
	Nitrogen	^{14}N	14	14.003074	99.636
		^{15}N	15	15.000109	0.364
A+2	Oxygen	^{16}O ^{17}O	16 17	15.994915 16.999132	99.757 0.038
		^{18}O	18	17.999161	0.205
	Silicon	^{28}Si	28	27.976927	92.223
		^{29}Si	29	28.976495	4.685
		^{30}Si	30	29.973770	3.092
	Sulphur	^{32}S	32	31.972071	94.99
		^{33}S	33	32.971459	0.75
		^{34}S	34	33.967867	4.25
		^{36}S	36	35.967081	0.01
	Chlorine	^{35}Cl	35	34.968853	75.76
		^{37}Cl	37	36.965903	24.24
	Potassium	^{39}K	39	38.963707	93.2581
		^{40}K	40	39.963998	0.0117
		^{41}K	41	40.961826	6.7302
	Bromine	^{79}Br	79	78.918337	50.69
		^{81}Br	81	80.916291	49.31

clusters in the mass spectrum. Those isotopic clusters can be used to distinguish metabolites of a specific compound, especially when it contains halogens such as chlorine or bromine. Indeed, such compound produces characteristic isotopic patterns easily detected in the mass spectrum even with low mass resolving power analyzers. As shown in Fig. 5.3, metabolites of vinclozolin, a molecule containing two chlorine atoms, can be easily detected thanks to the isotopic patterns characteristics of the presence of two chlorine atoms. The EC of metabolite can further be obtained from the accurate *m*/*z* value of characteristic ions. This example demonstrates the usefulness of stable isotopes and particularly that of the isotopic pattern annotation.

Hence, an approach based on characteristic isotopic patterns with very HRMS can be used for metabolite identification and especially for the characterization of xenometabolites (e.g. compounds containing chlorine or bromine atoms such as pesticides) in complex biofluids. Nevertheless, experimental isotopic patterns can be compared to simulated isotope distribution of theoretical EC.

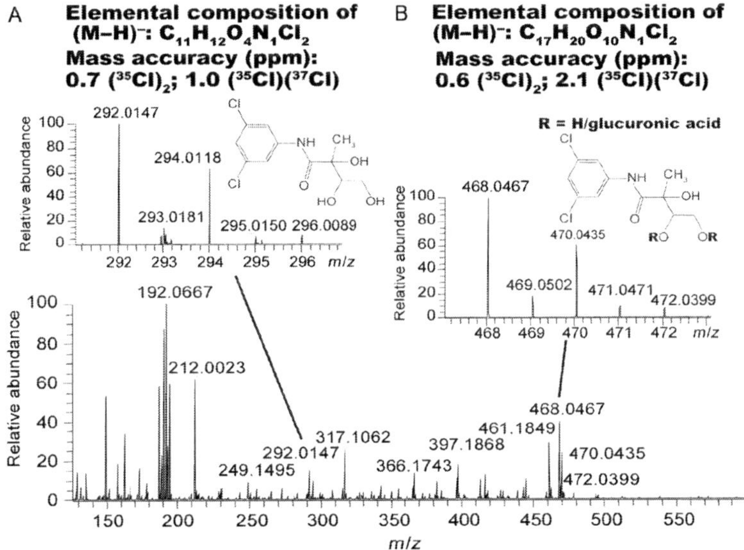

Figure 5.3 Negative ionization mass spectrum obtained from diluted rat urine (1/100). Experiments were conducted into a LTQ-Orbitrap mass spectrometer at a mass resolving power of 100,000 for *m/z* 400. The insets display the zoomed mass range of halogenated ion peaks (typical isotopic patterns) that is, (A) in the (*m/z* 291–297) range and (B) in the (*m/z* 467–473) range (personal communication from Rathahao-Paris, E., Bursztyka, J., Debrauwer, L., Jaeg, J.-P., & Cravedi, J.-P.). (For colour version of this figure, the reader is referred to the online version of this chapter.)

Theoretical isotopic distribution can be obtained for an EC $X_X Y_Y \ldots$ by expanding the following expression (Yergey, 1983):

$$(X_1 + X_2 + X_3 + \cdots)^x (Y_1 + Y_2 + Y_3 + \cdots)^y \cdots, \quad (5.2)$$

where $X_1, X_2, \ldots, Y_1, Y_2, \ldots$ are the individual natural isotopes of elements $X, Y, \ldots$ and $x, y, \ldots$ is the number of each element $X, Y, \ldots$ in the elemental formula.

For an EC of a given ion, absolute abundance (P) of a given permutation (i.e. a given element distribution) is calculated by resolving the following polynomial equation:

$$P = \frac{x!}{(x_1)!(x_2)!(x_3)!\ldots}(P_{X_1})^{x_1}(P_{X_2})^{x_2}(P_{X3})^{x_3}\ldots, \quad (5.3)$$

where P is the abundance of a given permutation, x is the total number of the element in the elemental formula, $x_1, x_2, x_3, \ldots$ are the number of each isotope ($x_1 + x_2 + x_3 + \cdots = x$) in a given permutation and $P_{X_1}, P_{X_2}, P_{X3}, \ldots$ are the abundances of each isotope.

And the molecular mass is equal to:

$$m = x_1 m_{X_1} + x_2 m_{X_2} + x_3 m_{X_3} + \cdots, \quad (5.4)$$

with m_{X_1} as the mass of the isotope X_1 of the element X.

The simulations resulted from a given EC are then displayed as a list of isotopic peaks within its abundance and mass (P_i, m_i). However, even the most powerful mass analyzers display a finite peak resolution and some isotopic peak overlapping arises, which must be accurately accessed.

The $A+1$ or $A+2$ isotopic clusters are composed of a mixture of isobaric species (i.e. ions having the same nominal mass but different exact masses). Each individual isotopic peak results from the contribution of specific isotope of a given element such as ^{2}H, ^{13}C or ^{15}N to the $A+1$ cluster. In order to resolve mass spectral peaks representative of isobaric species found for one individual $A+1$ or $A+2$ isotopic cluster, a very high mass resolving power instrument ($R_p > 100{,}000$) should be used. Each natural isotope is characterized by an exact isotopic mass (a non-integer value except for the ^{12}C isotope). The knowledge of exact isotopic masses enables determining the contribution of a specific isotope to an isotopic cluster. Indeed, the difference between the exact mass and the integer mass corresponds to the mass defect. There are positive and also negative mass defect values (Fig. 5.4A): for example, $1.007825 - 1 = +0.007825$ for ^{1}H and $16 - 15.994915 = -0.005085$ for ^{16}0. This mass defect is characteristic of

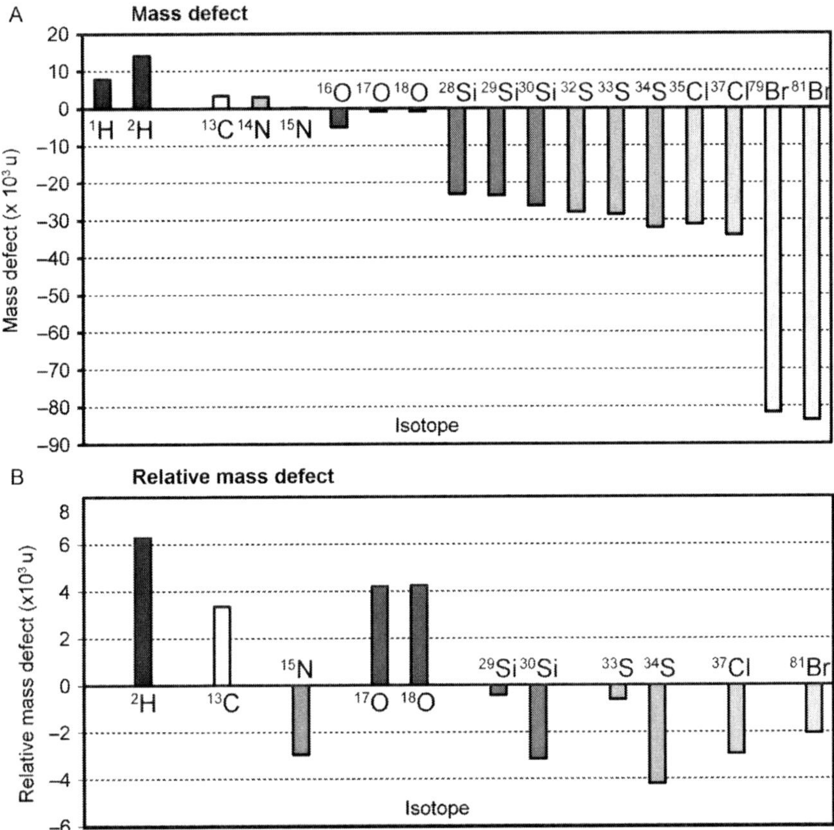

Figure 5.4 (A) Isotopic mass defects and (B) relative isotopic mass defects of the most common elements found in small-size organic compounds. (For colour version of this figure, the reader is referred to the online version of this chapter.)

each isotope and it can be used to identify its specific presence in the EC of molecules or charged species.

Alternatively, the use of the relative isotopic mass defect facilitates the identification of EC and then reveals the contribution of a given isotope to the $A+1$ or $A+2$ isotopic clusters (Thurman & Ferrer, 2010). The relative isotopic mass defect is defined as

$$\mathrm{R}\Delta m_{(A+n)} = m_{(A+n)} - m_A - n \tag{5.5}$$

when $m_{(A+n)}$ is the exact mass of the $A+n$ isotope, m_A the monoisotopic mass (*vide infra*) and n the difference between the integer masses of $m_{(A+n)}$ and m_A.

For example, the relative isotopic mass defect of the ^{2}H is obtained as follows:

$$\mathrm{R}\Delta m_{(A+1)2\mathrm{H}} = \left[m_{(2\mathrm{H})} - m_{(1\mathrm{H})}\right] - 1$$
$$R\Delta m_{(A+1)2\mathrm{H}} = [2.014102 - 1.007825] - 1$$
$$R\Delta m_{(A+1)2\mathrm{H}} = 0.0062677\,\mathrm{u}$$

The relative isotopic mass defect can be positive or negative as shown in Fig. 5.4B; its value is characteristic of an isotope for a given element.

An example for peak attribution from a mass spectrum is shown in Fig. 5.5. The simulated mass spectra of protonated glycine ($C_2H_5O_2N$) at the mass resolving power (R_p) of 100,000 and 500,000 display one main monoisotopic peak at m/z 76.0393 corresponding to $[M+H]^+$ species and several low-intensity isobaric peaks at m/z 77 and m/z 78 corresponding respectively to the $A+1$ and $A+2$ isotopic clusters. At a mass resolving power of 500,000 the peak width decreases, the number of detected isobaric peaks is then higher in both $A+1$ and $A+2$ isotopic clusters. By considering the relative isotopic mass defect, peak attribution can be easily achieved as shown in Fig. 5.5. For the $A+1$ isotopic cluster, the negative relative isotopic mass defect ($R\Delta m_{(A+1)} = -0.003$) is characteristic of the only ^{15}N isotope contribution (Fig. 5.4B). By taking into account that $R\Delta m_{(A+1)^{13}\mathrm{C}} < R\Delta m_{(A+1)^{17}\mathrm{O}} < R\Delta m_{(A+1)^{2}\mathrm{H}}$, the isotope attribution can be achieved for all detected peaks of the $A+1$ isotopic clusters having positive relative isotopic mass defects. For the $A+2$ isotopic patterns, the relative isotopic mass defect corresponds to the sum of the relative isotopic mass defect of the two involved elements. For example, the summed contribution of ^{13}C and ^{15}N atoms to the $A+2$ clusters could be identified if its relative isotopic mass defect corresponds to the value obtained in the following calculation:

$$R\Delta m_{(A+2)} = R\Delta m_{(A+1)^{13}\mathrm{C}} + R\Delta m_{(A+1)^{15}\mathrm{N}}$$
$$R\Delta m_{(A+2)} = \left(m_{(^{13}\mathrm{C})} - m_{(^{12}\mathrm{C})} - 1\right) + \left(m_{(^{15}\mathrm{N})} - m_{(^{14}\mathrm{N})} - 1\right)$$
$$R\Delta m_{(A+2)} = (13.003355 - 12.000000 - 1) + (15.000109 - 14.003074 - 1)$$
$$R\Delta m_{(A+2)} = 0.003355 - 0.002965$$
$$R\Delta m_{(A+2)} = 0.00039\,\mathrm{u}$$

Identification of all isotopes that contribute to the $A+1$ and $A+2$ isotopic clusters allows determination without ambiguity of the EC of a given chemical compound.

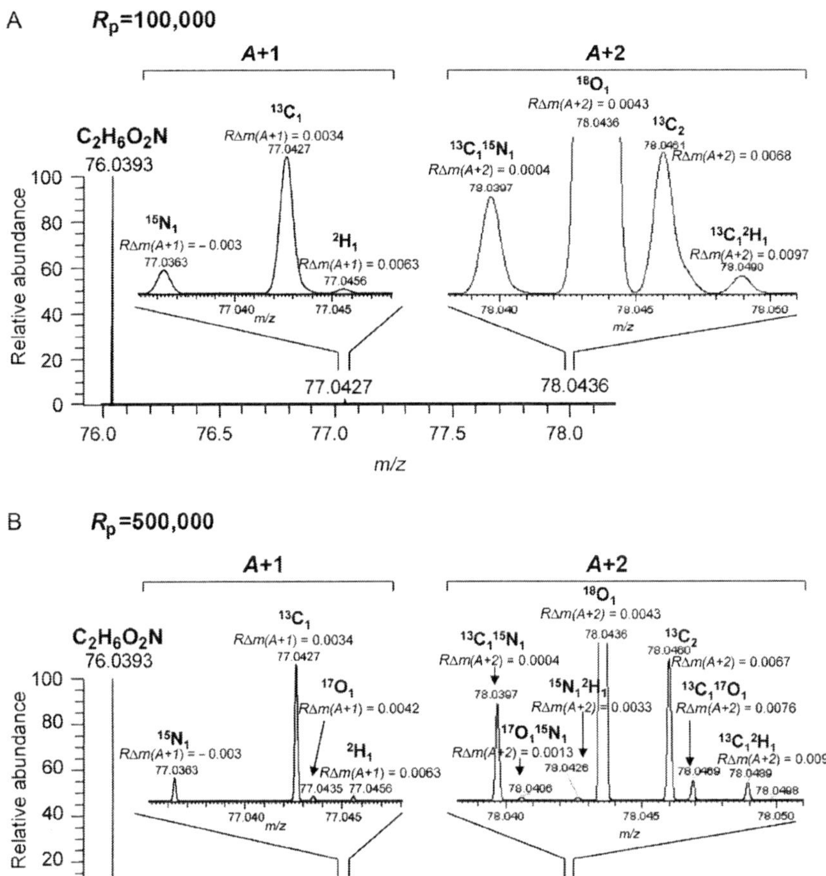

Figure 5.5 Simulated positive ionization mass spectra of the glycine ($C_2H_5O_2N$) at mass resolving power of (A) 100,000 and (B) 500,000. Attribution of elemental composition for each isotopic peak is achieved thanks to the relative isotopic mass defects (i.e. $R\Delta m_{(A+n)}$ values). Note the *m/z* value shifts when isobaric peaks are overlapped.

In the past, radioactive isotopes (e.g. ^{14}C or ^{3}H) have been widely used in kinetic studies of biochemical reactions as well as in metabolic studies for the elucidation of many metabolic pathways in cells (Rizzati et al., 2005), animals (Zalko et al., 2003) and plants (Pascal-Lorber, Rathahao, Cravedi, & Laurent, 2004). The decreasing in radioactivity level reflects the disappearance of radio-labelled precursors whereas the appearance of radioactivity in the produced metabolites facilitates their detections. However, its application in laboratory is not simple in terms of health and safety. The advances in mass spectrometry in terms of ultrahigh mass resolving power and high-accurate mass measurements allow the use of stable isotopes (e.g. ^{13}C and ^{15}N) in the replacement of radioactive isotopes for the study of metabolite pathways. In this way, the stable isotopes that are naturally present at very low abundances (e.g. 1.07% for ^{13}C and 0.364% for ^{15}N) are introduced in a biological system at artificially high abundances (e.g. 100% of ^{13}C or ^{15}N atoms). The application of stable isotopes has opened a new dimension in metabolomics study particularly in tracer-based metabolomics (i.e. a target approach focusing on metabolite distribution as described by Lee, Wahjudi, Xu, & Go, 2010) and in fluxomics, which is defined as the measurements of metabolic fluxes in order to characterize the metabolic networks in a cell (Winder, Dunn, & Goodacre, 2011).

2.3. Different concepts of mass

The *atomic mass* corresponds to the mass of an atom or its isotope. An ion or molecule corresponds to an association of a group of atoms. Its molecular mass can be determined by summing the atomic masses of all constituted atoms.

The *nominal mass* is the integer mass of the most abundant isotope of a chemical element (e.g. H=1; C=12; O=16; N=14). The nominal mass of an ion or molecule is the sum of the nominal masses of each chemical element contained in the EC. For example, we consider two molecules, glycine ($C_2O_2NH_5$) and 3-aminopropan-1-ol (C_3H_9ON), which have different chemical formulas (Table 5.2). After positive ionization, the resulting isobaric ions have the same nominal mass equal to 76 u.

The *average atomic mass* is the average weight of an atom, which is obtained by taking into account its natural isotopic composition (i.e. mass and percentage of each isotope). It can be found in the periodic table. The *average mass* of an ion or molecule is obtained by adding the average atomic masses of

Table 5.2 Nominal, average and monoisotopic masses of even electron species resulting from protonation of glycine and 3-aminopropan-1-ol

Chemical formula of even electron species	Nominal mass (u)	Average mass (u)[a]	Monoisotopic mass (u)[a]
$C_2H_6O_2N^+$	76	76.07463	76.03930
$C_3H_{10}ON^+$	76	76.08829	76.07569

[a]Ion masses were determined by taking into account that the mass of an electron is equal to 0.00055 u.

elements (e.g. H = 1.00794; C = 12.01100; O = 15.99940; N = 14.00674). For the previous example, the average masses are 76.07463 u for protonated glycine and 76.08829 u for protonated 3-aminopropan-1-ol (Table 5.2).

The *monoisotopic mass* of an ion or molecule is defined as an exact mass and is determined by summing each mass of the most abundant isotope (e.g. ^{1}H = 1.00782, ^{12}C = 12.00000, ^{16}O = 15.99491, ^{14}N = 14.00307) present in the chemical formula. For the same example, the calculated monoisotopic masses are 76.03930 u for protonated glycine and 76.07569 u for protonated 3-aminopropan-1-ol (Table 5.2). This example shows that two ions having different chemical formulas can have the same nominal mass but different monoisotopic masses, that is, they are isobaric ions.

The *exact mass* of an ion or a molecule is the calculated mass obtained by summing the exact masses of all isotopes of elements containing in the chemical formula.

The *accurate mass* is experimentally measured allowing the EC determination.

2.4. Mass measurement accuracy

The nature of the mass (e.g. nominal mass or monoisotopic mass) recorded by a mass spectrometer depends on the characteristics of the used instrument, especially on the *mass resolution* and *mass resolving power*. With low mass resolving power, only peaks differing by one mass unit can be separated and the recorded masses are then the nominal masses. Hence, mass spectrometers with insufficient mass resolving power do not allow distinguishing ions having the same nominal mass but different exact masses (i.e. isobaric ions). The increase in the mass resolving power reduces the peak width, allowing peaks differing by a small m/z increment to be resolved. Moreover, the accuracy of the mass measurement is improved with reduction of the peak width at a given m/z when the instrument is accurately calibrated in m/z. With sufficient high mass resolving power, the EC could be determined from the experimentally measured accurate mass. An error of the mass measurement is usually given when compared to a known compound based on the calculated monoisotopic mass. This mass measurement error is called *mass measurement accuracy*.

The *mass measurement accuracy* is the ability of the mass spectrometer to experimentally measure the accurate m/z value of a known compound reported from three to five decimals. It can be expressed either in milli-mass unit (mmu) or in ppm. The mass error in mmu corresponds to the difference between the theoretical (calculated) m/z of a compound and that measured by the mass spectrometer. The *mass measurement accuracy* expressed in ppm is a relative value and is defined as:

$$\frac{\Delta M}{M} \times 10^6 \tag{5.6}$$

where ΔM is the difference between the experimentally measured accurate m/z and the calculated exact m/z, and M the calculated exact m/z.

Note that for a given *mass measurement accuracy* in mmu, the mass error in ppm depends on the m/z values. For example, a 1 mmu mass error corresponds to 10 ppm at m/z 100 and to 2 ppm at m/z 500.

From an experimentally measured accurate mass, some ECs can be proposed. However, the number of possible ECs depends on the mass measurement accuracy and on the m/z value. As shown in Fig. 5.6, this number increases with increasing *mass measurement accuracy* for a given m/z value (*vide infra*; Grange, Genicola, & Socovol, 2002).

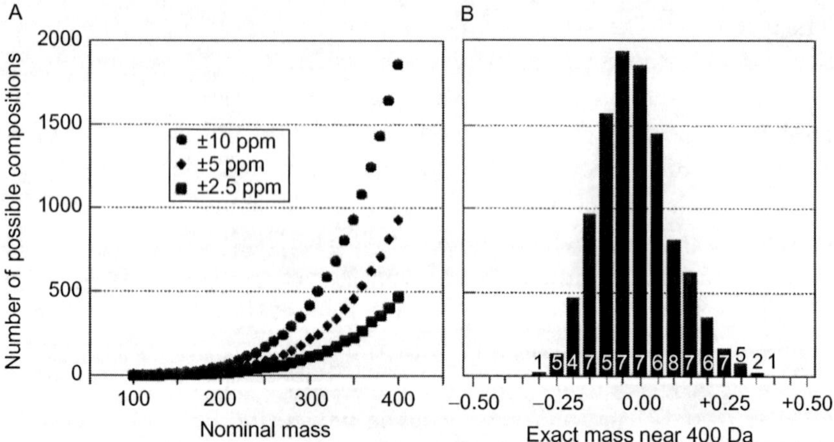

Figure 5.6 (A) Plots of the number of possible elemental compositions as a function of nominal masses with errors of 10, 5 and 2.5 ppm, by considering C, H, N, O, F, P and S atoms for the search procedure. Results are obtained by incrementing 10 u mass for each dot. (B) Number of possible compositions at nominal 400 u mass for different mass defect values (vertical bars are reported every 0.05 u intervals) with up to 10 ppm of error in the accurate mass. The numerals on/or above the vertical bars are the maximum number of remaining possible compositions after considering three accurate masses and two relative abundances measured from a single set of partial profiles acquired with 20,000 resolving power (10% valley). *Figure reproduced from Grange, Genicola, and Socovol (2002).*

3. PRINCIPLE OF FOURIER TRANSFORM MASS SPECTROMETERS

A huge evolution of spectroscopy has been done for *m/z* measurements of charged particles in vacuum since the first experiments of Thomson (1910). Nevertheless, the major instrumental innovations of the past 20 years have mainly contributed to the evolution of the ionization step including the development of atmospheric pressure sources and/or that of desorption/ionization methods. The best illustration of such advances is the Nobel Prize attribution in 2002 to Tanaka and Fenn, for the matrix-assisted laser desorption/ionization (MALDI) and electrospray ionization (ESI) desorption/ionization techniques in biomolecule analysis. Of course, technological improvements have been made for mass analyzers, although their principles of operation are fairly old. The last Nobel Prize in 1989 was shared by Hans Dehmelt and Wolfgang Paul for the development of the ion trap device in which relaxed ions are stored in a pseudo-potential well (March, 1997). Nevertheless, performances of instrumentation devices have been improved, for example, with the introduction by Comisarow and Marshall (1974) of FT treatment for ICR cell, the invention of the electrostatic mirror, that is, the reflectron mode for the TOF analyzer by Mamyrin, Karataev, Shmikk, and Zagulin (1973) or the alternatives in geometries of the radio frequency (RF) ion trapping devices, that is, two-dimensional ion trap (Douglas, Frank, & Mao, 2005). Exceptionally, a new ion trapping system allowing FT-MS experiments and thus ultrahigh resolution mass measurements has been presented in 1999 by Alexander Makarov termed as the Orbitrap analyzer.

Today, commercially available FT-MS instruments are based upon an ICR cell or Orbitrap technology. They have in common features the harmonic motion of the stored ion clouds into the ultrahigh vacuum of mass analyzers (10^{-10} mbar). It is then possible to accurately measure the harmonic motion frequency of spatially coherent ion packets by a common mode of detection, which is the time-domain signal. Indeed, frequencies of ion motions are the physical parameters that can be the most accurately measured than other physical parameters (e.g. ions' time arrival in TOF). For that, a time-domain signal as transient results from the detection of an oscillating current produced onto two opposed detection electrodes, when ions are approaching those electrodes. Afterwards, amplification and digitalization are achieved and a frequency-domain spectrum is then obtained by Fourier transformation of the digitized time-domain ICR signal (Comisarow & Marshall, 1974). Finally, a proportional relation between the ion cyclotron (or between the axial motion for Orbitrap cell) frequencies and ion m/z ratios results in ultrahigh-resolved mass spectrum using a simple mass scale calibration rule (see the principle of operations described in Fig. 5.7).

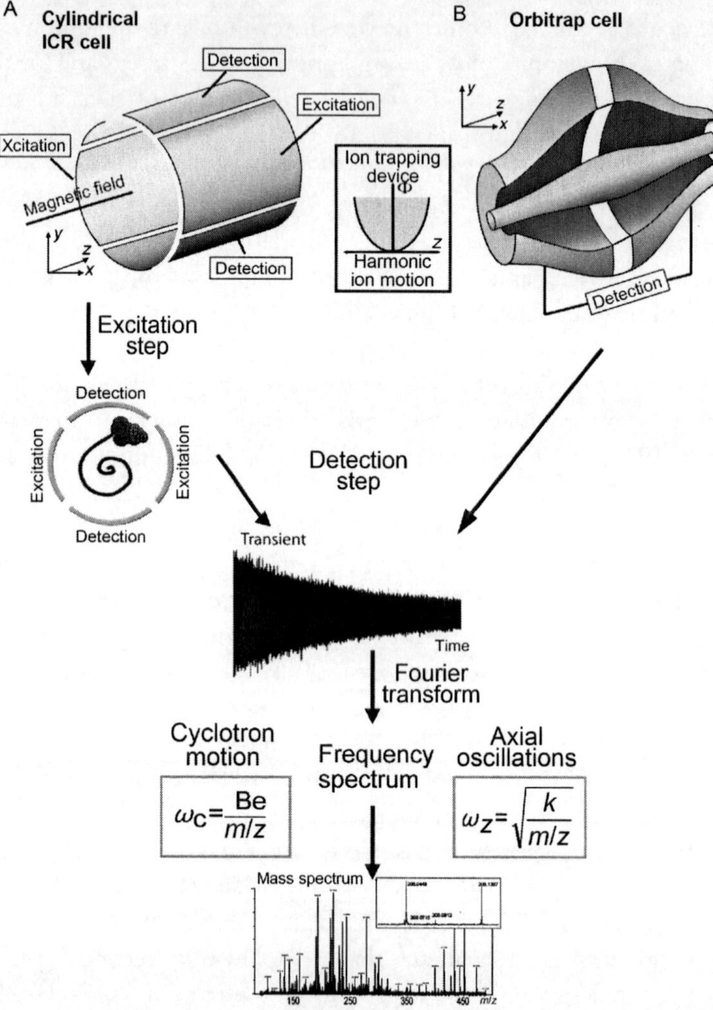

Figure 5.7 Principles of operation of both FT-MS instruments based upon (A) ICR cell or (B) electrostatic ion trapping (i.e. Orbitrap). Note that cylindrical geometry is reported here, for ICR cell, but various cell geometries have been developed along the FT-MS history. (For colour version of this figure, the reader is referred to the online version of this chapter.)

Hence, both these trapping devices are able to achieve ultrahigh-resolution mass spectrometry (HRMS). However, principles of ion trapping, geometries, and thus their respective analytical performances display major differences between both FT-MS analyzers. First distinguishable operating parameter is the ion storage conditions, that is, in ICR cell, ions are trapped in a uniform magnetic field combined with an electrostatic potential whereas Orbitrap analyzer uses a single electrostatic field to confine the ion packets.

3.1. Orbitrap-based instruments

The innovative Orbitrap analyzer as recently developed by Makarov (1999) is based upon an old principle of electrostatic ion trapping. The first trapping device introduced in 1923 by Kingdon (1923) had the ability to store charged particles into an electrostatic field within concentric cylinders (i.e. a thin wire central and a cylindric outer-electrode with end-cap electrodes enclosing the trapping volume). Later, the geometry of the trapping system was improved by Knight (1981) by modifying the external electrode geometry to produce a quadratic component of electric field into the axial direction without additional trapping plates. The Kingdon traps were initially trapping devices for charged particles (electrons or atomic ions) and were primarily employed in fundamental physics. Unfortunately, they were not considered as mass analyzers until the first attempts of Oksman (1995) and that of Gillig, Bluhm, and Russell (1996). For example, Gillig et al. (1996) tried such geometry cell for traditional FT-ICR experiments. Until then, applications such as measurements of lifetimes of the metastable ions, studies of electron capture process or academic studies on the orbital motion (astronomy) and even development of pumping system (Orbitron Ion pump) have been performed into the cylindrical and/or the ideal Knight geometry ion traps (Wineland, 1984).

3.1.1 Electrostatic ion trap analyzer

Principle of electrostatic ion trap as Orbitrap has been extensively described by Makarov (2000); Hardman and Makarov (2003); Hu et al. (2005) and Perry, Cooks, and Noll (2008). The ideal geometry of Orbitrap mass analyzer is constituted by an inner spindle-shaped electrode and an external electrode (in fact two bell-shaped outer electrodes, see Fig. 5.7). All electrode shapes were calculated so that a pure quadro-logarithmic potential exists inside the trapping cell, as expressed in the following equation:

$$V(r,z) = \frac{k}{2}\left(z^2 - \frac{r^2}{2}\right) + \frac{k}{2}(R_m)^2 \ln\left(\frac{r}{R_m}\right) + C, \qquad (5.7)$$

where r and z are cylindrical coordinates, R_m is termed as the characteristic radius, k the axial restoring force and C a constant (depending on cell dimensions and applied direct current (DC) voltage).

Under such a quadro-logarithmic potential, the stored ion adopts a spiral form motion. It orbits around the central electrode and at the same time oscillates along the z-axis (axial direction). Differential equations of ion motion according to the cylindrical coordinates were developed by Makarov (2000). The differential equation of axial ion motion is analog to that of a harmonic oscillator, whose solution is a sinusoidal function. Hence, the corresponding axial oscillation frequency is given by the following equation:

$$\omega_z = \sqrt{\frac{k}{m/z}} \tag{5.8}$$

where a direct proportional relation exists for a stored ion between its m/z ratio and its ω_z axial frequency, depending upon the k parameter as the axial restoring force.

This previous equation demonstrates that the axial motion is independent from the radial motion and initial conditions of injection and thus, gives a direct access to ultrahigh resolution and accurate mass measurements. This shows 'how the ideal Kingdon trap becomes a mass analyzer' (Hu et al., 2005). Indeed, the detection of the axial oscillation frequency is readily performed through time-domain signal acquisition at the external electrodes (in contrast to FT-ICR experiments, there is no excitation step before detection, *vide infra*). The external electrode is isolated at the centre (at $z=0$) by a ceramic plate in order to obtain two opposite detection electrodes allowing the measurement of the image current arisen from the ion axial oscillations.

In contrast, radial and angular frequencies (Hu et al., 2005; Makarov, 2000) characteristics of the ion circular motion around the central electrode vary for ions with slightly different initial positions and kinetic energies. This results in fast ion circular motion dephasing (few ms). Although radial/angular frequencies can be measured using detections plates, as demonstrated by Oksman (1995), they only allow low-resolution peak detection.

3.1.2 Hybrid instruments

The main difficulty of a mass spectrometer set-up based on Orbitrap cell is related to ion packet injection conditions. The limiting conditions are the initial kinetic energy (KE) of ions. SIMION simulations have been performed by Hu et al. (2005) for the Orbitrap cell depicted as a 360 ° electrostatic mass analyzer. The study has demonstrated that stable radial and axial motions imply nearly circular orbits, so that the initial KE should appropriately match the electric field, that is, at keV range value. Therefore, Makarov (2000) innovated optimal ion injection with the 'electrodynamic squeezing' process. It consists in the application of a fast DC voltage ramp (from 0 to −3500 V into 20–100 μs) during ion injection. Consequently, ion packet under the influence of the electrostatic force starts to adopt an orbit around the centre electrode. The continuous voltage raise causes a compression of the orbit motion, reducing the orbit radius, which simultaneously reduces the amplitude of axial oscillations. Then, stable orbits are reached and the detection step can start.

Additionally, adequate ion injection into the Orbitrap analyzer requires that ion packets must have narrow spatial (<few mm) and temporal distributions (<100–200 ns), to ensure the stability and coherency of the trapped ion motion (Hardman & Makarov, 2003). To obtain short pulses of ion packet injection, different home-built prototypes and finally, commercial Orbitrap-based instruments have been employed. These include: (i) electrostatic acceleration lenses for fast transmission of ion packets like those produced with a pulsed MALDI technique (Makarov, 2000), (ii) axial ejection from a linear quadrupole ion trap (Hardman & Makarov, 2003; Hu et al., 2005) of a prototype and (iii) a radial ejection from a 'C′-shaped' linear ion trap (in commercial LTQ-Orbitrap instruments, see Fig. 5.8; Olsen et al., 2005, 2007). The 'RF-only' C-trap transmits the ion packets from interface region into the Orbitrap cell, in an orthogonal fashion. The electrodes have a hyperbolic curved shape (the C-shaped electrodes), enclosed with the 'GateTrap' lenses. The first advantage of this geometry is that radial ion ejection results in a lower spatial dispersion than that obtained with an axial ion ejection (Douglas, Frank, & Mao, 2005). Above all, its particular geometry allows converging spatially ions at a focal point, out of the C-trap and coinciding with the entrance of the Orbitrap analyzer.

Hence, Thermofisher rapidly commercialized the solely available hybrid instruments based on Orbitrap devices, termed as LTQ-Orbitrap. This hybrid instrumentation consists of an atmospheric pressure source (API-MS) and combined two mass analyzers, that is, a linear ion trap (Douglas, Frank, & Mao, 2005; Schwartz, Senko, & Syka, 2002; as LTQ for linear trap quadrupole) and the Orbitrap cell (Olsen et al., 2005). The Orbitrap analyzer is characterized by its very high mass resolving power. However,

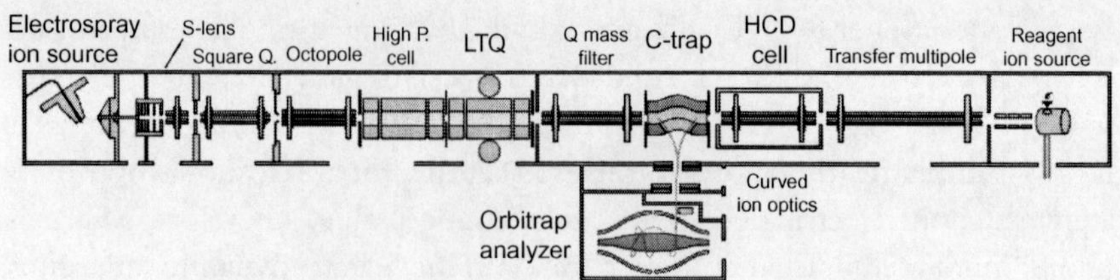

Figure 5.8 Schematic representation of LTQ-Orbitrap (Velos series, Thermofisher), equipped with ETD technology. This instrument includes a dual-Pressure Linear Ion Trap system (instead of a single LTQ cell), where the presence of combined high-pressure and low-pressure ion traps decouples ion accumulation and relaxation from scanning step. Briefly, externally produced ions (from API sources) are accumulated into LTQ system. They can be either radially ejected for low-resolution detection or further released into the curved C-trap (after axial ejection from the LTQ), which accumulates and stores the ions until they are sequentially transferred and analyzed into the high-resolution Orbitrap analyzer. Additional devices as the HCD cell, or Reagent Ion source can be implemented in the instrument to allow further MS/MS experiments based on 'in-beam CID' (HCD) or ETD mode (the latter technique consists in performing radical-induced processes inside the LTQ analyzer, after injection of precursor species and that of reagent anions). (For colour version of this figure, the reader is referred to the online version of this chapter.)

MS/MS and MS^n sequential experiments can only be performed into the LTQ system. The coupling with the linear ion trap analyzer allows also controlling the number of injected ions into the electrostatic ion trap using the automatic gain control (AGC) procedure. In addition, a detection system with a dynode and electron multiplier allows detecting ions at low-resolution conditions after radial ion ejection from the LTQ cell. Alternatively, ion packets can be transmitted through axial ejection from the LTQ to the Orbitrap analyzer through an octopole and the C-trap device. After the radial injection from the C-trap, a 'curved ion optics' accelerated ion beam is formed at keV KE before entrance into the Orbitrap (see Fig. 5.8).

Since the first commercial LTQ-Orbitrap version, new series of Orbitrap-based instruments with improved features have been proposed such as LTQ-Orbitrap-XL(-ETD), Discovery, Velos and recently, the Elite series. LTQ-Orbitrap-XL version has an additional mode of activation for MS/MS experiments (*vide infra*), that is, HCD mode for 'higher collision dissociation'. A collision cell placed in the axial direction of C-trap (opposite to atmospheric pressure ionization, API, source) allows conducting this non-resonant collision-induced dissociation (CID) activation mode (analog principle to the CID in a triple-quadrupole instrument, *vide infra*). The latest, Elite version, uses a new 'high field' Orbitrap cell (Michalski et al., 2012).

In fact, Makarov, Denisov, & Lange (2009) have proposed a new Orbitrap cell geometry with an inner electrode characterized by a higher radius. This geometry change provides larger orbit radii and consequently gives access to larger frequency values, leading to the increase in mass resolving power (until few hundreds of thousands) and in dynamic range. Note that Thermofisher company has also developed a bench-top instrument termed as Exactive, which does not include the LTQ cell.

3.1.3 Characteristics of Orbitrap-based instruments

The commercial Orbitrap mass spectrometers display the following performance characteristics (Hardman & Makarov, 2003; Makarov, et al., 2006; Makarov, Denisov, Lange, & Horning, 2006; Olsen et al., 2005): (i) mass resolving power up to 100,000 (in fact, a theoretical mass resolving power of 100,000 (FWHM) is given at m/z 400 for a maximum acquisition time of 1.9 s), (ii) mass measurement accuracy below 5 ppm and (iii) an in-spectrum linear dynamic range up to four orders of magnitude. The available mass resolving power is fixed by the time-domain signal duration, from 7500 to maximum 100,000 within 0.3–1.9 s scan duration is proposed in LTQ-Orbitrap instruments (Makarov, et al., 2006), exception for the Elite series equipped with 'high field cell' which offers a mass resolving power higher than 240,000 (Michalski et al., 2012). Actually, Orbitrap instrumentation displays lower resolution performances than that obtained routinely in FT-ICR instruments with high magnetic field strength (above 7 Tesla, T).

3.1.4 MS/MS experiments

Various activation modes are available on the commercial Orbitrap-based hybrid instruments, sometimes as optional devices. No activation technique can be currently performed inside the Orbitrap cell, although development for such experiment is on-going in Cooks' group (Hu, Cooks, & Noll, 2007; Hu, Makarov, Cooks, & Noll, 2006; Wu et al., 2006). All LTQ-Orbitrap hybrid instruments offer the possibility to operate MS/MS and sequential MS^n experiments into the LTQ via low-energy CID. Such resonant CID mode is obtained using an RF voltage applied on two opposite electrodes of the linear ion trap (i.e. radial dipolar excitation). Selection of the precursor ion can also be achieved by resonant ejection with an isolation window (e.g. ±1 u). However, resonant CID activation mode shows a limiting characteristic, that is the low-mass cut-off (LMCO). Indeed, fragment ions below an m/z value limit defined by the precursor ion q_z value (usually 0.25) into the stability diagram, are lost. This limit can be circumvented by using the pulsed quadrupolar dissociation mode. This latter low-energy resonant CID mode proceeds into the linear ion trap through three sequential stages constituted of precursor ion isolation, activation and product ion detection steps, in which the q_z position into the pseudo-potential well is quickly modified. By this way, the temporal sequential steps prevent the loss of low m/z product ions (no LMCO value).

Alternatively, non-resonant HCD activation mode (see Fig. 5.8) operates through fragmentation pathways induced by low-energy collisions with target gas into an external collision cell (Olsen et al., 2007). This activation mode is similar to the CID technique performed in a triple-quadrupole instrument. The precursor ions are trapped and performed round-trips along the collision cell. Such 'ion beam' CID conditions are characterized by extensive consecutive fragmentations (but still in a low-energy collisional regime), and allow low m/z product ion detection.

Furthermore, hybrid instruments can be optionally equipped with ETD (electron transfer dissociation) technology to access the multicharged ion dissociations. ETD mode is actually based upon ion–ion reactions conducted into the LTQ system, where a multicharged cation interacts with an anion radical reagent (produced in negative ion chemical ionization (CI) source, see Fig. 5.8; McAlister et al., 2008). ETD mode produces an unstable radical (charge-reduced) species that dissociates through non-ergodic fragmentation pathways. This mode of activation has been especially developed as a complementary tool for DNA, peptide or protein sequencing.

3.2. FT-ICR mass spectrometry

If Orbitrap cell is a recent mass analyzer concept, the principle of ion trapping upon a homogeneous static magnetic field for ion m/z measurement is much older. Lawrence and Livingston (1932) first proposed to filter ion m/z ratios according to their cyclotron motions. Following, Sommer, Thomas, and Hipple (1951) built the 'omegatron' instrument based upon the ICR, primarily for the measurement of the charge to mass ratio of the proton. They soon realized that such an instrument is suitable as a mass spectrometer. A real improvement has been brought with the introduction of FT data treatment, by analogy to NMR spectroscopy. Comisarow and Marshall (1974) have demonstrated the advantage of Fourier transform data treatment for simultaneous broadband measurements of ion cyclotron frequencies.

Today, this mass analyzer provides the highest peak resolution, that is, routinely hundreds of thousands in commercial high magnet FT-ICR instruments. A recent work has reported up to 20 millions of mass resolving power for reserpine ions (*m/z* 609) with the use of 7 T magnet using a new segmented cylindrical cell geometry (Nikolaev, Boldin, Jertz, & Baykut, 2011). FT-ICR displays also a sub-ppm mass accuracy, a wide mass range, various ion manipulation methods and also high-resolution ion isolation with the ability to perform multistage tandem mass spectrometry (MS^n) inside the cell using various activation modes (*vide infra*).

3.2.1 Ion cyclotron motion

Ion confinement using static magnetic field is a fairly old ion trapping principle. In spite of the different design and geometry (Guan & Marshall, 1995), ICR cells are always composed of different conductive electrodes that correlate with the different aspects of FT-ICR experiments, that is, (i) ion confinement under a homogeneous magnetic field combined with an electrostatic trapping potential (using the plates designed by T in Fig. 5.9); (ii) ion excitation and/or ejection procedure using an alternating (RF) electric potential applied on the E plates (Fig. 5.9) and finally (iii) the detection of image current produced by periodic cyclotron motion of the ions (on the D plates in Fig. 5.9).

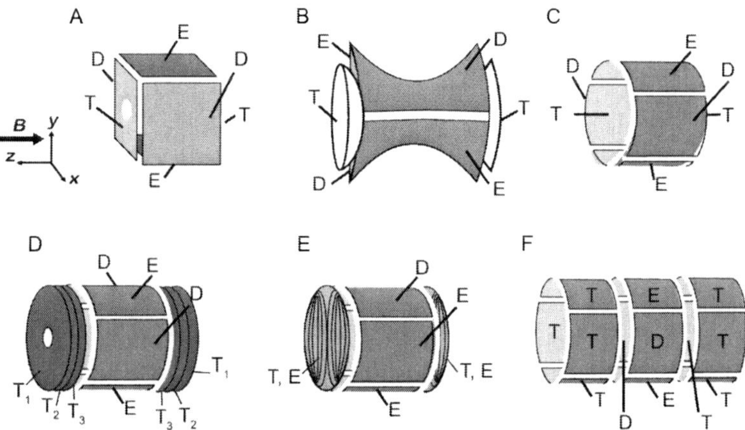

Figure 5.9 Scheme of ICR cell configurations, where E is the excitation; D the detection and T the trapping plates, based (A) cubic, (B) hyperboloidal (e.g. Penning trap), (C) cylindrical geometry electrodes or using (D) infinity, (E) open-ended and (F) segmented ICR configurations. *Figure adapted from Guan and Marshall (1995).*

Under a spatially uniform magnetic field B, a charged particle (of mass m and charge q moving at a constant velocity) undergoes the Lorentz force, which is perpendicular to both the particle velocity v and the magnetic field B direction. It adopts a helicoidal motion in the x–y plane, perpendicularly to B field direction (see Fig. 5.10; Marshall & Hendrickson, 2002; Marshall, Hendrickson, & Jackson, 1998). This periodic motion called cyclotron motion is characterized by the frequency of the ion rotation around its orbit, which is expressed (in rad/s or Hz) in the following equation:

$$\omega_c = \frac{qB}{m} \text{ or } \upsilon_c = \frac{1.535611 \times 10^7 B}{m/z} \qquad (5.9)$$

where q is the charge on the charged particle (coulombs, C), z is the charge number, ω_c is the cyclotron frequency (in rad/s), ν_c is the frequency in Hz and B is the magnetic strength (in Teslas, T).

An interesting feature of the previous equation is that ion at a given m/z rotates at the same orbit's radius, independently of its initial velocity (HRMS detection). Moreover, higher m/z ions have lower cyclotron frequencies, which explain the decreased resolution for higher m/z species after time-domain conversion. The equation shows that the magnetic field strength directly correlates with the available m/z range.

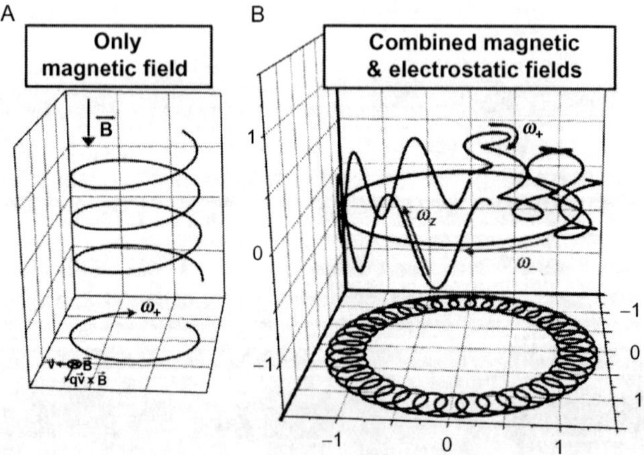

Figure 5.10 Schematic ion trajectories into a Penning trapping device (i.e. three dimensional and projection onto the x–y plane) (A) under the only magnetic field and (B) resulting from the combination of magnetic and electrostatic fields. *Figure adapted from Blaum (2006).* (For colour version of this figure, the reader is referred to the online version of this chapter.)

So confinement of ions in the x–y plane can be easily obtained within the magnetic field (see Fig. 5.10). Nevertheless, to ensure ion trapping in axial direction (z-axis), an electrostatic potential must be applied using two end-cap trapping plates (i.e. by application of low repulsing potential of 1V) to delimit axial motion and avoid ion escape. Ideally, a three-dimensional quadrupolar trapping potential allows harmonic ion oscillations into axial direction. Experimentally, it is only obtained in Penning cell geometry (Fig. 5.9), whereas other cell geometries (e.g. cubic and/or cylindrical) induce two-dimensional quadrupolar or dipolar trapping potential at least, at the centre of the ion trap. An alternative way to obtain harmonic axial oscillations is to segment the electrodes (open cells) of any shape to produce a nearly perfect quadrupolar electrostatic trapping potential (Guan & Marshall, 1995).

Incidentally, the presence of trapping potential renders much more complex the ion motion, which displays three components that is, the trapping and the cyclotron motions with the apparition of the 'magnetron' rotation perpendicular to the B field direction, at frequency, ω_- (see Fig. 5.10). Note that cyclotron frequency is modified from ω_c to ω_+. Expressions of trapping, reduced cyclotron and magnetron frequencies are:

$$\omega_z = \sqrt{\frac{2qV_{\text{trap}}\alpha}{ma^2}} \quad \text{as trapping oscillation frequency,} \tag{5.10}$$

where V_{trap} is the applied potential to the trapping electrodes (in volts, e.g. potential between the trapping plates and the side plates for a cubic trap; or between the ring and end-cap in hyperboloidal electrodes), a is the characteristic trap dimension and α is a factor determined by cell geometry.

$$\omega_+ = \frac{\omega_c}{2} + \sqrt{\left(\frac{\omega_c}{2}\right)^2 - \frac{\omega_z^2}{2}} \quad \text{as the 'reduced' cyclotron frequency,} \tag{5.11}$$

$$\omega_- = \frac{\omega_c}{2} - \sqrt{\left(\frac{\omega_c}{2}\right)^2 - \frac{\omega_z^2}{2}} \quad \text{as the magnetron frequency,} \tag{5.12}$$

3.2.2 FT-ICR sequential events

As the other ion trapping devices (see above Orbitrap principle of operation), FT-ICR experiment consists of a series of temporally separated events (Barrow, Burkitt, & Derrick, 2005; Marshall & Hendrickson, 2002; Marshall, Hendrickson, & Jackson, 1998). Four temporal events usually constitute a basic FT-ICR experiment: quench, ionization and/or ion trapping, excitation and detection. The quench step is employed to empty the ICR cell of any ions. Afterwards, the ionization step can occur internally (as electron bombardment) or externally. In the latter case, the ion packet should be transmitted along the *z*-direction into the ICR cell with a constant velocity, implying specific injection processes.

After the ion trapping event, the excitation step is performed to produce a spatially coherent ion packet motion at detectable orbital radius, close to the detection plates. The radius of the ion orbit is increased by applying a RF potential at the ion's cyclotron frequency on the opposite excitation plates (E plates in Fig. 5.9). The resulting post-excited radius of ion orbit, independently of its *m/z* ratio (i.e. of its cyclotron frequency) is expressed as:

$$r = \frac{V_{\text{p-p}} t_{\text{excit}}}{2dB} \tag{5.13}$$

where $V_{\text{p-p}}$ (peak-to-peak volts) is the excitation amplitude, t_{exc} (s), the excitation duration and d the distance between the excitation electrodes.

However, to obtain spatially coherent ion packet (all ions at the same orbit radius), different excitation processes can be performed using (i) rectangular pulses as 'off-resonant' technique, (ii) a frequency-sweep (chirp) method, which applies resonant excitation with a constant amplitude onto a ramp (broadband) of frequency range and/or using the (iii) SWIFT (stored-waveform inverse Fourier transform) technique. SWIFT procedure as proposed by Cody, Hein, Goodman, and Marshall (1987) applies an inverse Fourier transform to generate the specific time-domain waveform, which produces the desired excitation frequency-domain, when applying differentially on the two opposed electrodes.

3.2.3 FT-ICR characteristics

Mass resolving power in FT-ICR MS experiments routinely reaches hundreds of thousands in broadband (normal) mode, typically 300,000 at 9.4 T for 1 s time-domain signal. Alternative heterodyne mode is able to reach few millions but only, over a restricted *m/z* range (Marshall & Hendrickson, 2002).

From the first derivative of Eq. 5.9, the mass resolving power at the full-width at half-maximum peak height is expressed as:

$$\frac{m}{\Delta m_{50\%}} = \frac{qB}{m\Delta\omega_{50\%}} \tag{5.14}$$

This equation shows a direct proportionality to the magnetic field strength. So the most direct (but expensive) way for raising the peak resolution is the increase of magnetic field (currently available magnetic field: 7, 9.4, 12, 15 and a 21 T FT-ICR instrument is on-going construction). Note that increase of magnetic field strength produces also a raise of upper m/z limit, dynamic range and the number of ions that can be confined without coalescence, etc.

Furthermore, the duration of the time-domain signal (transient) and its shape directly impact the FT-ICR mass resolving power and the mass measurement accuracy (see Fig. 5.11). Indeed, a rapid transient signal damping due to ion-neutral collisions can occur, implying the need of ultrahigh vacuum conditions (10^{-10} mBar). The transient signal damping results in peak broadening and even in apparition of shoulder peaks. Consequently, it also induces decreasing mass measurement accuracy.

Additional phenomenon responsible of peak resolution lowering and limiting sensitivity is the space charge effects. Too many stored ions (i.e. more than 10,000 ions, Marshall, Hendrickson, & Jackson, 1998) cause losses of low-abundant peak resolution and mass measurement accuracy because of perturbing Coulombic interactions between ions. Peak coalescence (i.e. detection of a single peak due to overlapping peaks of closely spaced m/z) can eventually occur for very large populations of ion clouds with closely spaced m/z values. This effect explains the importance of controlling ion number

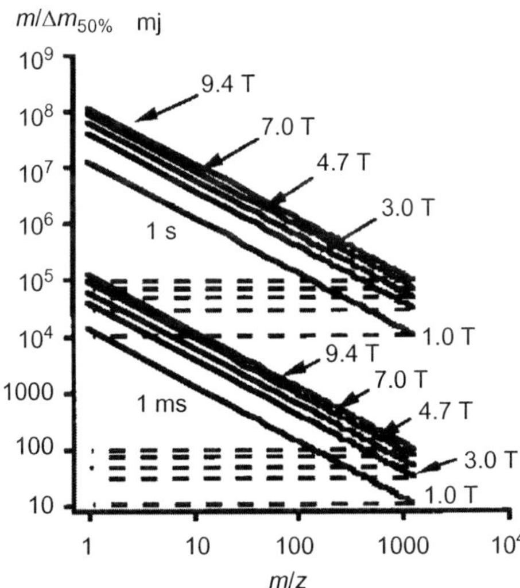

Figure 5.11 FT-ICR theoretical mass resolving powers at different magnetic field strengths (in Tesla) as a function of mass-to-charge ratio (m/z) for acquisition time of 1 ms and 1 s, respectively. The dashed lines indicate mass resolving power for magnetic field-swept instruments (e.g. magnetic sectors). *Figure reproduced from Marshall, Hendrickson, & Jackson, (1998).*

before injection into ICR cell. Then, regulation of the number of ions is eventually done into an external trapping device before ion packet injection into the ICR cell (Senko, Hendrickson, Emmett, Shi, & Marshall, 1997).

3.2.4 Hybrid instruments

Today, commercial FT-ICR devices are hybrid instruments in which ICR cells are mostly coupled with API sources, mainly dedicated to large-scale 'omics' applications. As already pointed out, external ionization step implies a carefully injection procedure. First, the ion source outside from the ICR magnet must be separated by several stages of differential pumping to finally reach the ultrahigh vacuum ICR cell. The ions should be transmitted through the magnetic fringe field by a suitable ion guide, that is, electrostatic Einzel lenses, RF-only multipole ion guides and/or through flight tube, etc. Additionally, external ion accumulation into an (multipolar) ion trapping device can be performed. Thermofisher proposed also an LTQ-FT hybrid instrument, in which ion accumulation (using automatic gain control, AGC) and eventual ion manipulation can be performed externally from the ICR into the LTQ (Peterman, Dufresne, & Horning, 2005).

3.2.5 MS/MS experiments into the ICR cell

For MS/MS experiments, FT-ICR offers variety of activation modes, which can be achieved inside the cell, in contrast to Orbitrap analyzer (Laskin & Futrell, 2005; Marshall, Hendrickson, & Jackson, 1998). First, resonant excitation process of ions provides not only ion detection (*vide supra*) but can be a way for precursor ion isolation and activation. Nevertheless, among the mostly used ICR activation modes, off-resonance irradiation collision-induced dissociation (SORI-CID; Peltz, Drahos, & Vékey, 2007) proceeds through sequential steps: (i) after isolation of precursor ions (resonant ejection), (ii) a relatively high-pressure collision gas is introduced into the ICR cell, (iii) then the selected ion is non-resonantly excited (i.e. ions undergo alternatively excitation and de-excitation by application of a frequency slightly different from the ion cyclotron frequency). (iv) Finally, ion orbits expand and contract their radius along the off-resonance excitation, so precursor ion and its prompt product ions undergo multiple low-energy collisions (slow heating process), resulting in the raise of ion internal energy until they dissociate. Other activation modes exist in the ultrahigh-vacuum ICR. Infrared multiphoton dissociation (IRMPD) technique (Eyler, 2009) used pulsed IR laser. Photon absorption by precursor ions results into slow heating process and thereby ion dissociation. The main advantage of IRMPD is the no requirement of collision gas, which avoids transient damping and subsequent loss of resolution and mass accuracy. Electron capture dissociation as ECD (or electron detachment for negatively charged species as EDD) uses low-energy electrons emitted from a cathode to the ICR cell to produce reduced charged species with prompt dissociations (analog fragmentation patterns to ETD mode as described by Cooper, Hakansson, & Marshall, 2005). Note that electron-induced dissociation (EID) processes can be performed for singly charged precursor ions allowing high energy fragmentation pathways (Lioe & O'Hair, 2007; Nguyen, Afonso, & Tabet, 2011).

4. ION PRODUCTION FOR METABOLITE ANALYSIS

Before being mass analyzed by the analyzer of the mass spectrometer, samples must be introduced and ionized in the ion source of the instrument. There are a lot of available ionization methods that can be coupled with FT-MS instruments. The choice of ionization technique to be used is essential, it is primarily directed by the physico-chemical properties of studied metabolites (e.g. volatility and thermal stability of the molecule but also by the deprotonation/protonation ability of chemical functional groups present in the molecule). In a less crucial manner, the sample status (e.g. solid samples as tissues or whole organisms or biofluids as blood and urine) can also justify the choice of a kind of ion source.

Furthermore, the appearance of the mass spectrum mainly depends on the used ionization technique and their operating conditions and especially correlated with the internal energy (E_{int}) of generated species (Vékey, 1996). The E_{int} of ionized molecules essentially comes from the energy acquired during the ionization processes since the thermal energy of molecule before ionization is usually very low compared to the deposed energy and can be neglected. The terms 'hard' and 'soft' ionization are related to the E_{int} range of ionized molecules. Indeed, 'hard' ionization indicates that molecular species will acquire a high E_{int} during the ionization step and will promptly decompose in the source, leading to a mass spectrum composed of intense fragment ion peaks and of very low abundance (or eventually absent) molecular species. The term 'soft' ionization implies that generated molecular species possess low E_{int} (a like thermal distribution) and will not (or weakly) quickly decompose.

In this present section, the ionization methods that can be used in the analysis of metabolites and in metabolomics applications will be presented. Their principles of operation are described in Fig. 5.12.

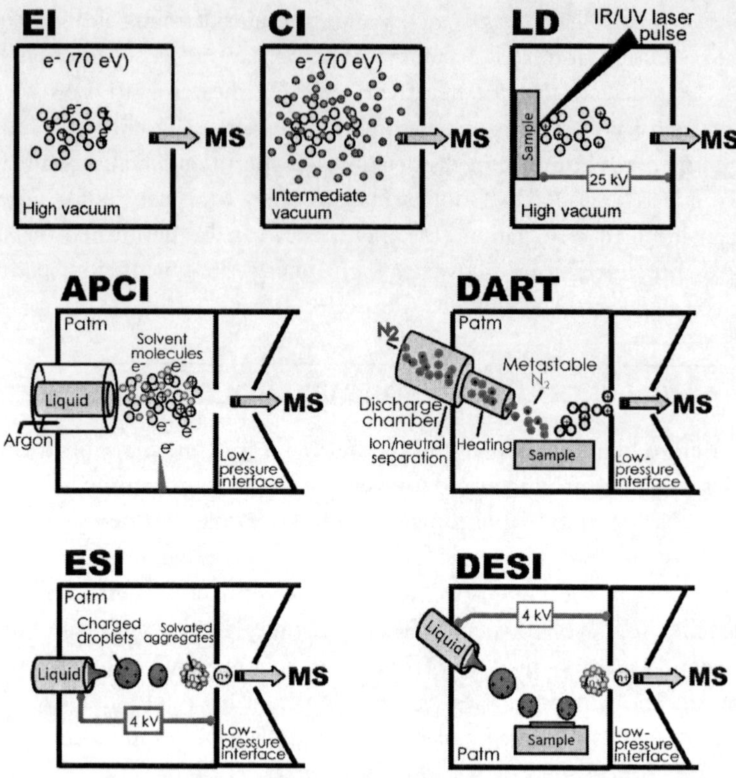

Figure 5.12 Scheme of ionization sources with their typical operating processes (EI, electron ionization; CI, chemical ionization; LD, laser desorption (and other laser-assisted techniques as MALDI, DIOS, etc.), APCI, atmospheric pressure chemical ionization (N_2 can also be used instead of argon); DART, direct analysis in real-time device; ESI, Electrospray and DESI, the desorption Electrospray Ionization). Note that internal ionization can be performed in FT-MS using electron bombardment. (For colour version of this figure, the reader is referred to the online version of this chapter.)

4.1. In-vacuum gas-phase ionization techniques

The in-vacuum gas-phase ionization techniques are used to ionize low mass (Molecular weight, MW < 1000 u), volatile, apolar or low polar and thermally stable metabolites.

Before the 1980s, electron ionization (EI; Hogness & Kvalnes, 1928) was the most commonly used technique in mass analysis. The ionization process takes place in the gas phase by interactions of molecules with accelerated electrons typically at 70 eV energy. $M^{+\cdot}$ molecular ions formed under EI conditions are odd-electron species, with a large E_{int} distribution that reaches 16 eV with the maximum close to 4 eV (Vékey, 1996). Only 0.8–2 eV energy range is enough to induce prompt ion decomposition. So, most molecular ions are extensively fragmented and EI mass spectra commonly display abundant fragment ion peaks, providing important structural information. Thanks to large ion E_{int} distribution, dependence of fragmentation ways upon instrumental conditions is limited, resulting in great mass spectrum reproducibility. Hence, EI mass spectral libraries are available such as National Institute of Standards and Technology (NIST, http://www.nist.gov/srd/nist1a.cfm) database (DB) which contents mass spectra of more than 200,000 compounds. Such DB is very useful for fast structural identification of metabolites. However, the weak abundance of molecular ions or eventually their absence makes difficult the identification of an unknown compound when its EI mass spectrum does not match the mass spectral data. Low-electron energy (5–20 eV) conditions are also used to reduce the fragmentation yield and to detect much abundant molecular ions but results in the loss of sensitivity (Vékey, 1996). However, the molecular ion does not often emerge.

Chemical ionization (CI) (Munson & Field, 1966) is a softer gas-phase ionization technique than EI. The CI processes involves (i) first the EI of a reagent gas (around 10^{-1} mbar) such as methane, isobutane, or ammonia (ii) followed by ion–molecule reactions between the reagent ions and its neutrals (iii) and/or the analyzed compound, leading to its ionization through adduct ion formation. Both positive and negative ions could be produced during the CI processes, for example, $[M+H]^+$ ions are produced in positive ionization mode and $[M-H]^-$ ions in negative ionization mode. The E_{int} acquired by the formed charged molecular species is proportional in positive mode to the difference between the gas-phase basicity (or the gas-phase acidity in the negative mode) of the molecule and that of the reagent gas. The exothermicity of ion–molecule reactions are usually less than 2 eV. Furthermore, the even-electron molecular species is more stable than the odd-electron ions due to the open-shell electronic structure of the latter species. Thus, the molecular species are then quite stable producing little fragmentations. CI can be used as a complementary ionization technique, especially when the radical molecular ions cannot be observed in the EI mass spectrum. It is generally applied for the determination of the compound molecular mass. However, the low reproducibility of CI mass spectra prevents the use of library search.

EI and CI are well-established gas-phase approaches covering all information required for structural identification of almost any volatile compounds. Indeed, stable molecular species from CI provide the compound MW, whereas fragmentation patterns of $M^{+\cdot}$ under EI give insights into the compound structure. Since, alternative gas-phase ionization techniques have been developed such as Penning ionization (Faubert, Paul, Giroux, & Bertrand, 1993), which involves bombardment of sample molecules by metastable species generated from N_2 or rare gases such as helium, neon, argon or krypton. This technique enables to control the E_{int} acquired by the generated radical molecular ions and eventually to produce selective ionization by the choice of the appropriate gas. Indeed, the E_{int} distribution is relatively narrow since the maximum E_{int} value corresponds to the ($EE–IE_M$) difference where EE is the excitation energy of the molecules yielding a metastable state and IE_M the compound ionization energy. Penning ionization mass spectra are then constituted of both molecular ions and fragment ions. This ionization mode has been successfully used in metabolomics applications (Dumas et al., 2002).

Those gas-phase ionization techniques can be combined with online gas chromatography/mass spectrometry (GC/MS). The main advantage of this coupling is related to the high separation power of GC column allowing the analysis of complex mixtures. However, the GC/MS approach is well suited to ionization of non-polar, volatile and thermally stable compounds. For polar and thermally labile metabolites, chemical derivatization procedure must be performed to reduce the polarity and improve the thermal stability and the volatility of the metabolites. Commonly, a silylation reagent such as *N,O*-bis(trimethylsilyl) trifluoroacetamide (BSTFA), *N*-methyl-*N*-(trimethylsilyl) trifluoroacetamide (MSTFA), or *N*-methyl-*N*-(*tert*-butyl-dimethylsilyl) trifluoroacetamide (MTBSTFA) is employed to derivatize functional groups such as hydroxyl, amine, and thiol groups (Fiehn, Kopka, Trethewey, & Willmitzer, 2000). To stabilize the open-chain forms of reducing sugars or to avoid the decarboxylation of α-keto acids such as pyruvic acid, oximation reaction using hydroxylamine or alkoxyamine hydrochlorides is also commonly performed (Fiehn et al., 2000).

4.2. Desorption/ionization methods in vacuum conditions

Other classes of sources in vacuum conditions appeared during the 1960s allowing desorption/ionization processes of polar and thermally labile (bio)-molecules (Busch, 1995). Some of them are based upon surface bombardment, that is, condensed-phase samples as eventually deposited and dried specimens onto a target using either ^{252}Cf radiation (PD as plasma desorption, Macfarlane & Torgerson, 1976) and neutral atom beam (FAB for fast atom bombardment, Barber, Bordoli, Sedgwick, & Tyler, 1981) or ion beam (SIMS as secondary ion mass spectrometry, Benninghoven,

1969; Castaing & Slodzian, 1962) surface impact as well as laser irradiation (**LD** as Laser Desorption; Opauszky, 1982; Tabet & Cotter, 1984). Today, laser-based techniques are mostly used in biological area. Primarily, laser irradiation of condensed-phase samples has been conducted using IR pulsed laser allowing desorption of preformed species. However, LD technique is limited in terms of mass range and in application areas. In the 1980s, two groups showed simultaneously that addition (by analogy with FAB conditions) into the samples of either glycerol mixed with Cobalt (Tanaka et al., 1988) or small organic aromatic molecules termed as matrix; (Karas & Hillenkamp, 1988) extended the mass range availability of laser-based ionization approaches to macromolecules. This was the first step of MALDI technique, which became a preeminent technique in the analysis of a wide variety of large mass compounds (from 600 u to more than 10^6 u) including polymers (Montaudo, Samperi, & Montaudo, 2006) and proteins (Domon & Aebersold, 2006). However, MALDI performances are strongly limited by the presence of abundant matrix species at low m/z range. Few publications have reported the possibility of detection of small size metabolites in biosamples (Guo, Zhang, Zou, Guo, & Ni, 2002).

Hence, matrix-free laser-assisted methods (Peterson, 2007) have been developed to small size compound analysis, as alternative approaches to MALDI technique. Principle of these 'softer' LD techniques is based upon the development of new surfaces used for depositing biological samples. For example, graphite targets (Sunner, Dratz, & Chen, 1995) or the use of nanostructured material as carbon nanotubes (Xu et al., 2003) act as an efficient way for laser energy transfer to the condensed-phase samples. A successful matrix-free preparation based on the modified porous silicon surface has been reported by Wei, Buriak, and Siuzdak (1999). This method, called desorption ionization on porous silicon (DIOS) appears well suited for analysis of small size compounds (i.e. for masses lower than 2000 u). Matrix-free techniques have successfully been applied to the analysis of low-molecular mass compounds in various biological samples (Budimir, Blais, Fournier, & Tabet, 2007; Peterson, 2007).

They are particularly well suited to direct analysis of condensed-phase samples such as tissue sections. Such ability of laser-based techniques (e.g. SIMS and desorption electrospray ionization (DESI) techniques) promotes development of imaging mass spectrometry (IMS, see Miura, Fujimura, & Wariishi, 2012). Indeed, an MS image of a tissue section or whole organism can be generated by plotting intensity of one m/z value (from a recorded mass spectrum) over (x,y) coordinate location, that is, the laser impact position onto the tissue surface, resulting in a two-dimensional ion density map. IMS offers a good technique for biological applications and clinical diagnosis. It can be operated by detection of drugs and/or their metabolites onto the sample surface (Li, Shrestha, & Vertes, 2008; Liu, Guo, & He, 2007). Figure 5.13 shows the high performances of MALDI-FT-ICR to specifically and spatially detect olanzapine and its metabolites on rat kidney (Cornett, Frappier, & Caprioli, 2008). Indeed, at low *m/z* range, MALDI displays a large number of peaks originating not only from endogenous compounds (of the sample) but also matrix-related species. Only the high-resolution detection is able to distinguish in a single run many isobaric species (here, hydroxyl-methyl olanzapine metabolite from three isobaric species at nominal *m/z* 329 value) and therefore to generate a lot of chemically specific MS images into a single experiment and within the simultaneous EC determination of all isobaric molecules.

4.3. Atmospheric pressure ionization approaches

Atmospheric pressure ionization (API) techniques refer to an ionization process that takes place at atmospheric pressure (Covey, Thomson, & Schneider, 2009). Those techniques considered as 'soft' ionization methods cover a large application range from low to highly polar, even ionic

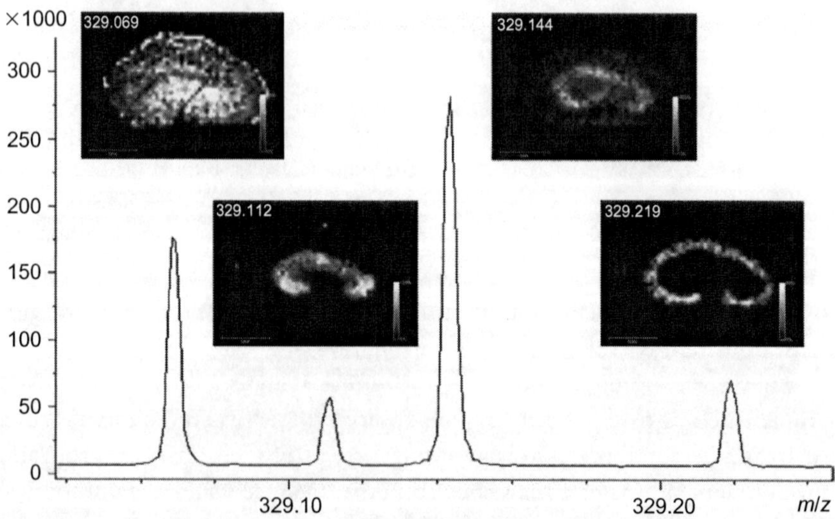

Figure 5.13 MS images and related MALDI-FT-ICR mass spectrum (i.e. mass range zoom at nominal *m/z* 329) of hydroxyl-methyl olanzapine and isobaric species. *Figure reproduced from Cornett, Frappier, and Caprioli (2008).*

compounds and also thermally labile molecules. When using gas-phase ionization in vacuum conditions, it is not possible to directly analyze aqueous solutions. The development of API techniques and low-pressure interface allows overcoming the challenge to convert polar metabolites present in the aqueous solutions and/or in biofluids into intact gas-phase molecular species. Moreover, they enable the coupling of liquid chromatography to the mass spectrometry (LC/MS).

The E_{int} of generated ions under API conditions generated from API is very narrow. For example, it has been shown that the E_{int} distribution of the ions formed by ESI is similar to thermal distribution (Naban-Maillet et al., 2005). The API mass spectra are then dominated by the even-electron molecular ions. Nevertheless, the E_{int} distribution of these species can be slightly modified during their transport from the API source into the high-vacuum region of the mass analyzer through an intermediate low-pressure interface (energy-controlled desolvation pathways). This variation depends on the source and interface geometry as well as on the setting parameters such as pressure, temperature, applied voltages, etc. According to the used operating conditions, API techniques include gas-phase ionization processes as well as ionization methods based on ion desorption phenomenon.

4.3.1 API gas-phase ionization techniques

The gas-phase ionization processes at atmospheric pressure are appropriate for analyzing metabolites characterized by low to medium polarity, having some volatility and limited MW (<1000 u). Those techniques as atmospheric pressure chemical ionization (APCI), atmospheric pressure photoionization (APPI) and direct analysis in real-time (DART) techniques (*vide infra*) imply very different ionization processes (see Fig. 5.12). However, all ionization methods display under adequate operating conditions the production of singly charged species such as $[M+H]^+$ protonated molecules in positive ionization mode or $[M-H]^-$ deprotonated molecules in negative ionization mode and/or adduct species. Other species such as $M^{+\cdot}$ ions may be eventually formed depending on the physical–chemical properties of studied molecules.

APCI (Huang, Wachs, Conboy, & Henion, 1990) presents analogies to conventional CI technique. During the ionization processes, reagent ions are produced from air and solvent molecules using a corona discharge and they subsequently react with gaseous sample molecules to generate molecular species via CI processes in the gas phase.

APPI (Robb, Covey, & Bruins, 2000) utilizes an ultraviolet lamp that emits photons at an energy level (10 eV for krypton) high enough for the direct photoionization of apolar molecules or for indirect ionization of most polar molecules. The last ionization process involves ionization of dopant or solvent molecules following by ion–molecule reactions between generated reagent ions and sample molecules (analog to APCI processes).

DART (Cody, Laramée, & Durst, 2005) is one 'ambient' ionization technique (Chen, Gamez, & Zenobi, 2009), which does not require sample preparation, allowing direct analysis of samples in solid, liquid or gas forms. Molecular species are produced by direct interaction of the sample with metastable species (generated by application of electric potential to a gas such as N_2, Ne or He) and subsequent gas-phase Penning processes and ion–molecule reaction cascades (close to APCI processes).

4.3.2 API desorption/ionization methods

Desorption/ionization API methods are suited for the ionization of non-volatile compounds, which possess a wide range of polarity (i.e. from medium to highly polar or ionic molecules) and a large MW range (from small size organic molecule to non-covalent macromolecules). The capacity of those techniques to produce multiply charged ions (e.g. $(M+nH)^{n+}$ species) from large size molecules allows extending the mass range of MS instruments. Those desorption/ionization methods are first ESI (Yamashita & Fenn, 1984) and also ambient desorption/ionization techniques such as DESI (Takáts, Wiseman, Gologan, & Cooks, 2004) and extractive electrospray ionization (EESI; Chen, Venter, & Cooks, 2006) and easy sonic spray ionization (Haddad, Catharino, Marques, & Eberlin, 2008).

The ESI technique introduced by John Fenn (Yamashita & Fenn, 1984; awarded by a Nobel Prize in 2002) is a very soft ionization method. The ESI processes (Kebarle & Verkerk, 2009) involve the production of ions from solutions at low concentrations. The sample solution is injected through a capillary tube into the ion source using a syringe pump or a chromatography system (LC/MS). Under the electric field and a nebulizer gas (e.g. nitrogen), a fine mist of highly charged droplets is produced. Those droplets diminish in size due to solvent evaporation. The increase of charge density on the droplet surface results in successive Coulombic explosions. Droplet size becomes more and more small, leading finally to desorption of highly charged solvated aggregates and/or molecular species. The ESI produced ions are then transferred to the high vacuum of mass spectrometer, within an intermediate low-pressure interface.

DESI and EESI are 'ambient' ionization techniques that can be performed with little or no sample preparation (Takáts et al., 2004). The DESI ionization processes present analogies with the conventional ESI mode. They involve the direct impact on the sample surface by a spray of charged solvent droplets generated by an ESI probe, producing desorbed ions and/or progeny droplets containing sample molecules. DESI is then a desorption/ionization technique particularly suited for solid-phase sample allowing surface analysis, whereas EESI is used for the analysis of sample in liquid or eventually gas form. EESI proceeds through two steps: (i) sample molecules are first nebulized to produce fine droplets, which are then carried by a gas stream (N_2) into the ion source, (ii) and then are ionized with charged solvent droplets produced by ESI probe. Mass spectra generated by both DESI and EESI techniques are very similar to that produced by conventional ESI approach (Bedair & Sumner, 2008).

4.4. Which ionization technique for high-throughput metabolomics?

In the earlier metabolomics studies, the GC/MS using EI mode has been widely employed (Pasikanti et al., 2008). GC/MS offers high chromatographic resolution as well as the ability to identify compounds from public libraries. Because of the polar character of most of metabolites, the use of GC/MS in metabolomics approach requires chemical derivatization (*vide supra*). GC/MS has some intrinsic drawbacks (i.e. time-consuming in sample processing like extraction and derivatization and chromatographic separation steps) to be fully recognized as a high-throughput fingerprinting research.

Since their first developments, API techniques appear as the most suited ionization methods for the analysis of polar and thermally labile metabolites. The comparison of different API techniques such as APCI, APPI and ESI is achieved by analyzing metabolite standards as displayed in Fig. 5.14. The results show that ionization yields of each molecule depends on the used technique, on the ionization polarity (An et al., 2010; Nordstrom, Want, Northen, Lehtio, & Siuzdak, 2008) and also to the intrinsic molecule properties. ESI source is able to ionize a large range of molecules. In addition, complementary molecular species can be detected in both positive and negative ESI mode for most of metabolites. This is related to the compound gas-phase properties. APCI displays ionization yield similar to that produced by ESI with the exception for some metabolites (e.g. cysteine), which show a better sensitivity in APCI. APPI is suitable for ionization of apolar or low-polar metabolites such as steroids (see Fig. 5.14). Thus, by cumulating data

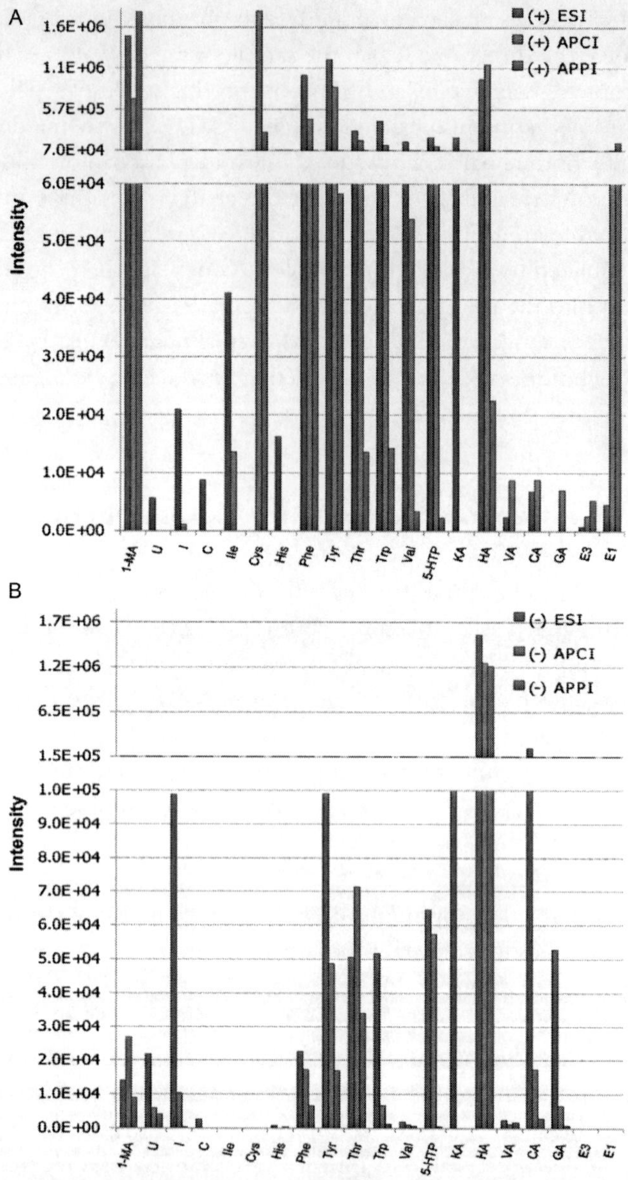

Figure 5.14 Average peak areas (within six LC–MS experiment repetitions) of representative metabolite standards using ESI, APCI and APPI techniques in both (A) positive and (B) negative ion modes. Typical metabolite standards are: 1-methyladenosine (1-MA), uridine (U), inosine (I), cytidine (C), L-isoleucine (Ile), L-cysteine (Cys), L-histidine (His), phenylalanine (Phe), L-tyrosine (Tyr), threonine (Thr), L-tryptophan (Trp), L-valine (Val), 5-hydroxy-L-tryptophan (5-HTP), kynurenic acid (KA), hippuric acid (HA), vanillic acid (VA), caffeic acid (CA), galic acid (GA), estriol (E3) and estrone (E1). LC–MS approach was conducted on a Q-TOF hybrid instrument. *Figure reproduced from An et al. (2010).* (For colour version of this figure, the reader is referred to the online version of this chapter.)

from various API modes more analytes can be detected increasing 'metabolite coverage'. Alternatively, dual sources and their ability to simultaneously perform two complementary ionization modes allow a gain in analysis time, which is an important aspect in high-throughput metabolomics. Additionally, to gain time, ambient API techniques such as DESI (Kauppila et al., 2007) and EESI (Chen et al., 2006; Chen, Wortmann, Zhang, & Zenobi, 2007) have been tested for global and targeted metabolomics. Since those techniques can be used without any sample treatment, they enable to perform high-throughput analysis and appear to be promising methods for metabolomics investigations.

The ability of API techniques to combine mass spectrometry with liquid chromatography (LC/MS) allows the development and the use of LC/MS in the analysis of metabolites and in metabolomics applications, especially using ESI technique (Roux, Lison, Junot, & Heilier, 2011; Wilson et al., 2005). The main advantage of LC/MS approach is the non-requirement to the derivatization step, facilitating the metabolite characterization through the m/z measurements and/or MS/MS experiments. Reverse-phase (RP) chromatography is the most commonly used separation technique for analyzing metabolites from biofluids as blood and urine. This technique involves the use of a hydrophobic (apolar) stationary phase and polar organic solvent as mobile phase. Hence, the most polar metabolites are first eluted since retention increases with decreasing polarity of metabolites. However, separation of all metabolites is never attained in all cases, because of large chemical diversity of metabolites (i.e. large scale of polarity). The main drawback of the RP chromatography is the insufficient retention on RP column of very polar compounds, which are eluted in the dead volume. Extensive research works have been developed for testing other physicochemical properties for stationary phase. For example, hydrophilic interaction liquid chromatography, which employs a polar stationary phase and polar organic solvent, is able to separate the most polar metabolites, which are usually un-retained onto a RP column (Idborg, Zamani, Edlund, Schuppe-Koistinen, & Jacobsson, 2005). The use of polar organic solvent allows overcoming the drawback of normal phase chromatography related to the insolubility of hydrophilic compounds in apolar mobile phase.

Nevertheless, LC/MS requires optimization of chromatographic conditions and results in long analysis time, which is incompatible with high-throughput metabolomics and serial analysis of biosamples (Bedair & Sumner, 2008). Chromatography performances can be improved with ultra-performance liquid chromatography (UPLC). The UPLC technology principle is to run separation using shorter column packed with particles smaller than 2 μm and higher flow rate for increased run speed, with superior peak capacity, higher resolution and sensitivity. In such conditions, UPLC system generates chromatographic peak width in the order of 2–3 s within a run of few minutes. This technique offers gain in time that is very attractive for high-throughput metabolomics studies. However, UPLC coupling with FT-MS instruments, which are based upon a time-domain transient (*vide supra*) as detection principle, presents a dilemna. Indeed, gain by performing fast chromatographic steps is in contrast to FT-MS optimal operating conditions where better peak resolution and mass measurement accuracy are directly proportional to transient duration.

Another system of coupling is the capillary electrophoresis CE-ESI-MS, which has been successfully applied for metabolomics (Bedair & Sumner, 2008; Ramautar et al., 2009). CE principle is based upon a separation mechanism completely different from that of LC, by which charged species (i.e. cationic, anionic and/or zwitterionic species) are separated in an electric field through an electrolyte buffer. It appears to be particularly well suited for the separation of polar and ionic compounds (e.g. well separation of polar amino acids as demonstrated by Desiderio, Iavarone, Rossetti, Messana, & Castagnola, 2010) and for the application to the analysis of polar metabolites in complex biological samples such as urine, plasma and whole blood. In the same objective, alternative ion-mobility coupled to mass spectrometry can be a promising approach especially for distinguishing structural isomeric compounds (Cuyckens et al., 2011).

A recent emerging analytical approach is the DIMS as an attractive method to perform high-throughput metabolome fingerprinting. The direct analysis of samples without the use of chromatographic separation has been achieved using pyrolysis mass spectrometry (Dumas et al., 2002; Goodacre & Kell, 1996) or ESI coupled to mass spectrometry (ESI-MS). (Bedair & Sumner, 2008; Castrillo, Hayes, Mohammed, Gaskell, & Oliver, 2003; Goodacre, York, Heald, & Scott, 2003; Kaderbhai, Broadhurst, Ellis, Goodacre, & Kell, 2003; Koulman et al., 2007). Those studies were initially performed on low-resolution instruments. DIMS analysis is susceptible to ion suppression effects, which are produced from competitive ionization of interfering compounds present in the biological matrix. The use of nano-ESI represents an alternative ionization technique to reduce the ionization suppression thanks to its higher ionization efficiency (Hop, Chen, & Yu, 2005). Alternatively HRMS offers better detection specificity. (Dunn, Overy, et al., 2005; Madalinski et al., 2008). With its highest dynamic range and mass resolving power, the FT-MS instrument appears as the technological solution for DIMS analysis by suppressing or at least minimizing its limitations. Indeed, a large number of compounds present in a wide concentration range in a complex mixture can be distinguished in a single FT-MS analysis. Hence, direct sample introduction method by using ESI coupled to FT-MS instrument appears as a very promising tool for high-throughput metabolomics analysis.

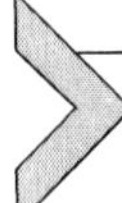

5. FT-MS APPLICATIONS FOR COMPLEX MIXTURE ANALYSIS

Global metabolomics analysis aims at recovering the sample composition, that is, obtain exhaustive metabolite detection and identification from a biological sample in given physiological conditions (Bedair & Sumner, 2008; Dettmer et al., 2007; Dunn, 2008; Dunn, Bailey, et al., 2005; Lindon & Nicholson, 2008; Villas-Bôas et al., 2005). High-throughput metabolite fingerprinting does not require structural characterization of metabolites to support extensive biochemical insights, but needs only the detection of a large number of metabolites, as many as possible. So, extensive works have been performed in development of complementary analytical platforms. Thanks to its sensitivity and specificity, mass spectrometer can be considered as a universal analytical tool, allowing detection of a large variety of molecules present in complex mixture in trace amounts. However, some compounds cannot be unambiguously identified, especially when the sample contains two or more molecules having the same nominal molecular mass. The overlapping between mass spectral peaks occurs if the mass resolving power of the analyzer is not high enough. Additionally, missing signals eventually occur because of the limited analyzer dynamic range. Yet, low-resolution mass spectrometer does not allow differentiation between well-known occurring metabolic processes such as (i) the methylation that corresponds to the addition of a CH_2 group and (ii) the hydroxylation (addition of one oxygen atom) followed by the oxidation (loss of H_2) because both metabolic processes lead an identical nominal mass shift of 14 u from the precursor. Therefore, ultra-high mass resolving power and high mass measurement accuracy FT mass spectrometers are required to perform DIMS analysis of complex mixtures since they can resolve at least partially such critical approach features with the advantage of EC determination ability (*vide infra*). FT-MS also offers characteristics and analytical performances in LC/MS approach for global as well as for targeted metabolomics studies (Junot et al., 2010).

5.1. DIMS approach using HRMS

DIMS approach is a promising analytical tool for high-throughput analyses since it enables the production of a mass spectrum from metabolites present in complex biological mixtures, in a few minutes. Despite various drawbacks mainly due to the matrix effects, such approach has been applied in several studies (Castrillo et al., 2003; Goodacre et al., 2003; Kaderbhai et al., 2003; Koulman et al., 2007). To suppress or at least to strongly minimize DIMS limitations, FT-MS instrument appears as the technological solution, with its highest dynamic range, mass resolving power and mass measurement accuracy. The best metabolome coverage can be obtained with FT-MS analyzers since it enables separation of very thin mass spectral peaks from isobaric species. Moreover, its powerful dynamic range allows detection of very low-abundant species from a single analysis (three or four orders of magnitude of relative peak intensity, see Marshall et al., 1998; Marshall & Hendrickson, 2002; Makarov, Denisov, Lange, et al., 2006). The relative peak intensities should correlate with the metabolite concentration but it is not always the case. Indeed, signal suppression effects exist during the MS experiments (ionization discrimination, detector saturation effects, etc.). When accurately calibrated in *m/z* value, ultrahigh-resolution mass spectrometer provides measurement of accurate *m/z* values, which are the keys for the determination of the ECs. The knowledge of compound EC is crucial for the structural elucidation of unknown biomarker metabolites in metabolomics studies.

To perform high-throughput analysis of biological samples, FIA approach using an ESI source coupled to an FT-MS analyzer seems to be an ideal analytical tool. Uneven performances of FT-MS instrumentations exist, that is, Orbitrap technology versus ICR cell. Orbitrap-based instruments display a limited mass resolving power compared to recent FT-ICR instruments equipped with high magnetic field (>7 T). A theoretical R_p value of 100,000 at *m/z* 400 is reached for the first generation of LTQ-Orbitrap hybrid instruments, which means a maximum time-domain signal duration of 1.9 s (Hardman and Makarov), whereas FT-ICR instrument routinely displays several hundreds of thousands of peak resolution at *m/z* until 1000 for acquisition of 1 s (Marshall & Hendrickson, 2002; Marshall et al., 1998). Additionally, the implementation of higher magnetic field enables enhanced number of detected peaks and consequently more complete metabolome coverage (Han et al., 2008). Higher magnetic field FT-ICR analyzers also allow a better mass measurement accuracy at sub-ppm (*vide supra*). Figure 5.15 shows the ESI mass spectrum acquired from human plasma extracts with a 12 T FT-ICR instruments (Han et al., 2008), where 570 distinct monoisotopic peaks with a signal-to-noise (S/N) ratio higher than 3 are collected (after data processing) within the *m/z*

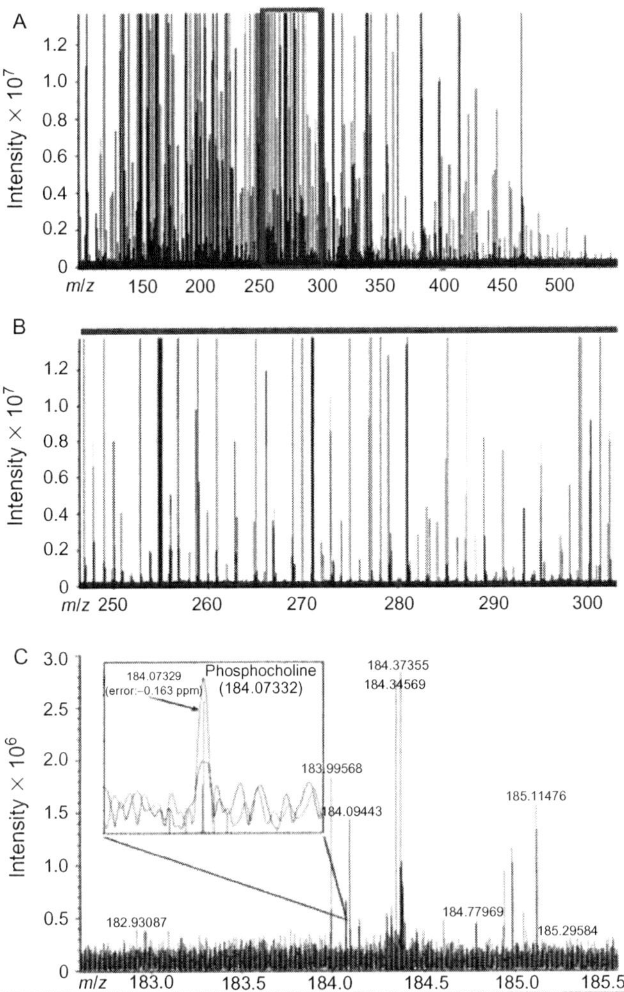

Figure 5.15 (A) Overlaid positive ESI-FT-ICR mass spectrum of four aqueous human plasma extracts (3 min of acquisition, using an internal calibration procedure and spiked standards). (B) Expansion of the *m/z* 250–300 range showing the mass resolution of 12 T FT-ICR instrument. (C) Identification of relatively low-intensity ion signal at *m/z* 184.07329, which is attributed to phosphocholine within an error of 0.163 ppm. *Figure reproduced from Han et al. (2008).* (For colour version of this figure, the reader is referred to the online version of this chapter.)

90–600 mass range. The powerful analyzer dynamic range allows the detection of very low-intensity peaks such as the peak at *m/z* 184.07329 in Fig. 5.15, which is subsequently annotated as the $C_5H_{15}N_1O_4P_1$ formula using an automatic search algorithm and identified as the phosphocholine species after DB query (*vide infra*).

So, the application of DIMS approach using an FT-MS instrument has been reported for numerous global metabolomics studies as investigated by Madalinski et al. (2008), Oikawa et al. (2006) and Takahashi et al. (2008). For example, Lucio et al. (2010) employed the FT-ICR instrument in non-targeted metabolome analysis of plasma samples from individuals displaying different insulin sensitivities. Similarly, Liger-Belair et al. (2009) used the DI-HRMS approach to record the metabolomics fingerprints of champagne wine and its aerosols. Differential metabolomics analysis can also use DI-HRMS approach for metabolite profiling, for example, to resolve metabolic activities of Arabidopsis cells under light/dark conditions (Nakamura et al., 2007) or to screen the metabolic differences between pre- and post-perfusion grafts of patients undergoing orthotopic liver transplantation (Hrydziuszko et al., 2010).

Another important advantage of FT-MS is the low ppm range of the mass error and the stability over time of the mass measurement for serial analysis of complex mixtures. Indeed, accurate mass measurements facilitate EC determination (Erve, Gu, Wang, DeMaio, & Talaat, 2009; Herniman, Langley, Bristow, & O'Connor, 2005; *vide infra*). However, small and short drift of mass calibration is also a critical factor for metabolite profiling, where reproducibility is an important feature for subsequent data processing (Bristow & Webb, 2003; Han et al., 2008; Thompson et al., 2003). Note that dynamic range performance directly impacts the accuracy of low-intensity peak measurements. The specifications of LTQ-Orbitrap are mass accuracy up to 2 ppm with internal calibration and 5 ppm with external calibration (Makarov, Denisov, Kholomeev, et al., 2006; Makarov, Denisov, Lange, & Horning, 2006). Xu et al. (2010) have tested DIMS analysis of mixture of 137 commercial authentic compounds of metabolites using an LTQ-Orbitrap hybrid instrument and have reported accurate mass measurements using external calibration with errors in the ppm range within 24 h, and a mass accuracy below 5 ppm over a 1-week period. A real improvement is obtained by internal calibration procedure but it is not always easy to find the ideal internal standard. Moreover, eventual interaction can occur between internal standard and matrix compounds. Alternative 'lock mass' function enabled in LTQ-Orbitrap instruments allows a real-time internal recalibration of MS and also MS^n data (Olsen et al., 2005) eventually using the background ions as standards. This may really improve further HRMS data processing (Koulman et al., 2009; Werner, Croixmarie, et al., 2008).

Optimal operating conditions can be applied by the scientists for improving analytical performances but sometimes, one parameter is optimized at the expense of the other. For example, enhanced metabolome coverage can be obtained by the raise of the number of stored ions (enhanced dynamic range). However, space charge effects may happen within peak broadening leading to inaccuracy mass measurement and eventually to peak coalescence (*vide supra*). Although modern FT-MS instruments allow an external ion accumulation before ICR cell injection, the AGC procedure should be carefully controlled to ensure the number of ions transferred to the ICR cell, as demonstrated by Southam, Payne, Cooper, Arvanitis, and Viant (2007). Since the number of metabolite ions to be detected is anyways limited in mass spectrum using full scan acquisition mode, 'Stitching' method has been proposed by Southam et al. (2007) to obtain enhanced analytical performances. They employed 12 T FT-ICR instruments for the acquisition of multiple selected ion monitoring (SIM) windows (within a small m/z range of 30 u and overlapping SIM windows) such as their sum constitutes the whole mass range (from m/z 100 to m/z 500). Using this method, an enhanced number of detected peaks (more than 3000) is obtained within a small average mass error of 0.18 ppm. Figure 5.16 compares the mass spectra acquired in broadband acquisition mode (using two AGC conditions, Fig. 5.16A and B) with the resulting mass spectrum using the 'stitching' procedure (Fig. 5.16C).

5.2. FT-MS instrumentation in LC/MS approach

Even if we mainly discussed about very high-throughput metabolomics analysis using DI-HRMS approach, FT-MS instrumentation has been extensively combined with separation technique (mostly LC technique coupled with API sources). The separation technique offers the essential advantage of isomeric compound resolution, compared to the single DIMS scenario approach. Additionally, LC/MS eliminates (or at least minimizes) the signal ion suppression and matrix effects unavoidable in mass spectrometry during ionization step. Such phenomena are highly dependent on compound structure and biological matrix. However, the separation technique suffers of its own limitations as time-consuming and less reproducible data format (i.e. retention time shift). In the specific case of LC/FT-MS

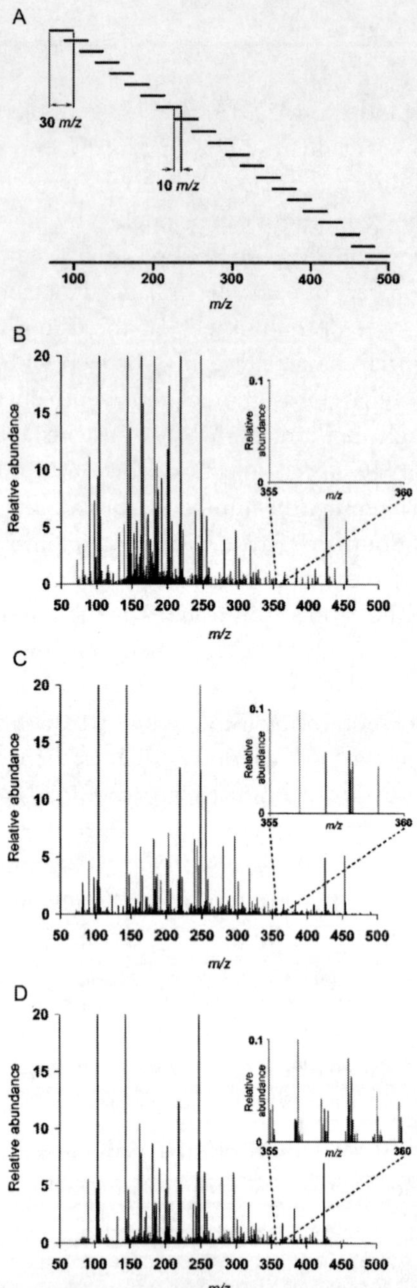

Figure 5.16 (A) Schematic of the optimized SIM-stitching method, which consists in acquisition of adjacent 30 u SIM windows, overlapping by 10 u and acquired for 15 s, to cover a total scan range of *m/z* 70–500. The AGC setting of 1.10^5 ions was applied for ion transmission to the ICR analyzer (7 T). Below are represented results obtained for a fish liver extract, that is, (B) the full FT-ICR mass spectrum with AGC target value of 1.10^5, (C) that with AGC target of 5.10^5 and (D) the mass spectrum resulting from optimized SIM-stitching method. All mass spectra were normalized to the largest peak, and the zoom of *m/z* 355–360 range is displayed in insets. The total acquisition time for each experiment was 5.5 min. *Figure reproduced from the works of Southam et al. (2007).*

approach, the space charge effect phenomenon is limited and, thus, allows reaching higher peak resolution and accurate mass measurements into the ICR cell. Nevertheless, a compromise must be found between optimal LC conditions for high-throughput analysis and the fact that the most accurate mass measurements are obtained through long time-domain signal acquisition (Fig. 5.11). Indeed, chromatographic peak width of 10 s means below 10 points for FT-MS acquisition of 1 s. Generally suitable sampling of data points across the chromatographic peak implies the decrease of the time-domain signal duration detrimental to achievable peak resolution (unless the access to higher magnet instruments). For example, most LC/MS experiments using Orbitrap-based instruments have employed short transients with mass resolving powers of 7500 or 15,000 (Kamleh, Hobani, Dow, & Watson, 2008; Kiefer, Portais, & Vorholt, 2008; Peterman, Duczak, Kalgutkar, Lame, & Soglia, 2006).

Thus, employment of FT-MS instruments in LC/MS approach remains an attractive analytical platform for metabolomics analysis owing to their respective performances (Wilson et al., 2005). For example, unresolved chromatography separation of metabolite peaks can be easily corrected by HRMS detection, especially for isobaric species (see Kiefer et al., 2008), while reduction in the chromatographic requirements is useful for reduction of the analytical time. UPLC, which displays improved chromatographic separation, peak capacity and sensitivity, has been successfully coupled to LTQ-Orbitrap instruments as reported by Dunn et al. (2008) for metabolite profiling of serum samples. Furthermore, another specificity of HRMS in metabolomics profiling is the high selectivity of metabolite detection by using the high-resolution extracted ion chromatography (EIC; Kamleh et al., 2008; Koulman et al., 2009; Peterman et al., 2006; Zhang, Yu, et al., 2009). Indeed, the extracted ion chromatographic traces based on accurate m/z values can be used to reveal the presence of isomeric metabolites (Peterman et al., 2006) or to facilitate targeted metabolomics analysis as applied by Koulman et al. (2009), who demonstrated the feasibility of simultaneous non-targeted and targeted metabolomics analysis using LC/HRMS for plasma samples spiked with phospholipids.

5.3. FT-MS instrumentation for metabolite quantification

An application less discussed here is the ability of FT-MS for metabolite quantitation. Yet, metabolite quantification is indispensable in metabolomics investigations for evidencing metabolic changes and for a better understanding of biological processes. This can be achieved in two manners, that is, relative quantification generally used in untargeted metabolomics or absolute quantification usually employed in targeted metabolomics. Relative quantification consists to normalize the signal intensity (or the peak area) of a metabolite of interest to that of a reference compound, which can be a molecule introduced in the samples or another metabolite considering to have a constant concentration. Thus, the measured metabolite levels are expressed as the peak ratios corresponding to their apparent relative concentrations since the ionization yield of each compound should be different. Absolute quantification implies the analysis of a reference standard at different concentrations to perform a calibration curve, indispensable for the determination of the absolute metabolite concentration.

The best and unrivalled method for such investigation is the selected reaction monitoring (SRM) mode using a triple quadruple tandem mass spectrometer (de Hoffmann & Stroobant, 2007). SRM scanning proceeds through the tracking of a given fragmentation way for each analyte of interest. The first quadrupole filter selects the precursor ion, the second (RF-only) quadrupole works as a collision cell and the third quadruple as a mass filter specifically detects diagnostic product ion. When multiple metabolites need to be simultaneously analyzed, this process is repeated for each metabolite in a cyclic manner. SRM method provides high selectivity and sensitivity with an incomparable limit of detection (LOD), for example, a LOD lower than 1 μg/mL has been obtained for 144 cellular metabolites by Bajad et al. (2006). However, development of quantitation SRM-based method including optimization of chromatography conditions and determination of the MS/MS parameters for each metabolite of interest is time-consuming. The application of SRM to metabolomics is then limited because of a relatively large number of metabolites to quantify, which exceeds the possible number of MS/MS experiments per analysis. Another drawback of SRM is that it is not possible to screen metabolites that are not identified. Post-targeted metabolomics analysis cannot be performed, that is, acquired MS data cannot be re-examined for the search of novel biomarker metabolite.

So, the ideal approach is to record mass spectra and to screen species of interest from EICs. The FT-MS instrumentation is well suited for such purpose since the selectivity of detection is ensured by the very high-mass resolving power and the accurate mass measurements. Moreover, quantification of a huge number of metabolites can be performed in a single run without the need of the pre-selection procedure like SRM transitions.

Hence, the full scan FT-MS becomes very attractive for the quantification of targeted and post-targeted metabolites in the metabolome scale.

Recently, the application of single MS acquisition approach for targeted metabolomics has been reported using the FT-MS instrumentation with or without the chromatographic system (Han et al., 2008; Kiefer et al., 2008; Madalinski et al., 2008; Zhang, Yu, et al., 2009). For example, an LC/MS experiment using the LTQ-Orbitrap has been achieved by Kiefer et al. (2008) allowing the quantification of 22 metabolites having concentrations ranging from 0.13 to 55.6 μM in cell extracts. Although most of peaks were not chromatographically separated, they were unambiguously identified thanks to the high accurate mass measurements, except for structural isomers such as leucine/isoleucine and glucose 6-phosphate/fructose 6-phosphate. The use of the FIA and the FT-MS instrument has also been reported by Madalinski et al. (2008), who have studied the cadmium toxicity in the yeast. They demonstrated that despite the occurrence of matrix effects, the DIMS method using LTQ-Orbitrap allowed a fast quantification of sulphur metabolites with acceptable precision. The obtained results were similar to that provided by performing common LC/MS/MS experiments onto a triple-quadrupole instrument (see Madalinski et al., 2008 and the works of Zhang, Yu, et al., 2009). Han et al. (2008) have also reported comparable results of DI/FT-ICR-MS and LC/MS (using an ion trap) techniques for the absolute quantification of choline in human plasma with sub-pmol sensitivity.

A major difficulty encountered in metabolite MS quantification is its inaccuracy (due to the non-proportionality between detected signal and concentration of metabolite and/or to missing data). This is due to the presence of interfering species that cause the matrix effects (Annesley, 2003). Indeed, ion signal suppression or enhancement due to the matrix effects affects the quality of metabolite quantification leading to an inaccurate determination of metabolite concentration, that is, underestimation or overestimation of measurements (Rathahao et al., 2005). The introduction of internal standard in the sample to be analyzed is a solution to overcome the problem of inaccurate quantification by controlling the detected signal variations during the analysis and the matrix effects. The stable isotope-labelled form of a metabolite represents the best internal standard for absolute quantification since it should display the same ionization efficiency than that of the corresponding targeted metabolite. The isotope dilution method is usually used (Sargent, Harte, & Harrington, 2002). The method proposed by Wu et al. (2005) utilizes uniformly ^{13}C-labelled metabolites from a cell extract as internal standards to quantify the microbial metabolome by LC/MS/MS technique. ^{15}N metabolic labelling has also been used for absolute quantification of nitrogen-containing metabolites in cells (Lafaye, Labarre, Tabet, Ezan, & Junot, 2005). However, quantitative metabolomics involves identification and quantification of a large number of metabolites (i.e. as many metabolites as possible) but the stable isotope-labelled form of metabolites is expensive, often commercially unavailable or cannot be chemically synthesized. An alternative way is the introduction of stable isotope-labelled tags to metabolites using chemical labelling reaction. A dimethylation reaction by using commercially available ^{13}C-labelled formaldehyde as labelling reagent has been described by Guo, Ji, and Li (2007) for quantitative analysis of amine-containing metabolites. Absolute quantification of 20 amino acids and 15 amines in human urine was performed using this labelling strategy. Similarly, Guo and Li (2009) reported the quantification of more than 90 metabolites present in human urine samples in a total run of 12 min by using isotope dansylation labelling and fast LC/FT-ICR approach.

6. METABOLITE IDENTIFICATION FROM HRMS DATA

A very large number of up to thousands of metabolites can be simultaneously measured in a single MS analysis. Because of the wide diversity of their chemical classes (i.e. sugars, amino acids, steroids, etc.) and the heterogeneity of their structure, metabolite identification (i.e. structural identification) from complex mixture represents one of the major challenges in metabolomics investigations. Of course, high mass resolving power mass spectrometer and accurate mass measurements facilitate such global metabolomics analysis. Nevertheless, identification of metabolites of interest from global, high-throughput metabolomics implies several levels of investigation (Han et al., 2008; Oishi et al., 2009):

- *Metabolite peak detection* constitutes the first critical step. After extraction of mass spectral data into the desired format, the resulting *m/z* value list and linked informations (as the peak intensity or eventually retention time for LC/MS or GC/MS experiments) can be preprocessed using statistical analysis (e.g. multivariate analysis) to detect the putative biomarker metabolite ions.
- The *EC determination* represents the following analytical step. This procedure is achieved for the ions of interest detected in the previous step,

usually using automatic search algorithm. The efficiency of algorithm correlates the correct annotation of peaks, for example, adduct species or eventual product ions recognition, but also the restrictive rules implemented into the search algorithm. Complementary tools have also been developed to facilitate identification of metabolites by improving visualization of HRMS data.

– The final step is the *metabolite structural elucidation*. Getting a single chemical composition is not sufficient for metabolite identification. Supplementary efforts are needed to resolve the metabolite structure. The most direct search way is the mass spectral DB query from the m/z values of metabolite ions of interest to provide putative identification. However, there are still many metabolites that cannot be identified through mass-based search. So additional physico-chemical (and/or biological information) is required, such as informative MS/MS and MS^n experiments. This latter approach can employ complementary activation modes and more rarely ion–molecule reactions (Habicht, Duan, Vinueza, Fu, & Kenttämaa, 2010).

Note that a new trend in metabolomics investigations is to perform metabolite identification in an exhaustive manner or at least the most exhaustive as possible. In the other words, the extracted m/z list is processed for peak annotation of all signals.

6.1. Mass spectral data

Extraction of mass spectral data provided from the raw mass spectrum needs automatic algorithm software and adapted data processing. Almost commercial instruments proposed such software. Home-made algorithms have also been developed for improvements of data processing (e.g. peak realignment to correct the shift in retention time) and peak annotations. Preprocessing is generally performed for all types of spectral data (LC/MS, GC/MS, MS and even sequential MS^n experiments). For example, suppression of background chemical noise, peak picking and mass spectral deconvolution are typical processes. MS data can be eventually pre-treated with centring, scaling and/or normalization processes. Optionally, different spectral alignment tools can be used to align or compare multiple GC/MS, LC/MS, LC and even NMR data sets (Katajamaa & Orešič, 2005; Smith, Want, O'Maille, Abagyan, & Siuzdak, 2006).

Afterwards, automatic m/z list extraction should be performed. Instrument performances and above all, the mass measurement accuracy should be taken into account (number of digits of accurate *m/z* values obtained from high or very high mass measurement accuracy (sub-ppm)). In addition, MS data interpretation should be accurately performed to correctly attribute the number and the type of charge. So, instrumental set-up is important and above all, the used source (*vide supra*). EI technique as energetic in-vacuum ionization mode often displays mass spectra without radical molecular ions. Alternatively, ESI sources allow soft ionization of polar systems thanks to the quasi-thermal energy distribution characteristic of the technique and the possibility to control desolvation process into the interface region. According to the ESI source operating parameters and/or sample conditions, redundant signals can be produced for one compound. Often Na^+, K^+ or $NH_4{}^+$ adduct species are often produced concomitantly with (or without) the protonated molecules. Eventually, non-specific multimer species (dimer, trimer) and/or fragment ions produced during controlled in-source fragmentations can be present in the mass spectrum.

Signal redundancy is resolved in LC/MS experiments by correlating retention time information with peak *m/z* values. For example, autocorrelation matrices as initially developed in NMR approach can highlight redundant signals in LC/MS experiments (Werner, Croixmarie, et al., 2008). In contrast, DIMS experiments display very complex mass spectral data. Nevertheless, HRMS and accurate mass measurements allow direct identification of adduct ions by searching specific mass differences, for example, 21.98195 u mass shifts between protonated and sodiated species. Therefore, redundant signals as adduct species and also isotopic peaks can be annotated or eventually subtracted in a preprocessing step (Draper et al., 2009) from the accurate *m/z* list before subsequent EC generation. Data preprocessing programs have been also developed to automatically annotate product ions resulting from characteristic small size neutral losses such as H_2O, NH_3, SO_3 and isobaric HPO_3 groups, etc., of a predicted mass shift list.

Finally, a decrease in the number of signals significantly reduces the number of *m/z* values which have to be forwarded for high-throughput EC determination step. Eventually, *a priori* multivariate analysis can be performed in differential metabolomics study to reveal the putative biomarker metabolites (Hrydziuszko et al., 2010; Nakamura et al., 2007; Oikawa et al., 2006). However, automatic data preprocessing, pretreatment can also be a source of errors due to artefact signals annotation and/or missing signals of potential metabolites of interest.

6.2. Elemental composition determination

EC determination provides important information indispensable for the structural elucidation of metabolites. It is usually performed based on accurate *m/z* values provided by FT-MS instrument.

6.2.1 Number of possible elemental formulae

Determining the EC from an accurate *m/z* value is less straightforward than often believed (Koch, Dittmar, Witt, & Kattner, 2007; Kujawinski & Behn, 2006). Manual formula assignment is not easy and extremely time-consuming. Therefore, automatic data postprocessing is important and critical for an efficient exploitation of the FT-MS complex data. Formula generator softwares are available in commercial FT-MS instruments. Typically, under formula computation, atoms are combined iteratively until the resulting exact molecular mass matches the experimental accurate mass (i.e. within a given mass measurement accuracy, see Fig. 5.6). A list of all possible ECs is then generated.

Nevertheless, the number of possible ECs for a given accurate *m/z* value is directly related to the accuracy of mass measurements, which constitutes the first limiting factor to correctly assess the chemical composition (*vide supra*). Even data with ultrahigh mass resolution and mass measurement accuracy below ppm are generally insufficient to obtain a single EC without ambiguity. In fact, Kind and Fiehn (2006) proved the impossibility to assign one unique chemical formula for masses upper 126 u in an exhaustive manner with only a mass error of 1 ppm (with a restricted C, H, O, N and P set of elements). Moreover, the number of possible chemical formulae increases exponentially with the increasing ion *m/z* values (see Fig. 5.6). Additional information or restrictive parameters should be implemented into the search procedure to reduce the number of possible combinations (i.e. list of candidate ECs) and most important to remove irrelevant results (Kind & Fiehn, 2007).

6.2.2 Basic chemical criteria

So, applying constraints is the most crucial step during the whole process of formula finding. First exclusion criterion is the restrictive choice of elements considered in the automatic search algorithm. However, it implies a basic knowledge of sample origins (i.e. plants, cells, animals and eventual sample preparation). Typical formulae as $C_cH_hN_nO_oS_s$ are postulated for petroleum samples. In the case of metabolome, large chemical diversity of compounds exists and generally, the most restricted set of elements takes into account C, H, O and N atoms. The presence of P and S (maximum number of two atoms) is also considered whereas metals are *a priori* excluded. Nevertheless, halogenated compounds could also be eventually detected in biological samples if xenometabolites are generated from metabolism of xenobiotics such as halogenated pesticides. Additionally, restriction of element content (i.e. maximal number of hydrogen, carbon atoms, etc.) and element number ratios can also facilitate the EC search procedure.

Physico-chemical rules are also implemented in automatic formula generator software. The most popular chemical rule is the *nitrogen rule*, which states that the odd mass of a molecule implies the presence of an odd number of ^{14}N atoms in this molecule. This is explained by the fact that the ^{14}N atom has an odd number of valences and an even mass. However, one must keep in mind the basic knowledge of mass spectrometry technique that is even electron species are almost produced by API techniques (exceptionally radical group losses are detected for N-containing compounds as observed by Fabre et al., 2000), whereas EI typical molecular ions are cation radical species. The number of rings plus double bonds (i.e. the aromatic character) can be also a restrictive parameter. Double bond equivalent (DBE) can be calculated from a theoretical EC, by taking into account the respective valence of elements:

$$\mathrm{DBE} = -\frac{A_1}{2} + \frac{A_3}{2} + A_4 + \frac{3}{2}A_5 + 2A_6 + 1 \quad (5.15)$$

where A_1 to A_6 are the number of atoms with valences state from 1 to 6 and $A_2 = 0$.

For example, DBE value of a $C_cH_hN_nO_o$ formula is equal to $(c - h/2 + n/2 + 1)$ expression with c, h and n the respective number of tetravalent C, monovalent H and trivalent N atoms. Note that other elements characteristics of metabolite structures as halogens (valence state equal to 1) can be easily implemented in DBE calculations. Other elements such as sulphur (valence states of 2, 4 and 6) and phosphorous (valence states of 3 and 5) render more complex the DBE calculation and even can be a source of errors (Kind & Fiehn, 2007). For example, large-size molecules containing several heteroatoms can display elements at different valence states. Finally, consistent DBE values with existing chemical compounds are above −0.5 and integer DBE values are characteristics of odd-electron ions whereas non-integer value signals even-electron species. Thus, integer values are automatically removed from the solution sets from compounds ionized by API techniques.

6.2.3 Isotope abundance patterns

Additionally, isotopic patterns can be a useful filter in calculating ECs. Kind and Fiehn (2006) showed that 95% of potential formulae are removed for masses above 500 u by implementing data-filtering processing using isotope patterns (within 3 ppm of mass error and 2% of relative error for isotope abundance). Manually or automatically, the comparison between simulations and experimental data (*vide supra*) is very helpful for MS data interpretation and EC determination (Kind & Fiehn, 2006; Koch et al., 2007; Miura, Tsuji, Takahashi, Wariishi, & Saito, 2010; Stoll, Schmidt, & Thurow, 2006). However, to include isotope pattern information into the EC calculations, the main interrogation is about the reliability of isotopic peak intensity and distribution, detected in the mass spectra above all, for the most complex DIMS data.

Of course, ultrahigh resolution instruments are of special interest allowing highly resolved isotopic peaks but, with a finite resolution and eventual isotopic peak overlapping. The isotopic contribution of isotopic peaks (e.g. ^{15}N, ^{13}C, ^{34}S and ^{18}O element substitutions) is assigned by detecting the mass difference between the monoisotopic and each isotopic peak. For example, the mass difference between $^{13}C_1{}^{12}C_{n-1}$ peak and the monoisotopic peak is 1.0034 u. Of course, calculation of the mass defect and/or the relative mass defect values of isotopic peak clusters provides the isotopic peak identification (*vide supra*). Afterwards, the area ratio between the monoisotopic and other isotopic peaks gives an estimation of the number of a given element (Koch et al., 2007; Miura et al., 2010). The restriction on the number of elements can then be used to reduce the number of possible chemical formulae during the EC search procedure.

The most critical parameter is the error in experimental isotopic ion abundances. Few studies on the accuracy of relative isotope abundance (RIA) have been reported for DIMS experiments on complex mixture samples. Moreover, the RIA values vary for each instrumental set-up and the RIA uncertainties correlate with operating principle of the analyzer. Grange, Zumwalt, and Sovocool (2006) have reported the RIA measurement error of $<20\%$, using an orthogonal accelerated time-of-flight (oa-TOF) mass spectrometer coupled to ESI technique. RIA errors certainly arise from the limited TOF analyzer dynamic range. In an FT/ICR instrument (7 T), a deviation of carbon content between the assigned formula and the calculated value (C_{dev}) of about 1.6 C atoms was found by Koch et al. (2007) in natural organic matter samples. Similar evaluation of the LTQ-Orbitrap hybrid instrument, in terms of mass measurement accuracy and

RIA measurement has been performed by Erve et al. (2009) onto 10 natural products and by Xu et al. (2010) who tested 137 commercial authentic compounds of metabolites. Xu et al. (2010) observed experimental errors systematically below 20% of their theoretical values, but above a signal threshold.

Indeed, absolute peak intensities appear as the principal limiting factor of RIA measurement accuracy. The largest RIA measurement errors of low-intensity peaks may arise from the insufficient dynamic range of Orbitrap analyzer, even if it is the most powerful. Xu et al. (2010) also assumed in the LTQ-Orbitrap hybrid instrument, that RIA errors could result from a variable ion trapping efficiency related to the detection mass window. Recently, Weber, Southam, Sommer, and Viant (2011) have published the accuracy of RIA measurements of ^{13}C isotopic peaks from PEG standards spiked in biological samples, using both LTQ-FT-ICR (mass resolving power of 100,000–750,000) and LTQ-Orbitrap ($R_p = 100,000$) ultrahigh mass resolving power mass spectrometers. They have similarly demonstrated that the quality of RIA measurements is limited by the S/N ratios as exemplified in Fig. 5.17. The authors concluded on the less accurate RIA values onto the Orbitrap-based values. Note that the RIA ratio measurements display large uncertainties for compounds characterized by a molecular mass below 200 u, because of the low abundances of related ^{13}C isotopic ions, as observed in Fig. 5.17. Changes in the mass resolving power values of both FT-Orbitrap (Erve et al., 2009; Weber et al., 2011; Xu et al., 2010) and FT-ICR instruments (Stoll et al., 2006; Weber et al., 2011) can impact also the RIA measurement accuracies. Indeed, longer signal acquisition times of ions necessary for raising the wished peak resolution implies to record a greater amount of noise, and thus results in an overall reducing of the S/N ratio (*vide supra*).

In fact, most of publications have reported the determination of ^{13}C-isotopic peaks with lower intensities compared to those provided by the simulations. Systematic underestimation of isotopic peak intensity measurements from FT-MS instruments may originate from different phenomena. The used Fourier deconvolution algorithm can be responsible for discriminating low-abundance ions. Moreover, relative abundances of trapped ions of closely spaced cyclotron frequencies can be biased. In addition, space charge effects could eventually occur, leading to ion–ion interactions into the ion cloud and even to peak coalescence (Marshall et al., 1998). Now, to limit and even to suppress the space charge effects, the number of stored ions is externally controlled in most recent hybrid

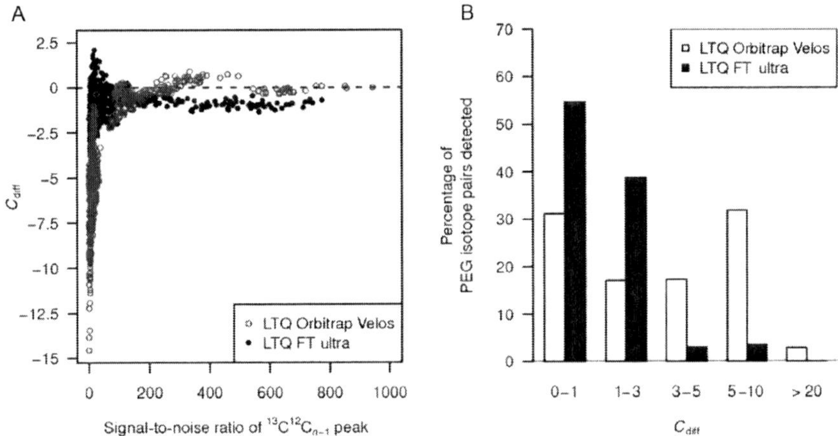

Figure 5.17 Comparison of the RIA measurement accuracy (i.e. through calculation of C_{diff} values) obtained on the LTQ-FT(-ICR) Ultra and LTQ-Orbitrap Velos high-resolution mass spectrometers (from Thermofisher). Only measurements of $^{13}C^{12}C_{n-1}$-containing PEG peaks (standards spiked in a biological matrix) were used in these plots. (A) C_{diff} values plotted against the signal-to-noise ratios of the $^{13}C^{12}C_{n-1}$-PEG peaks. (B) Distribution of C_{diff} values of PEG isotopic peaks detected in the *m/z* 70–590 range (with an S/N ratio greater than 3.5). C_{diff} was calculated by subtracting the theoretical number of carbons from the empirically calculated value as expressed in the following equation: $C_{diff} = \frac{100 \times \text{relative abudance } ^{13}C^{12}C_{n-1} \text{ PEG peak}}{1.10 \times \text{relative abundance } ^{12}C_n} \text{ PEG peak} - \text{theoretical no. of carbons } PEG_{parent}$. The authors used SIM-stitching scanning mode with direct infusion (DIMS approach) on an FT-MS instrument as demonstrated by Southam et al. (2007), that is, they collected and 'stitched' multiple overlapping SIM scans to reconstitute the whole mass spectrum enhancing the metabolome coverage. *Figure reproduced from the works of Weber et al. (2011).*

FT-MS instruments thanks to external ion accumulation systems (Senko et al., 1997).

Finally, Kind and Fiehn (2007) have described a typical automatic algorithm for EC determination, based on seven heuristic rules. Among them, (1) restriction on the number of elements and (2) Valence rules are basic constraints used in formula generator. Additionally, the authors filter ECs on the basis of (3) isotopic patterns, but also on restricted (4) H/C and (5) heteroatom ratios and also on (6) element content probabilities. Optionally, they have also included (7) the presence of trimethylsilylated compounds for particular GC/MS applications using EI technique (derivatization step). These rules have been implemented in an automated Excel® script, used for EC searching with accurate *m/z* values. It must be underlined that the reported RIA errors are considerably always greater than the value considered by Kind and Fiehn (2007) for testing isotope ratios, one of the seven-rule algorithm for filtering of molecular formulas.

6.3. Graphical methods and data mining for HRMS data

Detection and identification of metabolites from HRMS complex data is very challenging. Moreover, for MW of compounds exceeding 500 u, no single elemental formula can be easily determined even with the previously cited constraints. In fact, a very large number of possible elemental formulae exist for MW above 500 u and the decrease in the mass measurement accuracy for higher *m/z* value renders too complex and too long the data processing (see Fig. 5.6). To facilitate the EC determination of unknown metabolite species particularly at high *m/z* values in complex mixture, filtering tools can be implemented in data processing by exploiting the compound characteristics, that is, its (relative) mass defect and it EC, which are revealed by HRMS detection (Sleno, 2012). A valuable data-processing method is to selectively identify the *m/z* ratios of metabolite ions according to their common substructure (as 'functional group relationships'). For example, 'homologous series' of compounds connected through a specific mass shifts, such as 14.01565 u, 36.4 mu or 0.99525 u, can be evidenced in HRMS experiments. Indeed, the mass difference of 14.01565 u is attributed to the addition of a $^{12}CH_2$ group, the 36.4 mu mass shift is due to the exchange of $^{12}CH_4$ versus ^{16}O, and the exchange of $^{12}CH_2$ group with ^{14}N atom results in a mass difference of 0.99525 u. Hence, graphical methods allowing extraction of metabolite ions of interest according to a given 'functional group relationships' constitute a way to help in compound identification. They allow extending EC determination of annotated low *m/z* compounds to unknown high *m/z* compounds, additionally being a useful tool for representation of complex HRMS data.

6.3.1 Kendrick mass analysis

Among the possible data-processing approaches, the Kendrick mass analysis (Kendrick, 1963) is a well-suited method for petroleum complex mixture analysis (Hughey, Hendrickson, Rodgers, Marshall, & Qian, 2001). In this approach, high-resolution mass spectra are rescaled from the IUPAC mass scale (based on the ^{12}C atomic mass as exactly 12 u) to the Kendrick mass scale, according to the following equation:

$$\text{Kendrick mass} = \text{IUPAC mass} \times (14/14.01565). \qquad (5.16)$$

Then, homologous series of compounds with the same heteroatom composition and the same number of insaturations, but different numbers of CH_2 units have identical Kendrick mass defect, defined by:

$$\text{Kendrick mass defect} = \text{nominal Kendrick mass} - \text{exact Kendrick mass}. \quad (5.17)$$

A Kendrick plot is obtained by plotting the Kendrick mass defects versus the nominal Kendrick masses. This compact mass spectral diagram allows visual resolution of up to thousands of peaks as homologous series. Consequently, the EC assignment of higher m/z values is facilitated based on the identified lower m/z peaks of homologous series. Compounds containing the same heteroatom composition and the same number of insaturations (DBE values) are found on a single horizontal line with separation by unit of 14 u nominal Kendrick mass and the same Kendrick mass defect value. On the other hand, compounds of different heteroatom compositions are displaced vertically from those of other classes, because of their different Kendrick mass defects. In addition, every additional ring or double bond not only reduces the nominal Kendrick mass by 2 u but also increases the Kendrick mass defect.

Werner, Heilier, et al. (2008) have reported a Kendrick plot for 7000 compounds extracted from the KEGG DB. They observed (see Fig. 5.18A) in the region of mass defect values between 0.928 and 0.956, the homologous series of saturated and mono-unsaturated carboxylic fatty acids (Fig. 5.18B). If small compound metabolites display a large chemical diversity, a limited range of metabolites belongs to the same class (heteroatom content) and the same DBE value, with variable aliphatic group size. So, HMRS data representation based upon the CH_2 group relationships offers limited interest in metabolomics studies, except for specific field of lipidomics (He et al., 2007; Hughey et al., 2001).

6.3.2 van Krevelen diagrams

Another alternative way to visual analysis of HRMS data is the van Krevelen diagram, which has been introduced as a graphical method for analyzing elemental data of coal samples. Such a diagram is obtained by plotting element ratios, for example, atom number ratio of hydrogen to carbon (H/C) versus that of oxygen to carbon (O/C; Kim, Kramer, & Hatcher, 2003; Wu, Rodgers, & Marshall, 2004). Three-dimensional plots can be performed

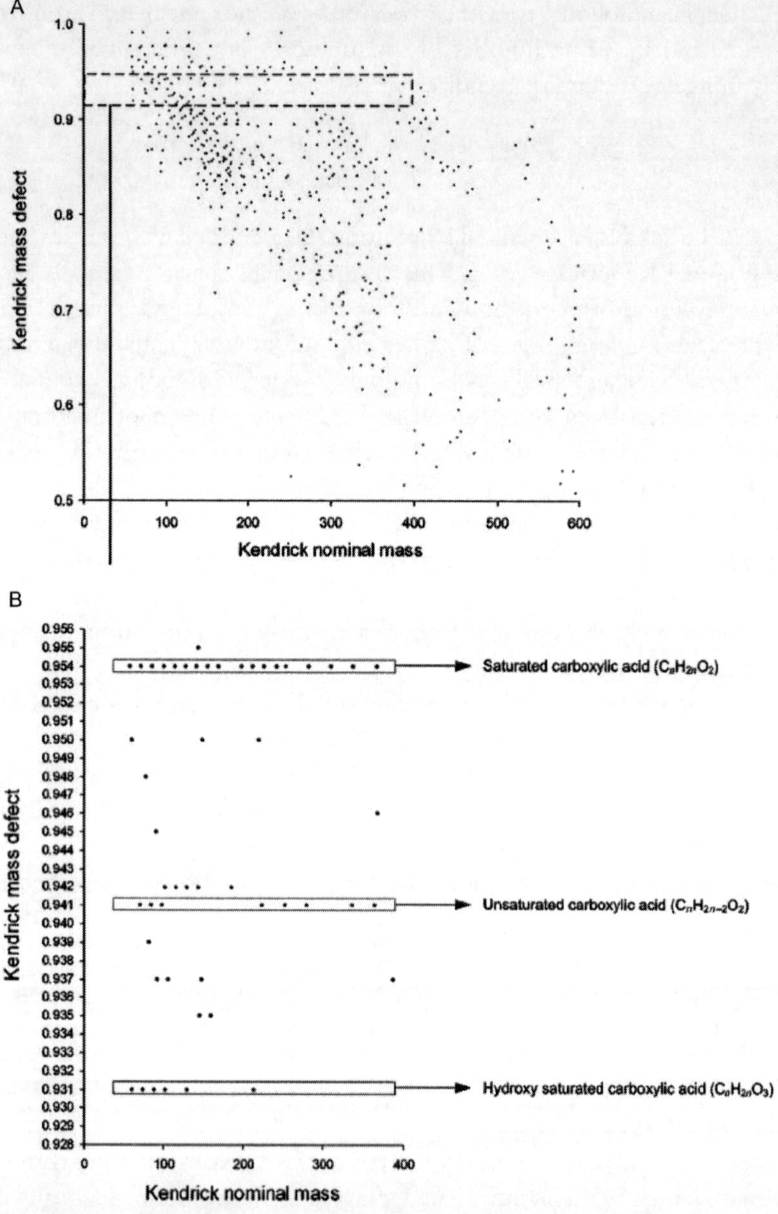

Figure 5.18 (A) Kendrick plot constructed from 7000 compounds contained in the KEGG database. The authors limited the plot to (i) compound exhibiting even molecular weights within the 50–1000 u range and for (ii) formulas containing C, H, N, O, P and S atoms. (B) Zoom of the [0.928–0.956] Kendrick mass defect range. *This figure is reproduced from Werner, Heilier, et al. (2008).* (For colour version of this figure, the reader is referred to the online version of this chapter.)

by displaying additional element ratio (e.g. N/O ratio) or the signal relative abundance scale. Alternatively, two-dimensional density plots can be drawn. This attractive graphical procedure allows graphically distinguishing all the compound classes of interest. Indeed, heteroatom classes may be separated on one axis (e.g. O/C ratio for distinction of compounds differing in O atom content), whereas H/C ratios separate compounds according to their insaturation character (DBE values). Homologous series composed of compounds with the same number of rings plus double bonds and the same heteroatom content, but different numbers of CH_2 units are then displayed on diagonal lines.

Van Krevelen diagrams have been applied to metabolomics, first for reference metabolites from DBs as demonstrated by Werner, Heilier, et al. (2008) and Ohta, Kanaya, and Suzuki (2010), but also in real complex samples as reported by Liger-Belair et al. (2009) for champagne wine and aerosols. To illustrate the usefulness of van Krevelen diagram for evidencing metabolic pathways, an example of endogenous estradiol metabolites is reported in Fig. 5.19. Figure 5.19A shows the metabolite pathways of estradiol, which has been identified by several groups, including Bellocq et al. (2010); Paris et al. (2000); Rathahao et al. (2000); Rizzati et al. (2005) and Yager and Liehr (1996). The Van Krevelen diagram produced by plotting the H/C versus the O/C ratios from ECs of estradiol and its metabolites is reported in Fig. 5.19B. It displays the interesting features of diagonals/horizontal lines of series of metabolites resulting from known chemical/enzymatic reactions such as oxidation (+O mass shift) and/or glycosylation processes ($+C_6H_8O_2$ diagonal lines in Fig. 5.19B). This figure demonstrates that such functional group relationships from an identified (or not) low m/z metabolite can be relevant not only for the confirmation of EC of metabolites at higher m/z value, but also in the analysis of metabolic pathways (Anari, Sanchez, Bakhtiar, Franklin, & Baillie, 2004).

Finally, such a 'formula extension approach' (that is data representation according to chemical properties of compound and its accurate m/z value) is potentially very attractive for global metabolomics studies (Kujawinski & Behn, 2006). Such graphical methods allow extending the EC determination by extrapolation from low m/z to higher m/z values, but also to link metabolites through metabolic pathway prediction. However, adequate functional group relationships (alternative to the use of CH_2 relationship in Kendrick mass approach) should be chosen to be most relevant for determination of metabolome composition, if possible in automatic fashion. In fact, similar Kendrick mass scale conversion could be obtained for series

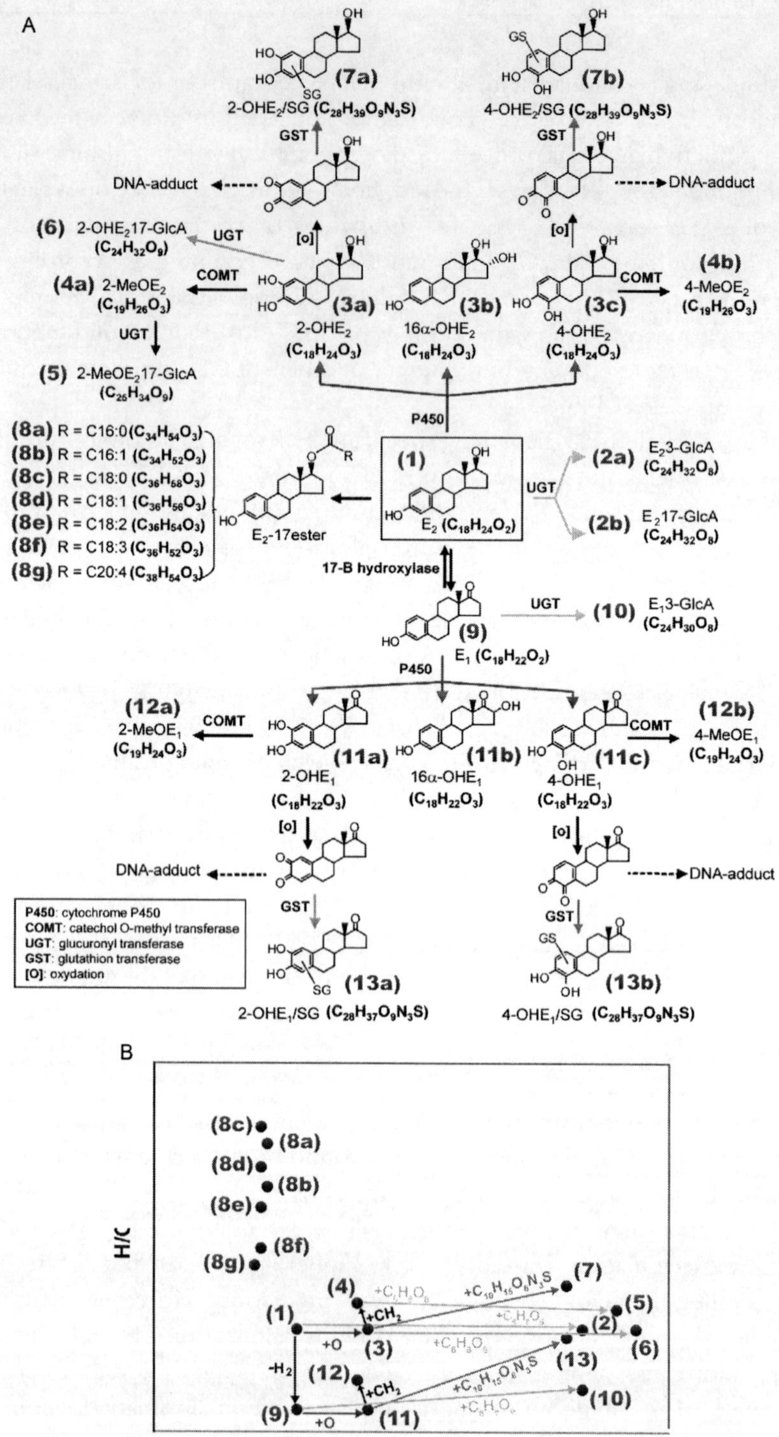
A
B
P450: cytochrome P450
COMT: catechol O-methyl transferase
UGT: glucuronyl transferase
GST: glutathion transferase
[O]: oxydation
2-OHE2/SG ($C_{28}H_{39}O_9N_3S$)
4-OHE2/SG ($C_{28}H_{39}O_9N_3S$)
DNA-adduct
(6) 2-OHE217-GlcA ($C_{24}H_{32}O_9$)
(4a) 2-MeOE2 ($C_{19}H_{26}O_3$)
(5) 2-MeOE217-GlcA ($C_{25}H_{34}O_9$)
(3a) 2-OHE2 ($C_{18}H_{24}O_3$)
(3b) 16α-OHE2 ($C_{18}H_{24}O_3$)
(3c) 4-OHE2 ($C_{18}H_{24}O_3$)
(4b) 4-MeOE2 ($C_{19}H_{26}O_3$)
(8a) R = C16:0 ($C_{34}H_{54}O_3$)
(8b) R = C16:1 ($C_{34}H_{52}O_3$)
(8c) R = C18:0 ($C_{36}H_{58}O_3$)
(8d) R = C18:1 ($C_{36}H_{56}O_3$)
(8e) R = C18:2 ($C_{36}H_{54}O_3$)
(8f) R = C18:3 ($C_{36}H_{52}O_3$)
(8g) R = C20:4 ($C_{38}H_{54}O_3$)
E2-17ester
(1) E2 ($C_{18}H_{24}O_2$)
(2a) E23-GlcA ($C_{24}H_{32}O_8$)
(2b) E217-GlcA ($C_{24}H_{32}O_8$)
17-B hydroxylase
(9) E1 ($C_{18}H_{22}O_2$)
(10) E13-GlcA ($C_{24}H_{30}O_8$)
(12a) 2-MeOE1 ($C_{19}H_{24}O_3$)
(11a) 2-OHE1 ($C_{18}H_{22}O_3$)
(11b) 16α-OHE1 ($C_{18}H_{22}O_3$)
(11c) 4-OHE1 ($C_{18}H_{22}O_3$)
(12b) 4-MeOE1 ($C_{19}H_{24}O_3$)
(13a) 2-OHE1/SG ($C_{28}H_{37}O_9N_3S$)
(13b) 4-OHE1/SG ($C_{28}H_{37}O_9N_3S$)
H/C
O/C
+O
$-H_2$
$+CH_2$
$+C_{10}H_{15}O_6N_3S$
$+C_6H_8O_6$

of compounds differing in oxidation or halogenation. For example, Taguchi, Nieckarz, Clement, Krolik, and Williams (2010) used a mass scale reflecting the substitution of ^{35}Cl for H (34.000000/33.960128 ratio) for the direct identification of halogenated dioxins and furans from vegetation exposed to an industrial fire.

6.3.3 Mass defect filters

A recent processing approach to simplify mass spectral data from HRMS analysis of biological complex samples is the mass defect filter (MDF). This post-acquisition processing method introduced by Zhang, Zhang, and Ray (2003) has been mainly applied for the detection and identification of drug metabolites in biological matrixes (Liang et al., 2011; Xie et al., 2012; Zhang, Zhang, Ray, & Zhu, 2009; Zhang, Zhu, Ray, Ma, & Zhang, 2008). It implies the removal of non-interesting signals (i.e. signals produced by interference compounds, eventually isobaric peaks), which are outside the pre-defined mass defect window. Consequently, MDF facilitates the search and identification of predicted and/or unknown metabolites ions within the filtered data. Most of biotransformation reactions imply the formation of metabolites with mass defects falling within a ± 50 mu window relative to the mass defect of a given parent compound, for example, the hydroxylation reaction results in the mass defect shift of -5 mu, the demethylation of -23 mu and the glucuronidation of $+32$ mu, excepted for the glutathione conjugation, which results in the mass defect change of $+68$ mu. Hence, an MDF with approximately ± 50 mu mass defect window (centred on the parent drug mass defect) can be applied to HRMS data for identification of exogenous (i.e. xenobiotic) as well as endogenous metabolites in biological samples. Figure 5.20 illustrates the efficiency of the MDF approach to evidence metabolites of the compound A in bile samples from dogs treated with ^{14}C-labelled compound A (Zhang et al., 2003). Without any data processing, the metabolite chromatographic peaks are not apparent in the unprocessed total ion chromatogram (TIC; Fig. 5.20A). Hence, common and/or unknown metabolite identification

Figure 5.19 (A) Metabolic pathways of estradiol established from published results of Bellocq et al. (2010); Paris, Rathahao, Debrauwer, Dolo, and Hamami (2000); Rathahao, Hillenweck, Paris, and Debrauwer (2000); Rizzati et al. (2005) and Yager and Liehr (1996) (B) Van Krevelen diagram is produced by plotting the H/C versus the O/C ratios from ECs of estradiol and its metabolites. (For colour version of this figure, the reader is referred to the online version of this chapter.)

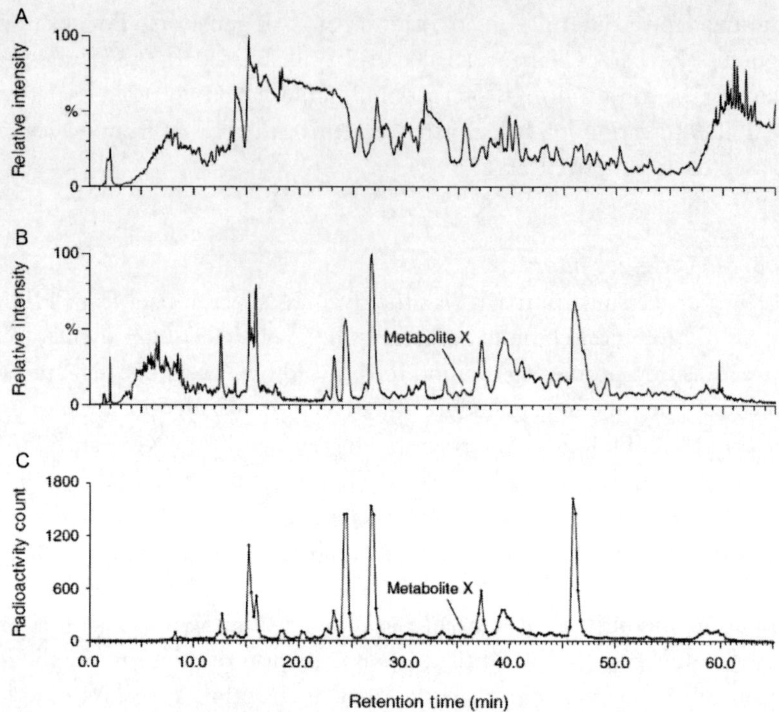

Figure 5.20 Metabolite profiles of compound A in dog bile: (A) total ion chromatogram of unprocessed LC/MS data; (B) total ion chromatogram of processed LC/MS data (with a mass defect filter designed to retain metabolite species with mass defects between 0.006 and 0.106 u from the protonated compound A mass defect (*m/z* 529.056 u)) and the (C) corresponding radioactivity chromatogram. *Figure reproduced from Zhang et al. (2003).*

from such complex data is very challenging. In contrast, the processed TIC by applying an MDF (Fig. 5.20B) is very similar to the metabolite profile from radiochromatogram, which provides specific detection of all ^{14}C-labelled metabolites (Fig. 5.20C). This shows the selectivity of the MDF tool to extract signals of all metabolites as done by common radiodetection, thus facilitating metabolite identification.

The MDF approach is very similar to the traditional extracted ion chromatogram procedure using neutral loss or precursor ion scan. Indeed, they enable selective detection of metabolites, even uncommon compounds. However, under MS/MS conditions the neutral loss and/or the precursor ion scanning modes imply a preliminary knowledge of fragmentation patterns. So, such MS/MS experiments are less selective than the MDF data postprocessing of full mass spectrum. Furthermore, it is possible to apply more than one MDF. Indeed, multiple MDFs approach is a screening strategy allowing the enhanced number of extracted metabolite ions (Zhu et al., 2006). In this strategy, various classes of metabolites can be detected from a given parent drug by using different filter templates such as dehalogenated metabolite filter, conjugate filter or glucuronide filter (Zhang, Zhang, et al., 2009). Additionally, to improve the selectivity of the MDF approach, the introduction of tags to metabolites can be an alternative way if the used tags are characterized by functional groups with substantially different mass defect value, compared to endogenous compounds. Indeed, tagging the compound of interest allows shifting signals into less noisy mass spectrum region. This mass defect labelling approach, currently used in proteomics (Hall et al., 2003), has been applied by LeBlanc, Shiao, Roy, and Sleno (2010) to biological samples for detection of glutathione (GSH) adducts. In this study, a GSH-analog containing a bromine atom has been used. Brominated tag reagent is a good candidate for such purpose because of its very characteristic mass defect and its specific isotopic pattern (^{79}Br and ^{81}Br with near equal natural abundances), which can serve as a secondary filtering tool.

6.4. Metabolite structural elucidation

The final level of investigation is the elucidation of metabolite structure from one unique EC. Exhaustive computation of all possible isomeric structures is excessively complex even for a single EC and almost impossible for a list of several chemical formula candidates. Isomer structure generator can be used (MOLGEN MS is available for EI data) but more than hundreds of billions of isomers can be constructed from a given molecular formula. Thus, the first search step should be limited on existing chemical structures or even more focused onto known natural products. For that, various web-free DBs are available.

6.4.1 HRMS data for DB query

Several kinds of DB can be used for resolving chemical compound structure from spectroscopic and/or biochemical data (Kind, Scholz, & Fiehn, 2009). General chemical DBs composed of synthetic and/or natural small size organic compounds as PubChem can be used. Alternatively, DBs including specific classes of compounds, for example, limited to metabolites such as KEGG (http://www.genome.jp/kegg/), KNApSAcK (http://kanaya.naist.jp/KNApSAcK) or restricted to specific compound classes such as as lipidomics in LipidMap (http://www.lipidmaps.org/) are developed. Alternatively, DBs can be exclusively dedicated to metabolite species of specific organisms, for example, Human Metabolome Database (HMDB, http://www.hmdb.ca/).

The DB query procedure can be performed not only with the molecular mass (nominal and/or accurate), a MW range or the whole mass spectrum, but also with the resolved EC. Better results are obtained from HRMS data. Mass spectral DBs (Tohge & Fernie, 2009) initially built from GC/MS data with reproducible EI mass spectra (e.g. Golm Metabolome Database; http://gmd.mpimp-golm.mpg.de) have been also developed from LC/MS, HRMS but also MS/MS experiments, for example, METLIN metabolite DB (http://metlin.scripps.edu; Smith et al., 2005), or MassBank (www.massbank.jp). However, such mass spectral data produced from API MS displays poor reproducibility, which limits strongly the constitution of universal DBs (in contrast to EI or NMR data). Nevertheless, proposed DB algorithm combined 'search tools' and such DB tools are still under development.

Nowadays, most of putative biomarker metabolites are still unidentified. Additional physicochemical and/or biological information are often required to get insights into metabolite structure (e.g. sample origins). More restrictive chemical information and even substructure knowledge for further isomer generator can be obtained from other spectroscopic approaches, for example, NMR and/or alternative MS approaches, such as H/D exchanges, labelled sample studies and sequential MS/MS experiments. For example, in-solution H/D exchanges are analyzed by mass spectrometry to reveal exchangeable protons produced on the analyzed compound. Such observations can be implemented in DB query or may be used to fit the fragmentation patterns provided from experimental MS/MS data. Nevertheless, most reliable chemical information is provided by the complementary MS/MS experiments.

6.4.2 MS/MS experiments for structural information

MS/MS experiments are a well-known approach to bring information about chemical groups or substructure present in the selected compound species by resolving the neutral loss and/or the product ions (elucidation of ion fragmentation pathways). Of course, mass spectral DB can be used in which MS/MS data are available, for example, NIST or METLIN... However, automated MS/MS data exploitation is still limited, as it suffers from lack of reliability between experimental and DB MS/MS data above all, for precursor species produced by API techniques. Nevertheless, general query approach looks for similarities between fragmentation patterns. MS/MS experiments can be subsequently performed on both putative biomarker metabolite and authentic compound of metabolite (if commercially available) to confirm MS data annotation. Nevertheless, constitution of universal DB with MS/MS data is limited by the low reproducibility of API mode, especially from fragmentation patterns changes related to the used activation modes (resonant or non-resonant CID in low-energy range, IRMPD or EID applied on singly charged precursor ions, Lioe & O'Hair, 2007). Changes in product ion abundances and types of ions can also be observed for different instrumental set-up and geometry (i.e. the kinetic shift).

Almost all MS/MS experiments are achieved using CID activation mode, which proceeds through collisions between accelerated precursor ions and target gas, leading to increasing precursor ion E_{int} above an energy threshold and to its dissociation. It is implemented in most of hybrid instruments (e.g. CID in a collision cell of triple-quadrupole instruments) and temporal stand-alone instruments (resonant or non-resonant activation of stored ions). However, collisionally activated fragmentation ways (CID mode) vary in terms of competitive, consecutive dissociation mechanisms or only into product ion abundances, according to the used instrumentation. First, the collision energy regime varies from keV to 2-200 eV kinetic energy range and results in different fragmentation patterns, for example, charge-remote process and homolytic cleavages are typically observed in high KE CID mode. Secondly, even in the sole low KE regime, fragmentation patterns are strongly dependant upon the analyzer and instrumental set-up (e.g. E_{lab}; gas pressure...). Large differences are observed between a low KE ion-beam filtered in a quadrupole, in which subsequently ion collisions occur with gas during the ion flight (as 'in-beam' CID) and the resonantly activated ions in ion trap or FT-MS instruments.

Nevertheless, FT-MS offers complementary activation modes (Marshall et al., 1998). Among them, IRMPD, which is an ergodic activation technique, presents fragmentation ways analogs to the CID technique (Sleno, Volmer, & Marshall, 2005), for example, charge directed dissociation processes, but this is more suited for higher mass compounds (Eyler, 2009). On the other hand, alternative ECD, EDD and ETD modes are also available in FT-MS instruments and can be performed onto multiple-charged species of high MW metabolites, which dissociate by radical-induced processes. Note that high-resolution isolation of precursor ions and subsequent *in situ* activation into ICR cell allow more specificity of MS/MS data.

Moreover, HRMS detection allows more relevant chemical information thanks to accurate mass measurements allowing distinction of isobaric species (e.g. CO vs. C_2H_4 losses; Sleno et al., 2005). Additionally, high mass resolving power of FT-MS instruments offers accurate isotope pattern determination, which can also be useful for determining chemical formula of product ions. For example, a CO_2 loss is not only determined by the detection of the product ion with a mass shift of 43.9898 u (which is different from isobaric CH_3CHO group loss) compared to that of the monoisotopic precursor ion, it further involves the lack of a carbon into product ion formula, which affects the product ion $A+1$ isotopic cluster peak.

Finally, interpretation of HRMS data and the wished metabolite identification should be automated to enhance metabolite coverage. The algorithm and the related elucidation strategy can be built from high-resolution MS/MS data under well-defined steps: (i) first small size neutral losses should be annotated with (ii) the presence of diagnostic product ions (e.g. m/z 175 for the glucuronide moiety; m/z 306 and m/z 272 for the glutathione moiety in negative ionization mode) (iii) ECs of series of productions (i.e. with the same heteroatom composition and substructure) can be deduced from their accurate m/z (or their mass defaults) values. The fragmentation patterns can be proposed from the comparison of those productions and/or by the observation of two complementary fragment ions in the MS/MS spectra involving the formation of an ion-dipole intermediate. Finally (iv) such results can be compared to DB MS/MS data.

7. CONCLUSION

The vast potential of FT mass spectrometers is demonstrated for various aspects of metabolomics, that is, global and targeted approaches, competing with the traditionally used spectroscopic or other MS approaches. Foremost, FT-MS instrument appears as the best well-suited analytical tool for very high-throughput metabolomics investigations, that is, it enables to perform a single and rapid measurement of a large number of metabolites in complex biological matrixes without prior separation step. Although the drawbacks of such DIMS approach could not be totally overcome, they are strongly diminished using FT-MS, allowing detection and identification of few hundreds metabolite peaks in a single scan acquisition but, it is still a non-exhaustive manner. Furthermore, FT-MS instruments provide measurements of accurate m/z values with errors in the sub-ppm range, greatly aiding the determination of ECs for small size molecules such as metabolites by limiting the number of EC returns for a given narrow accurate m/z value window. The EC determination represents a very important step indispensable for structural elucidation of metabolites and subsequent insights into biological processes. However, even if the mass measurement accuracy is taken into account together with the basic chemical rules to reject false ECs, one unique EC cannot be always obtained even for an accurate m/z value at sub-ppm error. So, the introduction of additional constraints, for example, by incorporating the isotope patterns, is now done for improving the EC determination.

Nowadays, a large number of metabolites are still unidentified. So, current limitations and issues must be addressed as the mass measurement accuracy and the dynamic range (even if it is the most powerful) to reach an exhaustive detection and annotation of metabolites. Hence, improvement of technology based upon FT-MS instrumentation are being developed such as the use of higher magnet (12, 15 and also 21 T magnets), which allows reaching higher peak resolutions and decreasing the transient duration (e.g. improving UPLC coupling). However, this means an extensive cost of instrumentation. So, optimal detection method conditions using commercial instruments were developed such as the 'stitching' acquisition method (Southam et al., 2007), to enhance the metabolome coverage. Alternatively, a novel promising ICR-based cell has been proposed by Nikolaev et al. (2011) allowing an extremely homogenous trapping field inside a dynamically harmonized infinity ICR cell under a modest 7 T magnetic field. Using this FT-ICR configuration, Nikolaev, Jertz, Grigoryev, and Baykut (2012) have reported a very high mass resolving power of 24 millions for the monoisotopic peak at m/z 609.28066 for protonated reserpine in heterodyne detection mode. Therefore, one unique EC can be obtained by perfectly tuning the FT-MS detection conditions (e.g. control the amount of stored ions) and using accurate EC determination constraints. Another critical challenging issue of FT-MS approaches is the isomer identification, unsolvable in DIMS approach; it still needs the further coupling of separation methods and/or eventual sequential MS^n experiments using various ion activation modes, for example, electronic excitation rather than vibrational dissociations.

Another feature of improvement is the HRMS data processing, which requires the help of complementary tools based on bioinformatics. Of course, existing MS DB and data-processing software exist. Additional efforts are taken to their automatic utilizations to allow exhaustive annotation of detected metabolite signals (Roux et al., 2012). Specific automatic software is required for very high-throughput global metabolomics analysis, in order to facilitate the HRMS data visualization (e.g. Kendrick mass graphics), the subsequent data filtering of the large MS data sets (e.g. the

MDF) and the final data interpretation, for example, by constitution of universal MS/MS DBs or elaboration of automated MS/MS data annotation (Benton, Wong, Trauger, & Siuzdak, 2008; Rojas-Chertó et al., 2011).

ACKNOWLEDGEMENTS

The authors wish to thank Alain Paris for his suggestions and helpful discussions on the biological aspects.

REFERENCES

An, Z., Chen, Y., Zhang, R., Song, Y., Sun, J., He, J., et al. (2010). Integrated ionization approach for RRLC-MS/MS-based metabonomics: Finding potential biomarkers for lung cancer. *Journal of Proteome Research, 9*, 4071–4081.

Anari, M. R., Sanchez, R. I., Bakhtiar, R., Franklin, R. B., & Baillie, T. A. (2004). Integration of knowledge-based metabolic predictions with liquid chromatography data-dependent tandem mass spectrometry for drug metabolism studies: Application to studies on the biotransformation of indinavir. *Analytical Chemistry, 76*, 823–832.

Annesley, T. M. (2003). Ion suppression in mass spectrometry. *Clinical Chemistry, 49*, 1041–1044.

Audi, G., Wapstra, A. H., & Thibault, C. (2003). The AME2003 atomic mass evaluation (II). Tables, graphs and references. *Nuclear Physics A, 729*, 337–676.

Bajad, S. U., Lu, W., Kimball, E. H., Yuan, J., Peterson, C., & Rabinowitz, J. D. (2006). Separation and quantitation of water soluble cellular metabolites by hydrophilic interaction chromatography-tandem mass spectrometry. *Journal of Chromatography. A, 1125*, 76–88.

Barber, M., Bordoli, R. S., Sedgwick, R. D., & Tyler, A. N. (1981). Fast atom bombardment of solids (F.A.B.): A new ion source for mass spectrometry. *Journal of the Chemical Society, Chemical Communications*, (7), 325–327.

Barrow, M. P., Burkitt, W. I., & Derrick, P. J. (2005). Principles of Fourier transform ion cyclotron resonance mass spectrometry and its application in structural biology. *Analyst, 130*, 18–28.

Bedair, M., & Sumner, L. W. (2008). Current and emerging mass-spectrometry technologies for metabolomics. *Trends in Analytical Chemistry, 27*(3), 238–250.

Bellocq, D., Molina, J., Rathahao-Paris, E., Canlet, C., Taché, S., Pierre, F., et al. (2010). Metabolic bioactivation of estradiol-17β (E2β) in mouse colon epithelial cells bearing ApcMin mutation. *Steroids, 75*, 665–675.

Benninghoven, A. (1969). Analysis of sub-monolayers on silver by secondary ion emission. *Physica Status Solidi, 34*(2), K169–K171.

Benton, H. P., Wong, D. M., Trauger, S. A., & Siuzdak, G. (2008). XCMS2: Processing tandem mass spectrometry data for metabolite identification and structural characterization. *Analytical Chemistry, 80*(16), 6382–6389.

Blaum, K. (2006). High-accuracy mass spectrometry with stored ions. *Physics Reports, 425*, 1–78.

Böhlke, J. K., De Laeter, J. R., De Bièvre, P., Hidaka, H., Peiser, H. S., Rosman, K. J. R., et al. (2005). Isotopic compositions of the elements, 2001. *Journal of Physical and Chemical Reference Data, 34*, 57–67.

Brenton, A. G., & Godfrey, A. R. (2010). Accurate mass measurement: Terminology and treatment of data. *Journal of the American Society for Mass Spectrometry, 21*, 1821–1835.

Bristow, A. W. T., & Webb, K. S. (2003). Intercomparison study on accurate mass measurement of small molecules in mass spectrometry. *Journal of the American Society for Mass Spectrometry, 14*, 1086–1098.

Budimir, N., Blais, J. C., Fournier, F., & Tabet, J. C. (2007). Desorption/ionization on porous silicon mass spectrometry (DIOS) of model cationized fatty acids. *Journal of Mass Spectrometry*, *42*(1), 42–48.

Busch, K. L. (1995). Desorption ionization mass spectrometry. *Journal of Mass Spectrometry*, *30*, 233–240.

Castaing, R., & Slodzian, G. J. (1962). Optique corpusculaire—Premiers essais de micro-analyse par emission ionique secondaire. *Microscopie*, *1*, 395–399.

Castrillo, J. I., Hayes, A., Mohammed, S. N., Gaskell, S. J., & Oliver, S. G. (2003). An optimized protocol for metabolome analysis in yeast using direct infusion electrospray mass spectrometry. *Phytochemistry*, *62*, 929–937.

Chen, H., Gamez, G., & Zenobi, R. (2009). What can we learn from ambient ionization techniques? *Journal of the American Society for Mass Spectrometry*, *20*, 1947–1963.

Chen, H., Venter, A., & Cooks, R. G. (2006). Extractive electrospray ionization for direct analysis of undiluted urine, milk and other complex mixtures without sample preparation. *Chemical Communications*, (19), 2042–2044.

Chen, H., Wortmann, A., Zhang, W., & Zenobi, R. (2007). Rapid in vivo fingerprinting of nonvolatile compounds in breath by extractive electrospray ionization quadrupole time-of-flight mass spectrometry. *Angewandte Chemie International Edition*, *46*, 580–583.

Cody, R. B., Hein, R. E., Goodman, S. D., & Marshall, A. G. (1987). Stored waveform inverse fourier transform excitation for obtaining increased parent ion selectivity in collisionally activated dissociation: Preliminary results. *Rapid Communications in Mass Spectrometry*, *1*, 99–102.

Cody, R. B., Laramée, J. A., & Durst, H. D. (2005). Versatile new ion source for the analysis of materials in open air under ambient conditions. *Analytical Chemistry*, *77*, 2297–2302.

Comisarow, M. B., & Marshall, A. G. (1974). Fourier transform ion cyclotron resonance spectroscopy. *Chemical Physics Letters*, *25*, 282–283.

Cooper, H. J., Hakansson, K., & Marshall, A. G. (2005). The role of electron capture dissociation in biomolecular analysis. *Mass Spectrometry Reviews*, *24*, 201–222.

Cornett, D. S., Frappier, S. L., & Caprioli, R. M. (2008). MALDI-FTICR imaging mass spectrometry of drugs and metabolites in tissue. *Analytical Chemistry*, *80*, 5648–5653.

Covey, T. R., Thomson, B. A., & Schneider, B. B. (2009). Atmospheric pressure ion sources. *Mass Spectrometry Reviews*, *28*, 870–897.

Cuyckens, F., Wassvik, C., Mortishire-Smith, R. J., Tresadern, G., Campuzano, I., & Claereboudt, J. (2011). Product ion mobility as a promising tool for assignment of positional isomers of drug metabolites. *Rapid Communication of Mass Spectrometry*, *25*, 3497–3503.

de Hoffmann, E., & Stroobant, V. (2007). *Mass spectrometry: Principles and applications* (3rd ed.). Chichester, UK: John Wiley and Sons Ltd.

Desiderio, C., Iavarone, F., Rossetti, D. V., Messana, I., & Castagnola, M. (2010). Capillary electrophoresis-mass spectrometry for the analysis of amino acids. *Journal of Separation Science*, *33*, 2385–2393.

Dettmer, K., Aronov, P. A., & Hammock, B. D. (2007). Mass spectrometry-based metabolomics. *Mass Spectrometry Reviews*, *26*, 51–78.

Domon, B., & Aebersold, R. (2006). Mass spectrometry and protein analysis. *Science*, *312*, 212–217.

Douglas, D. J., Frank, A. J., & Mao, D. (2005). Linear ion traps in mass spectrometry. *Mass Spectrometry Reviews*, *24*, 1–29.

Draper, J., Enot, D. P., Parker, D., Beckmann, M., Snowdon, S., Lin, W., et al. (2009). Metabolite signal identification in accurate mass metabolomics data with MZedDB, an interactive *m/z* annotation tool utilizing predicted ionisation behaviour 'rules'. *BMC Bioinformatics*, *10*, 227–239.

Dumas, M. E., Debrauwer, L., Beyet, L., Lesage, D., André, F., Paris, A., et al. (2002). Analyzing the physiological signature of anabolic steroids in cattle urine using pyrolysis/metastable atom bombardment mass spectrometry and pattern recognition. *Analytical Chemistry*, *74*, 5393–5404.

Dunn, W. B. (2008). Current trends and future requirements for the mass spectrometric investigation of microbial, mammalian and plant metabolomes. *Physical Biology*, *5*, 1–24.

Dunn, W. B., Bailey, N. J. C., & Johnson, H. E. (2005). Measuring the metabolome: Current analytical technologies. *Analyst*, *130*, 606–625.

Dunn, W. B., Broadhurst, D., Brown, M., Baker, P. N., Redman, C. W. G., Kenny, L. C., et al. (2008). Metabolic profiling of serum using Ultra Performance Liquid Chromatography and the LTQ-Orbitrap mass spectrometry system. *Journal of Chromatography. B*, *871*, 288–298.

Dunn, W. B., Overy, S., & Quick, W. P. (2005). Evaluation of automated electrospray-TOF mass spectrometry for metabolic fingerprinting of the plant metabolome. *Metabolomics*, *1*(2), 137–148.

Erve, J. C. L., Gu, M., Wang, Y., DeMaio, W., & Talaat, R. E. (2009). Spectral accuracy of molecular ions in an LTQ/Orbitrap mass spectrometer and implications for elemental composition determination. *Journal of the American Society for Mass Spectrometry*, *20*, 2058–2069.

Eyler, R. (2009). Infrared multiple photon dissociation spectroscopy of ions in penning traps. *Mass Spectrometry Reviews*, *28*, 448–467.

Fabre, N., Claparols, C., Richelme, S., Angelin, M.-L., Fouraste, I., & Moulis, C. (2000). Direct characterization of isoquinoline alkaloids in a crude plant extract by ion-pair liquid chromatography–electrospray ionization tandem mass spectrometry: Example of Eschscholzia californica. *Journal of Chromatography. A*, *904*, 35–46.

Faubert, D., Paul, G. J. C., Giroux, J., & Bertrand, M. J. (1993). Selective fragmentation and ionization of organic compounds using an energy-tunable rare-gas metastable beam source. *International Journal of Mass Spectrometry and Ion Processes*, *124*, 69–77.

Fiehn, O. (2001). Combining genomics, metabolome analysis, and biochemical modelling to understand metabolic networks. *Comparative and Functional Genomics*, *2*, 155–168.

Fiehn, O. (2002). Metabolomics—The link between genotypes and phenotypes. *Plant Molecular Biology*, *48*, 155–171.

Fiehn, O., Kopka, J., Trethewey, R. N., & Willmitzer, L. (2000). Identification of uncommon plant metabolites based on calculation of elemental compositions using gas chromatography and quadrupole mass spectrometry. *Analytical Chemistry*, *72*, 3573–3580.

Gillig, K. J., Bluhm, B. K., & Russell, D. H. (1996). Ion motion in a Fourier transform ion cyclotron resonance wire ion guide cell. *International Journal of Mass Spectrometry and Ion Processes*, *157/158*, 129–147.

Goodacre, R., & Kell, D. B. (1996). Pyrolysis mass spectrometry and its applications in biotechnology. *Current Opinion in Biotechnology*, *7*, 20–28.

Goodacre, R., Vaidyanathan, S., Dunn, W. B., Harrigan, G. G., & Kell, D. B. (2004). Metabolomics by numbers: Acquiring and understanding global metabolite data. *Trends in Biotechnology*, *22*, 245–252.

Goodacre, R., York, E. V., Heald, J. K., & Scott, I. M. (2003). Chemometric discrimination of unfractionated plant extracts analyzed by electrospray mass spectrometry. *Phytochemistry*, *62*, 859–863.

Grange, A. H., Genicola, F. A., & Socovol, G. W. (2002). Utility of three types of mass spectrometers for determining elemental compositions of ions formed from chromatographically separated compounds. *Rapid CommunIcation in Mass Spectrometry*, *16*, 2356–2369.

Grange, A. H., Zumwalt, M. C., & Sovocool, G. W. (2006). Determination of ion and neutral loss compositions and deconvolution of product ion mass spectra using an orthogonal acceleration time-of-flight mass spectrometer and an ion correlation program. *Rapid Communications in Mass Spectrometry*, *20*, 89–102.

Guan, S., & Marshall, A. G. (1995). Ion traps for Fourier transform ion cyclotron resonance mass spectrometry: Principles and design of geometric and electric configurations. *International Journal of Mass Spectrometry and Ion Processes, 146/147*, 261–296.

Guo, K., Ji, C., & Li, L. (2007). Stable-isotope dimethylation labeling combined with LC-ESI MS for quantification of amine-containing metabolites in biological samples. *Analytical Chemistry, 79*, 8631–8638.

Guo, K., & Li, L. (2009). Differential ^{12}C-/^{13}C-isotope dansylation labeling and fast liquid chromatography/mass spectrometry for absolute and relative quantification of the metabolome. *Analytical Chemistry, 81*, 3919–3932.

Guo, Z., Zhang, Q., Zou, H., Guo, B., & Ni, J. (2002). A method for the analysis of low-mass molecules by MALDI-TOF mass spectrometry. *Analytical Chemistry, 74*, 1637–1641.

Habicht, S. C., Duan, P., Vinueza, N. R., Fu, M., & Kenttämaa, H. I. (2010). Liquid chromatography/tandem mass spectrometry utilizing ion-molecule reactions and collision-activated dissociation for the identification of N-oxide drug metabolites. *Journal of Pharmaceutical and Biomedical Analysis, 51*, 805–811.

Haddad, R., Catharino, R. R., Marques, L. A., & Eberlin, M. N. (2008). Perfume fingerprinting by easy ambient sonic-spray ionization mass spectrometry: Nearly instantaneous typification and counterfeit detection. *Rapid Communication Mass Spectrometry, 22*, 3662–3666.

Hall, M. P., Ashrafi, S., Obegi, I., Petesch, R., Peterson, J. N., & Schneider, L. V. (2003). 'Mass defect' tags for biomolecular mass spectrometry. *Journal of Mass Spectrometry, 38*, 809–816.

Han, J., Danell, R. M., Patel, J. R., Gumerov, D. R., Scarlett, C. O., Speir, J. P., et al. (2008). Towards high-throughput metabolomics using ultrahigh-field Fourier transform ion cyclotron resonance mass spectrometry. *Metabolomics, 4*, 128–140.

Hardman, M., & Makarov, A. A. (2003). Interfacing the Orbitrap mass analyzer to an electrospray ion source. *Analytical Chemistry, 75*, 1699–1705.

He, H., Conrad, C. A., Nilsson, C. L., Ji, Y., Schaub, T. M., Marshall, A. G., et al. (2007). Method for lipidomic analysis: p53 expression modulation of sulfatide, ganglioside, and phospholipid composition of U87 MG glioblastoma cells. *Analytical Chemistry, 79*, 8423–8430.

Herniman, J. M., Langley, G. J., Bristow, T. W. T., & O'Connor, G. (2005). The validation of exact mass measurements for small molecules using FT-ICRMS for improved confidence in the selection of elemental formulas. *Journal of the American Society for Mass Spectrometry, 16*, 1100–1108.

Hogness, T. R., & Kvalnes, H. M. (1928). The ionization processes in methane interpreted by the mass spectrograph. *Physical Review, 32*, 942–945.

Hop, C., Chen, Y., & Yu, L. J. (2005). Uniformity of ionization response of structurally diverse analytes using a chip-based nanoelectrospray ionization source. *Rapid Communication of Mass Spectrometry, 19*, 3139–3142.

Hrydziuszko, O., Silva, M. A., Thamara, M., Perera, P. R., Richards, D. A., Murphy, N., et al. (2010). Application of metabolomics to investigate the process of human orthotopic liver transplantation: A proof-of-principle study. *OMICS A Journal of Integrative Biology, 14*(2), 143–150.

Hu, Q., Cooks, R. G., & Noll, R. J. (2007). Phase-enhanced selective ion ejection in an Orbitrap mass spectrometer. *Journal of the American Society for Mass Spectrometry, 18*, 980–983.

Hu, Q., Makarov, A. A., Cooks, R. G., & Noll, R. J. (2006). Resonant ac dipolar excitation for ion motion control in the Orbitrap mass analyzer. *Journal of Physical Chemistry A, 110*, 2682–2689.

Hu, Q., Noll, R. J., Li, H., Makarov, A., Hardman, M., & Cooks, R. G. (2005). The Orbitrap: A new mass spectrometer. *Journal of Mass Spectrometry, 40*, 430–443.

Huang, E. C., Wachs, T., Conboy, J. J., & Henion, J. D. (1990). Atmospheric pressure ionization mass spectrometry. Detection for the separation sciences. *Analytical Chemistry*, *62*(13), 713A–725A.

Hughey, C. A., Hendrickson, C. L., Rodgers, R. P., Marshall, A. G., & Qian, K. (2001). Kendrick mass defect spectrum: A compact visual analysis for ultrahigh-resolution broadband mass spectra. *Analytical Chemistry*, *73*, 4676–4681.

Idborg, H., Zamani, L., Edlund, P.-O., Schuppe-Koistinen, I., & Jacobsson, S. P. (2005). Metabolic fingerprinting of rat urine by LC/MS Part 1. Analysis by hydrophilic interaction liquid chromatography–electrospray ionization mass spectrometry. *Journal of Chromatography. B*, *828*, 9–13.

Junot, C., Madalinski, G., Tabet, J.-C. , & Ezan, E. (2010). Fourier transform mass spectrometry for metabolome analysis. *Analyst*, *135*, 2203–2219.

Kaderbhai, N. N., Broadhurst, D. I., Ellis, D. I., Goodacre, R., & Kell, D. B. (2003). Functional genomics via metabolic footprinting: Monitoring metabolite secretion by Escherichia coli tryptophan metabolism mutants using FT–IR and direct injection electrospray mass spectrometry. *Comparative and Functional Genomics*, *4*, 376–391.

Kamleh, M. A., Hobani, Y., Dow, J. A. T., & Watson, D. G. (2008). Metabolomic profiling of Drosophila using liquid chromatography Fourier transform mass spectrometry. *FEBS Letters*, *582*, 2916–2922.

Karas, M., & Hillenkamp, F. (1988). Laser desorption ionization of proteins with molecular masses exceeding 10,000 daltons. *Analytical Chemistry*, *60*(20), 2299–2301.

Katajamaa, M., & Orešič, M. (2005). Processing methods for differential analysis of LC/MS profile data. *BMC Bioinformatics*, *6*, 179–191.

Kauppila, T. J., Talaty, N., Kuuranne, T., Kotiaho, T., Kostiainen, R., & Cooks, R. G. (2007). Rapid analysis of metabolites and drugs of abuse from urine samples by desorption electrospray ionization-mass spectrometry. *Analyst*, *132*, 868–875.

Kebarle, P., & Verkerk, U. H. (2009). Electrospray: From ions in solution to ions in the gas phase, what we know now. *Mass Spectrometry Reviews*, *28*(6), 898–917.

Kendrick, E. (1963). A mass scale based on CH2 = 14.0000 for high resolution mass spectrometry of organic compounds. *Analytical Chemistry*, *35*(13), 2146–2154.

Kiefer, P., Portais, J.-C., & Vorholt, J. A. (2008). Quantitative metabolome analysis using liquid chromatography–high-resolution mass spectrometry. *Analytical Biochemistry*, *382*, 94–100.

Kim, S., Kramer, R. W., & Hatcher, P. G. (2003). Graphical method for analysis of ultrahigh-resolution broadband mass spectra of natural organic matter, the Van Krevelen diagram. *Analytical Chemistry*, *75*, 5336–5344.

Kind, T., & Fiehn, O. (2006). Metabolomic database annotations via query of elemental compositions: Mass accuracy is insufficient even at less than 1 ppm. *BMC Bioinformatics*, *7*, 234–244.

Kind, T., & Fiehn, O. (2007). Seven Golden Rules for heuristic filtering of molecular formulas obtained by accurate mass spectrometry. *BMC Bioinformatics*, *8*, 105–125.

Kind, T., Scholz, M., & Fiehn, O. (2009). How large is the metabolome? A critical analysis of data exchange practices in chemistry. *PLoS One*, *4*(5), e5440.

Kingdon, K. H. (1923). A method for the neutralization of electron space charge by positive ionization at very low gas pressures. *Physical Review*, *21*, 408–418.

Knight, R. D. (1981). Storage of ions from laser produced plasma. *Applied Physics Letters*, *38*(4), 221–223.

Koch, B. P., Dittmar, T., Witt, M., & Kattner, G. (2007). Fundamentals of molecular formula assignment to ultrahigh resolution mass data of natural organic matter. *Analytical Chemistry*, *79*, 1758–1763.

Koulman, A., Tapper, B. A., Fraser, K., Cao, M., Lane, G. A., & Rasmussen, S. (2007). High-throughput direct-infusion ion trap mass spectrometry: A new method for metabolomics. *Rapid Communication of Mass Spectrometry*, *21*, 421–428.

Koulman, A., Woffendin, G., Narayana, V. K., Welchman, H., Crone, C., & Volmer, D. A. (2009). High-resolution extracted ion chromatography, a new tool for metabolomics and lipidomics using a second generation Orbitrap mass spectrometer. *Rapid Communication of Mass Spectrometry, 23*, 1411–1418.

Kujawinski, E. B., & Behn, M. D. (2006). Automated analysis of electrospray ionization fourier transform ion cyclotron resonance mass spectra of natural organic matter. *Analytical Chemistry, 78*, 4363–4373.

Lafaye, A., Labarre, J., Tabet, J.-C., Ezan, E., & Junot, C. (2005). Liquid chromatography-mass spectrometry and ^{15}N metabolic labeling for quantitative metabolic profiling. *Analytical Chemistry, 77*, 2026–2033.

Laskin, J., & Futrell, J. H. (2005). Activation of large ions in FT-ICR mass spectrometry. *Mass Spectrometry Reviews, 24*, 135–167.

Lawrence, E. O., & Livingston, M. S. (1932). The production of high speed light ions without the use of high voltages. *Physical Review, 40*(1), 19–35.

LeBlanc, A., Shiao, T. C., Roy, R., & Sleno, L. (2010). Improved detection of reactive metabolites with a bromine-containing glutathione analog using mass defect and isotope pattern matching. *Rapid Communications in Mass Spectrometry, 24*, 1241–1250.

Lee, W.-N. P., Wahjudi, P. N., Xu, J., & Go, V. L. (2010). Tracer-based metabolomics: Concepts and practices. *Clinical Biochemistry, 43*, 1269–1277.

Li, Y., Shrestha, B., & Vertes, A. (2008). Atmospheric pressure infrared MALDI imaging mass spectrometry for plant metabolomics. *Analytical Chemistry, 80*, 407–420.

Liang, Y., Xiao, W., Dai, C., Xie, L., Ding, G., Wang, G., et al. (2011). Structural identification of the metabolites for strictosamide in rats bile by an ion trap-TOF mass spectrometer and mass defect filter technique. *Journal of Chromatography. B, 879*, 1819–1822.

Liger-Belair, G., Cilindre, C., Gougeon, R. D., Lucio, M., Gebefügi, I., Jeandet, P., et al. (2009). Unraveling different chemical fingerprints between a champagne wine and its aerosols. *PNAS, 106*(39), 16545–16549.

Lindon, J. C., & Nicholson, J. K. (2008). Analytical technologies for metabonomics and metabolomics, and multi-omic information recovery. *Trends in Analytical Chemistry, 27*, 194–204.

Lioe, H., & O'Hair, R. A. J. (2007). Comparison of collision-induced dissociation and electron-induced dissociation of singly protonated aromatic amino acids, cystine and related simple peptides using a hybrid linear ion trap–FT-ICR mass spectrometer. *Analytical and Bioanalytical Chemistry, 389*, 1429–1437.

Liu, Q., Guo, Z., & He, L. (2007). Mass spectrometry imaging of small molecules using desorption/ionization on silicon. *Analytical Chemistry, 79*, 3535–3541.

Lucio, M., Fekete, A., Weigert, C., Wägele, B., Zhao, X., Chen, J., et al. (2010). Insulin sensitivity is reflected by characteristic metabolic fingerprints—A Fourier transform mass spectrometric non-targeted metabolomics approach. *PLoS One, 5*(10), e13317.

Macfarlane, R. D., & Torgerson, D. F. (1976). Californium-252 plasma desorption mass spectroscopy. *Science, 191*, 920–925.

Madalinski, G., Godat, E., Alves, S., Lesage, D., Genin, E., Levi, P., et al. (2008). Direct introduction of biological samples into a LTQ-Orbitrap hybrid mass spectrometer as a tool for fast metabolome analysis. *Analytical Chemistry, 80*, 3291–3303.

Makarov, A. A. (1999). The Orbitrap: A novel high-performance electrostatic trap. In: *Proceedings of the 47th Conference on Mass Spectrometry and Allied Topics, Dallas, TX*.

Makarov, A. A. (2000). Electrostatic axially harmonic orbital trapping: A high-performance technique of mass analysis. *Analytical Chemistry, 72*, 1156–1162.

Makarov, A., Denisov, E., Kholomeev, A., Balschun, W., Lange, O., Strupat, K., et al. (2006). Performance evaluation of a hybrid linear ion trap/Orbitrap mass spectrometer. *Analytical Chemistry, 78*, 2113–2120.

Makarov, A., Denisov, E., & Lange, O. (2009). Performance evaluation of a high-field Orbitrap mass analyzer. *Journal of the American Society for Mass Spectrometry*, *20*(8), 1391–1396.

Makarov, A., Denisov, E., Lange, O., & Horning, S. (2006). Dynamic range of mass accuracy in LTQ Orbitrap hybrid mass spectrometer. *Journal of the American Society for Mass Spectrometry*, *17*, 977–982.

Mamyrin, B. A., Karataev, V. I., Shmikk, D. V., & Zagulin, V. A. (1973). The mass-reflectron, a new nonmagnetic time-of-flight mass spectrometer with high resolution. *Soviet Physics: Journal of Experimental and Theoretical Physics Letters*, *37*, 45.

March, R. (1997). An introduction to Quadrupole ion trap mass spectrometry. *Journal of Mass Spectrometry*, *32*(4), 351–369.

Marshall, A. G., & Hendrickson, C. L. (2002). Fourier transform ion cyclotron resonance detection: Principles and experimental configurations. *International Journal of Mass Spectrometry*, *215*, 59–75.

Marshall, A. G., Hendrickson, C. L., & Jackson, G. S. (1998). Fourier transform ion cyclotron mass spectrometry: A primer. *Mass Spectrometry Reviews*, *17*, 1–35.

McAlister, G. C., Berggren, W. T., Griep-Raming, J., Horning, S., Makarov, A., Phanstiel, D., et al. (2008). A proteomics grade electron transfer dissociation-enabled hybrid linear ion trap-orbitrap mass spectrometer. *Journal of Proteome Research*, 7(8), 3127–3136.

McNaught, A.D., Wilkinson, A. (1997). IUPAC Compendium of Chemical Terminology, 2nd ed.; Blackwell Science Publications: Oxford, U.K. IUPAC Compendium of Chemical Terminology—the Gold Book http://goldbook.iupac.org/ (2006) created by M. Nic, J. Jirat, B. Kosata; updates compiled by A. Jenkins. ISBN 0-9678550-9-8. http://dx.doi.org/10.1351/goldbook.

Michalski, A., Damoc, E., Lange, O., Denisov, E., Nolting, D., Müller, M., et al. (2012). Ultra high resolution linear ion trap Orbitrap mass spectrometer (Orbitrap Elite) facilitates top down LC MS/MS and versatile peptide fragmentation modes. *Molecular and Cellular Biology*, *11*(3). http://dx.doi.org/10.1074/mcp.O111.013698.

Miura, D., Fujimura, Y., & Wariishi, H. (2012). In situ metabolomic mass spectrometry imaging: Recent advances and difficulties. *Journal of Proteomics*, *75*, 5052–5060.

Miura, D., Tsuji, Y., Takahashi, K., Wariishi, H., & Saito, K. (2010). A strategy for the determination of the elemental composition by Fourier transform ion cyclotron resonance mass spectrometry based on isotopic peak ratios. *Analytical Chemistry*, *82*, 5887–5891.

Montaudo, G., Samperi, F., & Montaudo, M. S. (2006). Characterization of synthetic polymers by MALDI-MS. *Progress in Polymer Science*, *31*, 277–357.

Munson, M. S. B., & Field, F. H. (1966). Chemical ionization mass spectrometry. I. General introduction. *Journal American Chemistry Society*, *88*, 2621–2630.

Naban-Maillet, J., Lesage, D., Bossée, A., Gimbert, Y., Sztáray, J., Vékey, K., et al. (2005). Internal energy distribution in electrospray ionization. *Journal of Mass Spectrometry*, *40*(1), 1–8.

Nakamura, Y., Kimura, A., Saga, H., Oikawa, A., Shinbo, Y., Kosuke, K., et al. (2007). Differential metabolomics unraveling light/dark regulation of metabolic activities in Arabidopsis cell culture. *Planta*, *227*, 57–66.

Nguyen, V. H., Afonso, C., & Tabet, J. C. (2011). Concomitant EDD and EID of DNA evidenced by MS n and double resonance experiments. *International Journal of Mass Spectrometry*, *301*(1–3), 224–233.

Nikolaev, E. N., Boldin, I. A., Jertz, R., & Baykut, G. (2011). Initial experimental characterization of a new ultra-high resolution FTICR cell with dynamic harmonization. *Journal of the American Society for Mass Spectrometry*, *22*(7), 1125–1133.

Nikolaev, E. N., Jertz, R., Grigoryev, A., & Baykut, G. (2012). Fine structure in isotopic peak distributions measured using a dynamically harmonized Fourier transform ion cyclotron resonance cell at 7 T. *Analytical Chemistry*, *84*, 2275–2283.

Nordstrom, A., Want, E., Northen, T., Lehtio, J., & Siuzdak, G. (2008). Multiple ionization mass spectrometry strategy used to reveal the complexity of metabolomics. *Analytical Chemistry*, *80*, 421–429.

Ohta, D., Kanaya, S., & Suzuki, H. (2010). Application of Fourier-transform ion cyclotron resonance mass spectrometry to metabolic profiling and metabolite identification. *Current Opinion in Biotechnology*, *21*, 35–44.

Oikawa, A., Nakamura, Y., Ogura, T., Kimura, A., Suzuki, H., Sakurai, N., et al. (2006). Clarification of pathway-specific inhibition by Fourier transform ion cyclotron resonance/mass spectrometry-based metabolic phenotyping studies1[W]. *Plant Physiology*, *142*, 398–413.

Oishi, T., Tanaka, K., Hashimoto, T., Shinbo, Y., Jumtee, K., Bamba, T., et al. (2009). An approach to peak detection in GC-MS chromatograms and application of KNApSAcK database in prediction of candidate metabolites. *Plant Biotechnology*, *26*, 167–174.

Oksman, P. (1995). A Fourier transform time-of-flight mass spectrometer. A SIMION calculation approach. *International Journal of Mass Spectrometry and Ion Processes*, *141*, 67–76.

Olsen, J. V., de Godoy, L. M. F., Li, G., Macek, B., Mortensen, P., Pesch, R., et al. (2005). Parts per million mass accuracy on an Orbitrap mass spectrometer via lock mass injection into a C-trap. *Molecular & Cellular Proteomics*, *4*(12), 2010–2021.

Olsen, J. V., Macek, B., Lange, O., Makarov, A., Horning, S., & Mann, M. (2007). Higher-energy C-trap dissociation for peptide modification analysis. *Nature Methods*, *4*(9), 709–712.

Opauszky, I. (1982). Lasers in solid state mass spectrometry. *Pure and Applied Chemistry*, *54*(4), 879–887.

Paris, A., Rathahao, E., Debrauwer, L., Dolo, L., Hamami, S., & Hillenweck, A. (2000). Identification of long chain fatty acid 17b-estradiol-17-esters as non polar residues in perirenal fat in steers and heifers. In *Euroresidu IV, Veldhoven, Netherlands* (pp. 829–833).

Pascal-Lorber, S., Rathahao, E., Cravedi, J. P., & Laurent, F. (2004). Metabolic fat of [^{14}C]-2,4-dichlorophenol in macrophytes. *Chemosphere*, *56*, 275–284.

Pasikanti, K. K., Ho, P. C., & Chan, E. C. Y. (2008). Gas chromatography/mass spectrometry in metabolic profiling of biological fluids. *Journal of Chromatography. B*, *871*, 202–211.

Peltz, C., Drahos, L., & Vékey, K. (2007). SORI excitation: Collisional and radiative processes. *Journal of the American Society for Mass Spectrometry*, *18*, 2119–2126.

Perry, R. H., Cooks, R. G., & Noll, R. J. (2008). Orbitrap mass spectrometry: Instrumentation ion motion and applications. *Mass Spectrometry Reviews*, *27*, 661–699.

Peterman, S. M., Duczak, N., Kalgutkar, A. S., Lame, M. E., & Soglia, J. R. (2006). Application of a linear ion trap/orbitrap mass spectrometer in metabolite characterization studies: Examination of the human liver microsomal metabolism of the non-tricyclic anti-depressant nefazodone using data-dependent accurate mass measurements. *Journal of the American Society for Mass Spectrometry*, *17*, 363–375.

Peterman, S. M., Dufresne, C. P., & Horning, S. (2005). The use of a hybrid linear trap/FT-ICR mass spectrometer for on-line high resolution/high mass accuracy bottom-up sequencing. *Journal of Biomolecular Techniques*, *16*(2), 112–124.

Peterson, D. S. (2007). Matrix-free methods for laser desorption/ionization mass spectrometry. *Mass Spectrometry Reviews*, *26*, 19–34.

Psychogios, N., Hau, D. D., Peng, J., Guo, A. C., Mandal, R., Bouatra, S., et al. (2011). The human serum metabolome. *PLoS One*, *6*(2), e16957.

Ramautar, R., Somsen, G. W., & de Jong, G. J. (2009). CE-MS in metabolomics. *Electrophoresis*, *30*, 276–291.

Rathahao, E., Hillenweck, A., Paris, A., & Debrauwer, L. (2000). Investigation of the in vitro metabolism of 17ß-estradiol by LC-MS/MS using ESI and APCI. *Analusis*, *28*, 273–279.

Rathahao, E., Peiro, G., Martins, N., Alary, J., Gueraud, F., & Debrauwer, L. (2005). Liquid chromatography—Multistage tandem mass spectrometry for the quantification of dihydroxynonene mercapturic acid (DHN-MA), a urinary end-metabolite of 4-hydroxynonenal. *Analytical and Bioanalytical Chemistry, 381*, 1532–1539.

Rizzati, V., Rathahao, E., Gamet-Payrastre, L., Delous, G., Jouanin, I., Gueraud, F., et al. (2005). In vitro aromatic bioactivity of the weak estrogen E2alpha and genesis of DNA adducts. *Steroids, 70*, 161–172.

Robb, D. B., Covey, T. R., & Bruins, A. P. (2000). Atmospheric pressure photoionization: An ionization method for liquid chromatography-mass spectrometry. *Analytical Chemistry, 72*, 3653–3659.

Rojas-Chertó, M., Kasper, P. T., Willighagen, E. L., Vreeken, R. J., Hankemeier, T., & Reijmers, T. H. (2011). Elemental composition determination based on MS^n. *Bioinformatics, 27*(17), 2376–2383.

Roux, A., Lison, D., Junot, C., & Heilier, J.-F. (2011). Applications of liquid chromatography coupled to mass spectrometry-based metabolomics in clinical chemistry and toxicology: A review. *Clinical Biochemistry, 44*, 119–135.

Roux, A., Xu, Y., Heilier, J.-F., Olivier, M.-F., Ezan, E., Tabet, J.-C., et al. (2012). Annotation of the human adult urinary metabolome and metabolite identification using ultra high performance liquid chromatography coupled to a linear quadrupole ion trap-Orbitrap mass spectrometer. *Analytical Chemistry, 84*, 6429–6437.

Sargent, M., Harte, R., & Harrington, C. (2002). *Guidelines for achieving high accuracy in isotope dilution mass spectrometry (IDMS)*. Cambridge, UK: The Royal Society of Chemistry.

Schwartz, J. C., Senko, M. W., & Syka, J. E. P. (2002). A two-dimensional quadrupole ion trap mass spectrometer. *Journal of the American Society for Mass Spectrometry, 13*, 659–669.

Senko, M. W., Hendrickson, C. L., Emmett, M. R., Shi, S. D.-H., & Marshall, A. G. (1997). External accumulation of ions for enhanced electrospray ionization Fourier transform ion cyclotron resonance mass spectrometry. *Journal of the American Society for Mass Spectrometry, 8*, 970–976.

Sleno, L. (2012). The use of mass defect in modern mass spectrometry. *Journal of Mass Spectrometry, 47*, 226–236.

Sleno, L., Volmer, D. A., & Marshall, A. G. (2005). Assigning product ions from complex MS/MS spectra: The importance of mass uncertainty and resolving power. *Journal of the American Society for Mass Spectrometry, 16*, 183–198.

Smith, C. A., O'Maille, G., Want, E. J., Qin, C., Trauger, S. A., Brandon, T. R., et al. (2005). METLIN: A metabolite mass spectral database. *Therapeutic Drug Monitoring, 27*(6), 747–751.

Smith, C. A., Want, E. J., O'Maille, G., Abagyan, R., & Siuzdak, G. (2006). XCMS: Processing mass spectrometry data for metabolite profiling using nonlinear peak alignment, matching, and identification. *Analytical Chemistry, 78*(3), 779–787.

Sommer, H., Thomas, H. A., & Hipple, J. A. (1951). Measurement of e/M by cyclotron resonance. *Physical Review, 82*, 697–702.

Southam, A. D., Payne, T. G., Cooper, H. J., Arvanitis, T. N., & Viant, M. R. (2007). Dynamic range and mass accuracy of wide-scan direct infusion nanoelectrospray Fourier transform ion cyclotron resonance mass spectrometry-based metabolomics increased by the spectral stitching method. *Analytical Chemistry, 79*, 4595–4602.

Stoll, N., Schmidt, E., & Thurow, K. (2006). Isotope pattern evaluation for the reduction of elemental compositions assigned to high-resolution mass spectral data from electrospray ionization Fourier transform ion cyclotron resonance mass spectrometry. *Journal of the American Society for Mass Spectrometry, 17*, 1692–1699.

Sunner, J., Dratz, E., & Chen, Y.-C. (1995). Graphite surface-assisted laser desorption/ionization time-of-flight mass spectrometry of peptides and proteins from liquid solutions. *Analytical Chemistry, 67*(23), 4335–4342.

Tabet, J. C., & Cotter, R. J. (1984). Laser desorption time-of-flight mass spectrometry of high mass molecules. *Analytical Chemistry, 56*(9), 1662–1667.

Taguchi, V. Y., Nieckarz, R. J., Clement, R. E., Krolik, S., & Williams, R. (2010). Dioxin analysis by gas chromatography-Fourier transform ion cyclotron resonance mass spectrometry (GC-FTICR-MS). *Journal of the American Society for Mass Spectrometry, 21*, 1918–1921.

Takahashi, H., Kai, K., Shinbo, Y., Tanaka, K., Ohta, D., Oshima, T., et al. (2008). Metabolomics approach for determining growth-specific metabolites based on Fourier transform ion cyclotron resonance mass spectrometry. *Analytical and Bioanalytical Chemistry, 391*, 2769–2782.

Takáts, Z., Wiseman, J. M., Gologan, B., & Cooks, R. G. (2004). Mass spectrometry sampling under ambient conditions with desorption electrospray ionization. *Science, 306*, 471–473.

Tanaka, K., Waki, H., Ido, Y., Akita, S., Yoshida, Y., & Yoshida, T. (1988). Protein and polymer analyses up to *m/z* 100 000 by laser ionization time-of flight mass spectrometry. *Rapid Communication in Mass Spectrometry, 2*(20), 151–153.

Thompson, C. M., Richards, D. S., Fancy, S.-A., Perkins, G. L., Pullen, F. S., & Thom, C. (2003). A comparison of accurate mass techniques for the structural elucidation of fluconazole. *Rapid Communication of Mass Spectrometry, 17*, 2804–2808.

Thomson, J. J. (1910). Rays of positive electricity. *Philosophical Magazine, 20*, 752–767.

Thurman, E. M., & Ferrer, I. (2010). The isotopic mass defect: A tool for limiting molecular formulas by accurate mass. *Analytical and Bioanalytical Chemistry, 397*, 2807–2816.

Tohge, T., & Fernie, A. R. (2009). Web-based resources for mass-spectrometry-based metabolomics: A user's guide. *Phytochemistry, 70*, 450–456.

Vékey, K. (1996). Internal energy effects in mass spectrometry. *Journal of Mass Spectrometry, 31*, 445–463.

Villas-Bôas, S. G., Mas, S., Akesson, M., Smedsgaard, J., & Nielsen, J. (2005). Mass spectrometry in metabolome analysis. *Mass Spectrometry Reviews, 24*(5), 613–646.

Weber, R. J. M., Southam, A. D., Sommer, U., & Viant, M. R. (2011). Characterization of isotopic abundance measurements in high resolution FT-ICR and Orbitrap mass spectra for improved confidence of metabolite identification. *Analytical Chemistry, 83*, 3737–3743.

Wei, J., Buriak, J. M., & Siuzdak, G. (1999). Desorption/Ionization mass spectrometry on porous silicon. *Nature, 399*, 243–246.

Werner, E., Croixmarie, V., Umbdenstock, T., Ezan, E., Chaminade, P., Tabet, J.-C., et al. (2008). Mass spectrometry-based metabolomics: Accelerating the characterization of discriminating signals by combining statistical correlations and ultrahigh resolution. *Analytical Chemistry, 80*, 4918–4932.

Werner, E., Heilier, J.-F., Ducruix, C., Ezan, E., Junot, C., & Tabet, J.-C. (2008). Mass spectrometry for the identification of the discriminating signals from metabolomics: Current status and future trends. *Journal of Chromatography. B, 871*, 143–163.

Wilson, I. D., Plumb, R., Granger, J., Major, H., Williams, R., & Lenz, E. M. (2005). HPLC-MS-based methods for the study of metabonomics. *Journal of Chromatography. B, 817*, 67–76.

Winder, C. L., Dunn, W. B., & Goodacre, R. (2011). TARDIS-based microbial metabolomics: Time and relative differences in systems. *Trends in Microbiology, 19*, 315–322.

Wineland, D. J. (1984). Spectroscopy of stored ions. In B. N. Taylor & W. D. Phillips (Eds.), *Precision measurement and fundamental constants II* (pp. 83–92). United States: National Bureau of Standards, Special Publications 617.

Wu, L., Mashego, M. R., van Dam, J. C., Proell, A. M., Vinke, J. L., Ras, C., et al. (2005). Quantitative analysis of the microbial metabolome by isotope dilution mass spectrometry using uniformly C-13-labeled cell extracts as internal standards. *Analytical Biochemistry, 336*, 164–171.

Wu, G., Noll, R. J., Plass, W. R., Hu, Q., Perry, R. H., & Cooks, R. G. (2006). Ion trajectory simulations of axial ac dipolar excitation in the Orbitrap. *International Journal of Mass Spectrometry*, *254*, 53–62.

Wu, Z., Rodgers, R. P., & Marshall, A. G. (2004). Two- and three-dimensional van krevelen diagrams: A graphical analysis complementary to the kendrick mass plot for sorting elemental compositions of complex organic mixtures based on ultrahigh-resolution broadband Fourier transform ion cyclotron resonance mass measurements. *Analytical Chemistry*, *76*, 2511–2516.

Xie, T., Liang, Y., Hao, H., Jiye, A., Xie, L., Gong, P., et al. (2012). Rapid identification of ophiopogonins and ophiopogonones in Ophiopogon japonicus extract with a practical technique of mass defect filtering based on high resolution mass spectrometry. *Journal of Chromatography. A*, *1227*, 234–244.

Xu, Y., Heilier, J.-F., Madalinski, G., Genin, E., Ezan, E., Tabet, J.-C., et al. (2010). Evaluation of accurate mass and relative isotopic abundance measurements in the LTQ-Orbitrap mass spectrometer for further metabolomics database building. *Analytical Chemistry*, *82*, 5490–5501.

Xu, S., Li, Y., Zou, H., Qiu, J., Guo, Z., & Guo, B. (2003). Carbon nanotubes as assisted matrix for laser desorption/ionization time-of-flight mass spectrometry. *Analytical Chemistry*, *75*, 6191–6195.

Yager, J. D., & Liehr, J. G. (1996). Molecular mechanisms of estrogen carcinogenesis. *Annual Review of Pharmacology and Toxicology*, *36*, 203–332.

Yamashita, M., & Fenn, J. B. (1984). Another variation on the free-jet theme. Electrospray ion source. *Journal of Physical Chemistry*, *88*, 4451–4459.

Yergey, J. A. (1983). A general approach to calculating isotopic distributions for mass spectrometry. *International Journal of Mass Spectrometry and Ion Physics*, *52*, 337–349.

Zalko, D., Soto, A., Dolo, L., Dorio, C., Rathahao, E., Debrauwer, L., et al. (2003). Biotransformations of bisphenol A in a mammalian model: Answers and new questions raised by low-dose metabolic fate studies in pregnant CD1 mice. *Environmental Health Perspectives*, *111*, 309–319.

Zhang, N. R., Yu, S., Tiller, P., Yeh, S., Mahan, E., & Emary, W. B. (2009). Quantitation of small molecules using high-resolution accurate mass spectrometers—A different approach for analysis of biological samples. *Rapid Communications in Mass Spectrometry*, *23*, 1085–1094.

Zhang, H., Zhang, D., & Ray, K. (2003). A software filter to remove interference ions from drug metabolites in accurate mass liquid chromatography/mass spectrometric analyses. *Journal of Mass Spectrometry*, *38*, 1110–1112.

Zhang, H., Zhang, D., Ray, K., & Zhu, M. (2009). Mass defect filter technique and its applications to drug metabolite identification by high-resolution mass spectrometry. *Journal of Mass Spectrometry*, *44*, 999–1016.

Zhang, H., Zhu, M., Ray, K. L., Ma, L., & Zhang, D. (2008). Mass defect profiles of biological matrices and the general applicability of mass defect filtering for metabolite detection. *Rapid Communications in Mass Spectrometry*, *22*, 2082–2088.

Zhu, M., Ma, L., Zhang, D., Ray, K., Zhao, W., Humphreys, W. G., et al. (2006). Detection and characterization of metabolites in biological matrices using mass defect filtering of liquid chromatography/high resolution mass spectrometry data. *Drug Metabolism and Disposition*, *34*, 1722–1733.

CHAPTER SIX

Exploring Metabolome with GC/MS

Cyril Jousse*, Estelle Pujos-Guillot†,‡,1

*UBP-UFR Sciences, ICCF-UMR6296, Plateforme d'Exploration du Métabolisme, Aubière, France
†INRA, UMR 1019, Plateforme d'Exploration du Métabolisme, Nutrition Humaine, Saint Genès Champanelle, France
‡Clermont Université, UFR Médecine, UMR 1019 Nutrition Humaine, Clermont-Ferrand, France
[1]Corresponding author: e-mail address: estelle.pujos@clermont.inra.fr

Contents

Abstract

The rapidly expanding field of metabolomics has been driven by major advances in analytical tools such as mass spectrometry and the corresponding hyphenated methods such as gas chromatography (GC–MS) but also in chemometrics and bioinformatics. The metabolomic field has been mainly developed in plant sciences and GC/MS was applied very early as it allowed determining a large number of metabolites in a wide range of samples. A typical metabolomic experiment includes (i) sample preparation, (ii) derivatization, (iii) metabolic profiling by GC/MS, (iv) treatment of data set, (v) identification of key or discriminatory metabolites and eventually (vi) their quantification. Each of these steps needs to be optimized and technical choices are determined

Advances in Botanical Research, Volume 67
ISSN 0065-2296
http://dx.doi.org/10.1016/B978-0-12-397922-3.00006-X

depending on the scientific objective. This chapter gives a review of technologies and methods dedicated to GC/MS-based metabolomics, as well as the description of major advances in this area in recent years and perspectives for the future. Finally, the main application area of plant metabolomics, linked to phytochemistry, gene modifications and environmental disorders, are briefly presented.

1. INTRODUCTION

In recent years, the development of mass spectrometry (MS) significantly impacted biological research, including metabolomics. Because of its selectivity and sensitivity, MS has played an important role in the development of methods for metabolite profiling and fingerprinting.

A combination of techniques is often necessary to allow covering the different classes of metabolites with various physico-chemical properties and their wide concentration range in biological media (Büscher, Czernik, Ewald, Sauer, & Zamboni, 2009; Lenz & Wilson, 2006). Plants, in particular, contain a wide range of metabolites (200,000 according to Fiehn, 2002), which creates complexity when optimizing metabolomic protocols.

The use of gas chromatography (GC) is dedicated to the study of a set of biological molecules, that is, those that are volatile or could be derivatized. Metabolomic studies often used GC/MS techniques because of the widespread availability of instruments, broad coverage of metabolites, ease of use and the availability of extensive mass spectral libraries (Wilson, Hoggard, & Synovec, 2012).

Plant sciences and GC analyses have been strongly linked for several decades. Traditionally to get a vision of the physio-pathologic responses related to metabolic deregulation, single metabolites or classes of small molecules were measured using targeted analytical assays. Recent advances in analytical chemistry and computation now enable the simultaneous measurement of a large and diverse set of metabolites and the study of plant composition (Guy, Kopka, & Moritz, 2008) and the Physiologia Plantarum special issue 132 and plant tissue metabolic response to stimuli (Heinrich, 2008).

The metabolomic field has been firstly mainly developed in plant sciences and GC/MS technique was applied very early for the determination of a large number of metabolites in a wide range of plant species (Fiehn et al., 2000; Roessner, Wagner, Kopka, Trethewey, & Willmitzer, 2000). As with any other biological samples, metabolomics permit to evaluate physiological processes from genetic/enzymatic origins (refs.). It deals also with environmental adaptations to biotic and/or abiotic stress (Shulaev, Cortes, Miller, & Mittler, 2008). GC/MS and GC/FID have been selected for a long time as the techniques of choice in order to explore metabolic diversity of plant samples, exploring crude extracts and medicinal oils, and to validate purified drugs and goods (Noctor et al., 2007).

According to Fiehn et al. (2000), it is now considered as a 'mature and highly reproducible technique' and is suitable for compounds thermally stable or derivatizable and for small volatile non-polar ones (Daskalchuk, Ahiahonu, & Heath, 2006). As a result, metabolomics publications appeared early in this field with the number increasing at an exponential rate since about 2008 (Fig. 6.1).

This chapter is dedicated to the metabolome exploration using GC/MS. It will concern a review of technologies and methods dedicated to GC/MS-based metabolomics, as well as the description of major advances in this area in recent years and perspectives for the future.

A typical metabolomic experiment includes (i) preparation of the sample, (ii) analysis of samples via GC/MS, (iii) treatment of data set and (iv) the identification of key or discriminatory metabolites (Fig. 6.2).

The data acquisition and retrieval may be achieved by a targeted analysis where specific pathways are analyzed or an untargeted analysis where no prior selection of pathways is performed. Both have advantages and disadvantages and the approach taken depends ultimately on the study in question. Two alternative strategies have been carried out in MS-based

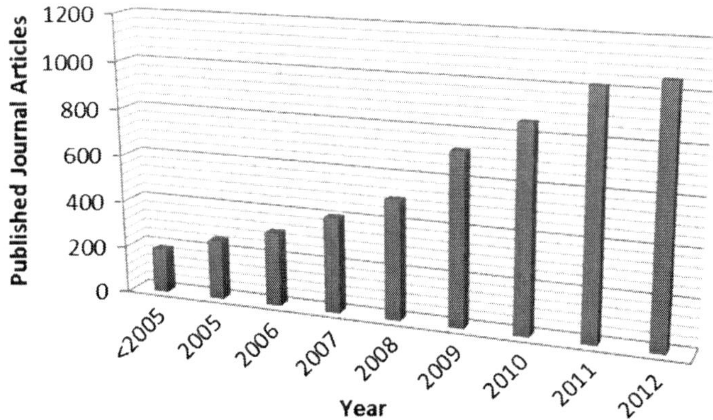

Figure 6.1 PubMed literature search result documents the continuously growing research areas of, 'metabolomics and GC/MS or metabolome or metabolite profiling or metabolite profile' based on numbers of publications. (For colour version of this figure, the reader is referred to the online version of this chapter.)

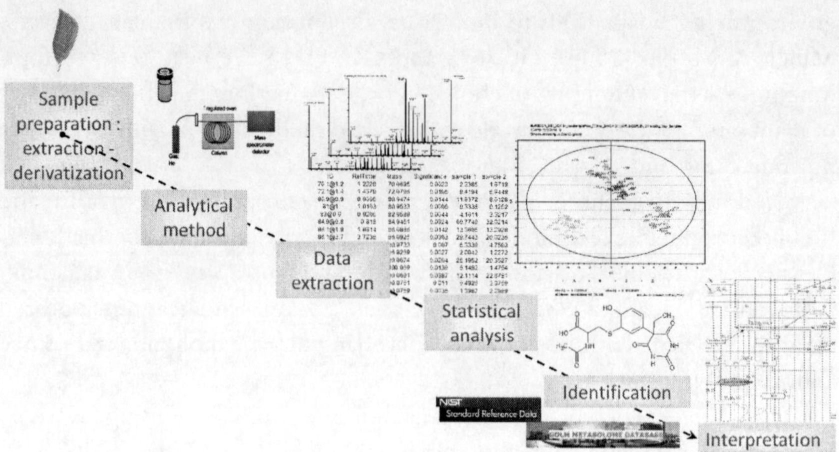

Figure 6.2 Metabolomic workflow. (For colour version of this figure, the reader is referred to the online version of this chapter.)

metabolomics: metabolic fingerprinting, which consists in determining metabolic signatures associated with a particular stress response or phenotype, and metabolic profiling, which aims at a simultaneous measurement of a set of metabolites in a sample (Shulaev et al., 2008).

This chapter well describes the different steps and finally presents the application of GC/MS metabolomics to plant study.

2. SAMPLE EXTRACTION FOR METABOLOMIC ANALYSIS

The extraction from a biological sample is a crucial step, independently of metabolomics. However, this approach requires different constraints: Firstly, the extract has to be representative of the *in vivo* conditions of the biological system at the time of sampling (i.e. enzymatic activity has to be negligible). Secondly, repeatability of extraction has to be maximized to be able to perform differential analysis of all samples in a batch. Therefore, samples are generally frozen (at −20 or −80 °C) as quickly as possible and stored before analysis.

The objective of the extraction step before a metabolomic analysis is to maximize the analytical coverage and to limit ionization suppression and source clogging by removing interferents (like proteins, salts, photosynthetic pigments, etc.). Therefore, the sample preparation should be as simple, universal and robust as possible. The efficiency of different methods of deproteinization using organic solvents, heat treatment or acid treatment for lowering the pH was evaluated based on the quantity of remaining proteins and/or the number of metabolites. Alternative approaches based on sample fractionation were also evaluated. For the extraction of free polar or non-polar metabolites, a methanol–water or methanol–chloroform–water extraction protocol is, respectively, used. Optimization of the protocol often requires the adjustment of the ratio solvent/biomass and the duration of extraction (Villas-Bôas, Mas, Åkesson, Smedsgaard, & Nielsen, 2005). Gullberg, Jonsson, Nordström, Sjöström, and Moritz (2004) investigate how different chemical and physical factors affect the extraction of the metabolome from leaves of the plant *Arabidopsis thaliana*. Riedl et al. (2012) analyzed low molecular weight compounds from hydrophilic and lipophilic extracts of leaves from individual plants using GC/MS and demonstrated that metabolic fingerprintings are suitable for studying physiological processes of aquatic macrophytes and are potentially applicable for environmental monitoring.

3. GC/MS PROFILING

Metabolomic GC/MS profiling requires different steps: extraction of metabolites from the biological sample, derivatization for non-volatile compounds, separation by GC and analysis by MS including ionisation and detection. The objective is to provide non-biased data representative of sample metabolic complexity.

3.1. Volatile compounds analysis (Vuckovic, Risticevic, & Pawliszyn, 2011)

Metabolomic analysis of plant volatiles is based on entrapment or adsorption to collect and concentrate components of interest. Headspace and/or direct contact *in vivo* solid-phase micro extraction (SPME) improve metabolite coverage in comparison to that of traditional extraction approaches, especially for collection of lower-molecular-weight compounds. SPME enables solventless extraction by means of a fused-silica or stainless-steel fibre coated with a thin film of polymer, which acts as the sorbent/solvent during the extraction of compounds. It includes, firstly, the extraction of analytes from the sample into the SPME coating, and secondly, the removal of the analytes by thermal desorption. The extraction process itself can be performed either in headspace mode (of a cell culture or plant to study volatile emissions) or in direct extraction mode (to intact plants) depending on the volatility of the analytes to be studied.

SPME coating plays another crucial role in metabolomics because small molecules can diffuse through the outer biocompatible layer and enter the sorbent pores, whereas large biomolecules cannot. This means that once a metabolite partitions into the coating, it is protected from further enzymatic reactions thus ensuring the capture of short-lived and labile species.

Various commercial SPME fibre assemblies are available for *in vivo* headspace SPME metabolomic profiling of different plant compartments (such as living flowers, leaves and bracts) as well as the intercompartmental metabolome composition of grapefruit at different developmental stages (Flamini, Cioni, & Morelli, 2005).

The precision achievable with this technique is comparable to that of traditional extraction methods, and an advantage is the information obtained on biologically important unbound metabolites.

As example, plant communications were studied on *A. thaliana* (Rohloff & Bones, 2005) with SPME–GC/MS on intact plant and organs analyses (leaves, stems, inflorescences). More than 100 volatiles were reported, 59 for the first time on Arabidopsis. In this study, headspace SPME, using both polar and apolar columns, allows covering a wide range of volatile compounds.

3.2. Extending GC/MS metabolome coverage: derivatization

The major disadvantage of GC/MS could be a limited analytical coverage and the need to require long sample preparations. In particular, derivatization reactions are often necessary to reduce polarity, increase thermal stability and volatility (Dettmer, Aronov, & Hammock, 2007). Derivatization also improves compound ionization.

3.2.1 Principle and reagents

Many derivatizing reagents were explored in metabolomics approach. The active functional groups, such as carboxylic, amide, amino and hydroxyl groups, are either alkylated, acylated or silylated (Little, 1999; Wells, 1999). During silylation, the active hydrogen is replaced by an alkylsilyl group, primarily trimethylsilyl (TMS). These derivatives show an increased thermal stability and a higher volatility compared to their precursors. Moreover, they often show characteristic MS fragmentation caused by a bond cleavage at the Si atom. Silylation occurs through nucleophilic attack (SN2).

The silylation reaction is driven by a good leaving group, which means a leaving group with a low basicity, ability to stabilize a negative charge in the transitional state and little or no π backbonding between the leaving group and silicon atom. Silylation reagents will react with water and alcohols first.

Consequently, the silylation derivatization procedure needs a nonaqueous environment for the reaction (use of aprotic organic solvent), whereas some nonsilylating derivatization techniques, such as chloroformate derivatization, can be performed in the presence of water.

A lot of alternative reagents are commercially available for silylation. These reagents differ in their reactivity, selectivity, side reactions and the character of the reaction by-products from the silylation reagent. A lot of reagents are commercially available for silylation but BSTFA (*N,O*-bis-(trimethylsilyl)-trifluoroacetamide) and MSTFA (*N*-methyl-trimethylsilyltrifluoroacetamide) are being predominantly used in metabolomic applications. BSTFA is highly volatile and produces volatile by-products, which causes little interference with early eluting peaks. Moreover, it can be used as its own solvent. Alternatively, MSTFA is the most volatile of the TMS acetamides and it is therefore the most useful for the analysis of volatile trace materials where the derivatives may be near the reagent or by-product peak. Often 1% trimethylchlorosilane is added to the reagents as a catalyst. In addition to trimethylsilylation, *tert*-butyldimethylsilylation (*t*-BDMS) is being used for derivatization in metabolomics. *t*-BDMS derivatives are more resistant to hydrolysis and can be up to 10,000 times more stable than TMS derivatives.

Because of its ability to derivatize a wide range of metabolites (including carbohydrates, polyols, etc.), the most commonly used derivatization method in GC/MS metabolomics is based on a two-step protocol involving a methoximation reaction followed by a silylation (Kanani, Chrysanthopoulos, & Klapa, 2008). Carbonyl groups are often transformed into the corresponding oximes with hydroxylamine or alkoxyamines, which primarily stabilizes alpha-keto acids and locks sugars in open-ring conformation. Fiehn et al. (2000) found alkoxyamines superior to hydroxylamines for identification purposes.

In some cases, a combination of different derivatives is necessary to obtain a good analytical coverage of a molecular family. Bruni et al. (2002) optimized a GC profiling method on Theobroma seeds covering fatty acids, tocopherols and sterols by using esterification combined with silylation on unsaponifiable fractions.

3.2.2 Optimizing the derivatization step

One of the objectives of a metabolomic approach is to guarantee a good metabolome coverage with a limited analytical variability. Therefore, it is important to optimize the derivatization reaction to increase its efficiency and its reproducibility.

This step is particularly complex when hundreds of compounds with different physico-chemical properties are present in the biological media as derivatization kinetic will differ from one compound to another (Fig. 6.3).

The second point to consider is matrix interferences. Matrix effects are an important problem for quantification as they can modify artificially the analytical response and introduce bias. In method development and validation, it is important to evaluate them to be able to characterize method performances. Even if electron impact (EI) ionization mode is very robust compared to chemical ionization (CI) or atmospheric pressure processes, matrix effects still occur during the derivatization step (Gao, Pujos-Guillot, & Sébédio, 2010) and have to be controlled.

3.3. GC/MS methods in metabolomics

The analysis of a wide range of metabolites requires several analytical strategies (from metabolic target analysis to metabolic fingerprinting), depending on the biological question. However, the choice of analytical technologies and the optimization of GC and MS methods for a broad metabolome coverage are often the same.

3.3.1 Gas chromatography

Chromatographic separation is complex in metabolomics due to a high number of metabolites and multiple derivatization products. Therefore, a long analysis time may be needed to separate most of the metabolites before they enter into the mass spectrometer for detection. Different factors will influence chromatographic separation including injection conditions and column properties (length, stationary phase, internal diameter) and temperature programme.

3.3.1.1 Injection

In most of the metabolomic applications, between 0.5 and 2 μL of derivatized sample are introduced into a heated injector (temperature between 200 and 250 °C). As the most common derivatization produces TMS derivatives, it is preferable to use deactivated inlets. Sample can either be injected in split or splitless mode depending on the sensitivity required. The advantage of a splitless system is that a larger amount of the sample is introduced into the column with less thermal degradation. However, the split system is often preferred in metabolomics because metabolites are present in wide range concentrations and because it allows analyzing volatile compounds eluting near the solvent peak.

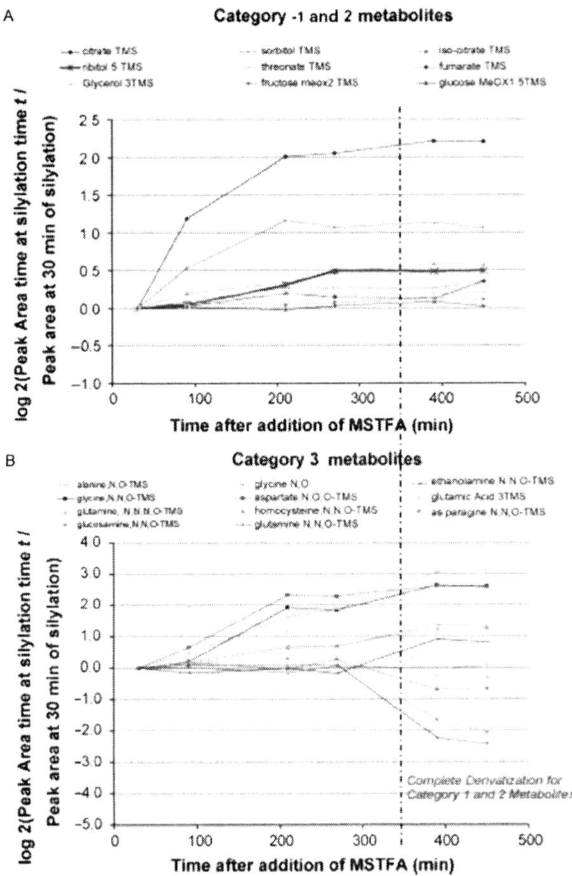

Figure 6.3 The profile over MSTFA silylation of the derivative marker ion peak area of (A) category 1 and category 2, and (B) category 3 metabolites. (For colour version of this figure, the reader is referred to the online version of this chapter.)

3.3.1.2 Column selection

Capillary GC columns made of fused silica are commonly used in GC/MS-based metabolomics. These columns can be used at very high temperature and allow obtaining good chromatographic resolution. These columns generally have a small internal diameter (0.25 mm) which limits the sample capacity to about 50–100 ng.

Columns with various polarity and chemical composition of stationary phases (DB-1–DB-50), and various lengths (10–60 m) have been used in metabonomic analysis. However, fused silica capillary columns with a 5% diphenyl cross-linked 95% dimethylpolysiloxane stationary phase

(0.25 μm film thickness), such as DB5-MS, HP-5MS or RTX-5MS, are the most typically used in metabolomic studies (Wiklund et al., 2007).

Optimization of GC parameters including flow rate, temperature programming and injection conditions (Wilson et al., 2012) is required to increase peak capacity production (i.e. peak capacity per separation run time). A compromise between the level of separation and length of analysis is often necessary in this selection.

To increase the number of separated metabolites in complex samples, comprehensive multidimensional GC × GC approaches are increasingly used (Koek, Muilwijk, van Stee, & Hankemeier, 2008; O'Hagan et al., 2007). Each peak eluting from a first non-polar column is transferred into a second polar column for further separation. This transfer is performed using a modulator: the eluent from the first column is divided into small segments, and cryo-trapped using CO_2 or N_2 cooled jets, and focussed. After a rapid heating, each fraction is then transferred into the second short column. This focussing step allows obtaining narrow injection bands and reduced peak widths, resulting in a higher sensitivity. The process is carried out very quickly as separation of a fraction has to be finished before the next segment reaches the modulator.

However, this approach has still much analytical variability especially in running large sample batches, and data processing is consequently more complex. No optimal solution is available yet and the data processing solutions are still under development, in particular concerning alignment software (Jeong, Shi, Zhang, Kim, & Shen, 2012).

3.3.2 Mass spectrometry

Initially, GC was the only separative method able to be coupled with MS. Even if GC/MS is considered as a mature technology, its application in metabolomics for high-throughput experiments requires fast acquisition rates, which are now available thanks to recent progresses.

3.3.2.1 Analyzer

GC is often coupled to single quadrupole analyzers that offer high sensitivity, good dynamic range but are inconvenient to operate with slower scan rates and lower resolution compared to Time of Flight systems. The introduction of rapid-scanning quadrupoles with scan speeds of up to 10,000 amu/s providing a less expensive alternative (Adahchour et al., 2005) have made them a popular and robust metabolomic platform. GC-triple quadrupole MS/MS allows to operate multiple reactions monitoring of analytes, which can overcome the challenging identification and quantification problems associated with co-eluting analytes in complex matrices.

However, GC coupled with Time-of-Flight Mass Spectrometry (GC/ToFMS) is increasingly used for metabolic profiling (Almstetter, Oefner, & Dettmer, 2012). Fast acquisition rates render ToFMS the ideal detector for metabolomic analysis of narrow chromatographic peaks. These fast scan speeds are particularly useful for an accurate deconvolution of overlapping peaks obtained from complex mixtures.

In particular, it has been demonstrated that combining 1D GC/ToFMS and GC × GC/ToFMS techniques was an optimal approach for metabolic phenotyping of natural variants (Kusano et al., 2007).

3.3.2.2 Ionisation source

EI ionization is the most commonly used method of ionization in metabolomics studies. As it is performed in a high-vacuum ion source where analytes in vapour state are bombarded with electrons at 70 eV, it provides many fragment ions. Fragmentation patterns are characteristic to a particular molecule and therefore this approach allows the use of spectral libraries.

CI has also been utilized in some specific metabolomic studies.

For example, this technique was combined to extraction, derivatization, dynamic headspace trapping and solvent desorption to optimize a robust profiling method for multiple phytohormones, VOCs and other metabolites (Schmelz, Engelberth, Tumlinson, Block, & Alborn, 2004).

As it is a softer ionization technique, it produces less fragmentation compared to EI. It is often used as it allows producing molecular ions of metabolites that do not give molecular ions in EI. This method is therefore complementary of EI experiments and can be used as a validation of candidate compounds given by library research. In particular, it provides molecular weight of the derivatives, as well as the fragmentation reaction laws.

Paolini, Costa, and Bernardini (2005) demonstrated that GC/MS was a powerful tool to determine complete profiles of essential oils. The combination of EI, positive and negative CI was found to be an effective method to study chemical structures of unknowns.

Recently, Wachsmuth et al. (2011) evaluated the performance of GC—atmospheric pressure chemical ionization-based methods in metabolomics. Its performance was found to be inferior to that of GC × GC/(EI)ToFMS in the quantitative analysis of potential biomarkers, but it exceeded the performance of the three GC-based techniques and proved to be a useful tool for metabolomics.

3.3.3 Source of variability and recommendations

In metabolomics, relative quantification based on mass intensities is usually applied to construct multivariate statistical models. To date, there are still no generally accepted assessment criteria for validation of metabolomic analysis. But it is really important to quantify analytical variability to confirm that the analytical method employed for a specific test is suitable for its intended use. Validations are generally focused on the robustness of sample pretreatment and on the instrument stability in terms of retention times, masses (essential for data alignment) and intensities (essential to detect biological variations).

Quality control samples are particularly useful to follow the analytical variations when running large sample sets and the development of procedures to correct batch effects are still of great importance (Almstetter et al., 2012).

As the usefulness of metabolomics highly depends on the quality of the acquired data as well as the level of detail used in describing the studies (associated metadata), the Metabolomics Society has established an initiative to propose minimum reporting standards (Fiehn, 2008).

As part of the META-PHOR project's (METAbolomics for Plants Health and OutReach: http://www.meta-phor.eu/), a GC–MS ring experiment based upon three complex matrices (melon, broccoli and rice) was launched in order to study the inter-laboratory reproducibility of GC/MS-based plant metabolomics (Allwood et al., 2009).

4. DATA PROCESSING

The main challenges for data processing in metabolomics are (i) the large amount of data to process, (ii) the alignment of chromatographic peaks along large sample sets, (iii) obtaining only one entry for each metabolite (Koek, Jellema, van der Greef, Tas, & Hankemeier, 2011) and (iv) achieving metabolite identification.

4.1. Extraction and alignment

Highly diverse and specialized software solutions for the GC/MS data preprocessing have been published. The objective of the preprocessing methods in metabolomic studies is to transform respective peak lists from series of chromatography data files into numeric data matrices. Each peak in the resulting three-dimensional data set is represented by retention time, mass to charge ratio (m/z) and its ion intensity in each sample. The matrix obtained is then exported for statistical analysis. Both automated peak extraction and the automated deconvolution of mass spectra allow the comprehensive analysis of GC/MS experiments.

The conventional approach in GC/MS is based on mass spectral deconvolution by processing the information present within single chromatograms. The most widely used tool may be the automated mass spectral deconvolution and identification system (AMDIS; http://chemdata.nist.gov/mass-spc/amdis/overview.html; Halket et al., 1999; Stein, 1999). For GC × GC-ToFMS data, two softwares are commonly used: the commercial ChromaTOF software (LECO Corporation, St. Joseph, MI, USA) and parallel factor analysis (PARAFAC, Harshman, 1970).

However, to be able to compare large numbers of samples and to improve data processing throughput, software projects, based on the comprehensive extraction of mass selective peak apex intensities, were initiated by academia. Typical examples of these tools are MetAlign (Lommen, 2009), MZmine (Katajamaa, Miettinen, & Orešič, 2006) or XCMS (Smith, Want, O'Maille, Abagyan, & Siuzdak, 2006).

Recently, alternative solutions were developed to combine both approaches: the Metab R package (Aggio, Villas-Bôas, & Ruggiero, 2011) was designed to automate the analysis of GC/MS datasets processed by AMDIS. New tools are still being developed, in particular, concerning the processing multidimensional data (Jeong et al., 2012).

4.2. Metabolite identification

Two different approaches are available for the identification of unknown: (1) the 'bottom-up' approach which consists a characterization from the spectra and validation using authentic standard of the predicted structure and (2) the 'top-down' characterization restricted to specific biological samples.

4.2.1 Spectral libraries

Metabolite identification using GC/MS-based metabolomics is performed by retention time comparison with pure standard compounds and/or comparison using the retention index of mass spectral library databases. Mass spectral matching alone was found insufficient for non-ambiguous identification, because of the existing multiple structural isomers present in highly complex biological samples (Strehmel, Hummel, Erban, Strassburg, & Kopka, 2008). Therefore, retention indices (RIs) are used to provide a second independent criterion. RIs are calculated by relating the retention time of the compound to retention times of standard *n*-alkanes analyzed under same analytical condition.

NIST 2005 is the latest edition of the most popular NIST spectrum library with a collection of 190,825 spectra and 121,112 RIs. It is available either alone or combined with the Wiley Registry (Wiley Registry 7/NIST 2005) for a complete mass spectral library solution. Although these libraries are extensive, they do not contain a large number of metabolites that are found in biological metabolic pathways. More than 10,000 metabolites have been described in plants and a GC/MS profile of a single extracts can contain hundreds of peaks. Most of the time, the majority of these peaks cannot be identified because many metabolites are not commercially available.

However, the distribution of pure reference substances in plant metabolomics community has allowed establishing comparability between laboratories using GC/MS in EI mode. A platform including mass spectra and RI has been established and is freely available in Golm (Hummel, Selbig, Walther, & Kopka, 2007; Kopka et al., 2005; Schauer et al., 2005). These large spectra libraries are often incomplete and the identification of unknown metabolites is still the major bottleneck in metabolomics (Fig. 6.4). The current developments to solve this problem concern algorithms to calculate RI and tools for *in silico* mass spectra prediction (Kumari, Stevens, Kind, Denkert, & Fiehn, 2011). One of the major challenges will also deal with the implementation of libraries with metabolites that may not be structurally identified (Styczynski et al., 2007). Mass spectral tags, that is, mass spectra characterized by a specific retention time and a repeated occurrence in biological samples (Kopka, 2006) were created to build new databases.

4.2.2 Characterization

Concerning the second alternative, it has been demonstrated that the elemental compositions of unknown metabolites can be determined from spectra after TBS derivatization (Fiehn et al., 2000). In this study, unknown peaks were characterized from both mass and isotopic ratio determination in Arabidopsis leaf extracts. *In vivo* stable isotope labelling is also an essential tool for 'top-down' characterization (Birkemeyer, Luedemann, Wagner, Erban, & Kopka, 2005). Fully labelled mass isotopomers provides unambiguous information on molecular ion and mass fragments and also a proof of the metabolic origin of compounds of interest.

Kumari et al. (2011) presented a comprehensive workflow to annotate GC–MS peaks without library spectra matching, just using accurate masses (determined using CI accurate mass GC-ToF) and matching predicted

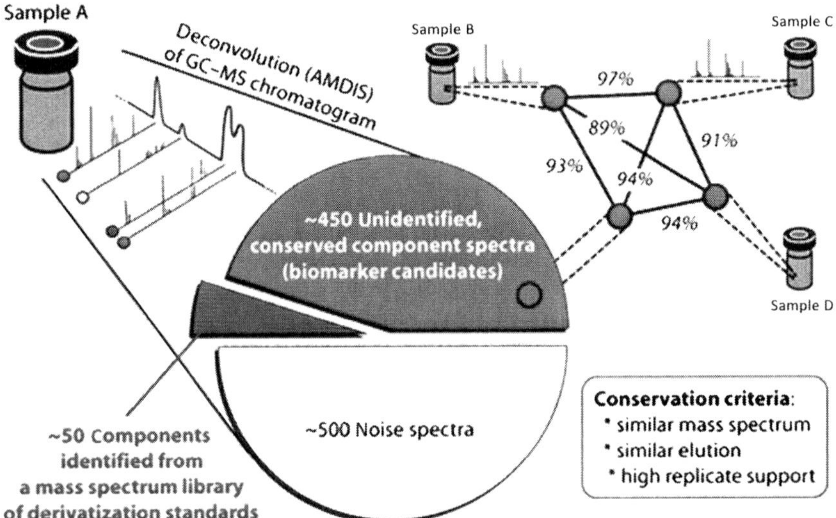

Figure 6.4 For the GC/MS vial labelled 'Sample A', we display the deconvolution and enumeration of components by AMDIS from the GC/MS output. Components labelled with blue circles did not match any spectra in a standard library but, as displayed in the upper right, had highly similar spectra occur in replicate samples. Note that though the relative amount of 'white' noise spectra may decrease with better deconvolution and peak enumeration, there will still remain a significant number of blue 'unidentifiable' spectra without an automated, systematic way to handle them. (For interpretation of the references to colour in this figure legend, the reader is referred to the online version of this chapter.)

against experimental RIs and mass spectra. Retention index prediction, as well as *in silico* fragmentation tools often allowed constraining hit lists of putative identifications.

5. QUANTIFICATION

In metabolomics, GC–MS is rather established as a semi-quantitative tool for metabolic fingerprinting or profiling approaches. However, standardisation is required to be able to compare results from diverse biological samples or experimental conditions, and achieve the same 'quantitative' level of transcriptome and proteome experiments.

Absolute quantification is still a challenging task using GC–MS technology because of the matrix effects occurring during derivatization and/or analysis (Koek et al., 2011). Isotope dilution MS is often the method of choice for the accurate determination of absolute metabolite concentrations (Vielhauer, Zakhartsev, Horn, Takors, & Reuss, 2011).

For this purpose, stable isotope labelling was developed to be used as a quantitative multiplex chemical standard. In particular, saturated *in vivo* labelling reference samples produced using 13C tracers have been used by (Birkemeyer et al., 2005). This standardised reference is generally mixed with non-labelled samples to determine recovery of all metabolic components using the appropriate isotopomer. The advantages of this approach are to obtain a better quantitative precision with an extended linear range (Wu et al., 2005), and a standardization of metabolomics experiments for the comparison between platforms and laboratories.

6. APPLICATIONS

Plant metabolomics was strongly linked with GC/MS (as well as NMR; Fan et al., 2001; Ritota et al., 2012) for historical reasons and because this technique allowed a wide analytical metabolome coverage, an easier identification process due to large databases, and robust quantifications. This section is dedicated to some current applications of plant metabolomics, linked to phytochemistry, gene modifications and environmental disorders (Fig. 6.5).

6.1. Metabolomics and phytochemistry

Drug discovery and chemotaxonomy (among others) are perhaps the oldest application fields to perform metabolic profiling, but they still are hot topics.

Because of the presence of more than 200,000 different plant compounds, metabolomics on medicinal plants is interesting for evidence-based development of new products (Shyur & Yang, 2008). All the platforms are presented in this review, with various advantages, but GC/MS technique is particularly emphasized because of its maturity and recent evolutions such as GC × GC/MS.

Oils (e.g. lipids and secondary metabolites) have been analyzed by GC/MS for decades, and lipidomics is now considered as a full topic among the metabolomic approaches. Phytochemical profiles of Condolia plants were used in order to identify compounds responsible of insecticidal effect (Cespedes et al., 2013). Among three different extracts, from conventional procedures, the hexane one was analyzed by GC/MS to determine long-chain *n*-alkanes composition in four plant populations. In this study, authors suggested synergistic effect on both insect growth regulators and insecticidal activity.

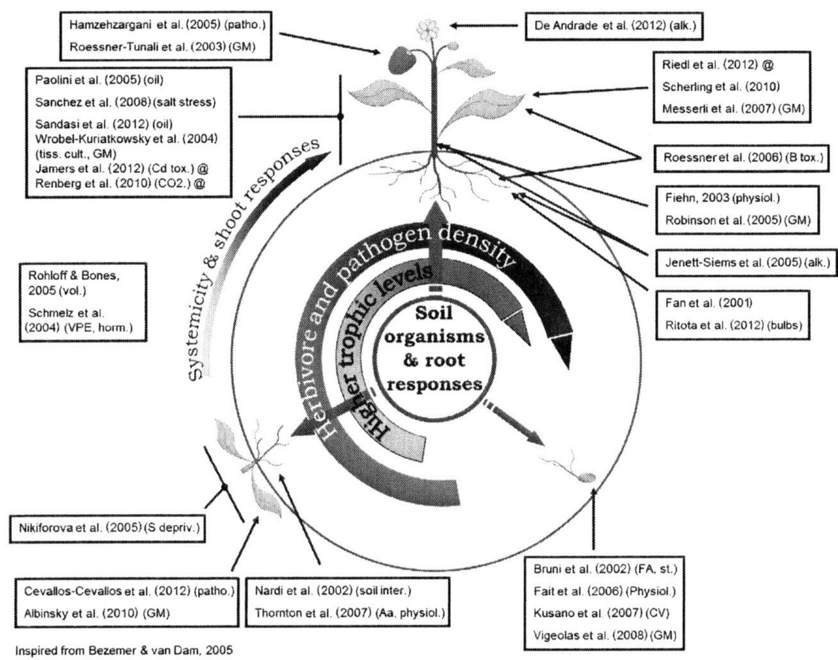

Figure 6.5 Examples of plant metabolomics studies. (For colour version of this figure, the reader is referred to the online version of this chapter.)

Oxylipins are important metabolites for plant defence, from fatty acid pathways. Analysis guidelines are often presented using GC/MS systems (Mueller, Mène-Saffrané, Grun, Karg, & Farmer, 2006). In spite of many drawbacks, authors indicated that GC was the most powerful platform. The alternative choices of each standard operating procedure steps were widely developed and chemical derivatization (which has to be the simplest) was discussed.

Alkaloids, well-known secondary metabolites, are sought in crude extracts of 105 Daffodils varieties (Torras-Claveria, Berkov, Codina, Viladomat, & Bastida, 2013). These data were correlated regarding to therapeutic activity and cultivars were screened to discover better pharmaceutical source. GC/MS single quad, as rapid device, with robust and specific methodologies (no derivatization step), was used for leaf and bulbs extracts analyses. In this study, no profile mining was conducted. Galanthamine and Sanguinine were the two compounds highlighted and compared to total alkaloid amount. This was the first time alkaloid contents and acetylcholinesterase inhibitory activities were determined on such a large Narcissus collection.

Volatile compounds are easily analyzed by GC/MS, as headspace and cold trapping increase analytical capabilities. VOC produced by Arabidopsis hairy-roots were analyzed by PTR/MS and GC/MS (Steeghs et al., 2004) using instruments either developed or in-house customized. Generally, root VOC production is under-documented, but these authors reached to draw cartography of root VOCs, which shows that some are constitutive and others are the result of biotic interactions. Such results must be linked to genomic investigations in Arabidopsis for a more integrative view of plant physiology.

Food quality and traceability is also a current hot topic. At the limit of plant metabolomics, phylogenic characterization of Tricholoma fruit bodies was conducted for traceability (Ding & Hou, 2012). Genomic evidences were confirmed with GC/MS profiling on volatile oils. Sixty-six molecules (from C6 to C37) were identified in ether extracts of samples from three different regions. At the end, authors discussed about nutritive, toxic and fragrance compositions.

6.2. Metabolomics and gene modifications

Artificial genetic mutants, even if mutations do not imply phenotypic modifications, are also evaluated in terms of metabolic profiles. *A. thaliana* L., the worldwide plant model, after a genome entirely sequenced in the late 1990s, still leaves so many questions across gene functionalities and relationships among omics (Quero, 2012).

GC/MS experiments are known to be faster and easier than mapping genes (Messerli et al., 2007). They offer as well, through statistical analysis and automated assignments, complementary metabolic information of genetic one. Within Populus, the plant model among trees, transgenic lines were explored in terms of metabolic profiles (Robinson, Gheneim, Kozak, Ellis, & Mansfield, 2005). Cell cultures and cambium were subjected to extraction and derivatization and then analyzed through GC/ion trap. Amino acid, benzene and carbohydrate abundances were evaluated as a first step of a global metabolism study in trees. Albinsky et al. (2010) used GC/ToF on genetically modified plants and explored metabolomic consequences, related to nitrogen metabolism. Rice and Arabidopsis mutants were screened, and primary and secondary metabolism interactions were supported by transcript profiling, which is a good example of a system biology study.

Agro-industry drive also research efforts on fruit and seed physiology. AtHXK1 overexpressing tomato plants were (Roessner-Tunali et al., 2003) evaluated through GC/MS profiling protocol, in terms of fruit composition and comparison of source and sink tissues. Primary and secondary metabolites were determined by GC/Quad (except starch, soluble sugars, hexophosphates, nucleotides and nucleotide sugars). With this study, authors highlighted the temporal resolution of metabolism. Correlation between genome, proteome and metabolome was investigated by Vigeolas et al. (2008) using developing Pea seeds. Pea albumin 2, resistant to digestion, was found to be the key for metabolic profiling on WT and variant PA2 lines. Sugars, TCA intermediates, amino acids and polyamines were highlighted by GC/MS analyses.

The use of genetically modified plants may allow developing the full potentialities of plant–pathogen interaction topics (Wróbel-Kwiatkowska et al., 2004). Experiments were performed on flax transgenic lines for fungi resistance and crop quality purposes. GC/Quad profiling of derivatized samples revealed important carbohydrate modifications, linked with organic acids, polyamines, fatty acids and other osmoprotectants, which participate to pathogenic resistance.

6.3. Metabolomics and environment

In their own biotope, plants are continuously challenged with biotic and non-biotic interactions. The pathogenetic issues are peculiarly supported by agro-industrial actors. However, pollution and global warming are also some of the current concerns in the scientific community.

Seed germination and vernalization imply different metabolic switches. On Arabidopsis, primary metabolism (mainly) of seed was analyzed through GC/Quad profiling protocol (Fait et al., 2006). Behaviour in CO_2 limiting environment was tested on the well-known unicellular green algae, *Chlamydomonas reinhardtii*. Using GC/ToF technique, Renberg et al. (2010) showed transient behaviour on amino acids, organic acids and lipids during acclimation to new environmental conditions. Authors suggested a metabolic pattern of carbon concentration mechanism after a 3-h induction.

Eco(toxico)logy is also actually a great challenge. As an example, cadmium toxicity was studied on the same plant model (Jamers, Blust, De Coen, Griffin, & Jones, 2013). Using gene transcription and metabolic profiles, authors observed a transient effect between 48 and 72 h of exposure. Water–MeOH–$CHCl_3$ extracts were obtained on sonicated frozen pellets and aqueous phases were submitted to NMR and GC/MS analyses.

Derivatized samples showed amino acids and A × P big modifications. This study revealed time and sensitivity dependences of omics approaches as well as its complementarities.

Boron concentration in soil is important for Barley physiology (Roessner et al., 2006). Through metabolomics on root and leaf tissues from tolerant and sensitive cultivars, GC/Quad analyses showed differences among tissues greater than between cultivars. Sulphur deprivation was observed by Nikiforova et al. (2005) on Arabidopsis using GC/MS and LC/MS. 1072 peaks are managed by GC/Quad and 78 ones were annotated through ICA (amino acids, organic acids, urea, amines, sugars and polyols). This revealed interdependency of sulphur, nitrogen, lipid, purine and photorespiratory pathways. Adaptation to salinity is explored through GC/ToF analyses on polar fractions of three plant models, Arabidopsis, Lotus and Oryza. It revealed that balances between amino acids and organic acids were a key response to salt stress (Sanchez, Siahpoosh, Roessner, Udvardi, & Kopka, 2008). Glycine supplying to Lolium roots at the place of NO_3^- implied proteins and amino acids profiles modifications (Thornton, Osborne, Paterson, & Cash, 2007). Concentrations and ^{15}N enrichment ratio were determined by GC/MS.

Plant fluids are, as for animals, source of evidences concerning physiology and adaptations. Root exudates have important functions for plant–soil relationship (Nardi et al., 2002), especially for metal mobilization. Saps are transportation media inside plant and reveal physiological status. In *Cucurbita* spp. (Fiehn, 2003), metabolomic profiles by GC/Quad revealed, through differences between plants at the same developmental stage, that plant vascular exudates are complex and versatile.

Metabolism is also a reflection of plant defence mechanisms in primary and secondary metabolic pools (Bezemer & van Dam, 2005). Plant–pathogen relations are complex because of a two living partner's relationship. But many reasons (e.g. food safety) drive scientific efforts in this interesting research field. Cevallos-Cevallos, Futch, Shilts, Folimonova, and Reyes-De-Corcuera (2012) were faced with an important challenge concerning a major Citrus disease, induced by *Candidatus liberibacter* and Psyllid species for spreading. In order to discover metabolic evidence of disease resistance, they compared metabolic profiles of healthy and inoculated plants on sensitive and tolerant varieties. Fresh leaves were used for GC/MS analyses, after extraction and derivatization. With GC/IT metabolomic profiles, Hamzehzarghani et al. (2005) experimented wheat Fusarium interactions among resistant and sensitive cultivars.

7. CONCLUSION

Towards a multi-platform approach, GC/MS has become a powerful metabolomic tool, due to major technical improvements, including multi-dimensional GC-coupled fast high-resolution mass analysers. It is considered as a mature technology fully part of the comprehensive metabolomic platform that uses complementary tools including LC/MS and NMR.

Recently, more sophisticated approaches using the multi-analytical platform were developed. It was first combined with LC/MS or NMR to ensure a broader metabolome coverage of metabolite diversity in plants (Fan et al., 2001; Scherling, Roscher, Giavalisco, Schulze, & Weckwerth, 2010). Sandasi, Kamatou, and Viljoen (2012) presented the use of an untargeted approach to differentiate two Salvia species by using GC/MS combined with mid-infrared spectroscopy on essential oils.

During the identification step, these multi-platform approaches are of great interest to accumulate structural proofs of metabolite identities. Jenett-Siems et al. (2005) studied tropane alkaloid family using a combination of GC/MS and NMR experiments and therefore were able to contribute to taxonomic efforts. De Andrade et al. (2012) combined GC/MS, GC/MS/MS and LC/MS to perform mass spectral studies of bioactive alkaloids from Narcissus.

Even if today a variety of algorithms are available for GC/MS data processing, the full procedure of obtaining knowledge from raw data still awaits complete automation and biomarker identification remains an analytical issue.

At last, there is a real need to develop quantitative analytical procedures compatible with high-throughput production of data to achieve the goal of metabolomics of gaining an understanding of biological systems.

After the study of simple model systems, researchers are currently dealing with more complex ones. In that way, the temporal and spatial characterization of the responses to perturbations, at an individual, population or ecosystem scale, is today a significant challenge to study functional metabolomics and understand the ecological basis of interactions between metabolites. Metametabolomics is consequently a logical perspective in community evaluation (ecology, phylogeny). Highlighted modern concerns are also sources for scientific topics. Drug and food mixes (e.g. vine varieties, cheeses, honeys) are investigated for traceability and safety (e.g. Datura, a source for news pharmaceuticals through tropane alkaloids production or a false positive doping problem for race horses, as weed in fodder?).

REFERENCES

Adahchour, M., Brandt, M., Baier, H.-U., Vreuls, R. J. J., Batenburg, A. M., & Brinkman, U. A. T. (2005). Comprehensive two-dimensional gas chromatography coupled to a rapid-scanning quadrupole mass spectrometer: Principles and applications. *Journal of Chromatography. A, 1067*, 245–254.

Aggio, R., Villas-Bôas, S. G., & Ruggiero, K. (2011). Metab: An R package for high-throughput analysis of metabolomics data generated by GC-MS. *Bioinformatics (Oxford, England), 27*, 2316–2318.

Albinsky, D., Kusano, M., Higuchi, M., Hayashi, N., Kobayashi, M., Fukushima, A., et al. (2010). Metabolomic screening applied to rice FOX arabidopsis lines leads to the identification of a gene-changing nitrogen metabolism. *Molecular Plant, 3*, 125–142.

Allwood, J. W., Erban, A., de Koning, S., Dunn, W. B., Luedemann, A., Lommen, A., et al. (2009). Inter-laboratory reproducibility of fast gas chromatography–electron impact–time of flight mass spectrometry (GC–EI–TOF/MS) based plant metabolomics. *Metabolomics, 5*, 479–496.

Almstetter, M. F., Oefner, P. J., & Dettmer, K. (2012). Comprehensive two-dimensional gas chromatography in metabolomics. *Analytical and Bioanalytical Chemistry, 402*, 1993–2013.

Bezemer, T. M., & van Dam, N. M. (2005). Linking aboveground and belowground interactions via induced plant defenses. *Trends in Ecology & Evolution, 20*, 617–624.

Birkemeyer, C., Luedemann, A., Wagner, C., Erban, A., & Kopka, J. (2005). Metabolome analysis: The potential of in vivo labeling with stable isotopes for metabolite profiling. *Trends in Biotechnology, 23*, 28–33.

Bruni, R., Medici, A., Guerrini, A., Scalia, S., Poli, F., Romagnoli, C., et al. (2002). Tocopherol, fatty acids and sterol distributions in wild Ecuadorian Theobroma subincanum (Sterculiaceae) seeds. *Food Chemistry, 77*, 337–341.

Büscher, J. M., Czernik, D., Ewald, J. C., Sauer, U., & Zamboni, N. (2009). Cross-platform comparison of methods for quantitative metabolomics of primary metabolism. *Analytical Chemistry, 81*, 2135–2143.

Cespedes, C. L., Molina, S. C., Muñoz, E., Lamilla, C., Alarcon, J., Palacios, S. M., et al. (2013). The insecticidal, molting disruption and insect growth inhibitory activity of extracts from Condalia microphylla Cav. (Rhamnaceae). *Industrial Crops and Products, 42*, 78–86.

Cevallos-Cevallos, J. M., Futch, D. B., Shilts, T., Folimonova, S. Y., & Reyes-De-Corcuera, J. I. (2012). GC–MS metabolomic differentiation of selected citrus varieties with different sensitivity to citrus huanglongbing. *Plant Physiology and Biochemistry, 53*, 69–76.

Daskalchuk, T., Ahiahonu, P., Heath, D., & Yamazaki, Y. (2006). The use of non-targeted metabolomics in plant science. In K. Saito, R. A. Dixon, & L. Willmitzer (Eds.), *Plant metabolomics. Biotechnology in agriculture and forestry*, Vol. 57, (pp. 311–325). Heidelberg: Springer.

De Andrade, J. P., Pigni, N. B., Torras-Claveria, L., Berkov, S., Codina, C., Viladomat, F., et al. (2012). Bioactive alkaloid extracts from Narcissus broussonetii: Mass spectral studies. *Journal of Pharmaceutical and Biomedical Analysis, 70*, 13–25.

Dettmer, K., Aronov, P. A., & Hammock, B. D. (2007). Mass spectrometry-based metabolomics. *Mass Spectrometry Reviews, 26*, 51–78.

Ding, X., & Hou, Y. L. (2012). Identification of genetic characterization and volatile compounds of Tricholoma matsutake from different geographical origins. *Biochemical Systematics and Ecology, 44*, 233–239.

Fait, A., Angelovici, R., Less, H., Ohad, I., Urbanczyk-Wochniak, E., Fernie, A. R., et al. (2006). Arabidopsis seed development and germination is associated with temporally distinct metabolic switches. *Plant Physiology, 142*, 839–854.

Fan, T. W.-M., Lane, A. N., Shenker, M., Bartley, J. P., Crowley, D., & Higashi, R. M. (2001). Comprehensive chemical profiling of gramineous plant root exudates using high-resolution NMR and MS. *Phytochemistry, 57*, 209–221.

Fiehn, O. (2002). Metabolomics—The link between genotypes and phenotypes. *Plant Molecular Biology, 48*, 155–171.

Fiehn, O. (2003). Metabolic networks of Cucurbita maxima phloem. *Phytochemistry, 62*, 875–886.

Fiehn, O. (2008). Extending the breadth of metabolite profiling by gas chromatography coupled to mass spectrometry. *Trends in Analytical Chemistry, 27*, 261–269.

Fiehn, O., Kopka, J., Dormann, P., Altmann, T., Trethewey, R. N., & Willmitzer, L. (2000). Metabolite profiling for plant functional genomics. *Nature Biotechnology, 18*, 1157–1161.

Flamini, G., Cioni, P. L., & Morelli, I. (2005). Composition of the essential oils and in vivo emission of volatiles of four Lamium species from Italy: L. purpureum, L. hybridum, L. bifidum and L. amplexicaule. *Food Chemistry, 91*, 63–68.

Gao, X., Pujos-Guillot, E., & Sébédio, J.-L. (2010). Development of a quantitative metabolomic approach to study clinical human fecal water metabolome based on trimethylsilylation derivatization and GC/MS analysis. *Analytical Chemistry, 82*, 6447–6456.

Gullberg, J., Jonsson, P., Nordström, A., Sjöström, M., & Moritz, T. (2004). Design of experiments: An efficient strategy to identify factors influencing extraction and derivatization of Arabidopsis thaliana samples in metabolomic studies with gas chromatography/mass spectrometry. *Analytical Biochemistry, 331*, 283–295.

Guy, C., Kopka, J., & Moritz, T. (2008). Plant metabolomics coming of age. *Physiologia Plantarum, 132*, 113–116.

Halket, J. M., Przyborowska, A., Stein, S. E., Mallard, W. G., Down, S., & Chalmers, R. A. (1999). Deconvolution gas chromatography/mass spectrometry of urinary organic acids—Potential for pattern recognition and automated identification of metabolic disorders. *Rapid Communications in Mass Spectrometry, 13*, 279–284.

Hamzehzarghani, H., Kushalappa, A. C., Dion, Y., Rioux, S., Comeau, A., Yaylayan, V., et al. (2005). Metabolic profiling and factor analysis to discriminate quantitative resistance in wheat cultivars against fusarium head blight. *Physiological and Molecular Plant Pathology, 66*, 119–133.

Harshman, R. A. (1970). Foundations of the PARAFAC procedure: Models and conditions for an "explanatory" multimodal factor analysis. *UCLA Working Papers in Phonetics, 16*, 1–84.

Heinrich, M. (2008). Ethnopharmacy and natural product research—Multidisciplinary opportunities for research in the metabolomic age. *Phytochemistry Letters, 1*, 1–5.

Hummel, J., Selbig, J., Walther, D., & Kopka, J. (2007). The Golm Metabolome Database: A database for GC-MS based metabolite profiling. In J. Nielsen & M. C. Jewett (Eds.), *Metabolomics. Topics in current genetics*, Vol. 18, (pp. 75–95). Heidelberg: Springer.

Jamers, A., Blust, R., De Coen, W., Griffin, J. L., & Jones, O. A. H. (2013). An omics based assessment of cadmium toxicity in the green alga Chlamydomonas reinhardtii. *Aquatic Toxicology, 126*, 355–364.

Jenett-Siems, K., Weigl, R., Böhm, A., Mann, P., Tofern-Reblin, B., Ott, S. C., et al. (2005). Chemotaxonomy of the pantropical genus Merremia (Convolvulaceae) based on the distribution of tropane alkaloids. *Phytochemistry, 66*, 1448–1464.

Jeong, J., Shi, X., Zhang, X., Kim, S., & Shen, C. (2012). Model-based peak alignment of metabolomic profiling from comprehensive two-dimensional gas chromatography mass spectrometry. *BMC Bioinformatics, 13*, 27.

Kanani, H., Chrysanthopoulos, P. K., & Klapa, M. I. (2008). Standardizing GC-MS metabolomics. *Journal of Chromatography. B, Analytical Technologies in the Biomedical and Life Sciences, 871*, 191–201.

Katajamaa, M., Miettinen, J., & Orešič, M. (2006). MZmine: Toolbox for processing and visualization of mass spectrometry based molecular profile data. *Bioinformatics, 22*, 634–636.

Koek, M. M., Jellema, R. H., van der Greef, J., Tas, A. C., & Hankemeier, T. (2011). Quantitative metabolomics based on gas chromatography mass spectrometry: Status and perspectives. *Metabolomics*, 7, 307–328.

Koek, M. M., Muilwijk, B., van Stee, L. L., & Hankemeier, T. (2008). Higher mass loadability in comprehensive two-dimensional gas chromatography-mass spectrometry for improved analytical performance in metabolomics analysis. *Journal of Chromatography. A, 1186*, 420–429.

Kopka, J. (2006). Current challenges and developments in GC–MS based metabolite profiling technology. *Journal of Biotechnology, 124*, 312–322.

Kopka, J., Schauer, N., Krueger, S., Birkemeyer, C., Usadel, B., Bergmuller, E., et al. (2005). GMD@CSB.DB: The Golm Metabolome Database. *Bioinformatics, 21*, 1635–1638.

Kumari, S., Stevens, D., Kind, T., Denkert, C., & Fiehn, O. (2011). Applying in-silico retention index and mass spectra matching for identification of unknown metabolites in accurate mass GC-TOF mass spectrometry. *Analytical Chemistry, 83*, 5895–5902.

Kusano, M., Fukushima, A., Kobayashi, M., Hayashi, N., Jonsson, P., Moritz, T., et al. (2007). Application of a metabolomic method combining one-dimensional and two-dimensional gas chromatography-time-of-flight/mass spectrometry to metabolic phenotyping of natural variants in rice. *Journal of Chromatography. B, Analytical Technologies in the Biomedical and Life Sciences, 855*, 71–79.

Lenz, E. M., & Wilson, I. D. (2006). Analytical strategies in metabonomics. *Journal of Proteome Research, 6*, 443–458.

Little, J. L. (1999). Artifacts in trimethylsilyl derivatization reactions and ways to avoid them. *Journal of Chromatography. A, 844*, 1–22.

Lommen, A. (2009). MetAlign: Interface-driven, versatile metabolomics tool for hyphenated full-scan mass spectrometry data preprocessing. *Analytical Chemistry, 81*, 3079–3086.

Messerli, G., Nia, V. P., Trevisan, M., Kolbe, A., Schauer, N., Geigenberger, P., et al. (2007). Rapid classification of phenotypic mutants of arabidopsis via metabolite fingerprinting. *Plant Physiology, 143*, 1484–1492.

Mueller, M., Mène-Saffrané, L., Grun, C., Karg, K., & Farmer, E. E. (2006). Oxylipin analysis methods. *The Plant Journal, 45*, 472–489.

Nardi, S., Sessi, E., Pizzeghello, D., Sturaro, A., Rella, R., & Parvoli, G. (2002). Biological activity of soil organic matter mobilized by root exudates. *Chemosphere, 46*, 1075–1081.

Nikiforova, V. J., Kopka, J., Tolstikov, V., Fiehn, O., Hopkins, L., Hawkesford, M. J., et al. (2005). Systems rebalancing of metabolism in response to sulfur deprivation, as revealed by metabolome analysis of arabidopsis plants. *Plant Physiology, 138*, 304–318.

Noctor, G., Bergot, G., Mauve, C., Thominet, D., Lelarge-Trouverie, C., & Prioul, J.-L. (2007). A comparative study of amino acid measurement in leaf extracts by gas chromatography-time of flight-mass spectrometry and high performance liquid chromatography with fluorescence detection. *Metabolomics, 3*, 161–174.

O'Hagan, S., Dunn, W. B., Knowles, J. D., Broadhurst, D., Williams, R., Ashworth, J. J., et al. (2007). Closed-loop, multiobjective optimization of two-dimensional gas chromatography/mass spectrometry for serum metabolomics. *Analytical Chemistry, 79*, 464–476.

Paolini, J., Costa, J., & Bernardini, A.-F. (2005). Analysis of the essential oil from aerial parts of Eupatorium cannabinum subsp. corsicum (L.) by gas chromatography with electron impact and chemical ionization mass spectrometry. *Journal of Chromatography. A, 1076*, 170–178.

Quero, A. (2012). Approche métabolomique de l'amélioration de la tolérance au stress hydrique chez le lin (Linum usitatissimum) par application de l'acide beta-amino butyrique (BABA) ou d'oligosaccharides. PhD thesis. UPJV, Amiens.

Renberg, L., Johansson, A. I., Shutova, T., Stenlund, H., Aksmann, A., Raven, J. A., et al. (2010). A metabolomic approach to study major metabolite changes during acclimation to limiting CO2 in Chlamydomonas reinhardtii. *Plant Physiology, 154*, 187–196.

Riedl, J., Kluender, C., Sans-Piché, F., Heilmeier, H., Altenburger, R., & Schmitt-Jansen, M. (2012). Spatial and temporal variation in metabolic fingerprints of field-growing Myriophyllum spicatum. *Aquatic Botany, 102*, 34–43.

Ritota, M., Casciani, L., Han, B. Z., Cozzolino, S., Leita, L., Sequi, P., et al. (2012). Traceability of Italian garlic (Allium sativum L.) by means of HRMAS-NMR spectroscopy and multivariate data analysis. *Food Chemistry, 135*, 684–693.

Robinson, A. R., Gheneim, R., Kozak, R. A., Ellis, D. D., & Mansfield, S. D. (2005). The potential of metabolite profiling as a selection tool for genotype discrimination in populus. *Journal of Experimental Botany, 56*, 2807–2819.

Roessner, U., Patterson, J. H., Forbes, M. G., Fincher, G. B., Langridge, P., & Bacic, A. (2006). An investigation of boron toxicity in barley using metabolomics. *Plant Physiology, 142*, 1087–1101.

Roessner, U., Wagner, C., Kopka, J., Trethewey, R. N., & Willmitzer, L. (2000). Simultaneous analysis of metabolites in potato tuber by gas chromatography–mass spectrometry. *The Plant Journal, 23*, 131–142.

Roessner-Tunali, U., Hegemann, B., Lytovchenko, A., Carrari, F., Bruedigam, C., Granot, D., et al. (2003). Metabolic profiling of transgenic tomato plants overexpressing hexokinase reveals that the influence of hexose phosphorylation diminishes during fruit development. *Plant Physiology, 133*, 84–99.

Rohloff, J., & Bones, A. M. (2005). Volatile profiling of Arabidopsis thaliana—Putative olfactory compounds in plant communication. *Phytochemistry, 66*, 1941–1955.

Sanchez, D. H., Siahpoosh, M. R., Roessner, U., Udvardi, M., & Kopka, J. (2008). Plant metabolomics reveals conserved and divergent metabolic responses to salinity. *Physiologia Plantarum, 132*, 209–219.

Sandasi, M., Kamatou, G. P. P., & Viljoen, A. M. (2012). An untargeted metabolomic approach in the chemotaxonomic assessment of two Salvia species as a potential source of α-bisabolol. *Phytochemistry, 84*, 94–101.

Schauer, N., Steinhauser, D., Strelkov, S., Schomburg, D., Allison, G., Moritz, T., et al. (2005). GC-MS libraries for the rapid identification of metabolites in complex biological samples. *FEBS Letters, 579*, 1332–1337.

Scherling, C., Roscher, C., Giavalisco, P., Schulze, E. D., & Weckwerth, W. (2010). Metabolomics unravel contrasting effects of biodiversity on the performance of individual plant species. *PLoS One, 5*, e12569.

Schmelz, E. A., Engelberth, J., Tumlinson, J. H., Block, A., & Alborn, H. T. (2004). The use of vapor phase extraction in metabolic profiling of phytohormones and other metabolites. *The Plant Journal, 39*, 790–808.

Shulaev, V., Cortes, D., Miller, G., & Mittler, R. (2008). Metabolomics for plant stress response. *Physiologia Plantarum, 132*, 199–208.

Shyur, L.-F., & Yang, N.-S. (2008). Metabolomics for phytomedicine research and drug development. *Current Opinion in Chemical Biology, 12*, 66–71.

Smith, C. A., Want, E. J., O'Maille, G., Abagyan, R., & Siuzdak, G. (2006). XCMS: Processing mass spectrometry data for metabolite profiling using nonlinear peak alignment, matching, and identification. *Analytical Chemistry, 78*, 779–787.

Steeghs, M., Bais, H. P., de Gouw, J., Goldan, P., Kuster, W., Northway, M., et al. (2004). Proton-transfer-reaction mass spectrometry as a new tool for real time analysis of root-secreted volatile organic compounds in arabidopsis. *Plant Physiology, 135*, 47–58.

Stein, S. E. (1999). An integrated method for spectrum extraction and compound identification from gas chromatography/mass spectrometry data. *Journal of the American Society for Mass Spectrometry, 10*, 770–781.

Strehmel, N., Hummel, J., Erban, A., Strassburg, K., & Kopka, J. (2008). Retention index thresholds for compound matching in GC-MS metabolite profiling. *Journal of Chromatography. B, Analytical Technologies in the Biomedical and Life Sciences, 871*, 182–190.

Styczynski, M. P., Moxley, J. F., Tong, L. V., Walther, J. L., Jensen, K. L., & Stephanopoulos, G. N. (2007). Systematic identification of conserved metabolites in GC/MS data for metabolomics and biomarker discovery. *Analytical Chemistry, 79*, 966–973.

Thornton, B., Osborne, S. M., Paterson, E., & Cash, P. (2007). A proteomic and targeted metabolomic approach to investigate change in Lolium perenne roots when challenged with glycine. *Journal of Experimental Botany, 58*, 1581–1590.

Torras-Claveria, L., Berkov, S., Codina, C., Viladomat, F., & Bastida, J. (2013). Daffodils as potential crops of galanthamine. Assessment of more than 100 ornamental varieties for their alkaloid content and acetylcholinesterase inhibitory activity. *Industrial Crops and Products, 43*, 237–244.

Vielhauer, O., Zakhartsev, M., Horn, T., Takors, R., & Reuss, M. (2011). Simplified absolute metabolite quantification by gas chromatography–isotope dilution mass spectrometry on the basis of commercially available source material. *Journal of Chromatography B, 879*, 3859–3870.

Vigeolas, H., Chinoy, C., Zuther, E., Blessington, B., Geigenberger, P., & Domoney, C. (2008). Combined metabolomic and genetic approaches reveal a link between the polyamine pathway and albumin 2 in developing pea seeds. *Plant Physiology, 146*, 74–82.

Villas-Bôas, S. G., Mas, S., Åkesson, M., Smedsgaard, J., & Nielsen, J. (2005). Mass spectrometry in metabolome analysis. *Mass Spectrometry Reviews, 24*, 613–646.

Vuckovic, D., Risticevic, S., & Pawliszyn, J. (2011). In vivo solid-phase microextraction in metabolomics: Opportunities for the direct investigation of biological systems. *Angewandte Chemie International Edition, 50*, 5618–5628.

Wachsmuth, C. J., Almstetter, M. F., Waldhier, M. C., Gruber, M. A., Nürnberger, N., Oefner, P. J., et al. (2011). Performance evaluation of gas chromatography-atmospheric pressure chemical ionization-time-of-flight mass spectrometry for metabolic fingerprinting and profiling. *Analytical Chemistry, 83*, 7514–7522.

Wells, R. J. (1999). Recent advances in non-silylation derivatization techniques for gas chromatography. *Journal of Chromatography. A, 843*, 1–18.

Wiklund, S., Johansson, E., Sjostrom, L., Mellerowicz, E. J., Edlund, U., Shockcor, J. P., et al. (2007). Visualization of GC/TOF-MS-based metabolomics data for identification of biochemically interesting compounds using OPLS class models. *Analytical Chemistry, 80*, 115–122.

Wilson, R. B., Hoggard, J. C., & Synovec, R. E. (2012). Fast, high peak capacity separations in gas chromatography–time-of-flight mass spectrometry. *Analytical Chemistry, 84*, 4167–4173.

Wróbel-Kwiatkowska, M., Lorenc-Kukula, K., Starzycki, M., Oszmiański, J., Kepczyńska, E., & Szopa, J. (2004). Expression of β-1,3-glucanase in flax causes increased resistance to fungi. *Physiological and Molecular Plant Pathology*, *65*, 245–256.

Wu, L., Mashego, M. R., van Dam, J. C., Proell, A. M., Vinke, J. L., Ras, C., et al. (2005). Quantitative analysis of the microbial metabolome by isotope dilution mass spectrometry using uniformly 13C-labeled cell extracts as internal standards. *Analytical Biochemistry*, *336*, 164–171.

CHAPTER SEVEN

The Lipid World Concept of Plant Lipidomics

Laetitia Fouillen*, Benoit Colsch†, René Lessire*,[1]
*University of Bordeaux, CNRS UMR 5200, and Metabolome Facility of Bordeaux, Functional Genomics Centre, INRA Centre of Bordeaux, Villenave d'Ornon, France
†Laboratoire d'Etude du Métabolisme des Médicaments, CEA- Centre d'étude de Saclay DSV/iBiTec-S/SPI/LEMM, Gif-sur-Yvette, France
[1]Corresponding author: e-mail address: rene.lessire@biomemb.u-bordeaux2.fr

Contents

Abstract

Lipidomics has emerged as a new field that allows using various approaches to the chemical structures and the quantitative composition of more than a hundred lipid molecular species constituting the cellular lipidome. The increase of the performance of lipidomic analysis has resulted in recent developments in electrospray ionization

Advances in Botanical Research, Volume 67
ISSN 0065-2296
http://dx.doi.org/10.1016/B978-0-12-397922-3.00007-1

mass spectrometry (ESI/MS) and rapid scanning tandem spectrometers that are capable of detecting and quantifying lipids at high sensitivity in an online high-performance chromatography. In this review, after a short description of the characteristic lipid classes of the plant kingdom, different approaches of 'lipidomics' will be addressed, including sample preparation (extractions, sample storage), MS analysis (ionization sources, shotgun lipidomics by direct infusion using tandem-in-space instruments and high-resolution systems and the use of separative methods before MS studies). Common fragmentation modes (MRM, CID including HCD) to determine molecular structures of lipid families in plants are also developed. The different principles of MS lipid analyses are briefly described and the different strategies using HPLC and MS/MS to quantify the different plant lipid molecular species are presented.

ABBREVIATIONS

Cer ceramide
CID collision-induced dissociation
DAGs diacylglycerols
ESI electrospray ionization
PC glycerophosphatidylcholine(s)
PE glycerophosphatidylethanolamine(s)
PG glycerophosphatidylglycerol(s)
PA glycerophosphatidic acid(s)
PI glycerophosphatidylinositol(s)
PS glycerophosphatidylserine(s)
LCB long-chain base
MALDI matrix-assisted laser desorption/ionization
MS mass spectrometry
Sph sphingomyelin(s)
TAG triacylglycerol(s)
TLE total lipid extract

1. INTRODUCTION

Lipids are essential cellular constituents of all plant cells that have many critical roles in cellular function. First of all, they are the components (phospholipids, glycolipids, sphingolipids, sterols) of cell membrane bilayers. They are involved in energy storage (triacylglycerols, TAGs), they participate to cell signalling as second messengers (lysolipids, diacylglycerols (DAGs), ceramides, phosphoinositides, phosphatidic acids) and they have a function of protective molecules (surface lipids). At the opposite of genomics and proteomics, which have known impressive developments, lipidomics has emerged later. Indeed, the first mass spectrometry (MS) analyses of a complex lipid mixture were reported in the 1990s (Han & Gross, 1994; Kim, Wang, & Ma, 1994) and actually with the advances in electrospray ionization mass spectrometry (ESI/MS), lipidomics is a rapidly expanding research field. Two of the reasons of this delay were the lack of sensitive analytical techniques and the high diversity of lipids. It was estimated that 9000–100,000 different molecular lipid species exist (Shevchenko & Simons, 2010). Consequently, a major difference with proteomics is that it is elusive to obtain a complete picture of the lipidome of a cell or tissue extracts in a single MS analysis due to the high molecular diversity of lipids. Nevertheless, MS analyses have revealed the complexity of the lipid composition of a cell compartment, which was modified in relation to the cellular environment. This has opened a new research field, which is to understand the relations between the cell biology and the lipidome. In addition, the term of 'lipidomics' has been used differently referencing to a global lipid composition (quantitative and qualitative) of a cell (tissue, organism) and its variations related to the state or environmental changes. The term has been used also to define quantitative analysis of the major lipids present in a sample. A general consensus was defined as 'the study of the composition, metabolism and biological role of lipids in cells at the level of molecular species' (German, Gillies, Smilowitz, Zivkovic, & Watkins, 2007; Wilson, 2003).

2. THE DIVERSITY OF PLANT LIPIDS

Cellular lipids were classified in three large groups including non-polar lipids, polar lipids and metabolites based on their relative head group polarities. This classification of lipids was ambiguous until the recent LIPIDMAPS project (http://www.lipidmaps.org/; Fahy et al., 2005, 2009) which has proposed eight lipid classes: fatty acids, glycerolipids, glycerophospholipids, sphingolipids, sterol lipids, prenol lipids, saccharolipids and polyketides.

Plant cells contain multiple subcellular organelles (e.g. plasma membrane, tonoplast, plastid, Golgi, endoplasmic reticulums, mitochondria) delineated by membranes, which represents 5–10% lipid/dry weight. These lipid membranes constitute barriers of cell compartments where essential functions take place as light harvesting, electron transport and photosynthesis. Lipids represent also the major form of energy storage in the seeds of many species, and epidermal cells synthesize cuticular lipids that constitute an important protective barrier against water loss, pathogens and environmental stress. Besides these abundant cellular lipids, minor lipids as phytohormones, oxylipids and eicosanoids are essential actors in cellular signalling.

2.1. Fatty acids

The first block constituting the more abundant lipids as glycerophospholipids, sphingolipids, sterolipids are the fatty acids. Generally, these membrane lipids contain fatty acids that are unbranched, mainly even carbon numbered (C14–C26) and possess *cis*-unsaturated double bond (Fig. 7.1). In plants, major

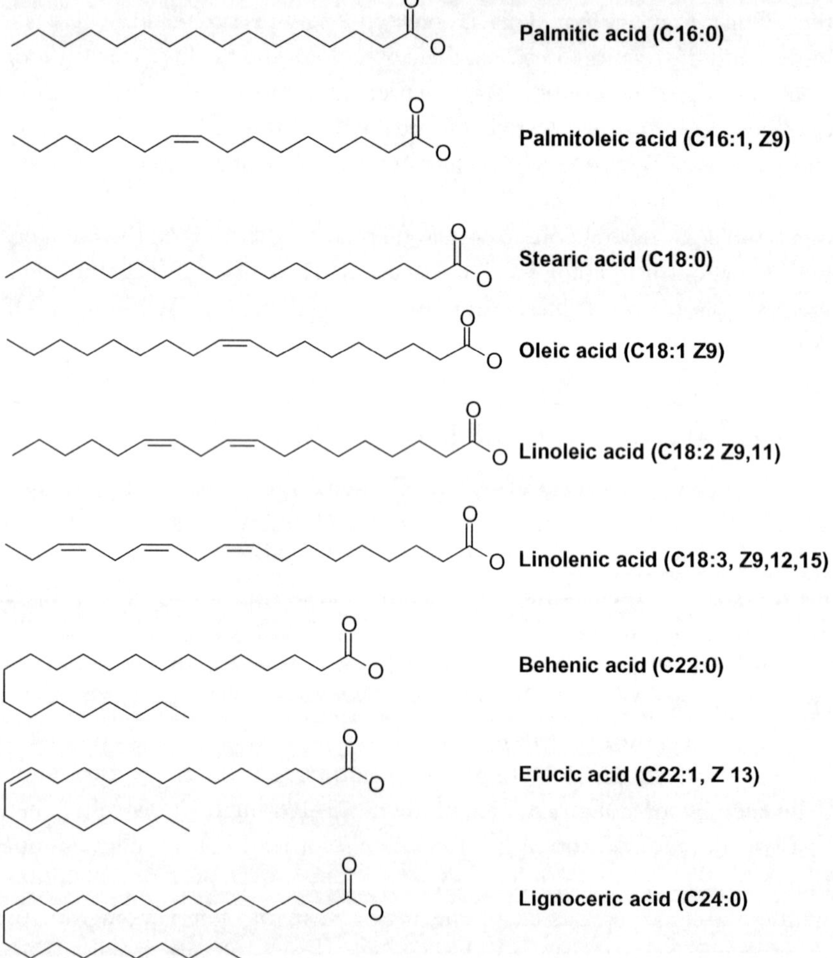

Figure 7.1 Most common fatty acids in plants.

fatty acids are: palmitic (C16:0), palmitoleic (C16:1, Z9), stearic (C18:0), oleic (C18:1, Z9), linoleic (C18:2, Z9,11) and linolenic (C18:3, Z9,11,13). They represent more than 80% of the total fatty acid amount of the cell. The localization of the unsaturation bond is conventionally indicated with the respect to the carboxyl function (Δ classification) or to the terminal methyl group (ω classification). In addition to these ubiquitous fatty acids, in plants the existence of unusual fatty acids is important particularly in seeds and in the surface lipids which cover the aerial parts. It has been estimated that more than 200 different fatty acids have been identified in the plant kingdom as: short chain (C8–C12), very long chain (C20–C32), branched, hydroxylated, conjugated polyunsaturated, acetylenic, cyclic, etc., fatty acids.

2.2. Membrane lipids

Polar lipids are amphiphilic and they have property of lipid to self-assembly in water to form a lipid bilayer. They are the major components of cell membranes. The most abundant class are the glycerophospholipids and among them the galactoglycerolipids which are the major constituent of photosynthetic thylakoid membranes. They consist of a diglyceride to which one (monogalactosyldiacylglycerols: MGDG) or two galactose sugars (digalactosyldiacylglycerols: DGDG) are attached (Fig. 7.2). Besides these two species, sulphoquinovosyldiacylglycerol is also found abundantly in higher plants' photosynthetic membranes. The phospholipids are the most common membrane lipids that are constituted of a glycerol backbone, which is most of the time acylated with a fatty acid having 16–18 carbon chain length (Fig. 7.2). The polar head consists of a phosphate group esterified at the sn-3 position of the glycerol and linked to a base (choline: Phosphatidylcholine PC, ethanolamine: Phosphatidylethanolamine PE), polyol (glycerol: phosphatidylglycerol PG, inositol: Phosphatidylinositol PI) or amino acid (serine: Phosphatidylserine PS).

The sphingolipids constitute a large class of polar lipids present in the membrane, at least 168 different exist in *Arabidopsis thaliana* (Markham, Li, Cahoon, & Jaworski, 2006; Pata, Hannun, & Ng, 2010; Sperling & Heinz, 2003). Plant sphingolipids can be divided into four classes: free long-chain bases (LCBs), ceramides (Cer), glycosylceramides and glycosyl inositol phosphoceramides (GIPCs). LCBs are mainly composed of 18 carbon atom chain with the presence of hydroxyl groups at C1 and C3 and an amine group at C2 (Fig. 7.3). According to the number and configuration of the double bonds of the acyl chain, nine different LCBs have been identified such as

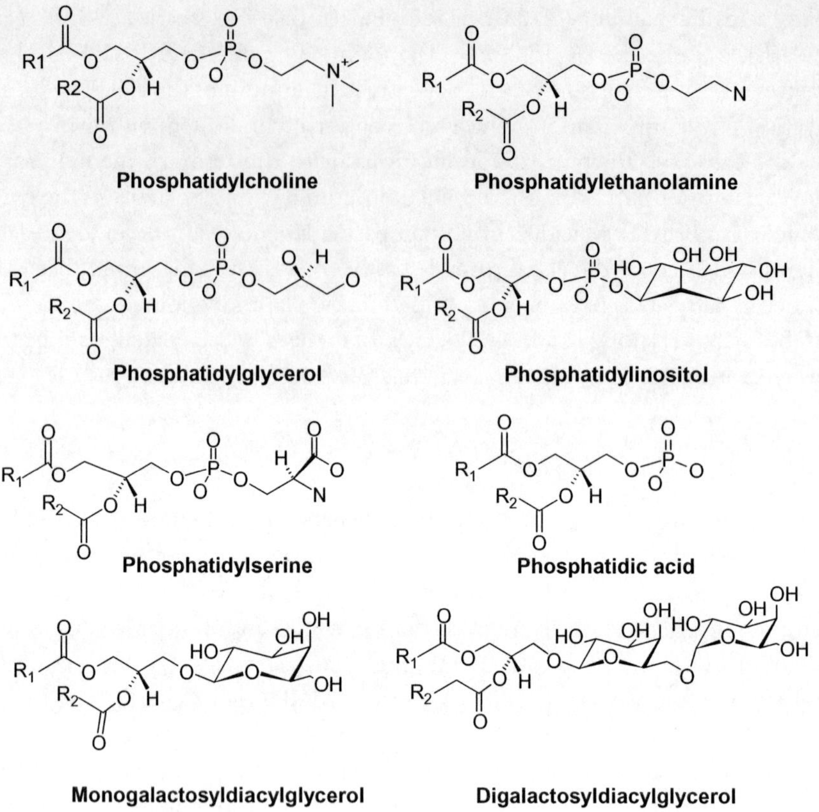

Phosphatidylcholine

Phosphatidylethanolamine

Phosphatidylglycerol

Phosphatidylinositol

Phosphatidylserine

Phosphatidic acid

Monogalactosyldiacylglycerol

Digalactosyldiacylglycerol

Figure 7.2 Glycerolipids.

sphinganine, sphingosine, phytosphingosine, sphingadienine and hydroxysphingenine (Pata et al., 2010). Ceramides are the *N*-acylated derivatives of an LCB by a fatty acid of 14–26 carbon atoms (Sperling & Heinz, 2003; Sperling, Franke, Lüthje, & Heinz, 2005; Fig. 7.3). A large variety of ceramides have been evidenced due to different modifications of the primary hydroxyl group of the LCB moiety by a phosphoryl group (ceramides-phosphates), either by one or several hexoses (glycosylceramides including glucosylceramides, galactosylceramides) or by an inositol-phosphate group inositol phospho-ceramide (IPC) or by both inositol phosphate and several hexoses (GIPCs; Sperling & Heinz, 2003; Sperling et al., 2005). The diversity of GIPCs in plants is very high and differs from plant to plant. It has been recently reported that 50 GIPC species have been characterized in *A. thaliana* and 120 in tobacco BY-2 cells (Buré et al., 2011).

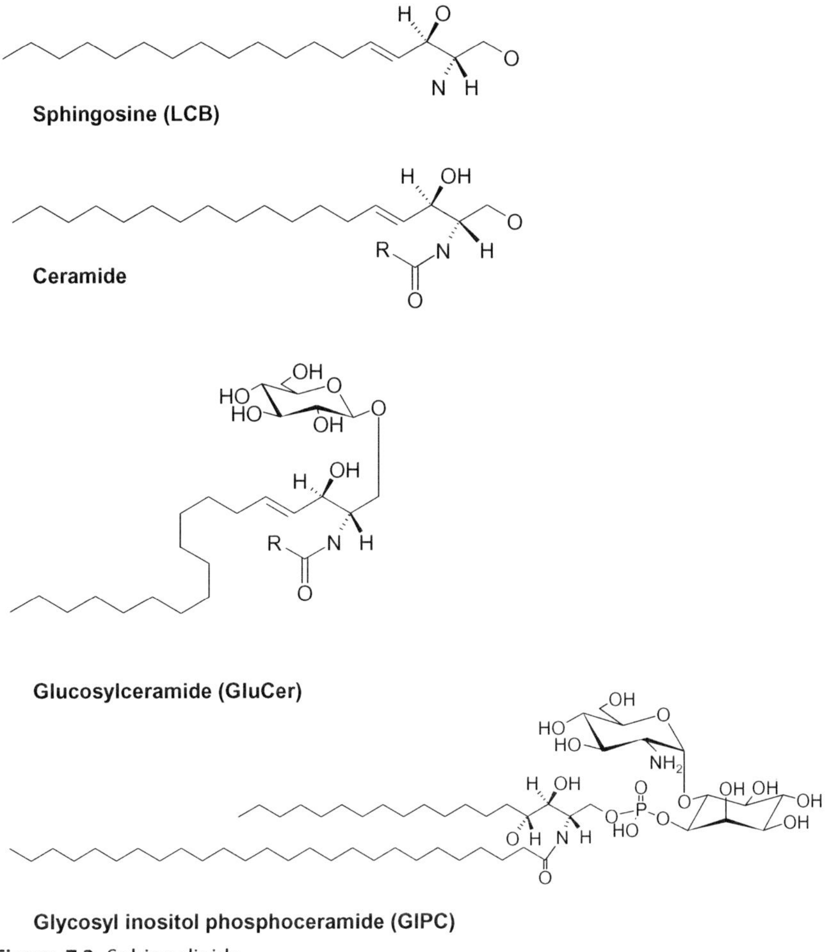

Sphingosine (LCB)

Ceramide

Glucosylceramide (GluCer)

Glycosyl inositol phosphoceramide (GIPC)

Figure 7.3 Sphingolipids.

Besides, these polar lipids more than 250 different types of phytosterols have been reported as membrane components in plant species. Representatives of these sterols are campesterol, β-sitosterol and stigmasterol (Fig. 7.4). Several sterol esters and glycosylated sterols can be found also in cell membranes, such as ergosteryl, stigmasteryl and β-sitosterol esters. Glycosylated steryl derivatives consist of one carbohydrate unit linked to the hydroxyl group of one sterol molecule: campesterol, stigmasterol, sitosterol, brassicasterol or dihydrositosterol and the sugar moiety is composed of glucose, xylose and even arabinose (Grille, Zaslawski, Thiele, Plat, & Warnecke, 2010). Acylated steryl glycosides are also a class of membrane

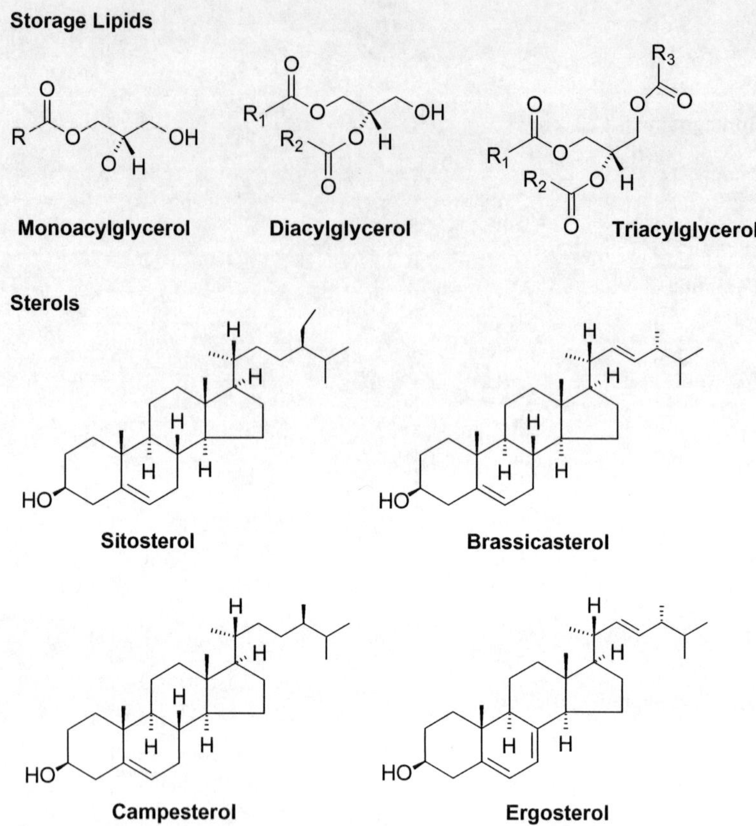

Figure 7.4 Neutral lipids and sterols in plants.

constituents. They are formed by acylation of a fatty acid to the primary alcohol group from the steryl glycoside molecule. The 6′-palmitoyl-β-D-glucoside of β-sitosterol is the major species (51%) detected in potato tubers while 6′-linoleoyl-β-D-glucoside of β-sitosterol is predominant (47%) in soybean extracts. In these products, other fatty acids were also detected (16:1, 18:1, 18:3).

2.3. Storage lipids

Most of plants synthesize and store neutral lipids as TAGs, which are the major compounds for long-term storage of energy. On the cellular level, the hydrophobic TAGs are stored in cytolytic and organelles as plastids in lipid droplets or oilbodies. In oilseeds, these TAGs constitute the source of vegetable oils used for human and animal nutrition. The oil droplets consist of a core of neutral lipids, mainly TAG and sterol esters, that are surrounded by a monolayer of polar lipids.

The TAGs are fatty acid triesters of glycerol and they differ accordingly to the nature of the fatty acid esterified to the different positions of the glycerol backbone (Fig. 7.4). Generally, the central position (sn-2) of the glycerol is bound to an unsaturated fatty acid mostly oleic acid (C18:1). In addition to TAGs and sterol esters, monoacylglycerols (MAGs) and DAGs are also minor components of lipid droplets. According to their denomination, one or two acyl groups can be linked to the glycerol, respectively.

Jojoba (*Simmondsia chinensis*) is unusual among higher plants in that its seed storage lipids are waxes instead of TAGs (Miwa, 1971). These waxes are esters of very long-chain fatty acids (C20–C24) and alcohols (C18–C22).

2.4. Surface lipids

Plants have three major extracellular hydrophobic barriers: cuticle, sporopollenin, and suberin (Samuels, Kunst, & Jetter, 2008). These three types of barriers that are essentially constituted of lipids are important for protection against various environmental stresses, such as water deficit, salinity, wounding and pathogen attack. The cuticle is a continuous lipophilic layer coating the aerial surfaces of land plants. It consists of waxes embedded within and overlaying a cutin matrix. Cutin is an esterified polymer of mainly C16 and C18 ω- and mid-chain hydroxy fatty acid esterified to glycerol skeleton (Molina, Bonaventure, Ohlrogge, & Pollard, 2006; Fig. 7.5). Waxes are mainly derivatives of very long-chain fatty acids (C20–C34) that include alkanes (C21–C33), aldehydes, primary and secondary alcohols, ketones and esters (Fig. 7.5; Samuels et al., 2008). Sporopollenin is a lipid- and phenolic-based polymer present in the outer exine layer of pollen walls (Morant et al., 2007). Similarly, suberin is a lipid- and phenolic-based polymer present in the cell walls of various external and internal tissue layers (Franke & Schreiber, 2007).

2.5. Other lipids

Isoprenoids and terpenes play also important biological functions. For example, In addition to give colours to the flowers in order to attract insects for pollinisation, carotenoids are antioxidants and protective agents. Another essential group is constituted by the phytohormones. They are grouped into several major classes including auxins, cytokinins, abscisic acid, gibberellins, ethylene, jasmonates, salicylic acid and brassinosteroids. They regulate plant

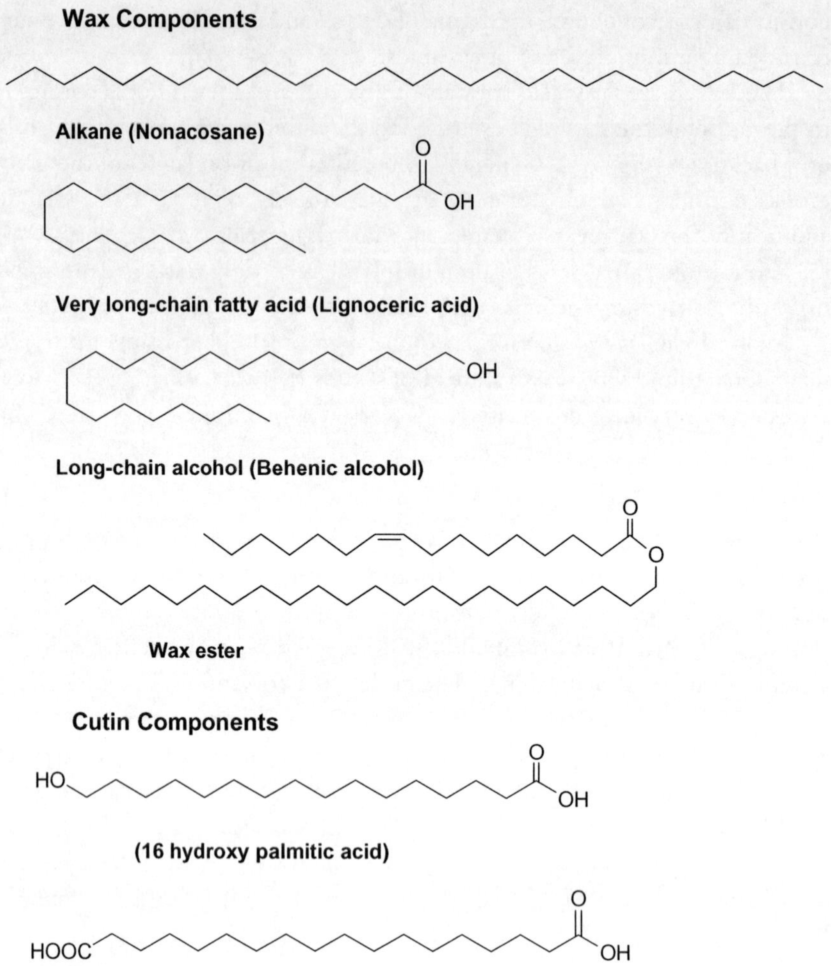

Wax Components

Alkane (Nonacosane)

O
OH

Very long-chain fatty acid (Lignoceric acid)

OH

Long-chain alcohol (Behenic alcohol)

O
O

Wax ester

Cutin Components

HO
O
OH

(16 hydroxy palmitic acid)

HOOC
O
OH

Figure 7.5 Surface lipids.

growth, development, and response to biotic and abiotic stress. Some lipids are only present at a low percent of total lipid mass in normal physiological conditions such as long-chain acyl-CoAs, lysolipids, DAGs, free fatty acids, sphingosine-3-phosphate, ceramides, for example. Most of them are involved in cellular or membrane signalling and act as second messengers.

3. LIPID EXTRACTION

The lipid extraction is very crucial and should be adapted to the lipid classes to be studied. The crucial points that have been identified concern: the separation of lipids from proteins, the choice of the extracting solvent and the minimization of lipid degradation (Wolf & Quinn, 2008). Usually, lipids are extracted from biological matrices using Folch method, which is a mixture of chloroform and methanol (Folch, Lees, & Sloane Stanley, 1957) or the modified method of Bligh and Dyer (1959).These methods are very popular since they are sufficient to get a good picture of the lipid profile but they are not sufficient to extract highly hydrophilic lipids such as GIPCs, and phosphoinositides. These methods are not enough selective to avoid the presence of water which can contain ions in sufficient concentration to extract contaminants as small molecules. So it is recommended to clean-up the total lipid extract (TLE) by an aqueous lower salt concentration solution (Wolf & Quinn, 2008). This aqueous solution can be acidic in order to improve the extraction of acid lipids. Addition of low LiCl concentration (50 mM) or NH_4Cl to the water phase results in better selectivity in ESI/MS for anionic lipid analyses as PE species (Han & Gross, 2005a). Disadvantages of these lipid extraction procedures are the presence of protein at the water-methanol/chloroform interface. If the protein amount is important, emulsions are facilitated and impairs lipid extraction and if the protein amount is important which facilitate emulsions. These latter are difficult to eliminate even by centrifugation, freezing, modification of the chloroform percentage or by adding ions. The most important consequence is that the chloroform phase contains a small proportion of water so solvent evaporation is longer and drying the lipid extract more difficult. Consequently, water contamination enhances lipid oxidation and vinyl ether bond decomposition (Wolf & Quinn, 2008). Solid phase extraction in which there is no need of water/solvent partition is an interesting alternative procedure (Ruiz-Gutiérrez & Pérez-Camino, 2000). Nevertheless, this method stays very common and can be modified according to the source of material. As an example, in the case of *A. thaliana* three different general lipid extractions are commonly used (Li-Beisson et al., 2010). The first one is recommended for low quantity of soft tissues and chloroform/methanol is the extracting solvent. The second hexane/isopropanol method is preferred for up to 1 g of fresh weight and the third which is a modified protocol of Bligh and Dyer method is more adapted to the leaves. In addition to the most common extraction methods described above, specific procedures have been developed for particular lipid classes or materials (see Cyberlipid Web site: www.cyberlipid.org). Consequently, a universal protocol allowing the extraction of lipids whatever the plant, the tissue and the amount of material does not exist and in most cases, the operator has to modify the conditions regarding the start sample. In addition, some other important lipid classes that are not recovered in the chloroform extract need specific protocols to be studied. This is the case for plant sphingolipids that have been exclusively represented by neutral sphingolipids, such as ceramide and monohexosylceramide, which are soluble in chloroform and resistant to mild-base hydrolysis. GIPCs, which have been characterized as (*N*-acetyl) glucosamine-glucuronic-inositolphosphoceramides with additional hexoses and pentoses at either the inositol or glucosamine residues, were omitted in most of the plant lipid characterization since they are not recovered in the chloroform phase because of their too high hydrophobicity. A multi-step complex protocol using a combination of propan-2-ol, hexane and water as solvents has been recently reported (Markham et al., 2006), which has allowed the first analysis and identification of the major sphingolipid classes of a plant *A. thaliana*. The picture of the GIPCs has been completed after optimization of their extraction showing the presence of GIPCs with up to six oses saccharide units in tobacco, whereas the *A. thaliana* species contain only two saccharide units (Buré et al., 2011). Some other minor lipids with high hydrophilic properties such as acyl-CoAs require a specific extraction procedure. For these particular molecules, a phosphate-buffered propanol–acetic acid extraction buffer (propanol–acetic acid) was developed (Mancha, Stokes, & Stumpf, 1975). This method, which involves several steps, is complicated and an alternative is to solubilize the acyl-CoAs and the lipids, directly in the reaction vessel, by 0.05 M Tris–HCl, pH 7.5/ $CHCl_3/CH_3OH$ (1/3/3, v/v/v) (Juguelin & Cassagne, 1984). This extraction mixture allows the quantitative analysis of acyl-CoAs in complex reaction mixtures. Plant surface lipids require also specific methods. Waxes are very simple to extract and a short time immersion (30 s) in $CHCl_3$, for example, is sufficient to get a complete recovery of the epicuticular lipids (Cassagne & Lessire, 1974; Kollatukudy, 1968; Wen & Jetter, 2009). Cutin is more difficult to analyze, since cutin is a polymer that has to be dissociate in its different constituents (Kolattukudy, 1970; Nawrath, 2002; Walton & Kolattukudy, 1972). The more efficient protocol consists of a first delipidation step followed by exhaustive solvent extractions: (1) boiling propanol, (2) $CHCl_3/CH_3OH$ (2/1), (3) $CHCl_3/CH_3OH$ (1/1) and (4) MeOH and by a depolymerization of the remaining residue using 6–5% sulphuric acid in MeOH for 3 h at 85 °C (Bonaventure, Beisson, Ohlrogge, & Pollard, 2004). Finally, the different cutin compounds are extracted by dichloromethane and derivatized using BSTFA before being subjected to GC/MS analysis.

Although different extraction methods have been developed in relation to the specific requirement of the different lipid classes, limitations still remain. Whatever the methodology used, it is rather difficult to appreciate if the lipid extraction is complete. Moreover, in the case of lipidomic studies, it seems to be very difficult to obtain a complete lipidome picture including all the lipid classes using a unique extraction method.

4. SAMPLE STORAGE

Lipid storage represents a crucial part before analysis. In the presence of oxygen, fatty acid double bonds can be oxidized and generated false lipids from existing species. The use of antioxidants such as butylated hydroxy-toluene (BHT, 2,6-di-tert-butyl-p-cresol) is recommended at a concentration that does not exceed a threshold of 0.00001 (w/w). It is also known that exposition to UV-light and elevated temperature favour isomerization as double-bond migration and diene conjugation (Khalil et al., 2010; Wolf & Quinn, 2008). So it is recommended that evaporation of solvent will be performed under nitrogen or argon gas at a temperature under 40 °C or using a SpeedVac and that the samples will be stored at −20 °C in TPFE-sealed opaque vessels under nitrogen (Wolf & Quinn, 2008). Under these conditions, glycerophospholipids were found to remain unchanged for up to 6 months storage at subzero temperatures. Sphingolipids are expected to be stable for much longer periods of time since they are generally more saturated (Wolf & Quinn, 2008).

5. PRINCIPLES OF MS LIPID ANALYSIS

The traditional methods as high-performance thin layer (HPTLC) or/and liquid chromatography (HPLC) used alone allowed only the determination of the lipid classes of a biological sample. Moreover, the combination of GC and one of these methods was necessary to quantify the fatty acids of each lipid class. These strategies were time-consuming and limitating since the quantification, most of the time based on an additional standard in the sample, is discutable and allow only to the global composition. The technological developments of MS have been an important factor for the emergence of news lipid analysis performances and strategies. The different aspects will be developed in the following parts.

5.1. Ionization

Lipids have a great chemical heterogeneity; their study requires the use of variety of ionization techniques. Lipidomics is mostly performed using atmospheric pressure ionization (API) techniques: mainly ESI followed by atmospheric pressure chemical ionization (APCI) and atmospheric pressure photoionization (APPI) but also using matrix-assisted laser desorption/ionization (MALDI) under vacuum. Soft ionization methods are able to preserve the lipid molecular structure after ionization without fragmenting them, meaning they can be analyzed intact (Fenn, Mann, Meng, Wong, & Whitehouse, 1989; Yamashita & Fenn, 1984).

ESI generates molecular ions in gas phase by desolvation mechanisms at atmospheric pressure, which allows easy coupling with direct infusion or liquid chromatography (Fenn et al., 1989; Whitehouse, Dreyer, Yamashita, & Fenn, 1985). The sample solution is sprayed at the end of a fine capillary needle held at a high electric potential (several kV), forming small charged droplets, which are directed into the mass analyzer by the applied electric field. During flight, the droplets are desolvated, either by passage through a curtain of heated inert gases or alternatively through a heated capillary, or both. Thus, during desolvation, the coulombic force between ions is dramatically increased. Once this force exceeds the surface tension of the solvent, the droplets explode to form smaller droplets. This cycle is iteratively repeated until molecular ions are generated prior to their entrance into the mass analyzer. Although many physicochemical features of the ionization and fragmentation process are still unclear, droplet surface tension and the spatial proximity of surface charges on sprayed droplets are critical determinants of the ionization process. The principles and mechanisms of ESI for lipids have been well summarized in many review articles and books (Duffin, Henion, & Shieh, 1991; Fenn, Mann, Meng, Wong, & Whitehouse, 1990; Murphy & Axelsen, 2011; Pulfer & Murphy, 2003).

APCI is a type of chemical ionization performed at atmospheric pressure, and can be used for the analysis of liquid chromatography effluents. The analytes are sprayed with high flow rates of nitrogen, creating an aerosol of fine droplets that readily evaporate in the ion source. This mixture of gas-phase analytes, solvent and atomizing gas is then subjected to a corona discharge created by raising a sharp metal needle to several thousand volts. The primary ions are produced resulting from the electrons from the corona discharge and the excess atomizing gas. These primary ions react with solvent molecules (such as H_2O, methanol and acetonitrile) in a complex sequence to ultimately form the solvated ions of highest proton affinity. Subsequently, the analyte ions are formed through gas-phase ion-molecule reactions between the analyte molecules and these solvated ions (Byrdwell, 2001; Horning et al., 1974).

APPI source is a recent ionization technique (Hanold, Fischer, Cormia, Miller, & Syage, 2004; Robb, Covey, & Bruins, 2000). In this approach, which is a modification of APCI, the liquid effluent is vaporized in a heated nebulizer to generate gas-phase analytes. The fundamental process in photoionization is the absorption of a high-energy photon, emitted from a UV lamp, by the molecule and subsequent ejection of an electron. In the more common dopant-assisted APPI mode, two steps are involved. In the first step, the ionization region is filled with the high-concentration dopant with a low ionization energy (e.g. toluene or acetone), and the dopant radical cations are produced by absorption of photons. In the second step, the analytes are then ionized by direct charge transfer from dopant radical cations or by the solvent-mediated ionization. During the latter, the solvent molecule is ionized by a proton transfer from the dopant radical cation, forming an ion, which in turn ionizes the analyte molecule by a second proton transfer reaction (Cai, Short, Syage, Potvin, & Curtis, 2007; Cai & Syage, 2006; Raffaelli & Saba, 2003).

MALDI represents alternative source of choice to analyze lipid molecular species (Fuchs, Süß, & Schiller, 2010). MALDI MS is based on the use of a matrix, which absorbs laser energy and leads quasimolecular ions (Hillenkamp and Peter-Katalinić, 2007). The matrix choice constitutes a crucial part to detect lipid species by MALDI-MS. The most common matrices used to analyze lipid species are usually a small organic acid such as 2,5-dihydroxybenzoic acid, α-cyano-4-hydroxy-cinnamic acid, but also 9-aminoacridine (9-AA) (Sun et al., 2008), 5-chloro-2-mercaptobenzothiazole and 2,4,6-trihydroxyacetophenone. More recently, the use of 2,6-dihydroxyacetophenone (DHA) was developed to limit salt adducts and increase signal intensity in positive ion mode (Colsch, Jackson, & Woods, 2011; Colsch & Woods, 2010; Delvolve, Colsch, & Woods, 2011). However, depending on the matrix choice, numerous studies were conducted to characterize lipid species using MS and MS/MS modes (Fuchs et al., 2010; Harvey, 1999; Schiller, 2004). Since 10 years, new developments using MALDI-MS were conducted to map biomolecules directly on tissue sections (Cornett, Reyzer, Chaurand, & Caprioli, 2007). At this time, numerous lipid species can be mapped by imaging MS (IMS; Fernández, Ochoa, Fresnedo, Giralt, & Rodríguez-Puertas, 2011; Goto-Inoue, Hayasaka, Zaima, & Setou, 2011; Horn et al., 2012; Jackson, Wang, & Woods, 2005).

The ionization efficiency of an analyte greatly depends on the electrical propensity of an individual analyte in its own microenvironment to lose or gain a charge (i.e. oxidize or reduce) in an electric field. Complete ionization of all classes of lipids in single ionization is really challenging due to chemical variety of lipid and their properties. Either the cationic or the anionic region in compounds that possess bonds with easily separable charges can be easily ionized in either positive- or negative-ion modes of ionization, respectively. The ionization efficiency for these ions is much higher relative to that for compounds that do not carry an inherent charge. Provided a sufficient dipole potential is present, ionization can also be achieved in the ion source through formation of adduct ions, with either a small anion (NH_3) or cation (Na^+, NH_4^+, Li^+) and thus is potentially suitable for analysis by ESI-MS, APCI-MS or MALDI-MS (Byrdwell, 2001; Duffin et al., 1991; Hsu & Turk, 1999, 2010; Pulfer & Murphy, 2003).

Another innovation of ESI in lipid analysis is ozone electrospray (OzESI). In this approach, ozone is introduced into the ion source together with the desolvation gas, permitting the analysis of double-bond positions in a broad range of common unsaturated lipids including acidic and neutral glycerophospholipids, sphingomyelins, and TAG (Thomas, Mitchell, & Blanksby, 2006; Thomas, Mitchell, Harman, Deeley, Murphy, & Blanksby, 2007; Thomas, Mitchell, Harman, Deeley, Nealon, & Blanksby, 2008).

5.2. Shotgun versus LC

Shotgun lipidomics is to perform a lipid extraction and analyze the crude extract directly by MS, mainly with the ESI ion source (Han & Gross, 2003; Han, Yang, Cheng, Fikes, & Gross, 2005; Schwudke et al., 2007). The exact analytical methodology employed depends on the type of mass spectrometer available; however, the goal is always to identify and quantify as many lipid molecular species as possible.

Using intra-source separation-based shotgun lipidomics, Han et al. develop a comprehensive profiling of the cellular lipidome by direct infusion of crude lipid extract without any previous chromatographic separation (Han & Gross, 2003; Han, Yang, Cheng, Ye, & Gross, 2004; Han et al., 2005). Sequentially changing experimental conditions such as adjusting pH of the lipid extract and employing both positive and negative ion modes can result in the preferential ionization of anionic lipids, weak anionic lipids and neutral lipids under different ESI experimental conditions. Intra-source separation can be used in combination with different MS/MS mode as multiple precursor ion scan (PIS) and neutral loss scan (NLS) using triple quadrupole mass spectrometer systems. The use of intra-source separation with its exploitation of selective ionization methods is expected to simplify the requirements for lipid purification prior to MS and give the unambiguous identification of many components in complex lipid extracts (Ejsing et al., 2006; Han, 2010; Hou et al., 2011).

Shevchenko and colleagues have developed a variant of shotgun lipidomics that they call 'top-down' lipidomics (Graessler et al., 2009; Schwudke et al., 2007). These analyses are performed on high-resolution mass spectrometers (HRMS), typically the LTQ-Orbitrap, and involve the acquisition of a mass spectrum at high resolution with exact mass (<4 ppm) measurement of peaks in the spectrum (Schuhmann et al., 2011). By accurately measuring the *m/z* of an ion its elemental composition can be determined and this is then exploited for lipid identification (Taguchi & Ishikawa, 2010; Wu, Rodgers, & Marshall, 2004).

The major advantages of direct infusion without previous chromatographic separation are the ease of implementation and high throughput. Crude lipid extracts can be analyzed without any extensive sample processing steps. Lipid extracts in the organic phase are usually cleaner and almost free of salts, which minimizes the possibility of ionization suppression by inorganic salt ions. Shotgun is capable of providing useful analyses of complex mixtures of lipid species that could have a dynamic range of several orders of magnitude. The major drawback of this technique is the ion suppression due to the ionization of all the analytes at a single time and the limited dynamic range of the ESI process. This leads to discrimination against compounds of low abundance lipids and/or with poor ionization characteristics. As an example, during sphingolipid analysis in positive ion mode, glycerophosphocholines (PC) molecules decrease the signal intensity of other species such as Sphingomyelins (SM) or Glycerophosphatidylethanolamines (PE). Otherwise, direct infusion can require a HRMS to identify more lipid species (mainly minor species) often not detected using low mass accuracy mass spectrometers (Taguchi & Ishikawa, 2010). Compared to direct infusion, coupling liquid chromatography to MS significantly reduces the complexity of the lipid mixture presented to the mass spectrometer for analysis. High abundant lipid species are limited to a relatively narrow chromatographic window and low abundance lipids species and isobaric compounds can be identified with less interference. Using LC prior to MS allow a far greater number of molecular species to be monitored. In addition, the chromatographic separation itself provides extra informations, that is, retention time in addition to *m/z* and help to perform MSn experiments more easily in complex mixtures, for lipid structure elucidation. However, due to the great diversity of the physico-chemical properties of lipid species, chromatographic separation is generally customized to analyze lipid families by class. However, global lipidomic approaches using HRLC–MS methods are developed with the addition of ammonium salts in mobile phases (Bird, Marur, Sniatynski, Greenberg, & Kristal, 2011a, 2011b; Hu et al., 2008) to detect neutral lipids such as glycerolipids (MAG, DAG, TAG) as $[M+NH_4]^+$ ions with other lipid classes. As a result, to achieve this quantitative measure for this large number of molecular species, multiple chromatographic separation techniques are normally required.

Several separations have been developed for different classes of lipids, like normal-phase separation (NL-LC/MS) or reverse-phase separation (RP-LC/MS) (Houjou, Yamatani, Imagawa, Shimizu, & Taguchi, 2005; Kalo, Ollilainen, Rocha, & Malcata, 2006; Mottram, Crossman, & Evershed, 2001). Mobile-phase gradients are also easily established in LC, thus allowing the collection of different lipid fractions for further examination depending of the ion source used, ESI or APCI ion source.

The use of HPTLC was performed too to analyze lipid families before MALDI-MS experiments. Lipids are separated by classes depending migration solvents, and analyzed directly on TLC plates (Fuchs et al., 2008; Rohlfing et al., 2007). A second approach consists of coupled TLC-Blot with MALDI-MS to quantify and image TLC bonds (Goto-Inoue et al., 2008).

5.3. MS mode

Species of some lipid classes have unique elemental compositions and could therefore be directly identified by their masses determined. Such lipidomics approaches require HRMSs, such as Orbitraps (Giavalisco et al., 2011; Koulman et al., 2009; Taguchi & Ishikawa, 2010). However, it is more common for lipids to be subjected to tandem MS analysis (MS/MS) for lipid structure elucidation and quantification in lipidomics research. The most commonly used approach for ion fragmentation in lipid analysis by MS/MS is collision-induced dissociation (CID), also termed collisionally activated dissociation. The most common types of tandem mass spectrometers are triple quadrupole (QqQ), ion trap (IT) and quadrupole time-of-flight (QTOF). Tandem MS can also include multiple stages of mass analysis, the so-called MSn approach that is practical only in IT instruments. MSn experiments enable the study of sequential multistage fragmentations of the precursor ions within the same run to characterize the structures of complex analytes such as lipids (Hsu & Turk, 2008a, 2008b; Larsen, Uran,

Jacobsen, & Skotland, 2001; McAnoy, Wu, & Murphy, 2005; Smith et al., 2008). However, using ITMS instruments, fragmentation of several lipid species require MS3 fragmentation due to the presence of low-mass cut off, which limits the detection of interest ions. Implementation of High-energy collisional dissociation (HCD) allows a complete fragmentation of molecules by MS/MS.

Tandem-in-space instruments such as triple quadrupole allow investigation of fragmentations of lipid species using three different scan modes for profiling lipidomics and one for target lipidomics (Fig. 7.6), described briefly above for triple quadrupole instruments: (i) a product ion scan for a selected

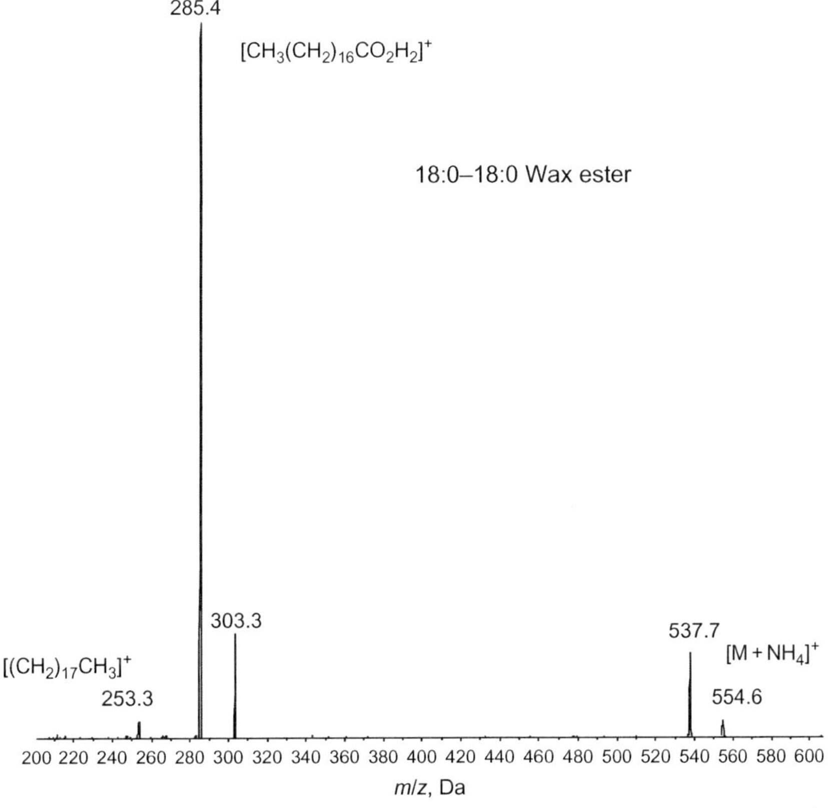

Figure 7.6 Positive ESI-MS/MS spectrum of 18:0–18:0 wax ester (*m/z* 554.6 $[M+NH_4]^+$). The major fragment ion *m/z* 285.4 is characteristic of the fatty acyl moiety of the 18:0–18:0 wax ester. The minor fragment at *m/z* 253.3 is coming from the fatty alcohol part of the wax ester. (For colour version of this figure, the reader is referred to the online version of this chapter.)

precursor ion, (Horai et al.) a PIS for a user-selected product ion and (iii) an NLS is performed to reveal precursors that fragment by ejecting a neutral fragment of molecular mass selected by the user. In addition, the target profiling multiple reaction monitoring (MRM) record signals arising from fixed user-selected *m/z* values for both precursor and product ions, and provides accurate and precise quantitative analyses currently possible.

To characterize an ion of interest, the mass spectrometer is set to product ion scan mode, the first quadrupole Q1 is fixed to transmit only the precursor ion with a user-selected *m/z* value. These ions enter the second quadrupole (Q2), referred as the collision cell, where they are fragmented by CID. The resulting product ions are displayed as a spectrum by the third quadrupole (Q3). In contrast of PIS mode, the PIS function focuses on sub-groups of ions carrying specific structural moieties that appear as the diagnostic fragment of a the specific class of lipid. In this mode, Q1 scans over the mass range and after the ions sequentially enter in Q2 for CID fragmentation, Q3 is set to transmit specific fragment ions with a user-selected *m/z* value.

In a neutral loss-scanning mode, both Q1 and Q3 are scanned in a synchronized manner with a constant *m/z* difference between the two analyzers. Because of the mass offset at any time, Q3 will transmit product ions with a fixed lower *m/z* value than the mass selected precursor ions transmitted by Q1. The resultant mass spectrum contains all the precursor ions that lose a neutral species of selected mass and used to detect sub-sets of lipids that contain a specific functional group. Precursor ion and NLSs have been used to detect sub-sets of lipids that contain a specific functional group, such as phospholipids (Ekroos, Chernushevich, Simons, & Shevchenko, 2002; Han, Yang, Yang, Cheng, & Gross, 2006).

In MRM mode, a series of user-specified precursor-fragment pairs are cycled to the detector by the triple quadrupole. Q1 is fixed to select one precursor ion, and Q3 is set to pass the fragment ion specified for that precursor. These transitions require user-defined input of the parent ion (Q1) and product ion (Q3). MRM excludes the observation of unexpected lipids. Applications of MRM in the quantification of specific lipid molecules, which yield known product ions, have been described (Bielawski, Szulc, Hannun, & Bielawska, 2006; Lembcke et al., 2005).

The careful choice of a particular scan mode that is appropriate for the class of lipids being studied provides excellent selectivity. In some cases, this added selectivity can help in the analysis of lipids that cannot be resolved chromatographically. However, the disadvantage of using specific scan modes is that several lipids can escape detection if an inappropriate scan mode is selected.

Although triple quadrupole mass spectrometers are unique in that they can perform product ion, precursor ion and neutral loss scanning, they also have several disadvantages; most notably, low resolving power and medium mass accuracy. A hybrid instrument, based on replacing Q3 with an IT, named QQ-IT (or Qtrap), can perform precursor and NLSs as well as multiple-stage MS/MS (MSn) (Buré et al., 2011; Hopfgartner et al., 2004; Markham & Jaworski, 2007). Alternatively, hybrid instruments in which the third quadrupole (Q3) is replaced by a TOF, named QQ-TOF and QTOF (Ekroos et al., 2002; Hou et al., 2011; Okazaki, Kamide, Hirai, & Saito, 2013), can provide very good mass accuracy and resolving power for product ions, but they cannot perform NLSs and MRM.

5.4. Quantification

Quantification still remains challenged in lipidomics area, in large part because it is very tedious to establish and carry out the quantitative analysis method. The quantitative purposes can imply measuring the absolute concentrations of particular lipid species, or the relative abundances of particular lipid species formed in two similar organisms under different physiological conditions or the relative abundances of different molecular species within a lipid class, in complex mixtures (Brugger, Erben, Sandhoff, Wieland, & Lehmann, 1997; Kerwin, Tuininga, & Ericsson, 1994; Liebisch, Drobnik, Lieser, & Schmitz, 2002; Zacarias, Bolanowski, & Bhatnagar, 2002).

Quantitative analyses of lipids are mostly performed on a case-by-case basis. The quantitative analysis of a lipidome is generally divided into global and targeted lipidomics. Global lipidomics allows the relative quantification of numerous molecular lipid species across multiple structural classes in TLEs. Targeted lipidomics, on the other hand, permits the quantitative analysis of a single or a few lipid species within a specific lipid class.

It is important to keep in mind that lipids species exhibit different behaviours using different types of mass spectrometers or using the same mass spectrometer under different experimental parameters (DeLong, Baker, Samuel, Cui, & Thomas, 2001; Larsen et al., 2001). Furthermore, the instrument response can vary between different classes of lipids (Koivusalo, Haimi, Heikinheimo, Kostiainen, & Somerharju, 2001).

In a mass spectrum, the relative signal intensities of the ions of different lipid species do not directly represent their molar abundances. The abundance of any molecular ion species typically carries information on concentration, but ion abundance is confounded by a number of features, including instrument response factors, ionization efficiency of the molecule, stability of the molecular ion species and the presence of other molecules that could cause ion suppression of the analyte of interest. A lipid species with a higher peak intensity than that of another could actually be present in a lower concentration. For example, it is impractical to use MS to directly estimate the relative abundances of different phospholipid subclasses due to the differences in the ionization efficiencies of their polar headgroups (DeLong et al., 2001; Koivusalo et al., 2001). Estimations of the relative abundances of lipid species is less problematic if the responses of all the lipid species are properly normalized with respect to the appropriate internal standards added to the lipid mixture to be analyzed (DeLong et al., 2001; Zacarias et al., 2002).

The absolute quantification of specific lipid species relies on the availability of internal standards preferably as isotopically labelled forms although closely related compounds (homologues or simple structural analogues) can be used. This approach is not available for most lipidomic studies because of the absence of internal standards for each and every lipid molecular species and, practically, the potential overlap of m/z between internal standards and endogenous lipids. The absence of reference standards for the different variants of structure observed in a molecular species series, for example, number and position of double bonds in fatty acyl chains, variety of fatty acyl chain length, ether analogs, even positional isomers. However, this strategy is applicable to a limited number of lipids in complex mixtures (Liebisch et al., 2002; Whitehead et al., 2007). The absolute concentration of a lipid species is then determined by spiking in known amounts of lipid standards labelled with stable isotopes (stable-isotope dilution) (Haroldsen, Clay, & Murphy, 1987; Liebisch et al., 2006). The isotopically labelled lipid standard and its normal counterpart exhibit identical responses in mass spectrometers as they have identical physicochemical properties if the isotopic labelling involves heavier atoms (usually 13C or deuterium). The central strategy of isotope dilution has been to use a stable isotope-labelled analog of the analyte as an internal standard and to measure the ratio of signal intensities of the analyte and internal standard rather than any absolute intensity. The ratio is then converted to analyte concentration by a calibration curve generated using reference standards. The absolute concentrations of lipids can also be determined without internal standards using standard addition methods. In brief, known amounts of standards of the target lipid are sequentially spiked in the complex mixture. The respective MS responses are used to determine by extrapolation the original concentration of the lipid species of interest (Liebisch et al., 2002; Whitehead et al., 2007).

An alternative approach to absolute quantification has been to express the abundance of molecular species within a single class of lipids as mole fraction using the abundance ratios of all molecular species in that class to normalize the data. In many cases, comprehensive lipid profiling comparison between different physiological states is required (Ekroos et al., 2002; Rouzer et al., 2006). This approach requires comparison of multiple mass spectrometric analyses using different scan modes. Uncontrolled fluctuations in signal intensity due to the lipid extraction and analytical processes can be an issue as each sample is individually processed. Fortunately, non-naturally occurring lipid standards, often as a mixture covering most lipid classes, can be spiked into the samples at different stages of sample processing, which facilitates the correction of signal intensity (Ejsing et al., 2006; Shaner et al., 2009).

5.5. Data processing

High-throughput methods quickly generate huge amounts of data that are difficult to process manually. Data processing thus becomes a central and rate-limiting step in lipidomic. As in other omics areas, bioinformatics tools will be needed to take the full advantage of information obtained by lipidomics studies. One key area for lipidomics studies is the data processing and lipid identification: methods and databases that facilitate turning raw instrumental data from lipidomics experiments into a final lipidomics dataset that can be analyzed and interpreted.

The complexity of lipid nomenclature represents a challenge as the virtual absence of comprehensive, integrated reference databases. Several public resources and international initiatives have been set up to find solutions to these problems (Table 7.1): LIPID MAPS consortium (http://www.lipidmaps.org/; Fahy et al., 2007; Schmelzer, Fahy, Subramaniam, & Dennis, 2007), European Lipidomics Initiative (http://www.lipidomics.net/; van Meer, Leeflang, Liebisch, Schmitz, & Goni, 2007) and similar developments in Japan (http://www.lipidbank.jp/; Taguchi et al., 2007).

The LIPID MAPS (LIPID metabolites and pathways strategy) consortium was created to identify and quantify all the major and many minor lipid

Table 7.1 Online resources for lipid classification and databases

Web site	Description	References
http://www.lipidmaps.org/	Classification scheme and extensive tools	Cotter, Maer, Guda, Saunders, and Subramaniam (2006), Fahy, Sud, Cotter, and Subramaniam (2007), and Sud et al. (2007)
http://www.cyberlipid.org/	Description of lipid classes, with references	
http://lipidlibrary.aocs.org/	Extensive highly referenced information	
http://lipidbank.jp/	Official database of Japanese Conference on the Biochemistry of Lipids	Taguchi, Nishijima, and Shimizu (2007)
http://www.massbank.jp/	High-resolution mass spectral database	Horai et al. (2010)
http://www.lipidat.tcd.ie/	Lipid thermodynamic and phase transition information	Caffrey and Hogan (1992)
http://www.chem.qmul.ac.uk/iupac/lipid/	IUPAC lipid nomenclature	

species in mammalian cells. They introduced a nomenclature that enables to represent a lipid compound by a unique 12-digit identifier (Fahy et al., 2009, 2007). The LIPID MAPS online provide several bioinformatics tools, like a tool enables the drawing of lipid structures, prediction of possible structures from MS data (Cotter et al., 2006; Fahy et al., 2007; Sud et al., 2007). The LipidBank has classified lipids into 17 categories covering a wide variety of animals and plants. The LIPIDAT is composed mostly of phospholipids and their thermodynamic data. The Lipid Library (http://lipidlibrary.aocs.org/) concentrates on fatty acyls and their derivatives.

Following the raw data acquisition from the MS instrument, either direct infusion techniques or liquid chromatography coupled to MS (LC/MS) methods, the data need to be processed. Over the past years, different methods and programmes have been developed for processing and identifying lipids from MS data. The increasing demand for better data processing has generated a number of software packages either from commercial

Table 7.2 Softwares for lipidomics data processing

Name	**Description**	**References**
TriglyAPCI	Interpretation of APCI mass spectra of triacylglycerols	Cvacka, Krafkova, Jiros, and Valterova (2006)
Lipid mass spectrum analysis (LIMSA)	Identification and quantification of lipids	Haimi, Chaithanya, Kainu, Hermansson, and Somerharju (2009) and Haimi, Uphoff, Hermansson, and Somerharju (2006)
LipidView (former lipid profiler)	Data processing tool for the molecular characterization and quantification of lipid species	Ejsing et al. (2006)
LipidQA	To identify and quantitate complex lipid molecular species	Song, Hsu, Ladenson, and Turk (2007)
LipidInspector		Schwudke et al. (2006)
FAAT (fatty acid analysis tool)	Analysis of Fourier transform mass spectral data from lipid extracts	Leavell and Leary (2006)
LipidX	Bottom-up and top-down shotgun lipidomics	Herzog et al. (2011) and Schuhmann et al. (2011)
LipID	Automated assignment of lipids	Hubner, Crone, and Lindner (2009)
LipidSearch		Houjou et al. (2005)
LipidAT	Automated analysis of raw MS data to display of the phospholipids differences	
SimLipid	Lipid identification and profiling	
LipidomeDB	Identification of lipids from direct infusion MS data	Zhou et al. (2011)

sources or the open source community. Table 7.2 provide non-exhaustive lists of commercially available and free software packages, respectively, that can be used to process lipidomics data. Many of the metabolomics data processing tools, which were extensively reviewed (Katajamaa & Oresic, 2007; Neumann & Bocker, 2010), can also be used in lipidomics MS-based applications.

6. PLANT LIPID ANALYSES

6.1. Fatty acids and related lipids

Most of analysis made for fatty acids and their derivatives such as fatty alcohol, fatty aldehydes and fatty esters are done using GC methods. However, for some specific cases, API sources and tandem MS can provide answers in term of characterization or/and sensibility.

Structural characterization of fatty acids by ESI source with CID fragmentation can be achieved by analyzing both negatively charged and positively charged (usually adduct ions with lithium) molecular species (Hsu & Turk, 1999; Kerwin, Wiens, & Ericsson, 1996). Fragmentation allows the precise localization of hydroxylation or insaturation on the fatty acid chain (Hsu & Turk, 2008a; Moe & Jensen, 2004; Moe, Strom, Jensen, & Claeys, 2004).

Identification and quantification of fatty aldehydes are challenging. They have to be derivatized to be analyzed by APCI-MS or ESI-MS. Different derivatization is used, all of them produced specific ions during CID fragmentation, which allow the characterization of fatty aldehydes (Berdyshev, 2011).

Wax ester (WE) molecular species can be analyzed by APCI-MS (Vrkoslav, Urbanova, & Cvacka, 2010) and ESI-MS (Fitzgerald & Murphy, 2007; Santos, Schreiber, & Graca, 2007). WE were detected as $[M+NH_4]^+$, $[M+H]^+$ or $[M+Li]^+$ adduct. The structure of the analytes was deduced from CID fragmentation spectra, fragmentation yielded the expected product ion characteristic of the fatty acyl moiety of the molecule (Fig. 7.6). MRM and neutral loss scanning are used. Direct-infusion (Santos et al., 2007) and chromatographic separation (with a non-aqueous reversed-phase column) (Medvedovici, Lazou, D'Oosterlinck, Zhao, & Sandra, 2002) are used for WE characterization and quantification.

6.2. TAG, DAG, MAG

The neutral lipids DAG, MAG and TAG are analyzed mainly by APCI and ESI sources, although they have no electrostatic charge in solution. MAGs, DAGs and TAGs are analyzed in positive ESI-MS and can form adduct ions with ammonium salts (mainly ammonium formate or acetate) $[M+NH_4]^+$ or with alkali-metal ions (Na^+, Li^+) and give mainly $[M+NH_4]^+$ ions (Duffin et al., 1991). Compared to ESI, APCI mass spectra of TAGs typically yield predominantly DAG fragment ions $[M+H\text{-}RCOOH]^+$ with some level of intact $[M+H]^+$ ions observed (Byrdwell, Neff, & List, 2001). The ionization efficiency is affected by the total number of carbons in fatty acyl chains and the degree of unsaturation. The sensitivity of each molecular species correlated with its unsaturation index and inversely correlated with its chain length (Byrdwell & Neff, 2002; Han & Gross, 2001).

The analysis of TAG molecular species analysis can be challenging if information about all unique molecular species is desired. This is because of the high number of species coming from the assembly of the fatty acyl groups on the glycerol framework. The analysis of MAGs, DAGs and TAGs by ESI-MS/MS was first reported by Duffin et al. (1991) who observed 'DAG fragments' after CID of $[M+NH_4]^+$ ions in a triple quadrupole mass spectrometer (Fig. 7.7). In general, CID mass spectra of $[M+X]^+$ ions ($X=NH_4$, Li, Na or Ag) generated preferential loss of the fatty acid at the sn-1 or sn-3 positions compared to the sn-2 position (Herrera, Potvin, & Melanson, 2010; Hvattum, 2001). This relative abundance of

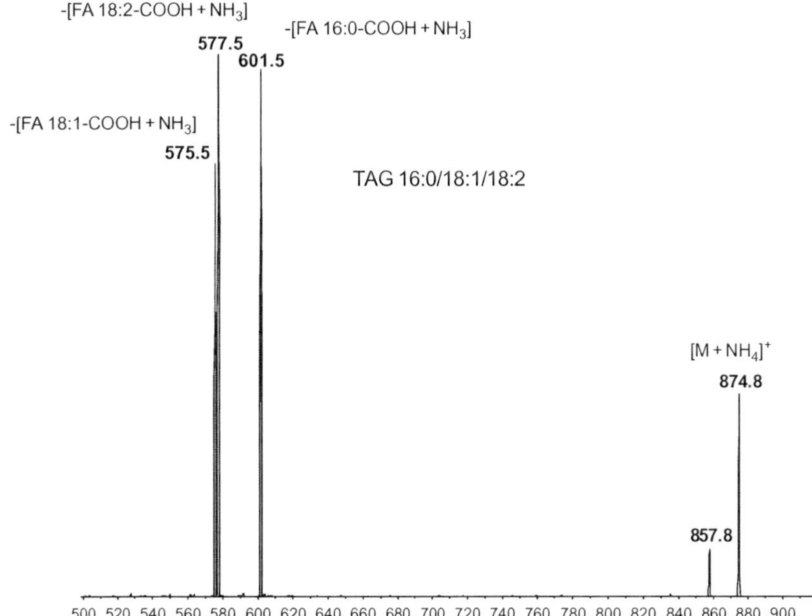

Figure 7.7 Positive ESI-MS/MS spectrum of 16:0/18:1/18:2 triacylglycerol (*m/z* 874.8 $[M+NH_4]^+$). The fragmentation yields three ions corresponding to the loss of the fatty acid. The less intense ion corresponds to the fatty acid 18:1 in the sn2 position. (For colour version of this figure, the reader is referred to the online version of this chapter.)

'DAG fragment' is used to determine the TAG regioisomers (Herrera et al., 2010; Hsu & Turk, 2010). Relying on in-source fragmentation for the generation of 'DAG fragments', APCI has been implemented successfully for the differentiation of TAG positional isomers: the least abundant DAG ion generated corresponded to the loss of the fatty acid from position sn-2 (Mottram, Woodbury, & Evershed, 1997). To determine the nature of the three groups in each TAG molecular species, a second phase of CID fragmentation (MS/MS/MS or MS3) can be used (Hsu & Turk, 2010; McAnoy et al., 2005) Each of these 'DAG ions', obtained from the first state of fragmentation, can then be further decomposed in a second stage of fragmentation to create 'MAG ions'. This ion reveals the nature of a second fatty acyl group. TAGs are also analyzed by APPI. APPI generates TAG mass spectra similar to those of APCI giving fragment ions $[M-RCOO]^+$ and/or protonated molecules $[M+H]^+$ as major ions. The intensity of $[M+H]^+$ increases as the degree of unsaturation increases. As with APCI spectra $[M-RCOO]^+$ fragment ion intensity can be related to the sn-2 substitution (Cai et al., 2007; Gomez-Ariza, Arias-Borrego, Garcia-Barrera, & Beltran, 2006).

TAGs are quantified using internal standards, usually with odd carbon chain length, and MRM or NLS, with and without chromatographic separation (Ejsing et al., 2009; Han & Gross, 2001). Typically, for chromatographic separation, reversed-phase high-performance liquid chromatography online with MS (RP-LC/MS) is employed. The separation is based on both the molecular size and the degree of unsaturation of molecules (Cozzolino & De Giulio, 2011; Laakso, 2002; Mottram et al., 2001, 1997). Silver ion is an alternative traditional approach commonly used in TAGs separations. The separation mechanism is based on specific silver ion/double bond interactions, and chromatographic retention increases with the number of double bonds (Dugo et al., 2005; Leveque, Heron, & Tchapla, 2010). Some studies used the normal-phase separation to identify MAG, DAG and TAG (Kalo, Kemppinen, Ollilainen, & Kuksis, 2003; Kalo et al., 2006). These approaches have been successfully employed to profile TAGs in a variety of vegetable oils (Byrdwell & Neff, 2002; Jakab, Jablonkai, & Forgacs, 2003; Leveque et al., 2010).

6.3. Glycerolipids

Glycerophospholipids consist of a glycerol phosphate backbone with a head group attached in the sn-3 position and two fatty acid moieties esterified in the remaining two positions (sn-1, sn-2 positions) of the glycerol.

Glycerophospholipids are mainly analyzed with ESI ion source in both positive and negative modes, although the polar head group generally drives a preference for either positive ion or negative ion formation. They can also form adduct with a variety of metal ions to yield both cations and anions (Han & Gross, 2005b; Pulfer & Murphy, 2003).

APCI has been successfully used in the analysis of phospholipids, although the vast majority of work has been done using ESI (Byrdwell, 1998). With APPI ion source, few studies have been done, in particular for PC study (Delobel, Touboul, & Laprevote, 2005).

The CID fragmentation patterns of glycerophospholipids normally reveal information on the lipid polar headgroup and its acyl chain moieties present on the glycerol backbone (Brugger et al., 1997; Han & Gross, 1994; Hsu & Turk, 2000a, 2000b, 2000c, 2001b; Kerwin et al., 1994) using QqQ and QTOF instruments. Each lipid class has characteristic headgroup MS/MS fragments, which is used as a diagnostic ion for each specific class of phospholipids Table 7.3. CID fragmentation of $[M-H]^-$ of PA molecule resulted in a product ion observed at m/z 153. Fragmentation of PE anions ($[M-H]^-$) generates a fragment at m/z 196, relating to the dehydrated glycerol phosphoethanolamine head group, and fragmentation of his cation $[M+H]^+$ found to yield an abundant ion characteristic of the PE polar head group that corresponded to the neutral loss of phosphoethanolamine

Table 7.3 Neutral loss of phospholipids head (Brugger et al., 1997; Pulfer & Murphy, 2003)

Phospholipid class	Polarity	Scan mode	Fragment
PA	Negative	Precursor of 153	phosphoryl glycerol–H_2O
PE	Negative	Precursor of 196	glycerol phosphoryl ethanolamine–H_2O
PE	Positive	Neutral loss of 141	phosphoryl ethanolamine
PI	Negative	Precursor of 241	phosphoryl inositol–H_2O
PS	Negative	Neutral loss of 87	serine-H_2O
PS	Positive	Neutral loss of 185	phosphoryl serine
PG	Negative	Precursor of 153	phosphoryl glycerol–H_2O
PC	Negative	Precursor of 184	phosphoryl choline

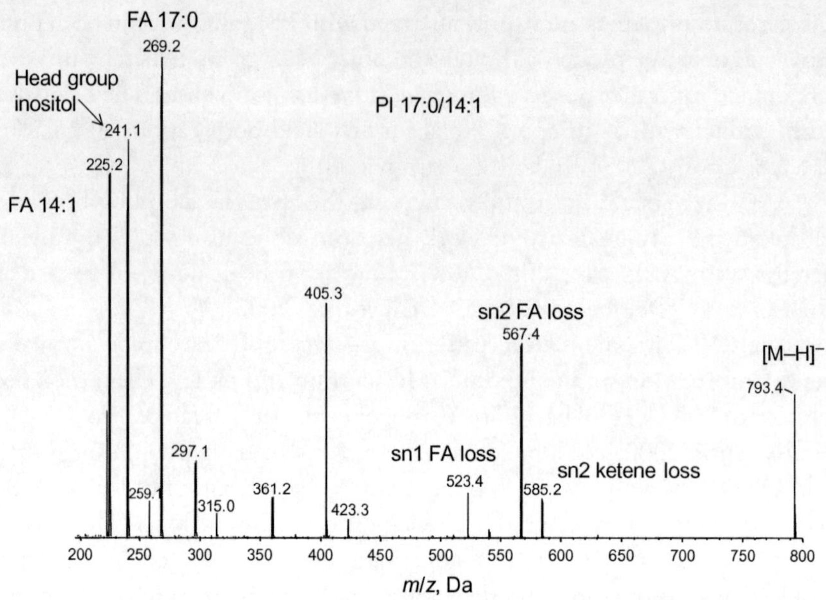

Figure 7.8 Negative ESI-MS/MS spectrum of 17:0/14:1 phosphoinositol (*m/z* 793.5 $[M-H]^-$). The major fragments ions correspond to the head group (phosphoryl inositol–H_2O) and the two fatty acids. (For colour version of this figure, the reader is referred to the online version of this chapter.)

(141 U). Fragmentation of PI anions $[M-H]^-$ appears to give a characteristic ions, independent of the fatty-acyl groups, at *m/z* 241. The $[M-H]^-$ species of PS yield abundant ions that correspond to the loss of the serine (87 U). CID fragmentation of positive protonated PC molecule ion ($[M+H]^+$) yields the phosphocholine ion at *m/z* 184, which is typical of phosphocholine lipids.

The acyl chain moieties present on the glycerol backbone are detected in negative CID fragmentation mode (Fig. 7.8). The most abundant fragment ion products of phospholipid molecular ions are the carboxylate ions derived from the sn-1 and sn-2 positions. The sn-2 carboxylate ion was more abundant than the sn-1 carboxylate ion, this observation or the ratio of intensity of the acyl group allows the assignment of their positions at the glycerol backbone (Hou et al., 2011; Hvattum, Hagelin, & Larsen, 1998; Vernooij, Brouwers, Kettenes-Van den Bosch, & Crommelin, 2002).

Phospholipids can be analyzed without chromatographic separation with the intra-source separation-based shotgun lipidomics (Ejsing et al., 2006; Han & Gross, 2005b; Hou et al., 2011). Chromatographic separation is used: reverse phase as well as normal phase. RP-LC/MS analysis, which is the most separation used, provides an approach that separates molecules by retention time based more on their fatty acyl chains (chain length and unsaturations) than the polarity of the headgroups (Houjou et al., 2005; Ogiso, Suzuki, & Taguchi, 2008; Retra et al., 2008; Taguchi & Ishikawa, 2010). The headgroup of a phospholipid predominantly determines the polar interactions that a phospholipid can participate in, and normal-phase chromatography (NP-LC/MS) hence fractionates a lipid extract into the lipid classes rather than into molecular species (Houjou et al., 2005; Pacetti, Boselli, Hulan, & Frega, 2005).

Quantification of phospholipids by LC–MS is a delicate issue. Addition of odd carbon chain length internal standards for each class can be used for quantification of lipid class and molecular species composition (Ejsing et al., 2006; Hou et al., 2011). Due to different ionization efficiencies, the relative signal intensities of the ions of different phospholipid classes do not directly represent their molar abundances, a correction factor is applied (Brugger et al., 1997; Koivusalo et al., 2001). One very important correction to consider is that of the carbon-13 natural abundance of molecular ion species (Han & Gross, 1994, 1996). Precursor ion scanning and 'neutral loss' scanning of polar head groups or fatty acids in phospholipids are widely used in detection, identification and quantification. MRM is used for quantification (Ekroos et al., 2002; Han et al., 2004; Taguchi et al., 2005).

MGDG and DGDG are analyzed in ESI-MS in positive mode mostly and yield adducts ions. The CID fragmentation yields two abundant specific fragment ions by the neutral loss of free fatty acid from the sn-1 or sn-2 position on the glycerol backbone, DGDG provides another fragment ion from the loss of fatty acid and one galactosyl moiety (Guella, Frassanito, & Mancini, 2003; Napolitano, Carbone, Saggese, Takagaki, & Pizza, 2007; Xu, Chen, Yan, Chen, & Zhou, 2010).

In general, galactoglycerolipids are analyzed directly or separated by reverse-phase chromatography before MS analysis (RP-LC/MS) and often analyzed with phospholipids. Internal standards are used for quantification (Cenzano, Cantoro, Hernandez-Sotomayor, Abdala, & Racagni, 2012; Devaiah et al., 2006; Finnie, Jeannotte, Morris, Giroux, & Faubion, 2010; Xu et al., 2010).

6.4. Sterols

Sterols analyses were mainly done with derivatization and GC analysis. However, sterols can be analyzed with soft ionization methods. Few studies concern the phytosterol, most of them are focused on mammalian sterols.

Phytosterol can be analyzed with ESI, APCI and APPI ionization source. Phytosterols formed $[M+H]^+$ and $[M+H-H_2O]^+$ ions in different ratio depending of the sterol during the APCI and APPI process (Lembcke et al., 2005; Lerma-Garcia, Ramis-Ramos, Herrero-Martinez, & Simo-Alfonso, 2008; Lerma-Garcia, Simo-Alfonso, Mendez, Lliberia, & Herrero-Martinez, 2010).

Free sterols were poorly ionized by ESI-MS, and were usually derivatized to improve ionization, with *N*-chlorobetainyl chloride (Wewer, Dombrink, vom Dorp, & Dormann, 2011), mono-(dimethylaminoethyl) succinyl (MDMAES) ester (Johnson, ten Brink, & Jakobs, 2001) or acetyl chloride (Liebisch et al., 2006). Derivatized sterols can then be analyzed in positive ESI mode by generation of protonated molecular ions or ammonium adducts. Fragmentation depends on the chemical adducts used for the derivatization, but provides a characteristic loss. Others sterols (steryl esters, glycosylated sterols) can form adducts with ammonium, lithium or sodium to be analyzed in ESI-MS (Bowden, Albert, Barnaby, & Ford, 2011; Bowden, Shao, et al., 2011; Schrick et al., 2012; Wewer et al., 2011). Fragmentation of adducts of steryl esters led to the neutral loss of the respective acyl groups (Kalo et al., 2006; Liebisch et al., 2006; Wewer et al., 2011). Glycosylated sterols differ from the type of sterol, the sugar (number, acylation) and the configuration of the linkage between the sterol and the sugar(s). Steryl glucoside was revealed with a scan for neutral loss of the hexose moiety and for the acyl steryl glucoside of the neutral loss of the hexose moiety acylated with a particular fatty acid (Schrick et al., 2012).

Usually chromatographic separation before MS is used to separate different sterol and is made with reverse-phase column (RP-LC/MS), the solvent depending on the ion source used (Lembcke et al., 2005; Lerma-Garcia et al., 2010; Wewer et al., 2011). Some studies used normal-phase column (NP-LC/MS) and separated sterol from others lipids, as neutral lipids, for multi-lipids analyses (Kalo et al., 2006; Rocha, Kalo, Ollilainen, & Malcata, 2010).

For the quantification, internal standards are used to correct the ionization difference (Brugger et al., 1997). They can be saturated sterol, which are absent from plants and can be used easily as internal standards (Schrick et al., 2012; Wewer et al., 2011), also synthetic molecule (Lembcke et al., 2005) or deuterated sterol (Liebisch et al., 2006) All sterols provides characteristic neutral loss during CID fragmentation, then neutral loss scanning mode and MRM scanning mode can be used to make accurate quantification (Cañabate-Díaz et al., 2007; Lembcke et al., 2005; Liebisch et al., 2006; Wewer et al., 2011).

6.5. Sphingolipids

Plant sphingolipids are mainly represented by four lipid classes including free LCBs, ceramides (Cer), glycosylceramides and glycosyl inositol phosphorylceramides (GIPCs). A hypothetical pathway of sphingolipid biosynthesis in plants was suggested by Warnecke and colleagues (Zäuner, Ternes, & Warnecke, 2010). Plant sphingolipids are involved in several biological processes such as cell signalling, membrane stability, permeability and apoptosis (Pata et al., 2010; Sperling & Heinz, 2003). Unlike mammalian LCBs, which are predominated by the sphingosine (d18:1^4) moiety, plant LCBs are more various and possess preferentially one unsaturation at C-8 such as sphingenine (d18:1^8), two unsaturations such as (4E,8E/Z)-sphinga-4,8-dienine (d18:$2^{4,8}$) or/and hydroxylations 4-hydroxysphingenine (t18:1^8), phytosphinganine (t18:0) (Sperling & Heinz, 2003). Using ESI-MS methods, LCBs can be analyzed in positive ion mode due to the presence of the amine group after separative techniques such as HTPLC and/or HPLC (Chigorno, Negroni, Nicolini, & Sonnino, 1997; Colsch et al., 2004). Ceramide backbones are formed by *N*-acetylation of fatty acids (FA) to the LCBs. FA are mainly hydroxylated in plants and their chain lengths vary from C_{14} to C_{26} (Fig. 7.9). By the addition of one or more hexose and an inositol phosphate group (IPC) on the ceramide polar group (OH), glycosylceramides and GIPCs can be formed respectively. They are structurally complex, having numerous possible combinations of fatty acids, desaturations, hydroxylations and head groups and have a broad range of cellular concentrations (Sperling & Heinz, 2003). They are mostly analyzed by ESI-MS, some used APCI-MS (Farwanah, Kolter, & Sandhoff, 2011; Markham et al., 2006; Pettus et al., 2003). In the presence of alkaline metal ions (LiOH), ceramides and glycosylceramides yield adducts ions under ESI source in positive and negative mode (Hsu & Turk, 2001a, 2002). CID fragmentation provides specific ions related to LCB and fatty acids moieties in both positive and negative ion mode, respectively. GIPCs can be analyzed by ESI-MS in positive mode, and yield adduct ions (with Na^+, NH_4^+ or Li^+) or in negative mode, given doubly charged precursor ions $[M-2H]^{2-}$. The fragmentation of the charged GIPC yield ions of the polar head, the ceramide, the inositol-phosphate moiety and the carbohydrate, which allows a complete characterization of all GIPC (Buré et al., 2011; Levery, Toledo, Straus, & Takahashi, 2001; Markham et al., 2006; Toledo et al., 2007).

For quantification of LCBs, ceramides, glycosylceramides, internal standards can be used, as synthetic standards with a C_{17} LCB or C_{12} fatty acid

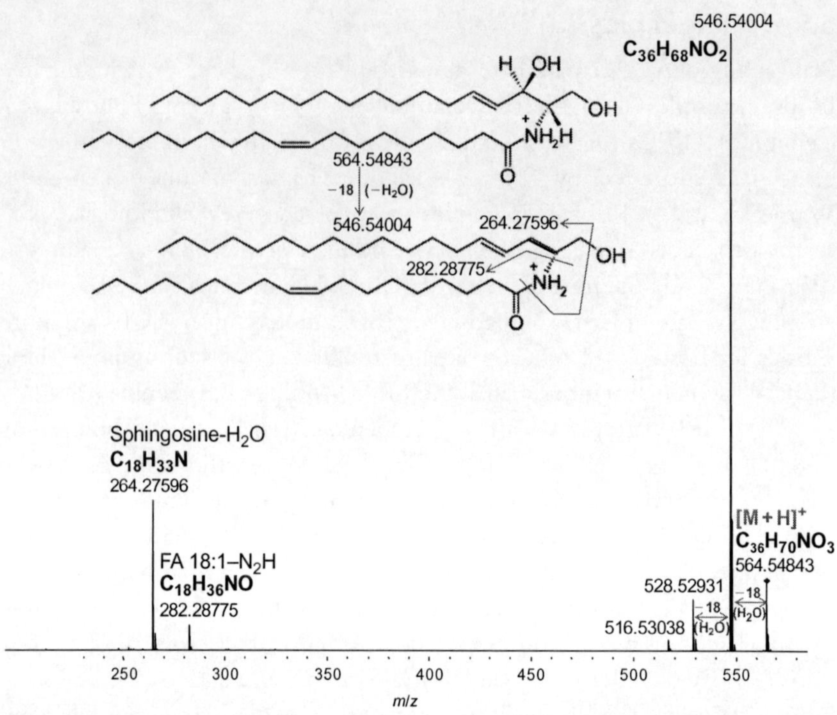

Figure 7.9 Positive ESI-MS/MS spectrum of Ceramide (d18:1/C18:0) (*m/z* 564.5, $[M+H]^+$). Cer fragmentation in positive ion mode reveals mainly the sphingoïd base (Sphingosine *m/z* 264.3) and the fatty acid group including the ceramide amine function (*m/z* 282.3). (For colour version of this figure, the reader is referred to the online version of this chapter.)

component, which can be found commercially and does not exist naturally in biological samples. Actually, no commercially available plant GIPC standard exists, so Markham et al. used animal sphingolipid (ganglioside) for their quantitation (Markham & Jaworski, 2007; Shaner et al., 2009). MALDI-MS was performed to analyze GIPC species using DHA matrix in negative mode (Buré et al., 2011).

Separation of sphingolipids by liquid chromatography has been performed either by reversed (RP-LC) or normal phase (NP-LC) chromatography. In reversed-phase chromatography, separations are based on the length, saturation and hydroxylation of LCB and/or fatty acid. In normal-phase chromatography, polarity of the head group drives the separation (Markham & Jaworski, 2007; Markham et al., 2006; Pettus et al., 2003; Shaner et al., 2009).

7. CONCLUSION

The recent advances in MS instrumentation concerning the sensitivity of the advanced tandem quadrupole and HRMSs have allowed remarkable progress in lipid analyses leading possible separation, identification and quantification of a large number of molecular species of the different lipid classes. The lipid profile or 'lipidome' of a tissue, a cell or an orgonite has gained in complexity, since the number of the different lipid species has been largely increased generating numerous data. Consequently, the interpretation of all these data has required 'omic technologies' as already used in genomics and proteomics assuming the emergence of a new field the 'lipidomic'. Moreover, new metabolites have been detected, which have to be identified and which raise the question of their biological function. Among them, the most promising are in particular the PI derivatives, the lysolipids and the cerebrosides, which appear to be important actors in cell signalling and host pathogen relationships. Plant lipidomics in addition to be a powerful method to investigate the lipidome has also emerged new research fields in plant pathology, physiology and cell biology.

ACKNOWLEDGEMENTS

The authors are thankful to Dr. P. Moreau for his critical reading and his comments. Dr. B. Colsch thanks the CEA Eurotalents programme funding from European commission.

REFERENCES

Berdyshev, E. V. (2011). Mass spectrometry of fatty aldehydes. *Biochimica et Biophysica Acta, 1811*, 680–693.

Bielawski, J., Szulc, Z. M., Hannun, Y. A., & Bielawska, A. (2006). Simultaneous quantitative analysis of bioactive sphingolipids by high-performance liquid chromatography-tandem mass spectrometry. *Methods, 39*, 82–91.

Bird, S. S., Marur, V. R., Sniatynski, M. J., Greenberg, H. K., & Kristal, B. S. (2011a). Lipidomics profiling by high-resolution LC-MS and high-energy collisionnal dissociation fragmentation: Focus on charaterization of mitochondrial cardiolipins and monolysocardiolipins. *Analytical Chemistry, 83*, 940–949.

Bird, S. S., Marur, V. R., Sniatynski, M. J., Greenberg, H. K., & Kristal, B. S. (2011b). Serum lipidomics profiling using LC-MS and high-energy collisionnal dissociation fragmentation: Focus on triglyceride detection and characterization. *Analytical Chemistry, 83*, 6648–6657.

Bligh, E. G., & Dyer, W. J. (1959). A rapid method of total lipid extraction and purification. *Canadian Journal of Biochemistry and Physiology, 37*, 911–917.

Bonaventure, G., Beisson, F., Ohlrogge, J., & Pollard, M. (2004). Analysis of the aliphatic monomer composition of polyesters associated with Arabidopsis epidermis: Occurrence of octadeca-cis-6, cis-9-diene-1,18-dioate as the major component. *The Plant Journal, 40*, 920–930.

Bowden, J. A., Albert, C. J., Barnaby, O. S., & Ford, D. A. (2011). Analysis of cholesteryl esters and diacylglycerols using lithiated adducts and electrospray ionization-tandem mass spectrometry. *Analytical Biochemistry, 417*, 202–210.

Bowden, J. A., Shao, F., Albert, C. J., Lally, J. W., Brown, R. J., Procknow, J. D., et al. (2011). Electrospray ionization tandem mass spectrometry of sodiated adducts of cholesteryl esters. *Lipids, 46*, 1169–1179.

Brugger, B., Erben, G., Sandhoff, R., Wieland, F. T., & Lehmann, W. D. (1997). Quantitative analysis of biological membrane lipids at the low picomole level by nano-electrospray ionization tandem mass spectrometry. *Proceedings of the National Academy of Sciences of the United States of America, 94*, 2339–2344.

Buré, C., Cacas, J.-L., Wang, F., Gaudin, K., Domergue, F., Mongrand, S., et al. (2011). Fast screening of highly glycosylated plant sphingolipids by tandem mass spectrometry. *Rapid Communications in Mass Spectrometry, 25*, 3131–3145.

Byrdwell, W. C. (1998). Dual parallel mass spectrometers for analysis of sphingolipid, glycerophospholipid and plasmalogen molecular species. *Rapid Communications in Mass Spectrometry, 12*, 256–272.

Byrdwell, W. C. (2001). Atmospheric pressure chemical ionization mass spectrometry for analysis of lipids. *Lipids, 36*, 327–346.

Byrdwell, W. C., & Neff, W. E. (2002). Dual parallel electrospray ionization and atmospheric pressure chemical ionization mass spectrometry (MS), MS/MS and MS/MS/MS for the analysis of triacylglycerols and triacylglycerol oxidation products. *Rapid Communications in Mass Spectrometry, 16*, 300–319.

Byrdwell, W. C., Neff, W. E., & List, G. R. (2001). Triacylglycerol analysis of potential margarine base stocks by high-performance liquid chromatography with atmospheric pressure chemical ionization mass spectrometry and flame ionization detection. *Journal of Agricultural and Food Chemistry, 49*, 446–457.

Caffrey, M., & Hogan, J. (1992). LIPIDAT: A database of lipid phase transition temperatures and enthalpy changes. DMPC data subset analysis. *Chemistry and Physics of Lipids, 61*, 1–109.

Cai, S. S., Short, L. C., Syage, J. A., Potvin, M., & Curtis, J. M. (2007). Liquid chromatography-atmospheric pressure photoionization-mass spectrometry analysis of triacylglycerol lipids—Effects of mobile phases on sensitivity. *Journal of Chromatography. A, 1173*, 88–97.

Cai, S. S., & Syage, J. A. (2006). Comparison of atmospheric pressure photoionization, atmospheric pressure chemical ionization, and electrospray ionization mass spectrometry for analysis of lipids. *Analytical Chemistry, 78*, 1191–1199.

Cañabate-Díaz, B., Segura Carretero, A., Fernández-Gutiérrez, A., Belmonte Vega, A., Garrido Frenich, A., Martínez Vidal, J. L., et al. (2007). Separation and determination of sterols in olive oil by HPLC-MS. *Food Chemistry, 102*, 593–598.

Cassagne, C., & Lessire, R. (1974). Studies on alkane biosynthesis in epidermis of Allium porrum L. leaves. Direct synthesis of tricosane from lignoceric acid. *Archives of Biochemistry and Biophysics, 165*, 274–280.

Cenzano, A. M., Cantoro, R., Hernandez-Sotomayor, S. M., Abdala, G. I., & Racagni, G. E. (2012). Lipid profiling by electrospray ionization tandem mass spectrometry and the identification of lipid phosphorylation by kinases in potato stolons. *Journal of Agricultural and Food Chemistry, 60*, 418–426.

Chigorno, V., Negroni, E., Nicolini, M., & Sonnino, S. (1997). Activity of 3- ketosphinganine synthase during differentiation and aging of neuronal cells in culture. *Journal of Lipid Research, 38*, 1163–1169.

Colsch, B., Afonso, C., Popa, I., Portoukalian, J., Fournier, F., Tabet, J. C., et al. (2004). Characterization of the ceramide moieties of sphingoglycolipids from mouse brain by

ESI-MS/MS: Identification of ceramides containing sphingadienine. *Journal of Lipid Research*, *45*, 281–286.

Colsch, B., Jackson, S. N., & Woods, A. S. (2011). Brain gangliosides' molecular microscopy, illustrating their distribution in hippocampal cell layers. *ACS Chemical Neuroscience*, *2*, 213–222.

Colsch, B., & Woods, A. S. (2010). Localization and imaging of sialylated glycosphingolipids in brain tissue sections by MALDI mass spectrometry. *Glycobiology*, *20*, 661–667.

Cornett, D. S., Reyzer, M. L., Chaurand, P., & Caprioli, R. M. (2007). MALDI imaging mass spectrometry: Molecular snapshots of biochemical systems. *Nature Methods*, *4*, 828–833.

Cotter, D., Maer, A., Guda, C., Saunders, B., & Subramaniam, S. (2006). LMPD: LIPID MAPS proteome database. *Nucleic Acids Research*, *34*, 507–510.

Cozzolino, R., & De Giulio, B. (2011). Application of ESI and MALDI-TOF MS for triacylglycerols analysis in edible oils. *European Journal of Lipid Science Technology*, *113*, 160–167.

Cvacka, J., Krafkova, E., Jiros, P., & Valterova, I. (2006). Computer-assisted interpretation of atmospheric pressure chemical ionization mass spectra of triacylglycerols. *Rapid Communications in Mass Spectrometry*, *20*, 3586–3594.

Delobel, A., Touboul, D., & Laprevote, O. (2005). Structural characterization of phosphatidylcholines by atmospheric pressure photoionization mass spectrometry. *European Journal of Mass Spectrometry*, *11*, 409–417.

DeLong, C. J., Baker, P. R., Samuel, M., Cui, Z., & Thomas, M. J. (2001). Molecular species composition of rat liver phospholipids by ESI-MS/MS: The effect of chromatography. *Journal of Lipid Research*, *42*, 1959–1968.

Delvolve, A. M., Colsch, B., & Woods, A. S. (2011). Highlighting anatomical sub-structures in rat brain tissue using lipid imaging. *Analytical Methods*, *3*, 1729–1736.

Devaiah, S. P., Roth, M. R., Baughman, E., Li, M., Tamura, P., Jeannotte, R., et al. (2006). Quantitative profiling of polar glycerolipid species from organs of wild-type Arabidopsis and a PHOSPHOLIPASE Dα1 knockout mutant. *Phytochemistry*, *67*, 1907–1924.

Duffin, K. L., Henion, J. D., & Shieh, J. J. (1991). Electrospray and tandem mass spectrometric characterization of acylglycerol mixtures that are dissolved in nonpolar solvents. *Analytical Chemistry*, *63*, 1781–1788.

Dugo, P., Kumm, T., Lo Presti, M., Chiofalo, B., Salimei, E., Fazio, A., et al. (2005). Determination of triacylglycerols in donkey milk by using high performance liquid chromatography coupled with atmospheric pressure chemical ionization mass spectrometry. *Journal of Separation Science*, *28*, 1023–1030.

Ejsing, C. S., Duchoslav, E., Sampaio, J., Simons, K., Bonner, R., Thiele, C., et al. (2006). Automated identification and quantification of glycerophospholipid molecular species by multiple precursor ion scanning. *Analytical Chemistry*, *78*, 6202–6214.

Ejsing, C. S., Sampaio, J. L., Surendranath, V., Duchoslav, E., Ekroos, K., Klemm, R. W., et al. (2009). Global analysis of the yeast lipidome by quantitative shotgun mass spectrometry. *Proceedings of the National Academy of Sciences of the United States of America*, *106*, 2136–2141.

Ekroos, K., Chernushevich, I. V., Simons, K., & Shevchenko, A. (2002). Quantitative profiling of phospholipids by multiple precursor ion scanning on a hybrid quadrupole time-of-flight mass spectrometer. *Analytical Chemistry*, *74*, 941–949.

Fahy, E., Subramaniam, S., Brown, H. A., Glass, C. K., Merrill, A. H., Murphy, R. C., et al. (2005). A comprehensive classification system for lipids. *Journal of Lipid Research*, *46*, 839–862.

Fahy, E., Subramaniam, S., Murphy, R. C., Nishijima, M., Raetz, C. R. H., Shimizu, T., et al. (2009). Update of the LIPID MAPS comprehensive classification system for lipids. *Journal of Lipid Research*, *50*, S9–S14.

Fahy, E., Sud, M., Cotter, D., & Subramaniam, S. (2007). LIPID MAPS online tools for lipid research. *Nucleic Acids Research, 35*, 606–612.

Farwanah, H., Kolter, T., & Sandhoff, K. (2011). Mass spectrometric analysis of neutral sphingolipids: Methods, applications, and limitations. *Biochimica et Biophysica Acta, 1811*, 854–860.

Fenn, J. B., Mann, M., Meng, C. K., Wong, S. F., & Whitehouse, C. M. (1989). Electrospray ionization for mass spectrometry of large biomolecules. *Science, 246*, 64–71.

Fenn, J. B., Mann, M., Meng, C. K., Wong, S. F., & Whitehouse, C. M. (1990). Electrospray ionization–principles and practice. *Mass Spectrometry Reviews, 9*, 37–70.

Fernández, J. A., Ochoa, B., Fresnedo, O., Giralt, M. T., & Rodríguez-Puertas, R. (2011). Matrix-assisted laser desorption ionization imaging mass spectrometry in lipidomics. *Analytical and Bioanalytical Chemistry, 401*, 29–51.

Finnie, S. M., Jeannotte, R., Morris, C. F., Giroux, M. J., & Faubion, J. M. (2010). Variation in polar lipids located on the surface of wheat starch. *Journal of Cereal Science, 51*, 73–80.

Fitzgerald, M., & Murphy, R. C. (2007). Electrospray mass spectrometry of human hair wax esters. *Journal of Lipid Research, 48*, 1231–1246.

Folch, J., Lees, M., & Sloane Stanley, G. H. (1957). A simple method for the isolation and purification of total lipids from animal tissues. *The Journal of Biological Chemistry, 226*, 11222–11228.

Fuchs, B., Schiller, J., Süß, R., Zscharnack, M., Bader, A., Müller, P., et al. (2008). Analysis of stem cell lipids by offline HPTLC–MALDI-TOF MS. *Analytical and Bioanalytical Chemistry, 392*, 849–860.

Fuchs, B., Süß, R., & Schiller, J. (2010). An update of MALDI-TOF mass spectrometry in lipid research. *Progress in Lipid Research, 49*, 450–475.

Franke, R., & Schreiber, L. (2007). Suberin—A biopolyester forming apoplastic plant interfaces. *Current Opinion in Plant Biology, 10*, 252–259.

German, J. B., Gillies, L., Smilowitz, J. T., Zivkovic, A. M., & Watkins, S. M. (2007). Lipidomics and lipid profiling in metabolomics. *Current Opinion in Lipidology, 18*, 66–71.

Giavalisco, P., Li, Y., Matthes, A., Eckhardt, A., Hubberten, H. M., Hesse, H., et al. (2011). Elemental formula annotation of polar and lipophilic metabolites using (13) C, (15) N and (34) S isotope labelling, in combination with high-resolution mass spectrometry. *The Plant Journal, 68*, 364–376.

Gomez-Ariza, J. L., Arias-Borrego, A., Garcia-Barrera, T., & Beltran, R. (2006). Comparative study of electrospray and photospray ionization sources coupled to quadrupole time-of-flight mass spectrometer for olive oil authentication. *Talanta, 70*, 859–869.

Goto-Inoue, N., Hayasaka, T., Sugiura, Y., Taki, T., Li, Y. T., Matsumoto, M., et al. (2008). High-sensitivity analysis of glycosphingolipids by matrix-assisted laser desorption/ionization quadrupole Ion trap time-of-flight imaging mass spectrometry on transfer membranes. *Journal of Chromatography B, 870*, 74–83.

Goto-Inoue, N., Hayasaka, T., Zaima, N., & Setou, M. (2011). Imaging mass spectrometry for lipidomics. *Biochimica et Biophysica Acta, 1811*, 961–969.

Graessler, J., Schwudke, D., Schwarz, P. E., Herzog, R., Shevchenko, A., & Bornstein, S. R. (2009). Top-down lipidomics reveals ether lipid deficiency in blood plasma of hypertensive patients. *PloS One, 4*, e6261.

Grille, S., Zaslawski, A., Thiele, S., Plat, J., & Warnecke, D. (2010). The functions of steryl glycosides come to those who wait: Recent advances in plants, fungi, bacteria and animals. *Progress in Lipid Research, 49*, 262–288.

Guella, G., Frassanito, R., & Mancini, I. (2003). A new solution for an old problem: The regiochemical distribution of the acyl chains in galactolipids can be established by electrospray ionization tandem mass spectrometry. *Rapid Communications in Mass Spectrometry, 17*, 1982–1994.

Haimi, P., Chaithanya, K., Kainu, V., Hermansson, M., & Somerharju, P. (2009). Instrument-independent software tools for the analysis of MS-MS and LC-MS lipidomics data. *Methods in Molecular Biology*, *580*, 285–294.

Haimi, P., Uphoff, A., Hermansson, M., & Somerharju, P. (2006). Software tools for analysis of mass spectrometric lipidome data. *Analytical Chemistry*, *78*, 8324–8331.

Han, X. (2010). Multi-dimensional mass spectrometry-based shotgun lipidomics and the altered lipids at the mild cognitive impairment stage of Alzheimer's disease. *Biochimica et Biophysica Acta*, *1801*, 774–783.

Han, X., & Gross, R. W. (1994). Electrospray ionization mass spectroscopic analysis of human erythrocyte plasma membrane phospholipids. *Proceedings of the National Academy of Sciences of the United States of America*, *91*, 10635–10639.

Han, X., & Gross, R. W. (1996). Structural determination of lysophospholipid regioisomers by electrospray ionization tandem mass spectrometry. *Journal of the American Chemical Society*, *118*, 451–457.

Han, X., & Gross, R. W. (2001). Quantitative analysis and molecular species fingerprinting of triacylglyceride molecular species directly from lipid extracts of biological samples by electrospray ionization tandem mass spectrometry. *Analytical Biochemistry*, *295*, 88–100.

Han, X., & Gross, R. W. (2003). Global analyses of cellular lipidomes directly from crude extracts of biological samples by ESI mass spectrometry: A bridge to lipidomics. *Journal of Lipid Research*, *44*, 1071–1079.

Han, X., & Gross, R. W. (2005a). Shotgun lipidomics: Electrospray ionization mass spectrometric analysis and quantitation of cellular lipidomes directly from crude extracts of biological samples. *Mass Spectrometry Reviews*, *24*, 367–412.

Han, X., & Gross, R. W. (2005b). Shotgun lipidomics: Multidimensional MS analysis of cellular lipidomes. *Expert Review of Proteomics*, *2*, 253–264.

Han, X., Yang, K., Cheng, H., Fikes, K. N., & Gross, R. W. (2005). Shotgun lipidomics of phosphoethanolamine-containing lipids in biological samples after one-step in situ derivatization. *Journal of Lipid Research*, *46*, 1548–1560.

Han, X., Yang, J., Cheng, H., Ye, H., & Gross, R. W. (2004). Toward fingerprinting cellular lipidomes directly from biological samples by two-dimensional electrospray ionization mass spectrometry. *Analytical Biochemistry*, *330*, 317–331.

Han, X., Yang, K., Yang, J., Cheng, H., & Gross, R. W. (2006). Shotgun lipidomics of cardiolipin molecular species in lipid extracts of biological samples. *Journal of Lipid Research*, *47*, 864–879.

Hanold, K. A., Fischer, S. M., Cormia, P. H., Miller, C. E., & Syage, J. A. (2004). Atmospheric pressure photoionization. 1. General properties for LC/MS. *Analytical Chemistry*, *76*, 2842–2851.

Haroldsen, P. E., Clay, K. L., & Murphy, R. C. (1987). Quantitation of lyso-platelet activating factor molecular species from human neutrophils by mass spectrometry. *Journal of Lipid Research*, *28*, 42–49.

Harvey, D. J. (1999). Matrix-assisted laser desorption/ionization mass spectrometry of carbohydrates. *Mass Spectrometry Reviews*, *18*, 349–450.

Herrera, L. C., Potvin, M. A., & Melanson, J. E. (2010). Quantitative analysis of positional isomers of triacylglycerols via electrospray ionization tandem mass spectrometry of sodiated adducts. *Rapid Communications in Mass Spectrometry*, *24*, 2745–2752.

Herzog, R., Schwudke, D., Schuhmann, K., Sampaio, J. L., Bornstein, S. R., Schroeder, M., et al. (2011). A novel informatics concept for high-throughput shotgun lipidomics based on the molecular fragmentation query language. *Genome Biology*, *12*, R8.

Hillenkamp, F., & Peter-Katalinić, J. (2007). MALDI MS: A practical guide to instrumentation, methods and applications methods and applications. *Applied Organometallic Chemistry*, *21*, 1067. http://dx.doi.org/10.1002/aoc.1309.

Hopfgartner, G., Varesio, E., Tschappat, V., Grivet, C., Bourgogne, E., & Leuthold, L. A. (2004). Triple quadrupole linear ion trap mass spectrometer for the analysis of small molecules and macromolecules. *Journal of Mass Spectrometry, 39*, 845–855.

Horai, H., Arita, M., Kanaya, S., Nihei, Y., Ikeda, T., Suwa, K., et al. (2010). MassBank: A public repository for sharing mass spectral data for life sciences. *Journal of Mass Spectrometry, 45*, 703–714.

Horn, P. J., Korte, A. R., Neogi, P. B., Love, E., Fuchs, J., Strupat, K., et al. (2012). Spatial mapping of lipids at cellular resolution in embryos of cotton. *The Plant Cell, 24*, 622–636.

Horning, E. C., Carroll, D. I., Dzidic, I., Haegele, K. D., Horning, M. G., & Stillwell, R. N. (1974). Atmospheric pressure ionization (API) mass spectrometry. Solvent-mediated ionization of samples introduced in solution and in a liquid chromatograph effluent stream. *Journal of Chromatographic Science, 12*, 725–729.

Hou, W., Zhou, H., Bou Khalil, M., Seebun, D., Bennett, S. A., & Figeys, D. (2011). Lysoform fragment ions facilitate the determination of stereospecificity of diacyl glycerophospholipids. *Rapid Communications in Mass Spectrometry, 25*, 205–217.

Houjou, T., Yamatani, K., Imagawa, M., Shimizu, T., & Taguchi, R. (2005). A shotgun tandem mass spectrometric analysis of phospholipids with normal-phase and/or reverse-phase liquid chromatography/electrospray ionization mass spectrometry. *Rapid Communications in Mass Spectrometry, 19*, 654–666.

Hsu, F. F., & Turk, J. (1999). Distinction among isomeric unsaturated fatty acids as lithiated adducts by electrospray ionization mass spectrometry using low energy collisionally activated dissociation on a triple stage quadrupole instrument. *Journal of the American Society for Mass Spectrometry, 10*, 600–612.

Hsu, F. F., & Turk, J. (2000a). Characterization of phosphatidylinositol, phosphatidylinositol-4-phosphate, and phosphatidylinositol-4,5-bisphosphate by electrospray ionization tandem mass spectrometry: A mechanistic study. *Journal of the American Society for Mass Spectrometry, 11*, 986–999.

Hsu, F. F., & Turk, J. (2000b). Charge-driven fragmentation processes in diacyl glycerophosphatidic acids upon low-energy collisional activation. A mechanistic proposal. *Journal of the American Society for Mass Spectrometry, 11*, 797–803.

Hsu, F. F., & Turk, J. (2000c). Charge-remote and charge-driven fragmentation processes in diacyl glycerophosphoethanolamine upon low-energy collisional activation: A mechanistic proposal. *Journal of the American Society for Mass Spectrometry, 11*, 892–899.

Hsu, F.-F., & Turk, J. (2001a). Structural determination of glycosphingolipids as lithiated adducts by electrospray ionization mass spectrometry using low-energy collisional-activated dissociation on a triple stage quadrupole instrument. *Journal of the American Society for Mass Spectrometry, 12*, 61–79.

Hsu, F.-F., & Turk, J. (2001b). Studies on phosphatidylglycerol with triple quadrupole tandem mass spectrometry with electrospray ionization: Fragmentation processes and structural characterization. *Journal of the American Society for Mass Spectrometry, 12*, 1036–1043.

Hsu, F.-F., & Turk, J. (2002). Characterization of ceramides by low energy collisional-activated dissociation tandem mass spectrometry with negative-ion electrospray ionization. *Journal of the American Society for Mass Spectrometry, 13*, 558–570.

Hsu, F. F., & Turk, J. (2008a). Elucidation of the double-bond position of long-chain unsaturated fatty acids by multiple-stage linear ion-trap mass spectrometry with electrospray ionization. *Journal of the American Society for Mass Spectrometry, 19*, 1673–1680.

Hsu, F. F., & Turk, J. (2008b). Structural characterization of unsaturated glycerophospholipids by multiple-stage linear ion-trap mass spectrometry with electrospray ionization. *Journal of the American Society for Mass Spectrometry, 19*, 1681–1691.

Hsu, F. F., & Turk, J. (2010). Electrospray ionization multiple-stage linear ion-trap mass spectrometry for structural elucidation of triacylglycerols: Assignment of fatty acyl groups on the glycerol backbone and location of double bonds. *Journal of the American Society for Mass Spectrometry*, *21*, 657–669.

Hu, C., Van Dommelen, J., Van Der Heijden, R., Spijksma, G., Reijmers, T. H., Wang, M., et al. (2008). RPLC-Ion-FTMS method for lipid profiling of plasma: Method validation and application to p53 mutant mouse model. *Journal of Proteome Research*, 7, 4982–4991.

Hubner, G., Crone, C., & Lindner, B. (2009). lipID—A software tool for automated assignment of lipids in mass spectra. *Journal of the American Society for Mass Spectrometry*, *44*, 1676–1683.

Hvattum, E. (2001). Analysis of triacylglycerols with non-aqueous reversed-phase liquid chromatography and positive ion electrospray tandem mass spectrometry. *Rapid Communications in Mass Spectrometry*, *15*, 187–190.

Hvattum, E., Hagelin, G., & Larsen, A. (1998). Study of mechanisms involved in the collision-induced dissociation of carboxylate anions from glycerophospholipids using negative ion electrospray tandem quadrupole mass spectrometry. *Rapid Communications in Mass Spectrometry*, *12*, 1405–1409.

Jackson, S., Wang, H.-Y., & Woods, A. (2005). In situ structural characterization of phosphatidylcholines in brain tissue using MALDI-MS/MS. *Journal of the American Society for Mass Spectrometry*, *16*, 2052–2056.

Jakab, A., Jablonkai, I., & Forgacs, E. (2003). Quantification of the ratio of positional isomer dilinoleoyl-oleoyl glycerols in vegetable oils. *Rapid Communications in Mass Spectrometry*, *17*, 2295–2302.

Johnson, D. W., ten Brink, H. J., & Jakobs, C. (2001). A rapid screening procedure for cholesterol and dehydrocholesterol by electrospray ionization tandem mass spectrometry. *Journal of Lipid Research*, *42*, 1699–1705.

Juguelin, H., & Cassagne, C. (1984). Assay of long-chain acyl-CoAs in a complex reaction mixture. *Analytical Biochemistry*, *142*, 329–335.

Kalo, P., Kemppinen, A., Ollilainen, V., & Kuksis, A. (2003). Analysis of regioisomers of short-chain triacylglycerols by normal phase liquid chromatography–electrospray tandem mass spectrometry. *International Journal of the American Society for Mass Spectrometry*, *229*, 167–180.

Kalo, P. J., Ollilainen, V., Rocha, J. M., & Malcata, F. X. (2006). Identification of molecular species of simple lipids by normal phase liquid chromatography–positive electrospray tandem mass spectrometry, and application of developed methods in comprehensive analysis of low erucic acid rapeseed oil lipids. *International Journal of the American Society for Mass Spectrometry*, *254*, 106–121.

Katajamaa, M., & Oresic, M. (2007). Data processing for mass spectrometry-based metabolomics. *Journal of Chromatography. A*, *1158*, 318–328.

Kerwin, J. L., Tuininga, A. R., & Ericsson, L. H. (1994). Identification of molecular species of glycerophospholipids and sphingomyelin using electrospray mass spectrometry. *Journal of Lipid Research*, *35*, 1102–1114.

Kerwin, J. L., Wiens, A. M., & Ericsson, L. H. (1996). Identification of fatty acids by electrospray mass spectrometry and tandem mass spectrometry. *Journal of the American Society for Mass Spectrometry*, *31*, 184–192.

Khalil, M. B., Hou, W., Zhou, H., Elisma, F., Swayne, L. A., Blanchard, A. P., et al. (2010). Lipidomics Era: Accomplishments and challenges. *Mass Spectrometry Reviews*, *29*, 877–929.

Kim, H. Y., Wang, T., & Ma, Y. C. (1994). Liquid chromatography/mass spectrometry of phospholipids using electrospray ionization. *Analytical Chemistry*, *66*, 3977–3982.

Koivusalo, M., Haimi, P., Heikinheimo, L., Kostiainen, R., & Somerharju, P. (2001). Quantitative determination of phospholipid compositions by ESI-MS: Effects of acyl chain length, unsaturation, and lipid concentration on instrument response. *Journal of Lipid Research*, *42*, 663–672.

Kolattukudy, P. (1970). Biosynthesis of a lipid polymer, cutin: The structural component of plant cuticle. *Biochemical and Biophysical Research Communications*, *41*, 299–305.

Kollatukudy, P. (1968). Further evidence for an elongation-decarboxylation mechanism in the biosynthesis of paraffins in leaves. *Plant Physiology*, *43*, 375–383.

Koulman, A., Woffendin, G., Narayana, V. K., Welchman, H., Crone, C., & Volmer, D. A. (2009). High-resolution extracted ion chromatography, a new tool for metabolomics and lipidomics using a second-generation orbitrap mass spectrometer. *Rapid Communications in Mass Spectrometry*, *23*, 1411–1418.

Laakso, P. (2002). Mass spectrometry of triacylglycerols. *European Journal of Lipid Science and Technology*, *104*, 43–49.

Larsen, A., Uran, S., Jacobsen, P. B., & Skotland, T. (2001). Collision-induced dissociation of glycero phospholipids using electrospray ion-trap mass spectrometry. *Rapid Communications in Mass Spectrometry*, *15*, 2393–2398.

Leavell, M. D., & Leary, J. A. (2006). Fatty acid analysis tool (FAAT): An FT-ICR MS lipid analysis algorithm. *Analytical Chemistry*, *78*, 5497–5503.

Lembcke, J., Ceglarek, U., Fiedler, G. M., Baumann, S., Leichtle, A., & Thiery, J. (2005). Rapid quantification of free and esterified phytosterols in human serum using APPI-LC-MS/MS. *Journal of Lipid Research*, *46*, 21–26.

Lerma-Garcia, M. J., Ramis-Ramos, G., Herrero-Martinez, J. M., & Simo-Alfonso, E. F. (2008). Classification of vegetable oils according to their botanical origin using sterol profiles established by direct infusion mass spectrometry. *Rapid Communications in Mass Spectrometry*, *22*, 973–978.

Lerma-Garcia, M. J., Simo-Alfonso, E. F., Mendez, A., Lliberia, J. L., & Herrero-Martinez,-J. M. (2010). Fast separation and determination of sterols in vegetable oils by ultra-performance liquid chromatography with atmospheric pressure chemical ionization mass spectrometry detection. *Journal of Agricultural and Food Chemistry*, *58*, 2771–2776.

Leveque, N. L., Heron, S., & Tchapla, A. (2010). Regioisomer characterization of triacylglycerols by non-aqueous reversed-phase liquid chromatography/electrospray ionization mass spectrometry using silver nitrate as a postcolumn reagent. *Journal of Mass Spectrometry*, *45*, 284–296.

Levery, S. B., Toledo, M. S., Straus, A. H., & Takahashi, H. K. (2001). Comparative analysis of glycosylinositol phosphorylceramides from fungi by electrospray tandem mass spectrometry with low-energy collision-induced dissociation of Li(+) adduct ions. *Rapid Communications in Mass Spectrometry*, *15*, 2240–2258.

Li-Beisson, Y., Shorrosh, B., Beisson, F., Andersson, M. X., Arondel, V., Bates, P. D., et al. (2010). *Acyl-lpid metabolism*. In C. R. Somerville & E. M. Meyerowitz (Eds.), *The arabidopsis book*. Rockville, MD, USA: ASPB Press. http://www.aspborg/publications/Arabidopsis/.

Liebisch, G., Binder, M., Schifferer, R., Langmann, T., Schulz, B., & Schmitz, G. (2006). High throughput quantification of cholesterol and cholesteryl ester by electrospray ionization tandem mass spectrometry (ESI-MS/MS). *Biochimica et Biophysica Acta*, *1761*, 121–128.

Liebisch, G., Drobnik, W., Lieser, B., & Schmitz, G. (2002). High-throughput quantification of lysophosphatidylcholine by electrospray ionization tandem mass spectrometry. *Clinical Chemistry*, *48*, 2217–2224.

Mancha, M., Stokes, G., & Stumpf, P. K. (1975). Fat metabolism in higher plants. The determination of acyl-acyl carrier protein and acyl coenzyme A in a complex lipid mixture. *Analytical Biochemistry*, *68*, 600–660.

Markham, J. E., & Jaworski, J. G. (2007). Rapid measurement of sphingolipids from Arabidopsis thaliana by reversed-phase high-performance liquid chromatography coupled to electrospray ionization tandem mass spectrometry. *Rapid Communications in Mass Spectrometry*, *21*, 1304–1314.

Markham, J. E., Li, J., Cahoon, E. B., & Jaworski, J. G. (2006). Separation and identification of major plant sphingolipid classes from leaves. *The Journal of Biological Chemistry, 281*, 22684–22694.

McAnoy, A. M., Wu, C. C., & Murphy, R. C. (2005). Direct qualitative analysis of triacylglycerols by electrospray mass spectrometry using a linear ion trap. *Journal of the American Society for Mass Spectrometry, 16*, 1498–1509.

Medvedovici, A., Lazou, K., D'Oosterlinck, A., Zhao, Y., & Sandra, P. (2002). Analysis of jojoba oil by LC-coordination ion spray-MS (LC-CIS-MS). *Journal of Separation Science, 25*, 173–178.

Miwa, T. (1971). Jojoba oil wax esters and derived fatty acids and alcohols: Gas chromatographic analyses. *Journal of the American Oil Chemists' Society, 48*, 259–264.

Moe, M. K., & Jensen, E. (2004). Structure elucidation of unsaturated fatty acids after vicinal hydroxylation of the double bonds by negative electrospray ionisation low-energy tandem mass spectrometry. *European Journal of Mass Spectrometry, 10*, 47–55.

Moe, M. K., Strom, M. B., Jensen, E., & Claeys, M. (2004). Negative electrospray ionization low-energy tandem mass spectrometry of hydroxylated fatty acids: A mechanistic study. *Rapid Communications in Mass Spectrometry, 18*, 1731–1740.

Molina, I., Bonaventure, G., Ohlrogge, J., & Pollard, M. (2006). The lipid polyester composition of Arabidopsis thaliana and Brassica napus seeds. *Phytochemistry, 67*, 2597–2610.

Morant, M., Jørgensen, K., Schaller, H., Pinot, F., Møller, B. L., Werck-Reichhart, D., et al. (2007). CYP703 is an ancient cytochrome P450 in land plants catalyzing in-chain hydroxylation of lauric acid to provide building blocks for sporopollenin synthesis in pollen. *The Plant Cell, 19*, 1473–1487.

Mottram, H. R., Crossman, Z. M., & Evershed, R. P. (2001). Regiospecific characterisation of the triacylglycerols in animal fats using high performance liquid chromatography-atmospheric pressure chemical ionisation mass spectrometry. *The Analyst, 126*, 1018–1024.

Mottram, H. R., Woodbury, S. E., & Evershed, R. P. (1997). Identification of triacylglycerol positional isomers present in vegetable oils by high performance liquid chromatography/atmospheric pressure chemical ionization mass spectrometry. *Rapid Communications in Mass Spectrometry, 11*, 1240–1252.

Murphy, R. C., & Axelsen, P. H. (2011). Mass spectrometric analysis of long-chain lipids. *Mass Spectrometry Reviews, 30*, 579–599.

Napolitano, A., Carbone, V., Saggese, P., Takagaki, K., & Pizza, C. (2007). Novel galactolipids from the leaves of Ipomoea batatas L.: Characterization by liquid chromatography coupled with electrospray ionization-quadrupole time-of-flight tandem mass spectrometry. *Journal of Agricultural and Food Chemistry, 55*, 10289–10297.

Nawrath, C. (2002). *The biopolymers cutin and suberin*. In C. R. Somerville & E. M. Meyerowitz (Eds.), *The arabidopsis book*. Rockville, MD, USA: ASPB Press. http://wwwaspborg/publications/Arabidopsis/.

Neumann, S., & Bocker, S. (2010). Computational mass spectrometry for metabolomics: Identification of metabolites and small molecules. *Analytical and Bioanalytical Chemistry, 398*, 2779–2788.

Ogiso, H., Suzuki, T., & Taguchi, R. (2008). Development of a reverse-phase liquid chromatography electrospray ionization mass spectrometry method for lipidomics, improving detection of phosphatidic acid and phosphatidylserine. *Analytical Biochemistry, 375*, 124–131.

Okazaki, Y., Kamide, Y., Hirai, M., & Saito, K. (2013). Plant lipidomics based on hydrophilic interaction chromatography coupled to ion trap time-of-flight mass spectrometry. *Metabolomics, 9*(Suppl. 1), 121–131.

Pacetti, D., Boselli, E., Hulan, H. W., & Frega, N. G. (2005). High performance liquid chromatography-tandem mass spectrometry of phospholipid molecular species in eggs from hens fed diets enriched in seal blubber oil. *Journal of Chromatography. A, 1097*, 66–73.

Pata, M. O., Hannun, Y. A., & Ng, C. K. (2010). Plant sphingolipids: Decoding the enigma of the Sphinx. *The New Phytologist, 185*, 611–630.

Pettus, B. J., Bielawska, A., Kroesen, B. J., Moeller, P. D., Szulc, Z. M., Hannun, Y. A., et al. (2003). Observation of different ceramide species from crude cellular extracts by normal-phase high-performance liquid chromatography coupled to atmospheric pressure chemical ionization mass spectrometry. *Rapid Communications in Mass Spectrometry, 17*, 1203–1211.

Pulfer, M., & Murphy, R. C. (2003). Electrospray mass spectrometry of phospholipids. *Mass Spectrometry Reviews, 22*, 332–364.

Raffaelli, A., & Saba, A. (2003). Atmospheric pressure photoionization mass spectrometry. *Mass Spectrometry Reviews, 22*, 318–331.

Retra, K., Bleijerveld, O. B., van Gestel, R. A., Tielens, A. G. M., van Hellemond, J. J., & Brouwers, J. F. (2008). A simple and universal method for the separation and identification of phospholipid molecular species. *Rapid Communications in Mass Spectrometry, 22*, 1853–1862.

Robb, D. B., Covey, T. R., & Bruins, A. P. (2000). Atmospheric pressure photoionization: An ionization method for liquid chromatography-mass spectrometry. *Analytical Chemistry, 72*, 3653–3659.

Rocha, J. M., Kalo, P. J., Ollilainen, V., & Malcata, F. X. (2010). Separation and identification of neutral cereal lipids by normal phase high-performance liquid chromatography, using evaporative light-scattering and electrospray mass spectrometry for detection. *Journal of Chromatography. A, 1217*, 3013–3025.

Rohlfing, A., Müthing, J., Pohlentz, G., Distler, U., Peter-Katalinić, J., Berkenkamp, S., et al. (2007). IR-MALDI-MS analysis of HPTLC-separated phospholipid mixtures directly from the TLC plate. *Analytical Chemistry, 79*, 5793–5808.

Rouzer, C. A., Ivanova, P. T., Byrne, M. O., Milne, S. B., Marnett, L. J., & Brown, H. A. (2006). Lipid profiling reveals arachidonate deficiency in RAW264.7 cells: Structural and functional implications. *Biochemistry, 45*, 14795–14808.

Ruiz-Gutiérrez, V., & Pérez-Camino, C. M. (2000). Update on solid-phase extraction for the analysis of lipid classes and related compounds. *Journal of Chromatography. A, 885*, 321–341.

Samuels, L., Kunst, L., & Jetter, R. (2008). Sealing plant surfaces: Cuticular wax formation by epidermal cells. *Annual Review of Plant Biology, 59*, 683–707.

Santos, S., Schreiber, L., & Graca, J. (2007). Cuticular waxes from ivy leaves (*Hedera helix* L.): Analysis of high-molecular-weight esters. *Phytochemical Analysis, 18*, 60–69.

Schmelzer, K., Fahy, E., Subramaniam, S., & Dennis, E. A. (2007). The lipid maps initiative in lipidomics. *Methods in Enzymology, 432*, 171–183.

Schrick, K., Shiva, S., Arpin, J. C., Delimont, N., Isaac, G., Tamura, P., et al. (2012). Steryl glucoside and acyl steryl glucoside analysis of Arabidopsis seeds by electrospray ionization tandem mass spectrometry. *Lipids, 47*, 185–193.

Schuhmann, K., Herzog, R., Schwudke, D., Metelmann-Strupat, W., Bornstein, S. R., & Shevchenko, A. (2011). Bottom-up shotgun lipidomics by higher energy collisional dissociation on LTQ Orbitrap mass spectrometers. *Analytical Chemistry, 83*, 5480–5487.

Schwudke, D., Hannich, J. T., Surendranath, V., Grimard, V., Moehring, T., Burton, L., et al. (2007). Top-down lipidomic screens by multivariate analysis of high-resolution survey mass spectra. *Analytical Chemistry, 79*, 4083–4093.

Schwudke, D., Oegema, J., Burton, L., Entchev, E., Hannich, J. T., Ejsing, C. S., et al. (2006). Lipid profiling by multiple precursor and neutral loss scanning driven by the data-dependent acquisition. *Analytical Chemistry, 78*, 585–595.

Shaner, R. L., Allegood, J. C., Park, H., Wang, E., Kelly, S., Haynes, C. A., et al. (2009). Quantitative analysis of sphingolipids for lipidomics using triple quadrupole and quadrupole linear ion trap mass spectrometers. *Journal of Lipid Research, 50*, 1692–1707.

Shevchenko, A., & Simons, K. (2010). Lipidomics: Coming to grips with lipid diversity. *Nature Reviews*, *11*, 593–598.
Smith, J. C., Hou, W., Whitehead, S. N., Ethier, M., Bennett, S. A., & Figeys, D. (2008). Identification of lysophosphatidylcholine (LPC) and platelet activating factor (PAF) from PC12 cells and mouse cortex using liquid chromatography/multi-stage mass spectrometry (LC/MS3). *Rapid Communications in Mass Spectrometry*, *22*, 3579–3587.
Song, H., Hsu, F. F., Ladenson, J., & Turk, J. (2007). Algorithm for processing raw mass spectrometric data to identify and quantitate complex lipid molecular species in mixtures by data-dependent scanning and fragment ion database searching. *Journal of the American Society for Mass Spectrometry*, *18*, 1848–1858.
Sperling, P., Franke, S., Lüthje, S., & Heinz, E. (2005). Are glucocerebrosides the predominant sphingolipids in plant plasma membranes? *Plant Physiology and Biochemistry*, *43*, 1031–1038.
Sperling, P., & Heinz, E. (2003). Plant sphingolipids: Structural diversity, biosynthesis, first genes and functions. *Biochimica et Biophysica Acta*, *1632*, 1–15.
Sud, M., Fahy, E., Cotter, D., Brown, A., Dennis, E. A., Glass, C. K., et al. (2007). LMSD: LIPID MAPS structure database. *Nucleic Acids Research*, *35*, 527–532.
Sun, G., Yang, K., Zhao, Z., Guan, S., Han, X., & Gross, R. W. (2008). Matrix-assisted laser desorption/ionization time-of-flight mass spectrometric analysis of cellular glycerophospholipids enabled by multiplexed solvent dependent analyte–matrix interactions. *Analytical Chemistry*, *80*, 7576–7585.
Taguchi, R., Houjou, T., Nakanishi, H., Yamazaki, T., Ishida, M., Imagawa, M., et al. (2005). Focused lipidomics by tandem mass spectrometry. *Journal of Chromatography B*, *823*, 26–36.
Taguchi, R., & Ishikawa, M. (2010). Precise and global identification of phospholipid molecular species by an Orbitrap mass spectrometer and automated search engine lipid search. *Journal of Chromatography. A*, *1217*, 4229–4239.
Taguchi, R., Nishijima, M., & Shimizu, T. (2007). Basic analytical systems for lipidomics by mass spectrometry in Japan. *Methods in Enzymology*, *432*, 185–211.
Thomas, M. C., Mitchell, T. W., & Blanksby, S. J. (2006). Ozonolysis of phospholipid double bonds during electrospray ionization: A new tool for structure determination. *Journal of the American Chemical Society*, *128*, 58–59.
Thomas, M. C., Mitchell, T. W., Harman, D. G., Deeley, J. M., Murphy, R. C., & Blanksby, S. J. (2007). Elucidation of double bond position in unsaturated lipids by ozone electrospray ionization mass spectrometry. *Analytical Chemistry*, *79*, 5013–5022.
Thomas, M. C., Mitchell, T. W., Harman, D. G., Deeley, J. M., Nealon, J. R., & Blanksby, S. J. (2008). Ozone-induced dissociation: Elucidation of double bond position within mass-selected lipid ions. *Analytical Chemistry*, *80*, 303–311.
Toledo, M. S., Levery, S. B., Bennion, B., Guimaraes, L. L., Castle, S. A., Lindsey, R., et al. (2007). Analysis of glycosylinositol phosphorylceramides expressed by the opportunistic mycopathogen Aspergillus fumigatus. *Journal of Lipid Research*, *48*, 1801–1824.
van Meer, G., Leeflang, B. R., Liebisch, G., Schmitz, G., & Goni, F. M. (2007). The European lipidomics initiative: Enabling technologies. *Methods in Enzymology*, *432*, 213–232.
Vernooij, E. A. A. M., Brouwers, J. F. H. M., Kettenes-Van den Bosch, J. J., & Crommelin, D. J. A. (2002). RP-hPLC/ESI MS determination of acyl chain positions in phospholipids. *Journal of Separation Science*, *25*, 285–289.
Vrkoslav, V., Urbanova, K., & Cvacka, J. (2010). Analysis of wax ester molecular species by high performance liquid chromatography/atmospheric pressure chemical ionisation mass spectrometry. *Journal of Chromatography. A*, *1217*, 4184–4194.
Walton, T. J., & Kolattukudy, P. E. (1972). Determination of structures of cutin monomers by a novel depolymerization procedure and combined gas chromatography and mass-spectrometry. *Biochemistry*, *11*, 1885–1896.

Wen, M., & Jetter, R. (2009). Composition of secondary alcohols, ketones, alkanediols, and ketols in Arabidopsis thaliana cuticular waxes. *Journal of Experimental Botany*, *60*, 1811–1821.

Wewer, V., Dombrink, I., vom Dorp, K., & Dormann, P. (2011). Quantification of sterol lipids in plants by quadrupole time-of-flight mass spectrometry. *Journal of Lipid Research*, *52*, 1039–1054.

Whitehead, S. N., Hou, W., Ethier, M., Smith, J. C., Bourgeois, A., Denis, R., et al. (2007). Identification and quantitation of changes in the platelet activating factor family of glycerophospholipids over the course of neuronal differentiation by high-performance liquid chromatography electrospray ionization tandem mass spectrometry. *Analytical Chemistry*, *79*, 8539–8548.

Whitehouse, C. M., Dreyer, R. N., Yamashita, M., & Fenn, J. B. (1985). Electrospray interface for liquid chromatographs and mass spectrometers. *Analytical Chemistry*, *57*, 675–679.

Wilson, J. F. (2003). Long-suffering lipids gain respect. *Scientist*, *17*, 34.

Wolf, C., & Quinn, P. J. (2008). Lipidomics: Practical aspects and applications. *Progress in Lipid Research*, *47*, 15–36.

Wu, Z., Rodgers, R. P., & Marshall, A. G. (2004). Characterization of vegetable oils: Detailed compositional fingerprints derived from electrospray ionization fourier transform ion cyclotron resonance mass spectrometry. *Journal of Agricultural and Food Chemistry*, *52*, 5322–5328.

Xu, J., Chen, D., Yan, X., Chen, J., & Zhou, C. (2010). Global characterization of the photosynthetic glycerolipids from a marine diatom Stephanodiscus sp. by ultra performance liquid chromatography coupled with electrospray ionization-quadrupole-time of flight mass spectrometry. *Analytica Chimica Acta*, *663*, 60–68.

Yamashita, M., & Fenn, J. B. (1984). Electrospray ion source. Another variation on the free-jet theme. *The Journal of Physical Chemistry*, *88*, 4451–4459.

Zacarias, A., Bolanowski, D., & Bhatnagar, A. (2002). Comparative measurements of multicomponent phospholipid mixtures by electrospray mass spectroscopy: Relating ion intensity to concentration. *Analytical Biochemistry*, *308*, 152–159.

Zhou, Z., Marepally, S. R., Nune, D. S., Pallakollu, P., Ragan, G., Roth, M. R., et al. (2011). LipidomeDB data calculation environment: Online processing of direct-infusion mass spectral data for lipid profiles. *Lipids*, *46*, 879–884.

Zäuner, S., Ternes, P., & Warnecke, D. (2010). Biosynthesis of sphingolipids in plants (and some of their functions). *Advances in Experimental Medicine and Biology*, *688*, 249–263.

CHAPTER EIGHT

Isotope Ratio Mass Spectrometry Technique to Follow Plant Metabolism: Principles and Applications of $^{12}C/^{13}C$ Isotopes

Jaleh Ghashghaie[*,1], **Guillaume Tcherkez**[†,‡]

[*]Laboratoire d'Ecologie, Systématique et Evolution (ESE), CNRS UMR8079, Université de Paris-Sud 11, Orsay, France

[†]Institut de Biologie des Plantes (IBP), CNRS UMR 8618, Université de Paris-Sud 11, Orsay, France

[‡]Institut Universitaire de France, Paris, France

[1]Corresponding author: e-mail address: jaleh.ghashghaie@u-psud.fr

Contents

Advances in Botanical Research, Volume 67
ISSN 0065-2296
http://dx.doi.org/10.1016/B978-0-12-397922-3.00008-3

Abstract

Isotope fractionations during biophysical and biochemical processes are at the origin of natural differences in heavy-to-light isotope ratios (e.g. $^{13}C/^{12}C$, $^{2}H/^{1}H$, etc.) among biosphere components (soil organic matter, vegetation and atmospheric CO_2), metabolites and atomic positions within molecules. The natural abundance in stable isotopes of biological common elements (C, H, O, N, S) is thus widely used as tracer or physiological marker to examine biochemical and ecophysiological mechanisms from the cellular to the ecosystemic scale. Since the pioneering studies on isotopic composition of plant organic matter in the 1980s, important methodological and technological advances have been made to improve the convenience, precision and rapidity of isotopic measurements. In this chapter, we explain the basics of isotopic composition and fractionation, and review the most important technical advances in plant biology. We focus on carbon stable isotopes since the natural ^{13}C-distribution has proved to be a powerful tool to investigate carbon primary metabolism and respiratory flux patterns.

ABBREVIATIONS

EA elemental analyser/analysis
GC gas chromatography
IRMS isotope ratio mass spectro-meter/metry
LC liquid chromatography
OAA oxaloacetate
OM organic matter
PEPC phospho*enol*pyruvate carboxylase (EC 4.1.1.31)
Rubisco ribulose-1,5-bisphosphate carboxylase/oxygenase (EC 4.1.1.39)
TCAP tricarboxylic acid pathway (Krebs cycle)

1. STABLE ISOTOPES AT NATURAL ABUNDANCE

Most of the elements have naturally more than one isotope, with the lightest form corresponding to the most abundant (e.g. ^{12}C, ^{14}N, ^{16}O, ^{1}H, etc.) and the heavy forms to rare isotopes (e.g. ^{13}C, ^{15}N, ^{17}O, ^{18}O, ^{2}H, etc.). The atomic mass difference between the heavy and light isotopes of a given element comes from the difference in the number of neutrons. Heavy isotopes have one, two, or more additional neutrons compared with the light isotope. Table 8.1 shows the natural abundance of some common isotopes involved in biological processes.

Both natural abundance of stable isotopes and labelling are used as tracers in a wide range of domains (e.g. hydrology, geophysics, geochemistry, forensics, criminology, medicine and different fields of biology, ecology

Table 8.1 Natural abundance of some heavy isotopes involved in biological processes

Heavy isotopes	^{13}C	^{15}N	^{18}O	^{2}H	^{34}S
Average natural abundance	1.1%	0.4%	0.2%	0.015%	4.2%

and agronomy). The atomic mass difference between heavy and light isotopes of the same element leads to differences in physical (e.g. coefficients for diffusion, dissolution, evaporation, etc.) and chemical properties (e.g. rate constants for enzymatic reactions) leading to isotope fractionations (e.g. during enzymatic reactions involving bond formation or cleavage). Fractionations are at the origin of the observed natural differences in heavy-to-light isotope ratios between different compartments of the biosphere (both organic and inorganic compartments), between metabolites, and also of the heterogeneous isotopic distributions within molecules (Table 8.1).

Since the atomic mass difference between heavy and light isotopes is low and heavy isotopes are naturally present in very small quantities (Table 8.1), changes in heavy-to-light isotope ratios due to isotope fractionations should be measured with high precision techniques, such as isotope ratio mass spectrometry (IRMS). If the natural abundance of the heavy isotopes were of the order of magnitude of a few %, the changes due to fractionations would not have been detectable (fractionations are in the ‰ range). The very small quantity in heavy isotopes thus allows one to carry out very precise quantification and calculations, useful for biological purposes.

In this chapter, we will focus on carbon isotopes since the natural ^{13}C/^{12}C distribution and IRMS have proved powerful to investigate carbon primary metabolism and respiratory flux patterns. In the following, a brief section of definitions and basics of photosynthetic fractionation (natural ^{13}C-abundance in plant material) are given. The major advances in following plant metabolism using natural ^{13}C-abundance are then summarised. The objective of the present review is not to provide a comprehensive picture of the isotopic field but rather to give basic knowledge in ^{12}C/^{13}C isotopics to starters and students.

2. DEFINITIONS

2.1. Isotope ratio

The "isotope ratio" is equal to the quotient of the content in the heavy isotope to that in the light isotope. For instance, carbon isotope ratio (R) is determined as follows:

$$R = {}^{13}\mathrm{C}/{}^{12}\mathrm{C}$$

Table 8.2 Standard isotope ratio (*R*) for some elements involved in biological processes

Element	Carbon	Nitrogen	Oxygen	Hydrogen	Sulphur
Isotope ratio	$^{13}C/^{12}C$	$^{15}N/^{14}N$	$^{18}O/^{16}O$	$^{2}H/^{1}H$	$^{34}S/^{32}S$
R values of the standard material	0.0112372	0.003677	0.002052	0.00015575	0.045005

Natural abundance of heavy isotopes being small, *R* values are also very small and thus very close to percentages. For example, *R* for carbon isotopes is around $R=1.1/98.9=0.011$, which is almost similar to the % of heavy carbon relative to the total carbon: $^{13}C/(^{12}C+^{13}C)=1.1\%$ (see Table 8.2 for some *R* values).

2.2. Isotope composition

In order to normalise isotope ratio values (fixed scale), an international standard is used as reference material. The isotopic composition of the samples for a given element is thus calculated relative to the isotope ratio of the corresponding standard. For example, the carbon isotope composition ($\delta^{13}C$) is calculated as follows:

$$\delta^{13}C=(R_S-R_{PDB})/R_{PDB}=(R_S/R_{PDB})-1,$$

where R_S and R_{PDB} are the isotope ratios of the sample and the standard PDB, respectively. The standard PDB is a belemnite fossil coming from geological formation Pee Dee in South Carolina, USA. The PDB is slightly ^{13}C-enriched ($R_{PDB}=0.0112372$) compared to almost all organic and inorganic materials but remains close to them so that $\delta^{13}C$ values of nearly all biological samples are very small and are thus expressed in per mil (‰, i.e. 10^{-3}) and are negative. This is not necessarily true for other isotopes (e.g. nitrogen, hydrogen and oxygen isotopes) for which negative or positive $\delta^{15}N$, $\delta^{18}O$ or $\delta^{2}H$ values may be found because the international standards (atmospheric N_2 and ocean water, respectively) are not particularly enriched. For carbon, the standard is now exhausted and the international atomic agency (AIEA, Vienna) uses a "secondary" standard Vienna-PDB (V-PDB, same isotope ratio) with respect to which all $\delta^{13}C$ values are expressed.

2.3. Isotope effect

The isotope effect (α) during an enzymatic reaction or a physical process transforming substrate A to product B is defined as follows:

$$\alpha = R_A/R_B, \tag{8.1}$$

where R_A and R_B are the isotope ratios of the substrate and the product of the reaction, respectively. Two types of isotope effects can be defined depending on the reaction considered: kinetic and equilibrium isotope effects. In kinetic effects (i.e. irreversible reactions), the isotope effect (α_k) corresponds to the ratio of the kinetic constant rates of the light over the heavy isotope (i.e. $\alpha_k = {}^{12}k/{}^{13}k$). The light "isotopologue" reacts generally (but not always) more rapidly than the heavy counterpart, thus $\alpha_k > 1$ and the product is depleted in ^{13}C compared with the substrate. For example, the isotope effect of the photosynthetic enzyme Rubisco is, in C_3 plants, around 1.029 (i.e. $^{12}CO_2$ goes 2.9% faster than $^{13}CO_2$) leading to a ^{13}C-depletion in the photosynthetic products compared with atmospheric CO_2. In equilibrium (thermodynamic) effects (i.e. reversible reactions), the isotope effect (α_e) corresponds to the ratio of equilibrium constant of the light over heavy isotopes (i.e. $\alpha_e = {}^{12}K/{}^{13}K$). The reaction being reversible, the isotope effect at full equilibrium (α_e) is generally lower than the kinetic one for usual reactions. Quite often, $\alpha_e < 1$, that is, the product is enriched in the heavy isotope compared with the substrate. For example, the equilibrium isotope effect (α_e) of the enzyme carbonic anhydrase (EC 4.2.1.1) catalysing the hydration of CO_2 into bicarbonate is about 0.991, leading to a ^{13}C-enrichment in HCO_3^- (product) of 9‰ compared with dissolved CO_2 (substrate).

When an irreversible enzymatic reaction occurs with an infinite or non-limiting amount of substrate, the isotope effect keeps its maximum value (kinetic effect, ${}^{12}k/{}^{13}k$). If the substrate is in limited amount, the isotope effect decreases and tends to unity as the substrate is consumed. In fact, at the end of the reaction, the product has the same isotope composition as the substrate initially provided. In other words, irreversible enzymatic reactions can fractionate between isotopes only if the substrate is not totally consumed. In equilibrium reactions, the isotope effect starts with the value of the kinetic effect of the forward reaction and then changes as soon as the reverse reaction occurs. When the equilibrium is reached, the net value of the isotope effect (α_e) depends upon the prevalence of the fractionation of the forward and backward reactions. Usually, reversible reactions that form bonds between carbon atoms do favour ^{13}C ($\alpha_e < 1$).

2.4. Isotope discrimination

In practice, (eco)physiologists use the isotope fractionation (or discrimination, Δ) rather than the isotope effect (α) in calculations. The discrimination is given by:

$$\Delta = \alpha - 1 \tag{8.2}$$

Isotope effects are only slightly lower or higher than 1 (see above), and therefore, discrimination values are very small and expressed in per mil (‰). When $\alpha > 1$ (general case for kinetic isotope effects), the discrimination is positive (i.e. ^{13}C-depletion in the product compared with the substrate). For example, the discrimination against ^{13}C during CO_2 fixation by Rubisco is positive ($\Delta = 1.029 - 1 = 0.029 = +29‰$), indicating indeed a ^{13}C-depletion in photosynthetic products. When $\alpha < 1$ (common case for equilibrium isotope effects), the discrimination is negative (i.e. ^{13}C-enrichment in products). For example, the discrimination during hydration of CO_2 is negative ($\Delta = 0.991 - 1 = -0.009 = -9‰$), indicating indeed a ^{13}C-enrichment in bicarbonate.

The isotope discrimination can also be expressed using δ-values (rearranging Eqs. (8.1) and (8.2)) as follows:

$$\Delta = (\delta_s - \delta_p)/(1 + \delta_p), \tag{8.3}$$

where δ_s and δ_p are the isotope composition of the substrate (source) material and the product of the reaction, respectively. δ-Values are negligible compared to 1 (expressed in per mil) and thus a common approximation is to neglect δ_p in the denominator: Δ is then simply close to the difference between source and product δ-values, that is, $\Delta \approx \delta_s - \delta_p$. For example, atmospheric CO_2 (photosynthetic carbon source) has a $\delta^{13}C$ value of around $-8‰$ and C_3 plant organic matter (OM) (product) is around $-28‰$ and the overall discrimination is thus:

$$\Delta \approx \delta_{CO_2} - \delta_{plant} = -8 - (-28) = +20‰$$

3. BASICS OF PHOTOSYNTHETIC $^{12}C/^{13}C$ DISCRIMINATION

Plants discriminate against ^{13}C during photosynthetic CO_2 assimilation. Therefore, plant OM is naturally ^{13}C-depleted compared to atmospheric CO_2. Photosynthetic carbon isotope discrimination is in average around 20‰ in C_3 and 4‰ in C_4 plants. Photosynthetic discrimination in C_3 plants occurs mainly during diffusion of CO_2 from the air into the leaves through stomata (4.4‰) and carboxylation by Rubisco (29‰). The photosynthetic carbon isotope discrimination has been extensively studied and robust models have been developed and validated for many C_3 and C_4 species (Farquhar, Ehleringer, & Hubick, 1989; Farquhar, O'Leary, & Berry, 1982). Variations in photosynthetic discrimination values in C_3 plants have been shown to be mainly due to changes in stomatal closure that are, in turn, caused by changes in environmental factors limiting CO_2 supply to Rubisco. In C_4 plants, the lower discrimination value compared to C_3 plants (≈4‰ compared to 20‰) is due to the involvement of the CO_2-concentrating mechanism (so-called CCM) involving the phospho*enol*pyruvate carboxylase (PEPC). Changes in the discrimination value of C_4 plants are mainly caused by changes in CO_2 leakage from the bundle sheath to mesophyll cells rather than stomatal closure (for a review, see Brugnoli & Farquhar, 2000). CAM plants have both C_3 and C_4 photosynthetic metabolism depending on environmental conditions, for example, during transition phases (light-to-dark and dark-to-light transitions). The discrimination may thus change accordingly between typical signatures of these two metabolic types but quite usually, the average discrimination is intermediate between C_3 and C_4 values. The average discrimination value of plants (in practice, the $\delta^{13}C$ value of total OM) is currently used to identify the prevalent photosynthetic pathway (C_3, C_4 or CAM).

The photosynthetic discrimination can be measured either using the $\delta^{13}C$ of bulk OM (integrated value during plant growth), $\delta^{13}C$ of sugars (integrates about 2–3 days of photosynthetic discrimination) or measured online during leaf CO_2 exchanges (instantaneous net photosynthetic discrimination under fixed conditions). However, the photosynthetic discrimination and thus the ^{13}C-content in photosynthetic products vary between plant species, plant developmental stages and environmental conditions. While the instantaneous discrimination value (net photosynthesis) is physiologically the most relevant, values obtained from organic materials should be viewed as average values and thus their biological significance is different.

Until recently, the $\delta^{13}C$ of leaf bulk OM have been considered to reflect properly net photosynthetic discrimination. However, additional discriminations have been shown to occur during post-carboxylation metabolic processes, which explain the observed inter-organ isotopic differences (Badeck, Tcherkez, Nogués, Piel, & Ghashghaie, 2005) and cause undesirable isotopic variations in leaf material. In fact, leaves have been shown to be generally ^{13}C-depleted compared to all other organs, suggesting that fractionating mechanisms do occur after photosynthate production (for a recent review, see Cernusak et al., 2009). For instance, there is a carbon isotope fractionation during respiration (reviewed by Ghashghaie et al., 2003; Werner & Gessler, 2011), that is, the $\delta^{13}C$ of carbon atoms lost by respiration is different from that in photosynthates. Many recent papers reported $\delta^{13}C$ values in respired CO_2 and showed some variability between species, organs and conditions (see below). It has progressively been made clear that investigating the metabolic origin of respiratory fractionation is an important tool to understand respiratory metabolic fluxes (Bathellier et al., 2009; Tcherkez et al., 2003). In what follows (Section 6), we will focus on the carbon isotope fractionation during respiratory processes and primary carbon metabolism, which are both exquisite examples of the use of IRMS techniques to disentangle metabolic flux patterns.

4. PRACTICAL ASPECTS OF EA-IRMS MEASUREMENTS

4.1. General principle

IRMS systems measure the isotope composition on pure gases (CO_2, for carbon) and thus both ordinary and reference samples should be in the form of carbon dioxide. CO_2 may originate from sampled air (measurement of atmospheric or respired CO_2) or the combustion of OM (metabolite of interest or total OM). In all instances, carbon dioxide should be purified from the background gas. Quite commonly, CO_2 analysed for biological purpose is separated using gas chromatography (GC) (or adsorption/desorption) and this requires a preparative device in front of the IRMS. The cryogenic purification of CO_2 and the subsequent direct injection in the mass spectrometer (dual-inlet) will not be detailed here for reasons that will become apparent below.

4.2. Isotope ratio mass spectrometry

In the source of the IRMS, CO_2 molecules are ionised by electron impact into CO_2^+. Ionised CO_2 is then directed (by focalisation plates) to a magnetic field that separates ions depending on their isotopic masses, due to slightly different diameters of their circular trajectory. CO_2^+ with a mass of 44 ($^{12}C^{16}O^{16}O$), 45 ($^{13}C^{16}O^{16}O$) and 46 ($^{12}C^{18}O^{16}O$) are sufficiently separated by the magnetic field to fall in separate collectors 1, 2 and 3 (Faraday cups). Collector 1 (smallest diameter of the trajectory) receives the lightest molecules (Fig. 8.1), etc. Note that for nitrogen isotopes

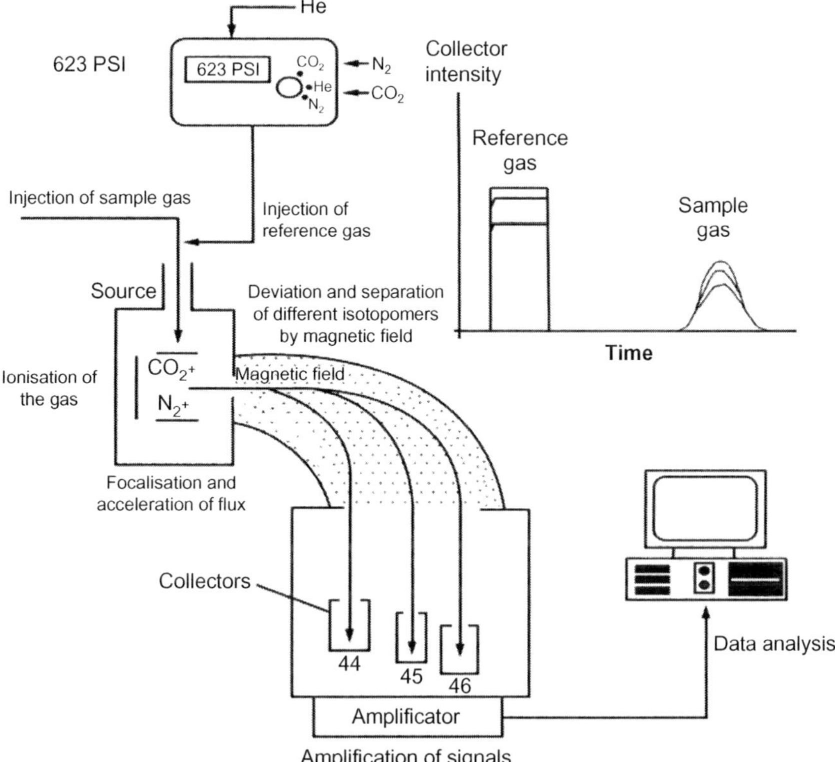

Figure 8.1 General principle of isotopic spectrometry. Stable Isotope Ratio Mass Spectrometer (IRMS) comprises an injection system for both reference and sample gases (e.g. CO_2 or N_2 for ^{13}C and ^{15}N, respectively). The carrier gas is helium (He). Sample and reference gas molecules are ionised by electron impact in the ion source. Ions (CO_2^+, N_2^+) are accelerated and enter a magnetic field, which separates the different isotopologues (since the mass difference causes a change in the circular trajectory) in individual collectors. Signals are amplified for detection. *Scheme from Deléens, Morot-Gaudry, Martin, Thoreux, and Gojon (1997).*

(analysed gas N_2), the masses of interest are 28 ($^{14}N^{14}N$), 29 ($^{14}N^{15}N$) and 30 ($^{15}N^{15}N$), and for oxygen isotopes in dioxygen (O_2), the masses of interest are 32 ($^{16}O^{16}O$), 34 ($^{18}O^{16}O$) and 36 ($^{18}O^{18}O$). Rather than changing the position of Faraday cups, the parameters of ion source, plates and the magnetic field are adjusted to adapt the analysis to the isotopic gas considered. The amounts of organic material required for a single (reliable) isotopic analysis of carbon is rather small, around 10^{-7}–10^{-5} mol of carbon dioxide (only a small fraction of which is effectively preleved by the source for the analysis: the source *per se* only requires 10^3–10^4 molecules to detect an isotopic signal). Therefore, the electrical signal induced by ion collection needs to be amplified for acquisition.

IRMS measurements are always relative, that is, compare the signal of the sample to a standard gas (cylinder of pure CO_2 of known isotope composition) injected prior to or after the sample. The isotope composition of the sample is then back-calculated from the imposed $\delta^{13}C$ of the reference gas ($\delta^{13}C_{ref}$) and the measured difference between the sample and the reference gas ($\delta^{13}C_{sr}$):

$$\delta^{13}C_{sample} = \left(\delta^{13}C_{ref} + 1\right) \times \delta^{13}C_{sr} + \delta^{13}C_{ref} \tag{8.4}$$

In other words, the $\delta^{13}C$ of the reference gas should be known to proceed IRMS analyses, using either an isotopically calibrated gas (very expensive) or a gas reverse-calibrated using a sample of known isotope composition (inexpensive), $\delta^{13}C_{ks}$, with $\delta^{13}C_{ref} = (\delta^{13}C_{ks} - \delta^{13}C_{sr})/(1 + \delta^{13}C_{sr})$ (derived from Eq. (8.4)).

The possible drawback of IRMS measurements is the occurrence of multi-isotopic species (or "clumped" isotopes) as well as different species with the same mass. For carbon, the IRMS systems acquire signals of mass 44, 45 (mainly $^{13}C^{16}O_2$) and 46 (mainly $^{12}C^{16}O^{18}O$) and calculate 45/44 and 46/44 ratios. However, the occurrence of ^{17}O in carbon dioxide gives also a mass of 45 ($^{12}C^{17}O^{16}O$) (this is often neglected because ^{17}O is rare) and ^{18}O gives a mass of 46 ($^{12}C^{18}O^{16}O$). ^{18}O cannot be neglected (Table 8.2) that is why the ratio 45/44 does not represent purely the $^{13}C/^{12}C$ ratio and thus a correction is needed (Craig correction, Craig & Boato, 1956). Note that the adulteration of samples by N_2O ($^{14}N_2^{16}O$, mass 44) and its isotopic species is not a problem any longer due to the use of chromatographic methods that separate CO_2 from possible N_2O traces.

4.3. Elemental analysis

Elemental analysis is generally used for measurement of %C and %N (and in the pyrolysis mode, %H and %O) contents of solid samples using combustion–reduction of OM and the quantitation of evolved CO_2 and N_2. Gases produced by the combustion in a O_2-rich helium atmosphere (i.e. CO_2, NO_2, SO_2, H_2O and the O_2 excess) are then reduced (with copper) and water vapour is trapped, yielding CO_2, N_2 and SO_2. Two rather different techniques of separation are now commercially available: gas chromatography and absorption/desorption (both under He flow). Gas chromatography allows the separation of sample gases with their retention time. Absorption/desorption takes advantage of temperature-dependent affinity of polymers in order to specifically trap or release gases.

4.4. Sample preparation

Sample preparation for EA usually includes drying (oven-drying or lyophilisation) and grinding to a fine, homogenous powder. The latter is weighed in tin (or silver) capsules that are placed into an autosampler (Fig. 8.2). Importantly, the complete combustion of samples is needed to avoid any fractionation, and this imperative requires an appropriate capsule (allowing a flash combustion) and a sufficient O_2 mole fraction during combustion. A rather small amount of dry material is needed for the $^{13}C/^{12}C$ isotope analysis of plant samples, which contain around 40% carbon. That is, each sample is of around 1 mg (this represents about 40 μmol of elemental C). The overall precision of EA-IRMS analyses is close to 0.2‰. Some analyses can be carried out on smaller samples, down to 0.1 mg (i.e. 4 μmol elemental C) with less precision (0.4‰). Note that this minimal value of about 1 μmol of elemental C may be problematic for small samples: it is usually not an issue for total OM but causes technical difficulties for the isotopic analysis of specific compounds. That is, EA-IRMS analyses require the compounds of interest to be purified. This may be achieved using HPLC, solvent–solvent extractions, etc. However, even for abundant plant metabolites such as sucrose, HPLC purification is tedious and this cannot be reasonably done on small starting material (typically less than 10 mg of dry plant matter). For less-abundant metabolites useful for flux patterns, other techniques should be exploited. This aspect (compound-specific IRMS analyses, GC-C- and LC-C-IRMS) is discussed below. Still, insoluble plant polymers cannot be isotopically analysed by GC-C- or LC-C-IRMS. So is the case of starch and cellulose, the analysis

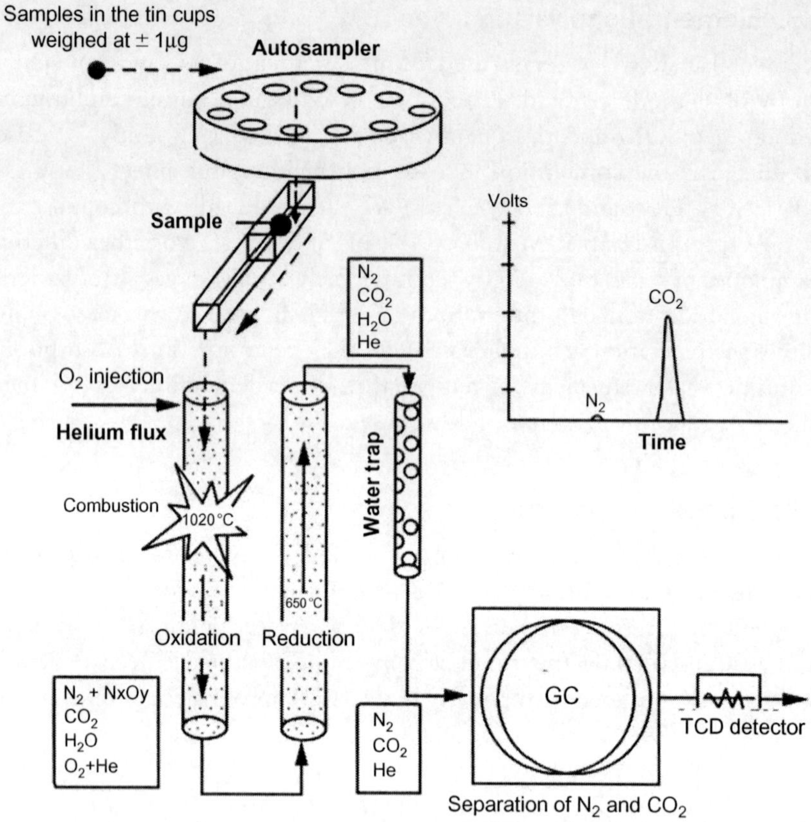

Figure 8.2 EA-IRMS principle. The elemental analyser (EA) is composed of an autosampler, oxidation and reduction tubes, a water vapour trap and a gas chromatography column (GC) (which may be replaced by a combination of differential adsorption columns). Tin capsules containing samples (dry powder) and placed in the autosampler fall in the oxidation–reduction tubes for complete, non-fractionating combustion and the evolved gases are directed by the carrier gas (He) to a water vapour trap and then to the GC for separation of CO_2 and N_2 peaks. Peaks are detected by the TCD and enter the source of the IRMS via a capillary. *Scheme from Deléens et al. (1997).*

of which requires specific, non-fractionating protocols (Loader, Robertson, & McCarroll, 2003; Richter et al., 2009; Wanek, Heintel, & Richter, 2001).

5. COMPOUND-SPECIFIC IRMS ANALYSES

5.1. Carbon dioxide gas exchange

Among key compounds useful for investigating plant metabolism with IRMS, CO_2 is certainly the most important in terms of papers and applications (reviewed by Bowling, Pataki, & Randerson, 2008). In fact, $^{13}C/^{12}C$ IRMS analyses of CO_2 allow one to determine the discrimination associated with net photosynthesis and thanks to calculations, individual CO_2 flux-component (isofluxes), namely, gross photosynthesis, photorespiration and day respiration can be inferred (for a recent example, see Tcherkez, Boex-Fontvieille, Mahé, & Hodges, 2012). A rough estimate of the photosynthetic fractionation may be obtained using a closed system: a leaf is placed in a closed chamber and after a certain time period in the light, the carbon isotopic composition of CO_2 inside the chamber is measured. The natural ^{13}C-enrichment in CO_2 is then used to calculate the photosynthetic fractionation (Δ). Such a value is apparent since gas exchange conditions are not constant during the experiment (e.g. CO_2 mole fraction; for a recent example, see Wingate, Seibt, Moncrieff, Jarvis, & Lloyd, 2007). More reliable is the use of open systems in which the $\delta^{13}C$ of inlet and outlet CO_2 are compared under constant conditions (CO_2 mole fraction, % O_2, etc.). The photosynthetic fractionation (Δ) is then calculated from mass balance (Evans, Sharkey, Berry, & Farquhar, 1986).

The carbon isotope composition of CO_2 evolved in the dark by plant organs is usually obtained with a closed system because this allows the build-up of respiratory CO_2 to reach a sufficient CO_2 mole fraction for IRMS determination. Most respiration rates are too small to use an open system. That is, the isotopic difference between inlet and outlet CO_2 is too low to allow an accurate mass balance calculation of the $\delta^{13}C$ in respired CO_2.

5.2. Carbon dioxide isotopic analysis

The IRMS analysis of CO_2 from gas exchange system involves either the cryogenic purification of CO_2 from air samples and the direct injection into the mass spectrometer (dual-inlet) or a coupling to gas chromatography. The former is now of little importance, due to the time needed for cryogenic purifications and dual-inlet analyses. Air samples are rather injected into a gas chromatograph (continuous flow) which is, in turn, coupled to the IRMS via a capillary. Similarly to EA-IRMS, the CO_2 peak separated by the GC specifically enters the mass spectrometer for the isotope analysis (Fig. 8.3). It should be noted that currently, other methods for analysing CO_2 based on laser technology (TDL spectrometry, Tunable Diode Laser) are available (Bowling, Sargent, Tanner, & Ehleringer, 2003). These methods take advantage of the differential absorption of infra-red (fixed wavelength) by

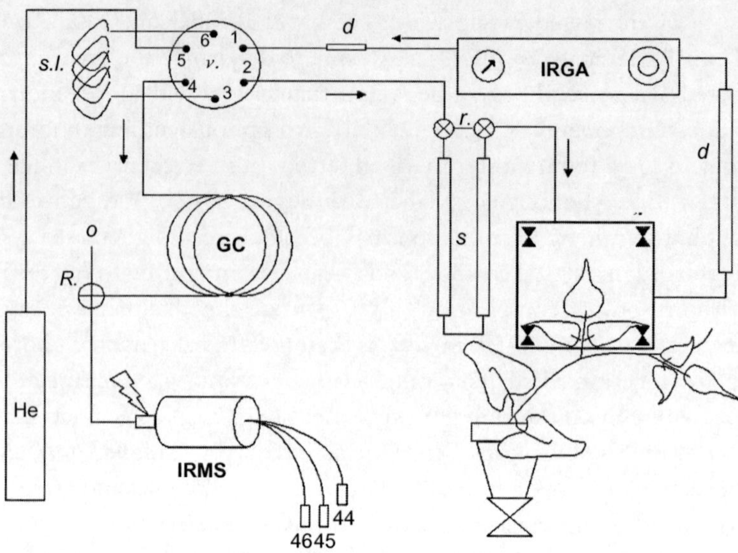

Figure 8.3 Online analysis of respired CO_2. The respiratory closed system is coupled to the IRMS via a six-ways valco-valve (*v.*) and a sampling loop (*s.l.*). The closed system includes a respiratory chamber, where attached leaves (in the present case) or roots should be placed. It further includes an infra-red gas analyser (IRGA) to monitor the CO_2 mole fraction during the experiment. Background CO_2 in the closed system is first removed using soda lime (*s*) columns, which are then isolated (*r*) from the closed system, allowing respired CO_2 to accumulate. Drying (*d*) columns included in the closed system removes water vapour coming from transpiration (thereby ensuring air dryness in the IRGA). Air circulation is maintained with a pump. When the CO_2 mole fraction is high enough, the valco-valve permutation allows the gas contained in the sampling loop to be carried by He and enters the GC. The CO_2 peak separated by the GC enters the ion source of the IRMS for the isotopic analysis. Commutating valve *R* isolates the sample line, that is, liberates the N_2 to the vent (*o*). Arrows indicate the direction of the flow. *Scheme from Tcherkez (2004).*

$^{12}CO_2$ and $^{13}CO_2$ isotopologues. TDL techniques are much more rapid (one $\delta^{13}C$ determination each 0.1 s) but appear to be noisier so that averaging remains necessary. For gas samples, IRMS techniques are mainly limited by the chromatographic step, which is a prerequisite to separate CO_2 from other gases. The maximum time resolution achievable by IRMS is thus of around 3 min, which is insufficient to catch up rapid respiratory kinetics (Barbour, McDowell, Tcherkez, Bickford, & Hanson, 2007).

5.3. GC-C- and LC-C-IRMS techniques

There is now a considerable body of studies on gas-chromatography-combustion-IRMS (GC-C-IRMS) in the scientific literature, simply because compound-specific analyses are of wide interest for forensics and fraud and adulteration detection. Other recent reviews address general aspects of GC-C-IRMS (Schmidt et al., 2004). Here, we shall concentrate on the specific analysis of plant samples in a metabolic perspective. GC-C- and LC-C-IRMS techniques take advantage of a preparative system (GC or LC) coupled to the IRMS via a combustion interface (which is rather a desolvation interface in the case of LC-C-IRMS). This allows both the separation of peaks and the online conversion of each of them into CO_2 for the isotopic analysis. Therefore, the $\delta^{13}C$ of each compound can be determined.

In practice, however, GC-C-IRMS requires quite large amounts of material (concentrated plant extract), a very good separation of compounds along chromatograms and needs the compounds of interest to be volatile. This is usually an issue in plant samples in which most compounds of interest (such as sugars, amino acids and organic acids) are ordinarily below 1 mmol L^{-1} (except glucose and sucrose under certain circumstances) and are not volatile. In the case of amino acids, several studies took advantage of protein hydrolysis, which yields concentrated extracts (reviewed in Werner & Schmidt, 2002). However, from a physiological perspective, the analysis of *free* amino acids is desirable since the turnover of their pools is rapid (less than 1 day). This, in turn, requires pre-purification protocols to enrich aqueous fractions of plant samples in the compounds of interest. Recently, a couple of studies have investigated the $\delta^{13}C$ in amino acids from alfalfa (Molero, Aranjuelo, Teixidor, Araus, & Nogués, 2011) using solid phase extraction. Volatilisation of plant metabolites is usually achieved using quantitative chemical derivatisation inspired from metabolomics GC–MS techniques, among which trimethylsilylation and isopropylsilylation are the most popular. It seems, however, that specific derivatisation protocols based on *N*-acetyl methyl ester derivatives rather than silylation give very good results for amino acids (Corr, Berstan, & Evershed, 2007). In addition to producing undesirable secondary products (with possible fractionation), the major drawback of chemical derivatisation is the addition of exogenous carbon (e.g. methyl or isopropyl groups) and thus a correction should be made on raw $\delta^{13}C$ values. In sugar derivatives, the number of added carbon atoms (n_a) is substantial (up to 18) due to the high number of substitutable (labile) hydrogen atoms (5 for glucose, 10 for sucrose). In this case, the associated isotopic correction is huge and causes large uncertainty in the $\delta^{13}C$ of endogenous carbons. Nevertheless, a recent comparison of GC-C-IRMS (with trimethylsilyl derivatives) and purification-EA-IRMS showed that although more variable, $\delta^{13}C$ values obtained with GC-C-IRMS agreed reasonably well with those obtained with the "classical" method requiring purification (Mauve et al., 2009). Still, uncertainty remains on whether the δ^{13}C values obtained by GC-C-IRMS are reliable since partial (incomplete) conversion of organic molecules into CO_2 might occur in the interface (furnace). Since each C-atom position has a different δ^{13}C value (intramolecular ^{13}C patterns, see below), the loss of a single C-atom position should change the observed δ^{13}C value of a given molecule. That is why particular types of calibration curves should be carried out with GC-C-IRMS using mixtures of labelled and unlabelled pure standard of known δ^{13}C so as to check whether the δ^{13}C response of the IRMS is linear and indicates a constant apparent number of added exogenous carbons n_a. Should n_a be found to be non-integer and the response poorly linear, this would suggest that some carbon atoms are lost in the furnace with a certain probability. Up to now, this has been done for sugars (Derrien, Balesdent, Marol, & Santaella, 2003) but not for amino acids or organic acids.

Quite recently (say, 5 years), IRMS suppliers have commercialised systems coupled to a HPLC (LC-C-IRMS) for the $^{13}C/^{12}C$ isotopic analysis of organic compounds that can be separated with water as a mobile phase (of course, there should be no carbon in the mobile phase; see examples in Guyon, Gaillard, Salagoïty, & Médina, 2011; Morrison, Taylor, & Preston, 2010). The interface between the HPLC and the IRMS withdraws the solvent and converts organic molecules to CO_2. This allows the direct analysis of δ^{13}C in sugars and other soluble organics. In that sense, LC-C-IRMS techniques are advantageous. However, three major difficulties remain. First, adapted HPLC conditions with good chromatographic separation are difficult to set up for compounds other than sugars, and often require extreme pH values. Second, the rather narrow range of possible chromatographic conditions may require pre-purification before injection (eliminates co-elution with undesirable compounds), derivatisation (improves separation) or alternatively, the use of ionic chromatography rather than HPLC (still awaiting extensive development, but see Morrison et al., 2010). Third, the isotopic analysis of nitrogen isotopes is not possible, since the breakdown of organic compounds into CO_2 cannot eliminate all dissolved N_2 from the liquid phase. Therefore, presently, LC-C-IRMS techniques are most adapted to the $^{13}C/^{12}C$ isotopic analysis of soluble sugars and the number of studies reporting analyses from wine, honey and plant samples is growing (for a review on LC-C-IRMS techniques, see Godin & McCullagh, 2011). LC-C-IRMS analyses of other compounds (amino acids and organic acids) are still relatively scarce (e.g. Abaye, Morrison, & Preston, 2011; Dunn, Honch, & Evershed, 2011; Hettmann & Gleixner, 2012; Krummen et al., 2004), but the comparison of GC-C-IRMS and LC-C-IRMS results on amino acids showed a relatively good agreement of $\delta^{13}C$ values (Dunn et al., 2011), indicating that LC-C-IRMS is a very promising technique.

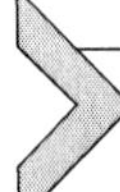

6. MAJOR ADVANCES IN METABOLIC PATTERNS PROVIDED BY IRMS TECHNIQUES

6.1. Carbon dioxide production by dark respiration

CO_2 evolved by respiration in darkness by plant organs is the product of complex respiratory processes and its origin may be traced using the carbon isotope composition ($\delta^{13}C_r$), which should reflect the carbon source oxidised by respiration (and possible associated fractionations). Due to possible isotopic variations in photosynthesis between plants or conditions, the $\delta^{13}C$ of respired CO_2 is usually expressed relative to source carbon (i.e. potential (but not necessarily actual) respiratory substrates, $\delta^{13}C_s$), that is, the respiratory $^{12}C/^{13}C$ discrimination (e_{dark}) is calculated as:

$$e_{dark} = \left(\delta^{13}C_s - \delta^{13}C_r\right)/\left(1 + \delta^{13}C_r\right)$$

$$e_{dark} \approx \delta^{13}C_s - \delta^{13}C_r$$

Quite generally, the isotope composition of soluble sugars (e.g. total soluble sugars or sucrose purified using HPLC) is used for $\delta^{13}C_s$. However, if more than one substrate type is used at the same time (e.g. a mixture of organic and fatty acids) or if substantial variations in sugars are observed, the carbon isotope discrimination during respiration can simply be expressed relative to total OM. It should be noted, however, that under fixed controlled conditions (including nutritional and isotopic growth conditions) and for a given plant species or tissue, $\delta^{13}C_r$ is rather constant and absolute values rather than e_{dark} may be exploited for metabolic purposes.

Pioneering investigations in the early 1970s have shown some variability in leaf $\delta^{13}C_r$ and e_{dark} (based on total OM) depending on species and environmental conditions (for a specific review, see Ghashghaie et al., 2003). Because of this variability (poorly understood at this time) and difficulties in identifying source carbon, this topic has not been investigated further for nearly 25 years. $\delta^{13}C_r$ has thus long been considered to be equal to $\delta^{13}C$ of photosynthetic products, thereby purely reflecting photosynthetic fractionation (see also Lin & Ehleringer, 1997). To the contrary, experimental data obtained in the last decade demonstrated that e_{dark} is not negligible and may vary so that $\delta^{13}C_r$ is an indicator of metabolic pathways and substrates used by leaf respiration, as explained below.

It has been shown that under typical conditions (20–25 °C, 21% O_2), dark-respired CO_2 evolved by C_3 leaves is ^{13}C-enriched compared with potential substrates, that is, leaf sucrose (Duranceau, Ghashghaie, Badeck, Deléens, & Cornic, 1999; Ghashghaie et al., 2001). In several crop leaves (bean, tobacco and sunflower), the $\delta^{13}C$ of leaf sucrose is naturally around −27‰ and $\delta^{13}C_r$ is around −21‰ so that $e_{dark} \approx -27 - (-21) = -6$‰. The negative value of e_{dark} indicates indeed the ^{13}C-enrichment in leaf-respired CO_2. The respiratory quotient[1] (RQ) measured on the same leaves is around 1, showing the use of carbohydrates as respiratory substrates under ordinary conditions (Tcherkez et al., 2003). Measurements on herbs and tree leaves detached and then incubated in vials gave similar results (Mortazavi et al., 2005; Mortazavi, Chanton, & Smith, 2006; Prater, Mortazavi, & Chanton, 2006; Werner, Unger, Pereira, Ghashghaie, & Máguas, 2007; Xu, Lin, Griffin, & Sambrotto, 2004; see also Werner & Gessler, 2011, for a recent review).

In order to elucidate the metabolic origin of the ^{13}C-enrichment in CO_2 and $\delta^{13}C_r$ values, the respiratory metabolism of leaves was modified by submitting the plants either to a continuous dark period (i.e. carbon deprivation) or to different temperatures (ranging from 10 to 30 °C) and the RQ was measured simultaneously (Tcherkez et al., 2003). A single linear relationship between RQ and $\delta^{13}C_r$ was observed, with ^{13}C-enriched CO_2 at high RQ (low temperature and no starving) and ^{13}C-depleted CO_2 at low RQ (high temperature and after a long period of darkness; Fig. 8.4). These original results have been since verified and exploited (for a review, see Bowling et al., 2008). In other words, leaf $\delta^{13}C_r$ appears to be an indicator of the nature of the respiratory substrate: the natural ^{13}C-enrichment indicates the consumption of sugars and the natural ^{13}C-depletion indicates the consumption of lipids.

However, there is a huge variability among species (ranging from −14‰ to −30‰ for C_3 and from −12‰ to −17‰ for C_4 species), with leaf-respired CO_2 being ^{13}C-depleted compared to plant OM in some cases (see reviews by Badeck et al., 2005; Ghashghaie et al., 2003; Werner & Gessler, 2011). Such

[1] The respiratory quotient is defined by the ratio of the rate of CO_2 evolution to the rate of O_2 consumption. This measures the oxygenation rate of the respiratory substrate and thus gives information on its nature: =1, carbohydrates; >1, gluconeogenesis (non-associated with lipid oxidation) or organic acids; <1, lipid oxidation.

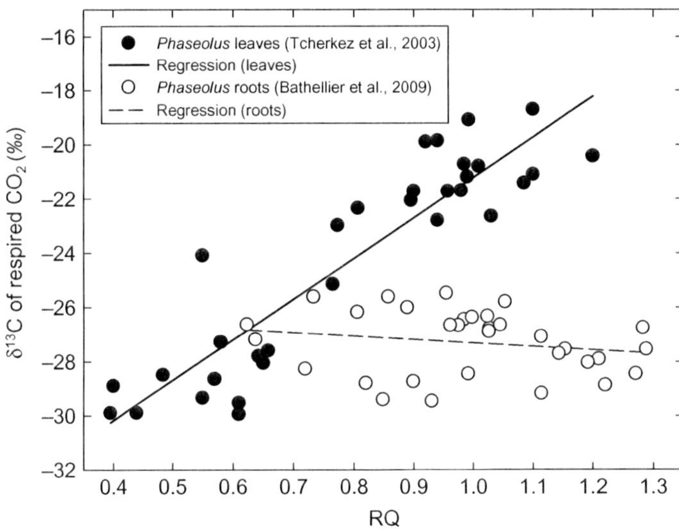

Figure 8.4 Variations of carbon isotope composition of respired CO_2 by attached leaves (closed symbols) and washed attached roots (open symbols) of *Phaseolus vulgaris* as a function of respiratory quotient (*RQ*). Regression lines are also presented. Data points correspond to individual measurements on different plants. Leaf data are from Tcherkez et al. (2003) and root ones from Bathellier et al. (2009).

variability probably reflects metabolic changes as well as different pathways (remobilisation of reserves, etc.) between species and conditions. Leaf $\delta^{13}C_r$ has also been shown to change a lot when plants are submitted to water deficit (Duranceau et al., 1999; Ghashghaie et al., 2001), but this mostly reflected changes in photosynthetic fractionation (stomatal conductance).

In roots, respired CO_2 has been shown to be relatively ^{13}C-depleted (compared to root sucrose) while under the same conditions and on the same plants, leaf-respired CO_2 was ^{13}C-enriched compared to sucrose (Badeck et al., 2005; Bathellier et al., 2008, 2009; Klumpp et al., 2005). Furthermore, in contrast to leaves, carbon deprivation (starvation under continuous darkness) did not cause any change in root $\delta^{13}C_r$. The metabolic origin of these differences between leaves and roots remains somewhat uncertain, but current hypotheses focus on changes in activity of enzymes catalysing (de)carboxylation with different isotope effects: pyruvate dehydrogenase (EC 1.2.4.1), the pentose phosphate pathway (PPP) and decarboxylases of the tricarboxylic acid pathway (TCAP) (isocitrate dehydrogenase and 2-oxoglutarate dehydrogenase). In other words, the current interpretation of $\delta^{13}C_r$ values involves intramolecular $\delta^{13}C$ patterns in metabolites and enzymatic isotope effects, as explained below.

6.2. Carbon dioxide produced in the light and in the light-to-dark transition

Classical gas exchange has shown for decades that respiratory leaf CO_2 evolution is inhibited in the light compared to the dark (for a review, see Atkin, Millar, Gärdestrom, & Day, 2000). Several isotopic techniques have now converged to a consistent picture of carbon metabolism feeding leaf day respiration (i.e. "mitochondrial", non-photorespiratory respiration occurring in the light). Isotope-sensitive gas analysers suggested that the turnover of the respiratory substrates is very slow in the light (Pinelli & Loreto, 2003). Recent IRMS measurements have further demonstrated that day respiration nearly exclusively oxidises carbon reserves and is thus divorced from current photosynthesis (Tcherkez, Mauve, Lamothe, Le Bras, & Grapin, 2011; Tcherkez et al., 2012). These recent findings highlight that (i) triose phosphates synthesised by photosynthesis and exported from the chloroplast to the cytoplasm are only marginally used by glycolysis and (ii) day/night cycle is crucial for leaf respiratory metabolism since carbon reserves accumulated in the dark are consumed in the light by day respiration (Gauthier et al., 2010). $\delta^{13}C_r$ associated with day respiration further show that day-evolved CO_2 is ^{13}C-depleted (up to 10‰ compared to leaf sucrose). This effect is presently believed to be caused by the large fractionation of respiratory decarboxylases which is, in turn, due to their low metabolic commitment (Tcherkez et al., 2012).

Upon darkening (light-to-dark transition), leaf respiration shows a peak lasting 5–20 min, the amplitude of which depends on irradiance during the light period (LEDR effect, *l*ight *e*nhanced *d*ark *r*espiration). New techniques (TDL spectrometry) have recently yielded high-resolution measurements of $\delta^{13}C_r$ during the light-to-dark transition and showed ^{13}C-enriched values, up to around −15‰, that is, nearly 15‰-enriched compared to OM (Barbour et al., 2007; Priault, Wegener, & Werner, 2009; Wegener, Beyschlag, & Werner, 2010). The LEDR-related isotopic effect is believed to stem from the abrupt consumption of (^{13}C-enriched) malate accumulated during the light period (Gessler et al., 2009; Werner, Buchmann, Siegwolf, Kornexl, & Gessler, 2011).

6.3. Primary carbon metabolism

As stated above, data on $\delta^{13}C$ values in individual metabolites (other than sugars) are relatively scarce and thus the current knowledge of isotopic metabolism at natural abundance is still fragmentary. It seems clear, however, that there are huge isotopic disparities among plant metabolites (Schmidt & Gleixner, 1998) due to enzymatic isotope effects. In fact, enzymes redistribute C-atom and fractionate between isotopes, causing substantial isotopic differences between C-atom positions and thus between metabolites. For example, aldolase (EC 4.1.2.13), which reversibly cleaves fructose-1,6-bisphosphate into two triose phosphates, fractionates against ^{13}C by around 3‰ and 4‰ in the C-3 and C-4 positions at equilibrium so that fructose-1,6-bisphosphate tends to be ^{13}C-enriched compared to triose phosphates (Gleixner & Schmidt, 1997). A comprehensive review of enzymatic isotope effects in plant primary metabolism has been published elsewhere (Tcherkez, Mahé, & Hodges, 2011). The compound-specific isotopic analysis of major plant metabolites has clearly shown that leaf starch is ^{13}C-enriched, fatty acids are ^{13}C-depleted and several organic acids are ^{13}C-enriched. A summary of metabolic information obtained from compound-specific $\delta^{13}C$ (including respired CO_2) is shown in Table 8.3. For example, the extent of the ^{13}C-enrichment in organic acids reflects the contribution of the PEPC (anaplerotic pathway). In fact, the PEPC fixes bicarbonate onto phospho*enol*pyruvate and yields oxaloacetate (OAA). Bicarbonate is naturally ^{13}C-enriched since atmospheric CO_2 is nearly 20‰-enriched compared to plant OM and furthermore, CO_2 hydration into bicarbonate fractionates against ^{12}C by $\approx$9‰. That is, the $\delta^{13}C$ of HCO_3^- fixed by the PEPC is thought to be of around -4‰. Combined to the isotope fractionation by PEPC catalysis itself ($\approx$3‰), the C-4 position in OAA evolved by the PEPC should be around -7‰. By contrast, the $\delta^{13}C$ of endogenous (non-anaplerotic) carbon is about -29‰. With a mass balance, one may calculate the plausible proportion of PEPC-derived molecules within OAA-derived metabolite pools, such as Asp: in tobacco leaves, the $\delta^{13}C$ of the C-4 position in Asp has been found to be near -5‰ and thus the PEPC should be responsible for $\geq$90% of Asp synthesis (Melzer & O'Leary, 1987).

6.4. The metabolic origin of ^{13}C-patterns

The enzymatic isotope effects that explain $\delta^{13}C$ values in metabolites and CO_2 are position specific since enzymatic catalysis primarily fractionates between isotopes at the level of C-atom positions involved in the chemical mechanisms (other positions being hardly affected). Therefore, the ^{13}C-distribution *within* metabolites is not homogeneous, with ^{13}C-enriched and ^{13}C-depleted C-atom positions. Conversely, if one has access to the

Table 8.3 Examples of metabolic commitments revealed by compound-specific IRMS analyses of plant samples

Metabolites analysed	Plant organ	Original finding	Biological significance	References
Starch, sucrose	Leaves	Starch ^{13}C-enriched, diel rhythm of $\delta^{13}C$ in sucrose (^{13}C-enriched in the dark)	Metabolic branching of triose phosphates and fractionation by aldolase (trioses exported by the chloroplast are ^{13}C-depleted)	Gessler, Tcherkez, Peuke, Ghashghaie, and Farquhar (2008) and Tcherkez, Farquhar, Badeck, and Ghashghaie (2004)
Asp, fumarate, citrate	Leaves	Aspartate ^{13}C-enriched particularly in the C-4 atom position, fumarate and others ^{13}C-enriched	Contribution of the PEPC in OAA synthesis	Hettmann and Gleixner (2012) and Melzer and O'Leary (1987)
Dark-respired CO_2, sucrose, glucose	Leaves	Respired CO_2 is ^{13}C-enriched (in the dark) compared to sucrose and glucose	Non-stoichiometric consumption of glucose molecules by respiration	Duranceau et al. (1999) and Tcherkez et al. (2003)
LEDR-respired CO_2	Leaves	CO_2 respired during the light-to-dark transition is considerably ^{13}C-enriched	Quantitative decarboxylation of ^{13}C-enriched malate by the malic enzyme	Barbour et al. (2007) and Gessler et al. (2009)
Free, non-proteinaceous Gly, Ser	Leaves, green algae	Gly and Ser are ^{13}C-enriched compared to other amino acids such as Glu	Photorespiration (glycine decarboxylase) liberates ^{13}C-depleted CO_2, thereby enriching in ^{13}C remaining glycine molecules	Abelson and Hoering (1961) and Tcherkez and Hodges (2008)
Respired CO_2, sucrose (and near-natural abundance labelling)	Leaves, roots	Respired CO_2 ^{13}C-depleted in roots compared to either leaf-respired CO_2 or root sucrose	Involvement of the pentose phosphate pathway to CO_2 production by roots	Bathellier et al. (2009)

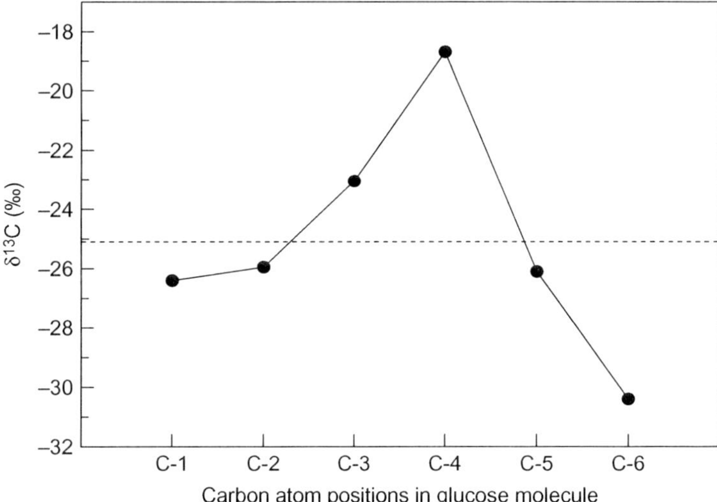

Figure 8.5 Non-statistical ^{13}C-distribution within glucose molecules measured on sugar beet syrup by Rossmann et al. (1991). The horizontal dashed line indicates δ^{13}C of the overall glucose.

intra-molecular ^{13}C-pattern in a given metabolite, this would give some information on the associated metabolism (see, e.g. the case of Asp above). The most important metabolite is certainly glucose because all the metabolites downstream inherit their carbon atoms from it. The first determination of the ^{13}C-distribution within plant glucose has been carried out by IRMS after (bio)chemical degradation. Glucose from both C_3 (sugar beet syrup) and C_4 (maize flour hydrolysate) has been shown to have a ^{13}C-enrichment in the C-3 and C-4 atom positions, while other positions (C-1, C-2, C-5 and C-6) are relatively ^{13}C-depleted compared to bulk glucose molecule (Fig. 8.5, adapted from Rossmann, Butzenlechner, & Schmidt, 1991). Recent data obtained with NMR-based analyses of glucose and the glucosyl moiety of sucrose from C_3, CAM and C_4 samples have shown rather similar patterns except in C-1 and C-2 (which are more enriched than with IRMS determinations of Gilbert, Silvestre, Robins, & Remaud, 2010; Gilbert, Silvestre, Robins, Tcherkez, & Remaud, 2011; Rossmann et al., 1991). The ^{13}C-pattern in fructose appears to be very different in C-1 and C-2, suggesting a crucial role of sugar interconversions (hexose isomerisation) (Gilbert et al., 2011). Very recently, the extensive examination (with NMR) of the intramolecular patterns in glucose and fructose in sugars from different origins as well as the experimental measurement of enzymatic isotope effects have highlighted clear relationships, sucrose metabolism and ^{13}C/^{12}C ratio within molecules glucose and fructose.

The heterogeneous ^{13}C-distribution in glucose explains some of the results described above. Along catabolism, CO_2 released by the pyruvate dehydrogenase (i.e. the C-1 of pyruvate that comes from the C-3 and C-4 positions of glucose) should be ^{13}C-enriched and the two other CO_2 molecules released by the TCAP should be ^{13}C-depleted. Any imbalance between the two isotopic types of CO_2 (i.e. non-stoichiometric CO_2 liberation) should cause an apparent respiratory fractionation. If dark CO_2 fixation by the PEPC and CO_2 production by the PPP are neglected, the isotope composition of dark-respired CO_2 should be close to −21‰ (which is the average of C-3 and C-4 positions in glucose) when pyruvate dehydrogenation predominates and close to −27‰ (average of C-1, C-2, C-5, and

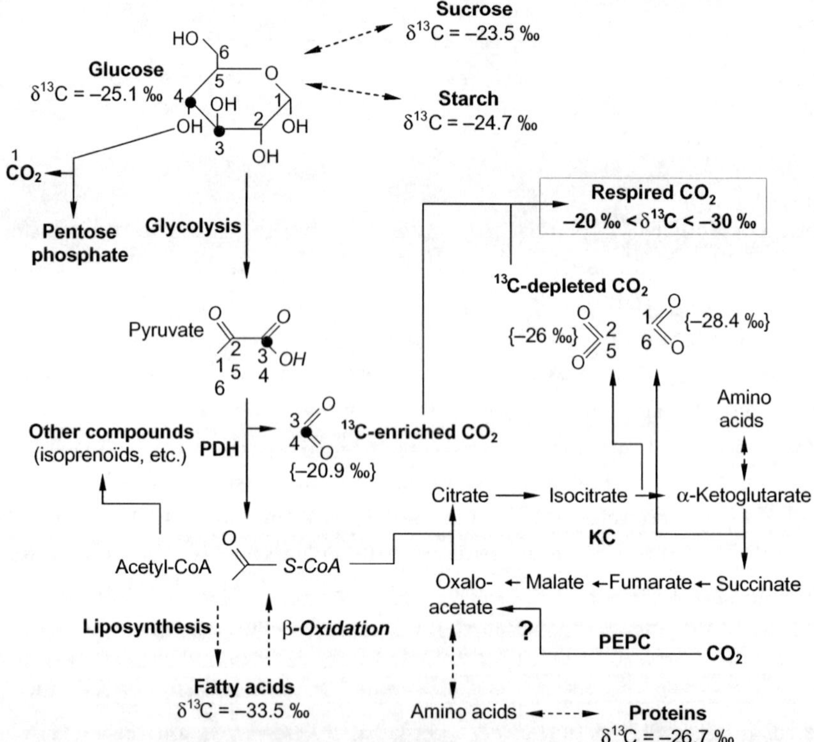

Figure 8.6 Carbon metabolism in darkness in relation to $\delta^{13}C$ of leaf respired CO_2. $\delta^{13}C$ values of metabolites are those measured in French bean leaves. $\delta^{13}C$ of CO_2 in brackets are derived from positional $\delta^{13}C$ values in natural glucose as given by Rossmann et al. (1991). PEPC, phospho*enol*pyruvate carboxylase; PDH, pyruvate dehydrogenase; KC, Krebs cycle. The symbol "?" points out the uncertainty about the amplitude of the PEPC reaction in darkness.. *Metabolic scheme from Tcherkez et al. (2003).*

C-6 positions in glucose) when the TCAP predominates. This interval fits well the observed variation range of evolved CO_2 under controlled conditions (Fig. 8.4). However, this does not explain $\delta^{13}C_r$ values (i) above −21‰, which could correspond to the use of ^{13}C-enriched substrates such as malate, and (ii) below −27‰ (down to −32‰), which correspond to the β-oxidation of fatty acids as demonstrated by the RQ (Fig. 8.6).

7. CONCLUSIONS AND PERSPECTIVES

In the past 10 years, considerable advances have been done in IRMS technology and applications: (i) the development of continuous-flow analyses (and also TDL spectrometry) of (respired) CO_2 not requiring cryogenic purification, (ii) a considerable number of isotopic measurements of $\delta^{13}C_r$ in leaves, other organs and soil either in the laboratory or in the field, (iii) protocols for GC-C-IRMS analyses of metabolites, (iv) the development of LC-C-IRMS techniques. However, data on compound-specific $\delta^{13}C$ in plant metabolites are still uncommon, impeding our understanding of plant metabolic isotopics. Although RQ data are not well developed currently, $\delta^{13}C_r$ is now used as a convenient tool to study respiratory metabolism at the organ or the plant level (e.g. differences between plant functional groups) and respiratory processes at the ecosystem level (Bowling et al., 2008; Werner et al., 2007; Wingate et al., 2007). For example, a significant difference in $\delta^{13}C_r$ between functional groups (fast growing herbs and slow growing Mediterranean shrubs) has been found (Priault et al., 2009). However, such observed values await clear metabolic explanations that will be eventually reached with high-throughput, reliable compound-specific isotopic analyses.

Beyond IRMS technology, crucial advances in plant isotopics will probably be provided by site-specific NMR analyses of the natural ^{13}C-abundance. The recent results obtained with plant sugars (described above) clearly show that intramolecular patterns give unique pieces of information on metabolic fluxes (for a review, see Gilbert et al., 2012). The NMR-based techniques applied to plant compounds is still made difficult by the large amounts required to carry out NMR analyses (≈ 10 g) and also by the complex chemical derivatisation required to obtain a fully resolved NMR spectrum (Gilbert et al., 2011). More adapted NMR sequences, polarisation transfer or probes with a better gain are thus needed in the near future to yield more intramolecular isotopic data at natural abundance.

REFERENCES

Abaye, D. A., Morrison, D. J., & Preston, T. (2011). Strong anion exchange liquid chromatographic separation of protein amino acids for natural ^{13}C-abundance determination by isotope ratio mass spectrometry. *Rapid Communications in Mass Spectrometry*, *25*, 429–435.

Abelson, H., & Hoering, T. C. (1961). Carbon isotope fractionation in formation of amino acids by photosynthetic organisms. *Proceedings of the National Academy of Sciences of USA*, *47*, 623–632.

Atkin, O. K., Millar, A. H., Gärdestrom, P., & Day, D. A. (2000). Photosynthesis, carbohydrate metabolism and respiration in leaves of higher plants. In R. C. Leegood, T. D. Sharkey, & S. von Caemmerer (Eds.), *Photosynthesis, physiology and metabolism* (pp. 153–175). London: Kluwer Academic Publisher.

Badeck, F. W., Tcherkez, G., Nogués, S., Piel, C., & Ghashghaie, J. (2005). Post-photosynthetic fractionation of stable carbon isotopes between plant organs—A widespread phenomenon. *Rapid Communications in Mass Spectrometry*, *19*, 1381–1391.

Barbour, M. M., McDowell, N. G., Tcherkez, G., Bickford, C. P., & Hanson, D. T. (2007). A new measurement technique reveals rapid post-illumination changes in the carbon isotope composition of leaf-respired CO_2. *Plant, Cell and Environment*, *30*, 469–482.

Bathellier, C., Badeck, F. W., Couzi, P., Harscoët, S., Mauve, C., & Ghashghaie, J. (2008). Divergence in $\delta^{13}C$ of dark respired CO_2 and bulk organic matter occurs during the transition between heterotrophy and autotrophy in *Phaseolus vulgaris* L. plants. *New Phytologist*, *177*, 406–418.

Bathellier, C., Tcherkez, G., Bligny, R., Gout, E., Cornic, G., & Ghashghaie, J. (2009). Metabolic origin of $\delta^{13}C$ of respired CO_2 from the roots of *Phaseolus vulgaris*. *New Phytologist*, *181*, 387–399.

Bowling, D. R., Pataki, D. E., & Randerson, J. T. (2008). Carbon isotopes in terrestrial ecosystem pools and CO2 fluxes. *New Phytologist*, *178*, 24–40.

Bowling, D. R., Sargent, S. D., Tanner, B. D., & Ehleringer, J. R. (2003). Tunable diode laser absorption spectroscopy for stable isotope studies of ecosystem-atmosphere CO_2 exchange. *Agricultural and Forest Meteorology*, *118*, 1–19.

Brugnoli, E., & Farquhar, G. D. (2000). Photosynthetic fractionation of carbon isotopes. In R. C. Leegood, T. D. Sharkey, & S. von Caemmerer (Eds.), *Photosynthesis: Physiology and metabolism*. (pp. 399–434). The Netherlands: Kluwer Academic Publishers.

Cernusak, L. A., Tcherkez, G., Keitel, C., Cornwell, W. K., Santiago, L. S., Knohl, A., et al. (2009). Viewpoint. Why are non-photosynthetic tissues generally ^{13}C enriched compared with leaves in C_3 plants? Review and synthesis of current hypotheses. *Functional Plant Biology*, *36*, 199–213.

Corr, R. T., Berstan, R., & Evershed, R. P. (2007). Optimisation of derivatisation procedures for the determination of $\delta^{13}C$ values of amino acids by gas chromatography/combustion/isotope ratio mass spectrometry. *Rapid Communications in Mass Spectrometry*, *21*, 3759–3771.

Craig, H., & Boato, G. (1956). Isotopes. *Annual Review of Physical Chemistry*, *6*, 403–432.

Deléens, E., Morot-Gaudry, J. F., Martin, F., Thoreux, A., & Gojon, A. (1997). Méthodologie ^{15}N. In J. F. Morot-Gaudry (Ed.), *Assimilation de l'azote chez les plantes: Aspects physiologique, biochimique et moléculaire* (pp. 265–280). Route de St Cyr Versailles Cedex, France: Edition INRA.

Derrien, D., Balesdent, J., Marol, C., & Santaella, C. (2003). Measurement of the 13C/12C ratio of soil-plant individual sugars by gas chromatography/combustion/isotope-ratio mass spectrometry of silylated derivatives. *Rapid Communications in Mass Spectrometry*, *17*, 2626–2631.

Dunn, P. J. H., Honch, N. J., & Evershed, R. P. (2011). Comparison of liquid chromatography–isotope ratio mass spectrometry (LC/IRMS) and gas chromatography–combustion–isotope ratio mass spectrometry (GC/C/IRMS) for the determination of collagen amino acid $\delta^{13}C$ values for palaeodietary and palaeoecological reconstruction. *Rapid Communications in Mass Spectrometry, 25*, 2995–3011.

Duranceau, M., Ghashghaie, J., Badeck, F. W., Deléens, E., & Cornic, G. (1999). $\delta^{13}C$ of CO_2 respired in the dark in relation to $\delta^{13}C$ of leaf carbohydrates *in Phaseolus vulgaris* L. under progressive drought. *Plant Cell and Environment, 22*, 515–523.

Evans, J. R., Sharkey, T. D., Berry, J. A., & Farquhar, G. D. (1986). Carbon isotope discrimination measured concurrently with gas exchange to investigate CO_2 diffusion in leaves of higher plants. *Australian Journal of Plant Physiology, 13*, 281–292.

Farquhar, G. D., Ehleringer, J. R., & Hubick, K. T. (1989). Carbon isotope discrimination and photosynthesis. *Annual Review of Plant Physiology and Plant Molecular Biology, 40*, 503–537.

Farquhar, G. D., O'Leary, M. H., & Berry, J. A. (1982). On the relationship between carbon isotope discrimination and the intercellular carbon dioxide concentration in leaves. *Australian Journal of Plant Physiology, 9*, 121–137.

Gauthier, P., Bligny, R., Gout, E., Mahé, A., Nogués, S., Hodges, M., et al. (2010). *In folio* isotopic tracing demonstrates that nitrogen assimilation into glutamate is mostly independent from current CO_2 assimilation in illuminated leaves of *Brassica napus*. *New Phytologist, 185*, 988–999.

Gessler, A., Tcherkez, G., Karyanto, O., Keitel, C., Ferrio, J. P., Ghashghaie, J., et al. (2009). On the metabolic origin of the carbon isotope composition of CO_2 evolved from darkened light-adapted leaves in *Ricinus communis*. *New Phytologist, 181*, 374–386.

Gessler, A., Tcherkez, G., Peuke, A., Ghashghaie, J., & Farquhar, G. D. (2008). Diel variations of the carbon isotope composition in leaf, stem and phloem sap organic matter in *Ricinus communis*. *Plant, Cell and Environment, 31*, 941–953.

Ghashghaie, J., Badeck, F. W., Lanigan, G., Nogués, S., Tcherkez, G., Deléens, E., et al. (2003). Carbon isotope discrimination during dark respiration and photorespiration in C_3 plants. *Phytochemistry Reviews, 2*, 145–161.

Ghashghaie, J., Duranceau, M., Badeck, F. W., Cornic, G., Adeline, M. T., & Deléens, E. (2001). $\delta^{13}C$ of CO_2 respired in the dark in relation to $\delta^{13}C$ of leaf metabolites: Comparison between *Nicotiana sylvestris* and *Helianthus annuus* under drought. *Plant, Cell and Environment, 24*, 505–515.

Gilbert, A., Silvestre, V., Robins, R. J., & Remaud, G. S. (2010). Impact of the deuterium isotope effect on the accuracy of ^{13}C NMR measurements of site-specific isotope ratios at natural abundance in glucose. *Analytical and Bioanalytical Chemistry, 398*, 1979–1984.

Gilbert, A., Silvestre, V., Robins, R. J., Tcherkez, G., & Remaud, G. S. (2011). A ^{13}C NMR spectrometric method for the determination of intramolecular $\delta^{13}C$ values in fructose from plant sucrose samples. *New Phytologist, 191*, 579–588.

Gilbert, A., Silvestre, V., Segebarth, N., Tcherkez, G., Guillou, C., Robins, R. J., et al. (2012). The intramolecular ^{13}C-distribution in ethanol reveals the influence of the CO_2-fixation pathway and environmental conditions on the site-specific ^{13}C variation in glucose. *Plant, Cell and Environment, 34*, 1104–1112.

Gleixner, G., & Schmidt, H. L. (1997). Carbon isotope effects on the fructose-1, 6-bisphosphate aldolase reaction, origin for non-statistical ^{13}C distribution in carbohydrates. *Journal of Biological Chemistry, 272*, 5382–5387.

Godin, J. P., & McCullagh, J. S. O. (2011). Review: Current applications and challenges for liquid chromatography coupled to isotope ratio mass spectrometry (LC/IRMS). *Rapid Communications in Mass Spectrometry, 25*, 3019–3028.

Guyon, F., Gaillard, L., Salagoïty, M. H., & Médina, B. (2011). Intrinsic ratios of glucose, fructose, glycerol and ethanol $^{13}C/^{12}C$ isotopic ratio determined by HPLC-co-IRMS: Toward determining constants for wine authentication. *Analytical and Bioanalytical Chemistry, 401*, 1551–1558.

Hettmann, E., & Gleixner, G. (2012). IRMS-LC/MS: $\delta^{13}C$ analysis of organic acids in plants. *Application Note #30075*, Thermo Electron Corporation, Bremen, Germany.

Klumpp, K., Schäufele, R., Lötscher, M., Lattanzi, F. A., Feneis, W., & Schnyder, H. (2005). C-isotope composition of CO_2 respired by roots: fractionation during dark respiration? *Plant Cell & Environment, 28*, 241–250.

Krummen, M., Hilkbert, A. H., Juchelka, D., Duhr, A., Schlueter, H. J., & Pesch, R. (2004). A new concept for isotope ratio monitoring LC/MS. *Rapid Communications in Mass Spectrometry, 18*, 2260–2266.

Lin, G., & Ehleringer, J. R. (1997). Carbon isotopic fractionation does not occur during dark respiration of C_3 and C_4 plants. *Plant Physiology, 114*, 191–194.

Loader, N. J., Robertson, I., & McCarroll, D. (2003). Comparison of stable carbon isotope ratios in the whole wood, cellulose and lignin of oak tree-rings. *Palaeogeography, Palaeoclimatology, Palaeoecology, 196*, 395–407.

Mauve, C., Jean Bleton, J., Bathellier, C., Lelarge-Trouverie, C., Guérard, F., Ghashghaie, J., et al. (2009). Kinetic $^{12}C/^{13}C$ isotope fractionation by invertase: Evidence for a small in vitro isotope effect and comparison of two techniques for the isotopic analysis of carbohydrates. *Rapid Communications in Mass Spectrometry, 23*, 2499–2506.

Melzer, E., & O'Leary, M. H. (1987). Anaplerotic CO_2 fixation by phospho*enol*pyruvate carboxylase in C_3 plants. *Plant Physiology, 84*, 58–60.

Molero, G., Aranjuelo, I., Teixidor, P., Araus, J. L., & Nogués, S. (2011). Measurement of ^{13}C and ^{15}N isotope labeling by gas chromatography/combustion/isotope ratio mass spectrometry to study amino acid fluxes in a plant–microbe symbiotic association. *Rapid Communications in Mass Spectrometry, 25*, 599–607.

Morrison, D. J., Taylor, K., & Preston, T. (2010). Strong anion-exchange liquid chromatography coupled with isotope ratio mass spectrometry using a Liquiface interface. *Rapid Communications in Mass Spectrometry, 24*, 1755–1762.

Mortazavi, B., Chanton, J. P., Prater, J. L., Oishi, A. C., Oren, R., & Katul, G. (2005). Temporal variability in ^{13}C of respired CO_2 in a pine and a hardwood forest subject to similar climatic conditions. *Oecologia, 142*, 57–69.

Mortazavi, B., Chanton, J. P., & Smith, M. C. (2006). Influence of ^{13}C-enriched foliage respired CO_2 on $\delta^{13}C$ of ecosystem-respired CO_2. *Global Biogeochemistry, 20*, 3029–3038.

Pinelli, P., & Loreto, F. (2003). $^{12}CO_2$ emission from different metabolic pathways measured in illuminated and darkened C_3 and C_4 leaves at low, atmospheric and elevated CO_2 concentration. *Journal of Experimental Botany, 54*, 1761–1769.

Prater, J. L., Mortazavi, B., & Chanton, J. P. (2006). Diurnal variation of the $\delta^{13}C$ of pine needle respired CO_2 evolved in darkness. *Plant, Cell and Environment, 29*, 202–211.

Priault, P., Wegener, F., & Werner, C. (2009). Pronounced differences in diurnal variation of carbon isotope composition of leaf respired CO_2 among functional groups. *New Phytologist, 181*, 400–412.

Richter, A., Wanek, W., Werner, R. A., Ghashghaie, J., Jäggi, M., Gessler, A., et al. (2009). Preparation of starch and soluble sugars from plant material for analysis of carbon isotope composition: A comparison of methods. *Rapid Communications in Mass Spectrometry, 23*, 2476–2488.

Rossmann, A., Butzenlechner, M., & Schmidt, H. L. (1991). Evidence for a non-statistical carbon isotope distribution in natural glucose. *Plant Physiology, 96*, 609–614.

Schmidt, H. L., & Gleixner, G. (1998). Carbon isotope effects on key reactions in plant metabolism and ^{13}C-patterns in natural compounds. In H. Griffiths (Ed.), *Stable isotopes: Integration of biological, ecological and geochemical processes. Environmental plant biology series.* (pp. 13–25). Oxford, UK: Bios Scientific Publishers.

Schmidt, T. C., Zwank, L., Elsner, M., Berg, M., Meckenstock, R. U., & Haderlein, S. B. (2004). Compound-specific stable isotope analysis of organic contaminants in natural environments: A critical review of the state of the art, prospects, and future challenges. *Analytical and Bioanalytical Chemistry, 378*, 283–300.

Tcherkez, G. (2004). Discrimination isotopique lors de la respiration à l'obscurité et marquage isotopique des photo-assimilats et des substrats de réserves : Outils pour étudier (i) les voies métaboliques et (ii) l'établissement et l'utilisation des réserves de la plante. PhD Thesis, Université de Paris-Sud 11, Faculté des Sciences d'Orsay, France.

Tcherkez, G., Boex-Fontvieille, E., Mahé, A., & Hodges, M. (2012). Respiratory carbon fluxes in leaves. *Current Opinion in Plant Biology, 15*, 308–314.

Tcherkez, G., Farquhar, G. D., Badeck, F., & Ghashghaie, J. (2004). Theoretical considerations about carbon isotope distribution in glucose of C_3 plants. *Functional Plant Biology, 31*, 857–877.

Tcherkez, G., & Hodges, M. (2008). How stable isotopes may help to elucidate primary nitrogen metabolism and its interaction with (photo)respiration in C_3 leaves. *Journal of Experimental Botany, 59*, 1685–1693.

Tcherkez, G., Mahé, A., & Hodges, M. (2011). $^{12}C/^{13}C$ fractionations in plant primary metabolism. *Trends in Plant Science, 16*, 499–506.

Tcherkez, G., Mauve, C., Lamothe, M., Le Bras, C., & Grapin, A. (2011). On the $^{13}C/^{12}C$ isotopic signal of day respired CO_2 in variegated leaves of *Pelargonium* x *hortorum*. *Plant, Cell and Environment, 34*, 270–283.

Tcherkez, G., Nogués, S., Bleton, J., Cornic, G., Badeck, F., & Ghashghaie, J. (2003). Metabolic origin of carbon isotope composition of leaf dark-respired CO_2 in French bean. *Plant Physiology, 131*, 237–244.

Wanek, W., Heintel, S., & Richter, A. (2001). Preparation of starch and other carbon fractions from higher plant leaves for stable carbon isotope analysis. *Rapid Communications in Mass Spectrometry, 15*, 1136–1140.

Wegener, F., Beyschlag, W., & Werner, C. (2010). The magnitude of diurnal variation in carbon isotopic composition of leaf dark respired CO_2 correlates with the difference between $\delta^{13}C$ of leaf and root material. *Functional Plant Biology, 37*, 849–858.

Werner, R. A., Buchmann, N., Siegwolf, R., Kornexl, B., & Gessler, A. (2011). Metabolic fluxes, carbon isotope fractionation and respiration—Lessons to be learned from plant biochemistry. *New Phytologist, 191*, 10–15.

Werner, C., & Gessler, A. (2011). Diel variations in the carbon isotope composition of respired CO_2 and associated carbon sources: A review of dynamics and mechanisms. *Biogeosciences, 8*, 2437–2459.

Werner, R. A., & Schmidt, H. L. (2002). The in vivo nitrogen isotope discrimination among organic plant compounds. *Phytochemistry, 61*, 465–484.

Werner, C., Unger, S., Pereira, J. S., Ghashghaie, J., & Máguas, C. (2007). Temporal dynamics in $\delta^{13}C$ of ecosystem respiration in response to environmental changes. In T. Dawson & R. Siegwolf (Eds.), *Isotopes as tracers of ecological change* (pp. 193–210). USA: Elsevier Academic Press, ISBN 978-012-373627-7.

Wingate, L., Seibt, U., Moncrieff, J. B., Jarvis, P. G., & Lloyd, J. (2007). Variations in ^{13}C discrimination during CO_2 exchange by *Picea sitchensis* branches in the field. *Plant, Cell and Environment, 30*, 600–616.

Xu, C., Lin, G., Griffin, K. L., & Sambrotto, R. N. (2004). Leaf respiratory CO_2 is ^{13}C-enriched relative to leaf organic components in five species of C_3 plants. *New Phytologist, 163*, 499–505.

CHAPTER NINE

High-Throughput Biochemical Phenotyping for Plants

Guillaume Ménard[*], Benoit Biais[*], Duyen Prodhomme[*], Patricia Ballias[*,†], Johann Petit[*], Daniel Just[*], Christophe Rothan[*], Dominique Rolin[†,‡], Yves Gibon[*,†,1]

[*]INRA, UMR 1332 Fruit Biology and Pathology, INRA Centre of Bordeaux, Villenave d'Ornon, France
[†]Metabolome Facility of Bordeaux, Functional Genomics Centre, INRA Centre of Bordeaux, Villenave d'Ornon, France
[‡]University of Bordeaux, UMR1332 Fruit Biology and Pathology, INRA Centre of Bordeaux, Villenave d'Ornon, France
[1]Corresponding author: e-mail address: yves.gibon@bordeaux.inra.fr

Contents

Abstract

There is an urgent need in low cost and fast technologies that enable the exploration of natural or induced biodiversity in plants. Biochemical phenotyping is often considered as particularly promising to identify given analytes that are linked to desirable plant phenotypes and that could be used as markers for plant performance. Over the past 15 years, metabolomics and other omics approaches have been intensively used to study plant phenotypes. Still, they do not enable routine screens of very large populations in

Advances in Botanical Research, Volume 67
ISSN 0065-2296
http://dx.doi.org/10.1016/B978-0-12-397922-3.00009-5

breeding programs, probably because this would represent unaffordable financial investments for equipment and specialised engineers. Along the last century, biochemical phenotyping and -screening have nevertheless been successfully exploited in medicine, pharmaceutical research and agro-industry. In medicine, over 6000 biochemical procedures exploiting a large variety of concepts and equipment are available for routine diagnosis and therapeutic monitoring. Pharmaceutical companies have spent massive investments to screen for biologically active molecules, whereas in the agro-industry increasingly sophisticated biochemical analysis has been developed to control the quality of raw and transformed products. Microplate technology is a well-established technology, which originated from the medical field and benefited from huge investments by the pharmaceutical industry. It offers the possibility to operate a wide range of analyses at very low costs per sample in plant research. To illustrate the potential of microplate technology as a high-throughput phenotyping tool, a screen for tomato metabolic mutants is shown where 24 enzymatic traits have been measured in more than 1500 samples within only 1 month with a relatively small investment.

1. INTRODUCTION

Today in agriculture, the challenge is to increase crop production in a sustainable way and in a changing world. Climate changes, in particular global warming due to greenhouse effects, pose a serious new challenge for agricultural production. Abiotic stresses such as drought, heat or salinity cause extensive losses to agricultural production worldwide, directly or indirectly by favouring pests and diseases. According to FAO recommendations (Ruane & Sonnino, 2011), biotechnologies can help to face these challenges in modern agriculture, assuming they are not the unique solution. They represent a broad range of technologies linked to fundamental biology and technological innovation, and are used to enhance the production of plant crops, livestock, forestry, fisheries and agro-industry. For crops, a combination of scientific strategies will be needed to maintain yield under climate changes, and in particular to significantly improve abiotic stress tolerance in the field. These will include more fundamental studies of stress responses, with extensive testing of new varieties in the laboratory, greenhouse and field.

Up to recently, new varieties were created by exploring natural diversity, selecting agronomical traits of interest and breeding into elite cultivars by using empirical approaches (Miflin, 2000). The selection of new varieties typically relied on the scoring of desired criteria such as biomass or harvest index. Then, breeders had to wait the next generation to see if their selection was efficient, a practise that was subject to environmental fluctuations and extremely time-consuming. In 1966, the International Rice Research Institute (IRRI) collected more than 10,000 rice varieties around the world, eventually creating the IR-8 rice variety, which has a stiff, strong and short straw and high yield, after more than 20 years of conventional breeding (Herdt, Capule, & International Rice Research Institute, 1983). Although a range of improvements including applications of biotechnology have considerably shortened the time it takes to bring new seeds on the market (it once took almost 25 years, it currently takes about 7–10 years) the time needed might become too slow regarding predictions for human demography and climate (Tester & Langridge, 2010). Then, successes of plant breeders have been based on relatively few improvements of relatively few genes (Gurian-Sherman, 2009). Finally, a survey of the world crop yields and a number of long-term experiments suggest that the rate of yield increases has slowed down over the last decade (Brisson et al., 2010; Finger, 2010; Lobell, Schlenker, & Costa-Roberts, 2011).

Over the past 20 years, considerable progress has been achieved in the development of technologies to both create and characterise genetic diversity. Integration of advances in biotechnology, genomic research and use of molecular markers with conventional plant breeding practises has created the foundation for molecular plant breeding, an interdisciplinary science that is revolutionising crop improvement. The implementation of large populations of mutants, cultivars and wild relatives has made an enormous genetic diversity available to researchers and breeders. Concurrently, the development of high-density technologies including next gene sequencing results in a dramatic acceleration of genotyping, and an exponentially increasing number of genomes are thus being fully sequenced. One consequence is that breeders will benefit from tools such as marker-assisted selection and high-density genetic mapping, which enable cost-effective tracking of genes/loci of interest in programs of crop improvement. However, the use of molecular markers to select traits of interest has not yet become the standard for breeders. Indeed, a large proportion of markers found in academic programs fail when transferred to the application domain (Xu & Crouch, 2008). Whereas major-effect genes have been successfully introgressed and pyramided, traits resulting from complex interactions such as epistasis and genotype by environment interactions remain challenging and are likely to require the implementation of accurate phenotype scoring (Furbank & Tester, 2011).

In contrast to genotyping, the evaluation of plant phenotypes, defined as sets of observable characteristics resulting from interactions between genotypes and environments, benefited from less methodological advances. In breeding programs, evaluations of phenotypes classically combine observations of visual traits (e.g. plant size, colour, presence or not of diseases), easy to score physiological indicators (e.g. growth rate, flowering date) or simple quantitative traits (e.g. yield). Whereas such rather empirical evaluations have proven very efficient, in particular when they are used to detect quantitative trait loci, it is likely that they will become insufficient in the next future. There are nevertheless opportunities to enhance our understanding of genotype–phenotype associations by studying phenotypes into much greater detail. On the one hand, a wide range of high-density phenotyping approaches including transcriptomics, proteomics and metabolomics, which enable the detection of many analytes per sample, have been developed during the last decade and have the potential to greatly facilitate gene discovery when combined with genomic tools (Lee et al., 2012). On the other hand, the development of high-throughput (HT) phenotyping approaches enabling the analysis of large numbers of samples or individuals, from whole plant traits to molecules, offers new opportunities for studying increasingly large populations of genotypes. Whereas high-density approaches are rather limited to studies with relatively small populations, HT methods are usually restricted to a few analytes. Both approaches might actually prove highly complementary, as in depth phenotyping of one or a few genotypes can provide robust markers of performance, which can then be used to screen large populations using cost-effective approaches (Gibon, Rolin, Deborde, Bernillon, & Moing, 2012). From this point of view, it seems logical that high-density approaches received much more attention in the recent past, thus revolutionising 'descriptive biology'. In contrast, the concept of HT plant phenotyping, also called 'plant phenomics' has only emerged recently in plant research (Furbank, 2009). Thus, various new tools are presently being adopted by an increasing number of research groups and seed companies to improve phenotyping. They include fully automated greenhouses, devices for whole plant imaging and new strategies for data management. Although plant phenomics rather focus on non-destructive and non-invasive techniques, there is room for biochemical traits. It is indeed expected that the integration of information being generated with high-density approaches will deliver valuable targets for crop improvement.

This chapter explores the high diversity of approaches of biochemical screening and phenotyping methodologies that have been developed in the medical, pharmaceutical, agro-industrial and breeding areas. It then highlights microplate technology, which appears as particularly appropriate for handling very large numbers of samples at reasonable costs. Finally, a screen for tomato metabolic mutants is described to exemplify the use of an HT phenotyping pipeline.

2. BIOCHEMICAL PHENOTYPING IN MEDICINE, AGRO INDUSTRY AND AGRICULTURE

Biochemical phenotyping is defined as the determination of the steady-state concentration of a broad spectrum of biological compounds including not only metabolites, but also RNA and proteins (Roessner, Willmitzer, & Fernie, 2002). The biochemical phenotype provides readouts of the metabolic state of the cells, tissues or individuals and as any phenotype it is the product of interactions between the genotype and the environment. Metabolic profiles provide information about the biochemistry, the physiological status and the environmental exposures that cannot necessarily be obtained from the genotype. Biochemical phenotypes have been exploited in many challenging purposes such as diagnosis or therapeutic monitoring in medicine, monitoring of product quality in agro-industry or survey of crop quality in agriculture.

2.1. Medical applications

The evolution of clinical chemistry and laboratory medicine along the last century provides a good example of how new insight from basic research in biology, biochemistry, engineering and informatics can be transferred into the daily routine to improve diagnosis and prevention. Over the last century, biochemical phenotyping played a major role in the strong transformation of medicine. Originally, biochemical parameters were determined only selectively and usually only during the course of therapy in order to confirm a tentative diagnosis. Led by the need to improve the ability to diagnose diseases, the biomedical industry developed a set of dedicated methods and instruments. All these technologies were based on the concept of biomarker, which defines as a characteristic that is objectively measured and evaluated as an indicator of a normal or pathogenic biological process, or a pharmacological response to a therapeutic intervention. In the second half of the twentieth century, enormous advances have been achieved in the fields of biochemistry, technology (mechanics, optics, hydraulics, electronics, data processing, etc.) and methodology. Greater sensitivity, more specificity and automation have shifted clinical chemistry from manual and artisanal activities to a profitable industry. For example, in 2007 more than 1 million biochemical tests have been prescribed in the Lariboisière-Fernand Widal hospital (Paris, France), a middle size hospital with 750 beds (Dozol et al., 2010). Apart from the need for standardised logistics, such sample throughput represents a challenge from the standpoint of the analytical instrumentation. Rapid mastery of this workload has been achieved in large part through advances in cuvette-based enzymatic colorimetry, which is the basis for many routine detection methods. Automation presupposed robotised pipetting systems, mechanical drive arrangements for the cuvettes and analogue/digital converters for the output signal. The first fully automated determinations of urea, glucose and calcium have been demonstrated at the International Congress on Clinical Chemistry in 1957 (Skeggs, 1957). A major breakthrough occurred in the early 1980s with the introduction of a new photometric principle: the diode array spectrophotometer, which enables simultaneous analyses at multiple wavelengths from a fixed spectrum with work cycles of less than 10 s. Table 9.1 lists metabolites that can be automatically measured by typical clinical chemistry analysers in a medical

Table 9.1 List of analytes measured by automated clinical chemistry analysers

Enzyme		*Specific proteins*	
Alanine aminotransferase	Creatine kinase	Apolipoprotein A	Immunoglobulin G
Alanine aminotransferase (P5P)	γ-Glutamyltransferase	Apolipoprotein B	Immunoglobulin M
Alkaline phosphatase (AMP)	Glucose hexokinase	β_2 Microglobulin	Lipoprotein (a)
Alkaline phosphatase (DEA)	Glucose oxidase	Complement 3 (C3)	Microalbumin
Amylase	Lactate dehydrogenase (L-P)	Complement 4 (C4)	Myoglobin
Aspartate aminotransferase	Lipase	C-reactive Protein (CRP)	Prealbumin
Aspartate aminotransferase (P5P)	Pancreatic amylase	CRP, wide range	Rheumatoid factor
Cholinesterase		CardioPhase® hsCRP	Transferrin

Table 9.1 List of analytes measured by automated clinical chemistry analysers—cont'd

Analyte		*Drugs of abuse/toxicology*	
Albumin (BCG)	Iron	Acetaminophen	Methadone
Albumin (BCP)	Lactate	Amphetamines	Nitrite specimen validity test
Ammonia	LDL cholesterol, direct	Barbiturates (serum)	Opiates
Bilirubin, direct	Magnesium	Barbiturates (urine)	Oxidant specimen validity test
Bilirubin, total	Phosphorus, inorganic	Benzodiazepines (serum)	pH specimen validity test
Calcium (Arsenazo)	Potassium	Benzodiazepines (urine)	Phencyclidine
Calcium (CPC)	Sodium	Cannabinoids (THC)	Propoxyphene
Carbon dioxide	Total iron-binding capacity	Cocaine metabolite	Salicylate
Chloride (ISE)	Total protein	Creatinine specimen validity test	Specific gravity specimen validity test
Cholesterol	Triglycerides	Ecstasy	Tricyclic antidepressants
Creatinine (Jaffé)	Urea nitrogen	Ethyl alcohol	
Creatinine (enzymatic)	Uric acid		
HDL cholesterol, direct			
Specific proteins		*Therapeutic drug monitoring*	
α_1-Acid glycoprotein	Ferritin	Carbamazepine	Phenobarbital
Antistreptolysin O	Haptoglobin	Digoxin	Phenytoin
α_1-Antitrypsin	Immunoglobulin A	Gentamicin Lithium	Theophylline Tobramycin

These compounds could all be analysed in a routine medicine diagnosis.

laboratory. Today over 6000 experimental procedures with large variety of equipment are available for the routine determination of medical analyses (Durner, 2010).

Among these, the enzyme-linked immune sorbent assay (ELISA) developed by Engvall and Perlmann (1971) has become very popular. It uses a heterogeneous solid-phase enzyme immunoassay to detect the presence of a substance in a liquid or wet sample. It became a widespread diagnostic tool in medicine and plant pathology, as well as a quality control check in various industries. Today more than 50,000 ELISA kits are available on the market, all using specific biomarkers (Crowther, 2010). Because ELISA tests can be performed to evaluate the presence of either an antigen or an antibody in a sample, it is used in routine for determining serum concentrations of antibodies raised against HIV and many other viral diseases. It has also found applications in the food industry, for example, the detection of potential food allergens originating from milk, peanuts, walnuts, almonds and eggs. Interestingly, ELISA tests can also be used to detect metabolites. Thus, in toxicology, they are used as a rapid presumptive screen for certain classes of drugs. Such tests are also used to detect and quantify phytohormones such as ABA (Nagamune et al., 2008). However, a large majority of metabolites is probably not detectable via ELISA tests, given they cannot be recognised by antibodies.

During the 1990s, rapid technical developments have been achieved in the field of mass spectroscopy (MS), leading to a range of widely used analytical techniques. Millington and Chace introduced the first routine application of metabolite screening by tandem mass spectrometry in the clinical laboratory and paved the way for metabolome investigations into clinical routine applications (Chace et al., 1993; Millington, Kodo, Terada, Roe, & Chace, 1991). The capacity to follow 30 metabolic parameters of amino acids and fatty acids metabolism for the detection of inherited metabolic disorders for newborns has been recognised as a big progress. This was the first multi-parametric metabolic approach, which proved to be a valuable and effective preventive diagnostic strategy (Schulze et al., 2003). The ongoing development of liquid chromatography (LC), gas chromatography (GC) or capillary electrophoresis coupled with different types of mass spectrometers still leads to enhanced power of identification, sensitivity and rate of analysis. However, it is unclear whether investigations of metabolomes, submetabolomes or even complete metabolite pathways will improve clinical routine diagnostics as compared to established targeted diagnostic assays. During the last decade, numerous targeted and non-targeted metabolome studies compared patients with healthy controls and encouraging results have been published, for instance, in research dealing with brain diseases (Urban et al., 2010), diabetes (Newgard et al., 2009), coronary heart diseases (Lewis, Asnani, & Gerszten, 2008) or cancer (Spratlin, Serkova, & Eckhardt, 2009; Suhre et al., 2010). A number of pre-analytical guidelines and requirements have nevertheless to be considered in the validation process for reproducible analyses; biological and technical factors have to be considered depending on the metabolites and sample types, and overall costs are relatively high. Certified commercial LC–MS/MS-based assays or platforms (e.g. Biocrates, Metabolon, Metanomics) with decreased analysis time became recently available, thanks to a new generation of mass spectrometers that require less expertise and maintenance while keeping high performances (Suhre et al., 2010), but the number of clinical applications remains limited, essentially to metabolic diseases. In conclusion, global metabolome analysis or metabolite profiling may help to identify new biomarkers or spotlight pathways of common diseases in human, but challenges remain for a broad use in clinical routine diagnostics.

The most fascinating field of medical diagnostics is perhaps *in situ* diagnostics, which recently became a reality in the management of diabetes (Yoo & Lee, 2010). Given the increasing importance of this disease now considered as epidemic (currently, there are more than 200 million patients worldwide), intensive research efforts have been initiated a long time ago to develop commercial blood glucose monitoring devices. A first generation of invasive kits allowing patients to measure their own blood samples with cost-effective glucose biosensors after simple finger pricking has been a huge success. Today the first non-invasive glucose monitoring devices that alleviate the inconvenience due to frequent skin pricking are coming on the market, with a potential market representing over US$ 1 billion per year (Vashist, 2012). For the moment, *in situ* biochemical monitoring is limited to glucose because it is an essential part of diabetes management. In the future, the challenge will be to follow major biomarkers associated to major diseases such as coronary heart disease or cancer.

2.2. Pharmaceutical research

Clinical chemistry and laboratory medicine are not the only fields where automation in analytical instrumentation has made strong progress. In the 1980s, a need to synthesise many chemical compounds rapidly and inexpensively spawned a new branch of chemistry known as combinatorial chemistry. The techniques of this rapidly growing field have initially been used to find new candidate drugs or pesticides for the pharmaceutical industry.

Combined to HT screening, combinatorial chemistry became an approach to drug discovery and has been associated to phenotyping approaches ranging from molecules (e.g. enzyme activities) to populations of individuals (e.g. insects). Such HT screenings targeting biological activity have also been applied to protein, peptide and gene libraries. To achieve such goals, assays have been developed and optimised for speed, efficiency, signal detection and low reagent consumption, eventually making a major contribution to the development of laboratory robotics, in particular microplate technology. These developments have allowed the screening of large libraries often composed of hundreds of thousands of compounds at rates exceeding 20,000 compounds per week (Martis, Radhakrishnan, & Badve, 2011). During the last decade, pharmaceutical companies have made massive investments in research and development, especially in screening technologies, but without getting a real burst in new active molecules. Concomitantly, the number of drugs approved by US Food and Drug Administration is decreasing. A huge proportion of putative new drugs actually failed during the validation process. For instance, out of 103 drugs approved in 2006, 91 failed during phase III of the tests and it is not known yet how many will succeed in phase III (Czerepak & Ryser, 2008). It is now clear that the first generation of HT strategies failed to provide a wide range of new active molecules. Hopefully, recently developed approaches combining HT, novel experimental design and fundamental knowledge about the biological activity of molecules will lead to a better output.

2.3. Agro-industry

Biochemical analysis is being used to control the quality of raw and transformed products (Ait Kaddour & Cuq, 2009; Cen & He, 2007) and to assess food safety (Davies, 2010). Today's consumer has become increasingly concerned about the quality and safety of food and negative effects of bio-industrial production. To deal with these challenges, companies around the world have increased the use of standard quality assurance systems and have looked for fast and efficient assessment methods. Since its first use by Norris (Norris, 1964) to measure grain moisture, near-infrared reflectance spectroscopy (NIRS) became popular and widespread in food- and agro-industry, which was the first to make intensive use of it. This non-destructive and rapid methodology uses the near-infrared region of the electromagnetic spectrum (from about 800 to 2500 nm) to generate complex spectra, which are usually difficult to assign to specific features or chemical components. Multivariate calibration techniques (e.g. principal components analysis, partial least squares or artificial neural networks) are necessary to extract the desired chemical information. However, with the development of informatics, statistics and chemometrics, NIRS is becoming a valuable biochemical phenotyping method to identify and track variations in composition during the processing of a wide variety of products such as fruits, oil, corn, wine, meat, fish or milk dairy products (Cen & He, 2007). This method indeed offers the rapid response required for in-line and intensive and even for '100% quality control' applications. In the beverage industry, researchers focused on the adulteration of beverages using NIRS techniques (Paradkar, Sivakesava, & Irudayaraj, 2002; Pontes et al., 2006). In the dairy industry, product quality is followed by NIR spectroscopy (Laporte & Paquin, 1999; Rodriguez-Otero, Hermida, & Cepeda, 1995). NIRS is also used in the discrimination and quantification of adulterated olive oils (Kasemsumran, Kang, Christy, & Ozaki, 2005). In fruit production, most applications of NIRS have focused on the non-destructive measurement of the soluble solid content of fruits. Other applications involving texture, dry matter, acidity or disorders of fruit and vegetables have also been reported (Nicolaï et al., 2007). In the production of wine, NIRS has been largely used for the determination of reducing sugar content during grape ripening, winemaking and ageing of white and red wines (Ali, Maltese, Choi, & Verpoorte, 2010). Following agricultural applications, many other areas are now embarking on an intensive use of NIRS technology, in particular petrochemical industries (including the polymer segment) where NIRS recently encountered the largest applicability (Morris et al., 2009).

2.4. Plant breeding

In the past, biochemical phenotyping has probably been insufficiently used to establish genotype–phenotype associations for efficient plant breeding (Houle, Govindaraju, & Omholt, 2010). In modern plant breeding, the better prediction of the phenotype that a particular genotype will produce is a primary goal of genomics-based breeding. Various available tools and techniques can be used to enhance our understanding of the various mechanisms linking the genotype and the phenotype. Thus, recent advances in sequencing and the development of 'omics' technologies such as transcriptomics, proteomics and metabolomics now enable a comprehensive understanding of many aspects of environmentally regulated biological processes, eventually opening a new way to phenotype plants at cellular and molecular levels.

Importantly, these 'omics' technologies provide an untargeted coverage of transcript, protein and metabolite networks, which opens up unprecedented possibilities for identifying the best molecular or biochemical markers of plant performance. In particular, it can be speculated that levels of given analytes are linked to desirable plant phenotypes. One important application is that such analytes could be used as markers for plant performance.

Metabolomics is often considered as particularly promising to identify such markers. Plant metabolomes indeed result from interactions between the genotype and the environment that include events resulting from epistasis, epigenetic- or post-translational processes. In modern plant breeding, the challenge of metabolomics is to find changes in the metabolic composition that are functionally correlated with the genotype and environment variables. Several classes of metabolites directly contribute to the colour, taste, aroma and scent of flowers and fruits, which makes them obvious markers for quality. Furthermore, a range of more or less specific metabolites, which accumulate or decrease in plants experiencing abiotic or biotic stresses, constitute potential indirect markers of plant performance. Finally, and as proposed by Rocha et al. (2010), combinations of metabolites can be associated to biomass production and quality. To promote plant metabolomics as a valuable functional genomics tool that provides a comprehensive characterisation of the biochemical phenotype of a plant, efforts have been taken to improve technologies and analytical tools for the determination of metabolites in complex plant tissues. Today most investigations of the plant metabolome tend to be based on either NMR spectroscopy or MS with or without hyphenation of chromatography or CE such as (LC–NMR, LC–SPE–NMR, LC–MS, GC–MS, GC–SPE–MS, CE–MS, Fourier Transform-MS). The use of metabolomics as a biomarker platform in plant breeding has hardly been exploited industrially, probably because such a strategy would need considerable financial investments to buy heavy equipment, to hire specialised engineers with strong expertise to develop then run dedicated routine procedures, to build up chemical libraries and metabolomics databases. Currently, the number of quantified metabolites in a given sample and in one shot is approximately 50 with proton NMR, 50–200 with GC–MS, >1000 with LC–high-resolution MS. Importantly, these approaches are often labelled 'high throughput' because they allow the determination of high numbers of *analytes* in a given sample. This might be misleading, because the expression 'high throughput' has originally been used to qualify methods allowing high numbers of *samples* to be processed. Thus, methodologies enabling both 'high density' (many analytes) and HT (many samples) are not readily available. In other words, 'omics' cannot be used to routinely screen thousands of individuals at reasonable costs. Today there is an urgent need for low cost and fast methodologies that enable the exploration of natural or induced biodiversity (Gibon et al., 2012). For this, targeted assays may be preferred due to their low needs in terms of labour and/or costs. The same reasoning actually applies to transcriptomics and proteomics, which also represent extensive costs per sample, but have the power to deliver markers of plant performance. Again, targeted assays that are amenable to HT are available, for example, real time quantitative PCR for transcripts, ELISA tests or enzyme activity assays for proteins.

Intuitively, it seems wise to pick HT phenotyping technologies that proved successful in the medical, pharmaceutical and agro-industrial fields. Spectacular advances realised in clinical chemistry and laboratory medicine are particularly appealing, although they represent costs per sample that might be too high regarding the usually very large size of populations used in breeding programs. A further complication comes from the wide metabolic diversity among crops as compared to humans or even mammals, which implies additional costs for optimisation. The possibility to use approaches based on NIRS and other imaging technologies such as spectral reflectance (Babar et al., 2006) as a way to phenotype crops, at relatively low costs and non-invasively, recently raised considerable interest. For example, NIRS has been shown to be as valuable as chemical analysis to 'fingerprint' barley seed (Jacobsen, Søndergaard, Møller, Desler, & Munck, 2005). It also appeared as particularly suitable for seed traits such as starch, protein and oil contents (Hacisalihoglu, Larbi, & Settles, 2010; Spielbauer et al., 2009), and has even been found to be reliable for the evaluation of specific fatty acids in single seeds (Niewitetzki, Tillmann, Becker, & Moellers, 2010). It is very likely that these technologies will enable the analysis of an increasing number of complex traits in crop species, thanks to the integration of ecophysiological, genomic and post genomic data (Montes, Melchinger, & Reif, 2007). However, such technologies are and will probably always be heavily dependent on the existence of good and suitable calibration, which means that an established and well-accepted supporting reference must be available to supply the analytical results required for the modelling step of imaging data. One consequence is that NIRS-based phenotyping appears restricted to features such as major storage compounds, water content or weight, but also to particular organs such as seeds that are relatively easy to calibrate. Finally, it is likely that the use of conventional sample-based methods will remain necessarily in crops phenotyping, either to complete phenotypic dataset or to enable a reliable calibration of the material under study.

Microplate technology probably represents a very good compromise for efficient although diversified analysis of large numbers of samples. This well-established technology, which has benefited from huge investments by the pharmaceutical industry, offers the possibility to operate a wide range of analyses at very low costs per sample. Paradoxically, its use for HT plant phenotyping remains rather limited.

3. SOLUTIONS FOR A MICROPLATE PIPELINE DEDICATED TO PLANT BIOCHEMICAL PHENOTYPING

The microplate, which represents over 100 millions of sales every year, has become one of the most familiar consumables in the laboratory. Used in combination with multichannel pipettes and a microplate reader, it already provides a significant increase in the analytical throughput, and ultimately a substantial decrease in costs per sample. However, this format has much more to offer, when used in combination with robots and laboratory information management systems (LIMS). Experimental processes, which traditionally took several months, can be completed within weeks or even days using technologies that are not particularly expensive, allowing more freedom and flexibility in research programs (more samples and/or more analytes), but without affecting precision and accuracy. Importantly and unlike most other phenotyping approaches, new staff can be trained within a few hours to produce data, a microplate platform being by nature highly accessible. Reciprocally, setting up a microplate platform probably requires expertise in both analytical biochemistry and robotics. Interestingly, while other technologies (NMR, GCMS, LCMS, etc.) are often proposed as highly sophisticated standalone solutions, microplate solutions rather result from the assembling of heterogeneous technologies and methodologies that have evolved in parallel during the last 60 years.

3.1. Sixty years of microplate history

During the 1950s, most biochemical assays were performed in glass tubes with volumes of several millilitres. In order to reduce assay volume and replace test tubes, Jonas Salk proposed a model of a handmade plastic plate in polymethylmethacrylate with 80 wells (250 μl each) (Salk, 1948). Few years later, Hungarian researchers were striving for a fast, economic and reliable test for the diagnosis of influenza. Dr. Gyula Takátsy had the idea to use a set of wire loops to fill several test tubes simultaneously. Rapidly, he replaced the wire loop with iron knitting needles, to be able to handle eight samples at the same time. This led him to create in 1951 the first 8 × 12-well plate in polymethylmethacrylate; the 96-well plate was born (Takátsy, 1955). In 1953, the American Linbro Company already began the massive production of a moulded 96-well microplate. Then, during the 1970s, the Londonian Centers for Disease Control started to use microplates for ELISA-based diagnostics. At that point, microplates used as matrices of wells being at the same time test tubes and cuvettes were becoming very common. In 1998, the Society for Biomolecular Screening, adopted standard rules for all microplates, the SBS standards, which considerably simplified the development of further microplate technology, in particular liquid handling. Strikingly, it is estimated that microplates really became adopted by the plant research community at that period, 40 years after their invention.

3.2. Liquid handling

Pipetting and handling microplates represents another critical step for achieving HT. Indeed pipetting microplates by hand might be exhausting and a lead to high technical error. Laboratory automation and the growing emergence of robotics to achieve tasks such as pipetting, mixing, gripping and moving labware around have transformed the typical workday in analytical biochemistry. The two main factors that have encouraged the development of laboratory automation are cost saving and low error. Today manufacturers offer systems ranging from semi-to-full automation, which are the result of the progress achieved in the second half of the twentieth century in technology (mechanics, optics, hydraulics, electronics and data processing) and biochemistry. When Takátsky created his 96-well microplate, he also created a new tool, the Microtitrator, which was designed for rapid and reliable serial dilution within a microplate (Takátsy, 1955). In 1962, Sever and colleagues, in collaboration with Cooke Engineering, marketed the Microtiter, a screw machine with loops and droppers, which strongly increased throughput and accuracy (Sever, 1962). In parallel, the hand automatic pipette was developed by several teams. In 1960, Dr. Hanns Schmitz designed the first fixed volume micropipette and few years later Dr. Heinrich Netheler, cofounder of Eppendorf, inherited the rights and diffused it in labs around the world. The next evolution came with Warren Gilson in 1974, who patented the mechanical basis for the popular adjustable pipette. At the same time, with the development of micro-scale motors and micro-processors, Clark Hamilton, the inventor of the first microlitre syringe, introduced an automated liquid aspirating and dispensing machine using motorised syringes, the MicroLab M, which was very accurate, but slow. Few years later, Hamilton and Tecan marketed the first automated liquid handling workstations, consisting in a single washable pipetting channel attached on a Cartesian arm, and equipped with the AMICA micro-processor for stopped flow fluidics control (Tecan Samplers 500 and Hamilton MicroLab 2200, 1985). The same year the Beckman Coulter Company marketed the Biomek 100, which was especially designed for microplates. In 1987, Hamilton introduced a 12-channel pipetting workstation using disposable tips, which has been used to screen blood samples for aids and hepatitis viruses. Then, in 1990, TomTec designed the first 96-channel head for microplates, which was equipped with pipette tips and reservoirs. Today, filling microplates of up to 9600 wells is no longer a bottleneck and solutions ranging for multichannel pipettes to fully automated robots are becoming affordable, thus offering the possibility to cope with a large range of throughputs.

3.3. Microplate readers

During the 1950s, the only available optical spectroscopy technology implied transferring the content of each assay tube or well into a suitable cuvette. During the late 1970s, Ruitenberg and colleagues, who worked on *Trichinella spiralis* (a parasitic nematode), dramatically improved the throughput of their experiments by converting a horizontal spectrophotometer into a vertical spectrophotometer. Their initial aim was to considerably reduce reagent amounts; their efforts eventually resulted in much easier sample handling (Ruitenberg, Brosi, & Steerenberg, 1976; Ruitenberg, Sekhuis, & Brosi, 1980). In parallel, it was sometimes necessary to obtain data at different wavelengths. To achieve this, batch analysers were initially employed at a given wavelength, to measure on up to 100 samples, after which the system was adjusted for the next wavelength. A major breakthrough occurred in the early 1980s with the diode array spectrophotometer making it possible to take measurements at as many wavelengths as desired from a fixed spectrum, with work cycles of less than 10 s. During the early 1980s, the processing of microplate data (ELISA tests) improved dramatically, once spectrophotometers could be connected to computers (Loomis & Stahl, 1986; Platt, Shore, Smithyman, & Kampfner, 1981). Today more than 15 companies provide a range of microplate readers, which offer UV–visible, fluorescence and luminometric assays and can be included in liquid handling stations for automated reading of microplates and even decision taking based on read data (i.e. adjustment of a sample dilution based on absorbance). It is also worth mentioning that real-time PCR is routinely performed on 96- or 384-well formats (Bustin, 2004; Schmittgen & Livak, 2008) in automated stations combining thermocycling and fluorimetric measurements, which can be integrated in a robotised pipetting station (Caldana, Scheible, Mueller-Roeber, & Ruzicic, 2007). Strikingly, to our knowledge, no microplate reader able to perform dual wavelength spectrophotometry has been developed so far. This technology, in which two different wavelengths are radiated simultaneously through one sample, enables an increase of the signal-to-noise ratio of up to 1000 times, making it the most sensitive technology to monitor changes in absorbance (Schmidt, 2004). It has been extensively used for the quantification of metabolic intermediates (e.g. Stitt, Lilley, Gerhardt, & Heldt, 1989). Although LC–MS/MS applications (Arrivault et al., 2009; Lunn et al., 2006), which can also be coupled to a microplate pipetting station, have surpassed this technology in terms of sensitivity and analyte coverage, the implementation of a 'dual wavelength microplate reader' might be of interest for targeting analytes of interest in very large numbers of samples.

3.4. Biochemical analysis

Any chemical or biochemical reaction can in principle be performed in microplates, thanks to a great diversity of formats and materials. Possibilities include the quantification of nucleic acids (PCR, Bustin, 2004), proteins (ELISA, Crowther, 2010), enzyme activities (Hadrich et al., 2011) and metabolites (Gibon, Vigeolas, Tiessen, Geigenberger, & Stitt, 2002) including polymers such as starch (Zhao, Mackown, Starks, & Kindiger, 2010) or cell wall components (Gomez, Whitehead, Barakate, Halpin, & Mcqueen-Mason, 2010). In particular, enzymatic analysis, which principles have essentially been developed during the twentieth century, enables the analysis of a large diversity of metabolites and enzyme activities, mainly via UV–visible absorbance assays. The famous book edited by Bergmeyer, Bergmeyer, and Grassl (1985), which proposes hundreds of protocols for the determination of metabolite levels and enzyme activities, provides a good example of this diversity. To our knowledge, there is no such book of protocols available for microplates, but we assume that many of the assays can be scaled down to be used in microplates, and automated. Advantages are multiple: decreased costs per sample, increased throughput and a dramatic acceleration of the optimisation or development of assays. Such scaling down is nevertheless not possible for analytes present at very low concentrations and that were routinely detected in cuvettes by using dual wavelength spectrophotometers (e.g. Stitt et al., 1989). In an average microplate absorbance reader, the theoretical limit of sensitivity would be 0.001 OD, which corresponds to 0.6 μM of NAD(P)H in a given well. In practise, sensitivity is typically of about 10 μM because of the background noise, whereas dual wavelength spectrophotometers have sensitivities in the range of 0.2–0.5 μM As mentioned above, this technology is not available for the microplate format. Instead, it is nevertheless possible to use highly sensitive assays. Many luminescence techniques are available for the study of various classes of enzyme properties and for the quantification of molecules (Lowry et al., 2008). However, the presence of various interfering molecules in plant extracts probably limits the use of fluorimetric and luminometric assays in phenotyping approaches. A further way to increase sensitivity is to use kinetic assays, in which the analyte is determined by measuring a reaction rate. Below saturation, reaction rates are proportional to substrate concentrations. If the initial rate is maintained (pseudo-zero order reaction) by recycling the substrate (cycling assays) or because the measured molecule is not the substrate (inhibition or activation tests) until the product becomes detectable, the sensitivity can be increased by a factor 100–10,000, which corresponds to sensitivities ranging from 1 to 100 nM. Such assays are usually operated in the UV–visible range and can therefore be performed in basic microplate readers (Gibon et al., 2002). Cycling assays, which have been discovered by Warburg, Christian, and Griese (1935), made popular by Lowry and Passonneau (1972), have been shown to be useful for the determination of a range of metabolites (Gibon et al., 2002) and enzyme activities (Gibon et al., 2004) in large to very large numbers of samples (e.g. Sulpice et al., 2010, 2007; Zhang et al., 2010).

3.5. Laboratory information management systems

The tabular nature of microplates represents a convenient format for data management. Microplate reader software usually enables the parameterisation of the reader and provides a range of options for data processing. At low to mid throughput, manipulating sample identifiers and processing raw data are easily performed in reader software and/or spreadsheet software. At HT, the implementation of a dedicated LIMS nevertheless represents an obligatory step. With the introduction of personal computers, the first generation of LIMS already offered an opportunity to utilise automated reporting tools. As personal computers became more powerful, LIMS adopted advantageous client/server architectures allowing better data processing and exchanges. Today, LIMS provide easy sample tracking and calculations as well as data storage. LIMS must allow non-experts to perform the experiments and process raw data instantly to obtain the first results. Connected to the web network, LIMS with flexible architecture provide smart data exchange interfaces, which allow the processing of data anywhere in the network.

3.6. Weighing and grinding

When the throughput has been optimised for every analytical step, sample handling inevitably becomes the next bottleneck. In most cases, biochemical analysis—unfortunately—requires to work on ground biological material. Furthermore, such material usually needs to be stored and manipulated at very low temperature prior to extraction. If sample preparation is manual, it usually represents the most time-consuming step in the workflow of an experiment, but also an important source of error. Samples can be ground to homogeneity by using a mechanical grinder (i.e. planetary ball mill, cryomill), but traditionally one by one. The grinding of up to 192 samples in parallel is nevertheless possible in microplate format, but it is restricted to very small and breakable samples such as leaf discs. Therefore, further technological developments are needed to enable the processing of large numbers of crop samples, which are usually of larger size and/or more difficult to break. If data are to be expressed on a weight basis, it is then necessary to weigh so-called aliquots. Automated 96-well solutions using a vibrating powder feeder that are also compatible with deeply frozen samples have emerged (Labman Ltd.). Furthermore, an all-in-one robotic solution designed to grind, weigh and analyse plant samples in 96-well microplates has recently been proposed by Gomez et al. (2010). Although this system is restricted to dry samples, it is representative of the versatility of microplate technology.

3.7. Harvest: The final bottleneck

The collection of samples is probably the last step to appear as a bottleneck when setting up an HT phenotyping pipeline. Producing and harvesting bacteria, yeasts, cells or leaves, roots or fruits for reproducible data is indeed very complex (Ap Rees & Hill, 1994; Rudell, 2010). Biochemical phenotypes are rather sensitive to the environment, which implies that perturbations need to be avoided or traced while harvesting. In particular, many metabolic-related traits (transcripts, activities of proteins, metabolites, etc.)

fluctuate during the day (Bläsing et al., 2005; Gibon et al., 2006). Given the more samples the longer it takes to harvest, the planning needs to be done carefully. Thus, knowing how much time one sample would require enables the prediction of harvest duration. For example, 1 min per sample would represent hours if hundreds of samples are to be harvested. One solution to minimise variations due to the time of harvest would then be to increase the number of operators adequately. Although using harvest automation as an alternative may sound ahead of time, it is not really a new challenge. Agriculture has solved this bottleneck a long time ago with, for example, combine harvesters (invented more than 150 years ago), but the challenge remains for screening purpose. Since two decades, technological research has focused on the automation of the most tedious and repetitive tasks in horticulture, in order to reduce the amount of human labour in the greenhouse or in the open field. Consistent efforts have been devoted to the development of autonomous agricultural robots, including guidance systems, greenhouse autonomous systems and fruit harvesting robots (Edan, 1995). Today, robust robots are able to autonomously navigate in different kinds of greenhouses. They have been used for many tasks such as spraying potentially toxic chemicals in the confined space of a hot and steamy glasshouse (Sammons, Furukawa, & Bulgin, 2005), moving pots and heavy loads (Belforte, Deboli, Gay, Piccarolo, & Ricauda Aimonino, 2006) or harvesting and transporting agricultural by-products. The robots reported in the literature are suited for harvesting apples, melon (Edan, 1995), tomatoes (Subrata et al., 1996), cucumbers (Arima & Kondo, 1999; Van Henten et al., 2002) and even cherry (Tanigaki, Fujiura, Akase, & Imagawa, 2008). However, their general use in commercial greenhouses remains limited. Indeed, operating robots in highly unstructured environments and under high humidity and temperature remains very difficult. Conversely, robots able to meet high-performance characteristics in terms of speed and success rate of the picking operation would also need to be cost-effective. Finally, designing robots for agricultural applications does not necessarily benefit from automation progress achieved in others industries (Van Henten et al., 2002). It can therefore be predicted that harvesting samples at HT will remain a major challenge for HT biochemical phenotyping.

4. EXAMPLE OF HT BIOCHEMICAL SCREEN

Among the different ways to induce genetic diversity in plants, the use of ethyl methyl sulfonate (EMS) is the most frequently used. It leads to populations of mutant with large numbers of point mutations in almost all species studied. In the last decade, EMS mutant collections have been developed in various plants species including rapeseed (Wu, Shi, Niu, Xing, & Xue, 2008), common bean (Porch et al., 2009), *Arabidopsis thaliana*, cotton, barley (Caldwell et al., 2004) and tomato (Emmanuel & Levy, 2002; Eshed and Zamir, 1995; Eshed et al., 1992; Meissner et al., 1997; Okabe et al., 2011; Watanabe et al., 2007). In our laboratory, an EMS mutant population has been obtained with the dwarf and determinate Micro-Tom variety (Scott & Harbaugh, 1989, 1995). Its small size (15 cm when grown in 130-ml pots) and short life cycle (70–90 days from sowing to fruit ripening) allow growing many individuals in a reduced space and obtaining advanced backcross generations in a relatively short time. The Bordeaux collection consists in a population of 8500 M2 families including 3500 already phenotyped on visual criteria such as shape, size, colour and number of flowers, fruits and seeds. Micro-Tom tomato mutants were obtained as described previously (Dan, Fei, & Rothan, 2007). Briefly, wild-type seeds (M0) were EMS-treated producing mutagenised seeds (M1), which were sown. M2 seeds collected from M1 plants were sown to give M2 plants (12 plants per M2 family) used for harvesting plant material for biochemical analyses.

Here, we have performed a forward enzymatic screen to uncover potential mutants carrying alterations in enzymes from central metabolism. For this a platform including two robotic stations, each equipped with a 96-channel pipette, a shaker, 12 temperature-controlled incubators and a gripper (Fig. 9.1) has been used in a HT workflow (Fig. 9.2). It enables the analysis of up to 50 enzymatic traits on up to 200 samples per day.

Plants were grown in a greenhouse under standard growth conditions (Letard, Erard, & Jeannequin, 1995). Leaf discs were sampled manually on 5-week-old plants and collected in SBS compliant storage tubes equipped with screw caps (Fig. 9.1A). In total, 3550 samples have been harvested in less than 4 days. Then, the samples were ground by using a ball mill type tissue lyser, directly in the tubes used for harvest to save time. This enabled the processing of up to 1000 samples per hour (Fig. 9.1B). Finally, samples were randomly distributed over 96 position racks.

Sample extraction was performed simultaneously. For all samples of a given rack, it took 1 min. For this, frozen tubes were opened with a Hamilton decapper (Fig. 9.1C) and 500 μl of extraction buffer were added immediately after the transfer to a microplate shaker on the robot desk. After extraction, samples were centrifuged and the supernatant containing the soluble enzymes was transferred to 96-well V-shaped plates (e.g. PCR plates). If needed, extract dilutions were performed, using a dedicated robotic method. Indeed, the amount of sample needed per well varies from 1 to

Figure 9.1 Overview of platform equipment. (A) Leaf discs were harvested and snap frozen in liquid nitrogen using micronic® tubes with screw caps. (B) Samples were milled in batch of 96 micronic® using a TissuLyser II (Qiagen®). (C) Automated decapper (Hamilton®) were used to cap/decap the tubes between each step of process. (D) An auxiliary arm iSwap® microplate carrier was used to transport the plate between the liquid handling workstation and the incubators (Hamilton®). (E) Each liquid workstation held 96 tips head Starlet® (Hamilton®). (F) The platform provided eight microplate readers MP96® and one microplate luminometer/fluorimeter reader Xenius® (SAFAS®). (G) Overview of the microplate workstations Starlet® (Hamilton®). (See Colour Insert.)

50 μg fresh weight, depending on the assay (Gibon et al., 2004). This procedure enables the extraction of up to 1000 samples per hour.

In parallel, assay mixes were prepared and loaded into standard polystyrene microplates using the second robot. These microplates were then

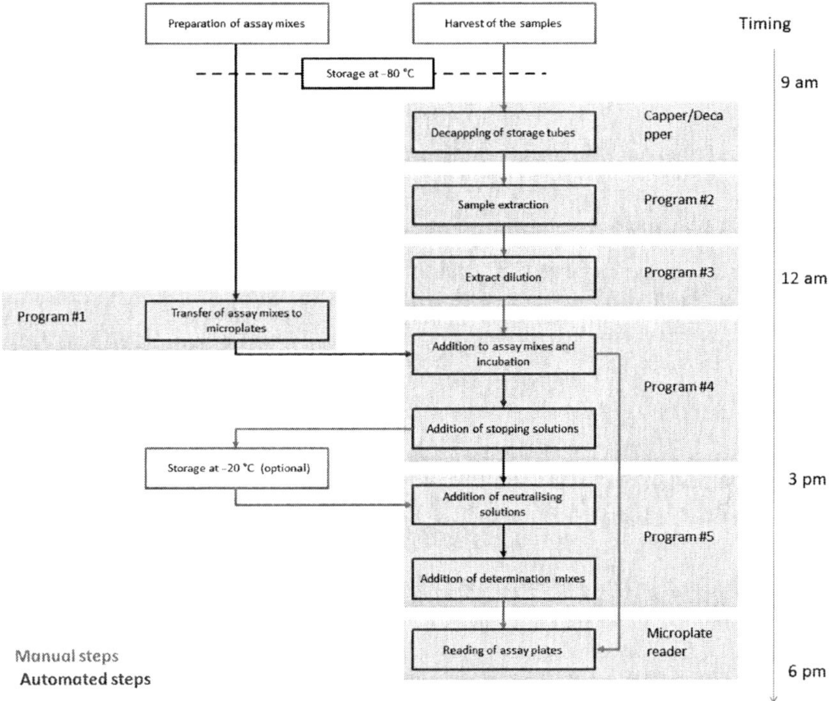

Figure 9.2 Workflow of a standard high-throughput experiment using the platform. Samples and reagent mixes could be stored at −80 °C and used on demand. The reagent mixes were optimised to allow long storage and defreeze on demand. The red boxes represent the non-automated steps and the blue boxes the automated steps. Robotised dedicated programs were developed for each step (in grey). Each program could be run independently but it could also be integrated in a complete workflow. The timing represents the standard workday to perform 5–10,000 analysis depending on the screen complexity. (For interpretation of the references to colour in this figure legend, the reader is referred to the online version of this chapter.)

placed on a shaker, loaded with samples at the appropriate dilution, and transported to the incubators (stopped assays) or placed into a stacker (direct assays performed in a microplate reader) by the gripper (Fig. 9.1D).

When enzyme activities are determined ‘directly’ in the microplate reader, measurements take 20–30 min on average (the number of microplate readers then limits the throughput). When enzymes are determined ‘indirectly’ (stopped assays), one to three additional pipetting steps are necessary (see Gibon et al., 2004 for details). While this apparently decreases the throughput, it actually offers more flexibility regarding time management. Indeed, large numbers of plates can be stored until the reaction product is determined. Stopped assays also enable substantial cost decreases, as the first step is typically run in only 20 µl of assay mix whereas a direct measurement requires at least 100 µl per well (Rogers & Gibon, 2009). Furthermore, the use of only a few types of determination mixes targeting NAD(H), NADP(H), glycerone-3P or glycerol-3P (direct or indirect products of the studied reaction) enables the determination of a wide range of enzyme activities. Standardising assay volumes and principles has several advantages. Firstly, it allows a higher throughput, as more than 10,000 enzyme activities could be determined within a day by using two robotic stations and nine microplate readers (Fig. 9.1E and F). Secondly, high precision is easier to achieve, as fewer processes need to be optimised. For this, each robotic method has been tested optically with bromophenol blue. Coefficients of variation found for each pipetting step are presented in Table 9.2A. During the screen, biological standards consisting in aliquots of a large sample of wild-type leaves were added in every batch to assess experimental error (from weighing of aliquots to microplate reading). Standard deviation found for enzymes measured during the screen is presented in Table 9.2B.

In total, 2000 mutants and wild-type plants have been assessed for 24 enzymatic traits, that is, the activities of 12 enzymes measured under saturating- or non-saturating substrate concentrations. The ratio between activities under non-saturated and saturated conditions has then been calculated and samples for which this ratio was below or above the mean + or − two times the coefficient of variation calculated for all samples were selected. As each plant was represented by two samples that were analysed separately, only those found with both samples being significant were validated.

Table 9.2 Schedule and cost synthesis for the high-throughput screen of an EMS mutant collection

	Variation co-efficient	
	Direct assay	**Stop assay**
(A)		
Loading sample	4.3	4.1
Loading assay mix	1.7	2.8
(B)		
Triose phosphate isomerase	8.2	–
Phosphofructokinase (PPi)	–	7.4
Hexokinase (fructose)	–	7.5

This high-throughput platform allowed scientist to perform more than 65,000 analysis in 4 months and for a very cheap investment, around 4000€. Every cost was calculated with VAT and taxes for salaries.

A family was selected when at least two members had been validated. At the end of the first screen 10 families were selected to perform a second screen (Fig. 9.3).

During the second screen, 12 plants of each selected family as well as wild-type plants were grown again to confirm the observed biochemical phenotypes. Again, two leaf samples were harvested on each individual and relevant enzyme traits assessed using the same criteria as above. Finally, two families with altered triose phosphate isomerase activities have been selected (Fig. 9.3). Several members of the P3A3 family showed higher non-saturated to saturated ratios for this activity, indicating a lower apparent

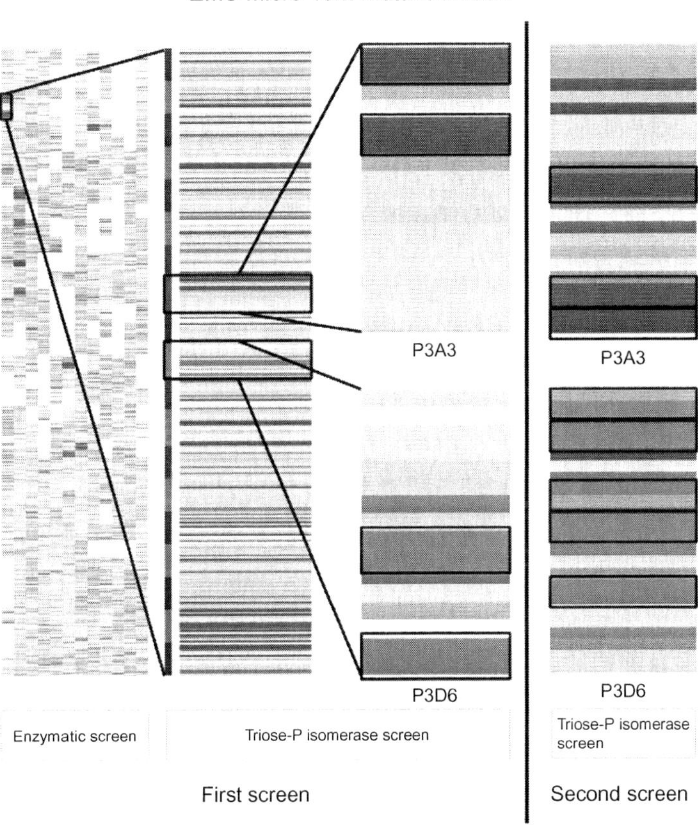

Figure 9.3 EMS mutant collection screen schema. An EMS mutant collection was screened to uncover potential enzymatic mutants. 150 M2 families (12 plants per family) from the collection were harvested and screened. The first screen allowed identifying two families with potential increased (red) or decreased (green) in triose phosphate isomerase activity. A second culture was conducted and screened in the process. This second screen confirmed the alterations initially uncovered. (See Colour Insert.)

Table 9.3 Variation coefficient analysis

Step	**Quantity**	**Material time**	**Human time**	**Costs**
Production	880 Plants	3 months	20 h	300€
Harvest	14,080 Leaf discs	16 h	16 h	300€
Extraction	2112 Extractions	3 h	16 h	200€
Assay preparation	1320 Microplates	16 h	40 h	100€
Assay	132,720 Assays	100 h	100 h	3000€
Calculation	66,360 Activity determinations	8 h	40 h	50€
Total	–	3 months 143 h	232 h	3650€

Each step of the programs was analysed and optimised to provide maximum reproducibility. (A) Variation co-efficient calculated using standard solution of bromophenol blue. (B) Variation co-efficient calculated using real sample and real determination. The triose phosphate isomerase, phosphofructokinase (PPi) and hexokinase (fructokinase activity) assays were performed on 192, 180 and 156 samples, respectively.

K_m. In the P3D6 family, individuals showed lower ratios, this time suggesting an increased apparent K_m.

To conclude, two potentially interesting mutations affecting an enzyme of central metabolism have been uncovered within only 5 months, which comprise the time needed for plant growth, sample processing, biochemical analyses and data handling. In particular, 24 enzymatic traits could be measured in more than 1500 samples within only 1 month. The entire experiment represented a relatively small investment, as the total cost (consumables, reagents, equipment depreciation, overheads), is estimated at less than 10,000€ (see Table 9.3 for details). The next steps will be to unravel the molecular bases of these mutations and to evaluate their consequences on the phenotype.

5. CONCLUDING REMARKS

Medicine, pharma- and agro-industry, as well as academic research continuously generate new methodologies and technologies that present opportunities for plant phenotyping. Among these, the microplate has been invented 60 years ago to speed up diagnosis in medical laboratories. It has then been improved and largely used to perform biochemical analyses, immunoassays or large-scale drug screens, eventually triggering the development of highly diversified technologies and methodologies. When the microplate finally hit plant sciences, many laboratories were already pioneering omics. Today, whereas improved analytical capabilities in transcriptomics, proteomics or metabolomics technologies combined with dedicated statistical and bioinformatics have come far, it appears that the microplate has not been fully exploited in plant sciences.

It can be envisioned that future plant research will combine high density and HT. On the one hand, high-density techniques will detect and interpret a broadening range of analytes and, on the other hand, HT targeted techniques will enable screens of increasingly large populations. Currently, microplate platforms probably represent the most accessible solution to achieve HT biochemical phenotyping of biodiversity. This technology offers numerous advantages including rapidity, robustness, sensitivity and low costs. Furthermore, when compared to high-density profiling techniques, a microplate platform able to process thousands of samples per week represents a rather moderate investment. However, building up such a platform requires expertise in analytical biochemistry and robotics, as constructors do not usually sell ready to run microplate solutions. In agriculture, the combined use of high-density techniques to identify potential biomarkers and microplate technology to validate them could benefit to breeding programs. Once information-rich analytes have been selected, fast, targeted and low-cost assays could be developed. If successful, such strategy will enable the prediction of crop performance with biochemical biomarkers.

Assuming that current microplate technology results from over 60 years of miniaturisation efforts, we can consider that micro-fluidics will be the next step in biochemical phenotyping. Micro-fluidics can be used to process many samples and assays in parallel while consuming extremely small amounts of sample and reagents (Fair, 2007; Sista et al., 2008). They have been shown to be suitable for enzyme kinetics (Miller and Wheeler, 2008), metabolite detection (Kraly, Holcomb, Guan, & Henry, 2009) or protein crystallisation (Lounaci et al., 2007, 2010) and commercial solutions are available, for example, for the quantification of transcripts via real-time PCR and more recently next-generation sequencing techniques. However, although micro-fluidics appeared more than 20 years ago, they are scarcely used in routine. Before this—still—emerging technology provides all the capabilities required for HT biochemical phenotyping, significant efforts will be needed. They will require the involvement of highly multidisciplinary teams (physicists, chemists, biochemists, biologists, programmers) but also consistent financial support. It can, nevertheless, be speculated that microplate technology is about to deliver the necessary proof of concept to convince research funding agencies and the industry to invest into micro-fluidics as a promising solution to improve plant phenotyping and ultimately make major contributions to breeding.

ACKNOWLEDGEMENTS

We wish to thank, Isabelle Atienza and Aurélie Honoré for technical support in the greenhouse, and Dr. Frédéric Delmas, Dr. Frédéric Gévaudant, Dr. Pierre Baldet, Dr. Martine Lemaire, Joanna Jorly and Jean-Philippe Mauxion and all the internship students for their help during sample collection.

REFERENCES

Ait Kaddour, A., & Cuq, B. (2009). In line monitoring of wet agglomeration of wheat flour using near infrared spectroscopy. *Powder Technology*, *190*, 10–18.

Ali, K., Maltese, F., Choi, Y., & Verpoorte, R. (2010). Metabolic constituents of grapevine and grape-derived products. *Phytochemistry Reviews*, *9*, 357–378.

Ap Rees, T., & Hill, S. A. (1994). Metabolic control analysis of plant metabolism. *Plant, Cell & Environment*, *17*, 587–599.

Arima, S., & Kondo, N. (1999). Cucumber harvesting robot and plant training system. *Journal of Robotics & Mechatronics*, *11*, 208–212.

Arrivault, S., Guenther, M., Ivakov, A., Feil, R., Vosloh, D., Van Dongen, J. T., et al. (2009). Use of reverse-phase liquid chromatography, linked to tandem mass spectrometry, to profile the Calvin cycle and other metabolic intermediates in Arabidopsis rosettes at different carbon dioxide concentrations. *The Plant Journal*, *59*, 824–839.

Babar, M. A., Reynolds, M. P., Van Ginkel, M., Klatt, A. R., Raun, W. R., & Stone, M. L. (2006). Spectral reflectance indices as a potential indirect selection criteria for wheat yield under irrigation. *Crop Science*, *46*, 578–588.

Belforte, G., Deboli, R., Gay, P., Piccarolo, P., & Ricauda Aimonino, D. (2006). Robot design and testing for greenhouse applications. *Biosystems Engineering*, *95*, 309–321.

Bergmeyer, H. U., Bergmeyer, J., & Grassl, M. (1985). *Methods of enzymatic analysis* (3rd ed.). Weinheim, Germany: Verlag Chemie.

Bläsing, O. E., Gibon, Y., Günther, M., Höhne, M., Morcuende, R., Osuna, D., et al. (2005). Sugars and circadian regulation make major contributions to the global regulation of diurnal gene expression in Arabidopsis. *The Plant Cell*, *17*, 3257–3281.

Brisson, N., Gate, P., Gouache, D., Charmet, G., Oury, F.-X., & Huard, F. (2010). Why are wheat yields stagnating in Europe? A comprehensive data analysis for France. *Field Crops Research*, *119*, 201–212.

Bustin, S. A. (2004). *A-Z of quantitative PCR*. La Jolla, CA: International University Line.

Caldana, C., Scheible, W.-R., Mueller-Roeber, B., & Ruzicic, S. (2007). A quantitative RT-PCR platform for high-throughput expression profiling of 2500 rice transcription factors. *Plant Methods*, *3*, 7.

Caldwell, D. G., Mccallum, N., Shaw, P., Muehlbauer, G. J., Marshall, D. F., & Waugh, R. (2004). A structured mutant population for forward and reverse genetics in Barley (*Hordeum vulgare* L.). *The Plant Journal*, *40*, 143–150.

Cen, H., & He, Y. (2007). Theory and application of near infrared reflectance spectroscopy in determination of food quality. *Trends in Food Science and Technology*, *18*, 72–83.

Chace, D. H., Millington, D. S., Terada, N., Kahler, S. G., Roe, C. R., & Hofman, L. F. (1993). Rapid diagnosis of phenylketonuria by quantitative-analysis for phenylalanine and tyrosine in neonatal blood spots by tandem mass-spectrometry. *Clinical Chemistry*, *39*, 66–71.

Crowther, J. R. (2010). *The ELISA guidebook* (2nd ed.). Clifton, NJ: Humana Press.

Czerepak, E. A., & Ryser, S. (2008). Drug approvals and failures: Implications for alliances. *Nature Reviews. Drug Discovery*, *7*, 197–198.

Dan, Y., Fei, Z., & Rothan, C. (2007). MicroTom—A new model system for plant genomics. *Genes, Genomes & Genomics*, *1*, 167–179.

Davies, H. (2010). A role for "omics" technologies in food safety assessment. *Food Control*, *21*, 1601–1610.

Dozol, A., Gana, I., Cocagne, N., Conilleau, B., Brignone, M., Moreau, A.-C., et al. (2010). Identifying, managing & monitoring laboratory test over utilization for inpatients. *Pratiques et Organisation des Soins*, *41*, 135–141.

Durner, J. (2010). Clinical chemistry: Challenges for analytical chemistry & the nanosciences from medicine. *Angewandte Chemie International Edition*, *49*, 1026–1051.

Edan, Y. (1995). Design of an autonomous agricultural robot. *Applied Intelligence*, *5*, 41–50.

Emmanuel, E., & Levy, A. A. (2002). Tomato mutants as tools for functional genomics. *Current Opinion in Plant Biology*, *5*, 112–117.

Engvall, E., & Perlmann, P. (1971). Enzyme-linked immunosorbent assay (ELISA). Quantitative assay of immunoglobulin G. *Immunochemistry*, *8*, 871–874.

Eshed, Y., & Zamir, D. (1995). An introgression line population of Lycopersicon pennellii in the cultivated tomato enables the identification and fine mapping of yield-associated QTL. *Genetics*, *141*, 1147–1162.

Eshed, Y., Abu-Abied, M., Saranga, Y., & Zamir, D. (1992). *Lycopersicon esculentum* lines containing small overlapping introgressions from *L. pennellii*. *Theoretical and Applied Genetics*, *80*, 1027–1034.

Fair, R. B. (2007). Digital microfluidics: Is a true lab-on-a-chip possible?. *Microfluidics and Nanofluidics*, *3*, 245–281.

Finger, R. (2010). Evidence of slowing yield growth—The example of Swiss cereal yields. *Food Policy*, *35*, 175–182.

Furbank, R. T. (2009). Plant phenomics: From gene to form and function. *Functional Plant Biology*, *36*, V–VI.

Furbank, R. T., & Tester, M. (2011). Phenomics—Technologies to relieve the phenotyping bottleneck. *Trends in Plant Science*, *16*, 635–644.

Gibon, Y., Blaesing, O. E., Hannemann, J., Carillo, P., Hohne, M., Hendriks, J. H. M., et al. (2004). A robot-based platform to measure multiple enzyme activities in Arabidopsis using a set of cycling assays: Comparison of changes of enzyme activities and transcript levels during diurnal cycles and in prolonged darkness. *The Plant Cell*, *16*, 3304–3325.

Gibon, Y., Rolin, D., Deborde, C., Bernillon, S., & Moing, A. (2012). New opportunities in metabolomics and biochemical phenotyping for plant systems biology. In U. Roessner (Ed.), *Metabolomics* (pp. 213–240). Rijeka: InTech Europe.

Gibon, Y., Usadel, U., Blaesing, O., Kamlage, B., Hoehne, M., Trethewey, R., et al. (2006). Integration of metabolite with transcript and enzyme activity profiling during diurnal cycles in Arabidopsis rosettes. *Genome Biology*, *7*, 1–23.

Gibon, Y., Vigeolas, H., Tiessen, A., Geigenberger, P., & Stitt, M. (2002). Sensitive and high throughput metabolite assays for inorganic pyrophosphate, ADPGlc, nucleotide phosphates, and glycolytic intermediates based on a novel enzymic cycling system. *The Plant Journal*, *30*, 221–235.

Gomez, L. D., Whitehead, C., Barakate, A., Halpin, C., & Mcqueen-Mason, S. J. (2010). Automated saccharification assay for determination of digestibility in plant materials. *Biotechnology for Biofuels*, *3*, 12.

Gurian-Sherman, D. (2009). *Failure to yield: Evaluating the performance of genetically engineered crops*. Cambridge: Union of Concerned Scientists.

Hacisalihoglu, G., Larbi, B., & Settles, A. M. (2010). Near-infrared reflectance spectroscopy predicts protein, starch, and seed weight in intact seeds of common bean (*Phaseolus vulgaris* L.). *Journal of Agricultural and Food Chemistry*, *58*, 702–706.

Hadrich, N., Gibon, Y., Schudoma, C., Altmann, T., Lunn, J. E., & Stitt, M. (2011). Use of TILLING and robotised enzyme assays to generate an allelic series of Arabidopsis thaliana mutants with altered ADP-glucose pyrophosphorylase activity. *Journal of Plant Physiology*, *168*, 1395–1405.

Herdt, R. W., Capule, C., & International Rice Research Institute (1983). *Adoption, spread, and production impact of modern rice varieties in Asia*. Los Baños: International Rice Research Institute.

Houle, D., Govindaraju, D. R., & Omholt, S. (2010). Phenomics: The next challenge. *Nature Reviews. Genetics*, *11*, 855–866.

Jacobsen, S., Søndergaard, I., Møller, B., Desler, T., & Munck, L. (2005). A chemometric evaluation of the underlying physical and chemical patterns that support near infrared spectroscopy of barley seeds as a tool for explorative classification of endosperm genes and gene combinations. *Journal of Cereal Science*, *42*, 281–299.

Kasemsumran, S., Kang, N., Christy, A., & Ozaki, Y. (2005). Partial least squares processing of near-infrared spectra for discrimination and quantification of adulterated olive oils. *Spectroscopy Letters*, *38*, 839–851.

Kraly, J. R., Holcomb, R. E., Guan, Q., & Henry, C. S. (2009). Review: Microfluidic applications in metabolomics and metabolic profiling. *Analytica Chimica Acta*, *653*, 23–35.

Laporte, M. F., & Paquin, P. (1999). Near-infrared analysis of fat, protein, and casein in cow's milk. *Journal of Agricultural and Food Chemistry*, *47*, 2600–2605.

Lee, J. M., Joung, J.-G., Mcquinn, R., Chung, M.-Y., Fei, Z., Tieman, D., et al. (2012). Combined transcriptome, genetic diversity and metabolite profiling in tomato fruit reveals that the ethylene response factor SlERF6 plays an important role in ripening and carotenoid accumulation. *The Plant Journal*, *70*, 191–204.

Letard, M., Erard, P., & Jeannequin, B. (1995). *Maîtrise de l'irrigation fertilisante: Tomate sous serre et abris, en sol et hors sol*. Paris: Centre technique interprofessionnel des fruits et légumes.

Lewis, G. D., Asnani, A., & Gerszten, R. E. (2008). Application of metabolomics to cardiovascular biomarker and pathway discovery. *Journal of the American College of Cardiology*, *52*, 117–123.

Lobell, D. B., Schlenker, W., & Costa-Roberts, J. (2011). Climate trends and global crop production since 1980. *Science*, *333*, 616–620.

Loomis, T. C., & Stahl, W. L. (1986). A rapid, flexible method for biochemical assays using a microtiter plate reader and a microcomputer. Application for assays of protein, Na, K-ATPase and K-p-nitrophenylphosphatase. *International Journal of Bio-Medical Computing*, *18*, 183–192.

Lounaci, M., Chen, Y., & Rigolet, P. (2010). Channel height dependent protein nucleation and crystal growth in microfluidic devices. *Microelectronic Engineering*, *87*, 750–752.

Lounaci, M., Rigolet, P., Abraham, C., Le Berre, M., & Chen, Y. (2007). Microfluidic device for protein crystallization under controlled humidity. *Microelectronic Engineering*, *84*, 1758–1761.

Lowry, M., Fakayode, S. O., Geng, M. L., Baker, G. A., Wang, L., Mccarroll, M. E., et al. (2008). Molecular fluorescence, phosphorescence, and chemiluminescence spectrometry. *Analytical Chemistry*, *80*, 4551–4574.

Lowry, O. H., & Passonneau, J. V. (1972). *A flexible system of enzymatic analysis*. New York, NY: Academic Press.

Lunn, J. E., Feil, R., Hendriks, J. H. M., Gibon, Y., Morcuende, R., Osuna, D., et al. (2006). Sugar-induced increases in trehalose 6-phosphate are correlated with redox activation of

ADP glucose pyrophosphorylase and higher rates of starch synthesis in Arabidopsis thaliana. *The Biochemical Journal*, *397*, 139–148.

Martis, E. A., Radhakrishnan, R., & Badve, R. R. (2011). High-throughput screening: The hits and leads of drug discovery—An overview. *Journal of Applied Pharmaceutical Science*, *1*, 2–10.

Meissner, R., Jacobson, Y., Melamed, S., Levyatuv, S., Shalev, G., Ashri, A., et al. (1997). A new model system for tomato genetics. *The Plant Journal*, *12*, 1465–1472.

Miflin, B. (2000). Crop improvement in the 21st century. *Journal of Experimental Botany*, *51*, 1–8.

Miller, E. M., & Wheeler, A. R. (2008). A digital microfluidic approach to homogeneous enzyme assays. *Analytical Chemistry*, *80*, 1614–1619.

Millington, D. S., Kodo, N., Terada, N., Roe, D., & Chace, D. H. (1991). The analysis of diagnostic markers of genetic disorders in human blood and urine using tandem mass-spectrometry with liquid secondary ion mass-spectrometry. *International Journal of Mass Spectrometry*, *111*, 211–228.

Montes, J. M., Melchinger, A. E., & Reif, J. C. (2007). Novel throughput phenotyping platforms in plant genetic studies. *Trends in Plant Science*, *12*, 433–436.

Morris, R. E., Hammond, M. H., Cramer, J. A., Johnson, K. J., Giordano, B. C., Kramer, K. E., et al. (2009). Rapid fuel quality surveillance through chemometric modeling of near-infrared spectra. *Energy & Fuels*, *23*, 1610–1618.

Nagamune, K., Hicks, L. M., Fux, B., Brossier, F., Chini, E. N., & Sibley, L. D. (2008). Abscisic acid controls calcium-dependent egress and development in *Toxoplasma gondii*. *Nature*, *451*, 207–211.

Newgard, C. B., An, J., Bain, J. R., Muehlbauer, M. J., Stevens, R. D., Lien, L. F., et al. (2009). A branched-chain amino acid-related metabolic signature that differentiates obese and lean humans and contributes to insulin resistance. *Cell Metabolism*, *9*, 565–566.

Nicolaï, B. M., Beullens, K., Bobelyn, E., Peirs, A., Saeys, W., Theron, K. I., et al. (2007). Nondestructive measurement of fruit and vegetable quality by means of NIR spectroscopy: A review. *Postharvest Biology and Technology*, *46*, 99–118.

Niewitetzki, O., Tillmann, P., Becker, H. C., & Moellers, C. (2010). A new near-infrared reflectance spectroscopy method for high-throughput analysis of oleic acid and linolenic acid content of single seeds in oilseed rape (*Brassica napus* L.). *Journal of Agricultural and Food Chemistry*, *58*, 94–100.

Norris, K. (1964). Design and development of a new moisture meter. *Agricultural Engineering*, *45*, 370–372.

Okabe, Y., Asamizu, E., Saito, T., Matsukura, C., Ariizumi, T., Bres, C., et al. (2011). Tomato TILLING technology: Development of a reverse genetics tool for the efficient isolation of mutants from Micro-Tom mutant libraries. *Plant & Cell Physiology*, *52*, 1994–2005.

Paradkar, M. M., Sivakesava, S., & Irudayaraj, J. (2002). Discrimination and classification of adulterants in maple syrup with the use of infrared spectroscopic techniques. *Journal of the Science of Food and Agriculture*, *82*, 497–504.

Platt, J. H., Shore, A. B., Smithyman, A. M., & Kampfner, G. L. (1981). A computerised ELISA system for the determination of total and antigen-specific immunoglobulins in serum and secretions. *Journal of Immunoassay*, *2*, 59–74.

Pontes, M., Santos, S., Araujo, M., Almeida, L., Lima, R., Gaiao, E., et al. (2006). Classification of distilled alcoholic beverages and verification of adulteration by near infrared spectrometry. *Food Research International*, *39*, 182–189.

Porch, T. G., Blair, M. W., Lariguet, P., Galeano, C., Pankhurst, C. E., & Broughton, W. J. (2009). Generation of a mutant population for TILLING common bean genotype BAT 93. *Journal of the American Society for Horticultural Science*, *134*, 348–355.

Rocha, M., Licausi, F., Araujo, W. L., Nunes-Nesi, A., Sodek, L., Fernie, A. R., et al. (2010). Glycolysis and the tricarboxylic acid cycle are linked by alanine aminotransferase during hypoxia induced by waterlogging of *Lotus japonicus*. *Plant Physiology*, *152*, 1501–1513.

Rodriguez-Otero, J. L., Hermida, M., & Cepeda, A. (1995). Determination of fat, protein, and total solids in cheese by near-infrared reflectance spectroscopy. *Journal of AOAC International*, *78*, 802–806.

Roessner, U., Willmitzer, L., & Fernie, A. R. (2002). Metabolic profiling and biochemical phenotyping of plant systems. *Plant Cell Reports*, *21*, 189–196.

Rogers, A., & Gibon, Y. (2009). Enzyme kinetics: Theory & practice. In J. Schwender (Ed.), *Plant Metabolic Networks* (pp. 71–103). Berlin, Heidelberg, New York: Springer.

Ruane, J., & Sonnino, A. (2011). Agricultural biotechnologies in developing countries and their possible contribution to food security. *Journal of Biotechnology*, *156*, 356–363.

Rudell, D. (2010). Standardizing postharvest quality and biochemical phenotyping for precise population comparison. *Hortscience*, *45*, 1307–1309.

Ruitenberg, E. J., Brosi, B. J., & Steerenberg, P. A. (1976). Direct measurement of microplates and its application to enzyme-linked immunosorbent assay. *Journal of Clinical Microbiology*, *3*, 541–542.

Ruitenberg, E. J., Sekhuis, V. M., & Brosi, B. J. (1980). Some characteristics of a new multiple-channel photometer for through-the-plate reading of microplates to be used in enzyme-linked immunosorbent assay. *Journal of Clinical Microbiology*, *11*, 132–134.

Salk, J. E. (1948). A plastic plate for use in tests involving virus hemagglutination and other similar reactions. *Science*, *108*, 749.

Sammons, P. J., Furukawa, T., & Bulgin, A. (2005). Autonomous pesticide spraying robot for use in a greenhouse. In: *Proceedings of the Australasian conference on robotics & automation, Sydney, Australia, 5–7 December.*

Schmidt, W. (2004). A high performance micro-dual-wavelength-spectrophotometer (MDWS). *Journal of Biochemical and Biophysical Methods*, *58*, 15–24.

Schmittgen, T. D., & Livak, K. J. (2008). Analyzing real-time PCR data by the comparative CT method. *Nature Protocols*, *3*, 1101–1108.

Schulze, A., Lindner, M., Kohlmuller, D., Olgemoller, K., Mayatepek, E., & Hoffmann, G. F. (2003). Expanded newborn screening for inborn errors of metabolism by electrospray ionization-tandem mass spectrometry: Results, outcome, and implications. *Pediatrics*, *111*, 1399–1406.

Scott, J. W., & Harbaugh, B. K. (1989). *Micro-Tom. A miniature dwarf tomato.* Circular—Agricultural Experiment Station, University of Florida, Gainesville, USA. pp. 1–6.

Scott, J. W., & Harbaugh, B. K. (1995). Micro-gold miniature dwarf tomato. *Hortscience*, *30*, 643–644.

Sever, J. L. (1962). Application of a microtechnique to viral serological investigations. *The Journal of Immunology*, *88*, 320–329.

Sista, R., Hua, Z., Thwar, P., Sudarsan, A., Srinivasan, V., Eckhardt, A., et al. (2008). Development of a digital microfluidic platform for point of care testing. *Lab on a Chip*, *8*, 2091–2104.

Skeggs, L. T., Jr. (1957). An automatic method for colorimetric analysis. *American Journal of Clinical Pathology*, *28*, 311–322.

Spielbauer, G., Armstrong, P., Baier, J. W., Allen, W. B., Richardson, K., Shen, B., et al. (2009). High-throughput near-infrared reflectance spectroscopy for predicting quantitative and qualitative composition phenotypes of individual maize kernels. *Cereal Chemistry*, *86*, 556–564.

Spratlin, J. L., Serkova, N. J., & Eckhardt, S. G. (2009). Clinical applications of metabolomics in oncology: A review. *Clinical Cancer Research*, *15*, 431–440.

Stitt, M., Lilley, R. M., Gerhardt, R., & Heldt, H. W. (1989). Metabolite levels in specific cells and subcellular compartments of plant leaves. In S. Fleischer & B. Fleischer (Eds.),

Methods in enzymology series: 174. Biomembranes part U: Cellular and subcellular transport: Eukaryotic (nonepithelial) cells (pp. 518–552). Amsterdam: Elsevier.

Subrata, I. D. M., Fujiura, T., Yamada, H., Hida, M., Yukawa, T., & Nakao, S. (1996). Cherry tomato harvesting robot using 3-D vision sensor (part 1). Recognition of cherry tomato. *Journal of the Japanese Society of Agricultural Machinery, 58*, 45–52.

Suhre, K., Meisinger, C., Doering, A., Altmaier, E., Belcredi, P., Gieger, C., et al. (2010). Metabolic footprint of diabetes: A multiplatform metabolomics study in an epidemiological setting. *PLoS One, 5*, e13953.

Sulpice, R., Trenkamp, S., Steinfath, M., Usadel, B., Gibon, Y., Witucka-Wall, H., et al. (2010). Network analysis of enzyme activities and metabolite levels and their relationship to biomass in a large panel of Arabidopsis accessions. *The Plant Cell, 22*, 2872–2893.

Sulpice, R., Tschoep, H., Von Korff, M., Bussis, D., Usadel, B., Höhne, M., et al. (2007). Description and applications of a rapid and sensitive non-radioactive microplate-based assay for maximum and initial activity of D-ribulose-1,5-bisphosphate carboxylase/oxygenase. *Plant, Cell & Environment, 30*, 1163–1175.

Takátsy, G. (1955). The use of spiral loops in serological and virological micro-methods. *Acta Microbiologica et Immunologica Hungarica, 50*, 369–383.

Tanigaki, K., Fujiura, T., Akase, A., & Imagawa, J. (2008). Cherry-harvesting robot. *Computers and Electronics in Agriculture, 63*, 65–72.

Tester, M., & Langridge, P. (2010). Breeding technologies to increase crop production in a changing world. *Science, 327*, 818–822.

Urban, M., Enot, D. P., Dallmann, G., Koerner, L., Forcher, V., Enoh, P., et al. (2010). Complexity and pitfalls of mass spectrometry-based targeted metabolomics in brain research. *Analytical Biochemistry, 406*, 124–131.

Van Henten, E. J., Hemming, J., Van Tuijl, B. A. J., Kornet, J. G., Meuleman, J., Bontsema, J., et al. (2002). An autonomous robot for harvesting cucumbers in greenhouses. *Autonomous Robots, 13*, 241–258.

Vashist, S. K. (2012). Non-invasive glucose monitoring technology in diabetes management: A review. *Analytica Chimica Acta, 750*, 1–12.

Warburg, O., Christian, W., & Griese, A. (1935). Wasserstoffübertragendes coferment seine Zusammensetzung und Wirkungsweise. *Biochemische Zeitschrift, 282*, 157–165.

Watanabe, S., Mizoguchi, T., Aoki, K., Kubo, Y., Mori, H., Imanishi, S., et al. (2007). Ethylmethanesulfonate (EMS) mutagenesis of *Solanum lycopersicum* cv. Micro-Tom for large-scale mutant screens. *Plant Biotechnology, 24*, 33–38.

Wu, G. Z., Shi, Q. M., Niu, Y., Xing, M. Q., & Xue, H. W. (2008). Shanghai RAPESEED Database: A resource for functional genomics studies of seed development and fatty acid metabolism of Brassica. *Nucleic Acids Research, 36*, D1044–D1047.

Xu, Y., & Crouch, J. H. (2008). Marker-assisted selection in plant breeding: From publications to practice. *Crop Science, 48*, 391–407.

Yoo, E.-H., & Lee, S.-Y. (2010). Glucose biosensors: An overview of use in clinical practice. *Sensors, 10*, 4558–4576.

Zhang, N., Gur, A., Gibon, Y., Sulpice, R., Flint-Garcia, S., Mcmullen, M. D., et al. (2010). Genetic analysis of central carbon metabolism unveils an amino acid substitution that alters maize NAD-dependent isocitrate dehydrogenase activity. *PLoS One, 5*, e9991.

Zhao, D., Mackown, C. T., Starks, P. J., & Kindiger, B. K. (2010). Rapid analysis of nonstructural carbohydrate components in grass forage using microplate enzymatic assays. *Crop Science, 50*, 1537–1545.

CHAPTER TEN

Two Computational Simplex Approaches to Graphical Highlighting Metabolic Phenotypes and Their Functional Origins: Correspondence Analysis and Weighted Metabolic Profiles Analysis

Nabil Semmar[*,†,1]

[*]UMR INRA 1260/INSERM 1025/Plateau BioMet/ Marseille, France
[†]Université Tunis El Manar. Institut Supérieur des Sciences Biologiques Appliquées de Tunis (ISSBAT), Tunis, Tunisia
[1]Corresponding author: e-mail address: nabilsemmar@yahoo.fr

Contents

Advances in Botanical Research, Volume 67
ISSN 0065-2296
http://dx.doi.org/10.1016/B978-0-12-397922-3.00010-1

Abstract

The metabolism is a complex system interacting with several intrinsic and extrinsic factors of biological organisms, viz. genome expressions, physiological states, environmental conditions, etc. This multifactorial interaction means that the metabolism works as a reactive and flexible system providing reliable biochemical pictures on the effects of different governing factors. Metabolic flexibility and reliability are linked to conservation laws, constraining the metabolism to a close system linking input (resources) to output (products) signals: any entering signal will be decomposed into weighted parts through different metabolic pathways. This gives to metabolic trends different functional degrees highlighted by different relative levels of metabolites. Sharing the same unit resource, the different metabolic pathways are statistically constrained to be regulated within a simplex space characterized by a unit sum of its components. Output metabolic responses and their inside regulatory processes can be analyzed by using two simplex-based approaches: correspondence analysis (CA) and weighted metabolic profiles analysis (WMPA), respectively. These two approaches are based on two opposite (complementary) principles consisting of decomposition and combination of metabolic variability. In CA, metabolic datasets are decomposed into extreme trends representing elementary components of metabolic polymorphism called metabotypes. In WMPA, iterated combinations between different metabolic components help to extract functional information on their generator backbone system.

1. INTRODUCTION

Metabolomics is an emergent multidisciplinary system, which combines chemical analysis methods with statistical tools to explore the variability of concentration or amount of datasets (*n* individuals by *p* metabolites) at biological population scales. This is carried out according to two complementary ways (Fig. 10.1; Fiehn, Kind, & Barupal, 2011; Fiehn & Weckwerth, 2003; Hollywood, Brison, & Goodacre, 2006; Redestig et al., 2011; Semmar, 2011; Steuer et al., 2007):

1. Identifying different metabolic phenotypes (called metabotypes) consisting of profiles in which one or more metabolites are highly regulated at the expense of others. The highlighted metabotypes provide frank metabolic trends that are separated by intermediate states leading to define the concept of chemical polymorphism in a biological population (Fig. 10.1B) and
2. Highlighting relationships between metabolites at the origin of metabotypes. This helps to analyze metabolic processes governing chemical polymorphism in the studied population (Fig. 10.1A).

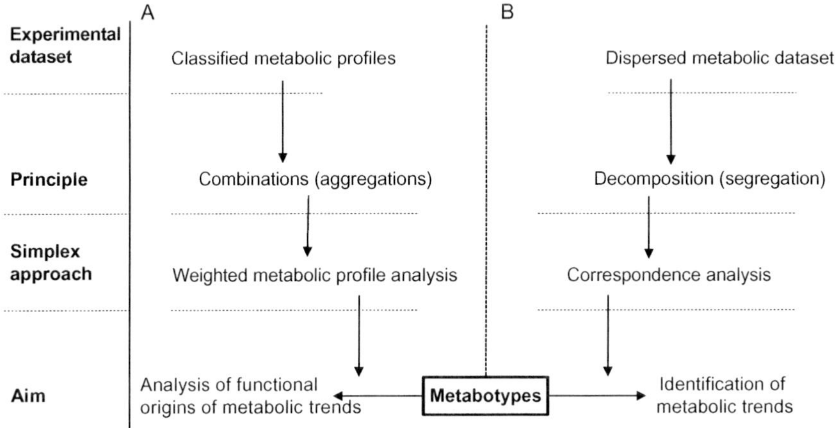

Figure 10.1 Two simplex approaches helping to statistically analyze metabotype structures and their functional origins, and consisting of correspondence analysis (CA) (B) and weighted metabolic profiles analysis (WMPA) (A), respectively.

The two questions can be conceptually extended by considering different intrinsic and extrinsic factors that could influence the metabolic variability in biological populations (Bowne, Bacic, Tester, & Roessner, 2011; Fernie & Keurentjes, 2011; Holmes, Wilson, & Nicholson, 2008; Rowan, 2011; Semmar, Nouira, & Farman, 2009; Van Dam & van der Meijden, 2011).

The first question concerns variability analysis of individual metabolic profiles leading to classify them into different metabolic trends (metabotypes). For that, the level of each metabolite is relativized according to all the others occurring in a same individual, leading to identify the most characteristic metabolites of different individuals in the population.

The second question concerns correlation analysis between metabolites tacking into account their variations the ones in relation to the others. Then, correlations can be interpreted in terms of metabolic variation ways favouring the development of metabolic trends observed under the first question.

These two questions can be statistically treated by using two simplex approaches that are based on two complementary (or inverse) computational principles:

1. Decomposition of crude dataset into homogeneous subsets representing background components (trends) of the system.

2. Combination of homogeneous subsets to extract backbone relationships governing the different trends' expressions and subsequent polymorphism at the population scale.

The first question can be treated by correspondence analysis (CA), which is a simplex topological multivariate analysis used to decompose or structure a dataset representing a biochemical system into decreasing and complementary variability parts called factors (Abdi & Williams, 2010a, 2010b; Bendixen, 2003; Frisvad, 1992; Greenacre, 1984, 2007; Renou, Lalanne-Cassou, Michelot, Gordon, & Doré, 1988; Savary, Madden, Zadoks, & Klein-Gebbinck, 1995; Semmar, 2011; Semmar, Jay, Farman, & Chemli, 2005). These decreasing variability factors make that the most general information of the dataset can be condensed into the first factors, whereas noise or residual information will be set aside along the last factors. Metabotypes provide background pictures on different polymorphic trends of a metabolic system. They can be defined as extreme profiles showing the highest regulation levels for some metabolites at expense of other metabolites and compared to all the individual profiles in population. Metabotypes can be used as reference metabolic patterns for the classification of all the (intermediate) profiles of the population.

Starting from a classified metabolic dataset into some well-defined metabotypes, a key question consists in analyzing the organization of the common metabolic backbone from which the different individual profiles were generated. To respond to such a question, correlations between metabolites can be analyzed by examining gradual variations between metabotypes. By considering that q metabotypes are issued from a same metabolic backbone, their statistical combinations usefully help to extract deep regulatory laws governing their variability. In this sense, statistical combinations between metabotypes give a multidimensional metabolic gradient from which regulatory relationships between metabolites can be highlighted. This can be carried out by applying a simplex mixture design which combines the q metabotypes according to gradual weights (Fig. 10.2). Therefore, average results issued from all the combinations provide a complete set of gradual variations between metabotypes. From such response set, correlations between metabolites can be analyzed to identify metabolic regulation ways favouring the development of observed metabotypes. This simplex approach is called weighted metabolic profile analysis (WMPA) (Mrabet & Semmar, 2010; Semmar, 2010a, 2011; Semmar, Jay, & Nouira, 2007).

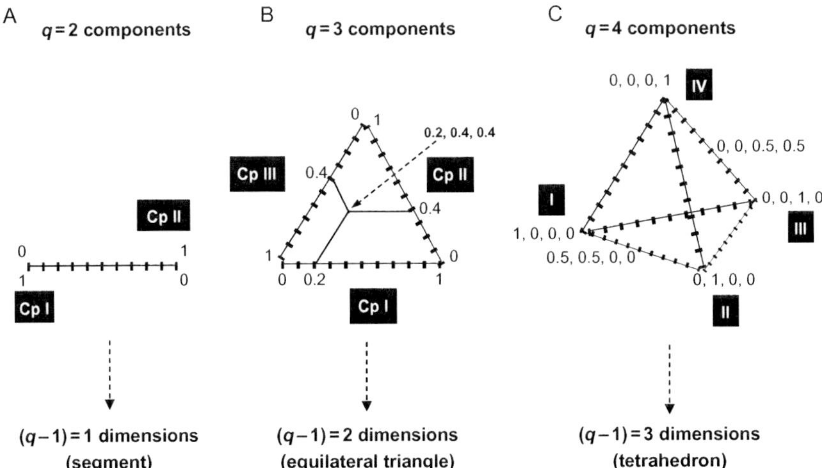

Figure 10.2 Simplex spaces with $(q-1)$ dimensions allowing to represent combinations between q additive components (Cp) (e.g. q metabolic regulations in a biological system). Simplex with $q=2$ (A), 3 (B) and 4 (C) components relatively varying in 1, 2 and 3 dimensions space, respectively.

2. GEOMETRICAL AND NUMERICAL BASES OF SIMPLEX SPACE

Numerically, simplex is used when q components of a system are governed by an equilibrium law controlling their relative levels. This implies the existence of intrinsic constraint in the system limiting the degree of freedom of the q components. Formally, any state point has q coordinates R_j obeying to the constraint:

$$\sum_{j=1}^{q} R_j = 1, \tag{10.1}$$

where R_j are the q relative levels of the q components ($j=1$ to q). In a given metabolic profile i containing p metabolites M_j with amounts x_j, the relative level of each metabolite M_j is given by:

$$R_j = \frac{x_j}{\sum_{j=1}^{p} x_j}. \tag{10.2}$$

Geometrically, all the points of simplex space combining q components (q coordinates) are included into a space with $(q-1)$ dimensions (Fig. 10.2).

The loss of one dimension ($q-1$) in geometrical space compared to the numerical one (q components) is due to the constraint (1). For example, with $q=2$ components, simplex space satisfying Eq. (10.1) is reduced to one dimension (a segment) (Fig. 10.2A); if $q=3$ or 4, the set of all additive states between components can be represented into an equilateral triangle or a tetrahedron corresponding to $(q-1)=2$ or 3 dimensions, respectively (Fig. 10.2B and C).

In metabolism, this constraint can be found by standardizing the amounts of different metabolites of a same metabolic profile by their sum (Fig. 10.3). In a chromatogram, this can be directly obtained by dividing each peak area by the total area of the chromatogram. This leads to the constraint of unity total regulation. In other words, the regulation of any metabolite takes a value depending on all the other metabolites' regulations, such as the sum of all the regulations is equal to 1.

This simplex rule is used as decomposition way to identify maximal and minimal regulation levels for some metabolites among the metabolic pool in a biological population. This simplex-based decomposition finds application in CA. From a large set of individual profiles containing several metabolites (e.g. HPLC dataset, NMR intensities dataset, etc.), CA explores the relative levels of metabolites in the ones in relation to the others to extract individual

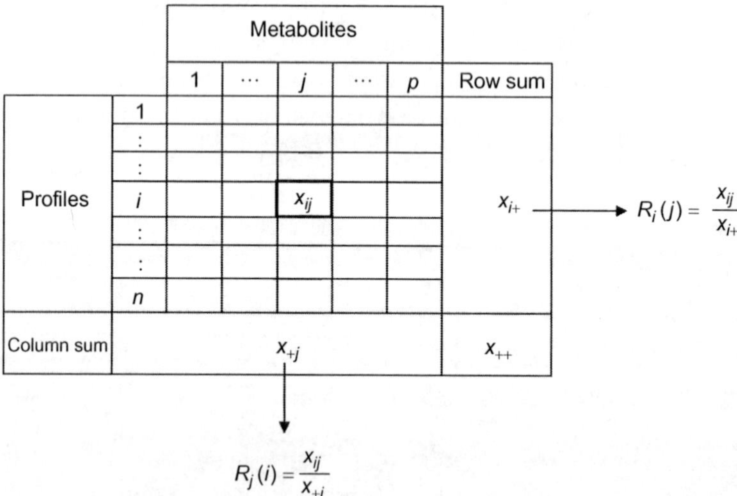

Figure 10.3 Double standardization of amount values x_{ij} of metabolites j in profiles i leading to identify the highest regulations within rows, $R_i(j)$, and within columns, $R_j(i)$. x_{i+}, x_{+j} and x_{++} represents the sums of row i, column j and the whole dataset, respectively.

profiles having the highest metabolic regulations for some compounds compared to the others.

For example, let us consider three HPLC (or NMR) profiles A, B, C giving the amounts (or intensities) of three metabolites M_1, M_2, M_3:

$$A=(6,10,4) \quad B=(6,12,5) \quad C=(3,5,2).$$

The profiles A and C will be identified to have the same relative levels for the three metabolites. By considering the sums of absolute levels $(6+10+4=20)$ and $(3+5+2=10)$, and by standardizing the values of each profile by corresponding sum, the two resulting relative profiles are found to be identical: $(6/20,\ 10/20,\ 4/20)=(3/10,\ 5/10,\ 2/10)=(0.3,\ 0.5,\ 0.2)$, such as $0.3+0.5+0.2=1$.

Apart from the decomposition way leading to identify different components of a simplex system (e.g. metabotypes), the components can be subjected to fusion or combination operations helping to extract central or backbone relationships governing their emergences. This ability is linked to a flexibility of simplex space to submit rotations, compressions and stretching resulting from iterative combinations of its components (Abdi & Williams, 2010a, 2010b; Duineveld, Smilde, & Doorhbos, 1993). Subsequently to this orientation flexibility and scale-linked variations, simplex approaches can be used to identify global and local regulation relationships between metabolites responsible for the chemical variability observed within and between metabolic profiles. More generally, simplex approaches provide structuration and filtration tools helping to extract background trends and backbone processes implied in the variability of metabolic system.

3. CA: PRINCIPLE AND METHOD

3.1. Aim

CA is a simplex-based method that can be applied on a metabolic dataset to extract metabolic profiles having the highest regulations for some metabolites compared to all the other metabolites and profiles of population. Characterization of a metabolic profile i by a metabolite j requires that such a metabolite has a high level compared to both:

- other metabolites' levels within the profile i;
- its levels within the other profiles of the studied population.

In metabolomics, such extreme metabolic profiles are called metabotypes. Beyond the extraction of extreme profiles, CA provides topological representation helping to analyze the regulation states of all the metabolic profiles the ones with respect to the others within a simplex space.

3.2. General principle of CA

Starting from a dataset containing the amounts of p metabolites (p columns) measured in n individuals (n rows), CA is a dual analysis that evaluates each value x_{ij} both along its row i and column j (Fig. 10.3). For that, the amount values are initially standardized by their row and column sums leading to two relative levels $R_i(j)$ and $R_j(i)$, respectively (Greenacre, 1984, 2007). The relativizations of amounts along (1) rows (metabolic profiles) and (2) columns (metabolites) lead to two analyses in which each value x_{ij} of the initial dataset will be:

1. considered within its row to be subjected to a between-columns analysis;
2. considered within its column to be subjected to a between-rows analysis.

A numerical example is given in Figs. 10.4 and 10.5.

Starting from a table of additive values containing the amounts of $p=3$ metabolites (M_1–M_3) in $n=4$ individuals (id_1–id_4), the aim consists in identifying the most original profiles in terms of high regulations of some metabolites. To respond to this question, all the amount or concentration values are standardized by their row sums on the one hand and by their column sums on the other hand. This dual standardization gives row and column profiles containing relative levels (Sections 3.3.1 and 3.3.2). The relative levels will be used to identify (profile, metabolite) pairs showing 'extreme' metabolic regulations in the studied population.

For example, metabolite M_2 is more regulated than the other metabolites in individual id_1 (Fig. 10.5A) and at the same time, it has a higher regulation in id_1 compared to its levels in the other individuals (Fig. 10.5B). Also, metabolite M_1 is well characteristic of id_2 because of a higher relative level compared to the other individuals (Fig. 10.5B); moreover, M_1 is more regulated than M_2, M_3 in id_2 (Fig. 10.5A). In id_4, metabolite M_3 has a high relative level compared to its negligible levels in $id_{1,2,3}$ (Fig. 10.5B); moreover, in profile id_4, M_3 shows higher relative levels than M_1 and M_2. Finally, id_3 shows intermediate metabolic state, because it is not characterized by a strict high regulation of any metabolite. Rather, id_3 can be characterized by a low regulation of M_1.

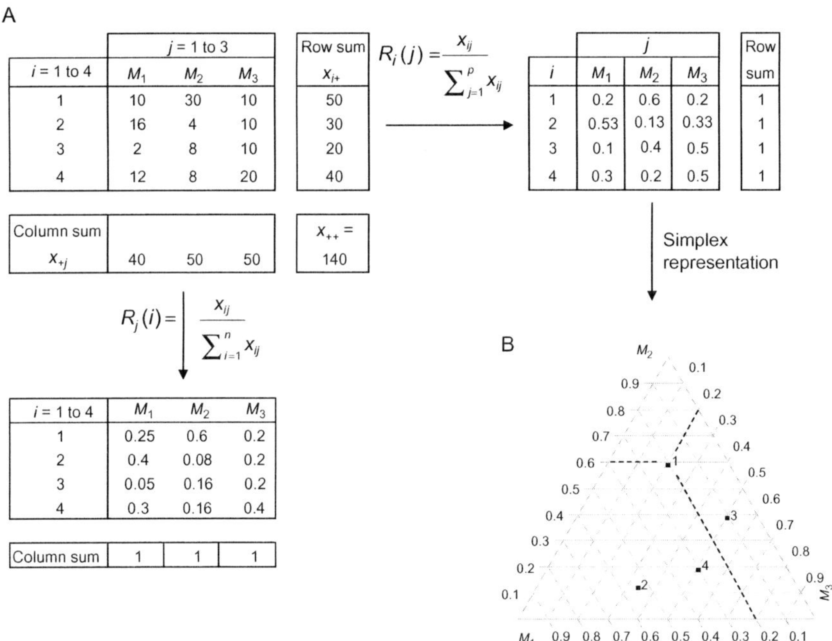

Figure 10.4 (A) A numerical example showing dual standardization of additive values (e.g. metabolites' amounts) according to their row and column sums giving row and column profiles from which affinities between individuals (rows) and metabolites (columns) can be highlighted. (B) Geometrical representation of the four row profiles *i* in a 2-dimension simplex space defined by unit sums of three metabolites' regulations ($\Sigma_{j=1}^{p=3} R_i(j) = 1$).

3.3. Dual calculation in CA

3.3.1 Between-profiles analysis

Between-profiles analysis evaluates how much individual profiles are close or distant between them in terms of relative levels (regulations) of p metabolites taking into account the variation scale of each metabolite in the population. The variation scales of metabolites are considered from the sum x_{+j} of all the values of a same metabolite j. By considering a given metabolite M_j, the square distance between two profiles i and i' can be calculated by the χ^2 formula (Greenacre, 1984, 2007; Legendre & Gallagher, 2001; Rencher, 2002):

$$d_j^2(i,i') = \frac{x_{++}}{x_{+j}} \left(\frac{x_{ij}}{x_{i+}} - \frac{x_{i'j}}{x_{i'+}} \right)^2, \tag{10.3}$$

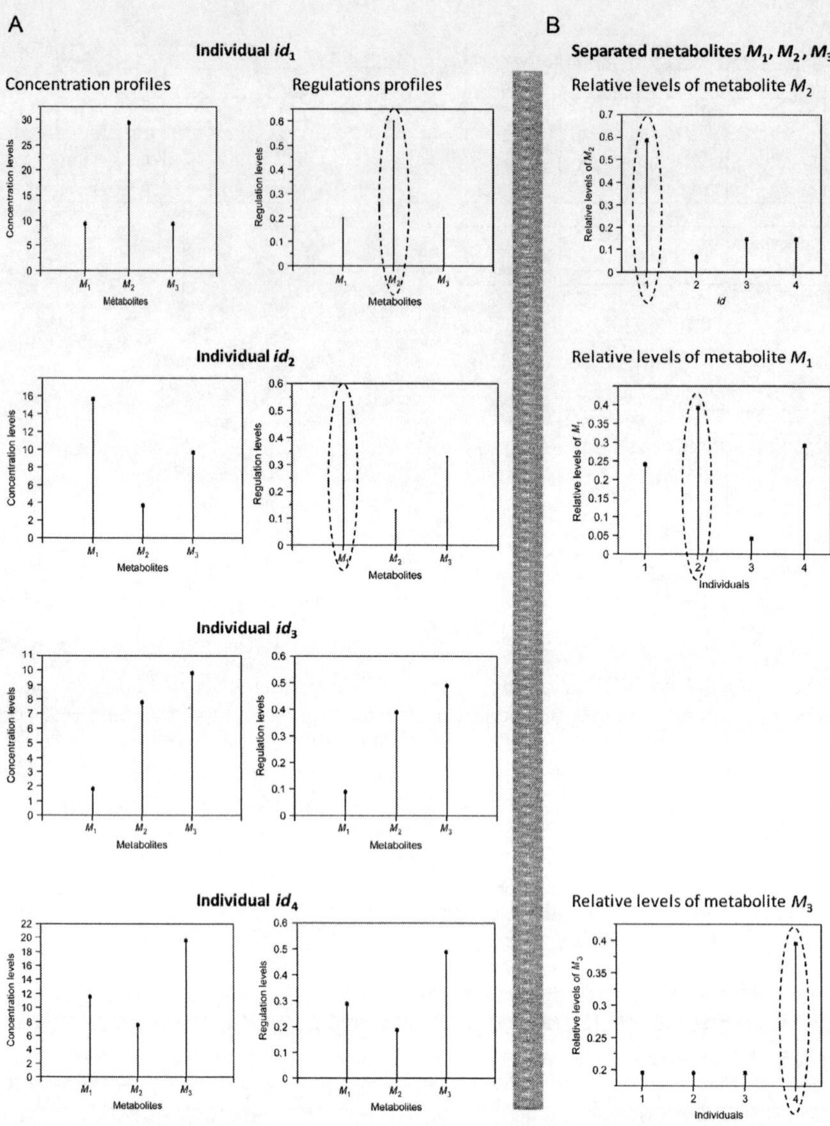

Figure 10.5 Standardizations of metabolites' concentrations (e.g. in mg/mL) (or amounts in mg/g DW) by row (A) and column (B) sums. Along rows, individual regulation profiles are obtained by dividing the concentration of each metabolite by the sum of all metabolites' concentrations of considered individual id_i (A). Along a column, the relative levels of a metabolite M_j are calculated by dividing its concentrations by the total sum of M_j in all the id_i (B). This dual standardization (in A and B) extracts the most characteristic metabolites of the different individual profiles. Metabolic profiles with extreme regulations for some metabolites are called metabotypes; For example, M_2 and M_1 are highly regulated in id_1 and id_2, respectively.

where

x_{ij} and $x_{i'j}$ are the amounts of metabolite M_j in profiles i and i', respectively.

x_{i+} and $x_{i'+}$ are the sums of amounts of all the metabolites of profiles i and i', respectively (Fig. 10.4A).

x_{+j} is the sum of all the amounts of metabolite M_j in all the profiles. In Eq. (10.3), it plays a weighting role for distance calculation between profiles.

x_{++} is the sum of all the amounts of all the (p) metabolites in all the (n) profiles of the dataset.

In Eq. (10.3) (distance formula), each amount x_{ij} of metabolite M_j in profile i is initially standardized by the sum x_{i+} of profile i (Fig. 10.4A). For a given metabolite M_j having two equal amounts x_{ij} and $x_{i'j}$ in two profiles i and i', the relative levels x_{ij}/x_{i+} and $x_{i'j}/x_{i'+}$ will depend on the sum x_{i+} and $x_{i'+}$: lower is the total amount x_{i+} in a profile i higher will be the relative levels x_{ij}/x_{i+} (with $j=1$ to p). Then for a given metabolite Mj, differences between its relative levels in profiles i and i' are calculated taking into account its cumulative value x_{+j} in the whole dataset.

By applying Eq. (10.3) for the p metabolites separately then by summing the p resulting square distances, one obtains the total square distance between two profiles i and i' (Eq. 10.4):

$$d^2(i,i') = \sum_{j=1}^{p} \frac{x_{++}}{x_{+j}} \left(\frac{x_{ij}}{x_{i+}} - \frac{x_{i'j}}{x_{i'+}} \right)^2. \tag{10.4}$$

This square distance corresponds to the χ^2 distance between two row profiles i and i'. Equations (10.3) and (10.4) show that the weighting by inverse of x_{+j} makes that lower is x_{+j} higher will be the relative level of a given amount x_{ij}. Consequently, a value x_{ij} of a metabolite M_j in a profile i will have higher relative importance when the metabolite sum x_{+j} is lower compared to the other metabolites. In other words, in a given profile i, two equal amounts x_{ij}, $x_{ij'}$ of two metabolites M_j, $M_{j'}$ (representing equal relative level in profile i) will contribute differently in distance calculation (Eq. 10.4) according to the sum x_{+j} and $x_{+j'}$ of columns j and j' at population scale (Fig. 10.5).

Based on the χ^2 distance, between-profiles analysis highlights different variability poles along which the different individuals (metabolic profiles) can be separated into different trends. In metabolic terms, such poles can be attributed to metabotypes providing a background picture on the chemical polymorphism in the studied population (Figs. 10.11 and 10.12).

3.3.2 Within-profiles or between-metabolites analysis

Within-profile analysis can be associated to a between-metabolites analysis because each profile is a set of regulated metabolites. It evaluates how much metabolites are regulated similarly or differently in the population taking into account the variation levels in each metabolic profile of the dataset. The variation levels in metabolic profiles are considered by reference to the sum x_{i+} of all the metabolites' amounts in a same profile i; this sum will be used as a weighting factor in distance calculation between metabolites (Eq. 10.5):

$$d^2(j,j') = \frac{x_{++}}{x_{i+}}\left(\frac{x_{ij}}{x_{+j}} - \frac{x_{ij'}}{x_{+j'}}\right)^2, \tag{10.5}$$

where

x_{ij} and $x_{ij'}$ are the amounts of metabolites M_j and $M_{j'}$ in profile i.

x_{+j} and $x_{+j'}$ are the sums of all the amounts of metabolites M_j and $M_{j'}$ in the dataset, respectively (Fig. 10.4A).

x_{i+} is the sum of amounts of all the metabolites in profile i.

x_{++} is the sum of the amounts of all the (p) metabolites in all the (n) profiles of the dataset.

Equation (10.5) gives square distance between two metabolites M_j and $M_{j'}$ in the same profile i: the relative state of each metabolite M_j is initially considered by dividing its amount x_{ij} by its cumulative sum x_{+j} in the population. Then at population scale, the distance between metabolites M_j and $M_{j'}$ is obtained by the sum of n distance terms corresponding to the n metabolic profiles i ($i=1$ to n). A difference between two relative values x_{ij}/x_{+j} and $x_{ij'}/x_{+j'}$ of two metabolites M_j and $M_{j'}$ will have higher importance when the total sum x_{i+} of profile i is lower (Eq. 10.6):

$$d^2(j,j') = \sum_{i=1}^{n} \frac{x_{++}}{x_{i+}}\left(\frac{x_{ij}}{x_{+j}} - \frac{x_{ij'}}{x_{+j'}}\right)^2. \tag{10.6}$$

Equation (10.6) gives the closeness or distance between the global regulation levels of two metabolites M_j and $M_{j'}$ in the population of n individuals. A numerical example is given in Fig. 10.6.

From amount dataset of $n=10$ individuals characterized by $p=5$ metabolites, profiles id_5 and id_6 initially show a same amount equal to 5 for metabolite M_3. By standardizing the values by row sums, between-profiles analysis shows that the value 5 does not correspond to same relative levels in id_5 and id_6: after division of $x_{53}=5$ by $x_{5+}=20$ and $x_{63}=5$ by $x_{6+}=40$, the same

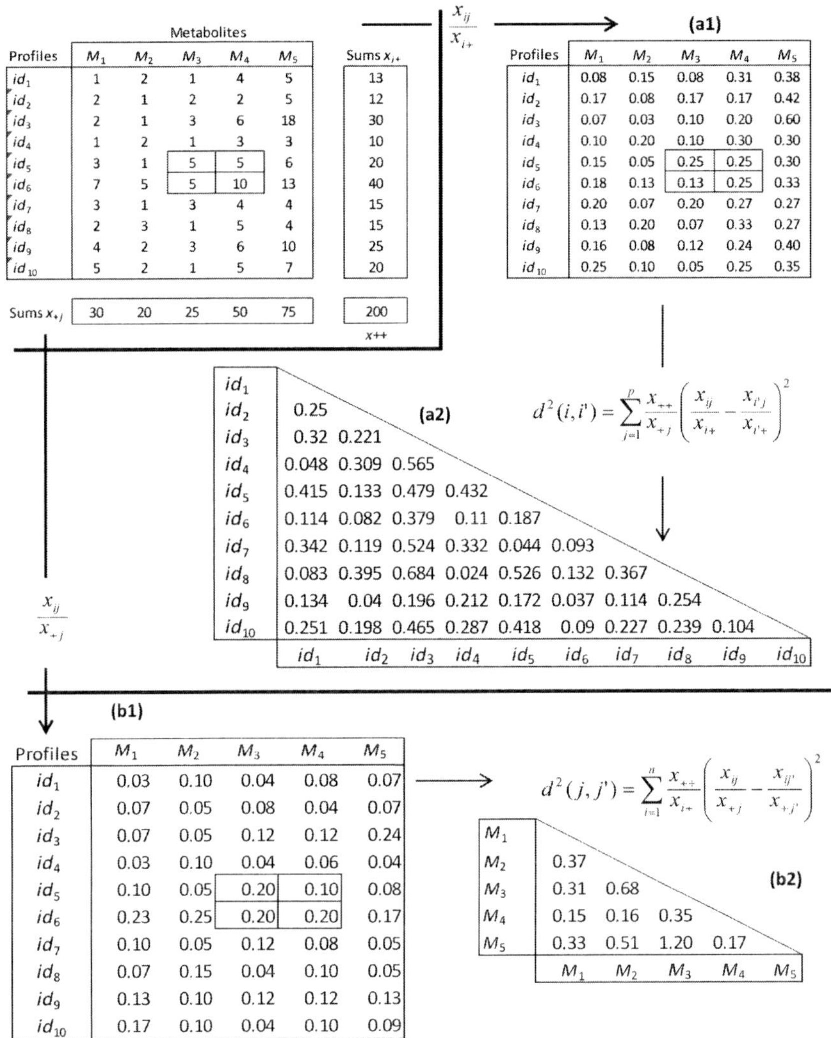

Figure 10.6 χ^2 distances calculations between row profiles (*a*1, *a*2) and between column profiles (*b*1, *b*2) of a metabolic dataset (10 individuals × 5 metabolites).

amount of 5 is found to correspond to regulation levels of 0.25 and 0.13 in id_5 and id_6, respectively (Fig. 10.6A1).

In between-metabolites analysis, the equal values of 5 for M_3 and M_4 in id_5 show different relative importances by reference to their column sums, viz. 0.2 (20%) and 0.1 (10%), respectively (Fig. 10.6B1).

χ^2 distances calculations between individual profiles show that id_4 and id_8 are the two closest cases in the dataset ($\chi^2 = 0.024$) followed by the pairs (id_6, id_9; $\chi^2 = 0.037$), (id_2, id_9; $\chi^2 = 0.04$) then (id_5, id_7; $\chi^2 = 0.044$) (Fig. 10.6A2).

The greatest χ^2 distances concern id_3 towards the other individual profiles, with a maximal χ^2 value for the pair (id_3, id_8; $\chi^2=0.684$). Interpretations of these distances are given in Section 3.4.4.

χ^2 distances calculations between metabolites (columns) give minimal values for the pairs (M_1, M_4; $\chi^2=0.15$), (M_2, M_4; $\chi^2=0.16$) and (M_4, M_5; $\chi^2=0.17$) (Fig. 10.6B2). However, the highest χ^2 values opposed M_2 to M_3 ($\chi^2=0.68$) and M_2 to M_5 ($\chi^2=0.51$). This indicates pairs of metabolites that may be regulated by some opposite ways in the population. More details will be graphically highlighted by CA algorithm (Section 3.4.4).

3.4. CA algorithm

3.4.1 Eigenvalues calculation

CA algorithm is performed infive steps through which data are transformed into factorial coordinates after eigenvalues and eigenvectors calculations (Fig. 10.7). The algorithm is applied separately on the n row- and the p column-profiles, leading to a dual analysis (Greenacre, 1984, 2007; Weller & Romney, 1990).

3.4.1.1 Weighted values matrix calculation

From a matrix X containing additive data x_{ij}, each value is standardized by the root square of product between its row and column sums:

$$x_{ij} \rightarrow \frac{x_{ij}}{\sqrt{x_{i+} \cdot x_{+j}}}. \tag{10.7}$$

This weighting or standardization (std) step gives a relative values' matrix S (n rows $\times$ p columns) (Fig. 10.7). The std makes that a high amount x_{ij} is considered all the more atypical as the product between its row and column sums is low. This is particularly the case of the value $x_{ij}=x_{53}=5$ metabolite M_3 in individual profile id_5, the standardization of which is equal to 0.22 (Fig. 10.7). This standardized value shows that $x_{53}=5$ (M_3 in id_5) is relatively more important than $x_{61}=7$ (M_1 in id_6) (std$=0.2$). Also, by comparing the equal amounts x_{53} and $x_{54}=5$ in the same profile id_5, the standardized level of M_3 in id_5 (std$=0.22$) appears to be more important than that of M_4 (std$=0.16$), because of the lower column sum x_{+3} (=25) compared to x_{+4} (=50). The highest standardized value (std$=0.38$) concerns metabolite M_5 in id_3 because of the high amount x_{35} (=18) in the whole dataset X (Fig. 10.7).

3.4.1.2 Square matrix calculation before diagonalization

In a second step, the matrix S is used to calculate two square matrices C and L from which eigenvalues (and eigenvectors) of the row- and column-systems will be calculated, respectively (Fig. 10.7A and B):

$$L = S \cdot S' \ (n \times n) \ \text{ and } \ C = S' \cdot S \ (p \times p).$$

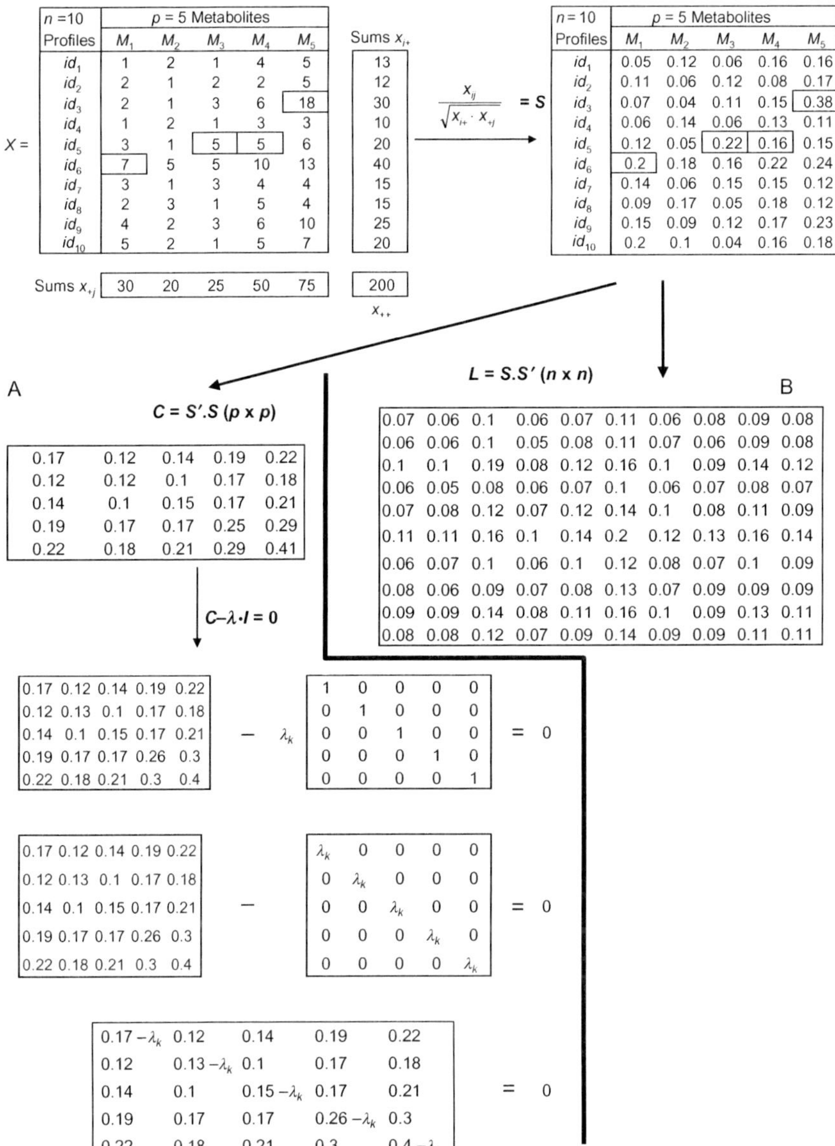

Figure 10.7 Standardization of a metabolic amount dataset X before eigenvalues' calculations from the roots of matrix determinant ($|C-\lambda.I|=0$). Squared columns (A) and rows (B) matrices used for dual analysis in CA.

The n diagonal elements of L are the sums of squared standardized values of each profile i. Out of the diagonal, the L-values are sums of products between the standardized values of different profiles taken two by two. Closer are the profiles, higher will be their product given by L.

The p diagonal elements of C are the sums of n squared standardized values of each metabolite j. Out of the diagonal, the C-values evaluate how much the metabolites are closely regulated: two metabolites regulated more similarly will have a higher product in the matrix C.

3.4.1.3 Eigenvalues calculation from square matrix

Initially, the number of eigenvalues (and eigenvectors) of a matrix X $(n \times p)$ is defined by the lowest dimension among n and p. Moreover, in simplex system, all the values of a same row or same column are dependent on each others by the sum unit constraint (Eq. 10.1). It results that if p is the lowest dimension of matrix X, we have $(p-1)$ eigenvalues to calculate. In the example given by Fig. 10.6 where $n=10$ individuals and $p=5$ metabolites, the system admits $p=5$ eigenvalues one of which is trivial (=0 or 1) because of the simplex rule governing the variability within rows and columns; subsequently, $(p-1)=4$ eigenvalues $(0<<1)$ will be calculated.

Eigenvalues (λ_k; $k=1$ to $p-1$) are additive and decreasing positive values representing decreasing information parts of the initial heterogeneous dataset (X). From the eigenvalues, CA calculates new variables, called principal components or factors, through which the total variability of the dataset is organized into well structured and complementary blocks. Under geometrical aspect, such components represent orthogonal axes along which total and heterogeneous information of the initial dataset X is decomposed into complementary and hierarchically decreasing homogeneous parts (Abdi & Williams, 2010a, 2010b; Rencher, 2002). This means that the first principal component F_1 of CA provides more information than all the others; 'residual' information not provided by F_1 will be partially treated by F_2, which provides a second structured variability block representing a greater information part than F_k $(k>2)$; etc.

Therefore, the fundamental concept of orthogonality between factors makes that total variability of dataset will be decomposed into homogeneous row and column blocks via calculations performed on the square matrices C $(p \times p)$ and L $(n \times n)$. Calculations of the eigenvalues λ_k for rows- and columns-systems are carried out by solving the equations (Fig. 10.7):

$$L = \lambda \cdot I, \text{ with} I : \text{identity matrix}(n \times n); \tag{10.8}$$

$$C = \lambda \cdot I, \text{ with} I : \text{identity matrix}(p \times p). \tag{10.9}$$

Identity matrix I represents a basis of orthogonal unitary vectors. For instance, with $p=4$, identity matrix consists of four orthogonal unitary vectors represented by its four columns:

$$\underset{4\times4}{I}=\begin{pmatrix}1&0&0&0\\0&1&0&0\\0&0&1&0\\0&0&0&1\end{pmatrix}.$$

To calculate the p eigenvalues λ_k ($k=1$ to p) of the system, we have to solve the determinant system:

$$|C-\lambda_k I|=0. \tag{10.10}$$

In the case where $n<p$, the eigenvalues will be extracted from the determinant system:

$$|L-\lambda_k I|=0. \tag{10.11}$$

Application on the numerical example given in Fig. 10.7 results in $p=5$ equations with 5 unknown λ_k the values of which are equal to 0.052 (λ_1), 0.036 (λ_2), 0.016 (λ_3), 0.002 (λ_4) and 1 (λ_5). The fifth eigenvalue (=1) is trivial leading to $(p-1)=4$ eigenvalues λ_1–λ_4. These four eigenvalues correspond to four structured variability blocks representing 49.1%, 34%, 14.8% and 2.1% of heterogeneous information mixed in the initial dataset (49.1+34+14.8+2.1=100%). These percentages are calculated from the ratios:

$$\%\text{explained variability}=\frac{\lambda_k}{\sum_{k=1}^{p-1}\lambda_k}. \tag{10.12}$$

3.4.2 Eigenvectors' calculation

Each of the $(p-1)$ eigenvalues is associated to an eigenvector with p or n coordinates, which are calculated from the squared matrix C or L, respectively. In other words, from the $(p-1)$ eigenvalues, the row and column analyses in CA imply the calculation of two sets of $(p-1)$ eigenvectors having n coordinates (row eigenvectors) and p coordinates (column eigenvectors). The calculation of eigenvector U_k is carried out by solving the system:

$$C\cdot U_k=\lambda_k\cdot I\cdot U_k \tag{10.13}$$

for column analysis, where $U_k=(u_{1k}, u_{2k}, \ldots, u_{pk})$ (p coordinates), and

$$L \cdot U_k = \lambda_k \cdot I \cdot U_k \tag{10.14}$$

for row analysis, where $U_k = (u_{1k}, u_{2k}, \ldots, u_{nk})$ (n coordinates).

3.4.2.1 Column eigenvector calculation

The $(p-1)$ column eigenvectors are calculated from the square matrix C (Fig. 10.7). They have p coordinates giving the roles of the p metabolites j ($j=1$ to p) in the considered column eigenvector U_k. From the numerical example of Figs. 10.6 and 10.7, the calculated column eigenvector matrix U_C is presented in Table 10.1.

The four eigenvectors (having five coordinates) represent four orthogonal bases defining the four principal components F_1–F_4 along which the variability of the n (=10) individuals (with five coordinates) will be structured. The first eigenvector U_1 provides the direction of the first principal component F_1 along which 49.1% variability of the dataset is explained; the second eigenvector U_2 gives the direction of the second factor F_2 explaining 34% total variability of the dataset; etc.

In U_1, the highest absolute coordinates concern the fifth and second values which are associated to metabolites M_5 and M_2, respectively; moreover, these two coordinates have opposite signs. This means that the 49.1% variability which will be observed along F_1 are mainly due to opposition between the regulations of metabolites M_5 and M_2 in the n (=10) individual profiles (Fig. 10.12A). Along F_2, the eigenvector U_2 shows a maximal coordinate for M_3 meaning that 34% variability is mainly due to the variability of M_3 between the n (=10) metabolic profiles (e.g. id_1 and id_4 vs. id_5 and id_7; Fig. 10.8). This metabolite shows some opposition to M_2 and M_5 because of opposite coordinates' signs in U_2 (Table 10.1). This will be more understood in the next steps after calculation and graphical visualization of factorial coordinates (Section 3.4.3).

Table 10.1 Matrix U_C containing $(p-1)$ column eigenvectors characterized by p (=5) coordinates corresponding to the p (=5) variables of the initial dataset

		$\lambda_1=0.052$	$\lambda_2=0.036$	$\lambda_3=0.016$	$\lambda_4=0.002$
	j	U1	U2	U3	U4
$U_C=$	1	−0.315	0.328	0.797	0.093
	2	−0.586	−0.419	−0.277	0.552
	3	0.083	0.766	−0.481	0.225
	4	−0.283	−0.109	−0.219	−0.781
	5 (=p)	0.686	−0.345	0.096	0.165

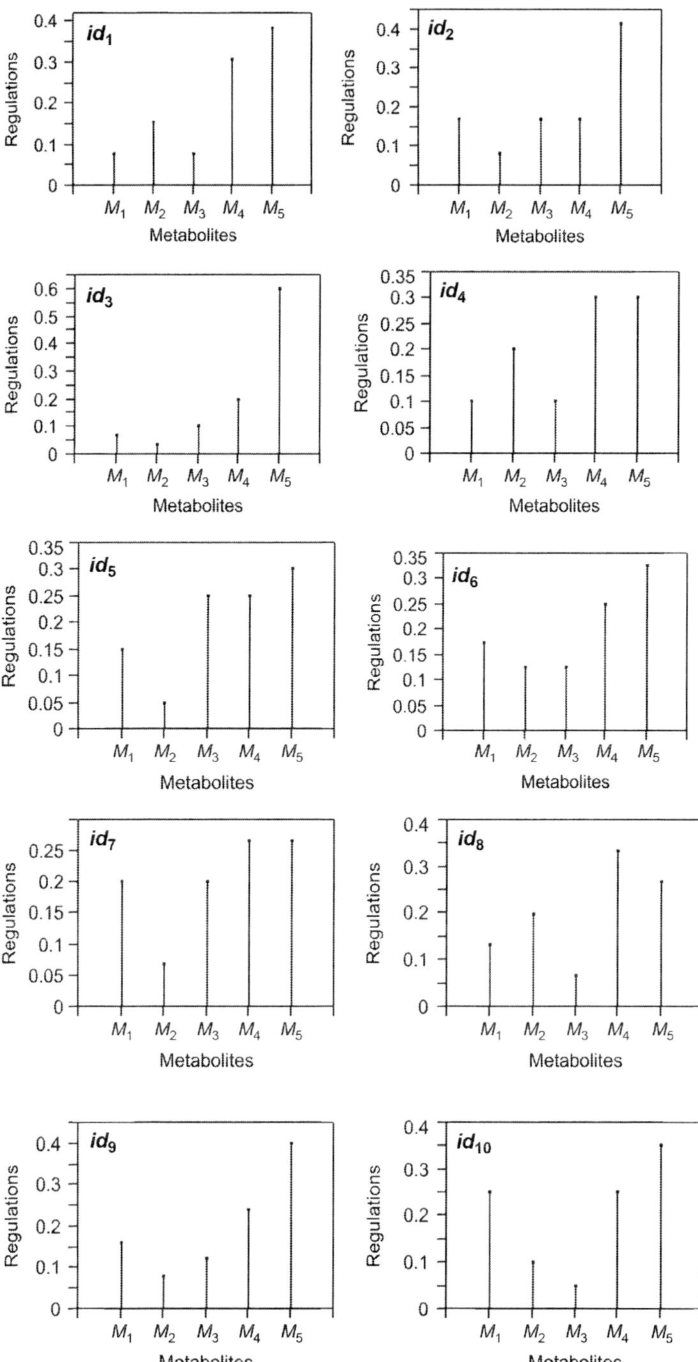

Figure 10.8 Regulation profiles of 10 individuals obtained by standardizing the amounts of the 5 metabolites by their sum within a same profile.

3.4.2.2 Row eigenvector calculation

In row analysis, $(p-1)$ eigenvectors with n coordinates are calculated. Each of the n coordinates gives the role of profile i ($i=1$–n) in the considered row eigenvector U_k. This calculation uses the square matrix L $(n\times n)$ (Fig. 10.7) by solving the Eq. (10.14). The four eigenvalues $\lambda_1=0.052$, $\lambda_2=0.036$, $\lambda_3=0.016$ and $\lambda_4=0.002$ give four row eigenvectors U_1, U_2, U_3, U_4 with $n=10$ coordinates that can be stored in a matrix U_L (Table 10.2):

The absolute values of the 10 coordinates of each eigenvectors quantify the variability states of the 10 individual profiles towards the p (=5) metabolites. For example, the highest coordinate of U_1 is the third value (0.787) which concerns the third metabolic profile id_3. In other words, id_3 presents a high variability compared to the other profile in the dataset due to marked variations between its metabolites (Fig. 10.8). The positive coordinate of id_3 in the first row eigenvector U_1 is opposite to four other coordinates associated to id_4 (−0.257), id_6 (−0.239), id_8 (−0.423) and id_{10} (−0.175) (Fig. 10.12). This globally means that on F_1, the metabolic regulations in id_3 seem to be opposite to those observed in $id_{4,\ 6,\ 8,\ 10}$. Taking into account the previous column analysis, the most influencing metabolites on F_1 are M_2 and M_5 (which are opposite) (Fig. 10.12). Therefore, the

Table 10.2 Matrix U_L containing $(p-1)$ (=4) row eigenvectors characterized by n (=10) coordinates corresponding to the n (=10) individual profiles of the initial dataset

		$\lambda_1=0.052$	$\lambda_2=0.036$	$\lambda_3=0.016$	$\lambda_4=0.002$
	Profiles *i*	***U1***	***U2***	***U3***	***U4***
$U_L=$	1	−0.082	−0.341	−0.316	−0.226
	2	0.129	0.156	0.069	0.745
	3	0.787	−0.309	−0.067	−0.112
	4	−0.257	−0.232	−0.338	0.231
	5	0.052	0.628	−0.347	−0.189
	6	−0.239	0.038	0.077	0.345
	7	−0.11	0.44	0.013	−0.314
	8	−0.423	−0.33	−0.21	−0.151
	9	0.095	0.023	0.15	−0.101
	10 (=n)	−0.175	−0.106	0.764	−0.205

opposition between id_3 and $id_{4,\ 6,\ 8,\ 10}$ is mainly linked to some opposition between the regulations of metabolites M_2 and M_5 in these five profiles (Fig. 10.8). This will be further illustrated after calculation of factorial coordinates (Section 3.4.3).

3.4.3 Factorial coordinates calculation

The final step in CA consists in calculating the factorial coordinates of rows and columns along the different principal components $F_1, \ldots, F_{p-1}$. This is carried out by multiplying each eigenvectors U_k representing the considered F_k by the matrix W_L or W_C containing the weighted row or column profiles (Figs. 10.9 and 10.10).

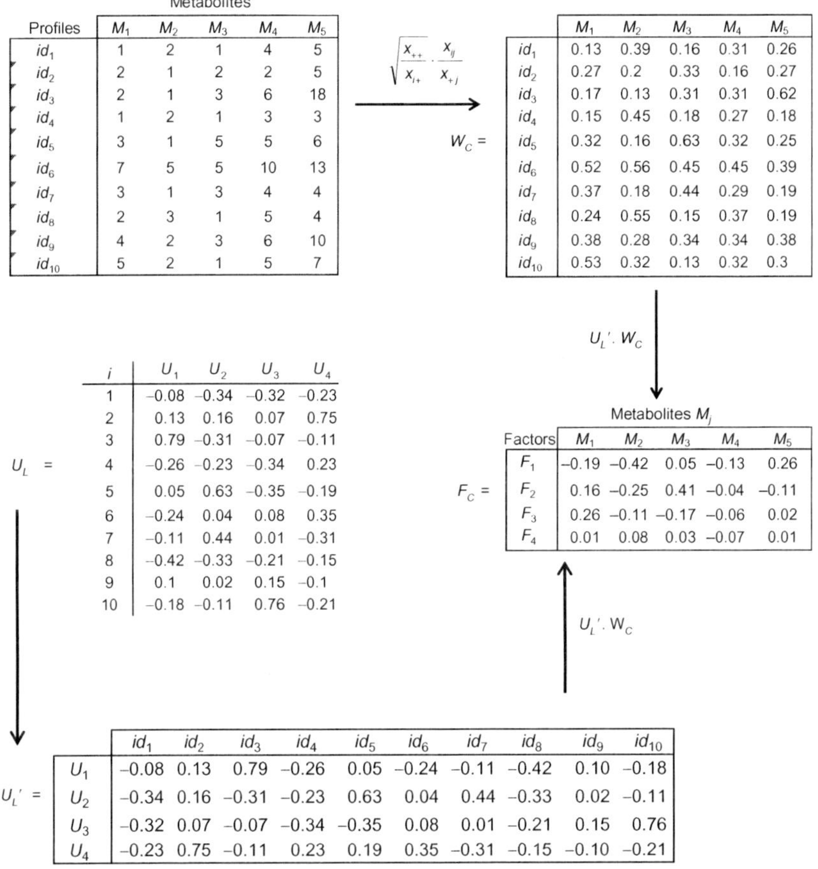

Profiles	M_1	M_2	M_3	M_4	M_5
id_1	1	2	1	4	5
id_2	2	1	2	2	5
id_3	2	1	3	6	18
id_4	1	2	1	3	3
id_5	3	1	5	5	6
id_6	7	5	5	10	13
id_7	3	1	3	4	4
id_8	2	3	1	5	4
id_9	4	2	3	6	10
id_{10}	5	2	2	5	7

$$\sqrt{\frac{x_{++}}{x_{i+}} \cdot \frac{x_{ij}}{x_{+j}}}$$

$W_C =$

	M_1	M_2	M_3	M_4	M_5
id_1	0.13	0.39	0.16	0.31	0.26
id_2	0.27	0.2	0.33	0.16	0.27
id_3	0.17	0.13	0.31	0.31	0.62
id_4	0.15	0.45	0.18	0.27	0.18
id_5	0.32	0.16	0.63	0.32	0.25
id_6	0.52	0.56	0.45	0.45	0.39
id_7	0.37	0.18	0.44	0.29	0.19
id_8	0.24	0.55	0.15	0.37	0.19
id_9	0.38	0.28	0.34	0.34	0.38
id_{10}	0.53	0.32	0.13	0.32	0.3

$U_L =$

i	U_1	U_2	U_3	U_4
1	−0.08	−0.34	−0.32	−0.23
2	0.13	0.16	0.07	0.75
3	0.79	−0.31	−0.07	−0.11
4	−0.26	−0.23	−0.34	0.23
5	0.05	0.63	−0.35	−0.19
6	−0.24	0.04	0.08	0.35
7	−0.11	0.44	0.01	−0.31
8	−0.42	−0.33	−0.21	−0.15
9	0.1	0.02	0.15	−0.1
10	−0.18	−0.11	0.76	−0.21

$F_C =$ ($U_L' \cdot W_C$)

Factors	M_1	M_2	M_3	M_4	M_5
F_1	−0.19	−0.42	0.05	−0.13	0.26
F_2	0.16	−0.25	0.41	−0.04	−0.11
F_3	0.26	−0.11	−0.17	−0.06	0.02
F_4	0.01	0.08	0.03	−0.07	0.01

$U_L' =$

	id_1	id_2	id_3	id_4	id_5	id_6	id_7	id_8	id_9	id_{10}
U_1	−0.08	0.13	0.79	−0.26	0.05	−0.24	−0.11	−0.42	0.10	−0.18
U_2	−0.34	0.16	−0.31	−0.23	0.63	0.04	0.44	−0.33	0.02	−0.11
U_3	−0.32	0.07	−0.07	−0.34	−0.35	0.08	0.01	−0.21	0.15	0.76
U_4	−0.23	0.75	−0.11	0.23	0.19	0.35	−0.31	−0.15	−0.10	−0.21

Figure 10.9 Columns' factorial coordinates matrix F_C calculated from the product between the weighted columns profiles matrix W_C and the row eigenvector matrix U_L.

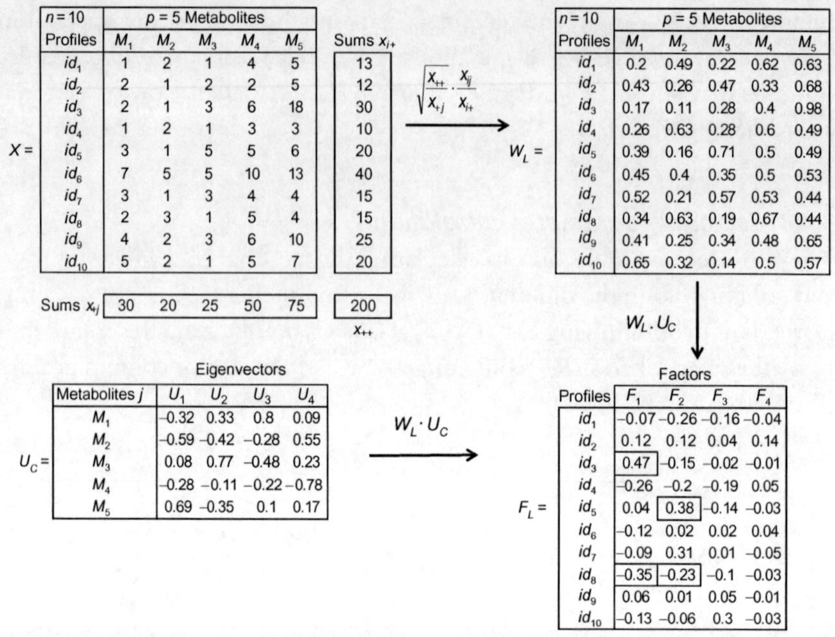

Figure 10.10 Matrix F_L of rows' factorial coordinates calculated from the product between the weighted row profiles matrix W_L and column eigenvector matrix U_C.

3.4.3.1 Columns' factorial coordinates calculation

The matrix F_C containing the factorial coordinates of the p (=5) metabolites along the $p-1$ factors F_k ($k=1$ to 4) is given by multiplying the transposed row eigenvector matrix U_L' by the weighted column profile matrix W_C (Fig. 10.9):

$$\underset{(p-1\times p)}{F_C} = \underset{(p-1\times n)}{U'_L} \cdot \underset{(n\times p)}{W_C} \tag{10.15}$$

where W_C has the general term:

$$W_C = \sqrt{\frac{x_{++}}{x_{i+}} \cdot \frac{x_{ij}}{x_{+j}}} \tag{10.16}$$

with

x_{ij} is the value (amount) of metabolite j in individual profile i given by the initial dataset X.

x_{i+} is the sum of values x_{ij} of row i ($j=1$ to p); $\sum_{j=1}^{p} x_{ij}$.

x_{j+} the sum of values x_{ij} of column j ($i=1$ to n); $\sum_{i=1}^{n} x_{ij}$.

x_{++} is the sum of all the values x_{ij} of the dataset X; $\sum_{i=1}^{n}\sum_{j=1}^{p} x_{ij}$.

In the matrix F_C (Fig. 10.9), metabolite M_2 shows the greatest variability along F_1 because of its higher coordinate (−0.42). It is opposite to M_5 having a high positive coordinate (0.26). Along F_2, M_3 with a high positive factorial coordinate (0.41) is opposite to M_2 with negative one (−0.25). Thus, M_2 shows some original variability which is opposite to either M_5 (along F_1) or M_3 (along F_2). Along F_3, another part of variability of M_2 ($F_{23} = -0.11$) is combined with that of M_3 ($F_{33} = -0.17$) to opposite to that of M_1 ($F_{13} = 0.26$). This indicates that in some metabolic profiles metabolites M_2 and M_3 share some regulation trend which is opposite to that of M_1. Metabolic profiles characterized by extreme regulations of metabolites will be identified after calculations of rows' factorial coordinates (Section 3.4.3). Columns' and rows' factorial coordinates will be considered together to graphically highlight the different extreme regulations of the metabolic system (Fig. 10.12).

3.4.3.2 Rows' factorial coordinates calculation

The matrix F_L containing the factorial coordinates F_{ik} of the n row profiles i along the $(p-1)$ factors F_k is calculated by the product between the weighted row profiles' matrix W_L and the column eigenvector matrix U_C (Fig. 10.10):

$$\underset{(n\times p-1)}{F_L} = \underset{(n\times p)}{W_L} \cdot \underset{(p\times p-1)}{U_C}, \tag{10.17}$$

where W_L has the general term:

$$W_L = \sqrt{\frac{x_{++}}{x_{+j}}} \cdot \frac{x_{ij}}{x_{i+}}. \tag{10.18}$$

In Eq. (10.18), the ratio x_{ij}/x_{i+} represents the relative value of variable j (metabolite j) in row i (individual profile i). A higher x_{ij} and a lower x_{i+} give a higher regulation (x_{ij}/x_{i+}) of metabolite M_j in profile i. Equation (10.18) shows also that high amounts x_{ij} of scarce metabolites (lower x_{+j} in dataset) will be relatively advantaged. Therefore in CA, an original metabolic state can be defined by a profile i showing a high regulation (high x_{ij}/x_{i+} due to high x_{ij} and/or low x_{i+}) for a metabolite M_j which is slightly (or rarely) produced in the population (low x_{+j}).

By applying Eq. (10.17), the row factorial coordinates of the n (=10) individuals along the $p-1$ (=4) factors F_k ($k=1$ to 4) can be calculated (Fig. 10.10).

High factorial coordinates indicate profiles with original metabolic states: id_3 is extreme along F_1 ($F_{31}=+0.47$); id_5 is extreme along F_2 ($F_{52}=0.38$); id_8 shows relatively high values both along F_1 and F_2 ($F_{81}=-0.35$, $F_{82}=-0.23$). Interpretation of these results will be presented in Section 3.4.4.

3.4.4 Graphical visualization and interpretation rules of factorial coordinates

From the matrix F_L and F_C, scatter plots are visualized to graphically analyze (i) behaviours of individual profiles, (ii) regulations trends of metabolites, and (iii) affinities between profiles and metabolites. The resulting factorial plots are of two types, viz. plots of factorial coordinates of rows (1) and columns (2), which are dually associated between them. From these dual factorial plots, three graphical criteria are considered to interpret three essential points (i–iii) (Fig. 10.11):

a. Distance of points from origin along a factor F_k: farther is a point from the origin and along F_k higher is its weight on F_k. Such a point contributes significantly to the meaning of F_k. In CA, the axes' origin (G) is the barycentre of the dataset

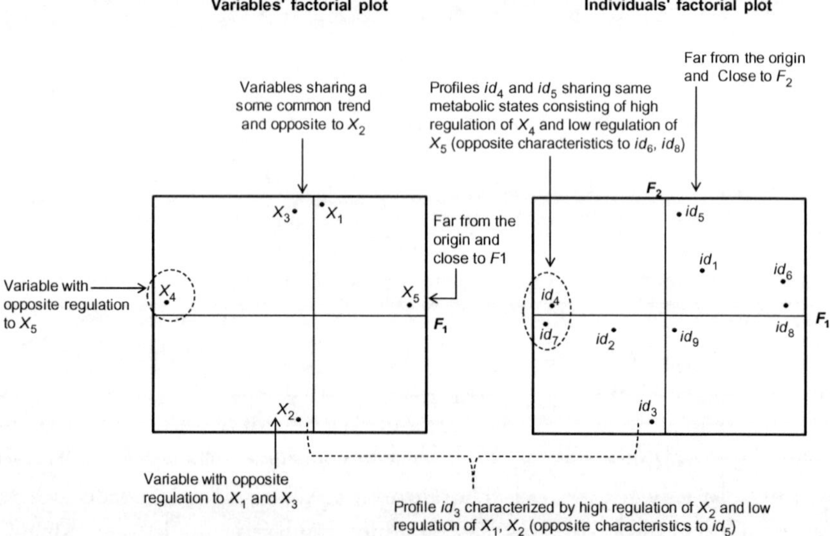

Figure 10.11 Graphical rules based on proximities and distances of projected points help to identify and characterize extreme states in correspondence analysis. Extreme states correspond to extreme points close to different factors, viz. id_4, id_7 opposite to id_6, id_8 (along F_1), and id_3 opposite to id_5 (along F_2). Variable points projected near extreme individual points indicate atypically high relative levels of considered variables in associated profiles. For instance, variable X_4 has high relative levels in profiles id_4 and id_7.

b. Closeness of points to a factor F_k: a close point to axis F_k has a variability that can be well explained by the global meaning of F_k.

c. Proximities between row and/or column points in factorial space. Points projected in a same close subspace are indicative of different facts that are all the more valid as the points are far from G and close to axes F_k:

- Two close individual profiles mean they share common metabolic trends.
- Two close metabolite points indicate they are regulated by a similar way in population.
- Opposite row or column points are indicative of inverse metabolic regulations between profiles or metabolites.
- Proximity between a profile point (*id i*) and metabolite point (M_j) indicates that *id i* is characterized by a high regulation of M_j.

A profile point *id i* that projects far from the origin and close to a factor F_k indicates an extreme metabolic state called metabotype (MTp) (Fig. 10.5). A metabotype is identified by some atypically high or low regulations for some metabolites M_j compared to the other metabolites. Metabolites with high or low regulations in a profile *i* will be projected in the same or opposite subspaces than *id i*, respectively (Fig. 10.11).

From the numerical example given in Figs. 10.6 and 10.7, the factorial plot F_1F_2 of variables highlights three poles defined by high regulations of M_2, M_5 and M_3, respectively (Fig. 10.12A). The opposition between metabolites M_2 and M_5 along F_1 indicates two metabolic trends in which M_2 and M_5 are inversely regulated. However, from the individuals' factorial plot F_1F_2, the points are projected in a way helping to identify three metabotypes combining different regulation levels of metabolites M_2, M_3, M_5 (Fig. 10.12B): along F_1, *MTp* I is represented by metabolic profiles of id_4, id_8 showing higher regulation of M_2 at the expense of M_5, compared to all the other profiles (Figs. 10.8 and 10.13). The extreme metabolic state is inversed in id_3 in which M_5 and M_2 reach their maximal and minimal regulation levels, respectively (*MTp* II) (Figs. 10.8 and 10.13).

Along F_2, the profiles id_5 and id_7 show affinity for metabolite M_3 at the expense of M_2 (projected in opposite subspace) (Fig. 10.12A and B); they define *MTp* III and are opposite to id_1, id_4 and id_8 showing high regulations of M_2 at the expense of M_3 (Figs. 10.8 and 10.13).

The other profiles represent intermediate metabolic states more or less close to one MTp compared to the others (Fig. 10.12A and B):

- The projection of id_{10} in the quadrat $F_1^-F_2^-$ containing id_1, id_4, id_8 indicates some common point with them, that is, its affinity for M_2 at the expense of M_3: these four profiles have the maximal ratios M_2/M_3

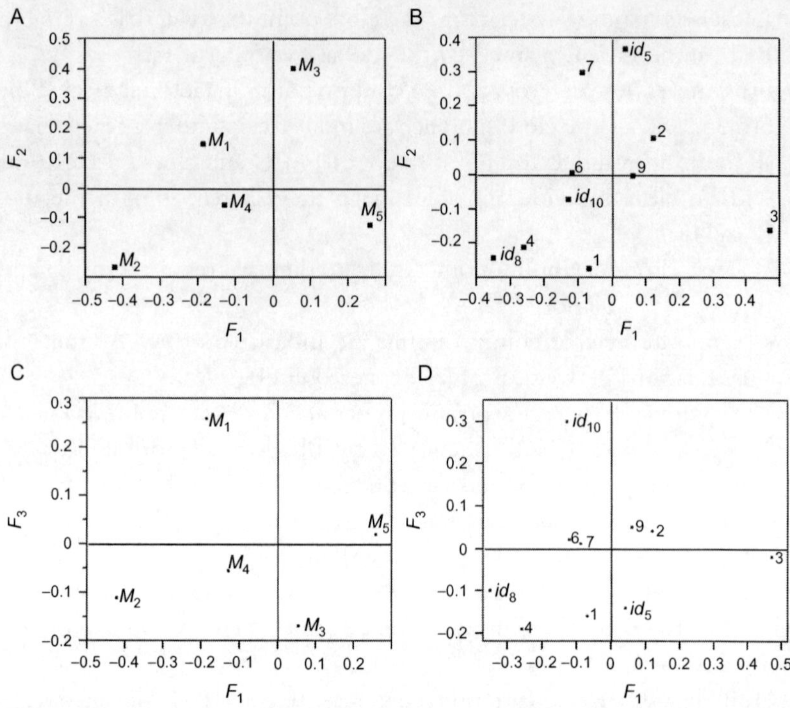

Figure 10.12 Correspondence analysis plots F_1F_2 (A, B) and F_1F_3 (C, D) giving the factorial coordinates of variables (metabolites) (A, C) and individuals (metabolic profiles) (B, D) of dataset given in Fig. 10.6.

(=3 in id_8 and =2 in $id_{1, 4, 10}$). Further metabolic characterizations of id_{10} are given by the factor F_3 (Fig. 10.12C and D): along F_3, the profile id_{10} shows affinity for metabolite M_1, and is opposite to id_5 which has affinity for M_3. This factorial observation helps to identify id_{10} and id_5 as the metabolic profiles having the maximal and minimal ratios M_1/M_3, respectively (Fig. 10.13). This is in favour of consideration of id_{10} as fingerprint of a fourth MTp, but the limited dataset associated to projection of id_{10} near the origin of F_1F_2 requires some precaution in interpretations. A larger dataset could highlight higher variability status of the metabolic ratio M_1/M_3 leading to a confirmed MTp. This numerical example shows that *id*10 is well described by its regulatory deficiency of M_3, a common point shared by *MTp* I and *MTp* II.

- In F_1F_2, the projection of id_2 in direction of id_5 indicates some affinity for metabolite M_3. On this basis *id*2 seems to be closer to *MTp* III. With *id*9, *id*2 projects in the same quadrat $F_1^+F_2^+$ characterized by some opposition to metabolites M_2 and M_4. This helps to identify id_2 and id_9 as low regulatory profiles of M_2, M_4 (Fig. 10.13).

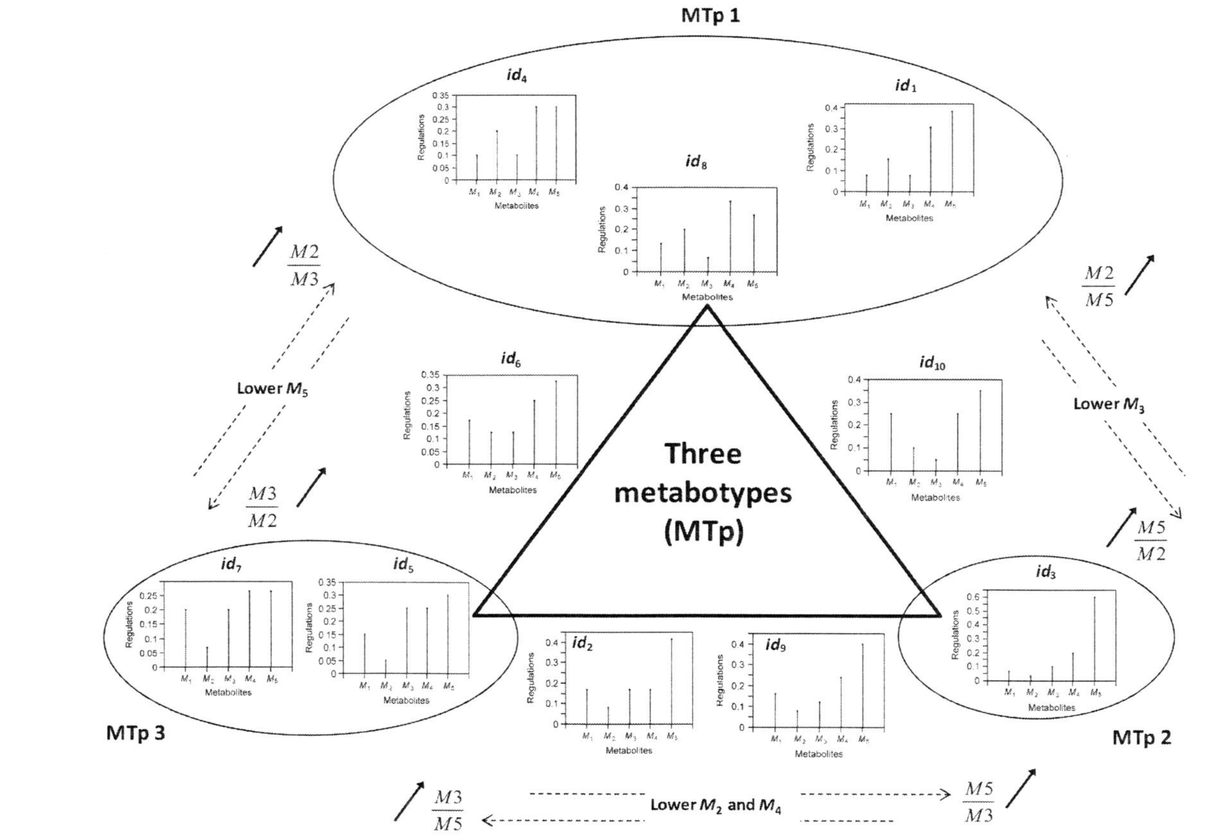

Figure 10.13 Extreme and intermediate metabolic profiles highlighted by CA and organized around of three metabotypes (MTp).

- The id_6 (non-extreme because near the centre) occupies an intermediate position between id_8 (representing *MTp* I associated to M_2) and id_5 (representing *MTp* III associated to M_3) (Fig. 10.12A and B). On this basis, id_6 provides intermediate metabolic picture between *MTp* I and *MTp* III (Fig. 10.13). Also, id_6 projects in the quadrat $F_1^- F_2^+$ which contains M_1 and opposites to M_5 (Fig. 10.12A and B). This indicates some affinity of id_6 for M_1 at the expense of M_5 (Fig. 10.13).

3.4.5 Help tools for interpretation of factorial results

3.4.5.1 Absolute contributions of factorial points

Graphical interpretations of CA results are particularly robust for extreme points projecting both along factors F_k and far from the origin G. Sometimes, a point can seem to be close to a factor F_k and/or distant from G just by simple projection effect on factorial plane. In other words, two closely projected points on the factorial plane can be distant in the multidimensional space.

To check the most far points (interpretable points) along a factor F_k, one calculates their absolute contributions (CtAb) to F_k. These contributions are calculated for row (individual) and column (metabolite) profiles (Eqs. 10.19 and 10.20) (Tables 10.3 and 10.4) (Escoufier & Pagès, 1991):

$$\mathrm{CtAb}_k(i) = \frac{1}{\lambda_k} \cdot \frac{x_{i+}}{x_{++}} \cdot F_{k(i)}{}^2, \text{ for row point } i \tag{10.19}$$

$$\mathrm{CtAb}_k(j) = \frac{1}{\lambda_k} \cdot \frac{x_{+j}}{x_{++}} \cdot F_{k(j)}{}^2, \text{ for column point } j \tag{10.20}$$

For example, the absolute contribution of id_3 to F_1 is calculated by:

$$\begin{aligned} \mathrm{CtAb}_1(i=3) &= \frac{1}{0.0525} \times \frac{30}{200} \times 0.4658^2 \\ &= 0.62, \text{ viz. } 62\% \text{ CtAb of } id_3 \text{ to } F_1. \end{aligned}$$

For example, the absolute contribution of metabolite M_5 to F_1 is:

$$\begin{aligned} \mathrm{CtAb}_1(j=5) &= \frac{1}{0.0525} \times \frac{75}{200} \times 0.2565^2 \\ &= 0.47, \text{ viz. } 47\% \text{ CtAb of } M_5 \text{ to } F_1. \end{aligned}$$

The absolute contribution $\mathrm{CtAb}_k(i)$ quantifies the part of point i in the explained variance along factor F_k, that is, the effect of point i in the construction of F_k. Higher CtAb_k value concern points which have higher coordinates along the considered factor F_k in the multidimensional space.

Table 10.3 Absolute contribution values CtAb(i) (in %) of individual profiles i to the three first factors $F1$–$F3$ given by CA applied on the dataset given in Fig. 10.10

id i	*F1*	*F2*	*F3*
1	0.66	11.64	9.99
2	1.67	2.42	0.47
3	61.99	9.54	0.44
4	6.58	5.36	11.44
5	0.27	39.49	12.03
6	5.73	0.14	0.59
7	1.21	19.32	0.01
8	17.9	10.9	4.38
9	0.89	0.05	2.23
10	3.06	1.11	58.36

Table 10.4 Absolute contribution values CtAb(j) (in %) of metabolites M_j to the three first factors $F1$–$F3$ given by CA applied on the dataset given in Fig. 10.10

M_j	*F1*	*F2*	*F3*
1	9.91	10.76	63.46
2	34.37	17.51	7.68
3	0.68	58.66	23.11
4	8.02	1.19	4.81
5	47	11.86	0.92

3.4.5.2 Relative contributions of factorial points

Complementary to absolute contributions, relative contributions (CtRel) can be calculated for row and column points to quantify how much they are close to (i.e. well explained by) considered factor F_k in the multi-dimensional factorial space (Eqs. 10.21 and 10.22) (Tables 10.5 and 10.6):

$$\mathrm{CtRel}_k(i) = \frac{F_k(i)^2}{\sum_{j=1}^{p}\left[\frac{x_{++}}{x_{+j}}\left(\frac{x_{ij}}{x_{i+}} - \frac{x_{+j}}{x_{++}}\right)^2\right]}, \quad \text{for row point } i \qquad (10.21)$$

Table 10.5 Relative contribution values $CtRel_k(i)$ (in %) of individual profiles id_i on the three first factors F_k ($k=1-3$) given by CA applied on the dataset of Fig. 10.10

id i	F1	F2	F3
1	5.59	67.45	25.16
2	28.71	28.84	2.45
3	90.1	9.61	0.19
4	47.11	26.61	24.66
5	0.87	87.12	11.52
6	88.06	1.53	2.74
7	8.06	89.16	0.03
8	66.59	28.13	4.91
9	54.43	2.15	40.83
10	14.18	3.58	81.41

Table 10.6 Relative contribution values $CtRel_k(j)$ (in %) of metabolites M_j on the three first factors F_k ($k=1-3$) given by CA applied on the dataset of Fig. 10.10

M_j	F1	F2	F3
1	27.14	20.43	52.32
2	68.59	24.24	4.61
3	1.4	83.81	14.34
4	62.42	6.42	11.27
5	84.5	14.79	0.49

$$\mathrm{CtRel}_k(j) = \frac{F_k(j)^2}{\sum_{i=1}^{n}\left[\frac{x_{++}}{x_{i+}}\left(\frac{x_{ij}}{x_{+j}} - \frac{x_{i+}}{x_{++}}\right)^2\right]}, \quad \text{for column point } j \tag{10.22}$$

Relative contribution value $\mathrm{CtRel}_k(i)$ can be interpreted as the dispersion of profile i explained by considered factor F_k. Closer is point i to F_k in multidimensional space higher will be $\mathrm{CtRel}_k(i)$. For example, Eq. (10.21) gives a $\mathrm{CtRel}_1(3)$ equal to 0.90 meaning that individual profile id_3 has dispersion, which is explained at 90% by factor F_1 (Fig. 10.10 → Eq. 10.21 → Table 10.5):

$$CtRel_1(i=3)=$$

$$=\frac{0.466^2}{\left[\frac{200}{30}\times\left(\frac{2}{30}-\frac{30}{200}\right)^2\right]+\left[\frac{200}{20}\times\left(\frac{1}{30}-\frac{20}{200}\right)^2\right]+\left[\frac{200}{25}\times\left(\frac{3}{30}-\frac{25}{200}\right)^2\right]+\left[\frac{200}{50}\times\left(\frac{6}{30}-\frac{50}{200}\right)^2\right]+\left[\frac{200}{75}\times\left(\frac{18}{30}-\frac{75}{200}\right)^2\right]}$$

$$=0.90.$$

The remaining 10% of variability of id_3 are shared between the other factors F_k $(k > 1)$ (Fig. 10.12B). Other calculations give CtRel values equal to 87% for id_5 to F_2, 67% for id_8 on F_1, 28% for id_8 on F_2, 47% for id_4 to F_1, 68% for id_1 to $F2$.

For variables (metabolites), the highest relative contributions show that the highest dispersions of M_2, M_4 and M_5 are explained on F_1 (68.6%, 62.4%, 84.5%), whereas those of M_3 and M_1 are mainly explained on F_2 (83.8%) and F_3 (52.3%), respectively (Table 10.6).

4. WEIGHTED METABOLIC PROFILES ANALYSIS

4.1. Aim and background

WMPA aims to extract correlations between metabolites from statistical combinations between metabotypes. This variability mixing-based approach provides a gradient of metabolic profiles from which regulatory relationships between metabolites can be analyzed.

Metabolic variability contained within and between metabotypes is gradually combined by varying the weights of metabotypes the ones at the expense of the others. The set of combinations provides a multiway space from which relationships between metabolites can be analyzed to identify metabolic regulation ways governing the development of metabotypes (Semmar, 2010a, 2011, 2007).

4.2. Mixture design parameters

WMPA is based on a complete set of mixtures between q metabolic components (Fig. 10.14A and B) which can be defined by q metabotypes. Statistical mixtures between metabotypes can be carried out by using an experimental design called Scheffé's mixture matrix (Fig. 10.14B) (Cornell, 2002; Murthy & Das, 1968; Scheffe, 1958, 1963; Thompson, 1981). This design gives the proportions of the q components (metabotypes)

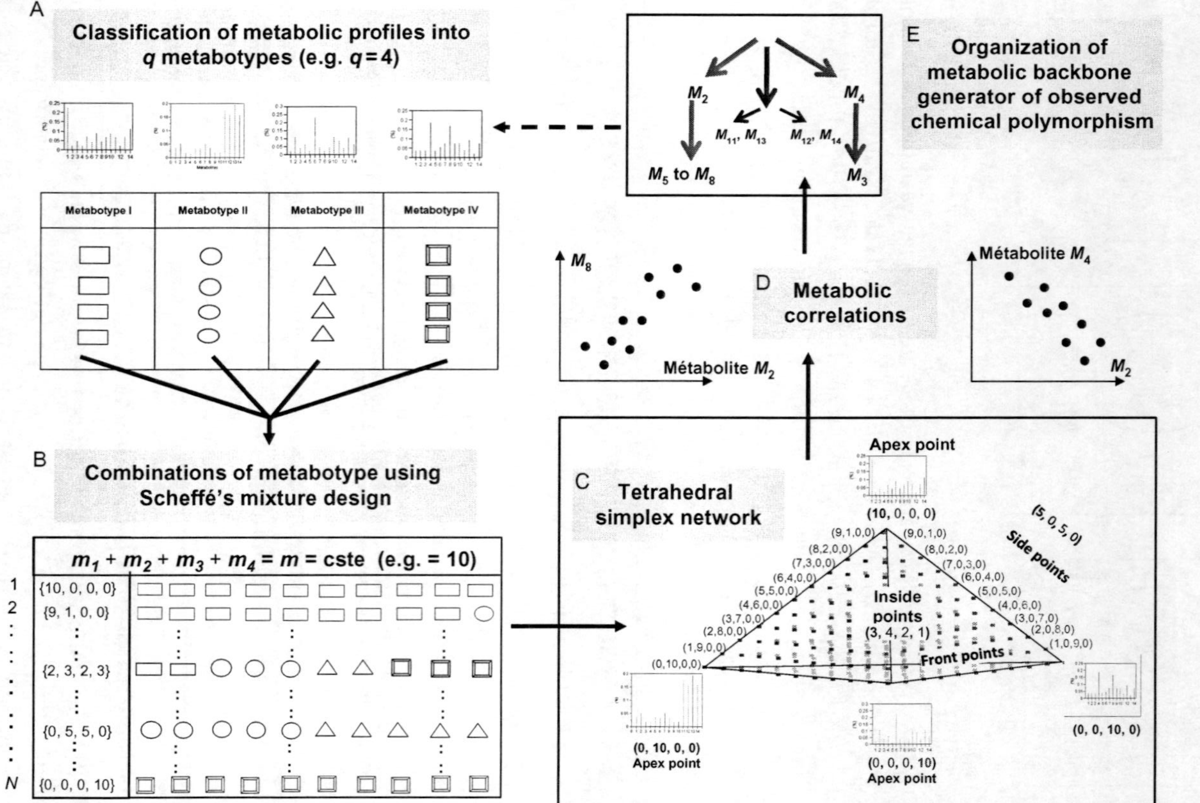

Figure 10.14 Numerical and geometrical simplex bases of the weighted metabolic profiles analysis (WMPA). (For colour version of this figure, the reader is referred to the online version of this chapter.)

in the different possible mixtures. These proportions are initially defined by reference to a constant total number of profiles to mix in each mixture, noted m.

Starting from a well-defined number q of components (Fig. 10.14A), and by fixing m, the total number N of all the mixtures to carry out is given by the combinatory formula:

$$N = \frac{(m+q-1)!}{(q-1)!m!}. \tag{10.23}$$

For example, with $q=4$ metabotypes and by fixing m to 10 metabolic profiles per mixture, the total number of mixtures N is equal to:

$$N = \frac{(10+4-1)!}{(4-1)!10!} = \frac{13!}{3! \times 10!} = 286.$$

The N mixtures of Scheffé's matrix differ the ones from the others by the q proportions m_g ($g=1$ to q) of the q metabotypes (Fig. 14B). For example, with $q=4$ and $m=10$, the mixture (4,2,3,1) consists of 4, 2, 3 and 1 individuals issued from metabotypes I, II, III and IV, respectively. The complete set of N mixtures allows transitions between extreme (pur) metabolic states implying the expression of only one metabotype (viz. (10,0,0,0), (0,10,0,0), (0,0,10,0), (0,0,0,10)) and all the possible intermediate states combining 2, 3 ($q-1$) or 4 (q) metabotypes.

4.3. Geometrical significance of simplex mixture design

Under geometrical aspect, the complete set of all the mixtures between q metabotypes generates a simplex space with q apexes (Figs. 10.2 and 10.14C). Under metabolomics aspect, each apex is exclusively associated to only one metabolic trend generating a well-defined metabotype. The q apexes represent pure or extreme metabolic states due to the expression of only one metabotype excluding the ($q-1$) others.

Between the q apexes, the ($N-q$) remaining points of the simplex network represent metabolic mixtures or combination states due to the contribution of more than one metabolic trend among the q ones. For example, with $q=4$ metabotypes, the N points are organized into a tetrahedric simplex network with four apexes (Figs. 10.2 and 10.14C). Between the four apexes, side, face and interior points to the network represent mixtures implying contributions of 2, 3 and 4 metabolic trends, respectively.

4.4. Elementary operations of WMPA

Before mixtures, the metabolic profiles (chromatograms) must be standardized into relative values by dividing each peak amount (or area) by the total area of the chromatogram. This consists in transforming the amounts x_{ij} of metabolites j of each profile i into relative values R_{ij} by dividing x_{ij} by the row sum x_{i+}. Each individual profile contains p metabolites leading to p relative levels having a unit-equal sum.

Taking into account the parameters q and m, each of the N mixtures is carried out by randomly sampling m (e.g. =10) metabolic profiles from the q (e.g. =4) metabotypes. Each mixture is applied by considering the q different weights m_g (g=1 to q) of the q components (metabotypes) given by the Scheffé matrix. The m profiles of each mixture are aggregated by calculating their average profile as a response parameter (Fig. 10.15). The average profile consists of p means $\bar{R}(j)$ of p metabolites j (j=1 to p); the mean of each metabolite j is calculated from its m values R_{ij} observed in the m random profiles (i) belonging to the q metabotypes.

In all, Scheffé's matrix of N mixtures expects a matrix response of N average metabolic profiles. The N random mixtures are characterized by

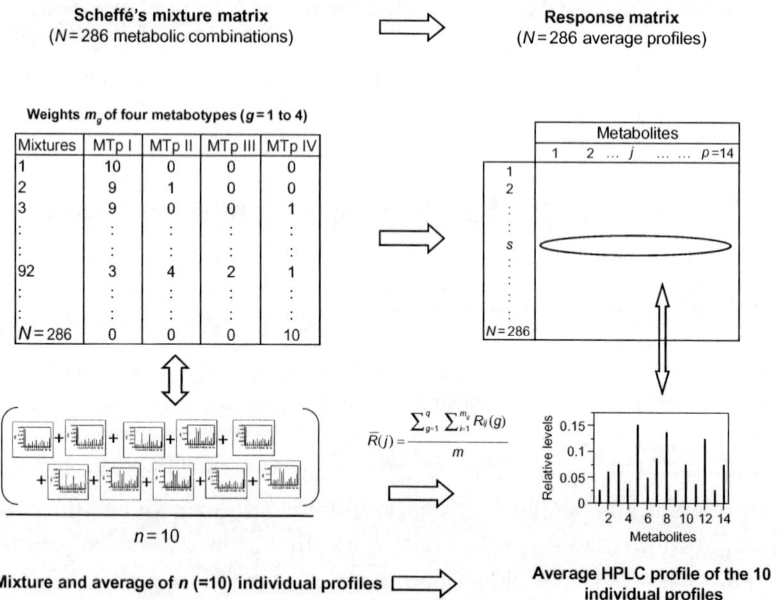

Figure 10.15 Elementary operation of WMPA algorithm consisting of N mixtures between m metabolic profiles issued from q metabotypes varying by their weights (m_g) (g=1 to q) and resulting in calculation of N average metabolic profiles.

different metabotypes' weights (m_g) ($g=1$ to q) the variations of which obey to the simplex constraint:

$$\sum_{g=1}^{q} m_g = m \text{ (constant)}. \tag{10.24}$$

4.5. Iterative operations

At the step where the m profiles of a given mixture are cumulated, their resulting average metabolic profile gives a combination picture of a limited number (m) of individuals; such a result underestimates the variability between and within metabotypes (Fig. 10.16B). To overcome this problem, the mixture design and its response matrix need to be iterated several times, and at each iterated mixture, 10 new profiles are randomly sampled from the classified dataset (Fig. 10.16C). This iteration can be considered as a bootstrap operation within metabotypes.

Therefore, by iterating k times the mixture design, one obtains k response matrices, each one containing N average metabolic profiles. Finally, the k response matrices are averaged to provide a single matrix containing N smoothed metabolic profiles (Fig. 10.16D).

From the final matrix of N smoothed metabolic profiles, scatter plots are visualized between metabolites to graphically identify metabolic relationships controlling the initially observed chemical polymorphism.

4.6. Interpretation of graphical results: Diversity of metabolic regulations

After smoothing, relationships between relative levels of metabolites can be graphically highlighted by scatter plots (Figs. 10.14D, 10.16E). Three types of relationships can be observed (Fig. 10.17):

- Global relationships consisting of thin monotonic clouds of points (Fig. 10.17A).
- Local relationships consisting of systematic variations included within global trends. A local variation can have a same or opposite sign than that of the global trend which includes it (Fig. 10.17B).
- Multidirectional relationships due to a low compression of the simplex space during iterations, and consisting of q superimposed metabotype-dependent relationships (Fig. 10.17C).

Global relationships help to identify different metabolic pathways: a positive global correlation between two metabolites can be interpreted as two metabolites belonging to a same metabolic pathway. A negative global correlation

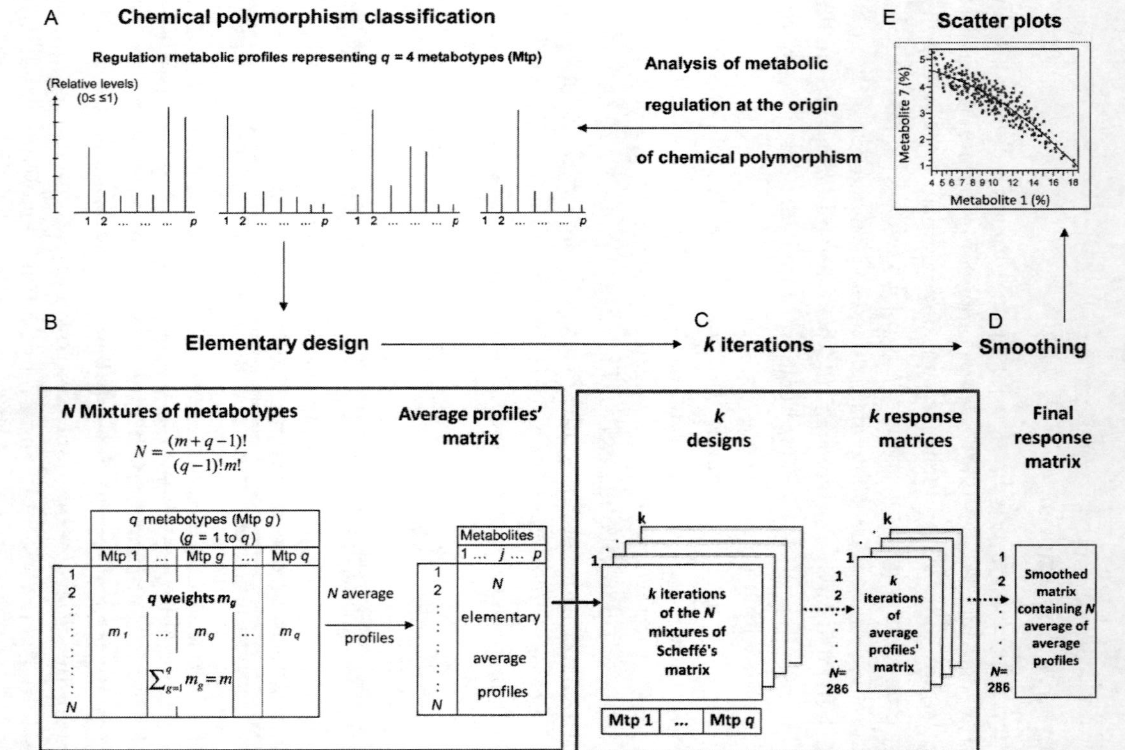

Figure 10.16 Iteration (bootstrapping) of the mixture design (B) giving several response matrices (C) from which a final matrix of smoothed metabolic profiles is calculated (D) to graphically analyze relationships between metabolites (E) helping to understand metabolic regulations responsible for the initially observed chemical polymorphism (A).

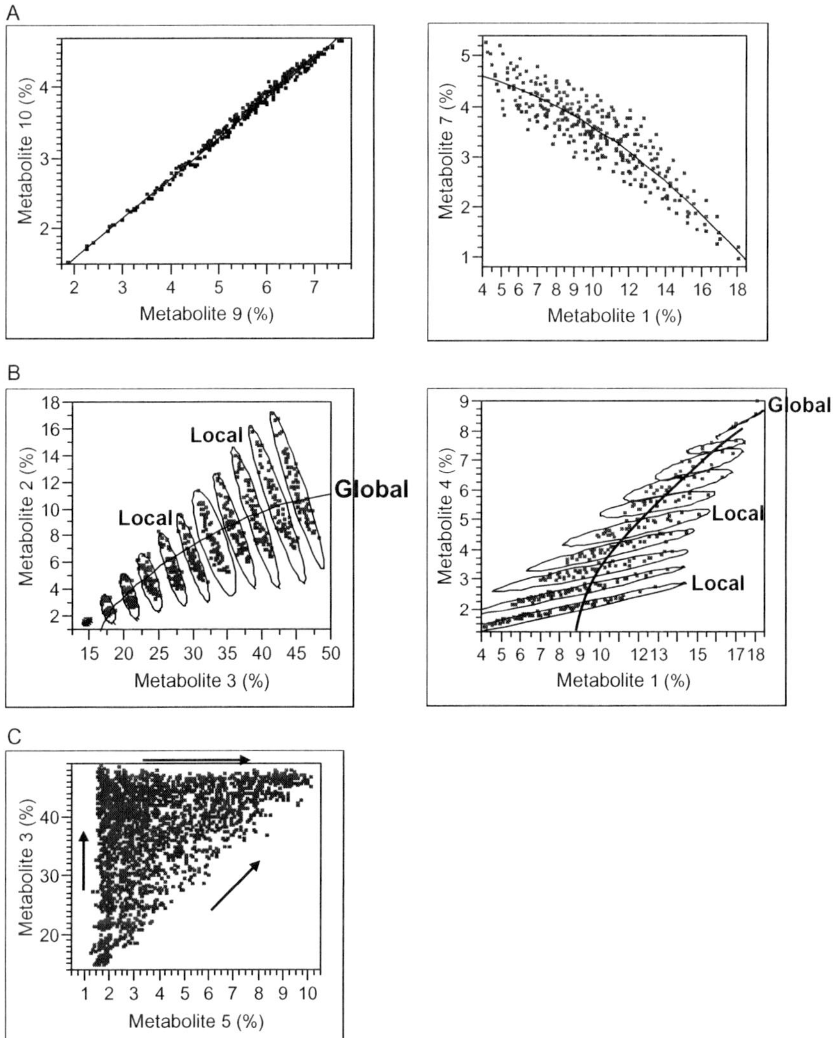

Figure 10.17 Scatter plots highlighting different scale- and metabotype-dependent relationships between metabolites given by the iterative WMPA algorithm. Legend: A: Global trends; B: Systematic or local variations; C: Multidimensional relationship.

can be indicative of two metabolites belonging to two competitive metabolic pathways; competitions can be for a same precursor or a same enzyme.

Local relationships are indicative of local processes within metabolic pathways. Positive local correlations can be indicative of common metabolic processes, which were shared by the two metabolites during their synthesis. A negative local correlation can be interpreted as a local competition between two metabolites within a same metabolic pathway; also, it can be the fingerprint of a previous competition occurring during the intermediate metabolism of the two concerned metabolites.

Multidirectional relationships are due to a weak compression of the simplex cloud of points by iterative WMPA algorithm. Such a cloud of points contains q superimposed metabolic relationships each one depending on one metabotype: the q metabotype-dependent relationships correspond to different variation ways or regulation modes favouring the development of the q metabotypes. To highlight each regulation mode within the multidirectional scatter plot, the weights of each metabotype need to be considered.

After projection of the N weights (of a given metabotype) on the N points of multidirectional scatter plot, the equal weight values can be delimitated by a confidence ellipse (Fig. 10.18). As the metabotype has $(m+1)$ weights, it results in $(m+1)$ confidence ellipses (0 to m). The ellipse of weight 0 indicates the absence of metabotype, whereas that of weight m indicates its full development. Subsequently, the succession of confidence ellipses highlights a trajectory that can be interpreted as metabolic regulation

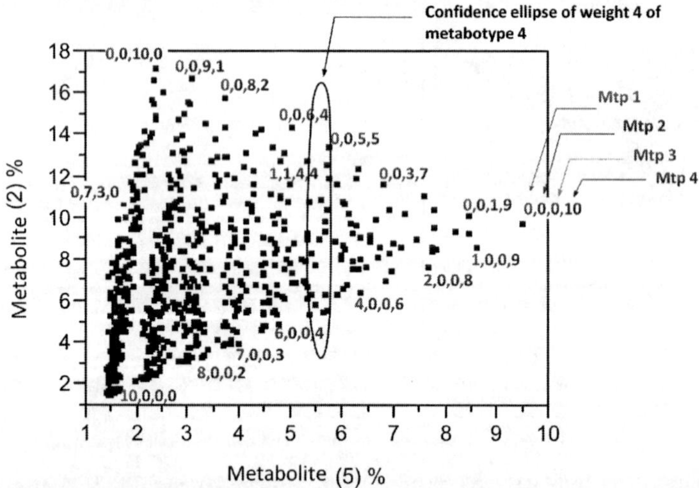

Figure 10.18 Projection of the N weights of a given metabotype on the N mixture points of simplex multidirectional scatter plot, and delimitation of equal weight values by a confidence ellipse. With $(m+1)$ weights of each metabotype, $(m+1)$ successive ellipses can be visualized to highlight a trajectory favouring the development of considered metabotype. In the figure $m=10$. (For colour version of this figure, the reader is referred to the online version of this chapter.)

way favouring the development of the considered metabotype. The approach will be illustrated from an application case on four metabotypes of *Astragalus caprinus* (Fabaceae) based on the flavonoids in the leaves.

4.7. Interest and application field of WMPA

In metabolomics, WMPA can be applied on classified HPLC dataset to extract correlations between unknown metabolites (HPLC peaks) helping to evaluate how and how much they are functionally linked. This can be applied independently from the structural identification of metabolites.

WMPA can be applied on a metabolic dataset, which is initially stratified according to any criterion having an implicit effect on metabolism: as metabolism is a crossing point between several intrinsic and extrinsic factors, stratification criteria influencing metabolism are numerous including genetic, chemical structural, physiological, ecological or clinical types, etc. Then, iterative combinations between classified metabolic profiles will provide smoothed results which give explicit pictures on the implicit effects of the considered stratification factor on metabolic variability.

When the metabolic profiles are classified into different metabotypes, WMPA provides results highlighting how metabolites were regulated by the ones at the expenses or in favour of the others leading to the development of observed metabotypes. The approach is useful in chemotaxonomy helping to statistically analyze the quantitative expression and qualitative organization of the high structural diversity of secondary metabolism in plant world.

Beyond this pure intrinsic metabolic analysis, classification of metabolic profiles into different groups representing different ecological conditions leads to smoothed results which may help to understand the dependence of metabolism on the considered environment factor. This helps to better understand plant biochemical responses in relation to extrinsic or intrinsic factors used as stratification criteria *a priori*. Highlighted links between metabolic flexibility and ecological plasticity provide useful information for a better management of natural area (e.g. natural parks, agricultural fields, etc.).

Also, WMPA can be extended to animal and human studies to analyze metabolic processes linked to controlled laboratory factors or clinical subpopulations (Semmar, 2010a). For instance, the metabolic profiles can be classified into different clinical groups including control group and different disease levels. In such a case, WMPA provides smoothed results that could help to identify metabolic trajectories favouring disease disappearance, appearance or amplification. The same remark remains valid concerning a physiological stratification: in such a case, WMPA can be applied to identify variation ways of metabolism depending on different physiological states considered *a priori*.

5. APPLICATION EXAMPLE COMBINING CA AND WMPA

Analysis of metabolic polymorphism and its functional origins are illustrated here by application of CA and WMPA on a dataset of flavonoids analyzed in the leaves of 404 individual plants *A. caprinus* (Fabaceae).

5.1. Identification of metabotypes by CA

After sampling of 404 individual plants in Tunisia (Semmar, Jay, & Chemli, 2001; Semmar et al., 2005), and HPLC analysis of flavonoids in the leaves, 14 major flavonoids were highlighted. After amount quantification of flavonoid peaks by means of internal standard, the dataset (404 plants × 14 flavonoids) was subjected to a CA and a cluster analysis which highlighted four metabotypes (Semmar et al., 2005; Fig. 10.19).

MTp I was characterized by high regulations of flavonoids *11–14* that are acylated diglycosides of 7-*O*-methylated flavonols (rhamnazin, rhamnetin or rhamnocitrin) (*ADGR 11–14*). This is shown in factorial plots of CA by extreme metabolite points *11–14* along F_1. On the individual plot, *MTp I* was represented by a well-separated plant group having high regulations of flavonoids *11–14* compared to the other metabolites (Fig. 10.19).

Metabotype II was characterized by a high regulation of tetraglycoside of quercetin 1 (*TGQ 1*) favouring its 4′-*O*-methylated derivatives, isorhamnetin tetraglycoside (TG-isoR 4) (Fig. 10.19). These flavonoids projected along the negative extremity of F_2 by opposing to flavonoid 5 (a kaempferol derivative). In individual plot, the factor F_2 shows a continuum of *Mtp II, III, IV*, meaning some gradual variations between them. The most representative profiles of *Mtp II* are those projected at the extreme of F_2^-; in these profiles, *TGQ 1* is highly regulated at the expense of the other metabolites. Beyond the extremity of F_2^-, less extreme profiles of *Mtp II* show some expressions of other metabolites (*TGK 2, TGK 3, ATGK6–10*) (Fig. 10.21A–D). This gradient between *Mtp II, III, IV* is also shown by the factorial coordinates along factor F_3 (Fig. 10.20). The F_3 shows intermediate profiles belonging to *Mtp II* in which flavonoid 5 is relatively well regulated (Fig. 10.21C). Also, some intermediate profiles of *Mtp II* tend to F_1^- which characterizes *Mtp I* (Figs. 10.19, 10.20). This indicates that some

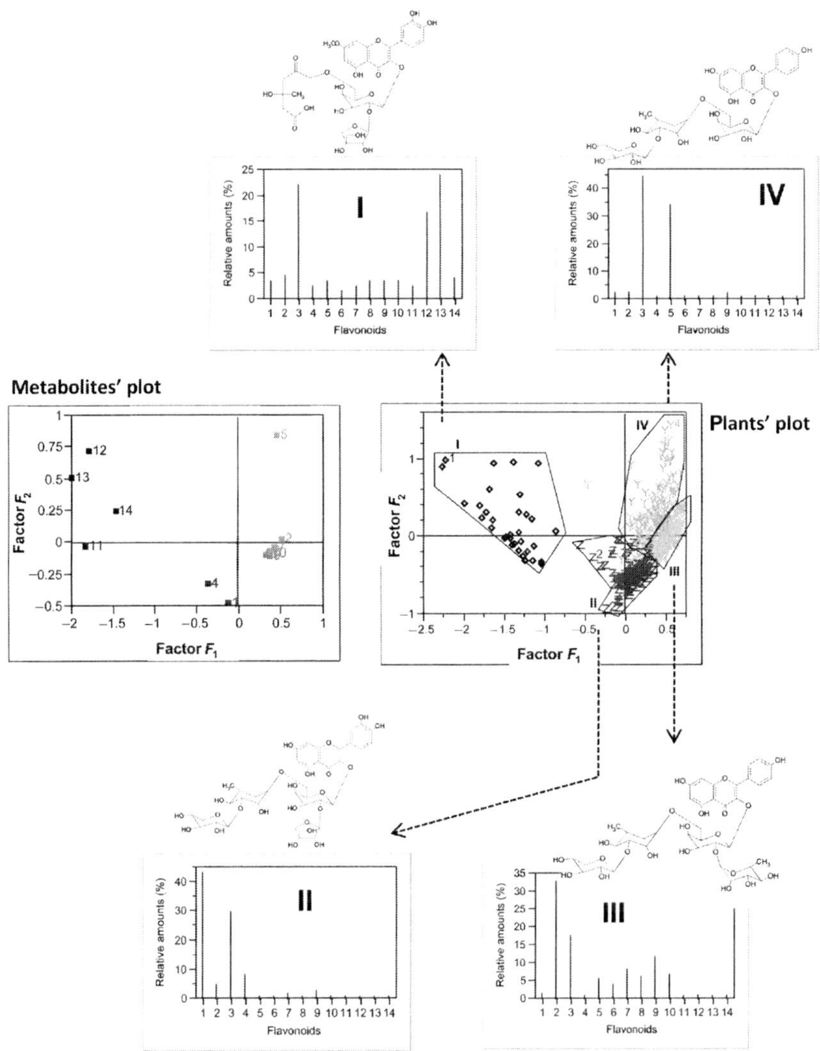

Figure 10.19 Factorial plots F_1F_2 of flavonoids (at left) and plants (at right) given by CA applied on the amount dataset (14 flavonoids × 404 individual plants) (Semmar et al., 2005). (For colour version of this figure, the reader is referred to the online version of this chapter.)

profiles associated to *Mtp II* can manifest some slight expressions of flavonoids *11–14* (Fig. 10.21D).

Metabotype III, highlighted along F_1^+, showed high regulations of kaempferol tetraglycoside *2* (*TGK 2*) with its acylated derivatives *6–10* (*ATGK 6–10*) (Fig. 10.19). It is relatively homogeneous compared to the

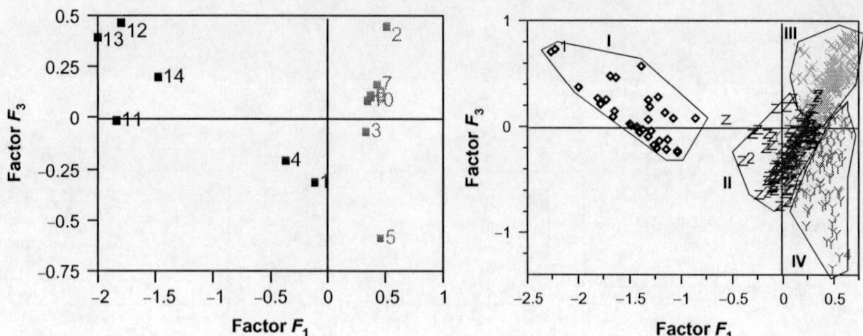

Figure 10.20 Factorial plot F_1F_3 of CA showing a continuum between flavonoid metabolic trends II, III, IV in *Astaragalus caprinus* (Semmar et al., 2005). (For colour version of this figure, the reader is referred to the online version of this chapter.)

other metabotypes; this is indicated on factorial plot by the small area occupied by the representative profiles of *Mtp III*. This homogeneity could be due to the greater number of metabolites characterizing *Mtp III*, viz. *TGK 2* and *ATGK 6–10*. The strong opposition of *Mtp III* to *Mtp I* along F_1 indicates that increase in metabolic regulations of *TGK 2* and *ATGK 6–10* is not favourable to the expression of *ADGR 11–14* at the population scale, and vice versa. This suggests the existence of different metabolic pathways in the studied system. Individuals' factorial plots show that *Mtp III* shares a continuum with *Mtps II* and *IV* meaning the existence of gradual intermediate metabolic states between them (Figs. 10.19, 10.20). The profiles of such intermediate states contain high regulations of *TGK 2* with good expressions of other metabolites characterizing other metabotypes (*TGQ1* for *MTp II*; Figs. 10.21A and 10.21C) (*tGK 5* for *MTp IV*; Fig. 10.21E).

Metabotype IV was distinguished by high regulations of triglycoside of kaempferol 5 (*tGK 5*). Along F_2, *tGK 5* has extreme coordinate and shares the quadrat $F_1^+F_2^+$ with *TGK 2* (Fig. 10.19). This reveals the existence of some profiles of *Mtp IV* in which *TGK 2* may be well regulated (Fig. 10.21F). Along F_3, *tGK 5* has also an extreme coordinate; it is opposite to *TGK 2* and shares the quadrat $F_1^+F_3^-$ with *TGK 3* (Fig. 10.20). This indicates metabolic profiles characterized by high regulations of *TGK 5* and *TGK 3* against low regulation of *TGK 2* (Fig. 10.21G). Such variability in metabolic regulations of *tGK 5* in the presence of the two other kaempferol glycosides (*TGK 2*, *TGK 3*) (along F_2 and F_3) could be responsible for the lesser homogeneity of *Mtp IV* (higher occupied area along F_2 and F_3) (Figs. 10.19 and 10.20). Responses concerning the 'why' of such a metabolic behaviour will be provided by applying WMPA on the dataset (404 × 14) classified into four metabotypes.

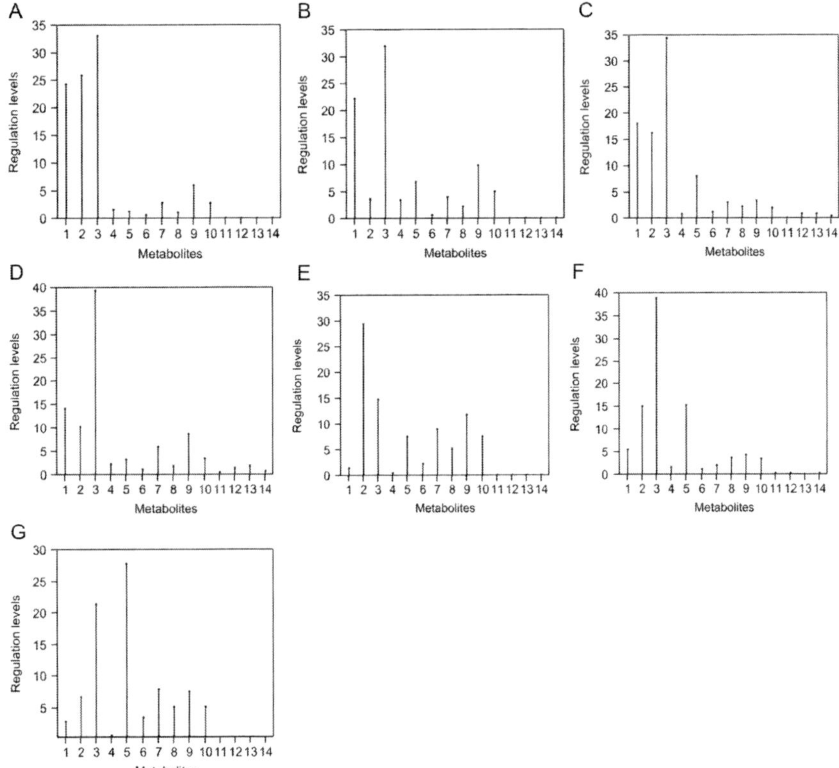

Figure 10.21 Different profiles showing different intermediate metabolic states between the four phenolic metabotypes of *Astragalus caprinus*. The labels A–G are used in manuscript to associate paragraphs to corresponding graphics.

Finally, metabolite 3 (kaempferol tetraglycoside; *TGK 3*) is projected near the origin because it is generally well regulated in all the profiles, and on this basis it does not characterize any metabolic trend.

5.2. Analysis of metabolic origins of metabotypes using WMPA

After classification of the 404 metabolic profiles of *A. caprinus* into q (=4) metabotypes using CA and cluster analysis, the next metabolomics question consists in analyzing how the p (=14) metabolites (flavonoids) vary the ones relatively to the others to generate the observed chemical polymorphism (four metabotypes).

For that, statistical combinations between the four metabotypes were applied *in silico* using Scheffé's mixture design ($q = 4$, $m = 10$, $N = 286$). After

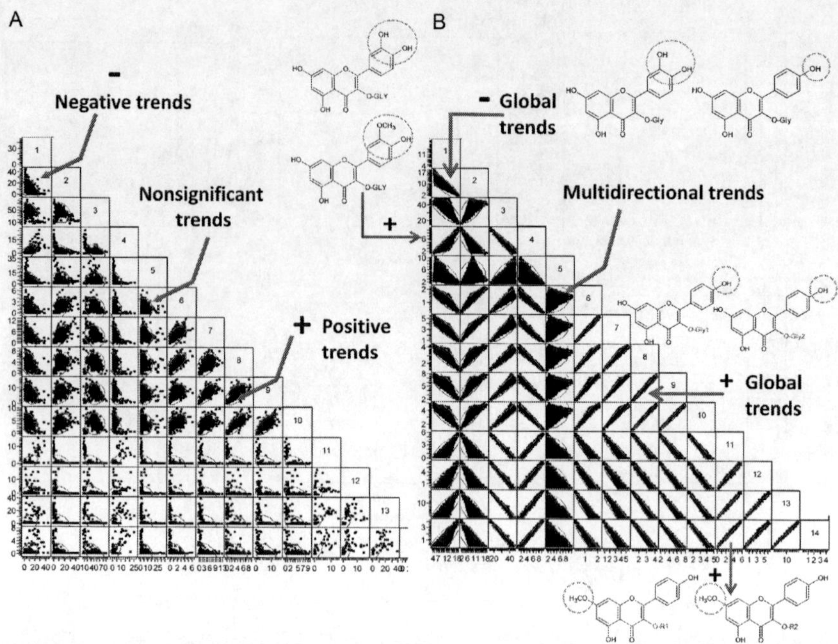

Figure 10.22 Two scatter plot matrices showing metabolic trends detected by Spearman correlations (A) and smoothed metabolic relationships extracted by WMPA algorithm (B). (For colour version of this figure, the reader is referred to the online version of this chapter.)

$k = 50$ iterations of the mixture design, a final response matrix containing $N = 286$ smoothed metabolic profiles was obtained (Fig. 10.16). From this matrix (286 smoothed profiles × 14 flavonoids), scatter plots were visualized between flavonoids to graphical analysis of their regulation ways (Fig. 10.22B).

The smoothed correlations were compatible with non-parametric Spearman correlations calculated on the 404 experimental HPLC profiles (Fig. 10.22A): Spearman's correlation coefficient is a robust statistics calculated on ranked data rather than measured values (Pearson's correlation coefficient). By opposition to Pearson statistic, Spearman's does not assume normal distributions for variables nor linearity between them; these properties make that Spearman statistic can be reliably used to detect trends between variables from dispersed (sparse, noised) data (Altam, 1991; Hauke & Kossowski, 2011; Spearman, 1904; Zar, 1999). Positive and negative trends initially detected by Spearman correlations were conserved after smoothing by WMPA (Fig. 10.22).

Beyond such general validation results, WMPA extracted different scale- and metabotype-dependent relationships hardly or not observed from the sparse scatter plots of crude data. Scale-dependent relationships included global and local relationships.

Global relationships are those conserving the significant Spearman correlations (Fig. 10.22A and B). Positive global relationships were observed between flavonoids having common aglycon characteristics, viz. kaempferol glycosides between them, quercetin glycosides between them, and the 7-*O*-methylated flavonoids between them (Fig. 10.22B). Independently from the knowledge of chemical structures, positive global correlations usefully helped to identify three metabolic poles confirming the Spearman's trends (Semmar et al., 2001).

Negative global relationships opposed kaempferol and quercetin derivatives that are known to belong to two metabolic pathways competing for a common precursor, the dihydrokaempferol (Fig. 10.23) (Ashihara,

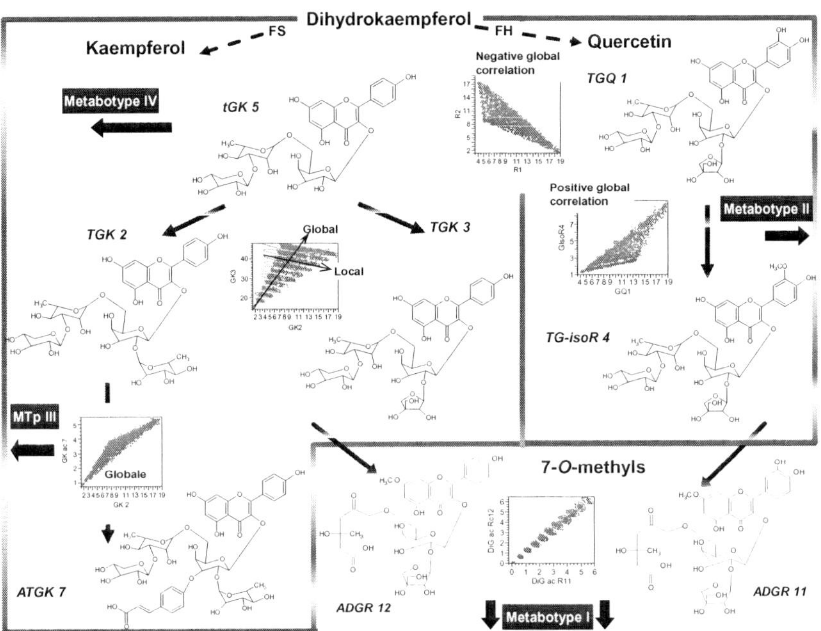

Figure 10.23 Metabolic pathways deduced from different smoothed relationships between flavonoids of *Astragalus caprinus* highlighted by WMPA. *TGK*, *TGQ* and *TG-isoR*: tetraglycoside of kaempferol, quercetin and isorhamnetin, respectively; *ATGK*: acylated tetraglycoside of kaempferol. *ADGR*: acylated diglycosides of rhamnetin or rhamnocitrin. (For colour version of this figure, the reader is referred to the online version of this chapter.)

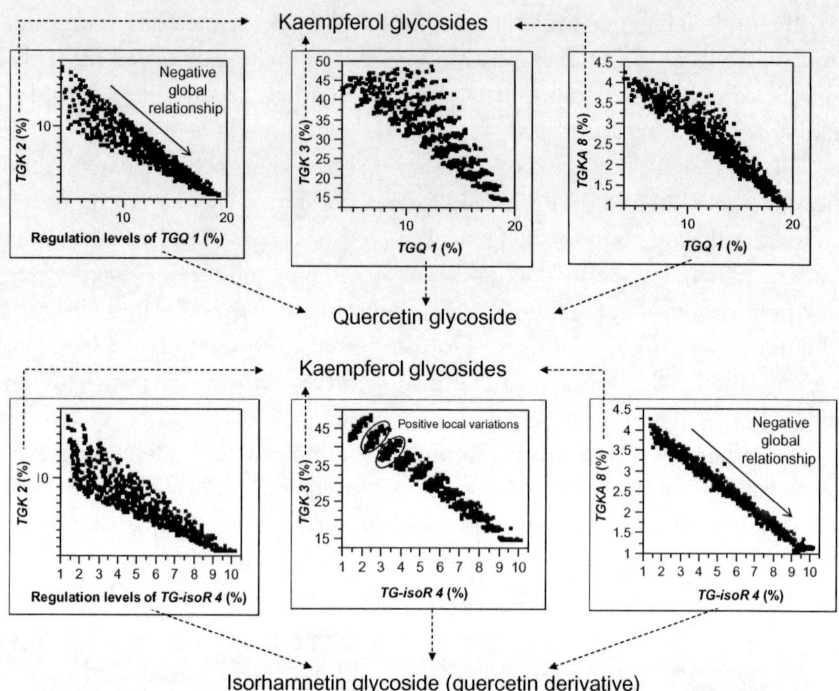

Figure 10.24 Smoothed scatter plots highlighted by WMPA algorithm showing negative global relationships between kaempferol and quercetin derivatives. A positive local relationship is highlighted between *TGK 3* and *TG-isoR 4*. *TGK, TGQ, TG-isoR*: tetraglycoside of kaempferol, quercetin and isorhamnetin, respectively; TGKA: acylated tetraglycoside of kaempferol.

Deng, Mullen, & Crozier, 2010; Gebhardt et al., 2005; Martens, Preub, & Matern, 2010; Schijlen, Ric de Vos, van Tunen, & Bovy, 2004; Semmar et al., 2007). The dihydrokaempferol is converted into kaempferol or quercetin according to its interaction with a flavonol synthase or flavonol hydroxylase, respectively (Fig. 10.23). Moreover, the metabolic competitions between the two pathways can be amplified for glycosylations (Osmani, Bak, & Møller, 2009).

Local relationships consisted of positive or negative systematic variations that manifest within some positive or negative global relationships (Figs. 10.23–10.25).

A negative local relationship was observed between kaempferol tetraglycosides *TGK 2* and *TGK 3* which have a positive global correlation (Fig. 10.23). The positive global correlation is compatible with the belonging of *TGK 2* and *TGK 3* to a same metabolic pathway, viz. kaempferol pathway. However, their negative local correlation could indicate a

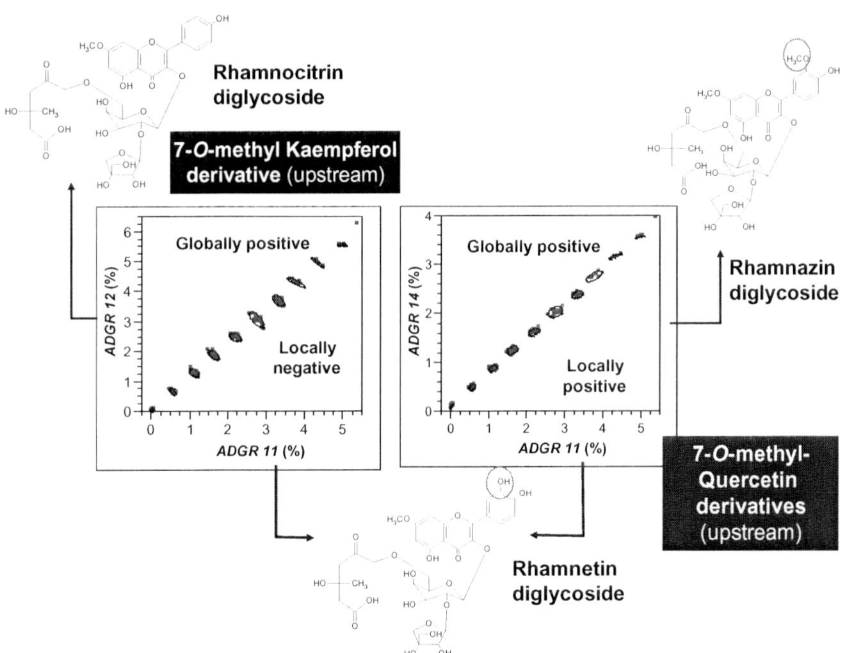

Figure 10.25 Smoothed scatter plots given by WMPA and showing (i) positive global relationships between 7-*O*-methylated flavonols in addition to (ii) positive or negative local relationships depending on their common or different biosynthetic pathway(s), respectively, in upstream. *ADGR 11, ADGR 12, ADGR 14* means acylated diglycosides of rhamnetin (11), rhamnocitrin (12) and rhamnazin (14).

metabolic competition between them for the common precursor, the kaempferol triglycoside *tGK 5*. The *tGK 5* can provide *TGK 2* or *TGK 3* if it is rhamnosylated or apiosylated, respectively (Fig. 10.23).

Although they were globally negatively correlated, *TGK3* and *TG-isoR 4* showed a positive local relationship that could memorize some common (locally shared) metabolic process (Fig. 10.24): the chemical structures show that these two metabolites have an apiose substituted on the 2-position of galactose (Fig. 10.23). Also, among all the kaempferol glycosides, *TGK 3* is the only apiosylated compound. The smoothed scatter plot shows that the positive local relationship between *TGK 3* and *TG-isoR 4* manifests at the high regulations of *TGK 3* and low regulations of *TG-isoR 4* (Fig. 10.24). This could indicate that the synthesis of *TGK 3* could favour that of *TG-isoR 4* from a certain high regulation level of *TGK 3*. This hypothesis could be confirmed by the disappearance of the positive local relationship at low regulations of *TGK 3*.

The 7-*O*-methylated flavonoids showed positive global relationships that could indicate some synchronic metabolism between them (Fig. 10.23: *ADGR 12* versus *ADGR 11*). However, they differed by positive or negative local relationships (Fig. 10.25): positive local relationships concerned the diglycosides of rhamnetin (*ADGR 11*) and rhamnazin (*ADGR 14*) that derive initially from quercetin (upstream of metabolic network) (Bate-Smith, 1962; Davies & Schwinn, 2006; Semmar, 2010b; Valant-Vetschera & Wollenweber, 2006; Williams, 2006). However, a negative local relationship was shown between rhamnetin (rhamnazin) and rhamnocitrin (*ADGR 12*) that derive initially from quercetin and kaempferol, respectively. Thus, such a negative local relationship between methylation-convergent metabolites showed the ability of WMPA to extract fingerprints on some memorized metabolic processes that are not directly observed in the crude data.

Apart from the global and local relationships, WMPA highlighted some multidirectional scatter plots, which concerned particularly the *tGK 5*. From the metabolic card (Fig. 10.23), the upstream localization of this flavonoid between two sub-pathways, those of *TGK 2* and *TGK 3*, could explain its multidirectional relationships; these relationships appeared as not significant by Spearman correlations. The multidirectional aspect of such cloud of points is due to the existence of several (=q) relationships leading to the q metabotypes within a same variation space. This is illustrated by the multidirectional plot of *TGK 2* versus *tGK 5* (Figs. 10.18 and 10.26):

By projecting the N weights of each metabotype and by delimiting the equal weight values by confidence ellipses, one obtains 11 successive ellipses ($m+1=10+1=11$). The succession of these 11 ellipses (from weights 0 to 10) highlighted a metabolic trajectory through which the considered metabotype developed (Fig. 10.26). For a given metabotype, the confidence ellipse of weight = 0 represents the absence of metabotype, and that of weight = 10 represents its maximal development.

By considering separately the four metabotypes I–IV in the same plot *TGK 2* versus *tGK 5*, the four corresponding trajectories highlight four variation ways leading to the development of the four metabotypes:

- Simultaneous diminutions of *TGK 2* and *tGK 5* led to the development of *Mtp I*.
- A strong decrease in regulations of *tGK 5* in presence of stable levels of *TGK 2* favoured *Mtp II*.
- A strong increase in *TGK 2* regulations against decrease in tGK5 levels led to the development of *Mtp III*.

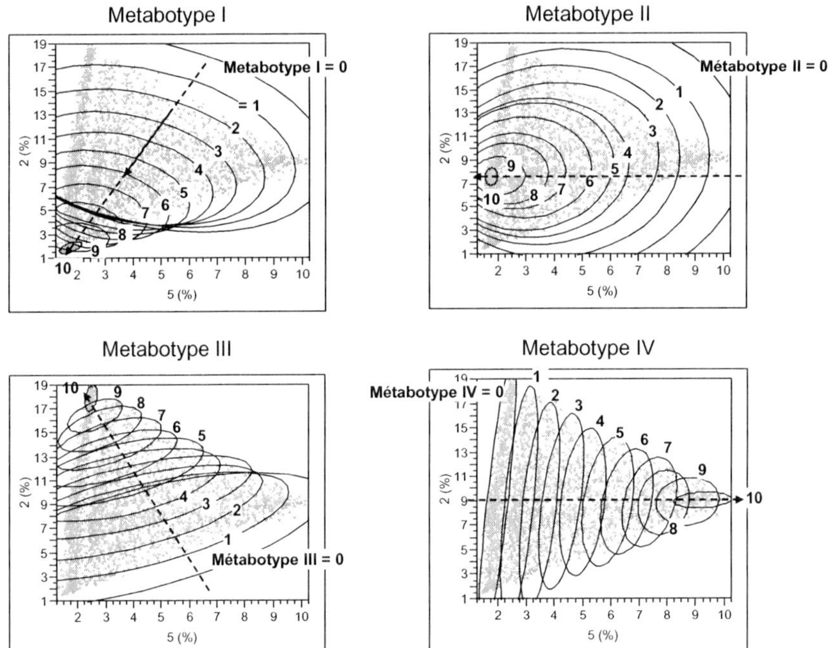

Figure 10.26 Four metabotype-dependent relationships between kaempferol triglycoside 5 (*tGK 5*) and kaempferol tetraglycoside 2 (*TGK 2*) highlighted by WMPA as four trajectories defined by successive weight confidence ellipses (0–10) of the four separated metabotypes.

- An increase in *tGK 5* regulations in presence of stable levels of *TGK 2* favoured *Mtp IV*.

These metabotype-dependent relationships highlighted by WMPA within a same metabolic variation space show the interest of this simplex approach to extract implicit metabolic processes governing explicit chemical polymorphism initially observed in a plant species.

REFERENCES

Abdi, H., & Williams, L. J. (2010a). Correspondence analysis. In N. J. Salkind, D. M. Dougherty, & B. Frey (Eds.), *Encyclopedia of research design* (pp. 267–278). Thousand Oaks, CA: Sage.

Abdi, H., & Williams, L. J. (2010b). Principal component analysis. *Wiley Interdisciplinary Reviews: Computational Statistics*, *2*, 433–459.

Altam, D. G. (1991). *Practical statistics for medical research*. London: Chapman & Hall.

Ashihara, H., Deng, W. W., Mullen, W., & Crozier, A. (2010). Distribution and biosynthesis of flavan-3-ols in *Camellia sinensis* seedlings and expression of genes encoding biosynthetic enzymes. *Phytochemistry*, *71*(5–6), 559–566.

Bate-Smith, E. C. (1962). The phenolic constituents of plants and their taxonomic significance. I. Dicotyledons. *Journal of the Linnean Society of London, Botany, 58*(371), 95–173.
Bendixen, M. (2003). A practical guide to the use of correspondence analysis in marketing research. *Marketing Bulletin, 14*, 16–38.
Bowne, J., Bacic, A., Tester, M., & Roessner, U. (2011). Abiotic stress and metabolomics. In R. D. Hall (Ed.), *Annual plant reviews, biology of plant metabolomics*, Vol. 43, (pp. 61–77). Oxford, UK: Willey-Blackwell.
Cornell, J. A. (2002). *Experiments with mixture designs, models and analysis of mixture data*. New York: John Willey & Sons.
Davies, K. M., & Schwinn, K. E. (2006). Molecular biology and biotechnology of flavonoid biosynthesis. In O. M. Andersen & K. R. Markham (Eds.), *Flavonoids: Chemistry, biochemistry and applications* (pp. 143–218). Boca Raton: CRC Press, Taylor & Francis Group.
Duineveld, C. A. A., Smilde, A. K., & Doorhbos, D. A. (1993). Comparison of experimental designs combining process and mixture variables. Part I. Design construction and theoretical evaluation. *Chemometrics and Intelligent Laboratory Systems, 19*, 295–308.
Escoufier, B., & Pagès, J. (1991). Presentation of correspondence analysis and multiple correspondence analysis with the help of examples. In J. Devillers & W. Karcher (Eds.), *Applied multivariate analysis in SAR and environmental studies* (pp. 33–83). Dordrecht: Kluwer Academic Publishers.
Fernie, A. R., & Keurentjes, J. J. B. (2011). Genetics, genomics and metabolomics. In R. D. Hall (Ed.), *Annual plant reviews, biology of plant metabolomics*, Vol. 43, (pp. 219–260). Oxford, UK: Willey-Blackwell.
Fiehn, O., Kind, T., & Barupal, D. K. (2011). Data processing, metabolomic databases and pathway analysis. In R. D. Hall (Ed.), *Annual plant reviews, biology of plant metabolomics*, Vol. 43, (pp. 367–406). Oxford, UK: Willey-Blackwell.
Fiehn, O., & Weckwerth, W. (2003). Deciphering metabolic networks. *European Journal of Biochemistry, 270*, 579–588.
Frisvad, J. C. (1992). Chemometrics and chemotaxonomy: A comparison of multivariate statistical methods for the evaluation of binary fungal secondary metabolite data. *Chemometrics and Intelligent Laboratory Systems, 14*, 253–269.
Gebhardt, Y., Witte, S., Forkmann, G., Lukačin, R., Matern, U., & Martens, S. (2005). Molecular evolution of flavonoid dioxygenases in the family Apiaceae. *Phytochemistry, 66*(11), 1273–1284.
Greenacre, M. J. (1984). *Theory and applications of correspondence analysis*. London: Academic Press.
Greenacre, M. J. (2007). *Correspondence analysis in practice* (2nd ed.). Boca Raton, FL: Chapman & Hall/CRC.
Hauke, J., & Kossowski, T. (2011). Comparison of values of Pearson's and Spearman's correlation coefficients on the same sets of data. *Quaestiones Geographicae, 30*(2), 87–93.
Hollywood, K., Brison, D. R., & Goodacre, R. (2006). Metabolomics: Current technologies and future trends. *Proteomics, 6*, 4716–4723.
Holmes, E., Wilson, I. D., & Nicholson, J. K. (2008). Metabolic phenotyping in health and disease. *Cell, 134*, 714–717.
Legendre, P., & Gallagher, E. D. (2001). Ecologically meaningful transformations for ordination of species data. *Oecologia, 129*, 271–280.
Martens, S., Preub, A., & Matern, U. (2010). Multifunctional flavonoid dioxygenases: Flavonol and anthocyanin biosynthesis. *Phytochemistry, 71*, 1040–1049.
Mrabet, Y., & Semmar, N. (2010). Mathematical methods to analysis of topology, functional variability and evolution of metabolic systems based on different decomposition concepts. *Current Drug Metabolism, 11*, 315–341.
Murthy, J. S., & Das, M. N. (1968). Design and analysis of experiments with mixtures. *Annals of Mathematical Statistics, 39*, 1517–1539.

Osmani, S. A., Bak, S., & Møller, B. L. (2009). Substrate specificity of plant UDP-dependent glycosyltransferases predicted from crystal structures and homology modeling. *Phytochemistry, 70*(3), 325–347.

Redestig, H., Szymanski, J., Hirai, M. Y., Selbig, J., Willmitzer, L., Nikoloski, Z., et al. (2011). Data integration, metabolic networks and systems biology. In R. D. Hall (Ed.), *Annual plant reviews, biology of plant metabolomics*, Vol. 43, (pp. 261–316). Oxford, UK: Willey-Blackwell.

Rencher, A. C. (2002). *Methods of multivariate analysis*. New York: Wiley.

Renou, M., Lalanne-Cassou, B., Michelot, D., Gordon, G., & Doré, J. C. (1988). Multivariate analysis of the correlation between noctuidae subfamilies and the chemical structure of their sex pheromones or male attractants. *Journal of Chemical Ecology, 14*, 1187–1215.

Rowan, D. D. (2011). Volatile metabolites. *Metabolites, 1*, 41–63.

Savary, S., Madden, L. V., Zadoks, J. C., & Klein-Gebbinck, H. W. (1995). Use of categorical information and correspondence analysis in plant disease epidemiology. In J. A. Callow, J. H. Andrews, & I. C. Tommerup (Eds.), *Advances in botanical research* (pp. 213–240). London: Academic Press Ltd..

Scheffe, H. (1958). Experiments with mixtures. *Journal of the Royal Statistical Society: Series B, 20*, 344–360.

Scheffe, H. (1963). Simplex centroid designs for experiments with mixtures. *Journal of the Royal Statistical Society: Series B, 25*, 235–263.

Schijlen, E. G. W. M., Ric de Vos, C. H., van Tunen, A. J., & Bovy, A. G. (2004). Modification of flavonoid biosynthesis in crop plants. *Phytochemistry, 65*(19), 2631–2648.

Semmar, N. (2010a). A New mixture design-based approach to graphical screening of potential interconnections and variability processes in metabolic systems. *Chemical Biology & Drug Design, 75*(1), 91–105.

Semmar, N. (2010b). *Chemotaxonomical analyses of herbaceous plants based on phenolic and terpenic patterns*. New York: Nova Science Publishers.

Semmar, N. (2011). *Computational metabolomics*. New York, USA: Nova Science Publishers.

Semmar, N., Jay, M., & Chemli, R. (2001). Chemical diversification trends in *Astragalus caprinus* (Leguminosae) based on the flavonoid pathway. *Biochemical Systematics and Ecology, 29*(7), 727–738.

Semmar, N., Jay, M., Farman, M., & Chemli, R. (2005). Chemotaxonomic analysis of *Astragalus caprinus* (Fabaceae) based on the flavonic patterns. *Biochemical Systematics and Ecology, 33*, 187–200.

Semmar, N., Jay, M., & Nouira, S. (2007). A new approach to graphical and numerical analysis of links between plant chemotaxonomy and secondary metabolism from HPLC data smoothed by a simplex mixture design. *Chemoecology, 17*, 139–156.

Semmar, N., Nouira, S., & Farman, M. (2009). Variability and ecological significance of secondary metabolites in terrestrial biosystems. In J. B. Aronoff (Ed.), *Handbook of nature conservation* (pp. 1–90). New York, USA: Nova Science Publishers.

Spearman, C. (1904). The proof and measurement of association between two things. *The American Journal of Psychology, 15*, 72–101.

Steuer, R., Nesi, A. N., Fernie, A. R., Gross, T., Blasius, B., & Selbig, J. (2007). From structure to dynamics of metabolic pathways: Application to the plant mitochondrial TCA cycle. *Bioinformatics, 23*, 1378–1385.

Thompson, D. R. (1981). Designing mixture experiments—A review. *Transactions of ASAE, 24*(4), 1077–1086.

Valant-Vetschera, K. M., & Wollenweber, E. (2006). Flavones and flavonols. In O. M. Andersen & K. R. Markham (Eds.), *Flavonoids: Chemistry, biochemistry and applications* (pp. 617–748). Boca Raton: CRC Press, Taylor & Francis Group.

Van Dam, N. M., & van der Meijden, E. (2011). A role of metabolomics in plant ecology. In R. D. Hall (Ed.), *Annual plant reviews, biology of plant metabolomics*, Vol. 43, (pp. 87–108). Oxford, UK: Willey-Blackwell.

Weller, S. C., & Romney, A. K. (1990). *Metric scaling: Correspondence analysis*. Thousand Oaks, CA: Sage.

Williams, C. A. (2006). Flavone and flavonol O-glycosides. In O. M. Andersen & K. R. Markham (Eds.), *Flavonoids: Chemistry, biochemistry and applications* (pp. 749–856). Boca Raton: CRC Press, Taylor & Francis Group.

Zar, J. H. (1999). *Biostatistical analysis*. New Jersey: Prentice Hall.

CHAPTER ELEVEN

Strategies for Data Handling and Statistical Analysis in Metabolomics Studies

Marianne Defernez, Gwénaëlle Le Gall[1]

Analytical Science Unit, Institute of Food Research, Norwich Research Park, Colney, Norwich, United Kingdom

[1]Corresponding author: e-mail address: gwenaelle.legall@IFR.AC.UK

Contents

Advances in Botanical Research, Volume 67
ISSN 0065-2296
http://dx.doi.org/10.1016/B978-0-12-397922-3.00011-3

Abstract

Metabolomics is classically defined as the holistic detection of metabolites of a system and usually involves the following multistep workflow: sample preparation, profile recording, data processing and pretreatment, data analysis, metabolite identification and data interpretation. In this chapter, we focus on the later part of the workflow: the preprocessing, pretreatment and data analysis. Thus we will present techniques and approaches that are commonly used for the analysis of metabolomics data. More importantly, we show that the data analysis does not sit in isolation but is instead intimately linked to the experimental steps that have taken place upstream of it. We will demonstrate that this interaction can be used in a beneficial way, by exploring how the knowledge of the experimental steps can inform the correct implementation of statistical techniques and conversely how a better understanding of these interactions can help us to improve the experimental aspects.

1. INTRODUCTION

Metabolomics is classically defined as the holistic detection of metabolites of a system (macromolecular level, organ, subunit cellular levels, etc.) (Patti, Tautenhahn, & Siuzdak, 2012). Some researchers using targeted technological platforms classify their studies as metabolomics (Bennett, Yuan, Kimball, & Rabinowitz, 2008; Sabatine et al., 2005; Sreekumar et al., 2009), but it is generally widely accepted that an omic approach is untargeted (Theodoridis, Gika, Want, & Wilson, 2012) and that significant data handling and possibly novel data mining approaches will have to be used/developed (Hendriks et al., 2011; Sugimoto, Kawakami, Robert, Soga, & Tomita, 2012) to obtain meaningful outputs. Studies that adhere to the principles defined by the concept of 'metabolomics'—under our definition—usually involve a multistep workflow, as schematised in Fig. 11.1.

In this Chapter, we focus on the later part of the workflow: the data analysis. Thus we will present techniques and approaches that are commonly used for the analysis of metabolomics data. More importantly, we show that the data analysis does not sit in isolation but is instead intimately linked to the experimental steps that have taken place upstream of it. We will demonstrate that this interaction can be used in a beneficial way, by exploring how the knowledge of the experimental steps can inform the correct implementation of statistical techniques, and conversely how a better understanding of these interactions can help us to improve the experimental aspects.

These interactions are directly linked to the nature of the data itself, and we will first present what these data typically consist of and what the

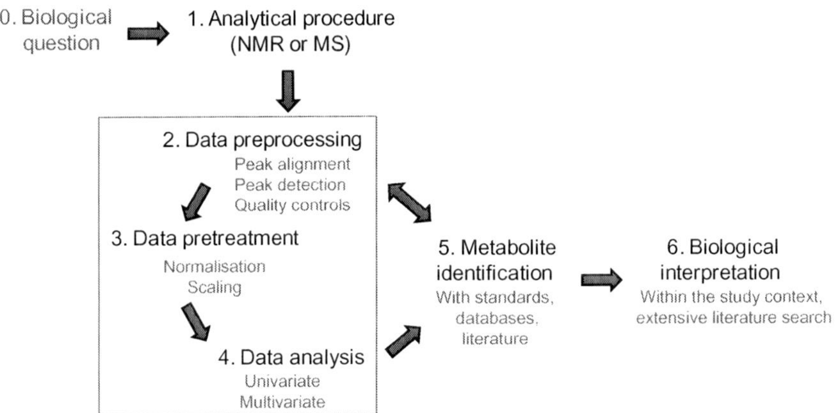

Figure 11.1 General workflow for performing a metabolomics study—all the steps from experimental to interpretation of data analysis interact with each other: study design and analytical methods are chosen (1), methods of data extraction are selected (2), data are prepared for statistical analysis (3), statistical analysis is applied on the processed data (4), metabolites are carefully identified (5) and results are commented (6). This chapter mainly deals with steps 2–4.

important characteristics are, namely that they are multivariate measurements collated with the view to provide comparable data across large series of samples (Section 2). From this, we will examine what methods may be appropriate to handle such data, including at the steps of data preprocessing and pretreatment (taking place between data acquisition and the statistical analysis *per say*) (Sections 3 and 6), and in terms of the data analysis itself (Section 7). Again in doing this, we will demonstrate how the different steps of the workflow are linked and have bearing on one another. Consideration will also be given to how we can check that experimental aspects do not have a detrimental effect that would affect the interpretation we may make of the results (Section 8), and finally, we will present robust ways of implementing the various techniques that are available, given the data are multidimensional and under-determined (Section 9).

2. GENERAL CHARACTERISTICS OF THE DATA

Metabolomics data mainly originate from two main analytical platforms: ^{1}H nuclear magnetic resonance (NMR) spectroscopy and mass spectrometry (MS). Instead of purifying a biological extract in order to accurately measure a small set of metabolites (usually, a class of molecules, e.g. amino acids) as is the case when applying traditional analytical chemistry, the profiling approaches aim at detecting holistically signals arising from numerous metabolites in the hope to cover a good part of the metabolome.

The two platforms are complementary to each other. Proton NMR profiling which is very reproducible and robust, and requires minimal sample preparation (most often a dilution)—and is hence conducive to high-throughput analysis—detects any molecule bearing a carbon backbone with protons attached to it. This platform would be the ultimate non-targeted technology if it was not for its lack of sensitivity (concentrations have to be >1 μM). Proton NMR can detect all kinds of molecules regardless of polarity (Fig. 11.2), which signifies that some of the compounds difficult to extract or analyse by MS have the potential to be easily detected by this platform (intact high- and low-density lipoproteins and the N-acetyl groups of albumin in blood plasma, sugars, phenolic compounds, organic and amino acids and osmolytic molecules such as myo-inositol or choline, phosphocholine or betaine seen in many biological matrices). On the other hand, MS is widely used (both hyphenated to gas and liquid chromatography) due to the versatility of the methods available, the wide coverage of metabolites detected (with the use of more than one platform), the sensitivity and lower costs (purchase and maintenance) associated with this type of platform (Theodoridis et al., 2012). MS is much less reproducible though, being particularly affected when parts involved in the molecules' separation need replacing (column, septum, etc.), and skilful parameter optimisation is required for the recording of quality metabolomic data. Therefore, the development of a successful analytical strategy is crucial to overcome the problems inherent to hyphenated LC/MS-based techniques (Dunn, Broadhurst, Begley, et al., 2011; Theodoridis et al., 2012).

The sample extraction is, in general, minimal, and this can have implications on the detection of signals. Problems such as signal line broadening by NMR or ion suppression by MS are recurrent mainly due to the fact that the sample is crude and numerous signals are detected simultaneously. These statements are important as the presence of artefacts, noise and redundant and ill-characterised signals is inherent to non-targeted metabolite approaches. Nevertheless, this type of analytical approach has the potential to highlight new links between metabolic pathways/interactions and offers a great insight into the metabolism of systems. Additionally, the constant technological advances extend the detecting possibilities even more. Thus it is not surprising that there has been a surge of publications on metabolomics over recent years (over 4000 studies are now listed in PubMed). Yet, most scientists having developed or used non-targeted metabolomics

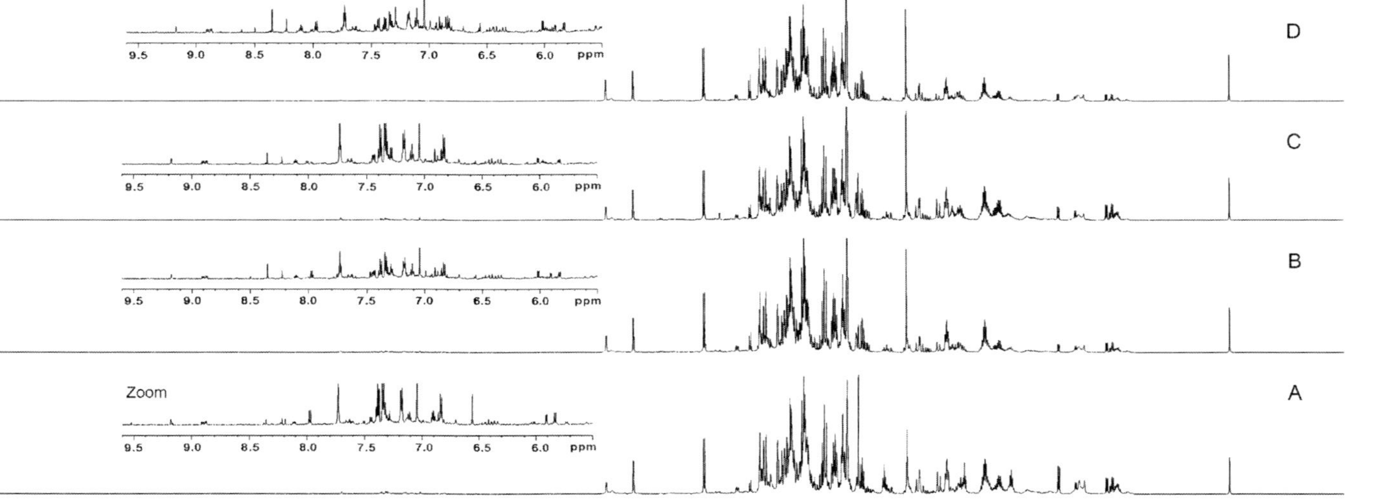

Figure 11.2—*Cont'd*

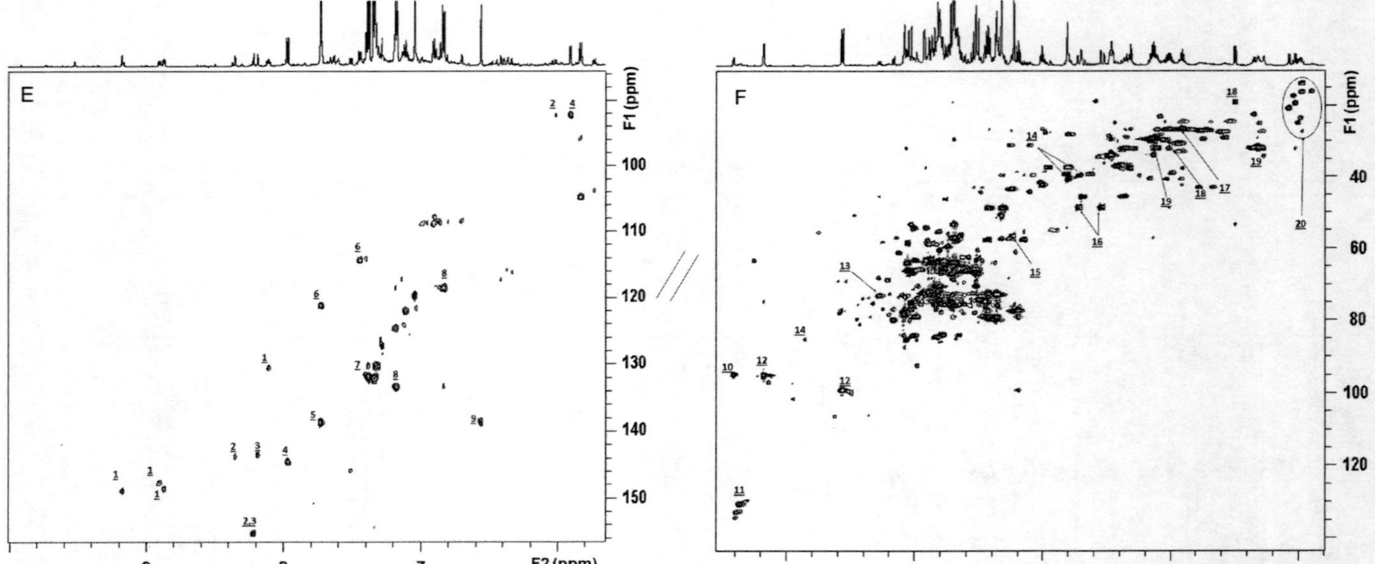

Figure 11.2 Data = metabolite profiles. ^{1}H NMR extracts of broccoli. (A) Raw fresh broccoli, (B) raw frozen broccoli, (C) cooked from fresh broccoli and (D) cooked from frozen broccoli (600 MHz, D_2O/CD_3OD, 27 °C, cryoprobe). The one-dimensional NMR spectra form complex profiles made of signals arising from metabolites. Small differences in levels can be observed by eye (see region around 1.80 ppm), but for a systematic spectral comparison, multivariate analysis is essential. (E and F) Metabolite identification in metabolite profiles. $^{1}H-^{13}C$ HSQC two-dimensional (2D) NMR spectrum of a broccoli extract (raw fresh). (E) Region covering 0.7–5.5 ppm. (F) Region covering 5.7–9.6 ppm (600 MHz, 70% D_2O/30% CD_3OD, 27 °C, cryoprobe). The HSQC 2D NMR spectrum displays the correlation $^{1}H-^{13}C$ signals allowing the identification of metabolites using the $^{1}H-^{13}C$ values to interrogate the human metabolome database (http://hmdb.ca). The second dimension gives specific carbon chemical shifts associated with the metabolite chemical groups, and this specificity removes/reduces the assignment ambiguity of 1-D information even in crowded region (1.5–3.0 ppm). Key to metabolites: 1, trigonelline; 2, adenosine; 3, adenine; 4, uridine; 5, xanthine; 6, tryptophan; 7, phenylalanine; 8, tyrosine; 9, fumarate; 10, sucrose; 11, lipids; 12, glucose; 13, malate; 14, glucosinolates; 15, choline; 16, citrate; 17, gamma-aminobutyrate (GABA); 18, glutamate; 19, glutamine and 20, branched amino acids (isoleucine, leucine and valine).

approaches by NMR or MS will have encountered limitations in terms of both signal detection (quality of the peak, threshold, etc.) and peak identification. It is important to mention these limitations and propose solutions to apprehend the best non-targeted metabolomics data. Manual preprocessing is cumbersome and involves multiple fastidious steps that may lead to errors, and for this reason, many research groups over the world have developed or proposed new tools to perform automatic data preprocessing (see Sections 3–5). But even after careful and thoughtful use of preprocessing tools and processing techniques (statistical analysis), it is still possible to be misled and it remains of upmost importance to check the raw data and have a strong analytical knowledge to be able to make a critical judgement on the outcomes following those two essential steps in the metabolomics workflow. This is in a way 'the price to pay' to have access to non-targeted metabolomics and the promises it holds. The next three sections introduce the concept of preprocessing and give an insight on recently developed packages (most of them being freely accessible) and approaches used to prepare the metabolomics data to statistical analysis.

3. NMR DATA RECORDING AND PREPROCESSING: THE NECESSARY STEPS PRIOR TO STATISTICAL ANALYSES

3.1. Data recording

A comprehensive description of the application of proton NMR (^{1}H NMR) can be found in Ross, Schlotterbeck, Dieterle, and Senn (2007). NMR is a versatile technique that exploits the magnetic properties of certain atomic nuclei. Isotopes with an odd number of protons and/or neutrons possess both a magnetic moment and an angular momentum; such isotopes have a non-zero nuclear spin described by a quantum number which may have values 1/2, 1, 3/2, etc., determined by the nuclear structure. Some examples of isotopes to which NMR can be applied are ^{1}H, ^{13}C, ^{15}N (all spin 1/2) and ^{2}H and ^{14}N (spin 1); NMR cannot be applied to spin zero isotopes such as ^{12}C or ^{16}O. In the presence of a magnetic field, the nucleus may occupy one of $2I+1$ energy levels, where I is the spin quantum number. For the proton (spin 1/2), there are thus two energy states, simply described as up (magnetic moment parallel to the applied magnetic field) or down (opposed to the magnetic field). The population (determined by the Boltzmann distribution) of the two states is slightly unbalanced with more nuclei occupying the lower energy level (spin up) than the higher one (spin down). The energy level difference (and hence population difference) increases with the strength of the external applied magnetic field. Application of electromagnetic (EM) radiation of the correct frequency allows a transfer of spins between the two levels to occur. This happens when the frequency of the rotating magnetic field associated with the EM radiation matches the natural precession frequency of the nuclear spins in the fixed magnetic field (the resonance condition). For the magnets most commonly used in NMR, the radiation is in the radiofrequency range of the spectrum typically between 20 and 1000 megahertz (MHz). The radiofrequency energy is most often applied as a short pulse which perturbs the spins away from their equilibrium condition; after the pulse, the system goes back to its normal state over a period of a few seconds, and during this period (the free induction decay), we get structural information by measuring the natural precession frequencies of the atoms in the molecules. Then, a Fourier transform converts the signals from the time domain (s) to the frequency domain (Hz). We obtain an NMR spectrum with the horizontal axis expressed in parts per million (ppm) from a zero reference point usually provided by an added internal standard. The location of a signal in ppm is obtained by measuring the distance between the signal and the reference (in Hz) and dividing that by the operating frequency of the NMR instrument (in MHz) so that the signal location may be expressed in units that are independent of the field strength of the NMR machine used.

Because each chemically different atom in a molecule has a slightly different environment (the neighbouring electrons shield the proton and its magnetic moment to different extents depending on the neighbouring atoms, type of chemical bond, etc.), each chemically distinct hydrogen will have a different precession frequency and hence signal location along the frequency axis. NMR spectroscopists know how to 'read' information based on the location, multiplicity and intensity of the peak groups in the spectrum. We can deduce the structure(s) of the molecule(s) contained in the NMR sample based on this information. Standard 'high-resolution' ^{1}H NMR measurements can be applied to any solution state samples (e.g. biofluids or solid tissue extracts). Sample preparation is kept to a minimum (simple dilution of fluids with a deuterated solvent or tissue extraction involving homogenisation in one or more solvents, followed by centrifugation, separation of the extract and possibly lyophilisation before reconstituting the sample in a deuterated solvent). Once the samples are prepared, NMR recording typically takes 10–15 min per sample allowing high-throughput analysis under automation (up to several thousand samples). NMR is a particularly repeatable analytical technique, and any rerun of a sample will virtually produce an identical NMR spectrum to the first (provided that there is no chemical/biological spoilage over time, sodium azide is usually added to the solvent as an antibacterial agent in the case of aqueous biological samples). Spectra are automatically phased and baseline corrected with minimal manual checking required using the software attached to the modern NMR instruments.

3.2. Data preprocessing

Due to pH and other factors, heavy shifts or displacement of signals can occur along the axis of a ^{1}H NMR spectrum (when comparing one spectrum to another). It is crucial to address this problem as the next step after spectral recording is usually to apply statistics to compare systematically the signals across the spectra in order to detect any differences in signal intensity among groups of samples.

There are several options for the analysis. One is to bin the data (also referred to as 'bucket'), meaning that the signal is added up over small chemical shift intervals. This may simply be done with regular sized bins (say, 0.04 ppm, which would then give ~300 to 500 variables depending on the spectral width used), or alternatively with varying-sized bins (also referred to as intelligent binning which is our favoured approach) (Fig. 11.3). The later approach can be taken in an attempt to provide variables that correspond to signals from a given compound (in which case it is similar to carrying out peak integration), although it is inevitable that in crowded regions it will simply not be possible to achieve this. Nevertheless, binning can have another advantage, that of mitigating the effect of peak position variations in the series of spectra. We will elaborate on this in Section 8 (registration effects). The majority of the NMR researchers in metabolomics use this approach (Ward, Baker, Llewellyn, Hawkins, & Beale, 2011; Ward et al., 2010).

Another approach is to apply a peak fitting based on a spectral database as performed in the software called Chenomx (Weljie, Newton, Mercier, Carlson, & Slupsky, 2006) which also like the binning process provides a solution to the peak registration problem.

A third option is to work on the whole spectrum (with the exception of unstable and/or uninformative regions and in particular the water region and the signal-free high- and low-frequency extremities); if spectra are digitised with 8192 points, we may end up with ~6000 points, that is, 6000 variables after removal of the aforementioned spectral regions. We used this approach

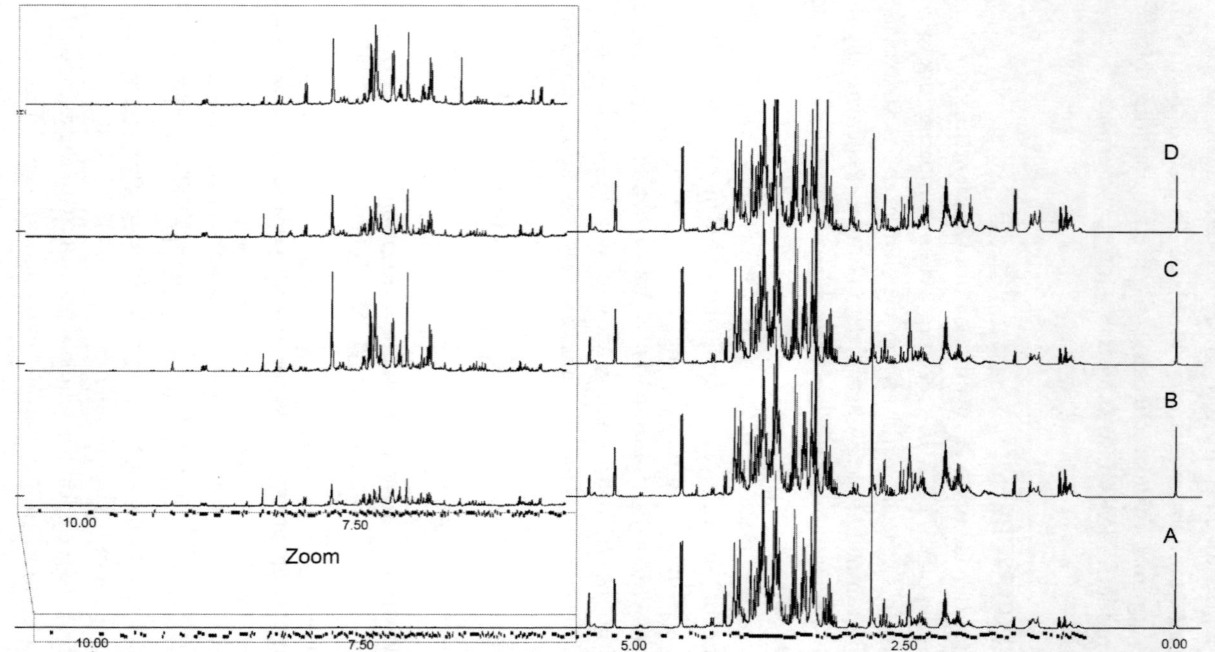

Figure 11.3 Typical ^{1}H NMR spectra of broccoli extracts in AMIX software. (A) Fresh raw broccoli. (B) Frozen raw broccoli. (C) Fresh steamed broccoli. (D) Frozen steamed broccoli. Variable buckets were drawn in AMIX (see black segments below spectrum A) using knowledge on the metabolite identity after careful scrutiny and peak assignment of 2D NMR spectra (see Fig. 11.2). Grouping of signals is regarded as important in our laboratory and we try to draw buckets around whole multiplets (rather than creating multiple buckets for peaks that are known to belong to one compound). Conversely, we also take into the overlapping of metabolite signals (it often happens that peaks from several metabolites arise in the same region). Identification is done methodically and recorded in a metabolite table list which will be the base of the identification of markers (some buckets will become markers after statistical analysis).

with some success on several datasets including extracts of orange juice (Le Gall, Puaud, & Colquhoun, 2001), tomato (Le Gall et al., 2003) or the plant model *Arabidopsis thaliana* (Le Gall, Metzdorff, Pedersen, Bennett, & Colquhoun, 2005). Our approach consisted in adapting the partial linear fit (PLF, Vogels, Tas, Venekamp, & van der Gref, 1996) to our datasets by creating in-house macros in Matlab (M. Defernez, personal communication). We have described briefly our procedure in one of our articles on green tea extracts (Le Gall, Colquhoun, & Defernez, 2004). This set of data particularly suffered from bad peak registration due to large shifts of pH between samples which heavily affected the patterns of scores when applying principal component analysis (PCA) on the unaligned data (Defernez & Colquhoun, 2003). We showed that using the PLF approach it was possible to align the main shifted peaks to gain meaningful information from the data recorded. Other alignment methods have been published, discussed and reviewed (Savorani, Tomasi, & Engelsen, 2010; Smolinska, Blanchet, Buydens, & Wijmenga, 2012; Veselkov et al., 2009; Vogels et al., 1996) and only a handful are described in the following section. Recursive segment-wise peak alignment (RSPA) is an alignment protocol published recently (Veselkov et al., 2009) based on fast Fourier transform (FFT) that aligns test segments to a reference and further subdivide these segments into subsegments to bring flexibility to the procedure. The principle of interval correlation optimised shifting algorithm (icoshift) (Savorani et al., 2010) is quite similar to other published methods as it is using FFT to boost the simultaneous alignment across the dataset. MacKinnon et al. (2012) applied the original version of RSPA and icoshift to two sets of data and also used two modified versions of the softwares (with improvement towards the reference spectrum and variable-length segments, respectively). The authors judged that the newly implemented procedures performed at least as well as the original ones. Another ingenious protocol has been introduced in 2011 (Beneduci, Chidichimo, Dardo, & Pontoni, 2011). It consists in simply adding HCl to the samples in order to reach a pH comprised between 1.2 and 2.0 where the effect of pH influence on peak chemical shifts is not more seen.

Recently, user-friendly processing solutions such as Metabolab or MetaboHunter have been proposed to the metabolomics research community. Metabolab (Ludwig & Gunther, 2011) is a graphical interface created in Matlab for performing some of the preprocessing of NMR spectra and providing tools for peak identification. MetaboHunter (Tulpan, Léger, Belliveau, Culf, & Cuperlović-Culf, 2011) is a Web server application aiming at identifying metabolites in ^{1}H NMR spectra in an automated manner. The software utilises access to databases such as HMDB. Similarly, two processing platforms were released in 2009. 'Metabolomics Package' (Izquierdo-García et al., 2009) is a novel R-package graphic user interface that provides preprocessing of raw spectra and an array of multivariate analysis tools and Automics (Wang et al., 2009) is an integrated open source software proposing preprocessing and nine statistical methods including the main ones (PCA, partial least squares (PLS), cluster analysis, etc.) Another very popular interface for processing of NMR data is MetaboAnalyst (Xia & Wishart, 2011) also released in 2009. This server provides like the others some preprocessing functions (especially in the version 2; Xia, Mandal, Sinelnikov, Broadhurst, & Wishart, 2012) and a variety of data analysis tools as well as access to multiple databases and pathway mapping. The same authors also proposed another semi-automated software called MetaboMiner (Xia, Bjorndahl, Tang, & Wishart, 2008) to identify metabolites in 2D NMR spectra.

4. LC/MS DATA RECORDING AND PREPROCESSING: THE NECESSARY STEPS PRIOR TO STATISTICAL ANALYSES

4.1. Data recording, QC and study design

Each step in a metabolomics flowchart is critical and this is particularly so for the step involving the data recording. It is essential in the recording strategy to avoid the introduction of a technical bias or at least to minimise it. For example, running samples in clusters—running samples from group 1 first followed by samples from group 2—would create an inevitable bias because the ion detection changes between each LC/MS run; the total intensity detection gradually decreases due to the source getting dirty and not ionising the molecules as well as earlier in the run. It is paramount to (1) know when to clean the source (or minimise the source contamination by injecting small sample volume) and (2) to randomise the sample order so the prejudicial ion intensity change observed for LC/MS profiling is equally applied to the different groups of sample studied.

It is now accepted that a pool sample or a quality control (QC) needs to be recorded at regular time intervals (every 5–10 samples) to monitor changes in the total intensity current (TIC) and in ion detection, in general. It is also advisable to run a blank on several occasions to assess the carry-over. QC samples determine the quality of signal detection during an analytical batch by providing data to establish the technical precision within each analytical block (the US Food and Drug Administration (FDA) recommends <15–20% of the QC mean) and by providing reference data for the use of non-linear signal correction within and between blocks to compensate for the analytical 'drifts' and 'dips' in signal intensity often encountered during the data recording of long runs (>50 samples) (Dunn, Broadhurst, Begley, et al., 2011; Zelena et al., 2009).

Moreover, a QC sample is essential for the conditioning of the analytical system at the beginning of a sample run (between 3 and 7 QC injections are usually needed to equilibrate the platform; Sangster, Major, Plumb, Wilson, & Wilson, 2006).

Last but not least, this chapter will not cover in detail the importance of the design of a metabolomics study experiment (for this, please see Dunn, Broadhurst, Atherton, Goodacre, & Griffin, 2011; Dunn, Broadhurst, Begley, et al., 2011). Carefully establishing the correct number of samples is paramount to ensuring the significance of uni- and multivariate statistics and like in any scientific study, knowledge of the study's biological context and of the tools available to evaluate the number of samples needed (Broadhurst & Kell, 2006; Chadeau-Hyam et al., 2010; Freedman, Back, & Bernstein, 2001; Guo, Graber, McBurney, & Balasubramanian, 2010) is critical to the success of the outcomes.

Data recording is usually performed on accurate mass spectrometers such as time-of-flight (TOF) hybrids such as the quadrupole-TOF and Orbitrap equipments (Allwood & Goodacre, 2010; Roux, Lison, Junot, & Heilier, 2011). It is limiting to perform untargeted metabolite profiling using a unit mass spectrometer like a quadrupole as marker identification is much more hypothetical.

4.2. Data preprocessing

Analysis of sets of untargeted chromatographic data requires a data preprocessing step so as to make this data comparable across samples. This is because there are variations in retention times (RTs), and the first approaches to deal with this challenge were developed at a time where detection of chromatographic peaks by ultraviolet (UV) or flame ionisation detector (FID) was the norm. The move towards using MS as detection method led to radically different strategies for data pretreatment being developed, though the purpose was still to make the data comparable across samples regardless of instrumental variability.

4.2.1 Liquid chromatography with detection by UV or FID

With detection based on UV or FID, the data are two-dimensional: RT (or retention index, RI) in dimension 1, ion current or ion intensity in dimension 2. Two main issues arise with this type of data, namely, variation in overall intensity and variation in RT. The first is addressed by applying a signal intensity correction (on the second dimension); it is performed for both GC and LC methods by ratioing the signal of each metabolite peaks with an appropriate internal standard (usually a molecule in known amount having an identical behaviour to that of the metabolite under study under the particular chromatographic conditions used; Bennett, Mellon, & Kroon, 2004; Kern, Bennett, Mellon, Kroon, & Garcia-Conesa, 2003; Le Gall et al., 2003).

The second issue has been the object of numerous investigations. Since the late 1980s, several alignment methods were created to overcome the shifts in time domain observed between samples of chromatographic nature. Dynamic time warping (DTW), parametric time warping (PTW), semi-parametric time warping (STW), correlation optimised warping (COW), fuzzy warping and peak alignment by a genetic algorithm were all developed to propose a solution to this recurrent problem of registration (Nederkassel, Daszykowski, Eilers, & Heyden, 2006; Tistaert, Dejaegher, & Heyden, 2011). Hendriks, Cruz-Juarez, De Bont, and Hall (2005) describe PTW, DTW and COW as three alignment programmes that stretch and compress data points along the time domain differing only by the smoothness of the alignment function whilst Tomasi, Savorani, and Engelsen (2011) define two main types of correction function: compression/expansion (C/E) or insertion/deletion (I/D) and categorise PTW, STW and COW as C/E approaches and DTW as I/D. Several publications drew comparisons between DTW and COW (Hendriks et al., 2005; Pravdova, Walczak, & Massart, 2002) and concluded that both methods led to satisfactory peak alignment. Nederkassel et al. (2006) compared DTW, PTW and COW using two large sets of HPLC/UV chromatograms of plant extracts. Each algorithm needs implementation (e.g. optimisation of parameters) and the way the authors proceeded for each method is detailed in the publication. They concluded that COW and STW performed the best on their data and that STW resulted in a faster alignment. COW (Nielsen, Carstensen, & Smedsgaard, 1998) is a popular choice and one of the most effective algorithms (it preserves the area and shape of peaks; Veselkov et al., 2009), but due to the iterative nature of the peak alignment routine, the method can be time-consuming and not suitable for the large datasets often produced by metabolomics approaches (Nederkassel et al., 2006; Tistaert et al., 2011). Dynamic programming methods such as DTW are also time-consuming (Zhang, Chen, & Liang, 2011) and sometimes prone to signal 'over-warps' (Zhang, Liang, et al., 2012). PTW and STW are fast because unlike for COW and DTW the warping function is explicitly modelled (Veselkov et al., 2009), but the principle of those methods can result in the preferential alignment of the large peaks sometimes at the detriment of smaller peaks (Veselkov et al., 2009) and PTW has shown some limitation when dealing with complex chromatographic shifts (Nederkassel et al., 2006). Another set of alignment methods are based on FFT cross correlation and seem particularly suited for large datasets. However, FFT can introduce artefacts and remove peak points (Zhang, Liang, et al., 2012). In the past 4 years, a string of publications proposed new developments essentially based on existing methods. COW has been combined with a component detection algorithm (CODA) to align TIC chromatograms from LC/MS data (Christin et al., 2008). In this algorithm, some of the information in the MS dimension is used to define a reference chromatogram using CODA and COW is then applied on the data. The authors compared COW–CODA and COW using three different sets of data and concluded that COW–CODA had a superior alignment power especially with highly variable chromatograms such as those recorded from urine samples. The purpose of the rest of the algorithms developed was to shorten the computing time of the chromatographic alignment. Zhang et al. (2011) proposed in 2011 a method called peak alignment using differential evolution (DE), a C/E method they have called alignDE. Another method called new supervised alignment method based on linear interpolation initially involved the input of the user (peaks to align have to be selected) but then guaranteed a fast automatic alignment of HPLC/diode array detector data (Struck, Wiczling, Waszczuk-Jankowska, Kaliszan, & Markuszewski, 2012). The last two warping methods mentioned in the present review are I/D models both based on FFT cross correlation. Multiscale peak alignment is using a continuous wavelet transform to detect peaks in chromatograms (Zhang, Liang, et al., 2012), and the icoshift initially developed for NMR (Savorani et al., 2010) was implemented for HPLC/Corona charged aerosol detection data (Tomasi et al., 2011) and showed better performance compared to COW. Authors of both methods claimed to provide the speedy alignment of two-dimensional datasets—an advantage which seems particularly suited for the analysis of large sets of complex extracts. No single method is, however, perfect and none can account, for example, for the change in order of peaks (Savorani et al., 2010). Thus new developments possibly involving the better characterisation of the peaks that need the alignment are needed in order to overcome the limitation associated with the peak registration observed in the metabolic profile datasets using liquid chromatography. Once the alignment is performed, the resulting table of data is usually a two-dimensional data matrix formed by a peak-table comprising RT and peak intensity.

4.2.2 Untargeted liquid chromatography with MS detection

The continuum of technological advances in MS seen in the past 15 years has led to the natural increase of the use of MS as a mode of detection over the more classically used ones (UV, FID, etc.), and in combination with LC or GC, these approaches produce the three-dimensional data that scientists in metabolomics are now familiar with: for each sample, the raw chromatography-MS data are a three-dimensional matrix of RT in dimension 1, ion current or ion intensity in dimension 2 (this second dimension represents the sum of all the ions in dimension 3 for a given 'slice of time' that is a scan—typically 2–20 ms—Theodoridis et al., 2012) and a series of mass over charge (m/z) ion peaks (typically covering a range of 80–1000 m/z) along dimension 3 (Fig. 11.4). Thus adding an extra dimension not only resulted in a steep improvement in the number of metabolite detected but also created the need to deconvolute the MS peaks—the peak deconvolution of a time window is the thorough characterisation of all the metabolites detected within this time frame on the basis of their m/z ions (molecular ions, adducts, fragments, etc.). This need is also driven by the increased use of MS to profile crude extracts and the complexity in comparing many metabolites that comes with this, which means that correcting for peak registration and variations in signal intensity using peak deconvolution and peak picking tools is now the critical step before data pretreatment (see next section).

There are two main preprocessing strategies to handle data generated by LC/MS or UPLC/MS platforms: either create time windows of defined size (in dimension 1, i.e. in seconds) and sum the signals of ions in the third dimension for each window (slicing data) or establish a pipeline of steps leading to the detection of a selection of individual peaks from the third dimension for small time windows along the first dimension (Katajamaa & Oresic, 2007; Smith, Want, O'Maille, Abagyan, & Siuzdak, 2006). The second approach is more challenging yet this is the preferred one as (1) it is conceptually attractive to make use of the extra dimension afforded by the MS to identify and distinguish co-eluting compounds and (2) it gives direct access

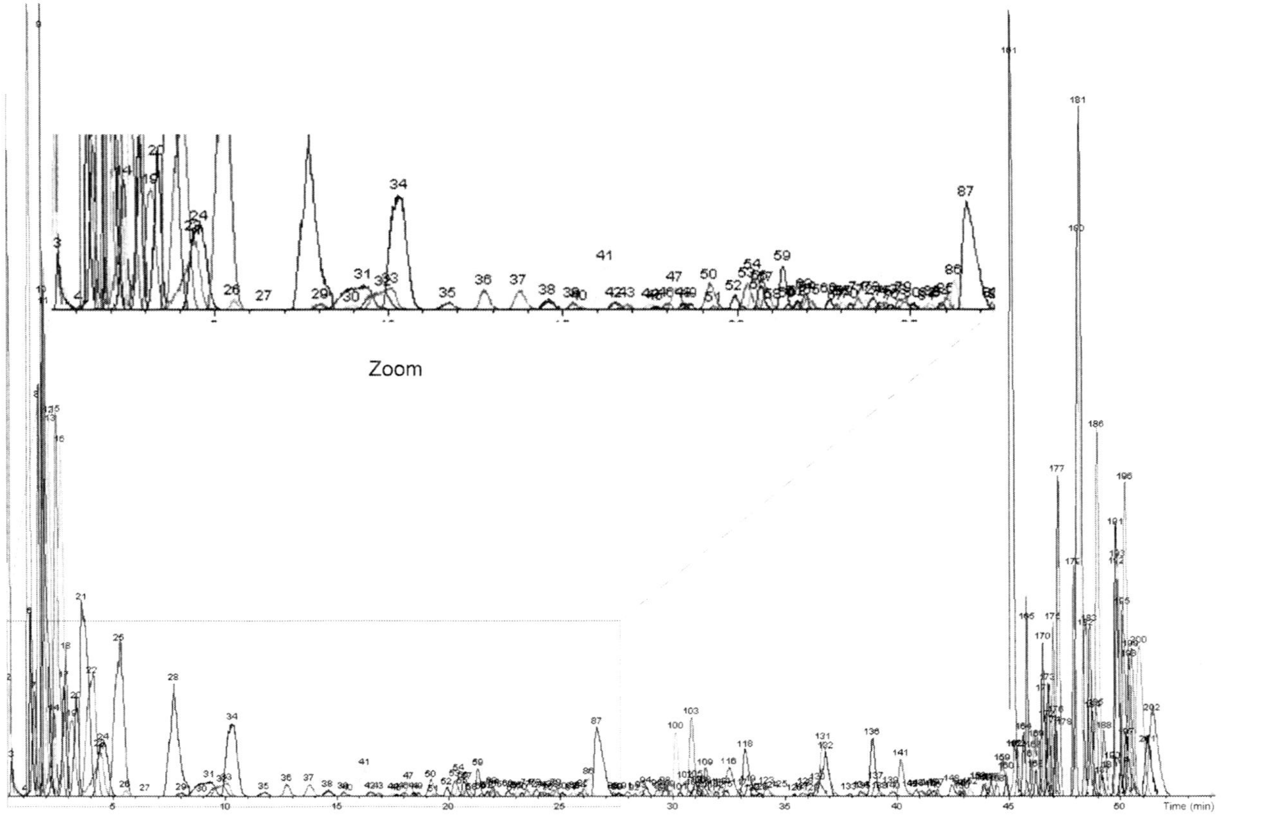

Figure 11.4 Typical LC/MS chromatogram displayed in the Bruker DataAnalysis software using the function 'dissect extracted chromatogram'. The function scans for any rise and fall of ions and will group ions for individual compounds numbered in the plot. (For colour version of this figure, the reader is referred to the online version of this chapter.)

to the information responsible for any sample groups separation demonstrated by application of uni- or multivariate analyses. A number of software packages now exist to tackle this challenge. Broadly described, the preprocessing workflow includes several steps (not all compulsory) which are the use of noise filtering, baseline correction, peak detection, peak alignment, peak integration, filling for missing data, automatic identity recognition (or the attempt to do so), normalisation and the use of interface user tools (uni- and multivariate statistics, visual implementations such as network analysis or access to selected peaks from the raw data).

One possibility is to use the vendors' software packages (e.g. MarkerLynx, Markerview from AB SCIEX, MassHunter from Agilent, Sieve from ThermoFischer Scientific, etc.), and a number of publications refer to the use of these softwares to deconvolute untargeted LC/MS data (Bruce et al., 2009; Denery, Nunes, & Dickerson, 2010; Grata et al., 2008; Hanhineva et al., 2011; Kind et al., 2012; Lutz, Lutz, & Lutz, 2006; Nylund, Weinberger, Rempt, & Pohnert, 2011; Sadygov, Maroto, & Hühmer, 2006; Sangster, Wingate, Burton, Teichert, & Wilson, 2007; Spagou et al., 2011; t'Kindt, De Veylder, Storme, Deforce, & Van Bocxlaer, 2008; t'Kindt, Morreelm, Deforce, Boerjan, & Van Bocxlaer, 2009; Vaclavik, Lacina, Hajslova, & Zweigenbaum, 2011). This option is in some sense the most convenient as the software package is integrated with the MS instrument and offers access to information and tailored tutorials. A limitation is that this type of software package cannot be used directly with data from other MS equipments, though it is sometimes possible to translate raw MS data from one format to another via the use of, for example, netCDF or mzXML formats (Katajamaa & Oresic, 2007).

Another (lengthy) option is to create a list of m/z signals selected manually using a LC/MS chromatogram representative of the dataset (e.g. a pool sample) and to automatically measure the intensity of each selected peak using vendor softwares such as Xcalibur from Thermo Scientific, Masslynx from Waters, massHunter from Agilent or Analyst from AB SCIEX. For instance, Shepherd and colleagues (Shepherd et al., 2010) have used this approach, carrying out the first step (peak detection) with Automated Mass Spectral Deconvolution and Identification System (AMDIS) (see Section 5 on GC–MS for which it was originally developed) followed by automatic peak integration in Xcalibur to analyse extracts of potato by LC/MS. In our own work, we have used the same procedure and compared it with the use of XCMS (see below). Both approaches provided variables that correspond to RT/(m/z) features, but typically, a package such as XCMS produces more variables than the manual procedure—this, however, does not necessarily mean more information. The outcome of the data analysis was remarkably similar (Fig. 11.5) though it was not possible to draw a direct link between the two sets of variables (i.e. the information in the loadings differed).

The need for less labour-intensive, more comprehensive and free solution led to several groups putting efforts towards developing their own

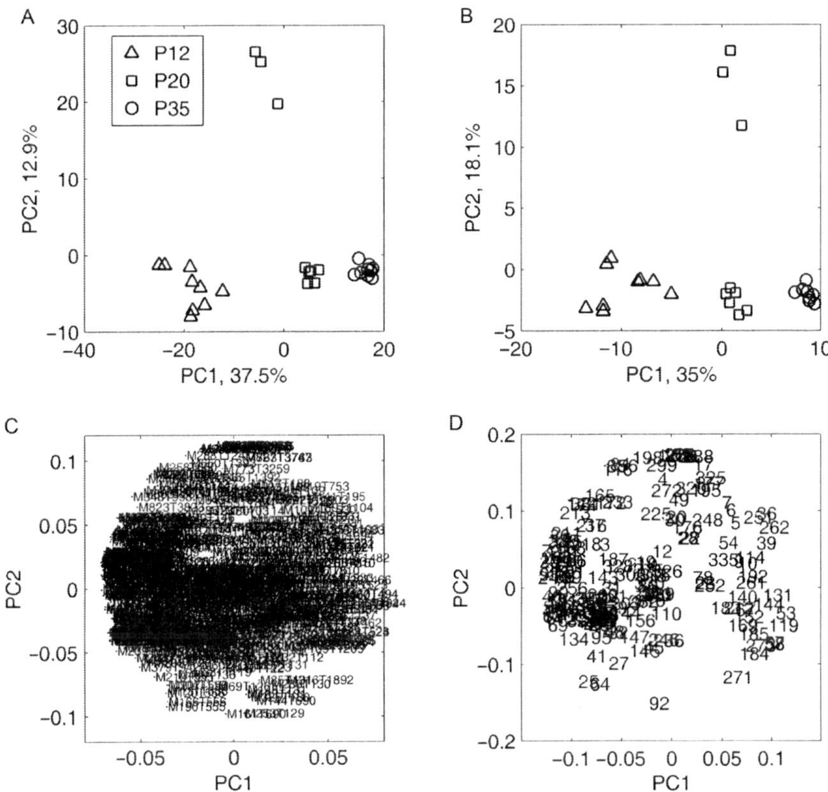

Figure 11.5 Principal component analysis (PCA) applied onto two processed 2-dimensional datasets arising from the same analytical untargeted LC/MS sample set. The sample set consists in the profiling of three biological replicates of tomato mesocarp tissue (run in triplicate) using reverse-phase HPLC/MS on a triple quadrupole mass spectrometer (Le Gall et al., unpublished). The samples were harvested at 12, 20 and 35 days post-anthesis. (A) PCA scores after the automated signal extraction procedure (using XMCS software) based on 604 mass/RT features (after removal of redundant peaks) and (B) PCA scores after the manual signal extraction procedure based on 176 mass/RT features. (C) and (D) are the respective plots of the corresponding loadings. The figure shows that the automated and manual procedures give similar results.

solutions. In 2005, Trygg, Moritz and colleagues proposed a semi-automatic processing implemented in Matlab that consists of selecting signals in the *m/z* channel using criteria that include minimum intensity above the 'noise threshold level' and a peak shape that satisfies the quality required. The data collated are then compressed by summing the spectral dimension (Jonsson, Bruce, et al., 2005). This group and others have successfully used this approach to analyse untargeted LC/MS and GC/MS datasets (Bruce et al., 2008; Hoffman et al., 2010; Jonsson, Johansson, et al., 2005).

In 2006, an open access software package called XCMS (Smith et al., 2006) became available to the community of researchers in the field of metabolomics. XCMS which proposes an all automated peak detection, alignment and integration with filling for missing data was an instant hit as the software was relatively user-friendly. Although accessed through the command-line package 'R', the application only required a limited number of commands with possible amendments to parameters. As the original version, new options for these parameters have been developed, for instance, for feature detection the algorithm *centwave* was added to that offered originally (*matchedFilter*) (Patti et al., 2012). *Centwave* has the ability to handle multiple peak widths whilst the user is restricted to one peak width parameter for *matchedFilter*. In 2008, Tautenhaun and colleagues (Tautenhahn, Böttcher, & Neumann, 2008) compared these two algorithms as well as *centroidPicket* from MZmine (Katajamaa & Oresic, 2005) using two test datasets and concluded based on those results that *centwave* was performing slightly better than the other two algorithms. Siuzdak's group who has developed XCMS also recently proposed two new programs based on XCMS. MetaXCMS (Tautenhahn et al., 2010) now allows the comparison of multiple groups of samples recorded on separate occasions but sharing identical sample preparation and recording conditions. Also, a Web-based version of the package now circumvents the need for command-line: XCMSonline (Tautenhahn, Patti, Rinehart, & Siuzdak, 2012) proposes a solution for the entire untargeted metabolomics workflow (data processing, alignment, statistical analysis and metabolite assignment). Some add-ups can also be used in conjunction with XCMS to automatically assign the ions detected. Links to PUTMEDID and CAMERA software packages can be implemented in R in combination with XCMS. CAMERA (Kuhl, Tautenhahn, Böttcher, Larson, & Neumann, 2012) looks for molecular masses and tries to assign many possible combinations of adducts such as $[M+H]^+$ and $[M+Na]^+$ and isotopes using the XCMS data table as input. No compounds are named. CAMERA is an annotation system that provides possible assignment for adducts (including fragments). The software called PUTMEDID_LCMS within the framework TAVERNA (Brown et al., 2011) allocates putative compound assignments to a selection of metabolite features using a comprehensive database of candidate compounds found in biological matrices. This software has a sophisticated automatic annotation process that often gives multiple outputs for one metabolite feature. It includes a tracking system to group the m/z ions suspected to belong to a same molecule although fragment ions are not included in the search. An alternative (and laborious) identification process is to manually check the metabolite features listed as potential markers in the XCMS table (based on the t-test) by searching online databases such as METLIN (Smith et al., 2005), HMDB (Wishart et al., 2009) or MZedDB (Draper et al., 2009). For this, the analyst needs to decide based on the ion features found in representative raw LC/MS chromatograms if the quality of the feature ion is good enough (e.g. not noise) and hypothesise which m/z feature might be the molecular ion to then interrogate the online databases. MZedDB is particularly flexible and proposes a multitude of adducts including fragments representing the loss of a water molecule $[M+H-H_2O]^+$ or ammonia $[M+H-NH_3]^+$, two fragments that can often be observed in metabolomics.

XCMS is now one of the most used deconvolution softwares (over 45,000 downloads as of 2011 (Patti et al., 2012) and more than 463 citations, 22/10/12 based on Web of Knowledge). Many research groups rely on XCMS for the deconvolution of their LC/MS data (including us), and the usage of this popular software reaches all areas of research (plant, mammalian metabolism, etc.) (Arbona, Iglesias, Talón, & Gómez-Cadenas, 2009; Boyard-Kieken et al., 2011; Dunn, Broadhurst, Begley, et al., 2011; Dunn et al., 2008; Eliasson et al., 2012; Fujimura et al., 2011; Kenny et al., 2010; Masson, Alves, Ebbels, Nicholson, & Want, 2010). Breitling and colleagues (Scheltema, Jankevics, Jansen, Swertz, & Breitling, 2011) have recently proposed a workflow integrating the feature detection of XCMS and a newly designed alignment and identification routine they have developed, mzMatch. Similar to XCMS, mzMatch provides a set-up comprising multiple (some optional) steps including a mass chromatogram extraction, matching, noise filtering, normalisation and alignment. The graphical interface enables a rapid visualisation at every step of the analysis including inspection of raw data. The software was successfully used on extracts from the parasite *Tryponosoma brucei brucei* to align and annotate peaks (Creek et al., 2011). Recently, the authors published IDEOM, an Excel interface aimed at automatically removing the many false-identifications after deconvolution of LC/MS data using the mzMatch pipeline rather than applying a removal by manual curation (Creek, Jankevics, Burgess, Breitling, & Barrett, 2012). Likewise, Melamud, Vastag, and Rabinowitz (2010) developed a pipeline similar to that of XCMS and Mzmine for the purpose of deconvoluting targeted and untargeted LC/MS data and annotating *m/z* features, but the effort was focused on making the graphical interface user-friendly. The workflow called Metabolomic Analysis and Visualisation Engine proposes access to the data at any stage of the processing and includes an optional pathway-based visualisation of isotope-labelling data.

MZmine (Katajamaa & Oresic, 2005) also provides broad screening of biomolecular components across multiple samples. The software has the ability to filter noise, detects selectively LC/MS peaks and aligns them, and proposes a normalisation step with a series of options (total unit, peak height or based on internal standards) (Katajamaa & Oresic, 2005). The authors are particularly concerned with the peak detection step as many features in biological matrices are at trace levels. At the time of MZmine release, they were particularly keen to progress further on the feature detection. In the following year, they released an amended version of MZmine which could process larger batches of samples (Katajamaa, Miettinen, & Oresic, 2006). The peak picking routine was also improved by refining the step of filling the gaps in the dataset using the raw data following the detection of missing values. Subsequently, the authors released a new version of MZmine called MZmine 2 (cited 43 times 22/10/12) which is based on a new alignment technique named RANSAC (random sample consensus algorithm) (Pluskal, Castillo, Villar-Briones, & Oresic, 2010). The motivation behind MZmine 2 was to propose a friendly interface and a flexible and modular platform for MS-based data processing. In a similar manner to XCMS, it is now possible to use MZmine 2 to make an attempt at identifying metabolite features through links to online resources KEGG (Kanehisa & Goto, 2000), METLIN (Smith et al., 2005) and HMDB (Wishart et al., 2009). MZmine is widely used by the metabolomics community with more than 363 citations (22/10/12) (Ivanisevic, Thomas, Lejeusne, Chevaldonne, & Perez, 2011; Katajamaa & Oresic, 2005; Katajamaa et al., 2006; Pluskal, Nakamura, Villar-Briones, & Yanagida, 2010; Toffali et al., 2011; Zhang et al., 2011; Zhao et al., 2012).

MetAlign (Vorst et al., 2005) was probably with MSFACTs (Duran, Yang, Wang, & Sumner, 2003) and MET-IDEA (Broeckling, Reddy, Duran, Zhao, & Sumner, 2006), one of the first deconvolution softwares to provide an alternative to vendors' software packages. At the time, access to the package was through a fee-based membership, but now the membership is free and information on downloading and details on the algorithm have been published (Lommen, 2009). This software package proposes noise filtering, baseline correction, peak detection, peak alignment and peak integration. The baseline correction in MetAlign is a distinctive step that not many deconvolution softwares offer. It is widely used by the metabolomics community using non-targeted GC/MS approaches (see Section 5). The peak alignment routine of MetAlign can also be integrated in a workflow like the one proposed by Tikunov and colleagues who aimed at removing the signal redundancy per metabolite often found in dataset outputs after deconvolution (Tikunov, Laptenok, Hall, Bovy, & de Vos, 2011) using a fuzzy clustering approach. Similar to XCMS and Mzmine, Metalign is also a popular choice regularly cited in various metabolomics research publications (Kobayashi et al., 2012; Kusano et al., 2011; Kuzina et al., 2009; Matsuda et al., 2012; Redestig et al., 2011; Tikunov et al., 2005; Zamboni et al., 2010).

The processing time for the three main freely accessible deconvolution softwares (XCMS, MZmine and metAlign) depends in part on the number of samples processed but also on the capacity of the application performance of the PC used. The three softwares have evolved simultaneously with the recent change in the standard desktop PC configuration (introduction of mainstream 64-bit operating systems) and all three now recommend the efficient use of powerful systems with 64-bit CPU or large random access memory of 4 GB or more to process the data (Lommen & Kools, 2012; http://mzmine.sourceforge.net/download.shtml; nSlaves =2 command in Patti et al., 2012) in order to decrease the processing time of the deconvolution or increase the number of samples studied (up to 1000 for MetAlign; Lommen & Kools, 2012) and similar numbers for MetaXCMS (Tautenhahn et al., 2010) and Mzmine 2 (Pluskal, Castillo, et al., 2010).

4.2.3 Dealing with other, specific, LC-based data types

(1) Processing targeted LC/MS data is easier as only a selection of predefined metabolites is detected; thus, this methodology bypasses the deconvolution step. Usually, the m/z ions are detected using a selection ion monitoring (SIM) approach where only a narrow range of m/z is recorded or the multiple reaction monitoring which is a highly specific tandem MS/MS. Both approaches are particularly suited for the quantitation of metabolites in complex extracts. The detection sensitivity is also usually superior compared to the full-scan approach (used for untargeted MS metabolomics) which typically covers the range 80–1000 m/z. Full-scan (untargeted) and tandem MS

(targeted) approaches are, in fact, complementary (Lu et al., 2010) and having both platforms is of benefit to any metabolomics research unit. This dual tactic has been chosen by several research groups who wished to use the advantages that each platform has to offer (metabolite identity is known with the targeted approach, metabolites are quantified and the untargeted platform allows the unbiased screening of complex extracts) (Bajad et al., 2006; Evans, DeHaven, Barrett, Mitchell, & Milgram, 2009; Lu et al., 2010). When used comprehensively, this procedure can cover the detection of several hundreds of primary and secondary metabolites (Boer, Crutchfield, Bradley, Botstein, & Rabinowitz, 2010; Oliver et al., 2011; Xu et al., 2012).

(2) Recently, a series of LC/MS data processing tools were published to address solutions to specific types of datum, such as capillary electrophoresis (CE/MS), Fourier transform (FT/MS) and isotopic data.

The need for processing specific data led to the creation of MathDAMP and P-BOSS, two softwares aimed at processing CE data sets. CE data sets are prone to large amount of redundancy and one way to deal with this is to reduce the redundancy in the dataset with the software called P-BOSS (Morohashi et al., 2007) or to use a software that bypasses the feature detection step like MathDAMP (Baran et al., 2006). MathDAMP aims at grouping features along the *m/z* dimension, performing a peak alignment and providing an output intuitively easy to understand based on a colour-coded graphical display that reveals which *m/z* values differ in intensity between groups of samples.

The problem encountered with high-resolution (HR) MS measurements is different to that with lower mass resolution. As very HR data are usually associated with lower background noise, the authors of apLCMS (Yu, Park, Johnson, & Jones, 2009) proposed a software that leaned towards sensitivity rather than specificity thus reducing the risk of merging separate MS features. Specificity is often set to a high threshold with lower resolution data to avoid detecting false features from the noise. The philosophy behind apLCMS was to develop an adaptive scheme that separates features which *m/z* values are very close to each other. This software was successfully applied to mitochondrial extracts from mouse liver extracted by two LC/FTMS approaches based on an anion exchange and a reverse-phase chromatography platforms (Roede, Park, Li, Strobel, & Jones, 2012). Similarly, Wei and colleagues developed a package specifically dedicated to HR-MS data called MetSign (Wei et al., 2011) which was applied on direct-infusion FTMS data. The software was able to deconvolute MS features quickly and also provided the analysis of stable isotope-labelled data and temporal data and access to putative assignment by matching the experimental data (including isotopic peaks) with theoretical data of metabolites in the MetSign database (itself created from a collation of databases, namely, KEGG, LIPID MAPS (www.lipidsmap.org) and HMBD). The authors successfully applied MetSign to a set of mouse liver samples yielding a total of 800 metabolite peaks (Shi et al., 2012). Since the first paper, the same developers published an optimised deconvolution method to deal (better) with LC/FTMS datasets (Wei et al., 2012). Krishnan et al. (2012) recently proposed a binning process to reduce the size of HR datasets, an essential step prior to the deconvolution process of data associated with variable mass accuracy along the mass chromatogram *m/z* axis. They based their approach on a so-called minimal distance approach to distinguish metabolite-related signals from noise. The authors intended to provide further tools in the future such as a Web-based LC/HR-MS interface or feeding knowledge related to isotopic distribution or peaks or the *m/z* variation over scan in order to improve the binning procedure they have implemented.

(3) Recently, Schuhmacher and colleagues published an automatic extraction aimed at dealing with the tracking of labelled metabolites in *in vivo* stable isotopic labelling biological samples (Bueschl et al., 2012). They proposed MetExtract a software that proceeds to an automatic matching of ^{12}C and uniformly ^{13}C-labelled isotopologues followed by a comparison of their intensity ratios and the calculation of a Person's correlation coefficient. The peaks' characteristics (intensity, RT, etc.) are then listed in an output table specifying the number of atoms predicted from the labelling element (e.g. carbon) with a putative assignment (i.e. adduct, fragment, polymer ions, etc.).

Lipidomic approaches are in essence similar to non-targeted aqueous-based metabolomics. They are MS profiling techniques based on either a hyphenated liquid chromatography system or a direct infusion approach (shotgun lipidomics). The preprocessing of a lipidomics LC/MS dataset can be achieved using either XCMS or MZmine (Lankinen et al., 2009; Massodi, Eiden, Koulman, Spaner, & Volmer, 2010). Interestingly, two recently developed softwares dedicated to lipidomics were published in 2011 (Oresic, 2011). The first software called Lipid Data Analyser provides classical LC/MS data deconvolution with a peak detection based on a three-dimensional algorithm which can track low levels of minor lipid species (Hartler et al., 2011). The second software called LipidXplorer (Herzog et al., 2012, 2011) identifies and quantifies lipidic entities from shotgun experiments.

(5) The alternative to the usage of one of those deconvolution softwares is to employ Web-based tools to perform preprocessing and other steps such as statistical analysis and metabolite assignment. As highlighted earlier in this chapter, XCMSonline was recently released to provide a solution for the entire untargeted metabolomics workflow (Tautenhahn et al., 2012). Other Web-tools such as metaP-Server (Kastenmüller, Römisch-Margl, Wägele, Altmaier, & Suhre, 2011) or MetaboAnalyst (Xia & Wishart, 2011) aim at performing data analysis covering statistical analysis and sample group comparisons whilst another subset of Web tools such as TICL (Antonov, Dietmann, Wong, & Mewes, 2009), MetDB (Neuweger et al., 2008), MetExplore (Cottret et al., 2010) and Pathos (Leader, Burgess, Creek, & Barrett, 2011) would focus on mapping metabolic pathways.

4.2.4 Conclusions

Package freely available are listed in Table 11.1. Castillo, Gopalacharyulu, Yetukuri, and Oresic (2011), Sugimoto et al. (2012) and Katajamaa and Oresic (2007) also provided a detailed comparison of a sub-selection of softwares.

The output after preprocessing using one of the softwares described is often a table of processed (mainly aligned and integrated) peaks readable in a spread sheet programme, for example, Microsoft Excel or similar which can then be transferred to a multivariate analysis software ready for pretreatment.

5. GC/MS DATA RECORDING AND PREPROCESSING: THE NECESSARY STEPS PRIOR TO STATISTICAL ANALYSES

5.1. Data recording

The two platforms LC/MS and GC/MS have the communal principle of using chromatographic columns to separate substances on the basis of their chemical properties. For the LC/MS platform reviewed above, the compounds are separated on the basis of their polarity; for the GC/MS platform, the separation occurs due to differences in volatility among the chemicals analysed and their interaction with the liquid phase coating the inside of the GC chromatographic column. Like for LC/MS, the compounds are eluted one by one out of the column (sometimes co-eluted) and then detected by a mass spectrometer usually a single quadrupole or a TOF instrument. Running QC samples and randomising the order of the samples also apply to the recording of GC/MS data (Begley et al., 2009), but whereas it is

Table 11.1 A selection of preprocessing packages freely available

Software	Feature detection	Other preprocessing steps
XCMS (Smith et al., 2006)	Combination of chromatographic slices of chosen *m/z* within the time domain followed by integration with a filter function	Matching across samples using *m/z* bins and alignment with a Kernel density estimator to identify groups of peaks of similar retention times. Selected data are then fitted with a local regression fitting method loess. Filling for missing peaks with raw data
Mzmine (Katajamaa & Oresic, 2005)	Noise reduction through data smoothing using filters and feature detection using either the local maxima method or the recursive threshold method	Peak alignment using a master peak list followed by a second step that fills the gaps for massing values. Normalisation methods are proposed: linear normalisation (to average peak height, total raw signal, etc.) or normalisation using multiple internal standard compounds. Note that the newer version of the software MZmine2 offers a modular pipeline and a new alignment method called random consensus aligner (RANSAC)
metAlign (Lommen, 2009)	Data smoothing is performed by digital filters. MetAlign is characterised by a background subtractive step followed by the extraction of mass peak information	Use of a sophisticated alignment of the chromatographic axis using high S/N peaks ('landmark peaks') and iterative fine alignment by increment of the number of landmark peaks. Missing data are awarded a value three times the estimated local noise
MathDAMP (Baran et al., 2006)	Raw dataset is binned along the *m/z* dimension to a preferred resolution. Baseline correction and smoothing filters are optional steps	Alignment of datasets along the time domain is achieved using the DTW method. Data are normalised to a reference dataset and visualised on a colour-coded 2D plot (time versus *m/z*)

Continued

Table 11.1 A selection of preprocessing packages freely available—cont'd

Software	Feature detection	Other preprocessing steps
apLCMS (Yu et al., 2009)	Removal of noise and feature identification based on a kernel density algorithm followed by a run filter to estimate if the detected feature is real or noise	Use of non-parametric method based on kernel density estimators to align features and achieve fine-tune intensity grouping
mzMatch (Scheltema et al., 2011)	Feature detection from XCMS (CentWave) or that of Mzmine	Customizable pipeline providing filtering, peak-matching, gap-filling and annotation of mass spectra

possible to perform non-targeted LC/MS without resorting to the use of internal standards, the gold standards for GC/MS is to use (1) at least one internal standard to ratio each m/z peak used for data analysis and (2) a series of standards usually alkanes or fatty acid methyl esters (FAMEs) added to the samples to realign the peaks along the time axis (once realigned the peak RT becomes RI). This renders the automatic matching of the peaks defined by their mass to charge ratio (m/z) and their RI much easier. These peak characteristics are called sometime mass spectral tags (Kopka, 2006; Luedemann, Strassburg, Erban, & Kopka, 2008). Contrary to LC/MS, there are several mature databases available to assign compounds detected by GC/MS; thus a large portion of metabolites can be identified. The most comprehensible is called National Institute of Standard and Technology (NIST), but others dedicated to metabolomics exist such as the Golm database (Kopka et al., 2005) or the Fiehn Libraries (Kind et al., 2009).

5.2. Data preprocessing

As explained in Section 4.2 (LC–MS data preprocessing), application of an MS-based profiling approach results in outputs that take the form of a three-dimensional table. Similar to LC–MS, the preprocessing of this type of data aims to detect peaks via deconvolution and peak integration so that a two-dimensional table (intensity for each sample of 'features' corresponding to $\{RT/m/z\}$ pairs) is produced that can then be fed into a statistical analysis. The metabolite identification for GC/MS metabolomic data is somewhat different to that of LC/MS data tough, mainly due to the inherent differences in detection of the two methods. With GC/MS, it is possible to obtain reproducible mass spectra and to consult very large databases (NIST (National Institute of Standards and Technology), the Wiley spectral libraries, the Golm Metabolome database (Kopka et al., 2005), the BinBase mass spectral database (Fiehn, Wohlgemuth, & Scholz, 2005) and the FiehnLib libraries (Kind et al., 2009)) to identify the metabolites based on characteristic recognisable fragment ions. Consequently, researchers have concentrated efforts towards the automation and the accuracy of this process. A multitude of scripts and packages has been developed and pipelines described, putting different emphasis and integrating to different extent the peak identification, integration and annotation.

The first free package called AMDIS was made available to the GC–MS community in 1999 (Halket et al., 1999; Stein, 1999). AMDIS offers a powerful deconvolution approach that allows the creation of outputs—deconvoluted m/z peaks—that can then be used to perform multivariate analysis in order to tease out the difference in metabolic phenotypes between groups of samples. This tool has been widely used by the GC-based metabolomic community and until today it remains a popular choice for preprocessing GC/MS metabolomics data (Jiye et al., 2005; Birkemeyer, Kolasa, & Kopka, 2003; Lohse et al., 2005; Villas-Bôas, Delicado, Akesson, & Nielsen, 2003). A classical way of using AMDIS is to create a list of m/z peaks in order to integrate the selected metabolites using another software package (e.g. a vendor's package like Xcalibur) (Dobson et al., 2010; Shepherd et al., 2010). Some authors have reported that AMDIS has a tendency to produce too many false positive hits, that is, artefacts (O'Callaghan et al., 2012). This led to the creation of several software packages that use AMDIS as a starting point in their flowchart. Styczynski et al. proposed in 2007 SpectConnect a freely available software implementation that can detect components from GC/MS chromatograms without the use of a reference library or manual curation (Styczynski et al., 2007). The authors were able to gain information on additional peaks (compared with AMDIS) tracked by the SpectConnect library which consists of peaks not identified but detected across the samples. This popular approach has been applied to green tea (Ku et al., 2010), algae (Barupal, Kind, Kothari, Lee, & Fiehn, 2010) and yeast (Usaite et al., 2009). Behrends, Tredwell, and Bundy (2011) produced a script called GC/MS Assignment Validator and Integrator that combines AMDIS outputs and raw data to automatically backfill missing values. Aggio, Villas-Bôas, and Ruggiero (2011) combine AMDIS and some preprocessing steps written in R-package and based on XCMS to automate the pipeline for analysis of metabolomics GC/MS data. Some research groups have also used AMDIS as a base for a procedure that quantifies metabolites in an absolute manner; in a procedure called MetaQuant (Bunk et al., 2006), they use the flexibility of AMDIS to refine this essential step in GC/MS metabolomics. Liebeke, Brözel, Hecker, and Lalk (2009) used MetaQuant in conjunction with Chemstation and the NIST library (Liebeke et al., 2009). The same group also developed a complete pipeline of steps called MetaboliteDetector that includes peak smoothing, deconvolution, integration and identification of GC/MS peaks (Hiller et al., 2009). The authors compared their software package to the manual approach using AMDIS followed by Xcalibur and concluded that their approach led to similar if not better results using an automation process that lasted 45 min compared to the manual approach which took 20 h. MetaboliteDetector was successfully applied to cell work analysed by GC/MS (Abu Dawud, Schreiber, Schomburg, & Adjaye, 2012). Finally, Li et al. (2012) proposed an elegantly designed pipeline that consists of running a pooled extract by full-scan GC/MS and using an algorithm written in Visual C++ that uses AMDIS to deconvolute peaks to automatically evaluate which peaks to include in a metabolite table based on them passing a number of criteria. Once the peaks are included in a table list, the individual samples are run using a GC/MS-SIM approach which improves the detection limit compared with the full-scan approach.

ChemStation (Agilent) is a software package regularly used to select peaks for quantification in metabolomics and a popular choice for data analysis of GC/MS peaks (Choe et al., 2012; Duran et al., 2003). Another software regularly used called Masslab (ThermoQuest) can perform both peak identification (using RT or RI) and quantification (Fiehn et al., 2000; Roessner, Wagner, Kopka, Trethewey, & Willmitzer, 2000; Roessner et al., 2001) sometimes in combination with the NIST library (Desbrosses, Kopka, & Udvardi, 2005) or the Golm database (Lombardo et al., 2011; Semel, Schauer, Roessner, Zamir, & Fernie, 2007). Xcalibur (Thermo Scientific) is a vendor's software similar to the ones previously cited that can automatically integrate multiple GC/MS peaks manually preselected for analysis. Researchers used this software after peak detection in AMDIS (Dobson et al., 2008, 2010; Shepherd et al., 2010). Xcalibur can be used with the NIST and Golm databases (Bowne et al., 2012). AnalyzerPro from SpectralWorks proposes a deconvolution option in its setting which has been used by several research groups (Lowe et al., 2008; Lu, Dunn, Shen, Kell, & Liang, 2008; O'Callaghan et al., 2012). ChromaTOF from LECO is the most popular vendor software with its user-friendly fully integrated automated deconvolution algorithm. Numerous laboratories used this software (Castillo, Mattila, Miettinen, Orešič, & Hyötyläinen, 2011) in conjunction with libraries such as NIST or Golm (Begley et al., 2009; Biais et al., 2009; Huege et al., 2007; Ma et al., 2010; Sanchez et al., 2008) or in-house libraries (Jozefczuk et al., 2010; Krall, Huege, Catchpole, Steinhauser, & Willmitzer, 2009; Lisec, Schauer, Kopka, Willmitzer, & Fernie, 2006). The software called metAlign (Vorst et al., 2005) is also used for dealing with GC/MS datasets. Carreno-Quintero et al. (2012) applied it onto GC/TOF-MS data profiling of potato tuber extracts using the NIST and Wiley libraries for identification and Rohloff et al. (2012) used MeAlign and TagFinder to study cold responses by applying untargeted GC/MS on extracts of leaves and roots of woodland strawberry.

Several articles refer to the comparison of AMDIS and ChromaTOF sometimes with other available softwares. Lu et al. (2008) described a comparative evaluation of AMDIS, ChromaTOF and AnalyzerPro for deconvolution of metabolomics data. As expected, AMDIS tended to produce numerous artefacts but so did ChromaTOF (false positives). On the other hand, AnalyzerPro omitted peaks (false negatives) thus highlighting the need for caution in treating the output data. A possible solution the authors pointed out was to amalgamate the results of the three softwares which would involve some careful manual inspection of each set of outputs (a recommended practice that should be applied to any data after preprocessing). Manually characterising the peaks or checking manually the identity tags automatically allocated after deconvolution and identification remains a valid process to handle the data analysis of untargeted GC/MS metabolomics data as it is often possible to end up with false peak annotations. The same authors published later another comparative paper this time focussing on the comparison between their new non-deconvolution method and MZmine (Lu, Gan, Zhang, & Liang, 2011). The authors showed using PCA that the scores of the sample groups were separated slightly better using their approach compared with MZmine, but on the whole, results appeared comparable since the loadings contained similar information. Lee, Choi, Cho, and Kim (2010) also compared ChromaTOF and AMDIS using a GC/MS dataset produced from plant extracts. Both softwares gave satisfactory results but slightly different metabolite lists. In this work, AMDIS coupled to SpectConnect appeared in this work to be more convenient to use. Kind, Tolstikov, Fiehn, and Weiss (2007) compared XCMS and MZmine applied on a set of urinary samples analysed by GC/MS. The authors concluded that once the parameters were optimised both softwares gave similar results. Koh,

Pasikanti, Yap, and Chan (2010) compared MetAlign, MZmine and TagFinder using a set of human urine samples recorded by a GC/MS approach. They reportedly found that MZmine performed the best in terms of alignment accuracy, but altogether the software took twice as long to process that data compared with MetAlign.

Some packages are specifically geared towards metabolite identification. Using both RT information and the MS features improves the reliability of the compounds' annotation. To maximise the peak matching, standards—typically either FAMEs or alkanes—are used so that the ions can be referenced by their RI, which are independent of the chromatographic conditions, rather than their RTs (Kind et al., 2009; Kopka, 2006; Luedemann et al., 2008). Thus it is now possible to automatically characterise GC/MS peaks using this approach and several groups propose similar approaches. Recently, several software packages offered access to the automatic identification of GC/MS peaks (TagFinder, TargetSearch and SetupX). TagFinder (Luedemann et al., 2008) includes a peak deconvolution package (based on ChromaTOF) and identification of peaks based on RIs from an alkane series. This pipeline has been cited almost 100 times since 2006 and is widely used (Calingacion et al., 2012; Moing et al., 2011; Scherling, Roscher, Giavalisco, Schulze, & Weckwerth, 2010; Steinbrenner, Gómez, Osorio, Fernie, & Orians, 2011). TargetSearch (Cuadros-Inostroza et al., 2009) uses a peak apex approach to selectively measure peak intensities and like TagFinder converts RT to RI in order to proceed to automatic peak identification but uses a FAMEs series to calculate the RIs instead of alkanes. Both TagFinder and TargetSearch are often utilised in combination with ChromaTOF for peak deconvolution (Hische et al., 2012; Szymanski et al., 2009).

SetupX (Scholz & Fiehn, 2007) is a data housing repository that integrates the metabolic annotation BinBase (Fiehn et al., 2005) and proposes to delineate the dimensions of a study and to help defining the semantic of analyses realised using a GC/MS metabolomics approach. The ideas behind such repository tools are to harmonise procedures for metabolomics studies and to have access to large datasets in order to further the identification of metabolites in GC/MS metabolomics datasets of multiple origins (plants, animals and microorganisms) (Fiehn et al., 2005). Both databases (BinBase and FiehnLib) are regularly used by the metabolomics community (Shin, Lee do, Liu, Fiehn, & Kim, 2010).

Earlier work proposed a fast-matching algorithm to align GC/FID peaks (Johnson, Wright, Jarman, & Synovec, 2003), whilst more recently, Clifford et al. (2009) published a novel variation of a warping method entitled variable penalty DTW which aims at aligning TIC chromatograms. The authors were focused on minimising the production of artefacts during the warping of a very large batch of samples (712 files). The latter method was applied onto GC/MS data from wine extracts (Berna, Trowell, Clifford, Cynkar, & Cozzolino, 2009). Moritz et al. proposed in 2004 a strategy for dealing with the data analysis of GC/MS data that does not involve deconvolution (Jonsson et al., 2004). The method written in Matlab bypasses the problem of peak registration by summing the peak intensity of small portion of the chromatogram in the time domain. Those two-dimensional structures (*m/z* vs. samples) are then further compressed using alternating regression and the data submitted to multivariate analysis. This paper has so far cited 168 times (December 2012) and is regularly used (Lu et al., 2012). In the following year, the authors published another method similar to the previous that makes use of a hierarchical multivariate curve resolution approach (Jonsson, Johansson, et al., 2005; Jonsson et al., 2006). This popular development (which has 161 citations, December 2012) has also been regularly cited (Jiye et al., 2005; Kimbara et al., 2012; Lu et al., 2012; Qiu et al., 2009; Zheng et al., 2011). Once the data are preprocessed, peaks can be selected in ChromaTOF and identified with the NIST, Golm or in-house libraries (Jiye et al., 2005; Zheng et al., 2011).

MSFACTS (Duran et al., 2003) has been released in 2003 and is probably the first freely available preprocessing software conceived to deal with large datasets such as GC/MS metabolomics data. The method is designed to deal with one-dimensional data typically lists of peaks characterised by intensity (sum of some *m/z* ions) and classified according to their RT. Usually, this peak list would come from a third-part software like ChemStation. MSFACTs has been cited by most of the deconvolution softwares published afterwards and was used mainly for dealing with GC/MS datasets (Hovell, Pereira, Arruda, & Rezende, 2010; Tarpley, Duran, Kebrom, & Sumner, 2005). Another programme from the same research group called MET-IDEA (Broeckling et al., 2006) was proposed in 2006 to deal with large batch automatic listing of ion intensity for peaks preselected by a preprocessing software such as AMDIS or Xcalibur. This software is regularly used by the metabolomics community for both GC/MS and LC/MS datasets (Booth et al., 2011; Wone, Donovan, & Hayes, 2011; Zhou et al., 2012).

Recently, additional deconvolution softwares appeared in the literature. MetaboliteDetector described earlier in the text is one software published in 2009 (Hiller et al., 2009) that proposes the full workflow needed to analyse metabolomics GC/MS data. Similar to this development, another pipeline was proposed in 2010 called Automated Data Analysis Pipeline (ADAP) which includes automated peak detection, deconvolution, alignment and library search (Jiang et al., 2010). The authors successfully used their software on a series of blood plasma samples (Su et al., 2012) and they have recently updated ADAP (version 2) and published the changes operated on the pipeline particularly improving the peak detection and addressing problems of peak co-elution (Ni et al., 2012). CompExtractor is another in-house program published in 2012 that uses the macro modules of Agilent ChemStation to measure extracted peaks (thus no deconvolution is used) in an automated manner (Choe et al., 2012). The authors compared their programme with AMDIS-SpectConnect and concluded that both approaches could successfully handle the data analysis of three of their experimental datasets. Similarly, Furbo and Christensen (2012) proposed an algorithm they have called automated peak extraction and quantification based on parallel factor analysis which extracts and quantifies peaks in GC/MS chromatograms. The procedure is fast but relies on having relatively simple profiles with co-eluting peaks that differ in their mass spectra. The method can handle the detection of peaks expected to be present in the chromatograms and also unknown peaks. Another processing toolkit called PyMS has been released in 2012 to perform large sample batch processing (that includes peak detection and deconvolution followed by dynamic programming peak alignment and quantitation) (O'Callaghan et al., 2012). This novel software package was compared to XCMS, AnalyzerPro and AMDIS. PyMS and AnalyzerPro achieved overall the most consistent results with manual analyses.

Finally, the MetabolomeExpress Project (Carroll, Badger, & Harvey Millar, 2010) is an online comprehensive workflow for data storage, data deconvolution, peak detection and integration, peak matching and access to an array of statistics to analyse the outputs (hierarchical clustering analysis, PCA, correlation analysis, etc.). The tools proposed were compared with established methods (use of ChemStation for peak detection, AMDIS combined with the Golm database) and considered as good as the existing tools if not better.

6. DATA PRETREATMENT

6.1. Normalisation

Normalising the intensity of peaks in ^{1}H NMR spectra is often a necessary step to be able to compare profiles across samples. Several normalisation procedures related to NMR metabolomics data have been published mostly to normalise peaks in urinary NMR spectra because this fluid is probably the material that suffers the most from peak registration. However, the methods developed on that type of data are applicable to other data sets.

The most classical normalisation method was inspired from work on urine in clinical chemistry (Ross et al., 2007). The idea is that creatinine clearance is considered constant; thus it is possible to correct for differences in metabolite concentration in urine using creatinine as a reference (ratioing all the signals with the signal of creatinine). However, normalising urinary ^{1}H NMR spectra to a house-keeping metabolite such as creatinine has to be taken with caution in the context of metabolomics as it may be that creatinine excretion is not constant (Craig, Cloarec, Holmes, Nicholson, & Lindon, 2006). Another similar approach is to ratio the NMR peaks with that of the reference added to the samples usually (TSP (sodium 3-(trimethylsilyl)-propionate-d_4) or DSS (sodium 2,2-dimethyl-2-silapentane-5-sulphonate)) (Weljie et al., 2006), but this is not appropriate all the time (this will not compensate dilution effects).

One commonly used method of normalisation is to set each variable in the NMR spectrum to a unit total intensity by expressing each data point as a fraction of the total spectral integral (Craig et al., 2006). This approach provides a good compromise between accounting for dilution and other sources of variation. The only downside of the total intensity approach is that if a/some large peak(s) occur(s) in the spectrum, the normalisation of the sample will be skewed and the sample will end up being an outlier on the PC plot (if applying PCA) (Craig et al., 2006; Ross et al., 2007). The total intensity approach was, for example, preferred to the normalisation using creatinine in the work by Schnackenberg et al. (2007). An alternative to total intensity is probabilistic quotient normalisation (PQN), a method published in 2006 with 156 citations (December 2012) (Dieterle, Ross, Schlotterbeck, & Senn, 2006). This popular variation in normalising is based on the calculation of a most probable dilution factor deduced from the distribution of quotients of the amplitudes of a test spectrum by those of a reference spectrum. Other researchers (Torgrip, Aberg, Alm, Schuppe-Koistinen, & Lindberg, 2008) have developed a normalisation algorithm based on the construction of a histogram of the spectral intensities (or bins) called histogram matching (HM) to estimate the dilution factor. They have compared it to total intensity and PQN normalisation approaches and concluded that all the three methods performed satisfactorily and that HM was more accurate. Zhang et al. (2009) also compared a set of urines by ^{1}H NMR applying no normalisation, creatinine and total intensity normalisations and a new normalisation called variance stabilising normalisation (VSN) initially developed for the analysis of gene expression microarrays. They also determined if log transformation on the data was improving the separation of the samples using statistical analysis. They have concluded that VSN and log transformation were beneficial to the data analysis. Finally, Kohl et al. (2012) compared multiple normalisation approaches on a set of urinary data (creatinine-normalised, PQN, quantile, cyclic locally weighted scatter plot smoothing (cyclic loess), cubic-spline normalisation, etc.) and concluded that for their dataset quantile and cubic-spline normalisation were the best performing methods.

Classically normalising LC/MS and GC/MS data consist in ratioing the peaks detected by MS with that from one reference standard except that this approach is not easily applicable to metabolomics data (for LC/MS) as all peaks do not vary in the same way. Sysi-Aho, Katajamaa, Yetukuri, and Oresic (2007) proposed an alternative to the use of a reference standard which consists in using a combination of standards and applying a modulated normalisation based on a linear combination of the standard responses. They named their optimised normalisation NOMIS. Veselkov et al. (2011) compared four different normalisation techniques (loess, quantile, median fold change and total intensity) on a set of urinary data and settle for the median fold change normalisation. Dunn and colleagues (Dunn, Broadhurst, Begley, et al., 2011) use the loess approach to smooth the discrepancies between 'blocks of recording' of LC/MS data over a long period.

Finally, Lehallier, Ratel, Hanafi, and Engel (2012) proposed a systematic ratio normalisation to improve both sample set discrimination and biomarkers identification on GC/MS data.

6.2. Mean-centring

Mean-centring is applied on a variable basis; for a given variable, the mean value over the whole data set is calculated and is then subtracted from each measurement. This is often an integral part of further analyses (like PCA) and simply shifts the reference used to express the data—it does not alter the relative scale of the measurements.

6.3. Autoscaling and pareto scaling

Unlike mean-centring, autoscaling (alternatively know as variance-scaling) and pareto scaling alter the relative scale of the different measurements (variables) and are thus more likely to have profound effects on the outcomes of analyses (van den Berg, Hoefsloot, Westerhuis, Smilde, & van der Werf, 2006). In autoscaling, each measurement is divided by the standard deviation of the variable over the dataset (so that for variance-scaled data each variable has a variance of 1), and in pareto scaling, each measurement is divided by the square root of the standard deviation (so that the scaled data are a bit closer to the original one).

6.4. Log transform

Some researchers have used log transformation as a mean to correct for heretodasticity (variables with sub-populations that have different variabilities from others) (van den Berg et al., 2006). Parsons, Ludwig, Günther, and Viant (2007) applied generalised logarithm (glog), autoscaling, pareto scaling and unscaling to three datasets (1-D NMR spectra) and concluded that sample classification accuracies were the highest with glog transformation for two datasets. We also applied log transformation on urinary NMR spectra to overcome a strong skewness in the distribution of NMR peaks (Rahmioglu et al., 2011).

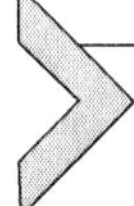

7. BASIC METHODS FOR STATISTICAL ANALYSIS: POSSIBLE APPROACHES

One of the characteristics of metabolomics data is that it is high-dimensional, that is to say there are usually many more variables than there are observations ($n \ll d$). In addition, there is likely to be redundancy, and the data are intrinsically noisy. There are essentially two routes that can be chosen in terms of analysis, being univariate and multivariate approaches.

7.1. Univariate

With univariate methods, each variable is analysed separately, with, for example, traditional statistical techniques such as ANOVA or *t*-tests. Although applicable in the same fashion as with any other type of data, there are a few issues which are related to the nature of the experiments: (1) the data may not be normally distributed; (2) in tests such as *t*-test, which are frequently used to establish whether there are significant differences between groups, there are consequences to carrying out *t*-tests on many variables, as opposed to on just one *t*-test, on the expected false positive rate and (3) there can be ‘registration’ issues. We will examine each of these issues in more detail below.

The first consideration is that many statistical tests assume a normal distribution of the data, which is not always the case (Ebbels, Holmes, Lindon, & Nicholdon, 2004). There are several options to deal with this, the simplest being to use the parametric tests but apply a transformation to the data prior to the test, such as a log or more generally a power transformation (Box & Cox, 1964). Another alternative is to use rank-based methods such as Mann–Whitney instead, as these do not require the data to be normally distributed.

The second consideration is that of the effect of carrying out multiple tests on the chance of making a type I error (i.e. in a *t*-test to wrongly reject the null hypothesis 'there is no difference between groups'). In a single test setting, the critical *p*-value to 0.05 means that there is a 5% chance of doing so. When carrying out multiple tests, there is the same chance of making a type I error for each test individually, but this means the overall likelihood of wrongly rejecting the null hypothesis ('no difference') is much higher. One possible way to ensure this does not happen is to lower the critical *p*-value for each test with the aim of obtaining the same overall chance of type I error, with a popular choice being the Bonferroni adjustment (Bonferroni-corrected *p*-value of 0.05/nb_met). Given the number of variables that are typically being handled, this can greatly increase the stringency of the test, with the danger that significant differences are not recognised. Also, this correction assumes that each test is independent, whereas correlations between variables are to be expected in metabolomics—a further argument for this test being too stringent. Another approach, which is getting more popular, is to use the false discovery rate (Benjamini & Hochberg, 1995). The purpose here is to state what proportion of tests deemed to be statistically significant are actually false positive (as opposed to false negative) and set the significance level of the test so that this proportion is acceptable (Storey, 2002).

The last consideration is that of registration issues. In the section on multivariate methods, we will present this potential problem, as it can have quite subtle effects on the analysis. For univariate methods, it does not constitute a problem in terms of statistics, but it would lead to incorrect interpretation if left unnoticed.

7.2. Multivariate

The purpose of multivariate techniques is to consider all variables simultaneously (rather than separately), with methods available to serve different possible aims: data exploration, classification, relationship with other (univariate) measurements or (multivariate) datasets. We will present one of the most widely used techniques, namely, PCA; this will allow us to introduce the basic principle of many of these methods, as well as explain in which way the experimental aspects may affect such analyses.

PCA is a technique that reduces the dimensionality of the data whilst retaining most of the information. It does this by expressing the measurements in terms of the samples' coordinates on axes (principal components) that are linear combinations of the original variables, with the criterion for selecting the axes being that the variance of these coordinates is maximised on successive axes, and the constraints that the axes are orthogonal to one another.

The principle is explained graphically in Fig. 11.6. For this, let us assume that we only have measurements on three variables; in this case, we can visualise the measurements from all samples on three axes which correspond to these variables (we have also mean-centred the data for each variable). We can see that these axes may not be optimal for examining the variability in this dataset. Thus we define a new axis (the first PC axis), such that the projection of the samples on this axis (the coordinates, also called PC scores) has a maximum variance. Having done this, we can define a second axis which is orthogonal to the first and explains as much of the remaining variance as possible, and then a third axis with the same properties again. Effectively, we have defined the axes by rotation of the original set. In addition to the scores, the other main output is the loadings; we can see that each axis is in fact a linear combination of the original ones, with, for instance, variable 1 having a much higher weight in defining PC1, than variables 2 and 3. These weights are called loadings and are the link between the original measurements and the PC axes.

In reality, we have many more than three variables to start with and, whilst it cannot be demonstrated graphically, the same operation can be done mathematically. This is summarised by way of the equations shown in Fig. 11.7.

The input variables to this analysis depend on the experimental technique: For spectral data such as infrared (a signal as a function of wavelength), the digitised spectral trace would be used, whilst for NMR, this is an option, as is the use of data already reduced by, for example, binning or peak integration. For GC– and LC–MS, the data are somehow integrated (see Sections 4 and 5) prior to analysis. Outputs such as the loadings may be presented differently according to what the input variables were, as they correspond to one another; for spectral data, it makes sense that the link to the

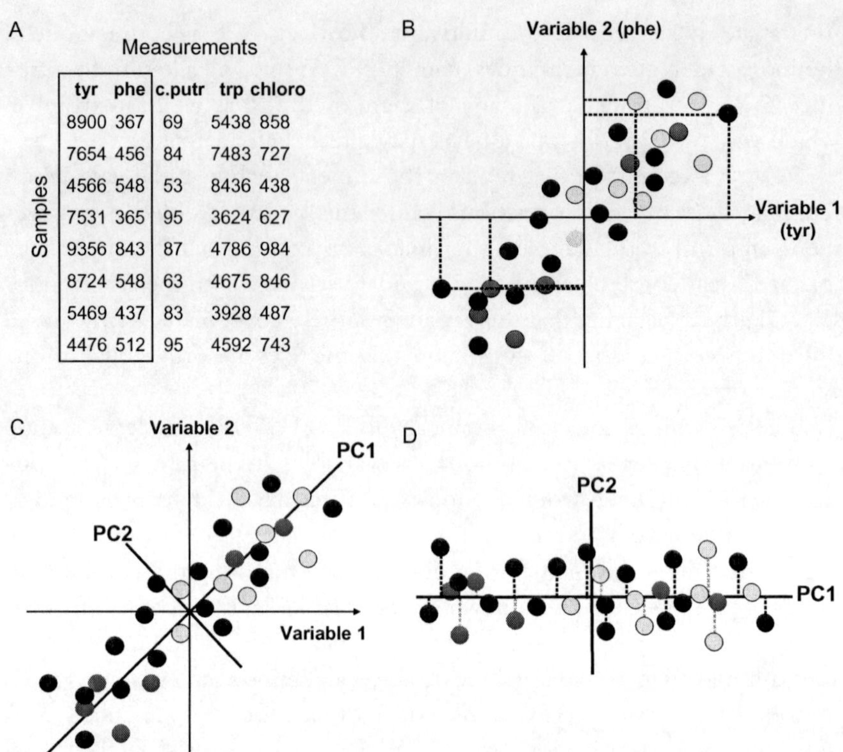

Figure 11.6 Graphical representation of PCA. The measurements (A) are shown as a set of points on axes that correspond to two of the variables (B). A principal component is then defined (C) such that the variance of the projection of the datapoints on this axis (i.e. the coordinates) is maximised, following which a second PC is defined which corresponds to the maximum remaining variance and is orthogonal to the first. The new axes are used to express the measurements' coordinates (D); these are referred to as PC scores.

spectral dimension is preserved and so, the loadings may be shown for a given PC, on an axis corresponding to the spectra (frequency, chemical shift) (Fig. 11.8). When there is no physical meaning to the order of the variables, the loadings for two PCs are often shown against one another in scatterplots (such as with LC–MS data, see, e.g. Fig. 11.5).

8. REPRODUCIBILITY AND EXPERIMENTAL ISSUES

8.1. General remarks

NMR is considered to be fairly reproducible, in that signal from given chemical moieties are expected to appear at the same chemical shift under

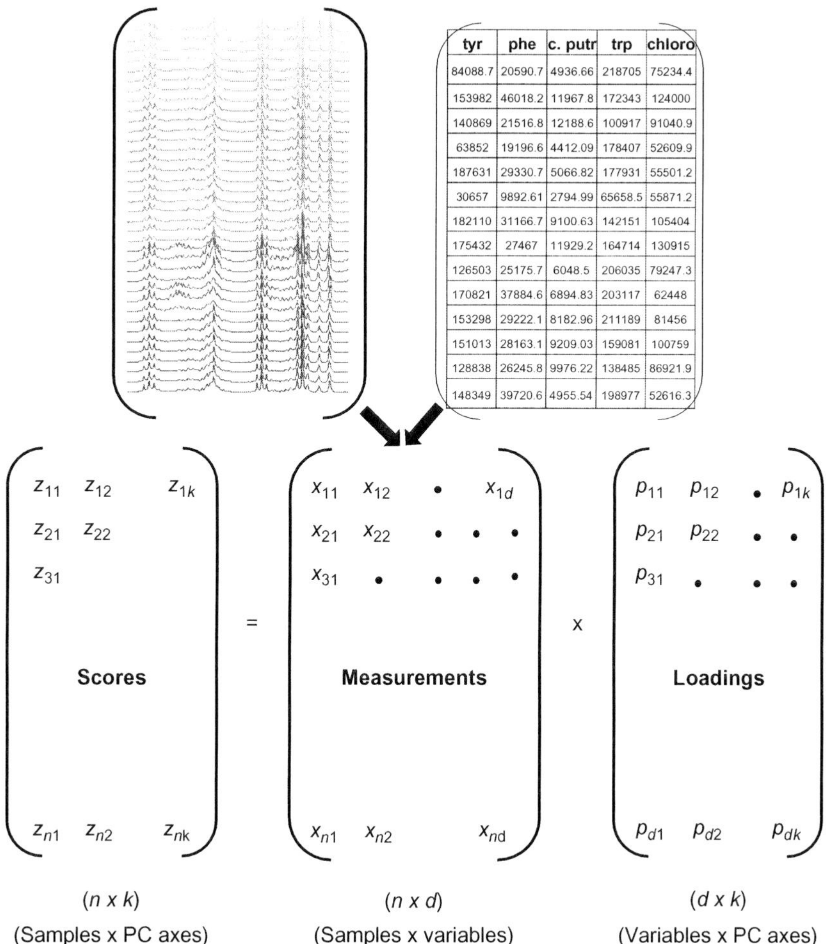

tyr	phe	c. putr	trp	chloro
84088.7	20590.7	4936.66	218705	75234.4
153982	46018.2	11967.8	172343	124000
140869	21516.8	12188.6	100917	91040.9
63852	19196.6	4412.09	178407	52609.9
187631	29330.7	5066.82	177931	55501.2
30657	9892.61	2794.99	65658.5	55871.2
182110	31166.7	9100.63	142151	105404
175432	27467	11929.2	164714	130915
126503	25175.7	6048.5	206035	79247.3
170821	37884.6	6894.83	203117	62448
153298	29222.1	8182.96	211189	81456
151013	28163.1	9209.03	159081	100759
128838	26245.8	9976.22	138485	86921.9
148349	39720.6	4955.54	198977	52616.3

Figure 11.7 Mathematical equations involved in PCA. The input data (measurements) can be, for example, the digitised NMR traces, or integrated values (features in LC- or GC–MS). (For colour version of this figure, the reader is referred to the online version of this chapter.)

a given chemical environment, and with the same structure (e.g. doublets and triplets). On a given instrument then at least and with a standardised protocol, it is anticipated that series of spectra will be consistent. With GC– and LC–MS, it is recognised that there is variability in both the chromatography (RTs, particularly for LC) and the spectrometry (fragmentation). This is why the data integration involves algorithms that take these parameters into account over the course of the experiment (see Sections 4 and 5). In the following section, we assume the data

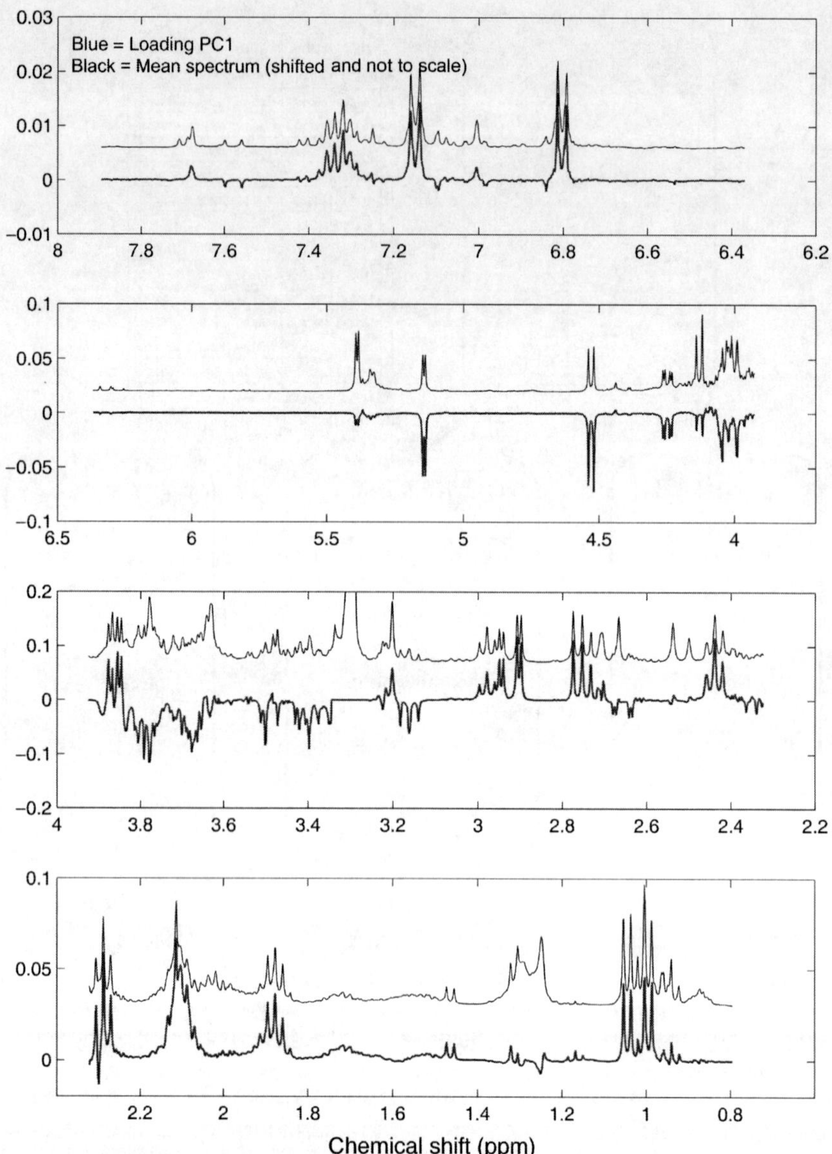

Figure 11.8 Loading representation in cases where the variable order has a physical meaning. (For colour version of this figure, the reader is referred to the online version of this chapter.)

integration was carried out carefully and present issues that may arise despite this, downstream in the workflow.

8.2. NMR

With NMR, the most important phenomenon that has been recognised as having a negative influence on multivariate analyses is that of peak registration. In the most severe cases, the effect on, for example, PCA is quite obvious, as was the case in one of our pieces of work on a tea dataset (Defernez & Colquhoun, 2003). Here, the distribution of the scores was clearly odd, and the loadings typical of situations where the position of some of the peaks is not fixed in the dataset but rather is distributed around a mean position. The first loading is derivative-like, with the zero-crossing point at the mean peak position, whilst the second loading appears at the mean position with dips at either side. Other, more subtle effects can also occur; to demonstrate this, examples are shown which use a reproducibility study on potato extracts that consisted in carrying out three repeat extractions and four NMR runs on six samples. In the first case, PCA shows that the repeats of each sample do cluster, but there is also clearly a spread of the scores on the second PC (Fig. 11.9A). When looking at spectra that lie at the extremities of this spread, we can see that there are differences in bandshape, with the base of the peak differing between the two (Fig. 11.9B). In a second, similar experiment (with three NMR runs), disparities between the scores of repeats only appear on the fourth PC (Fig. 11.10) and are again linked to subtle inconsistencies in bandshape.

It appears then that both peak position and bandshape affect the multivariate analyses, and we now examine the points in the workflow which can be influenced to minimise this. The first one is at the experimental level; the peak position being affected by factors such as the pH, it will be beneficial to control it as tightly as possible. The second level at which users can intervene is post-acquisition, with the first aspect being the mitigation of inconsistencies in bandshape by careful data pretreatment. In addition, mathematical treatment can be attempted to realign peaks in cases where it is not possible to control their position at the experimental level, such as in the case of chelation. An example is shown in Le Gall et al. (2004) of the result of applying an in-house written macro (based on Vogels et al., 1996) to the tea dataset mentioned above.

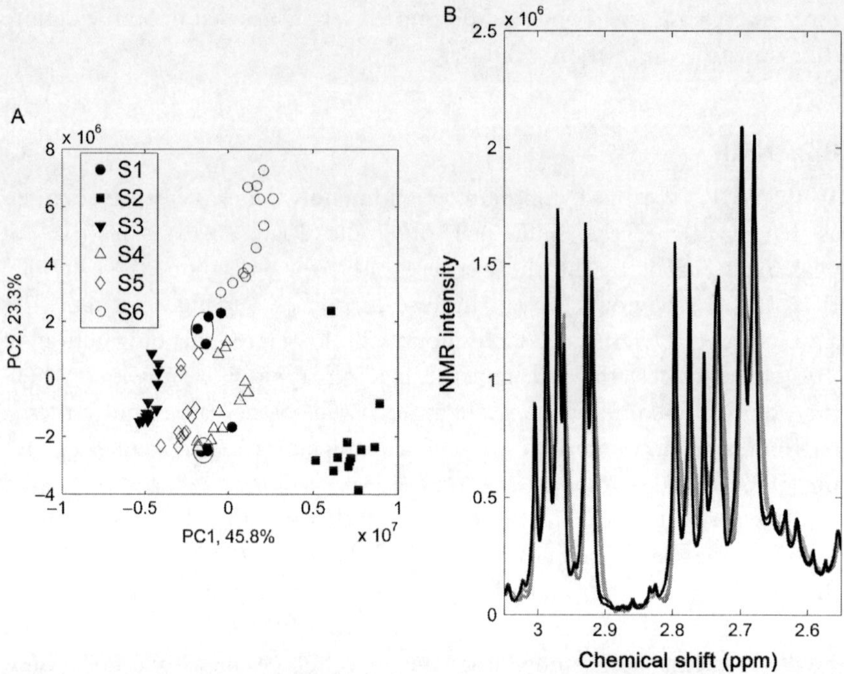

Figure 11.9 (A) PCA of a set of six potato samples extracted four times and ran by NMR three times, showing a discrepancy between observations on PC2, and (B) a portion of the spectra from the four spectra highlighted in (A) showing lineshape differences responsible for these discrepancies. (For colour version of this figure, the reader is referred to the online version of this chapter.)

We can see that the procedure is quite successful in correcting the peak positions, and this correction was crucial to be able to analyse this dataset. The correction is not always perfect though, and we can see an example of this when peaks overlap which do not shift in the same way (Fig. 11.11). Aligning all peaks in such cases would necessitate the use of deconvolution. Also, our experience suggests that the success of the alignment may depend on the sample type, with, for example, samples such as tomato being more difficult to align than potato; this is likely linked to the pH. If the outcome of using such tools is not satisfactory, the most widely used alternative is to deresolve the data by binning (or bucketing, see Section 3.2), with the option to use larger-sized bins for peaks that shift more, such as citric acid. Note that binning is also used as standard procedure even if no alignment problem is specifically identified.

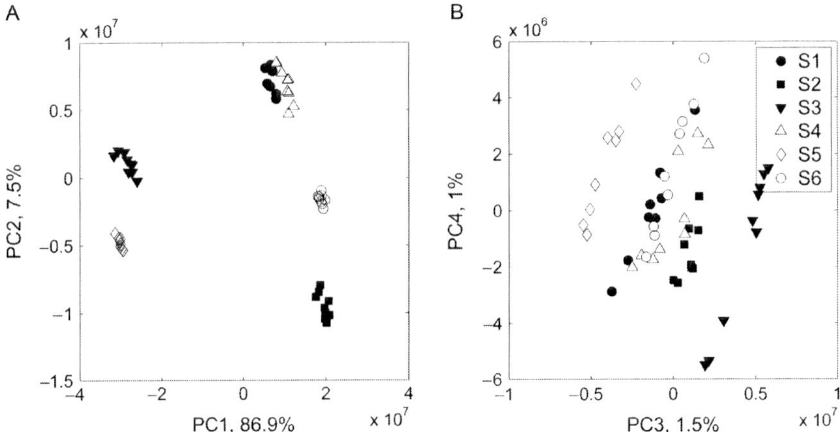

Figure 11.10 PCA scores (A) PC1 VS 2, (B) PC3 VS 4, for PCA of a set of six potato samples extracted three times and ran by NMR three times, showing a discrepancy between observations but only on PC4 (1% variance).

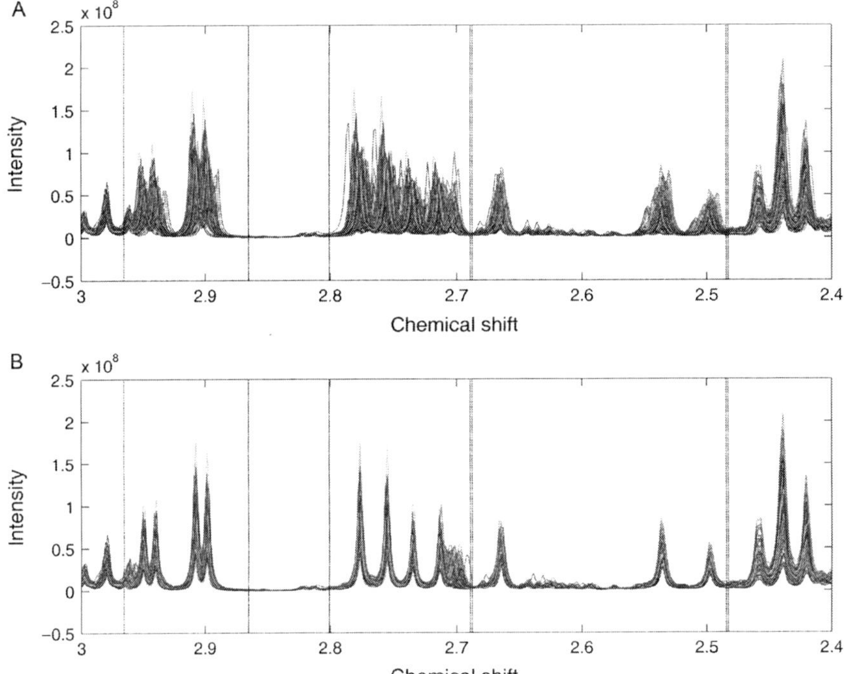

Figure 11.11 Set of potato NMR spectra (A) before alignment and (B) after alignment carried out separately for each of segment of the spectrum (marked by vertical lines). The alignment is based on PLF. (For colour version of this figure, the reader is referred to the online version of this chapter.)

8.3. GC- and LC-MS

With chromatography, registration issues are usually much more severe than with NMR and are often accompanied by additional problems such as the peak shape (particularly trailing peaks, or compounds whose order of elution may be swapped). As it is clear that even relatively small imperfections in the data can affect the analyses (see Section 8.2), attempts at analysing chromatograms as profiles are usually only successful for small datasets of limited complexity. These difficulties generally mean that the data are not used as profile but are instead integrated using the MS as a second dimension. Although the issues still have to be dealt with, it is deemed much more achievable to tackle them at this stage.

9. SUPERVISED METHODS FOR CLASSIFICATION PROBLEMS: POTENTIAL FOR STATISTICAL BIAS

9.1. Introduction to supervised methods

The purpose of many of the pieces of work in the metabolomics field is to highlight differences between samples belonging to different groups (for instance, types of mutants and variety). 'Unsupervised' techniques such as PCA, meaning those that highlight structure in the data based on the matrix of measurements alone, are a good starting point because they are not biased. However, the structure that they show may not be directly related to the groups of interest, as the biggest sources of variance may not be related to them. Thus it can also be useful to apply 'supervised' methods, which make use of the group membership information in the modelling step. For instance, there are a host of techniques which, like PCA, perform a rotation of axes, but where the criterion for choosing the axes is different and utilises the group membership information. PLS discriminant analysis is such a technique (the criterion is to maximise the covariance between the matrix of measurements, and that encoding group membership), as is canonical variates analysis (the criterion is to maximise the ratio of between-group variance over that of within-group variance). These techniques are undoubtedly very useful in that they can enable us to highlight characteristics that are unclear or obscured with unsupervised analyses. It is, however, very important to realise that they are prone to producing results that are over-optimistic because they are biased towards selecting features that are characteristic of the group membership for the dataset, whether this is information or noise. This is now well recognised and striking examples are shown in Defernez and Kemsley (1997).

9.2. Avoiding overfitting by using validation

The reason for overfitting is that the group information is used in the modelling step, which can lead to situations where we can find a model to describe the data structure very well, but this is specific to that particular data. The best way to avoid such situation is to ensure that were the models applied to new data, they would generalise to it. This is the purpose of model validation. In its simplest form (Fig. 11.12), it can be carried out by dividing the dataset into two subsets, a training (or modelling) and a test (or validation) set, preferably with similar proportions of samples from the different groups in both subsets. The model is then built using the training set and applied to the test set. It is the results from the test set classification which inform us on the model's ability to classify the samples.

There are possible variants to this scheme, most notably when the dataset size is such that subdividing it leads to unacceptably small sets. In such cases, leave-one-out (also called jack-knife) procedure is a suitable alternative; one sample is taken out as test set and classified using a model

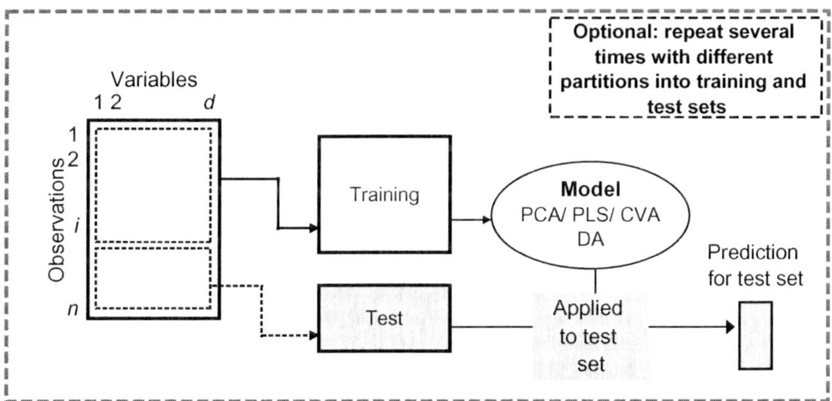

Figure 11.12 Schematic of the validation procedure. Both training and test sets should comprise observations from the different groups. The model is developed using the training set and is applied to a totally independent test set; the prediction for the test set gives the real measure of the usefulness of the model. An optional addition is to repeat this procedure and average the test set results.

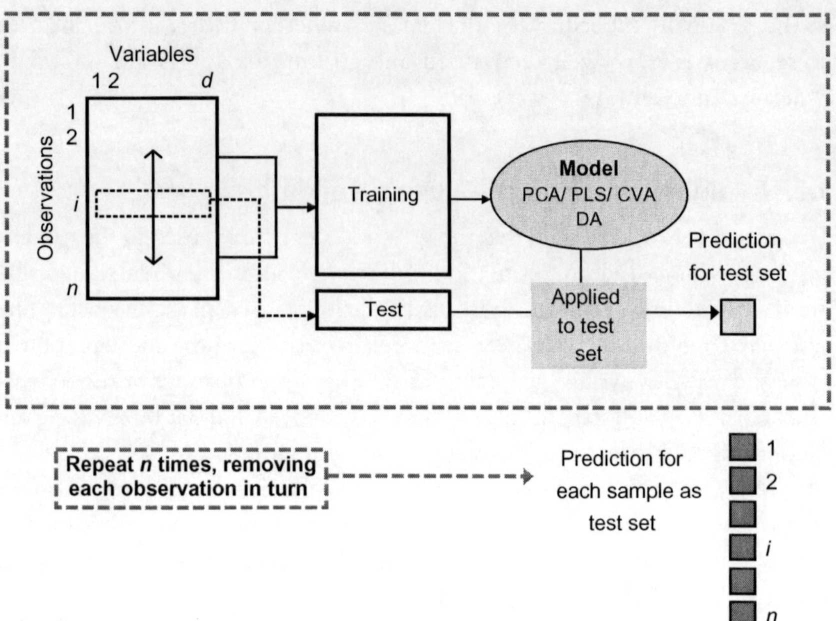

Figure 11.13 Schematic of the leave-one-out validation procedure. In this and the leave-segment-out procedure, each observation serves once as a test.

built on all other samples (Fig. 11.13), and this is repeated, in turn, until all samples have been taken out. The model's performance is then estimated by amalgamating the outcomes from all test sets. A similar procedure can be, if the dataset is somewhat larger, to take a few samples out at a time, still taking all samples out once at some point during the procedure. An alternative, again when the dataset is somewhat larger, is to simply carry out a validation by splitting the dataset into training and validation set, but repeating this procedure a number of times and average the results (so the difference here is that some samples may be tested several times, and some not) (Fig. 11.12).

Not as widely used, but worth considering is the use of a third dataset called tuning set. The idea here is to build the model with a training set, apply it to a tuning set, make decisions (such as the number of axes to retain) based on this set and then carry out a completely independent evaluation of the power of the model by applying it to a separate test set.

There are a few subtleties that are worth noting to diminish further the risk of overfitting, in particular, in cases where there are technical

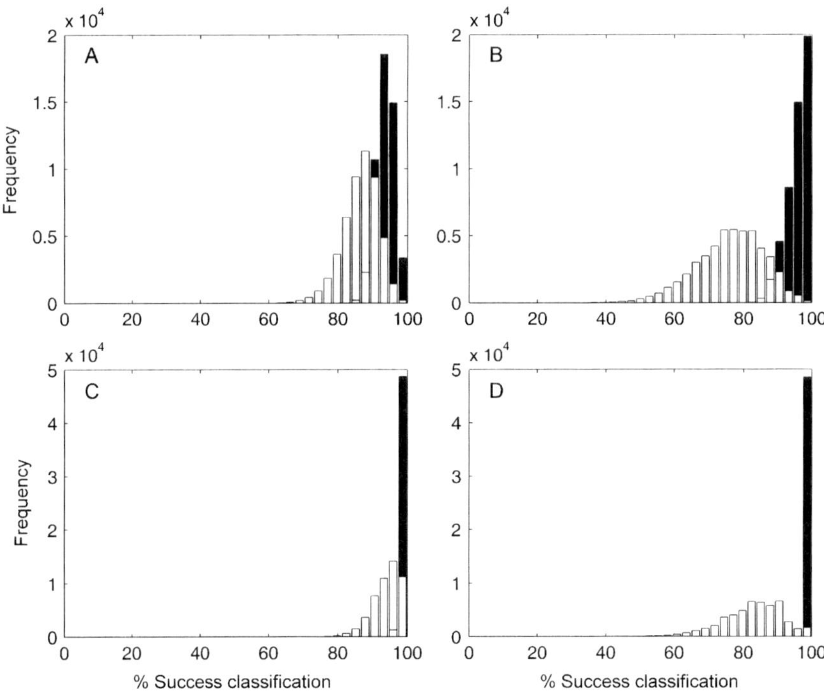

Figure 11.14 Percentage success for classification using PCA and discriminant analysis (A–B using 13 PCs, and C–D using 20 PCs in the DA), based on 50,000 repeats of the training/test set partitions with 38 out of 57 (group 1) and 36 out of 54 (group 2) spectra in the training set. (A) and (C) show the results when the spectra are completely randomly attributed to training and test sets, and (B) and (D) when they are randomly attributed but with triplicate measurements kept together in either training or test set.

replicates in the dataset. Figure 11.14 shows that if the technical replicates are split between training and test sets, the results will again be overfit. Instead, all technical replicates should be kept together either in the training or in the test set (so that if a leave-one out procedure is used, it means here a leave-one sample out, i.e. all replicates at once). This will ensure that the discrimination is based on biological replicates alone.

REFERENCES

Abu Dawud, R., Schreiber, K., Schomburg, D., & Adjaye, J. (2012). Human embryonic stem cells and embryonal carcinoma cells have overlapping and distinct metabolic signatures. *PLoS One*, *7*, e39896.

Aggio, R., Villas-Bôas, S. G., & Ruggiero, K. (2011). Metab: An R package for high-throughput analysis of metabolomics data generated by GC-MS. *Bioinformatics*, *27*, 2316–2318.

Allwood, J. W., & Goodacre, R. (2010). An introduction to liquid chromatography-mass spectrometry instrumentation applied in plant metabolomic analyses. *Phytochemical Analysis, 21*, 33–47.

Antonov, A. V., Dietmann, S., Wong, P., & Mewes, H. W. (2009). TICL—A webtool for network-based interpretation of compound lists inferred by high-throughput metabolomics. *FEBS Journal, 276*, 2084–2094.

Arbona, V., Iglesias, D. J., Talón, M., & Gómez-Cadenas, A. (2009). Plant phenotype demarcation using nontargeted LC–MS and GC–MS metabolite profiling. *Journal of Agricultural and Food Chemistry, 57*, 7338–7347.

Bajad, S. U., Lu, W., Kimball, E. H., Yuan, J., Peterson, C., & Rabinowitz, J. D. (2006). Separation and quantitation of water soluble cellular metabolites by hydrophilic interaction chromatography–tandem mass spectrometry. *Journal of Chromatography A, 1125*, 76–88.

Baran, R., Kochi, H., Saito, N., Suematsu, M., Soga, T., Nishioka, T., et al. (2006). MathDAMP: A package for differential analysis of metabolite profiles. *BMC Bioinformatics, 13*, 530–538.

Barupal, D. K., Kind, T., Kothari, S. L., Lee, D. Y., & Fiehn, O. (2010). Hydrocarbon phenotyping of algal species using pyrolysis-gas chromatography mass spectrometry. *BMC Biotechnology, 10*, 40–48.

Begley, P., Francis-McIntyre, S., Dunn, W. B., Broadhurst, D. I., Halsall, A., Tseng, A., et al. (2009). Development and performance of a gas chromatography-time-of-flight mass spectrometry analysis for large-scale nontargeted metabolomic studies of human serum. *Analytical Chemistry, 81*, 7038–7046.

Behrends, V., Tredwell, G. D., & Bundy, J. G. (2011). A software complement of AMDIS for processing GC–MS metabolomics data. *Analytical Biochemistry, 415*, 206–208.

Beneduci, A., Chidichimo, G., Dardo, G., & Pontoni, G. (2011). Highly routinely reproducible alignment of ^{1}H NMR spectral peaks of metabolites in huge sets of urines. *Analytica Chimica Acta, 685*, 186–195.

Benjamini, Y., & Hochberg, Y. (1995). Controlling the false discovery rate: A practical and powerful approach to multiple testing. *Journal of the Royal Statistical Society, Series B, 57*(1), 289–300.

Bennett, R. N., Mellon, F. A., & Kroon, P. A. (2004). Screening crucifer seeds as sources of specific intact glucosinolates using ion-pair high-performance liquid chromatography negative ion electrospray mass spectrometry. *Journal of Agricultural and Food Chemistry, 52*, 428–438.

Bennett, B. D., Yuan, J., Kimball, E. H., & Rabinowitz, J. D. (2008). Absolute quantitation of intracellular metabolite concentrations by an isotope ratio-based approach. *Nature Protocols, 3*, 1299–1311.

Berna, A. Z., Trowell, S., Clifford, D., Cynkar, W., & Cozzolino, D. (2009). Geographical origin of Sauvignon Blanc wines predicted by mass spectrometry and metal oxide based electronic nose. *Analytica Chimica Acta, 648*, 146–152.

Biais, B., Allwood, J. W., Deborde, C., Xu, Y., Maucourt, M., Beauvoit, B., et al. (2009). ^{1}H NMR, GC-EI-TOFMS, and data set correlation for fruit metabolomics: Application to spatial metabolite analysis in melon. *Analytical Chemistry, 81*, 2884–2894.

Birkemeyer, C., Kolasa, A., & Kopka, J. (2003). Comprehensive chemical derivatization for gas chromatography–mass spectrometry-based multi-targeted profiling of the major phytohormones. *Journal of Chromatography A, 993*, 89–102.

Boer, V. M., Crutchfield, C. A., Bradley, P. H., Botstein, D., & Rabinowitz, J. D. (2010). Growth-limiting intracellular metabolites in yeast growing under diverse nutrient limitations. *Molecular Biology of the Cell, 21*, 198–211.

Booth, S. C., Workentine, M. L., Wen, J., Shaykhutdinov, R., Vogel, H. J., Ceri, H., et al. (2011). Differences in metabolism between the biofilm and planktonic response to metal stress. *Journal of Proteome Research, 10*, 3190–3199.

Bowne, J. B., Erwin, T. A., Juttner, J., Schnurbusch, T., Langridge, P., Bacic, A., et al. (2012). Drought responses of leaf tissues from wheat cultivars of differing drought tolerance at the metabolite level. *Molecular Plant, 5*, 418–429.

Box, G. E. P., & Cox, D. R. (1964). An analysis of transformations. *Journal of the Royal Statistical Society, Series B, 26*(2), 211–252.

Boyard-Kieken, F., Dervilly-Pinel, G., Garcia, P., Paris, A. C., Popot, M. A., Le Bizec, B., et al. (2011). Comparison of different liquid chromatography stationary phases in LC–HRMS metabolomics for the detection of recombinant growth hormone doping control. *Journal of Separation Science, 34*, 3493–3501.

Broadhurst, D. I., & Kell, D. B. (2006). Statistical strategies for avoiding false discoveries in metabolomics and related experiments. *Metabolomics, 2*, 171–196.

Broeckling, C. D., Reddy, I. R., Duran, A. L., Zhao, X., & Sumner, L. W. (2006). MET-IDEA: Data extraction tool for mass spectrometry-based metabolomics. *Analytical Chemistry, 78*, 4334–4341.

Brown, M., Wedge, D. C., Goodacre, R., Kell, D. B., Baker, P. N., Kenny, L. C., et al. (2011). Automated workflows for accurate mass-based putative metabolite identification in LC/MS-derived metabolomic datasets. *Bioinformatics, 27*, 1108–1112.

Bruce, S. J., Jonsson, P., Antti, H., Cloarec, O., Trygg, J., Marklund, S. L., et al. (2008). Evaluation of a protocol for metabolic profiling studies on human blood plasma by combined ultra-performance liquid chromatography/mass spectrometry: From extraction to data analysis. *Analytical Biochemistry, 372*, 237–249.

Bruce, S. J., Tavazzi, I., Parisod, V., Rezzi, S., Kochhar, S., & Guy, P. A. (2009). Investigation of human blood plasma sample preparation for performing metabolomics using ultrahigh performance liquid chromatography/mass spectrometry. *Analytical Chemistry, 81*, 3285–3296.

Bueschl, C., Kluger, B., Berthiller, F., Lirk, G., Winkler, S., Krska, R., et al. (2012). MetExtract: A new software tool for the automated comprehensive extraction of metabolite-derived LC/MS signals in metabolomics research. *Bioinformatics, 28*, 736–738.

Bunk, B., Kucklick, M., Jonas, R., Münch, R., Schobert, M., Jahn, D., et al. (2006). MetaQuant: A tool for the automatic quantification of GC/MS-based metabolome data. *Bioinformatics, 23*, 2962–2965.

Calingacion, M. N., Boualaphanh, C., Daygon, V. D., Anacleto, R., Sackville Hamilton, R., et al. (2012). A genomics and multi-platform metabolomics approach to identify new traits of rice quality in traditional and improved varieties. *Metabolomics, 8*, 771–783.

Carreno-Quintero, N., Acharjee, A., Maliepaard, C., Bachem, C. W., Mumm, R., Bouwmeester, H., et al. (2012). Untargeted metabolic quantitative trait loci analyses reveal a relationship between primary metabolism and potato tuber quality. *Plant Physiology, 158*, 1306–1318.

Carroll, A. J., Badger, M. R., & Harvey Millar, A. (2010). The MetabolomeExpress Project: Enabling web-based processing, analysis and transparent dissemination of GC/MS metabolomics datasets. *BMC Bioinformatics, 11*, 376–388.

Castillo, S., Gopalacharyulu, P., Yetukuri, L., & Oresic, M. (2011). Algorithms and tools for the preprocessing of LC–MS metabolomics data. *Chemometrics and Intelligent Laboratory Systems, 108*, 23–32.

Castillo, S., Mattila, I., Miettinen, J., Orešič, M., & Hyötyläinen, T. (2011). Data analysis tool for comprehensive two-dimensional gas chromatography/time-of-flight mass spectrometry. *Analytical Chemistry, 83*, 3058–3067.

Chadeau-Hyam, M., Ebbels, T. M., Brown, I. J., Chan, Q., Stamler, J., Huang, C. C., et al. (2010). Metabolic profiling and the metabolome-wide association study: Significance level for biomarker identification. *Journal of Proteome Research, 9*, 4620–4627.

Choe, S., Woo, S. H., Kim, D. W., Park, Y., Choi, H., Hwang, B. Y., et al. (2012). Development of a target component extraction method from GC–MS data with an in-house program for metabolite profiling. *Analytical Biochemistry, 426*, 94–102.

Christin, C., Smilde, A. K., Hoefsloot, H. C., Suits, F., Bischoff, R., & Horvatovich, P. L. (2008). Optimized time alignment algorithm for LC–MS data: Correlation optimized warping using component detection algorithm-selected mass chromatograms. *Analytical Chemistry, 80*, 7012–7021.

Clifford, D., Stone, G., Montoliu, I., Rezzi, S., Martin, F. P., Guy, P., et al. (2009). Alignment using variable penalty dynamic time warping. *Analytical Chemistry, 81*, 1000–1007.

Cottret, L., Wildridge, D., Vinson, F., Barrett, M. P., Charles, H., Sagot, M. F., et al. (2010). MetExplore: A web server to link metabolomics experiments and genome-scale metabolic networks. *Nucleic Acids Research, 38*, W132–W137.

Craig, A., Cloarec, O., Holmes, E., Nicholson, J. K., & Lindon, J. C. (2006). Scaling and normalization effects in NMR spectroscopic metabonomic data sets. *Analytical Chemistry, 78*, 2262–2267.

Creek, D. J., Jankevics, A., Breitling, R., Watson, D. G., Barrett, M. P., & Burgess, K. E. (2011). Toward global metabolomics analysis with hydrophilic interaction liquid chromatography–mass spectrometry: Improved metabolite identification by retention time prediction. *Analytical Chemistry, 83*, 8703–8710.

Creek, D. J., Jankevics, A., Burgess, K. E., Breitling, R., & Barrett, M. P. (2012). IDEOM: An Excel interface for analysis of LC–MS-based metabolomics data. *Bioinformatics, 28*, 1048–1049.

Cuadros-Inostroza, A., Caldana, C., Redestig, H., Kusano, M., Lisec, J., Peña-Cortés, H., et al. (2009). TargetSearch—A Bioconductor package for the efficient preprocessing of GC–MS metabolite profiling data. *BMC Bioinformatics, 10*, 428–439.

Defernez, M., & Colquhoun, I. J. (2003). Factors affecting the robustness of metabolite fingerprinting using ^{1}H NMR spectra. *Phytochemistry, 62*, 1009–1017.

Defernez, M., & Kemsley, E. K. (1997). The use and misuse of chemometrics for treating classification problems. *Trends in Analytical Chemistry, 16*, 216–221.

Denery, J. R., Nunes, A. A., & Dickerson, T. J. (2010). Characterization of differences between blood sample matrices in untargeted metabolomics. *Analytical Chemistry, 83*, 1040–1047.

Desbrosses, G. G., Kopka, J., & Udvardi, M. K. (2005). *Lotus japonicus* metabolic profiling. Development of gas chromatography–mass spectrometry resources for the study of plant-microbe interactions. *Plant Physiology, 137*, 1302–1318.

Dieterle, F., Ross, A., Schlotterbeck, G., & Senn, H. (2006). Probabilistic quotient normalization as a robust method to account for dilution of complex biological mixtures. Application in ^{1}H NMR metabonomics. *Analytical Chemistry, 78*, 4281–4290.

Dobson, G., Shepherd, T., Verrall, S. R., Conner, S., McNicol, J. W., Ramsay, G., et al. (2008). Phytochemical diversity in tubes of potato cultivars and landraces using a GC–MS metabolomics approach. *Journal of Agricultural and Food Chemistry, 56*, 10280–10291.

Dobson, G., Shepherd, T., Verrall, S. R., Griffiths, W. D., Ramsay, G., McNicol, J. W., et al. (2010). A metabolomics study of cultivated potato (Solanum tuberosum) groups Andigena, Phureja, Stenotomum, and Tuberosum using gas chromatography–mass spectrometry. *Journal of Agricultural and Food Chemistry, 58*, 1214–1223.

Draper, J., Enot, D. P., Parker, D., Beckmann, M., Snowdon, S., Lin, W., et al. (2009). Metabolite signal identification in accurate mass metabolomics data with MZedDB, an interactive *m/z* annotation tool utilising predicted ionisation behaviour 'rules'. *BMC Bioinformatics, 10*, 227–242.

Dunn, W. B., Broadhurst, D. I., Atherton, H. J., Goodacre, R., & Griffin, J. L. (2011). Systems level studies of mammalian metabolomes: The roles of mass spectrometry and nuclear magnetic resonance spectroscopy. *Chemical Society Reviews*, *40*, 387–426.

Dunn, W. B., Broadhurst, D., Begley, P., Zelena, E., Francis-McIntyre, S., Anderson, N., et al. (2011). Procedures for large-scale metabolic profiling of serum and plasma using gas chromatography and liquid chromatography coupled to mass spectrometry. *Nature Protocols*, *6*, 1060–1083.

Dunn, W. B., Broadhurst, D., Brown, M., Baker, P. N., Redman, C. W., Kenny, L. C., et al. (2008). Metabolic profiling of serum using ultra performance liquid chromatography and the LTQ-Orbitrap mass spectrometry system. *Journal of Chromatography A*, *871*, 288–298.

Duran, A. L., Yang, J., Wang, L., & Sumner, L. W. (2003). Metabolomics spectral formatting, alignment and conversion tools (MSFACTS). *Bioinformatics*, *19*, 2283–2293.

Ebbels, T. M. D., Holmes, E., Lindon, J. C., & Nicholdon, J. K. (2004). Evaluation of metabolic variation in normal rat strains from a statistical analysis of (1)H NMR spectra of urine. *Journal of Pharmaceutical and Biomedical Analysis*, *36*, 823–833.

Eliasson, M., Rännar, S., Madsen, R., Donten, M. A., Marsden-Edwards, E., Moritz, T., et al. (2012). Strategy for optimizing LC–MS data processing in metabolomics: A design of experiments approach. *Analytical Chemistry*, *84*, 6869–6876.

Evans, A. M., DeHaven, C. D., Barrett, T., Mitchell, M., & Milgram, E. (2009). Integrated, nontargeted ultrahigh performance liquid chromatography/electrospray ionization tandem mass spectrometry platform for the identification and relative quantification of the small-molecule complement of biological systems. *Analytical Chemistry*, *81*, 6656–6667.

Fiehn, O., Kopka, J., Dörmann, P., Altmann, T., Trethewey, R. N., & Willmitzer, L. (2000). Metabolite profiling for plant functional genomics. *Nature Biotechnology*, *18*, 1157–1161.

Fiehn, O., Wohlgemuth, G., & Scholz, M. (2005). Setup and annotation of metabolomic experiments by integrating biological and mass spectrometric metadata. In B. Ludascher & L. Raschid (Eds.), *Data integration in the life sciences* (pp. 224–239). Berlin: Springer-Verlag.

Freedman, K. B., Back, S., & Bernstein, J. (2001). Sample size and statistical power of randomised, controlled trials in orthopaedics. *The Journal of Bone and Joint Surgery. British Volume*, *83*, 397–402.

Fujimura, Y., Kurihara, K., Ida, M., Kosaka, R., Miura, D., Wariishi, H., et al. (2011). Metabolomics-driven nutraceutical evaluation of diverse green tea cultivars. *PLoS One*, *6*, e23426.

Furbo, S., & Christensen, J. H. (2012). Automated peak extraction and quantification in chromatography with multichannel detectors. *Analytical Chemistry*, *84*, 2211–2218.

Grata, E., Boccard, J., Guillarme, D., Glauser, G., Carrupt, P. A., Farmer, E. E., et al. (2008). UPLC–TOF-MS for plant metabolomics: A sequential approach for wound marker analysis in *Arabidopsis thaliana*. *Journal of Chromatography A*, *871*, 261–270.

Guo, Y., Graber, A., McBurney, R. N., & Balasubramanian, R. (2010). Sample size and statistical power considerations in high-dimensionality data settings: A comparative study of classification algorithms. *BMC Bioinformatics*, *11*, 447.

Halket, J. M., Przyborowska, A., Stein, S. E., Mallard, W. G., Down, S., & Chalmers, R. A. (1999). Deconvolution gas chromatography/mass spectrometry of urinary organic acids—potential for pattern recognition and automated identification of metabolic disorders. *Rapid Communications in Mass Spectrometry*, *21*, 2965–2970.

Hanhineva, K., Rogachev, I., Aura, A. M., Aharoni, A., Poutanen, K., & Mykkänen, H. (2011). Qualitative characterization of benzoxazinoid derivatives in whole grain rye and wheat by LC–MS metabolite profiling. *Journal of Agricultural and Food Chemistry*, *59*, 921–927.

Hartler, J., Trötzmüller, M., Chitraju, C., Spener, F., Köfeler, H. C., & Thallinger, G. G. (2011). Lipid Data Analyzer: Unattended identification and quantitation of lipids in LC–MS data. *Bioinformatics*, *27*, 572–577.

Hendriks, M. M. W. B., Cruz-Juarez, L., De Bont, D., & Hall, R. D. (2005). Preprocessing and exploratory analysis of chromatographic profiles of plant extracts. *Analytica Chimica Acta*, *545*, 53–64.

Hendriks, M. M. W. B., Eeuwijk, F. A. V., Jellema, R. H., Westerhuis, J. A., Reijmers, T. H., Hoefsloot, H. C. J., et al. (2011). Data-processing strategies for metabolomics studies. *Trends in Analytical Chemistry*, *30*, 1685–1698.

Herzog, R., Schuhmann, K., Schwudke, D., Sampaio, J. L., Bornstein, S. R., Schroeder, M., et al. (2012). LipidXplorer: A software for consensual cross-platform lipidomics. *PLoS One*, 7, e29851.

Herzog, R., Schwudke, D., Schuhmann, K., Sampaio, J. L., Bornstein, S. R., Schroeder, M., et al. (2011). A novel informatics concept for high-throughput shotgun lipidomics based on the molecular fragmentation query language. *Genome Biology*, *12*, 88–112.

Hiller, K., Hangebrauk, J., Jäger, C., Spura, J., Schreiber, K., & Schomburg, D. (2009). MetaboliteDetector: Comprehensive analysis tool for targeted and nontargeted GC/MS based metabolome analysis. *Analytical Chemistry*, *81*, 3429–3439.

Hische, M., Larhlimi, A., Schwarz, F., Fischer-Rosinský, A., Bobbert, T., Assmann, A., et al. (2012). A distinct metabolic signature predicts development of fasting plasma glucose. *Journal of Clinical Bioinformatics*, *2*, 3–12.

Hoffman, D. E., Jonsson, P., Bylesjö, M., Trygg, J., Antti, H., Eriksson, M. E., et al. (2010). Changes in diurnal patterns within the *Populus* transcriptome and metabolome in response to photoperiod variation. *Plant, Cell & Environment*, *33*, 1298–1313.

Hovell, A. M., Pereira, E. J., Arruda, N. P., & Rezende, C. M. (2010). Evaluation of alignment methods and data pretreatments on the determination of the most important peaks for the discrimination of coffee varieties Arabica and Robusta using gas chromatography–mass spectroscopy. *Analytica Chimica Acta*, *678*, 160–168.

Huege, J., Sulpice, R., Gibon, Y., Lisec, J., Koehl, K., & Kopka, J. (2007). GC–EI-TOF-MS analysis of in vivo carbon-partitioning into soluble metabolite pools of higher plants by monitoring isotope dilution after $^{13}CO_2$ labelling. *Phytochemistry*, *68*, 2258–2272.

Ivanisevic, J., Thomas, O. P., Lejeusne, C., Chevaldonne, P., & Perez, T. (2011). Metabolic fingerprinting as an indicator of biodiversity: Towards understanding inter-specific relationships among Homoscleromorpha sponges. *Metabolomics*, 7, 289–304.

Izquierdo-García, J. L., Rodríguez, I., Kyriazis, A., Villa, P., Barreiro, P., Desco, M., et al. (2009). A novel R-package graphic user interface for the analysis of metabonomic profiles. *BMC Bioinformatics*, *10*, 363–372.

Jiang, W., Qiu, Y., Ni, Y., Su, M., Jia, W., & Du, X. (2010). An automated data analysis pipeline for GC–TOF-MS metabonomics studies. *Journal of Proteome Research*, *9*, 5974–5981.

Jiye, A., Trygg, J., Gullberg, J., Johansson, A. I., Jonsson, P., Antti, H., et al. (2005). Extraction and GC/MS analysis of the human blood plasma metabolome. *Analytical Chemistry*, 77, 8086–8094.

Johnson, K. J., Wright, B. W., Jarman, K. H., & Synovec, R. E. (2003). High-speed peak matching algorithm for the retention time alignment of gas chromatography data for chemometric analysis. *Journal of Chromatography A*, *996*, 141–155.

Jonsson, P., Bruce, S. J., Moritz, T., Trygg, J., Sjöström, M., Plumb, R., et al. (2005). Extraction, interpretation and validation of information for comparing samples in metabolic LC/MS data sets. *The Analyst*, *130*, 701–707.

Jonsson, P., Gullberg, J., Nordström, A., Kusano, M., Kowalczyk, M., Sjöström, M., et al. (2004). A strategy for identifying differences in large series of metabolomic samples analysed by GC/MS. *Analytical Chemistry*, *76*, 1738–1745.

Jonsson, P., Johansson, A. I., Gullberg, J., Trygg, J., A, J., Grung, B., et al. (2005). High-throughput data analysis for detecting and identifying differences between samples in GC/MS-based metabolomics analyses. *Analytical Chemistry, 77*, 5635–5642.

Jonsson, P., Johansson, E. S., Wuolikainen, A., Lindberg, J., Schuppe-Koistinen, I., Kusano, M., et al. (2006). Predictive metabolite profiling applying hierarchical multivariate curve resolution to GC–MS data—A potential tool for multi-parametric diagnosis. *Journal of Proteome Research, 5*, 1407–1414.

Jozefczuk, S., Klie, S., Catchpole, G., Szymanski, J., Cuadros-Inostroza, A., Steinhauser, D., et al. (2010). Metabolomic and transcriptomic stress response of *Escherichia coli*. *Molecular Systems Biology, 6*, 364–379.

Kanehisa, M., & Goto, S. (2000). KEGG: Kyoto encyclopedia of genes and genomes. *Nucleic Acids Research, 28*, 27–30.

Kastenmüller, G., Römisch-Margl, W., Wägele, B., Altmaier, E., & Suhre, K. (2011). metaP-server: A web-based metabolomics data analysis tool. *Journal of Biomedicine & Biotechnology, 2011*, Article ID 839862.

Katajamaa, M., Miettinen, J., & Oresic, M. (2006). MZmine: Toolbox for processing and visualization of mass spectrometry based molecular profile data. *Bioinformatics, 22*, 634–636.

Katajamaa, M., & Oresic, M. (2005). Processing methods for differential analysis of LC/MS profile data. *BMC Bioinformatics, 6*, 179–190.

Katajamaa, M., & Oresic, M. (2007). Data processing for mass spectrometry-based metabolomics. *Journal of Chromatography A, 1158*, 318–328.

Kenny, L. C., Broadhurst, D. I., Dunn, W., Brown, M., North, R. A., McCowan, L., et al. (2010). Robust early pregnancy prediction of later preeclampsia using metabolomics biomarkers. *Hypertension, 56*, 741–749.

Kern, S. M., Bennett, R. N., Mellon, F. A., Kroon, P. A., & Garcia-Conesa, M. T. (2003). Absorption of hydroxycinnamates in humans after high-bran cereal consumption. *Journal of Agricultural and Food Chemistry, 51*, 6050–6055.

Kimbara, J., Yoshida, M., Ito, H., Hosoi, K., Kusano, M., Kobayashi, M., et al. (2012). A novel class of *sticky peel* and *light green* mutations causes cutile deficiency in leaves and fruits of tomato (*Solanum lycopersicum*). *Planta, 236*, 1559–1570.

Kind, T., Meissen, J. K., Yang, D., Nocito, F., Vaniya, A., Cheng, Y. S., et al. (2012). Qualitative analysis of algal secretions with multiple mass spectrometric platforms. *Journal of Chromatography A, 1244*, 139–147.

Kind, T., Tolstikov, V., Fiehn, O., & Weiss, R. H. (2007). A comprehensive urinary metabolomics approach for identifying kidney cancer. *Analytical Biochemistry, 363*, 185–195.

Kind, T., Wohlgemuth, G., Lee, D. Y., Lu, Y., Palazoglu, M., Shahbaz, S., et al. (2009). FiehnLib: Mass spectral and retention index libraries for metabolomics based on quadrupole and time-of-flight gas chromatography/mass spectrometry. *Analytical Chemistry, 81*, 10038–10048.

Kobayashi, S., Nagasawa, S., Yamamoto, Y., Donghyo, K., Bamba, T., & Fukusaki, E. (2012). Metabolic profiling and identification of the genetic varieties and agricultural origin of *Cnidium officinale* and *Ligusticum chuanxiong*. *Journal of Bioscience and Bioengineering, 114*, 86–91.

Koh, Y., Pasikanti, K. K., Yap, C. W., & Chan, E. C. (2010). Comparative evaluation of software for retention time alignment of gas chromatography/time-of-flight mass spectrometry-based metabonomic data. *Journal of Chromatography A, 1217*, 8308–8316.

Kohl, S. M., Klein, M. S., Hochrein, J., Oefner, P. J., Spang, R., & Gronwald, W. (2012). State-of-the art data normalization methods improve NMR-based metabolomics analysis. *Metabolomics, 8*, S146–S160.

Kopka, J. (2006). Current challenges and developments in GC–MS based metabolite profiling technology. *Journal of Biotechnology, 124*, 312–322.

Kopka, J., Schauer, N., Krueger, S., Birkemeyer, C., Usadel, B., Bergmüller, E., et al. (2005). GMD@CSB.DB: The Golm Metabolome Database. *Bioinformatics*, *21*, 1635–1638.

Krall, L., Huege, J., Catchpole, G., Steinhauser, D., & Willmitzer, L. (2009). Assessment of sampling strategies for gas chromatography–mass spectrometry (GC–MS) based metabolomics of cyanobacteria. *Journal of Chromatography A*, *877*, 2952–2960.

Krishnan, S., Vogels, J. T., Coulier, L., Bas, R. C., Hendriks, M. W., Hankemeier, T., et al. (2012). Instrument and process independent binning and baseline correction methods for liquid chromatography–high resolution-mass spectrometry deconvolution. *Analytica Chimica Acta*, *740*, 12–19.

Ku, K. M., Choi, J. N., Kim, J., Kim, J. K., Yoo, L. G., Lee, S. J., et al. (2010). Metabolomics analysis reveals the compositional differences of shade grown tea (*Camellia sinensis* L.). *Journal of Agricultural and Food Chemistry*, *58*, 418–426.

Kuhl, C., Tautenhahn, R., Böttcher, C., Larson, T. R., & Neumann, S. (2012). CAMERA: An integrated strategy for compound spectra extraction and annotation of liquid chromatography/mass spectrometry data sets. *Analytical Chemistry*, *84*, 283–289.

Kusano, M., Redestig, H., Hirai, T., Oikawa, A., Matsuda, F., Fukushima, A., et al. (2011). Covering chemical diversity of genetically-modified tomatoes using metabolomics for objective substantial equivalence assessment. *PLoS One*, *6*, e16989.

Kuzina, V., Ekstrøm, C. T., Andersen, S. B., Nielsen, J. K., Olsen, C. E., & Bak, S. (2009). Identification of defense compounds in *Barbarea vulgaris* against the herbivore *Phyllotreta nemorum* by an ecometabolomic approach. *Plant Physiology*, *151*, 1977–1990.

Lankinen, M., Schwab, U., Erkkilä, A., Seppänen-Laakso, T., Hannila, M. L., Mussalo, H., et al. (2009). Fatty fish intake decreases lipids related to inflammation and insulin signalling—A lipidomics approach. *PLoS One*, *4*, e5258.

Le Gall, G., Colquhoun, I. J., & Defernez, M. (2004). Metabolite profiling using ^{1}H NMR spectroscopy for quality assessment of green tea, *Camellia sinensis* (L.). *Journal of Agricultural and Food Chemistry*, *52*, 692–700.

Le Gall, G., DuPont, M. S., Mellon, F. A., Davis, A. L., Collins, G. J., Verhoeyen, M. E., et al. (2003). Characterization and content of flavonoid glycosides in genetically modified tomato (*Lycopersicon esculentum*) fruits. *Journal of Agricultural and Food Chemistry*, *51*, 2447–2456.

Le Gall, G., Metzdorff, S. B., Pedersen, J., Bennett, R. N., & Colquhoun, I. J. (2005). Metabolite profiling of *Arabidopsis thaliana* (L.) plants transformed with an antisense chalcone synthase gene. *Metabolomics*, *1*, 181–198.

Le Gall, G., Puaud, M., & Colquhoun, I. J. (2001). Discrimination between orange juice and pulp wash by (1)H nuclear magnetic resonance spectroscopy: Identification of marker compounds. *Journal of Agricultural and Food Chemistry*, *49*, 580–588.

Leader, D. P., Burgess, K., Creek, D., & Barrett, M. P. (2011). Pathos: A web facility that uses metabolic maps to display experimental changes in metabolites identified by mass spectrometry. *Rapid Communications in Mass Spectrometry*, *25*, 3422–3426.

Lee, S., Choi, H. K., Cho, S. K., & Kim, Y. S. (2010). Metabolic analysis of guava (*Psidium guajava* L.) fruits at different ripening stages using different data-processing approaches. *Journal of Chromatography A*, *878*, 2983–2988.

Lehallier, B., Ratel, J., Hanafi, M., & Engel, E. (2012). Systematic ratio normalization of gas chromatography signals for biological sample discrimination and biomarker discovery. *Analytica Chimica Acta*, *733*, 16–22.

Li, Y., Ruan, Q., Li, Y., Ye, G., Lu, X., Lin, X., et al. (2012). A novel approach to transforming a non-targeted metabolic profiling method to a pseudo-targeted method using the retention time locking gas chromatography/mass spectrometry-selected ions monitoring. *Journal of Chromatography A*, *1255*, 228–236.

Liebeke, M., Brözel, V. S., Hecker, M., & Lalk, M. (2009). Chemical characterization of soil extract as growth media for the ecophysiological study of bacteria. *Applied Microbiology and Biotechnology*, *83*, 161–173.

Lisec, J., Schauer, N., Kopka, J., Willmitzer, L., & Fernie, A. R. (2006). Gas chromatography mass spectrometry-based metabolite profiling in plants. *Nature Protocols, 1*, 387–394.

Lohse, S., Schliemann, W., Ammer, C., Kopka, J., Strack, D., & Fester, T. (2005). Organisation and metabolism of plastids and mitochondria in arbuscular mycorrhizal roots of *Medicago truncatula*. *Plant Physiology, 139*, 329–340.

Lombardo, V. A., Osorio, S., Borsani, J., Lauxmann, M. A., Bustamante, C. A., Budde, C. O., et al. (2011). Metabolic profiling during peach fruit development and ripening reveals the metabolic networks that underpin each developmental stage. *Plant Physiology, 157*, 1696–1710.

Lommen, A. (2009). MetAlign: Interface-driven, versatile metabolomics tool for hyphenated full-scan mass spectrometry data preprocessing. *Analytical Chemistry, 81*, 3079–3086.

Lommen, A., & Kools, H. J. (2012). MetAlign 3.0: Performance enhancement by efficient use of advances in computer hardware. *Metabolomics, 8*, 719–726.

Lowe, R. G., Lord, M., Rybak, K., Trengove, R. D., Oliver, R. P., & Solomon, P. S. (2008). A metabolomics approach to dissecting osmotic stress in the wheat pathogen *Stagonospora nodorum*. *Fungal Genetics and Biology, 45*, 1479–1486.

Lu, W., Clasquin, M. F., Melamud, E., Amador-Noguez, D., Caudy, A. A., & Rabinowitz, J. D. (2010). Metabolomic analysis via reversed-phase ion-pairing liquid chromatography coupled to a stand-alone orbitrap mass spectrometer. *Analytical Chemistry, 82*, 3212–3221.

Lu, H. M., Dunn, W. B., Shen, H. L., Kell, D. B., & Liang, Y. Z. (2008). Comparative evaluation of software for deconvolution of metabolomics data based on GC–TOF-MS. *Trends in Analytical Chemistry, 27*, 215–227.

Lu, H., Gan, D., Zhang, Z., & Liang, Y. (2011). Sample classification of GC–TOF-MS metabolomics data without the requirement for chromatography deconvolution. *Metabolomics, 7*, 191–205.

Lu, J., Zhou, J., Bao, Y., Chen, T., Zhang, Y., Zhao, A., et al. (2012). Serum metabolic signatures of fulminant type 1 diabetes. *Journal of Proteome Research, 11*, 4705–4711.

Ludwig, C., & Gunther, U. L. (2011). Metabolab—Advanced NMR data processing and analysis for metabolomics. *BMC Bioinformatics, 12*, 366–371.

Luedemann, A., Strassburg, K., Erban, A., & Kopka, J. (2008). TagFinder for the quantitative analysis of gas chromatography–mass spectrometry (GC–MS)-based metabolite profiling experiments. *Bioinformatics, 24*, 732–737.

Lutz, U., Lutz, R. W., & Lutz, W. K. (2006). Metabolic profiling of glucuronides in human urine by LC–MS/MS and partial least-squares discriminant analysis for classification and prediction of gender. *Analytical Chemistry, 78*, 4564–4571.

Ma, Y., Liu, W., Peng, J., Huang, L., Zhang, P., Zhao, X., et al. (2010). A pilot study of gas chromatography/mass spectrometry-based serum metabolic profiling of colorectal cancer after operation. *Molecular Biology Reports, 37*, 1403–1411.

MacKinnon, N., Ge, W., Khan, A. P., Somashekar, B. S., Tripathi, P., Siddiqui, J., et al. (2012). Variable reference alignment: An improved peak alignment protocol for NMR spectral data with large intersample variation. *Analytical Chemistry, 84*, 5372–5379.

Massodi, M., Eiden, M., Koulman, A., Spaner, D., & Volmer, D. A. (2010). Comprehensive lipidomics analysis of bioactive lipids in complex regulatory networks. *Analytical Chemistry, 82*, 8176–8185.

Masson, P., Alves, A. C., Ebbels, T. M., Nicholson, J. K., & Want, E. J. (2010). Optimization and evaluation of metabolite extraction protocols for untargeted metabolic profiling of liver samples by UPLC–MS. *Analytical Chemistry, 82*, 7779–7786.

Matsuda, F., Okazaki, Y., Oikawa, A., Kusano, M., Nakabayashi, R., Kikuchi, J., et al. (2012). Dissection of genotype-phenotype associations in rice grains using metabolome quantitative trait loci analysis. *The Plant Journal, 70*, 624–636.

Melamud, E., Vastag, L., & Rabinowitz, J. D. (2010). Metabolomic analysis and visualization engine for LC–MS data. *Analytical Chemistry, 82*, 9818–9826.

Moing, A., Aharoni, A., Biais, B., Rogachev, I., Meir, S., Brodsky, L., et al. (2011). Extensive metabolic cross-talk in lemon fruit revealed by spatial and developmental combinational metabolomics. *The New Phytologist, 190*, 683–696.

Morohashi, M., Shimizu, K., Ohashi, Y., Abe, J., Mori, H., Tomita, M., et al. (2007). P-BOSS: A new filtering method for treasure hunting in metabolomics. *Journal of Chromatography A, 1159*, 142–148.

Nederkassel, A. M., Daszykowski, M., Eilers, P. H., & Heyden, Y. V. (2006). A comparison of three algorithms for chromatograms alignment. *Journal of Chromatography A, 1118*, 199–210.

Neuweger, H., Albaum, S. P., Dondrup, M., Persicke, M., Watt, T., Niehaus, K., et al. (2008). MeltDB: A software platform for the analysis and integration of metabolomics experiment data. *Bioinformatics, 23*, 2726–2732.

Ni, Y., Qiu, Y., Jiang, W., Suttlemyre, K., Su, M., Zhang, W., et al. (2012). ADAP-GC 2.0: Deconvolution of coeluting metabolites from GC/TOF-MS data for metabolomics studies. *Analytical Chemistry, 84*, 6619–6629.

Nielsen, N. V., Carstensen, J. M., & Smedsgaard, J. (1998). Aligning of single and multiple wavelength chromatographic profiles for chemometric data analysis using correlation optimised warping. *Journal of Chromatography A, 805*, 17–35.

Nylund, G. M., Weinberger, F., Rempt, M., & Pohnert, G. (2011). Metabolomic assessment of induced and activated chemical defense in the invasive red alga *Gracilaria vermiculophylla. PLoS One, 6*, e29359.

O'Callaghan, S., Desouza, D. P., Isaac, A., Wang, Q., Hodkinson, L., Olshansky, M., et al. (2012). PyMS: A Python toolkit for processing of gas chromatography–mass spectrometry (GC–MS) data application and comparative study of selected tools. *BMC Bioinformatics, 13*, 115–132.

Oliver, M. J., Guo, L., Alexander, D. C., Ryals, J. A., Wone, B. W., & Cushman, J. C. (2011). A sister group contrast using untargeted global metabolomics analysis delineates the biochemical regulation underlying desiccation tolerance in *Sporobolus stapfianus. The Plant Cell, 23*, 1231–1248.

Oresic, M. (2011). Informatics and computational strategies for the study of lipids. *Biochimica et Biophysica Acta, 1811*, 991–999.

Parsons, H. M., Ludwig, C., Günther, U. L., & Viant, M. R. (2007). Improved classification accuracy in 1- and 2-dimensional NMR metabolomics data using the variance stabilising generalised logarithm transformation. *BMC Genomics, 8*, 234–249.

Patti, G. J., Tautenhahn, R., & Siuzdak, G. (2012). Metabolomics: The apogee of the omics trilogy. *Nature Reviews. Molecular Cell Biology, 13*, 263–269.

Pluskal, T., Castillo, S., Villar-Briones, A., & Oresic, M. (2010). MZmine 2: Modular framework for processing, visualizing, and analyzing mass spectrometry-based molecular profile data. *BMC Bioinformatics, 11*, 395–405.

Pluskal, T., Nakamura, T., Villar-Briones, A., & Yanagida, M. (2010). Metabolic profiling of the fission yeast *S. pombe*: Quantification of compounds under different temperatures and genetic perturbation. *Molecular BioSystems, 6*, 182–198.

Pravdova, V., Walczak, B., & Massart, D. L. (2002). A comparison of two algorithms for warping of analytical signals. *Analytica Chimica Acta, 456*, 77–92.

Qiu, Y., Cai, G., Su, M., Chen, T., Zheng, X., Xu, Y., et al. (2009). Serum metabolite profiling of human colorectal cancer using GC–TOFMS and UPLC–QTOFMS. *Journal of Proteome Research, 8*, 4844–4850.

Rahmioglu, N., Le Gall, G., Heaton, J., Kay, K. L., Smith, N. W., Colquhoun, I. J., et al. (2011). Prediction of variability in CYP3A4 induction through an integrative ^{1}H NMR metabonomics approach. *Journal of Proteome Research, 10*, 2807–2816.

Redestig, H., Kusano, M., Ebana, K., Kobayashi, M., Oikawa, A., Okazaki, Y., et al. (2011). Exploring molecular backgrounds of quality traits in rice by predictive models based on high-coverage metabolomics. *BMC Systems Biology*, *5*, 176–186.

Roede, J. R., Park, Y., Li, S., Strobel, F. H., & Jones, D. P. (2012). Detailed mitochondrial phenotyping by high-resolution metabolomics. *PLoS One*, *7*, e33020.

Roessner, U., Luedemann, A., Brust, D., Fiehn, O., Linke, T., Willmitzer, L., et al. (2001). Metabolic profiling allows comprehensive phenotyping of genetically or environmentally modified plant systems. *The Plant Cell*, *13*, 11–29.

Roessner, U., Wagner, C., Kopka, J., Trethewey, R. N., & Willmitzer, L. (2000). Simultaneous analysis of metabolites in potato tuber by gas chromatography–mass spectrometry. *The Plant Journal*, *23*, 131–142.

Rohloff, J., Kopka, J., Erban, A., Winge, P., Wilson, R. C., Bones, A. M., et al. (2012). Metabolite profiling reveals novel multi-level cold responses in the diploid model *Fragaria vesca* (woodland strawberry). *Phytochemistry*, *77*, 99–109.

Ross, A., Schlotterbeck, G., Dieterle, F., & Senn, H. (2007). NMR spectroscopy techniques for application to metabonomics. In J. C. Lindon, J. K. Nicholson, & E. Holmes (Eds.), *The Handbook of Metabonomics and Metabolomics* (pp. 55–112). Oxford: Elsevier.

Roux, A., Lison, D., Junot, C., & Heilier, J. F. (2011). Applications of liquid chromatography coupled to mass spectrometry-based metabolomics in clinical chemistry and toxicology: A review. *Clinical Biochemistry*, *44*, 119–135.

Sabatine, M. S., Liu, E., Morrow, D. A., Heller, E., McCarroll, R., Wiegand, R., et al. (2005). Metabolomic identification of novel biomarkers of myocardial ischemia. *Circulation*, *112*, 3868–3875.

Sadygov, R. G., Maroto, F. M., & Hühmer, A. F. (2006). ChromAlign: A two-step algorithm procedure for time alignment of three-dimensional LC–MS chromatographic surfaces. *Analytical Chemistry*, *78*, 8207–8217.

Sanchez, D. H., Lippold, F., Redestig, H., Hannah, M. A., Erban, A., Krämer, U., et al. (2008). Integrative functional genomics of salt acclimatization in the model legume *Lotus japonicus*. *The Plant Journal*, *53*, 973–987.

Sangster, T., Major, H., Plumb, R., Wilson, A. J., & Wilson, I. D. (2006). A pragmatic and readily implemented quality control strategy for HPLC–MS and GC–MS-based metabonomic analysis. *The Analyst*, *131*, 1075–1078.

Sangster, T. P., Wingate, J. E., Burton, L., Teichert, F., & Wilson, I. D. (2007). Investigation of analytical variation in metabonomic analysis using liquid chromatography/mass spectrometry. *Rapid Communication in Mass Spectrometry*, *21*, 2965–2970.

Savorani, F., Tomasi, G., & Engelsen, S. B. (2010). icoshift: A versatile tool for the rapid alignment of 1D NMR spectra. *Journal of Magnetic Resonance*, *202*, 190–202.

Scheltema, R. A., Jankevics, A., Jansen, R. C., Swertz, M. A., & Breitling, R. (2011). PeakML/mzMatch: A file format, Java library, R library, and tool-chain for mass spectrometry data analysis. *Analytical Chemistry*, *83*, 2786–2793.

Scherling, C., Roscher, C., Giavalisco, P., Schulze, E. D., & Weckwerth, W. (2010). Metabolomics unravel contrasting effects of biodiversity on the performance of individual plant species. *PLoS One*, *5*, e12569.

Schnackenberg, L. K., Sun, J., Espandiari, P., Holland, R. D., Hanig, J., & Beger, R. D. (2007). Metabonomics evaluations of age-related changes in the urinary compositions of male Sprague Dawley rats effects of data normalization methods on statistical and quantitative analysis. *BMC Bioinformatics*, *8*, S3–S16.

Scholz, M., & Fiehn, O. (2007). Setup X—A public study design database for metabolomics projects. *Pacific Symposium on Biocomputing*, *12*, 169–180.

Semel, Y., Schauer, N., Roessner, U., Zamir, D., & Fernie, A. R. (2007). Metabolite analysis for the comparison of irrigated and non-irrigated field grown tomato of varying genotype. *Metabolomics*, *3*, 289–295.

Shepherd, L. V. T., Alexander, C. A., Sungurtas, J. A., McNicol, J. W., Stewart, D., & Davies, H. V. (2010). Metabolomic analysis of the potato tuber life cycle. *Metabolomics, 6*, 274–291.

Shi, X., Wahlang, B., Wei, X., Yin, X., Falkner, K. C., Prough, R. A., et al. (2012). Metabolomic analysis of the effects of polychlorinated biphenyls in non-alcoholic fatty liver disease. *Journal of Proteome Research, 11*, 3805–3815.

Shin, M. H., Lee do, Y., Liu, K. H., Fiehn, O., & Kim, K. H. (2010). Evaluation of sampling and extraction methodologies for the global metabolic profiling of *Saccharophagus degradans*. *Analytical Chemistry, 82*, 6660–6666.

Smith, C. A., O'Maille, G., Want, E. J., Qin, C., Trauger, S. A., Brandon, T. R., et al. (2005). METLIN: A metabolite mass spectral database. *Therapeutic Drug Monitoring, 27*, 747–751.

Smith, C. A., Want, E. J., O'Maille, G., Abagyan, R., & Siuzdak, G. (2006). XCMS: Processing mass spectrometry data for metabolite profiling using nonlinear peak alignment, matching, and identification. *Analytical Chemistry, 78*, 779–787.

Smolinska, A., Blanchet, L., Buydens, L. M., & Wijmenga, S. S. (2012). NMR and pattern recognition methods in metabolomics: From data acquisition to biomarker discovery: A review. *Analytica Chimica Acta, 750*, 82–97.

Spagou, K., Wilson, I. D., Masson, P., Theodoridis, G., Raikos, N., Coen, M., et al. (2011). HILIC-UPLC–MS for exploratory urinary metabolic profiling in toxilogical studies. *Analytical Chemistry, 83*, 382–390.

Sreekumar, A., Poisson, L. M., Rajendiran, T. M., Khan, A. P., Cao, Q., Yu, J. D., et al. (2009). Metabolomic profiles delineate potential role for sarcosine in prostate cancer progression. *Nature, 467*, 910–914.

Stein, S. E. (1999). An integrated method for spectral extraction and compound identification from gas chromatography/mass spectrometry data. *Journal of the American Society for Mass Spectrometry, 10*, 770–781.

Steinbrenner, A. D., Gómez, S., Osorio, S., Fernie, A. R., & Orians, C. M. (2011). Herbivore-induced changes in tomato (*Solanum lycopersicum*) primary metabolism: A whole plant perspective. *Journal of Chemical Ecology, 37*, 1294–1303.

Storey, J. D. (2002). A direct approach to false discovery rates. *Journal of the Royal Statistical Society, Series B, 64*, 479–498.

Struck, W., Wiczling, P., Waszczuk-Jankowska, M., Kaliszan, R., & Markuszewski, M. J. (2012). New supervised alignment method as a preprocessing tool for chromatographic data in metabolomics studies. *Journal of Chromatography A, 1256*, 150–159.

Styczynski, M. P., Moxley, J. F., Tong, L. V., Walther, J. L., Jensen, K. L., & Stephanopoulos, G. N. (2007). Systematic identification of conserved metabolites in GC/MS data for metabolomics and biomarker discovery. *Analytical Chemistry, 79*, 966–973.

Su, M., Zheng, X. Y., Zhang, T., Pei, L., Qang, F., Zheng, X., et al. (2012). Integrated profiling of metabolites and trace elements reveals a multifaceted malnutrition in pregnant women from a region with a high prevalence of congenital malformations. *Metabolomics, 8*, 831–844.

Sugimoto, M., Kawakami, M., Robert, M., Soga, T., & Tomita, M. (2012). Bioinformatics tools for mass spectroscopy-based metabolomic data processing and analysis. *Current Bioinformatics*, 7, 96–108.

Sysi-Aho, M., Katajamaa, M., Yetukuri, L., & Oresic, M. (2007). Normalization method for metabolomics data using optimal selection of multiple internal standards. *BMC Bioinformatics, 8*, 93–109.

Szymanski, J., Jozefczuk, S., Nikoloski, Z., Selbig, J., Nikiforova, V., Catchpole, G., et al. (2009). Stability of metabolic correlations under changing environmental conditions in *Escherichia coli*—A systems approach. *PLoS One, 4*, e7441.

t'Kindt, R., De Veylder, L., Storme, M., Deforce, D., & Van Bocxlaer, J. (2008). LC–MS metabolic profiling of *Arabidopsis thaliana* plant leaves and cell cultures: Optimization of pre-LC–MS procedure parameters. *Journal of Chromatography A, 871*, 37–43.

t'Kindt, R., Morreelm, K., Deforce, D., Boerjan, W., & Van Bocxlaer, J. (2009). Joint GC–MS and LC–MS platforms for comprehensive plant metabolomics: Repeatability and sample pre-treatment. *Journal of Chromatography A, 877*, 3572–3580.

Tarpley, L., Duran, A. L., Kebrom, T. H., & Sumner, L. W. (2005). Biomarker metabolites capturing the metabolite variance present in a rice plant developmental period. *BMC Plant Biology, 5*, 8–19.

Tautenhahn, R., Böttcher, C., & Neumann, S. (2008). Highly sensitive feature detection for high resolution LC/MS. *BMC Bioinformatics, 9*, 504–519.

Tautenhahn, R., Patti, G. J., Kalisiak, E., Miyamoto, T., Schmidt, M., Lo, F. Y., et al. (2010). metaXCMS: Second-order analysis of untargeted metabolomics data. *Analytical Chemistry, 83*, 696–700.

Tautenhahn, R., Patti, G. J., Rinehart, D., & Siuzdak, G. (2012). XCMSonline: A web-based platform to process untargeted metabolomics data. *Analytical Chemistry, 84*, 5035–5039.

Theodoridis, G. A., Gika, H. G., Want, E. J., & Wilson, I. D. (2012). Liquid chromatography–mass spectrometry based global metabolite profiling: A review. *Analytica Chimica Acta, 711*, 7–16.

Tikunov, Y. M., Laptenok, S., Hall, R. D., Bovy, A., & de Vos, R. C. (2011). MSClust: A tool for unsupervised mass spectra extraction of chromatography–mass spectrometry ion-wise aligned data. *Metabolomics, 8*, 714–718.

Tikunov, Y., Lommen, A., de Vos, C. H., Verhoeven, H. A., Bino, R. J., Hall, R. D., et al. (2005). A novel approach for nontargeted data analysis for metabolomics. Large scale profiling of tomato fruit volatiles. *Plant Physiology, 139*, 1125–1137.

Tistaert, C., Dejaegher, B., & Heyden, Y. V. (2011). Chromatographic separation techniques and data handling methods for herbal fingerprints: A review. *Analytica Chimica Acta, 690*, 148–161.

Toffali, K., Zamboni, A., Anesi, A., Stocchero, M., Pezzotti, M., Levi, M., et al. (2011). Novel aspects of grape berry ripening and post-harvest withering revealed by untargeted LC-ESI–MS metabolomics analysis. *Metabolomics*, 7, 424–436.

Tomasi, G., Savorani, F., & Engelsen, S. B. (2011). icoshift: An effective tool for the alignment of chromatographic data. *Journal of Chromatography A, 1218*, 7832–7840.

Torgrip, R. J. O., Aberg, K. M., Alm, E., Schuppe-Koistinen, I., & Lindberg, J. (2008). A note on normalization of biofluid 1D ^{1}H-NMR data. *Metabolomics, 4*, 114–121.

Tulpan, D., Léger, S., Belliveau, L., Culf, A., & Cuperlović-Culf, M. (2011). MetaboHunter: An automatic approach for identification of metabolites from ^{1}H NMR spectra of complex mixtures. *BMC Bioinformatics, 12*, 400–421.

Usaite, R., Jewett, M. C., Oliveira, A. P., Yates, J. R., 3rd., Olsson, L., & Nielsen, J. (2009). Reconstruction of the yeast Snf1 kinase regulatory network reveals its role as a global energy regulator. *Molecular Systems Biology, 5*, 319–330.

Vaclavik, L., Lacina, O., Hajslova, J., & Zweigenbaum, J. (2011). The use of high performance liquid chromatography–quadrupole time-of-flight mass spectrometry coupled to advanced data mining and chemometric tools for discrimination and classification or red wines according to their variety. *Analytica Chimica Acta, 685*, 45.51.

van den Berg, R. A., Hoefsloot, H. C., Westerhuis, J. A., Smilde, A. K., & van der Werf, M. J. (2006). Centering, scaling, and transformations: Improving the biological information content of metabolomics data. *BMC Genomics*, 7, 142–156.

Veselkov, K. A., Lindon, J. C., Ebbels, T. M., Crockford, D., Volynkin, V. V., Holmes, E., et al. (2009). Recursive segment-wise peak alignment of biological ^{1}H NMR spectra for improved metabolic biomarker recovery. *Analytical Chemistry, 81*, 56–66.

Veselkov, K. A., Vingara, L. K., Masson, P., Robinette, S. L., Want, E., Li, J. V., et al. (2011). Optimized preprocessing of ultra-performance liquid chromatography/mass spectrometry urinary metabolic profiles for improved information recovery. *Analytical Chemistry*, *83*, 5864–5872.

Villas-Bôas, S. G., Delicado, D. G., Akesson, M., & Nielsen, J. (2003). Simlutaneous analysis of amino and nonamino organic acids as methyl chloroformate derivatives using gas chromatography–mass spectrometry. *Analytical Biochemistry*, *322*, 134–138.

Vogels, J. T. W. E., Tas, A. C., Venekamp, J., & van der Gref, J. (1996). Partial linear fit: A new NMR spectroscopy preprocessing tool for pattern recognition applications. *Journal of Chemometrics*, *10*, 425–438.

Vorst, O., DeVos, C. H. R., Lommen, A., Staps, R. V., Visser, R. G. F., Bino, R. J., et al. (2005). A non-directed approach to the differential analysis of multiple LC–MS-derived metabolic profiles. *Metabolomics*, *1*, 169–180.

Wang, T., Shao, K., Chu, Q., Ren, Y., Mu, Y., Qu, L., et al. (2009). Automics: An integrated platform for NMR-based metabonomics spectral processing and data analysis. *BMC Bioinformatics*, *10*, 83–95.

Ward, J. L., Baker, J. M., Llewellyn, A. M., Hawkins, N. D., & Beale, M. H. (2011). Metabolomic analysis of Arabidopsis reveals hemiterpenoid glycosides as products of a nitrate ion-regulated, carbon flux overflow. *Proceedings of the National Academy of Sciences of the United States of America*, *108*, 10762–10767.

Ward, J. L., Baker, J. M., Miller, S. J., Deborde, C., Maucourt, M., Biais, B., et al. (2010). An inter-laboratory comparison demonstrates that [H]-NMR metabolite fingerprinting is a robust technique for collaborative plant metabolomic data collection. *Metabolomics*, *6*, 263–273.

Wei, X., Shi, X., Kim, S., Zhang, L., Patrick, J. S., Binkley, J., et al. (2012). Data preprocessing method for liquid chromatography–mass spectrometry based metabolomics. *Analytical Chemistry*, *84*, 7963–7971.

Wei, X., Sun, W., Shi, X., Koo, I., Wang, B., Zhang, J., et al. (2011). MetSign: A computational platform for high-resolution mass spectrometry-based metabolomics. *Analytical Chemistry*, *83*, 7668–7675.

Weljie, A. M., Newton, J., Mercier, P., Carlson, E., & Slupsky, C. M. (2006). Targeted profiling: Quantitative analysis of ^{1}H NMR metabolomics data. *Analytical Chemistry*, *78*, 4430–4442.

Wishart, D. S., Knox, C., Guo, A. C., Eisner, R., Young, N., Gautam, B., et al. (2009). HMDB: A knowledgebase for the human metabolome. *Nucleic Acids Research*, *27*, D603–D610.

Wone, B., Donovan, E. R., & Hayes, J. P. (2011). Metabolomics of aerobic metabolism in mice selected for increased maximal metabolic rate. *Comparative Biochemistry and Physiology. Part D, Genomics & Proteomics*, *6*, 399–405.

Xia, J., Bjorndahl, T. C., Tang, P., & Wishart, D. S. (2008). MetaboMiner—Semi-automated identification of metabolites from 2D NMR spectra of complex biofluids. *BMC Bioinformatics*, *9*, 400–421.

Xia, J., Mandal, R., Sinelnikov, I. V., Broadhurst, D., & Wishart, D. S. (2012). MetaboAnalyst 2.0—A comprehensive server for metabolomics data analysis. *Nucleic Acids Research*, *40*, W127–W133.

Xia, J., & Wishart, D. S. (2011). Web-based inference of biological patterns, functions and pathways from metabolomics data using MetaboAnalyst. *Nature Protocols*, *6*, 743–752.

Xu, Y. F., Zhao, X., Glass, D. S., Absalan, F., Perlman, D. H., Broach, J. R., et al. (2012). Regulation of yeast pyruvate kinase by ultrasensitive allostery independent of phosphorylation. *Molecular Cell*, *48*, 52–62.

Yu, T., Park, Y., Johnson, J. M., & Jones, D. P. (2009). apLCMS—Adaptive processing of high-resolution LC/MS data. *BMC Bioinformatics*, *25*, 1930–1936.

Zamboni, A., Di Carli, M., Guzzo, F., Stocchero, M., Zenoni, S., Ferrarini, A., et al. (2010). Identification of putative stage-specific grapevine berry biomarkers and omics data integration into networks. *Plant Physiology*, *154*, 1439–1458.

Zelena, E., Dunn, W. B., Broadhurst, D., Francis-McIntyre, S., Carroll, K. M., Begley, P., et al. (2009). Development of a robust and repeatable UPLC–MS method for the long-term metabolomic study of human serum. *Analytical Chemistry*, *81*, 1357–1364.

Zhang, Z. M., Chen, S., & Liang, Y. Z. (2011). Peak alignment using wavelet pattern matching and differential evolution. *Talanta*, *83*, 1108–1117.

Zhang, Z. M., Liang, Y. Z., Lu, H. M., Tan, B. B., Xu, X. N., & Ferro, M. (2012). Multiscale peak alignment for chromatographic datasets. *Journal of Chromatography A*, *1223*, 93–106.

Zhang, S., Zheng, C., Lanza, I. R., Nair, K. S., Raftery, D., & Vitek, O. (2009). Interdependence of signal processing and analysis of urine ^{1}H NMR spectra for metabolic profiling. *Analytical Chemistry*, *81*, 6080–6088.

Zhao, Y., Xie, Z. H., Niu, Y. G., Shi, H. M., Chen, P., & Yu, L. L. (2012). Chemical compositions, HPLC/MS fingerprinting profiles and radical scavenging properties of commercial *Gynostemma pentaphyllum* (Thunb.) Makino samples. *Food Chemistry*, *134*, 180–188.

Zheng, X., Xie, G., Zhao, A., Zhao, L., Yao, C., Chiu, N. H., et al. (2011). The footprints of gut microbial—Mammalian co-metabolism. *Journal of Proteome Research*, *10*, 5512–5522.

Zhou, J., Zhang, L., Chang, Y., Lu, X., Zhu, Z., & Xu, G. (2012). Alteration of leaf metabolism in Bt-transgenic rice (*Oryza sativa* L.) and its wild type under insecticide stress. *Journal of Proteome Research*, *11*, 4351–4360.

CHAPTER TWELVE

Qualitative Modelling of Metabolic Networks

Fabien Jourdan[1]

INRA, UMR 1331 TOXALIM (Research Center in Food Toxicology), Metabolism of Xenobiotics (MeX), Toulouse, France

[1]Corresponding author: e-mail address: Fabien.Jourdan@toulouse.inra.fr

Contents

Abstract

Global and untargeted approaches such as metabolomics require analysis of experimental data with no assumption regarding the organism's functions affected by any environmental or genetic stress. From a metabolism perspective, it means that the metabolic shift induced by the studied perturbation may span several pathways. The challenge is then to incorporate and analyze simultaneously all the metabolic reactions. This is achieved by using genome-scale metabolic networks since they gather in a single model all the anabolic and catabolic reactions a cell or an organism can perform. This information is generally retrieved by processing genome sequences and applying validation methods. In this chapter, we will discuss how they are built and raise some issues on databases and network exchanges between different platforms. Then, we will show how graph modelling can be used to model and process this qualitative

Advances in Botanical Research, Volume 67
ISSN 0065-2296
http://dx.doi.org/10.1016/B978-0-12-397922-3.00012-5

information. In particular, we will focus on the way to identify reaction paths between metabolites and how it can be used to create sub-networks based on metabolomics data. Finally, since these networks often contain thousands of reactions and metabolites, we will illustrate how visualization can be used to enhance this qualitative mining of metabolism.

1. INTRODUCTION

Metabolism is now studied at the organism or cellular level using global untargeted approaches such as metabolomic, transcriptomic and/or proteomic. Their first objective is to provide a qualitative description of metabolism by monitoring as many biochemical entities as possible (e.g. metabolome coverage in metabolomics). The second aim is the quantitative and dynamic measurement of an organism's metabolic response to an external stress (e.g. sulphur depletion for plants; Nikiforova et al., 2005). In this chapter, we focus on the qualitative and descriptive aspect of genome-scale metabolic studies. The quantitative aspect is discussed in the following chapter.

Metabolism is the sum total of all the catabolic and anabolic processes an organism can perform to sustain life. These metabolic functions, such as glycolysis, are achieved through a set of biochemical reactions. These reactions are connected since the product of a reaction can be the substrate of another one. Prior to the advent of global biochemical monitoring approaches, biochemistry studies generally focused on a few metabolic pathways. These pathways were first described in textbooks and are now stored in databases and displayed online via metabolic map representations such as the ones proposed by KEGG website (Kanehisa et al., 2008). But, as mentioned in Green and Karp (2006), pathways are often arbitrary and do not take into account the whole metabolic complexity and interactions. To overcome this limitation, all pathways can be gathered to create a single large web called the metabolic network. This network provides a perfect context for global untargeted approaches such as metabolomics since it does not require any *a priori* data on the metabolic function to be studied. Currently, the available knowledge on metabolism allows creating networks with up to two thousand reactions (Thiele & Palsson, 2010a). This complexity raises new challenges in systems biology, requiring the development of computational mining tools.

In order to model *in silico* the whole metabolic activity of an organism, it is necessary to generate an exhaustive list of all its metabolic reactions. This is challenging since this knowledge is spread over thousands of scientific publications and databases. Nevertheless, this work has been performed for model organisms such as *Saccharomyces cerevisiae* (Herrgård et al., 2008), *Escherichia coli* (Orth et al., 2011), *Homo sapiens* (Duarte et al., 2007) and *Arabidopsis thaliana* (De Oliveira Dal'Molin, Quek, Palfreyman, Brumbley, & Nielsen, 2010). Even for non-model organisms, it is now possible to build the first draft of a genome-scale metabolic network. It is achieved by using sequence homology between an annotated genome and an organism-independent repository of biochemical reactions (e.g. MetaCyc; Caspi et al., 2012). This first draft is then refined using a complex pipeline of computational tools, as described in Thiele and Palsson (2010a). Finally, reconstructions are validated by checking the stoichiometric consistency of the network (see the following chapter for more details). But for multicellular organisms, these methods provide a reference network that is not tissue dependent. In order to take into account the metabolic specificity of each tissue (e.g. human liver), approaches were developed basing reconstruction on experimental and bibliographic data (Jerby, Shlomi, & Ruppin, 2010). Nevertheless, such reconstruction will always require inputs from experts (Thiele & Palsson, 2010b). To facilitate and increase this collaborative effort, web platforms are currently being developed. This metabolic network reconstruction is discussed in the Section 2 of this chapter.

Reconstructed genome-scale metabolic networks allow raising new hypotheses regarding the organism's metabolism since they take into account all metabolic reactions simultaneously. This computational analysis can highlight, for instance, the structural weakness of metabolism that may correspond to potential drug targets (Rahman & Schomburg, 2006). To do so, graph mathematical formalism can be used. A graph is a set of elements (nodes) connected by links (edges, called arcs if they have a source and a target). This object is intensively studied in discrete mathematics (Tutte, 2001) and is used in several fields such as communication or sociology (Girvan & Newman, 2002). The challenge is to provide algorithms to automatically find interesting features in graphs when they contain thousands of nodes and edges. Moreover, in the field of metabolism, turning the network into a formal graph is not straightforward. Finally, particular care has to be taken when modelling ubiquitous compounds such as water or ATP since they are involved in a large number of reactions (Arita & Proc, 2004; Jeong, Tombor,

Albert, Oltvai, & Barabási, 2000). In fact, if they are considered in the model like the other metabolites, they will create shortcuts in the graphs and then will bias the algorithm results.

The first application of graph modelling is in drug discovery (e.g. see Cottret & Jourdan, 2010 for a review of the field of parasitology). But graph formalism can also be used to study the metabolic dialogue between a symbiotic organism and its host (Cottret et al., 2010). In fact, symbiotic organisms lost most of their biosynthetic functions during evolution. On the other hand, hosts depend on these organisms to provide important metabolites (e.g. amino acids). By querying the metabolic graph structure, it is possible to identify metabolites that cannot be synthesized or those that are created and not consumed by each organism. Moreover, the structural (topological) study of metabolic graphs can help in formulating hypotheses on metabolism evolution. This can be achieved by comparing metabolic networks (Zhu & Qin, 2005) or by looking for repeated patterns within the networks (Acuña et al., 2009). In the Section 3 of this chapter, we will describe metabolic graphs and discuss approaches to circumvent limitations due to ubiquitous compounds. We will also introduce some methods to computationally identify relevant patterns or elements within metabolic networks.

Genome-scale metabolic networks allow interpreting data generated using untargeted approaches on metabolites (metabolomic) and reactions (transcriptomic, proteomic and fluxomic) (Gehlenborg et al., 2010). Since these descriptive models may contain thousands of reactions and metabolites, visualization is a relevant method to get first clues on the processes leading to the experimental observations. Visualization of metabolic networks can be achieved by drawing each metabolic pathway and assembling all the pathways in a single poster-like view (as done in KEGG (Arakawa, Kono, Yamada, Mori, & Tomita, 2005) or BioCyc (Paley & Karp, 2006)). But in this case, the connectivity of each element is restricted to its pathway. For instance, pyruvate in the AraCyc database (Zhang et al., 2005) is connected to two reactions in glycolysis while it is connected to 34 reactions in the genome-scale metabolic graph corresponding to the same database. Graph visualization can be used to get a faithful image of this information. The challenge is then to compute a drawing of this large structure. Finally, using graph algorithms it is possible to highlight metabolic processes leading, for instance, to the expression of specific metabolites observed in metabolomics. This issue regarding metabolic network visualization is discussed in the Section 3 of this chapter.

2. RECONSTRUCTION, STORAGE AND SHARING

2.1. Metabolic network reconstruction

Most of the biochemical reactions stored in metabolic network databases are inferred on the basis of genomic information (Francke, Siezen, & Teusink, 2005; Lacroix, Cottret, Thébault, & Sagot, 2008; Reed, Famili, Thiele, & Palsson, 2006). In fact, once the genome of an organism is annotated, bioinformatics pipelines can be applied to go from the genes to the proteins (enzymes) and finally to metabolic reactions (DeJongh et al., 2007; Karp, Latendresse, & Caspi, 2011; Karp et al., 2009). These programs look for the enzymatic activities (encoded by the EC number) that can be performed by each or several proteins. Finally, these EC numbers are linked to metabolic reactions using databases such as ENZYME (Bairoch, 2000), or BRENDA (Schomburg et al., 2004).

Once the first draft of the metabolic reaction list has been built, it may still contain some false positive and false negative reactions. The first reason is that reactions are retrieved using sequence homology between the studied organism and an organism-independent database. This approach can lead to false annotations, especially when the organism is phylogenetically far from the model ones. This organism specificity also implies that some of its reactions may not have been described in any other organism and are thus not proposed in the reconstructed network. To help in correcting these errors, software such as Pathway Tools (Karp et al., 2009) offers algorithms that will propose potential reactions to be added (called pathway holes), especially the ones for which no gene has been identified but which will help to complete a metabolic pathway. The SEED (DeJongh et al., 2007) is improving this approach by taking into account stoichiometric constraints. The whole protocol, including other refinements, is described in Thiele and Palsson (2010a).

Nevertheless, automatic reconstructions still require a final step of manual pruning to add information such as reaction reversibility, missing gene/protein references (especially for pathway holes) and reactions not predicted during reconstruction and postprocessing. But this annotation effort has to be shared by several experts because of network complexity. It can be achieved by sending feedbacks to databases (KEGG, BioCyc) or by gathering experts corresponding to different organisms for a few days in 'jamborees' as described in Thiele and Palsson (2010b). This collaborative approach was, for instance, successfully applied to build the Yeast metabolic network

(Herrgård et al., 2008). Moreover, interactive tools are now available to manually curate metabolic network models (Forth, McConkey, & Westhead, 2010; Gille, Hübner, Hoppe, & Holzhütter, 2011).

The genome is the most reliable source of information to build metabolic networks. But only part of the genome is annotated, and some metabolic functions are thus missing in the reconstruction. It was recently proposed to use metabolomics data to complete network knowledge. In particular, high-resolution mass spectrometry data can be used to infer potential chemical transformations between metabolites. This method, called *ab initio* reconstruction (Breitling, Ritchie, Goodenowe, Stewart, & Barrett, 2006; Jourdan, Breitling, Barrett, & Gilbert, 2008), will link two metabolites (or spectrum peaks) if their exact mass difference is equal (up to a few parts per million) to the loss or addition of a cofactor (e.g. a carboxylation reaction will be associated to a mass difference of 43.98983 Da, which is the molecular weight of CO_2).

Most of the information on metabolism still remains in thousands of articles available in the literature. Text mining approaches are being developed to retrieve connectivity information from within publication databases such as Pubmed. To do so, abstracts are processed by tools such as PathBinder (Zhang, Berleant, Ding, Cao, & Syrkin Wurtele, 2009). This kind of approach is often called bibliomic.

Metabolic network reconstruction is a long and tedious task that will involve several laboratories and require using a large range of bioinformatics tools. Fortunately, for genome annotations, more and more genome-scale models are readily available. For plants, reconstruction has been applied to generate networks for *Arabidopsis* (De Oliveira Dal'Molin et al., 2010; Poolman, Miguet, Sweetlove, & Fell, 2009; Radrich et al., 2010), barley seeds (Grafahrend-Belau, Schreiber, Koschützki, & Junker, 2009), *Brassica napus* seeds (Hay & Schwender, 2011; Pilalis, Chatziioannou, Thomasset, & Kolisis, 2011), maize (Saha, Suthers, & Maranas, 2011) and *Chlamydomonas* (Boyle & Morgan, 2009; Cogne et al., 2011).

2.2. Sharing metabolic networks

The output of the reconstruction process is a reaction list. This list will then be used in systems biology as a knowledge source (database) or as the input of modelling programs (exchange files). For this latest objective, the simplest way is to write reaction lists into text files (generally provided in publication supplementary material). The main drawback of text-based file formats is that there is no predefined data structure. For instance, metadata columns (names, identifiers, cross references) may differ in their orders and names. Moreover, stoichiometric coefficients are often directly written in reaction equations, thus complicating the automatic extraction of this information. Finally, there is no logical link between data elements (e.g. metabolite M is involved in reaction R). These formats are relevant for manual search but cannot be used with several tools.

To standardize metabolic network electronic description, a more descriptive file format has been introduced: SBML (Systems Biology Markup Language) (Hucka et al., 2003). Here, we will focus only on the qualitative description of metabolism stored in SBML files, but this format has also been established to describe kinetic models. SBML follows Markup Language (ML) specifications, meaning that it is a way to electronically describe a document while keeping a readable file format. One of the most common examples of the ML format is HTML, which is used to describe in a single file the text, graphic design and structure of web pages. Each element of the document is opened (<tag>) and then closed (</tag>) using a specific tag written between brackets. As shown in Fig. 12.1 (inset 2), metabolite definition will be described between tags <species> and </species>. In addition to this formalism, the structure of the document is fixed, meaning that tags are hierarchically organized. For instance, as shown in Fig. 12.1 (inset 3), <listOfReactants> and <listOfProducts> will always have to be within the <reaction> tags. Within each tag, values are also defined with normalized labels (e.g. 'name', 'compartment', 'charge' for species). This format is becoming a standard and is used as input and output by many tools (in August 2012, 238 tools mentioned on the SBML website that they use this format).

BioPax is a more complete file format since it is able to handle hierarchical links between objects, and all the information, from genes to metabolites, can be stored in a single file. Nevertheless, it seems that because of its complexity, it is rarely used in metabolic network tools. Both formats and their specificity have been described in Strömbäck, Hall, and Lambrix (2007).

SBML is a handy way to describe metabolic processes and also gene regulation events. However, in the current version, information tags are missing. For instance, there are no predefined tags to specify whether a reaction belongs to a specific pathway. There are also no tags for cross references such as publication identifiers or links to experimental data. The way to circumvent this limitation is to add information within <notes> tags (see Fig 12.1, inset 3). But in this case, as for text files, there is no predefined structure. This limitation is

```
<sbml xmlns="http://www.sbml.org/sbml/level2" xmlns:html="http://www.w3.org/1999/xhtml" level="2" version="1">
  <model id="AraC4_2009" name="C4GEM_v2">
    <listOfCompartments>
      <compartment id="Extracellular" />
      <compartment id="Cytosol" outside="Extracellular" />
      <compartment id="Mitochondria" outside="Cytosol" />
      <compartment id="Plastid" outside="Cytosol" />
      <compartment id="Vacuole" outside="Cytosol" />
      <compartment id="Peroxisome" outside="Cytosol" />
    </listOfCompartments>
    <listOfSpecies>
     [...]
    <species id="Pyrophosphate_m" name="C00013_m" compartment="Mitochondria" charge="0" boundaryCondition="false" />
    <species id="Pyrophosphate_p" name="C00013_p" compartment="Plastid" charge="0" boundaryCondition="false" />
    <species id="Pyrophosphate_x" name="C00013_x" compartment="Peroxisome" charge="0" boundaryCondition="false" />
    <species id="Pyruvate_c" name="C00022_c" compartment="Cytosol" charge="0" boundaryCondition="false" />
    <species id="Pyruvate_m" name="C00022_m" compartment="Mitochondria" charge="0" boundaryCondition="false" />
    <species id="Pyruvate_p" name="C00022_p" compartment="Plastid" charge="0" boundaryCondition="false" />
     [...]
    </listOfSpecies>
    <listOfReactions>
     [...]
    <reaction id="R00006_c" name="acetolactate synthase, chloroplast / acetohydroxy-acid synthase (ALS)" reversible="false">
      <listOfReactants>
        <speciesReference species="Pyruvate_c" stoichiometry="2" />
      </listOfReactants>
      <listOfProducts>
        <speciesReference species="2-Acetolactate_c" stoichiometry="1.0" />
        <speciesReference species="CO2_c" stoichiometry="1.0" />
      </listOfProducts>
      <notes>
        <html:listOfGenes>
          <html:p>ENTRY: AT3G48560</html:p>
          <html:p>ID:</html:p>
          <html:p>ENZYME: acetolactate synthase, chloroplast / acetohydroxy-acid synthase (ALS)</html:p>
          <html:p>EC: 2.2.1.6</html:p>
          <html:p>PATHWAY: Pantothenate and CoA biosynthesis</html:p>
        </html:listOfGenes>
      </notes>
    </reaction>
     [...]
    </listOfReactions>
  </model>
</sbml>
```

Figure 12.1 Overview of SBML files. The whole document is included between the <sbml> and </sbml> tags while the model stands between <model> tags. Each model is divided into three sections. (1) The list of all compartments providing contextual information (e.g. the fact that mitochondria are located within the Cytosol is depicted by *outside='Cytosol'*). (2) The list of all species (metabolites) specifying their compartments (compartment), charge (charge) and whether they are considered in or out of the flux model (boundaryCondition). (3) The list of all reactions with, for each reaction, the list of reactants and products (with their corresponding stoichiometric value), and a 'notes' section for free text (giving details about gene, enzyme, EC and pathway). The reversibility of the reaction is an attribute of each reaction. (For the colour version of this figure, the reader is referred to the online version of this chapter.)

mainly due to the original purpose of SBML, which was focused on pathway kinetic models. With the increasing activity in genome-scale modelling, the structure will probably evolve in the future releases of SBML specifications.

2.3. Metabolic network databases and web portals

Several online repositories of metabolic reactions and pathways exist. Some of them, such as KEGG (Kanehisa et al., 2008) and BioCyc (Caspi et al., 2012; Latendresse, Paley, & Karp, 2012), tend to cover a wide range of organisms (1968 organisms in BioCyc and 2186 in KEGG on August 2012). Both of these well-known Web servers offer query facilities to mine their metabolic databases. They also provide online tools to interpret omics data in the context of metabolic pathways using visualization (Arakawa et al., 2005; Paley & Karp, 2006). The main difference between KEGG and BioCyc is the way each creates and updates its database. KEGG is a centralized server with all reactions, pathways and tools stored and developed in-house. It results in a better compatibility between metabolic databases of different organisms. But it is not straightforward to suggest curation on database entries. It is still possible to access KEGG online but until July 2011, it was not possible to freely download data. This licencing policy change is particularly affecting bioinformatics tools that are using KEGG's data. On the other hand, BioCyc is redirecting users toward Cyc-like pages that are maintained by specialists of each organism (e.g. S. Rhee, Department of Plant Biology, Carnegie Institution, USA for AraCyc; Zhang et al., 2005). To help these groups in their annotations, they provide a software suite called Pathway Tools allowing reconstruction, curation, and web page creation (Karp et al., 2009).

Development of SBML as a standard has resulted in the publication of many genome-scale models, which are not necessarily stored in KEGG or BioCyc. Databases have been designed to import, store and export these models. Among them, BIGG web server (Schellenberger, Park, Conrad, & Palsson, 2010) is providing around 40 highly curated metabolic reconstructions. MetExplore web server (Cottret et al., 2010) also proposes access to these models and allows uploading them when they are not in the database. Moreover, computational tools for graph and flux analysis can be applied. Finally, like KEGG and BioCyc, it allows embedding omics experimental data in the visual representation of metabolic networks. Nevertheless, these databases are oriented toward modellers while KEGG and BioCyc are more easily usable by biologists since they provide pathway representations and a large range of querying interfaces and cross references.

2.4. Database and model compatibility

Publicly available genomes and reconstruction tools allow building metabolic networks for any sequenced organisms. Thus, even for the same organism, several genome-scale models may be published. For instance, there are at least six databases or models for *A. Thaliana* (see Table 12.1). This multiplication of redundant reconstructions can be used to validate the network model of an organism by cross-checking reactions between all models. Moreover, networks can also be used to compare the metabolic capacities of different organisms, for instance, comparing Arabidopsis with other plants. These validations and comparisons can be achieved if models are described with the same conventions.

But we are currently facing a great challenge in terms of model compatibility. The main issue is the naming convention, in particular for compounds. In Radrich et al. (2010), authors mention that pyruvate has five different names in KEGG and nine (10 on August 2012) in AraCyc. Moreover, the identifier in KEGG is C00022 and PYRUVATE in AraCyc (see Table 12.2). Several efforts have been made to standardize compound

Table 12.1 *Arabidopsis thaliana* genome-scale models

Model/database	URL	References	Curated	SBML
Poolman2009	http://www.plantphysiol.org/content/151/3/1570/suppl/DC1	Poolman et al. (2009)	Yes	Yes
AraGEM	http://www.plantphysiol.org/content/152/2/579/suppl/DC1	De Oliveira Dal'Molin et al. (2010)	Yes	Yes
BMID000000140799	http://www.ebi.ac.uk/biomodels-main/BMID000000140799	Li et al. (2010)	No	Yes
Radrich2010	http://www.biomedcentral.com/1752-0509/4/114/additional	Radrich et al. (2010)	No	Yes
KEGG	http://www.genome.jp/kegg/	Kanehisa et al. (2008)	Yes	No
AraCyc	http://www.arabidopsis.org/biocyc/	Zhang et al. (2005)	Yes	No

Table 12.2 Metabolite identifier issue illustrated for pyruvate

Database/model	Formula	Names	DB identifier	SMILES	InChI	InChIKey	CHEBI	PubChem	CAS
Poolman2009	No	1	PYRUVATE	No	No	No	No	No	No
AraGEM	No	3	Pyruvate_*	No	No	No	No	No	No
BMID	$C_3H_3O_3$	1	C00022	No	No	No	15361	No	No
Radrich2010	No	1	Ath_C0020	No	No	No	No	No	No
KEGG	$C_3H_4O_3$	5	C00022	No	No	No	32816	3324	127-17-3
AraCyc	$C_3H_3O_3$	10	PYRUVATE	Yes	Yes	No	15361	107735	127-17-3

identifiers in a non-ubiquitous way: ChEBI (Chemical Entities of Biological Interest; De Matos et al., 2010), CAS (Chemical Abstracts Service) and PubChem identifier (Bolton, Wang, Thiessen, & Bryant, 2008). Recently, the InChI (International Chemical Identifier) code was introduced in order to provide, in the same string, atomic composition and stereochemistry. These strings can be very long for large molecules (e.g. the InChI entry for pyruvate in ChemSpider (Pence & Williams, 2010) is 1S/C3H4O3/c1-2(4)3(5)6/h1H3,(H,5,6)/p-1). In order to standardize these identifiers, an algorithm allows generating InChIKeys with a fixed number of 27 characters (e.g. the InChIKey for pyruvate in ChemSpider is LCTONWCANYUPML-UHFFFAOYSA-M). Currently, InChIKey is not stored in all databases (in neither AraCyc nor KEGG) and almost never in SBML files. Moreover, even when these identifiers are provided, there are still some differences between databases regarding the same compounds (e.g. see ChemSpider InChI and AraCyc InChI for pyruvate).

In Radrich et al. (2010), a semi-automatic method is proposed to overcome this identifier limitation when integrating metabolic networks coming from several sources, for instance, KEGG and AraCyc. This work cannot be achieved by the mere union of the two networks since there are several inconsistencies between the two (naming conventions, reaction definition). The method is based on a correspondence table between the models' reactions and metabolites, which is iteratively populated. For instance, if there is an established correspondence for a reaction between two models, then it provides clues regarding the correspondence of its substrate and product identifiers. This new information is then used in the next iteration of the integration process. This approach also allows ranking reactions according to their reliability regarding both reconstructions. Nevertheless, this method remains semi-automatic since some ambiguities cannot be solved without user specialist feedback.

Standardization efforts are currently underway. For instance, in MIRIAM (Juty, Le Novère, & Laibe, 2012; Le Novère et al., 2005), a minimal set of standardized information to store metabolic networks is proposed. For Arabidopsis, the TAIR initiative (Lamesch et al., 2012) is also providing resources to help in sharing metabolic knowledge. Finally, SBO (Systems Biology Ontology) (Courtot et al., 2011) is proposing an organized and controlled vocabulary to describe any systems biology network (not limited to metabolism). For metabolites, InChIKey seems to be the most reliable solution and will probably populate most of the metabolic network databases in the near future.

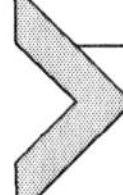

3. GRAPH FORMALISM FOR NETWORK MINING

3.1. From network to graphs

Once the network has been built and stored, it can be queried using web server tools. But the aim is also to use the structure of metabolic networks in order to interpret experimental data. In particular, the aim is to identify relevant reaction chains connecting biomarkers identified in metabolomics. This search cannot be focused on one metabolic pathway since the metabolic process shifted by the perturbation can span several of them. To perform this connectivity-based search, the most appropriate mathematical formalism is a graph (Tutte, 2001). A graph is a set of elements (nodes) and links (edges) connecting them. To model a biological network using a graph, it is necessary to associate nodes to biological bodies or functions and to define when they will be connected (Kepes, 2008; Lacroix et al., 2008). The model choice depends on the available data, the biological question and the algorithmic method that will be applied. Mainly three classes of graphs exist (Cottret & Jourdan, 2010; Lacroix et al., 2008). The first one is based on metabolites and will be called 'compound graphs'. The second focuses on enzymatic reactions, two variants being called 'enzymatic graphs' and 'reaction graphs'. The third class is the most descriptive since it will model both reactions and metabolites using 'bipartite graphs'.

Compound graphs are designed to model connections between metabolites; thus, the nodes in these networks are metabolites. Two nodes are connected if one is the input of a reaction and the other is the output of the same reaction. For instance, in Fig. 12.2 (inset 4), *myo*-inositol is connected to galactinol since reaction 2.4.1.123 has *myo*-inositol as substrate and produces galactinol. With this model, the notion of cofactors is lost since there is no modelling information on the fact that two metabolites are used as substrate in the same reaction. For instance, both galactinol and sucrose are required by reaction 2.4.1.82 to synthesize raffinose, but it is not shown in the compound graph. On the other hand, this network is well suited to interpret chemical data generated on metabolites (metabolomics, fluxomics). In fact, it will highlight the chemical transformation path, for instance, carbon chain modifications along glycolysis.

Reaction and enzyme graphs are designed to provide a network view of the interplay between metabolic reactions. In the reaction graph nodes are reactions that are connected if the output of one reaction is the input of the other. For instance, in Fig. 12.2 (inset 5), reactions 2.4.1.123 and 2.4.1.82

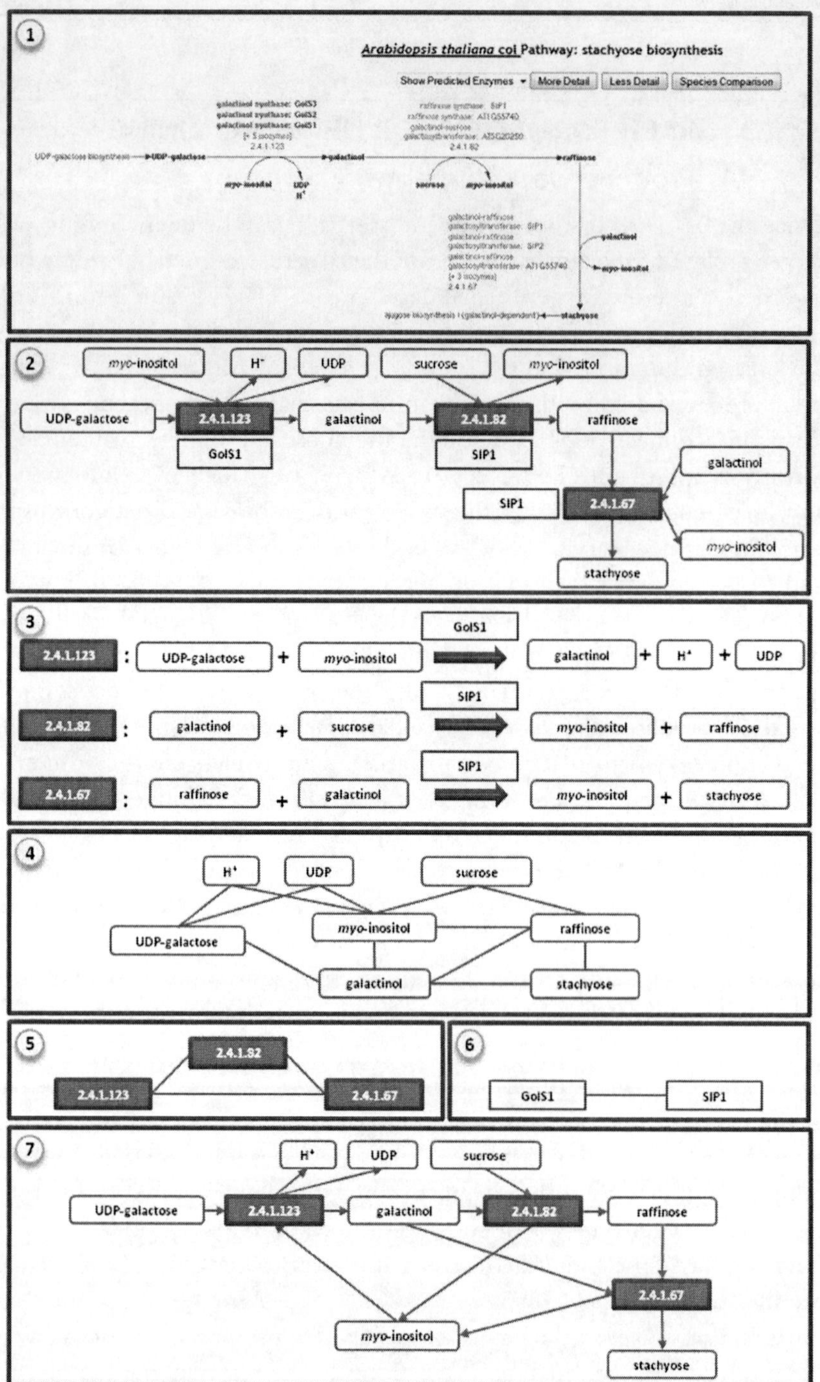

Figure 12.2 (1) The online pathway representation of stachyose biosynthesis as stored in the AraCyc database. (2) Same pathway with a simplification of enzyme catalyzing

are connected since galactinol is produced by the first reaction and is a substrate of the second. The process is the same for enzyme graph except that nodes are enzymes (see Fig. 12.2, inset 6). These two graphs can be different since a reaction can be performed by the same enzyme. For instance, in the stachyose biosynthesis pathway, enzyme SP1 catalyzes two reactions. The drawback of this method is that the connectivity information can differ and as for compound graphs, part of the original information is lost. This model is often used when studying organism evolution in order to highlight the connectivity between reactions and, in particular, to find a model linking genome evolution and metabolism evolution (Caetano-Anolles et al., 2008). More generally, these graphs allow establishing a link between genomic and metabolic information.

A bipartite graph is designed to model both reactions and metabolites into a single graph. It will contain two kinds of nodes (metabolites and reactions). A metabolite will be connected to a reaction if it is one of its substrates (resp. products). This graph is called bipartite since nodes can be divided into two sets (metabolites and reactions) such that there is no edge connecting two nodes of the same set. To keep this graph property true, spontaneous and transport processes are considered as reactions. These graphs are the most commonly used to display networks (e.g. KEGG maps). In fact, they allow analyzing simultaneously data on reactions and metabolites. But, in contrast to KEGG maps, a metabolite is represented only once. For instance, in stachyose biosynthesis, myo-inositol is involved as a substrate or a product in all reactions. In Fig. 12.2 (inset 1), corresponding to the AraCyc pathway, this metabolite is duplicated while in the bipartite model presented in Fig 12.2 (inset 7), it is a single node. The connectivity information is thus better conserved in bipartite graph models than in models where nodes are duplicated. This is generally the main difference between pathway- and graph-oriented studies.

Directionality in graphs consists in providing an orientation for each edge (arc in this case). An arc will go from a source node to a target node. Adding directionality to the previous graph models is a challenging problem. In fact, the directionality of reactions is not known for all them. Moreover, some

---

reactions (for simplicity, only one enzyme per reaction was kept). In this box and the following ones, the rounded rectangles are metabolites, blue rectangles are reactions (with the corresponding EC number) and red rectangles are enzymes catalyzing the closer reaction. (3) The reaction list corresponding to this pathway. (4) The compound graph. (5) Reaction graph. (6) Enzyme graph. (7) Bipartite graph. (For interpretation of the references to colour in this figure legend, the reader is referred to the online version of this chapter.)

reactions can occur in both directions and are called reversible. Let us assume that the reaction turning metabolite A into metabolite B is reversible. Then it will be modelled in the compound graph by two arcs: one from A to B and the other from B to A. The drawback of this modelling choice is that it creates cycles (e.g. path from A to B then B to A) that are not realistic from a biological point of view. The solution generally chosen consists in using undirected compound graphs and labelling edges (whether each edge is reversible or not). For reaction and enzyme graphs, labelling is performed on nodes (whether each reaction/enzyme is reversible or not). For bipartite graphs, labelling is applied to reaction nodes showing if the reaction is reversible or not.

Using MetExplore web server (Cottret, Wildridge, et al., 2010), we generated the bipartite and compound graphs for the AraCyc database. Both of them were filtered in the same way: known ubiquitous side compounds such as water were removed (see MetExplore website for a detailed list). Figure 12.3 (inset 1) shows the bipartite metabolic network on the left while Fig. 12.3 (inset 2) shows the compound network. In both networks, we selected all the nodes that were involved in the stachyose biosynthesis described in Fig. 12.2. Then we extracted from the genome-scale network all the edges connecting two of these nodes. For the bipartite graph (resp. compound graph), it resulted in the network shown on the right of Fig. 12.3 (inset 1; resp. Fig. 12.3, inset 2). In contrast to the pathway shown in Fig. 12.2, these sub-networks are taking into account the whole network. For the bipartite modelling, sub-network 12.3 (inset 2) is similar to the one in Fig. 12.2 (inset 7), except for H+, which was filtered during the graph reconstruction process. But the compound graphs in Fig. 12.2 (inset 4) and 12.3 (inset 2) are different. In particular, there is an extra link between *myo*-inositol and stachyose (red edge) in Fig. 12.3 (inset 2). This edge comes from the ajugose biosynthesis pathway (galactinol-dependent) shown in Fig. 12.3 (inset 3). In fact, verbascose synthase has stachyose as a substrate and *myo*-inositol as a product meaning that, in the compound graph, these two metabolites are linked. This example shows that even on a pathway, it is possible to gain new insights based on genome-scale studies. Moreover, it shows that compound graphs can help in analyzing metabolic data even if they focus on metabolites.

3.2. Network topology

Network topology either describes the spatial embedding of the graph as for wire networks (physical topology) or describes the network flux potentiality

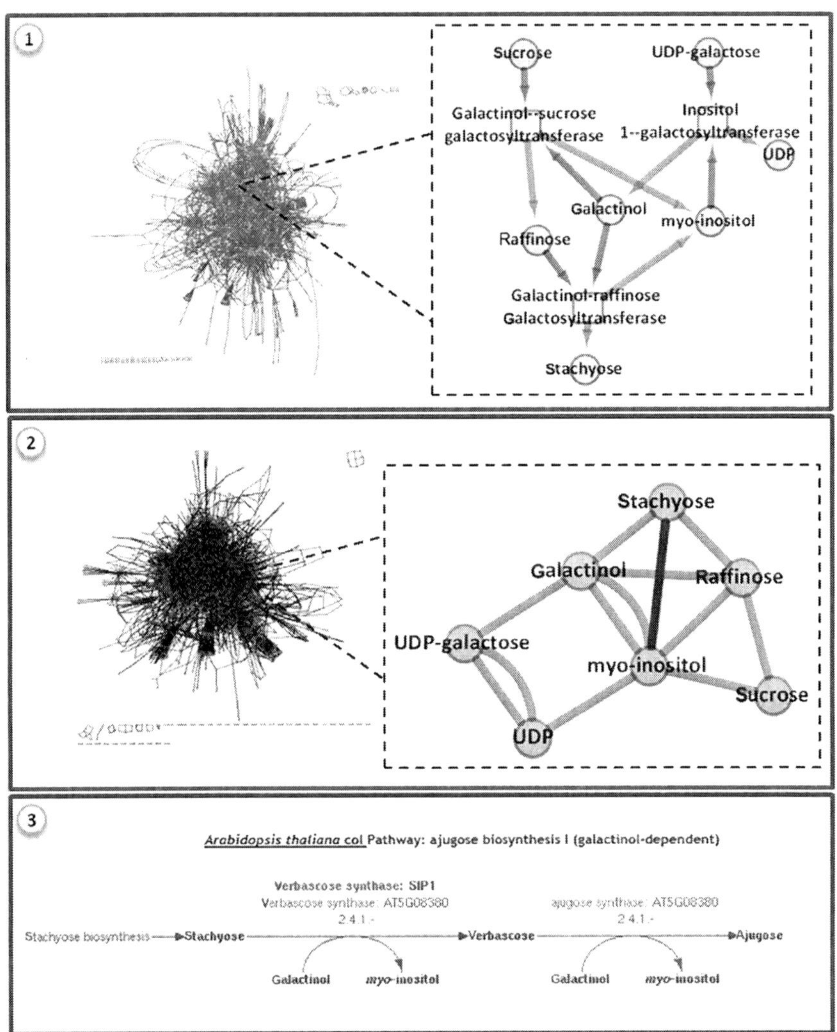

Figure 12.3 (1) Bipartite metabolic graph built with MetExplore based on AraCyc data. On the right, a sub-network containing all the metabolites and reactions involved in stachyose biosynthesis and all the edges connecting these nodes in the bipartite graph. (2) Compound graph built with MetExplore based on AraCyc data. On the right, a sub-network containing all the metabolites involved in stachyose biosynthesis and all the edges connecting these nodes in the compound graph. (3) The ajugose biosynthesis I (galactinol-dependent) as it is described on AraCyc website. (For the colour version of this figure, the reader is referred to the online version of this chapter.)

regardless of nodes and edges localization in space (logical topology). Even though localization of metabolic network elements is crucial (cell compartments), topology has been investigated only from a logical point of view. In such case, it will allow identifying weak elements in the network that can be used as potential drug targets or detecting repeated patterns corresponding to possible metabolic evolutionary scenarios.

Node degree is the number of edges connected to a node (we will consider the graph without directionality on edges). In Albert, Jeong, and Barabasi (2000), the authors measured the node degree distribution in metabolic networks. They identified that this distribution follows a power law (low number of highly connected nodes and a high number of poorly connected ones). This can be explained by the preferential attachment theory (Barabasi & Albert, 1999; Jeong et al., 2000), meaning that new nodes are preferentially connected to ones that are already highly connected (note that most of these observations find analogies in social sciences; Newman, Watts, & Strogatz, 2002). It allows raising the hypothesis that metabolism is evolving by adding functions preferentially around already highly active ones. Moreover, high degree nodes are of much importance since they are involved in many biochemical reactions. Therapeutically targeting these nodes may be relevant since they may have a strong impact on the organism. On the other hand, power law distribution means that, for instance, a random viral attack will have small chances of reaching an important node (Albert et al., 2000) since there are only a few of them.

These conclusions have been intensively discussed in the community (Arita & Proc, 2004). Indeed, the graph model chosen (compound, reaction, bipartite) may lead to different node degree distributions. Moreover, if water is considered as a node in the network, it will be connected to a large number of reactions (458 in AraGEM, 596 in Biomodel and 618 in AraCyc) and will thus be considered, following the previous reasoning scheme, as a potential weak point of the network (see the 10 most connected nodes in the five bipartite models in Table 12.3). Such molecules will also have strong impacts on the reaction graph since they will connect many reactions. With these molecules in the graph model, the AraGEM reaction graph contains 1609 reactions connected by 125622 edges; this number falls to a few thousands when they are removed.

Consequently, in order to get more realistic topological information and relevant results from graph algorithms, it is necessary to filter the networks. The first option consists in removing the most connected nodes as they often correspond to cofactors (see Table 12.3). But important metabolites that are

Table 12.3 Highly connected metabolites in genome-scale models

Poolman2009		AraGEM		BMID		Radrich2010		AraCyc	
Node	**Degree**	**Node**	**Degree**	**Node**	**Degree**	**Node**	**Degree**	**Node**	**Degree**
O_2	193	H_2O_c	458	H_2O	596	Water	756	H_2O	618
NADP	183	H+_c	200	protium(1)	535	O_2	263	O_2	226
NADPH	183	ATP_c	162	H( )	334	ATP	249	ATP	204
ATP	148	NADP+_c	127	NADP(3 −)	214	NADP	233	NADP	185
CO_2	145	NADPH_c	127	NADPH(4 −)	213	NADPH	233	NADPH	183
PPI	121	CO_2_c	126	O_2	185	CO_2	213	diphosphate	181
ADP	117	Orthophosphate_c	119	ATP(3-)	179	PPI	204	CO_2	180
Pi	116	Oxygen_c	118	lipid	160	ADP	161	H	141
NAD	103	Pyrophosphate_c	118	NAD(1 −)	141	NAD	149	Phosphate	140
NADH	100	ADP_c	114	CO_2	140	NADH	144	ADP	136

also highly connected, such as pyruvate, might be removed from the metabolic network. Moreover, there is no objective way to choose a threshold for node degree values to decide if a metabolite has to be removed or not. At last, some important metabolic processes (e.g. the formation of ATP) might disappear from the metabolic network (Lacroix et al., 2008). Another way consists in listing all the cofactor transformations (e.g. ATP → ADP + Pi) and to model them only once. More subtle methods split the reaction by computing the most probable exchanges of atom groups between subsets of substrates and products. In KEGG, such information is available in the RPAIR database (Oh, Yamada, Hattori, Goto, & Kanehisa, 2007).

3.3. Metabolic paths between metabolites

High-throughput methods such as metabolomics are designed mostly to identify biomarkers. These biomarkers can be used as signatures of a particular metabolic status. The challenge is then to understand the underlying processes explaining these metabolic changes. For metabolites, it means finding the chains of biochemical reactions that imply changes in metabolite concentrations. Since these biomarkers can be involved in several metabolic processes (lipid metabolism, amino acid metabolism), a genome-scale interpretation is required. Using metabolic networks, it will be possible to identify paths (set of reactions) that can lead from one biomarker to another. Unifying these paths in a single sub-network will provide clues to the metabolic shift induced by a perturbation. More generally, the aim will be to decompose the network into functional modules or clusters (Faust, Dupont, Callut, & Van Helden, 2010).

A common approach to analyze graphs consists in computing paths between elements. For instance, one may want to identify the set of reactions that can lead from one compound to another. In this case, a path is a sequence of nodes connected by edges. A commonly used notion is the shortest path, meaning that the sequence has to contain a minimal number of nodes. When the network is not filtered, these shortest paths will often go through highly connected nodes such as water or ATP since they behave like shortcuts in the network. A way to solve this problem consists in, as mentioned before, filtering the network in order to remove these shortcuts. Another option consists in changing the definition of a path by taking into account the degree of the nodes it is going through. It will result in weighted paths, which will be penalized if they go through compounds such as water. The method will then consist in selecting the lightest of these paths (Croes,

Couche, Wodak, & Van Helden, 2006). For instance, in Fig. 12.2 (inset 4), *myo*-inositol will be avoided by the lightest paths since its degree is five. This is relevant since this metabolite is mainly a side compound in the pathway.

Once paths are identified between each pair of biomarkers, it is possible to gather all these paths into a single sub-network. But as noticed in Antonov, Dietmann, Wong, and Mewes (2009), the resulting sub-network may be large. In fact, many paths with the same length may connect two nodes. For instance, in Fig. 12.2 (inset 4), *myo*-inositol can be linked to stachyose with a length two path via raffinose or via galactinol. To restrict the size of the sub-network, they propose to generate sub-networks unifying all paths of length one, then paths of length up to two and so on. They then compute the statistical relevance of each sub-network and select the most relevant one (Antonov et al., 2009). As mentioned previously, by using the shortest paths, it may lead to false positive results due to hub metabolites. Moreover, the authors do not take into account the availability of all the substrates to add a reaction to the sub-network. This notion is taken into account in Jourdan et al. (2010), but for a restricted path length.

In Jourdan et al. (2010), it was also shown that sub-network extraction can be used to improve the interpretation of spectral data such as the ones generated using high-resolution mass spectrometry. In fact, some metabolites may not be detected due to technical limitations or to relatively low abundances. Sub-networks connecting already identified compounds will provide hypotheses for the ones that should be in the spectral data. Based on these hypotheses, it would then be possible to perform a targeted search in the raw dataset.

3.4. Precursors and scopes of metabolites

A metabolic graph is also a powerful tool to investigate the cross talk between different organisms, in particular when studying symbiosis relationships. In Cottret, Milreu, et al. (2010), the authors decipher the metabolic exchange needed between a host (the leafhopper *Homalodisca coagulata*) and two endosymbiotic bacteria *Sulcia muelleri* and *Baumannia cicadellinicola*. Since genomes of the two bacteria are available, they built the corresponding metabolic networks. They used graph modelling to determine the list of metabolites that have to be imported by each organism (Borenstein, Kupiec, Feldman, & Ruppin, 2008). By comparing these lists, they detected that both organisms import almost two distinct sets of metabolites from their host (only three source metabolites in common). Moreover, in order to investigate the complementarity between the two metabolisms, they used another algorithm called PITUFO, which computes the minimum set of metabolites (seeds) required to produce a set of metabolites of interest (such as amino acids). The method starts from the end compounds and goes backward, checking each time that all the cofactors required for synthesizing the compound are present. The computation ends when metabolites cannot be created anymore; these metabolites are called precursors. PITUFO is original since it takes into account cycles occurring in the metabolic graph. A complementary approach was developed to compute the biosynthetic capacity of a set of metabolites. Once again, at each step, a reaction is added to this scope only if all the cofactors are available (Handorf, Ebenhöh, & Heinrich, 2005).

4. VISUALIZATION OF METABOLIC NETWORKS

4.1. Drawing metabolic graphs

Metabolism may contain thousands of elements (see Table 12.2). Mining these large datasets is a challenging problem (Gehlenborg et al., 2010; Suderman & Hallett, 2007). By using graphical representations, it is possible to exploit the human visual system to obtain new knowledge of the entire network. Since metabolism has been established by slowly gathering information on specific metabolic functions, graphical representations have been drawn manually and are available in different formats (paper, digital). But with the increasing number of sequenced organisms and the development of automatic reconstruction tools, it is necessary to automate these drawings. Moreover, it is not possible to manually draw genome-scale metabolic networks. The challenge is to provide algorithms that will automatically generate suitable representations (Michal, 1998).

In scientific visualization, such as medicine visualization, data points have a predefined position in the representation space (e.g. the location of a particular activated area of the brain). In contrast, information visualization deals with data with no predefined positions such as economic or social data (Card, Mackinlay, & Shneiderman, 1999; Ware, 2004). It requires embedding points in two or three dimensions. Graphs fall in this category and thus algorithms are required to decide the position of nodes and the drawing of edges (edge routing). This area of research is called graph drawing (Herman, Melancon, & Marshall, 2000; Kaufmann & Wagner, 2001; Tollis, Battista, Eades, & Tamassia, 1999). In the bibliography, series of aesthetic criteria are defined in order to measure the quality of representations. These criteria are mainly avoiding edge crossings, avoiding node overlapping and getting homogeneous length of edges. Optimizing simultaneously all these parameters is a computationally difficult problem (Tollis et al., 1999). This is the reason why heuristics (methods that do not necessarily provide the best solution) were introduced. The challenge in metabolic network drawings is to select the relevant algorithms and adapt them according to users' expectations (Michal, 1998).

The most commonly used algorithms are the force-based ones. This approach is quite popular since it can be applied to any kind of networks regardless of their topology. Moreover, they tend to optimize several of the aesthetic criteria of graph drawing. All these methods are based on the similar analogy: the graph is considered as a physical system where, for instance, nodes are steel balls and edges are springs (as was introduced in 1984 by Peter Eades; Eades, 1984). The hypothesis is that the simulation of this dynamical system (for instance, modelled by Hooke's law; Fruchterman & Reingold, 1991; Kamada & Kawai, 1989) will end up in an equilibrium state corresponding to a relevant drawing of the graph (see Kaufmann & Wagner, 2001, for a review). This method is implemented in most visualization tools and applied to other biological networks (protein interaction networks (Consortium, 2011) or gene networks). It may appear under different names in software, for example, 'organic' in the yFiles graph drawing library used in the Cytoscape network visualization tool (Shannon et al., 2003). Since forces are affected to edges, it is possible to parameterize the strength of each edge in order, for instance, to take into account experimental data. The problem is that it may interfere with the quality of aesthetic criteria. Finally, methods are developed to constrain the forces by defining the area where set nodes will be drawn, for instance, to mimic cell compartments (Barsky, Gardy, Hancock, & Munzner, 2007). The main drawback of the force-directed approach is that it is a non-deterministic method, meaning that the algorithm will not always create the same drawing given the same graph. In fact, the physical system used to model the graph can have several equilibrium positions corresponding to several drawings. Currently, methods are being developed to limit this variability.

Another drawback of force-directed algorithms is that they do not necessarily highlight metabolic patterns such as cycles or hierarchical structures. Nevertheless, methods for drawing graphs for these kinds of patterns (Sugiyama & Misue, 1991) exist. Since each of these structures can be drawn, the idea was then to arrange them in a two-dimensional space. Each cycle, for instance, was considered as a new meta-node and all these meta-nodes were then drawn using a force-directed algorithm (Wegner & Kummer, 2005). But metabolites and reactions can be involved in several of these structural modules. In Bourqui et al. (2007), a method was designed to maximize the number of these well-drawn structures and also to maximize the number of fully drawn metabolic pathways. It resulted in representations that mimic text-book drawing conventions. The remaining challenge lies in finding the way to connect these well-drawn pathways.

4.2. Visualizing metabolomics data

The distinction between metabolic networks and metabolic pathways led to the introduction of the metabolic graph concept. But the visualization of both levels is complementary since genome-scale metabolic network views will provide a global representation of the data while pathway views will focus on a particular function. This top-down approach is known in the visualization field as Shneiderman's mantra: 'Overview first, zoom and filter, then details-on-demand' (Shneiderman, 1996). Among the pathway-oriented visualizations, the KEGG ones are commonly used since they allow highlighting metabolites and reactions according to omics data (Arakawa et al., 2005). Figure 12.4 (inset 2) illustrates the use of KEGG on a sample metabolomics dataset where coloured nodes correspond to metabolites identified in metabolomics (seven of the most significant biomarkers identified in Fukushima, Kusano, Redestig, Arita, & Saito, 2011). The omics viewer (Paley & Karp, 2006) provided for BioCyc databases also allows highlighting data on metabolic maps but only at the overview level (when all the maps are displayed) and not for individual pathways. Figure 12.4 (inset 1) shows part of this representation. Note that more than seven metabolites are highlighted since the representation duplicates nodes when they appear in several pathways. A more organism-specific tool called MapMan (Usadel et al., 2009) has been developed, first for Arabidopsis (it is now possible to extend this tool to other organisms). MapMan is not a node link diagram; it is a visualization showing a set of enzymes and metabolites (see Fig. 12.4, inset 3). MapMan is a handy tool to import and visualize omics data in the context of plant metabolism, regulation network and genetic material (Pauwels et al., 2008).

Many tools are available for graph visualization purposes (Auber, 2004; Bastian, Heymann, & Jacomy, 2009; Ellson et al., 2001). Most of them are generic enough to adapt to any kind of graphs. But it can be challenging to adapt them to biological networks, in particular, to import and visualize omics data. In a more targeted objective, specialized bioinformatics visualization tools were designed for specific networks such as protein–protein

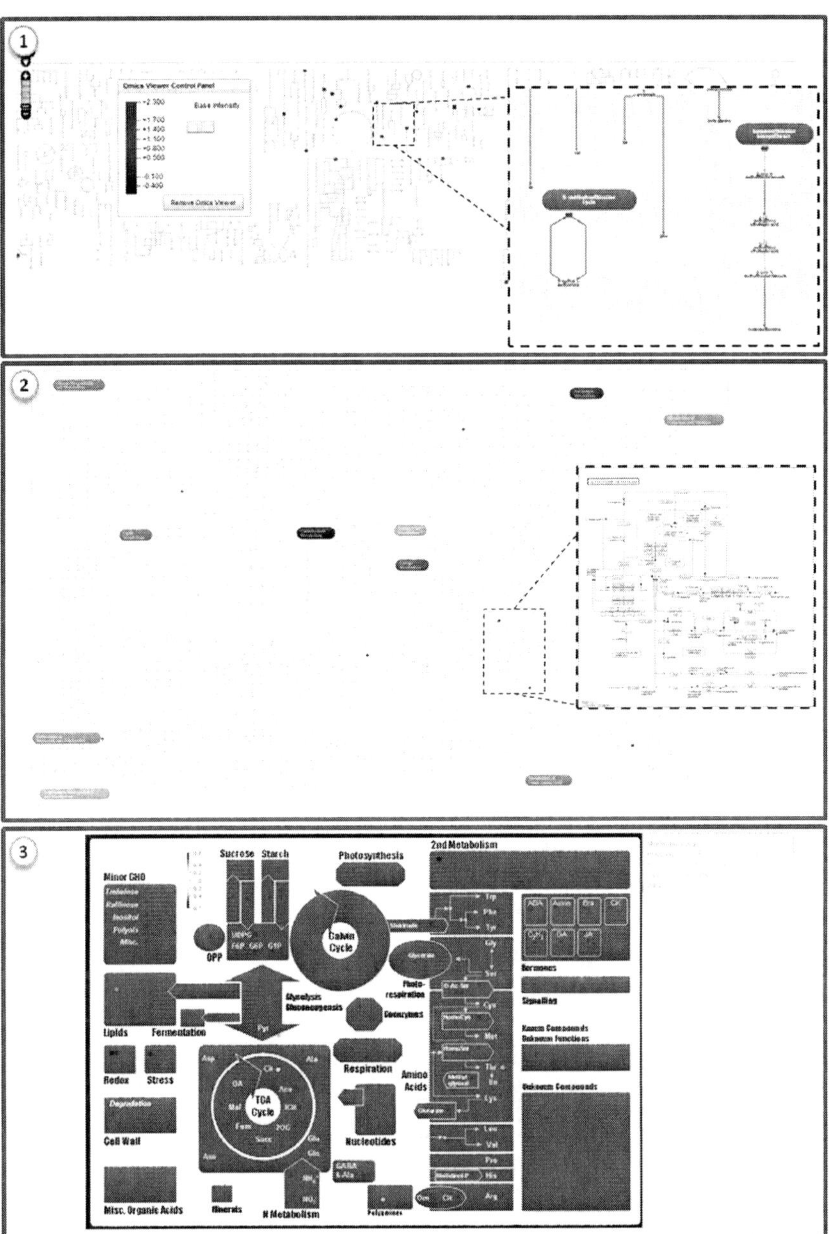

Figure 12.4 (1) Metabolic overview proposed by AraCyc. Metabolites found in metabolomics experiments are highlighted using the omics viewer and a computed colour scale (red to blue dots). The representation presents all metabolic pathways. It is possible to zoom in (inset on the right). (2) KEGG metabolism visualization. Metabolites found in experiments are coloured using a colour set by the user (red and blue in this

(Continued)

interaction networks (e.g. Proviz; Iragne, Nikolski, Mathieu, Auber, & Sherman, 2005) or metabolic networks (e.g. Systrip; Bourqui et al., 2007). Halfway between generic and specific tools stands software such as Vanted (Klukas & Schreiber, 2010) or Cytoscape (Shannon et al., 2003), which was developed to visualize any kind of biological networks. Cytoscape is a freely available platform developed by an international consortium of universities and companies (see the Cytoscape webpage for a detailed list). Its main advantage is that it allows easily creating and adding domain-specific plugins (software components) in the field of metabolic networks, for instance, plugins for reconstruction of high-resolution mass spectrometry-based networks (Jourdan et al., 2008), compartment networks (Barsky et al., 2007), multi-omics visualization (Enjalbert, Jourdan, & Portais, 2011) or graph topological analysis (Assenov, Ramírez, Schelhorn, Lengauer, & Albrecht, 2008). Cytoscape comes with various graph drawing algorithms, search functions and facilities to modify visual attributes on nodes and edges. By importing SBML files and data, it is then possible to create representations where metabolomics data are highlighted (see Fig. 12.5).

Figure 12.5 illustrates the whole graph-based analysis pipeline for metabolomics data. First, as introduced in Section 2, a relevant metabolic database is chosen. For *A. thaliana*, we proposed to use AraCyc, which was imported in the MetExplore web server using SBML files. MetExplore is then used to create a bipartite graph (Fig. 12.5, inset 1), as described in Section 3. This graph is created by filtering out a defined set of side compounds (water, proton, CO_2, etc.). By using the mapping function, we import a list of metabolites obtained during a metabolomics experiment (Fukushima et al., 2011). The resulting graph is then processed using Cytoscape since it allows visualizing the whole graph by applying a drawing algorithm (force-directed algorithm), as described in this section. Finally, as shown after step 4 in Fig. 12.5, the lightest path is computed between each pair of metabolites using a Cytoscape plugin developed by us. Then all these

Figure 12.4—Cont'd example). By clicking on a specific pathway, it is possible to open up a new page containing a diagram of the selected metabolic pathway. The same colouring scheme is used. (3) MapMan visualization where metabolites are represented by circles. A colour scale is computed based on experimental values. Each metabolic function is represented by a specific grey shape. If a metabolite identified in the metabolomics experiment is detected, then a coloured circle is painted. (For interpretation of the references to colour in this figure legend, the reader is referred to the online version of this chapter.)

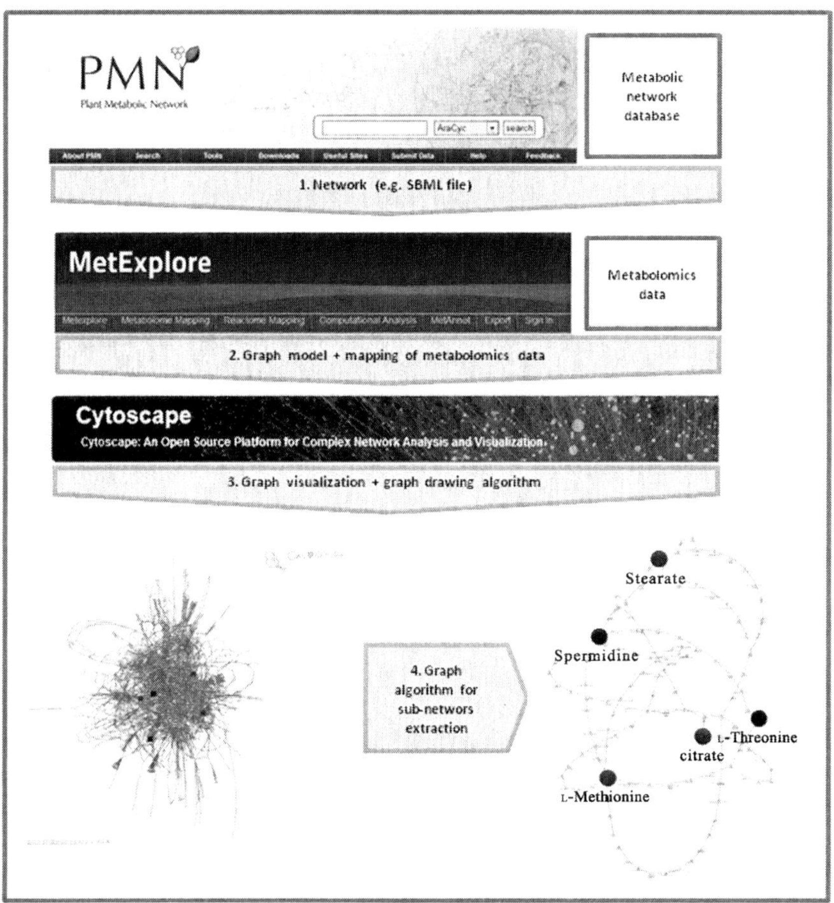

Figure 12.5 Bioinformatics pipeline used to interpret metabolomics data in the context of metabolic networks. (1) A database containing a metabolic network is chosen (AraCyc). The network is exported in a software readable format (SBML). (2) This file is imported in software in order to generate a graph model of the metabolic network. In addition, metabolomics data are mapped to graph nodes (using MetExplore mapping function). (3) Graph files are imported in biological network visualization software (Cytoscape). Using a graph drawing algorithm (force-directed), the network is visualized. Larger nodes are metabolites identified in metabolomics experiments. (4) The lightest path between each pair of identified metabolites is computed (using the lightest path algorithm). All these paths are merged in a sub-network (112 nodes and 126 edges) showing metabolic processes connecting identified metabolites. (See Colour Insert.)

paths are merged in a single sub-network. This sub-network has the advantage of containing less nodes and edges than the original one (132 nodes and 149 edges while the whole network contains 3969 nodes and 6622 edges). It is then easier to follow metabolic paths between identified metabolites. More importantly, since it is based on the genome-scale network, these paths can span several metabolic pathways. This sub-network then provides better connectivity information than predefined metabolic pathways.

5. CONCLUSION

Global and untargeted approaches such as metabolomics require interpreting generated data at the organism level. With regard to metabolism, it means deciphering the metabolic potential of an organism by listing all its metabolic reactions: its metabolic network. Methods and tools are now available to retrieve this information according to genome sequence and experimental data. The main challenge in terms of metabolic network reconstruction is the need for expert feedback once the automatic process is performed. This can be achieved by gathering groups of experts and providing them with tools to simultaneously and remotely work on the same database. This concept of collaborative reconstruction will be a bioinformatics issue in the next years. But all these efforts will have to take particular care to model compatibility. In fact, even for model organisms such as *A. thaliana*, there is no control vocabulary for reaction and metabolite identifiers. This constitutes a strong limitation as regards model comparison and investigation of multi-organism systems. Nevertheless, exchange formats such as SBML are being used more and more, allowing for the import and export of models between different tools and databases.

Until recently, studies on metabolism were, for experimental and computational reasons, performed mainly at the pathway level by focusing on a few functions. But these focused studies do not take into account all the biochemical links between metabolites. It is thus necessary to model the entire metabolic network in a single object called a graph. Building a graph based on a network database requires choosing between several modelling options for nodes and edges (compound, reaction, enzyme, bipartite graph). It is also necessary to filter the graph (removing side compounds) in order to get relevant results from computational approaches. These two issues have to be treated according to the biological question and the data that needs to be interpreted. Once this graph has been built, it is possible to use a large range of graph methods, for instance, to identify cascades of reactions between metabolites or to extract sub-networks affected by an environmental or genetic stress.

Finally, visualization is a relevant approach to mine these datasets and to orient further in-depth studies. Until recently, metabolism was visualized mainly at the pathway level. As for graph studies, they present only one part of the information and miss some connectivity data. This is the reason why genome-scale visualization methods and tools are now being developed. The main challenge is the drawing of these large networks since it is necessary to take into account text-book drawing conventions.

Network databases, graph modelling and visualization allow identifying processes involved in the organism response to a particular stress. But this approach does not take into account quantitative values. It constitutes a first step in metabolism modelling toward quantitative studies. These approaches, based on stoichiometric values of the model, are treated in the following chapter.

REFERENCES

Acuña, V., Chierichetti, F., Lacroix, V., Marchetti-Spaccamela, A., Sagot, M.-F., & Stougie, L. (2009). Modes and cuts in metabolic networks: Complexity and algorithms. *Biosystems*, *95*(1), 51–60. http://dx.doi.org/10.1016/j.biosystems.2008.06.015.

Albert, R., Jeong, H., & Barabasi, A. (2000). Error and attack tolerance of complex networks. *Nature*, *406*(6794), 378–382. http://dx.doi.org/10.1038/35019019.

Antonov, A. V., Dietmann, S., Wong, P., & Mewes, H. W. (2009). TICL—A web tool for network-based interpretation of compound lists inferred by high-throughput metabolomics. *The FEBS Journal*, *276*(7), 2084–2094. http://dx.doi.org/10.1111/j.1742-4658.2009.06943.x.

Arakawa, K., Kono, N., Yamada, Y., Mori, H., & Tomita, M. (2005). *KEGG-based pathway visualization tool for complex omics data. In Silico Biology*, *5*(4), 419–423. Retrieved from, http://www.ncbi.nlm.nih.gov/pubmed/16268787.

Arita, M., & Proc, D. A. R. (2004). The metabolic world of Escherichia coli is not small. *Proceedings of the National Academy of Sciences of the United States of America*, *101*(6), 1543–1547.

Assenov, Y., Ramírez, F., Schelhorn, S.-E., Lengauer, T., & Albrecht, M. (2008). Computing topological parameters of biological networks. *Bioinformatics (Oxford, England)*, *24*(2), 282–284. http://dx.doi.org/10.1093/bioinformatics/btm554.

Auber, D. (2004). Tulip—A huge graph visualization framework. In M. Jünger & P. Mutzel (Eds.), *Graph Drawing Software* (pp. 105–126). Berlin, Heidelberg: Springer Berlin Heidelberg.

Bairoch, A. (2000). The ENZYME database in 2000. *Nucleic Acids Research*, *28*(1), 304–305. http://dx.doi.org/10.1093/nar/28.1.304.

Barabasi, A., & Albert, R. (1999). *Emergence of scaling in random networks. Science (New York, NY)*, *286*(5439), 509–512. Retrieved from, http://www.ncbi.nlm.nih.gov/pubmed/10521342.

Barsky, A., Gardy, J. L., Hancock, R. E. W., & Munzner, T. (2007). Cerebral: A Cytoscape plugin for layout of and interaction with biological networks using subcellular localization annotation. *Bioinformatics (Oxford, England), 23*(8), 1040–1042. http://dx.doi.org/10.1093/bioinformatics/btm057.

Bastian, M., Heymann, S., & Jacomy, M. (2009). *Gephi: An open source software for exploring and manipulating networks.* In *International AAAI conference on weblogs and social media; Third international AAAI conference on weblogs and social media.* Retrieved from, http://citeseerx.ist.psu.edu/viewdoc/summary?doi=10.1.1.172.7704.

Bolton, E. E., Wang, Y., Thiessen, P. A., & Bryant, S. H. (2008). PubChem: Integrated platform of small molecules and biological activities. In W. Cornell (Ed.), *Annual reports in computational chemistry, 4,* (pp. 217–241).

Borenstein, E., Kupiec, M., Feldman, M. W., & Ruppin, E. (2008). Large-scale reconstruction and phylogenetic analysis of metabolic environments. *Proceedings of the National Academy of Sciences of the United States of America, 105*(38), 14482–14487. http://dx.doi.org/10.1073/pnas.0806162105.

Bourqui, R., Cottret, L., Lacroix, V., Auber, D., Mary, P., Sagot, M.-F., et al. (2007). Metabolic network visualization eliminating node redundance and preserving metabolic pathways. *BMC Systems Biology, 1,* 29. http://dx.doi.org/10.1186/1752-0509-1-29.

Boyle, N. R., & Morgan, J. A. (2009). Flux balance analysis of primary metabolism in Chlamydomonas reinhardtii. *BMC Systems Biology, 3*(1), 4. http://dx.doi.org/10.1186/1752-0509-3-4.

Breitling, R., Ritchie, S., Goodenowe, D., Stewart, M. L., & Barrett, M. P. (2006). Ab initio prediction of metabolic networks using Fourier transform mass spectrometry data. *Metabolomics, 2*(3), 155–164. http://dx.doi.org/10.1007/s11306-006-0029-z.

Caetano-Anolles, G., Sun, F.-J., Wang, M., Yafremava, L. S., Harish, A., Kim, H. S., et al. (2008). *Origins and evolution of modern biochemistry: Insights from genomes and molecular structure. Frontiers in Bioscience: A Journal and Virtual Library, 13,* 5212–5240. Retrieved from, http://www.ncbi.nlm.nih.gov/pubmed/18508583.

Card, S. K., Mackinlay, J. D., & Shneiderman, B. (1999). *Readings in information visualization: Using vision to think.* USA: Morgan Kaufmann. (p. 712); Retrieved from, http://dl.acm.org/citation.cfm?id=300679.

Caspi, R., Altman, T., Dreher, K., Fulcher, C. A., Subhraveti, P., Keseler, I. M., et al. (2012). The MetaCyc database of metabolic pathways and enzymes and the BioCyc collection of pathway/genome databases. *Nucleic Acids Research, 40*(Database issue), D742–D753. http://dx.doi.org/10.1093/nar/gkr1014.

Cogne, G., Rügen, M., Bockmayr, A., Titica, M., Dussap, C.-G., Cornet, J.-F., et al. (2011). A model-based method for investigating bioenergetic processes in autotrophically growing eukaryotic microalgae: Application to the green algae Chlamydomonas reinhardtii. *Biotechnology Progress, 27*(3), 631–640. http://dx.doi.org/10.1002/btpr.596.

Cottret, L., & Jourdan, F. (2010). Graph methods for the investigation of metabolic networks in parasitology. *Parasitology, 137*(9), 1393–1407. http://dx.doi.org/10.1017/S0031182010000363.

Cottret, L., Milreu, P. V., Acuña, V., Marchetti-Spaccamela, A., Stougie, L., Charles, H., et al. (2010). Graph-based analysis of the metabolic exchanges between two co-resident intracellular symbionts, Baumannia cicadellinicola and Sulcia muelleri, with their insect host, Homalodisca coagulata. *PLoS Computational Biology, 6*(9). http://dx.doi.org/10.1371/journal.pcbi.1000904.

Cottret, L., Wildridge, D., Vinson, F., Barrett, M. P., Charles, H., Sagot, M.-F., et al. (2010). MetExplore: A web server to link metabolomic experiments and genome-scale metabolic networks. *Nucleic Acids Research, 38*(Web Server issue), W132–W137. http://dx.doi.org/10.1093/nar/gkq312.

Courtot, M., Juty, N., Knüpfer, C., Waltemath, D., Zhukova, A., Dräger, A., et al. (2011). Controlled vocabularies and semantics in systems biology. *Molecular Systems Biology, 7,* 543. http://dx.doi.org/10.1038/msb.2011.77.

Croes, D., Couche, F., Wodak, S. J., & Van Helden, J. (2006). Inferring meaningful pathways in weighted metabolic networks. *Journal of Molecular Biology, 356*(1), 222–236. http://dx.doi.org/10.1016/j.jmb.2005.09.079.

De Matos, P., Alcántara, R., Dekker, A., Ennis, M., Hastings, J., Haug, K., et al. (2010). Chemical entities of biological interest: An update. *Nucleic Acids Research, 38*(Database issue), D249–D254. http://dx.doi.org/10.1093/nar/gkp886.

De Oliveira Dal'Molin, C. G., Quek, L.-E., Palfreyman, R. W., Brumbley, S. M., & Nielsen, L. K. (2010). *AraGEM, a genome-scale reconstruction of the primary metabolic network in Arabidopsis. Plant Physiology, 152*(2), 579–589. Retrieved from, http://www.plantphysiol.org/cgi/content/abstract/152/2/579.

DeJongh, M., Formsma, K., Boillot, P., Gould, J., Rycenga, M., & Best, A. (2007). Toward the automated generation of genome-scale metabolic networks in the SEED. *BMC Bioinformatics, 8*, 139. http://dx.doi.org/10.1186/1471-2105-8-139.

Duarte, N. C., Becker, S. A., Jamshidi, N., Thiele, I., Mo, M. L., Vo, T. D., et al. (2007). Global reconstruction of the human metabolic network based on genomic and bibliomic data. *Proceedings of the National Academy of Sciences of the United States of America, 104*(6), 1777–1782. http://dx.doi.org/10.1073/pnas.0610772104.

Eades, P. (1984). *A heuristic for graph drawing. Congressus Numerantium, 42*, 149–160. Retrieved from, http://www.citeulike.org/user/hansjoergmueller/article/570874.

Ellson, J., Gansner, E., Koutsofios, L., North, S., Woodhull, G., Description, S., et al. (2001). *Graphviz—Open source graph drawing tools. Lecture Notes in Computer Science, 2265*, 483–484. Retrieved from, http://citeseerx.ist.psu.edu/viewdoc/summary?doi=10.1.1.87.9389.

Enjalbert, B., Jourdan, F., & Portais, J.-C. (2011). Intuitive visualization and analysis of multi-omics data and application to Escherichia coli carbon metabolism. *PLoS One, 6*(6), e21318. http://dx.doi.org/10.1371/journal.pone.0021318.

Consortium, A. I. M. (2011). Evidence for network evolution in an Arabidopsis interactome map. *Science, 333*(6042), 601–607. http://dx.doi.org/10.1126/science.1203877.

Faust, K., Dupont, P., Callut, J., & Van Helden, J. (2010). Pathway discovery in metabolic networks by subgraph extraction. *Bioinformatics (Oxford, England), 26*(9), 1211–1218. http://dx.doi.org/10.1093/bioinformatics/btq105.

Forth, T., McConkey, G. A., & Westhead, D. R. (2010). MetNetMaker: A free and open-source tool for the creation of novel metabolic networks in SBML format. *Bioinformatics (Oxford, England), 26*(18), 2352–2353. http://dx.doi.org/10.1093/bioinformatics/btq425.

Francke, C., Siezen, R. J., & Teusink, B. (2005). Reconstructing the metabolic network of a bacterium from its genome. *Trends in Microbiology, 13*(11), 550–558. http://dx.doi.org/10.1016/j.tim.2005.09.001.

Fruchterman, T. M. J., & Reingold, E. M. (1991). Graph drawing by force-directed placement. *Software: Practice and Experience, 21*(11), 1129–1164. http://dx.doi.org/10.1002/spe.4380211102.

Fukushima, A., Kusano, M., Redestig, H., Arita, M., & Saito, K. (2011). Metabolomic correlation-network modules in Arabidopsis based on a graph-clustering approach. *BMC Systems Biology, 5*(1), 1. http://dx.doi.org/10.1186/1752-0509-5-1.

Gehlenborg, N., O'Donoghue, S. I., Baliga, N. S., Goesmann, A., Hibbs, M. A., Kitano, H., et al. (2010). Visualization of omics data for systems biology. *Nature Methods*, 7(3 Suppl.), S56–S68. http://dx.doi.org/10.1038/nmeth.1436.

Gille, C., Hübner, K., Hoppe, A., & Holzhütter, H.-G. (2011). Metannogen: Annotation of biological reaction networks. *Bioinformatics (Oxford, England)*, 27(19), 2763–2764. http://dx.doi.org/10.1093/bioinformatics/btr456.

Girvan, M., & Newman, M. E. J. (2002). Community structure in social and biological networks. *Proceedings of the National Academy of Sciences of the United States of America, 99*(12), 7821–7826. http://dx.doi.org/10.1073/pnas.122653799.

Grafahrend-Belau, E., Schreiber, F., Koschützki, D., & Junker, B. H. (2009). Flux balance analysis of barley seeds: A computational approach to study systemic properties of central metabolism. *Plant Physiology*, *149*(1), 585–598. http://dx.doi.org/10.1104/pp. 108.129635.

Green, M. L., & Karp, P. D. (2006). The outcomes of pathway database computations depend on pathway ontology. *Nucleic Acids Research*, *34*(13), 3687–3697. http://dx.doi.org/10.1093/nar/gkl438.

Handorf, T., Ebenhöh, O., & Heinrich, R. (2005). Expanding metabolic networks: Scopes of compounds, robustness, and evolution. *Journal of Molecular Evolution*, *61*(4), 498–512. http://dx.doi.org/10.1007/s00239-005-0027-1.

Hay, J., & Schwender, J. (2011). Metabolic network reconstruction and flux variability analysis of storage synthesis in developing oilseed rape (Brassica napus L.) embryos. *The Plant Journal: For Cell and Molecular Biology*, *67*(3), 526–541. http://dx.doi.org/10.1111/j.1365-313X.2011.04613.x.

Herman, I., Melancon, G., & Marshall, M. S. (2000). Graph visualization and navigation in information visualization: A survey. *IEEE Transactions on Visualization and Computer Graphics*, *6*(1), 24–43. http://dx.doi.org/10.1109/2945.841119.

Herrgård, M. J., Swainston, N., Dobson, P., Dunn, W. B., Arga, K. Y., Arvas, M., et al. (2008). A consensus yeast metabolic network reconstruction obtained from a community approach to systems biology. *Nature Biotechnology*, *26*(10), 1155–1160. http://dx.doi.org/10.1038/nbt1492.

Hucka, M., Finney, A., Sauro, H. M., Bolouri, H., Doyle, J. C., Kitano, H., et al. (2003). *The systems biology markup language (SBML): A medium for representation and exchange of biochemical network models. Bioinformatics (Oxford, England)*, *19*(4), 524–531. Retrieved from, http://www.ncbi.nlm.nih.gov/pubmed/12611808.

Iragne, F., Nikolski, M., Mathieu, B., Auber, D., & Sherman, D. (2005). ProViz: Protein interaction visualization and exploration. *Bioinformatics (Oxford, England)*, *21*(2), 272–274. http://dx.doi.org/10.1093/bioinformatics/bth494.

Jeong, H., Tombor, B., Albert, R., Oltvai, Z. N., & Barabási, A. L. (2000). The large-scale organization of metabolic networks. *Nature*, *407*(6804), 651–654. http://dx.doi.org/10.1038/35036627.

Jerby, L., Shlomi, T., & Ruppin, E. (2010). Computational reconstruction of tissue-specific metabolic models: Application to human liver metabolism. *Molecular Systems Biology*, *6*(1), 401. http://dx.doi.org/10.1038/msb.2010.56.

Jourdan, F., Breitling, R., Barrett, M. P., & Gilbert, D. (2008). MetaNetter: Inference and visualization of high-resolution metabolomic networks. *Bioinformatics (Oxford, England)*, *24*(1), 143–145. http://dx.doi.org/10.1093/bioinformatics/btm536.

Jourdan, F., Cottret, L., Huc, L., Wildridge, D., Scheltema, R., Hillenweck, A., et al. (2010). Use of reconstituted metabolic networks to assist in metabolomic data visualization and mining. *Metabolomics: Official Journal of the Metabolomic Society*, *6*(2), 312–321. http://dx.doi.org/10.1007/s11306-009-0196-9.

Juty, N., Le Novère, N., & Laibe, C. (2012). *Identifiers.org and MIRIAM Registry: Community resources to provide persistent identificationNucleic Acids Research*, *40*(Database issue), D580–D586. http://dx.doi.org/10.1093/nar/gkr1097.

Kamada, T., & Kawai, S. (1989). An algorithm for drawing general undirected graphs. *Information Processing Letters*, *31*(1), 7–15. http://dx.doi.org/10.1016/0020-0190(89)90102-6.

Kanehisa, M., Araki, M., Goto, S., Hattori, M., Hirakawa, M., Itoh, M., et al. (2008). KEGG for linking genomes to life and the environment. *Nucleic Acids Research*, *36*(Database issue), D480–D484. http://dx.doi.org/10.1093/nar/gkm882.

Karp, P. D., Latendresse, M., & Caspi, R. (2011). The pathway tools pathway prediction algorithm. *Standards in Genomic Sciences*, *5*(3), 424–429. http://dx.doi.org/10.4056/sigs.1794338.

Karp, P. D., Paley, S. M., Krummenacker, M., Latendresse, M., Dale, J. M., Lee, T. J., et al. (2009). Pathway tools version 13.0: Integrated software for pathway/genome informatics and, systems biology. *Briefings in Bioinformatics*, *11*(1), 40–79.

Kaufmann, M., & Wagner, D. (2001). M. Kaufmann & D. Wagner (Eds.), *Drawing graphs: Methods and models* (p. p. 312). *LNCS2025*. (p. p. 312). Springer. Retrieved from, http://www.springer.com/computer/theoretical+computer+science/book/978-3-540-42062-0.

Kepes, F. (2008). *Complex systems and interdisciplinary science. Biological Networks*. Singapore: World Scientific Publishing.532. Retrieved from, http://www.amazon.co.uk/BIOLOGICAL-NETWORKS-Complex-Systems-Interdisciplinary/dp/981270695X.

Klukas, C., & Schreiber, F. (2010). Integration of -omics data and networks for biomedical research with VANTED. *Journal of Integrative Bioinformatics*, 7(2), 112. http://dx.doi.org/10.2390/biecoll-jib-2010-112.

Lacroix, V., Cottret, L., Thébault, P., & Sagot, M.-F. (2008). An introduction to metabolic networks and their structural analysis. *IEEE/ACM transactions on computational biology and bioinformatics/IEEE, ACM*, *5*(4), 594–617. http://dx.doi.org/10.1109/TCBB.2008.79.

Lamesch, P., Berardini, T. Z., Li, D., Swarbreck, D., Wilks, C., Sasidharan, R., et al. (2012). *The Arabidopsis Information Resource (TAIR): Improved gene annotation and new tools. Nucleic Acids Research*, *40*(Database issue), D1202–D1210. Retrieved from, http://nar.oxfordjournals.org/cgi/content/abstract/40/D1/D1202.

Latendresse, M., Paley, S., & Karp, P. D. (2012). Browsing metabolic and regulatory networks with BioCyc. *Methods in Molecular Biology (Clifton, NJ)*, *804*, 197–216. http://dx.doi.org/10.1007/978-1-61779-361-5_11.

Le Novère, N., Finney, A., Hucka, M., Bhalla, U. S., Campagne, F., Collado-Vides, J., et al. (2005). Minimum information requested in the annotation of biochemical models (MIRIAM). *Nature Biotechnology*, *23*(12), 1509–1515. http://dx.doi.org/10.1038/nbt1156.

Li, C., Donizelli, M., Rodriguez, N., Dharuri, H., Endler, L., Chelliah, V., et al. (2010). BioModels Database: An enhanced, curated and annotated resource for published quantitative kinetic models. *BMC systems biology*, *4*, 92. http://dx.doi.org/10.1186/1752-0509-4-92.

Michal, G. (1998). On representation of metabolic pathways. *Biosystems*, *47*, 1–7.

Newman, M. E. J., Watts, D. J., & Strogatz, S. H. (2002). Random graph models of social networks. *Proceedings of the National Academy of Sciences of the United States of America*, *99*(Suppl. 1), 2566–2572. http://dx.doi.org/10.1073/pnas.012582999.

Nikiforova, V. J., Kopka, J., Tolstikov, V., Fiehn, O., Hopkins, L., Hawkesford, M. J., et al. (2005). Systems rebalancing of metabolism in response to sulfur deprivation, as revealed by metabolome analysis of Arabidopsis plants. *Plant Physiology*, *138*(1), 304–318. http://dx.doi.org/10.1104/pp.104.053793.

Oh, M., Yamada, T., Hattori, M., Goto, S., & Kanehisa, M. (2007). Systematic analysis of enzyme-catalyzed reaction patterns and prediction of microbial biodegradation pathways. *Journal of Chemical Information and Modeling*, *47*(4), 1702–1712. http://dx.doi.org/10.1021/ci700006f.

Orth, J. D., Conrad, T. M., Na, J., Lerman, J. A., Nam, H., Feist, A. M., et al. (2011). A comprehensive genome-scale reconstruction of Escherichia coli metabolism–2011. *Molecular Systems Biology*, 7, 535. http://dx.doi.org/10.1038/msb.2011.65.

Paley, S. M., & Karp, P. D. (2006). The Pathway Tools cellular overview diagram and Omics Viewer. *Nucleic Acids Research*, *34*(13), 3771–3778. http://dx.doi.org/10.1093/nar/gkl334.

Pauwels, L., Morreel, K., De Witte, E., Lammertyn, F., Van Montagu, M., Boerjan, W., et al. (2008). Mapping methyl jasmonate-mediated transcriptional reprogramming of metabolism and cell cycle progression in cultured Arabidopsis cells. *Proceedings of the*

National Academy of Sciences of the United States of America, *105*(4), 1380–1385. http://dx.doi.org/10.1073/pnas.0711203105.

Pence, H. E., & Williams, A. (2010). ChemSpider: An online chemical information resource. *Journal of Chemical Education*, *87*(11), 1123–1124. http://dx.doi.org/10.1021/ed100697w.

Pilalis, E., Chatziioannou, A., Thomasset, B., & Kolisis, F. (2011). An in silico compartmentalized metabolic model of Brassica napus enables the systemic study of regulatory aspects of plant central metabolism. *Biotechnology and Bioengineering*, *108*(7), 1673–1682. http://dx.doi.org/10.1002/bit.23107.

Poolman, M. G., Miguet, L., Sweetlove, L. J., & Fell, D. A. (2009). *A genome-scale metabolic model of Arabidopsis and some of its properties*. *Plant Physiology*, *151*(3), 1570–1581. Retrieved from, http://www.plantphysiol.org/cgi/content/abstract/151/3/1570.

Radrich, K., Tsuruoka, Y., Dobson, P., Gevorgyan, A., Swainston, N., Baart, G., et al. (2010). *Integration of metabolic databases for the reconstruction of genome-scale metabolic networks*. *BMC Systems Biology*, *4*(1), 114. Retrieved from, http://www.biomedcentral.com/1752-0509/4/114.

Rahman, S. A., & Schomburg, D. (2006). Observing local and global properties of metabolic pathways: "Load points" and "choke points" in the metabolic networks. *Bioinformatics*, *22*(14), 1767–1774. http://dx.doi.org/10.1093/bioinformatics/btl181.

Reed, J. L., Famili, I., Thiele, I., & Palsson, B. O. (2006). Towards multidimensional genome annotation. *Nature Reviews. Genetics*, 7(February), 130–141. http://dx.doi.org/10.1038/nrg1769.

Saha, R., Suthers, P. F., & Maranas, C. D. (2011). *Zea mays iRS1563: A comprehensive genome-scale metabolic reconstruction of maize metabolism. Andersen, M. R. (Ed.), PloS one*, vol. 6(7), e21784. Retrieved from, http://dx.plos.org/10.1371/journal.pone.0021784.

Schellenberger, J., Park, J. O., Conrad, T. M., & Palsson, B. Ø. (2010). BiGG: A Biochemical Genetic and Genomic knowledgebase of large scale metabolic reconstructions. *BMC Bioinformatics*, *11*, 213. http://dx.doi.org/10.1186/1471-2105-11-213.

Schomburg, I., Chang, A., Ebeling, C., Gremse, M., Heldt, C., Huhn, G., et al. (2004). BRENDA, the enzyme database: Updates and major new developments. *Nucleic Acids Research*, *32*(Database issue), D431–D433. http://dx.doi.org/10.1093/nar/gkh081.

Shannon, P., Markiel, A., Ozier, O., Baliga, N. S., Wang, J. T., Ramage, D., et al. (2003). Cytoscape: A software environment for integrated models of biomolecular interaction networks. *Genome Research*, *13*(11), 2498–2504. http://dx.doi.org/10.1101/gr.1239303.

Shneiderman, B. (1996). The eyes have it: A task by data type taxonomy for information visualizations. In: *Proceedings on 1996 IEEE Visual Languages* (pp. 336–343).

Strömbäck, L., Hall, D., & Lambrix, P. (2007). A review of standards for data exchange within systems biology. *Proteomics*, 7(6), 857–867. http://dx.doi.org/10.1002/pmic.200600438.

Suderman, M., & Hallett, M. (2007). Tools for visually exploring biological networks. *Bioinformatics (Oxford, England)*, *23*(20), 2651–2659. http://dx.doi.org/10.1093/bioinformatics/btm401.

Sugiyama, K., & Misue, K. (1991). Visualization of structural information: Automatic drawing of compound digraphs. *IEEE Transactions on Systems, Man, and Cybernetics*, *21*(4), 876–892. http://dx.doi.org/10.1109/21.108304.

Thiele, I., & Palsson, B. Ø. (2010a). A protocol for generating a high-quality genome-scale metabolic reconstruction. *Nature Protocols*, *5*(1), 93–121. http://dx.doi.org/10.1038/nprot.2009.203.

Thiele, I., & Palsson, B. Ø. (2010b). Reconstruction annotation jamborees: A community approach to systems biology. *Molecular Systems Biology*, *6*, 361. http://dx.doi.org/10.1038/msb.2010.15.

Tollis, I. G., Battista, G. Di, Eades, P., & Tamassia, R. (1999). *Graph drawing: Algorithms for the visualization of graphs*. USA: Prentice Hall. p. 397; Retrieved from, http://www.amazon.fr/Graph-Drawing-Algorithms-Visualization-Graphs/dp/0133016153.

Tutte, W. T. (2001). *Graph Theory (Cambridge Mathematical Library)*. Cambridge: Cambridge University Press. (p. 360); Retrieved from, http://www.amazon.com/Graph-Theory-Cambridge-Mathematical-Library/dp/0521794897.

Usadel, B., Poree, F., Nagel, A., Lohse, M., Czedik-Eysenberg, A., & Stitt, M. (2009). A guide to using MapMan to visualize and compare Omics data in plants: A case study in the crop species, Maize. *Plant, Cell & Environment*, *32*(9), 1211–1229. http://dx.doi.org/10.1111/j.1365-3040.2009.01978.x.

Ware, C. (2004). *Information visualization* (2nd ed.): *Perception for Design* (*Interactive Technologies*) (p. 486). Morgan Kaufmann. Retrieved fromhttp://www.amazon.com/Information-Visualization-Second-Edition-Technologies/dp/1558608192.

Wegner, K., & Kummer, U. (2005). A new dynamical layout algorithm for complex biochemical reaction networks. *BMC Bioinformatics*, *6*(1), 212. http://dx.doi.org/10.1186/1471-2105-6-212.

Zhang, L., Berleant, D., Ding, J., Cao, T., & Syrkin Wurtele, E. (2009). PathBinder—Text empirics and automatic extraction of biomolecular interactions. *BMC Bioinformatics*, *10*(Suppl. 1), S18. http://dx.doi.org/10.1186/1471-2105-10-S11-S18.

Zhang, P., Foerster, H., Tissier, C. P., Mueller, L., Paley, S., Karp, P. D., et al. (2005). MetaCyc and AraCyc. Metabolic pathway databases for plant research. *Plant Physiology*, *138*(1), 27–37. http://dx.doi.org/10.1104/pp. 105.060376.

Zhu, D., & Qin, Z. S. (2005). Structural comparison of metabolic networks in selected single cell organisms. *BMC Bioinformatics*, *6*, 8. http://dx.doi.org/10.1186/1471-2105-6-8.

CHAPTER THIRTEEN

Modelling Metabolic Networks—The Theories of Metabolism

Jean-Pierre Mazat*[,1], Bertrand Beauvoit†
*Université de Bordeaux, IBGC-CNRS UMR 5095, Bordeaux, France
†Bordeaux University, UMR1332 Fruit Biology and Pathology, INRA Centre of Bordeaux, Villenave d'Ornon
[1]Corresponding author: e-mail address: jpm@u-bordeaux2.fr

Contents

Abstract

There are several theoretical approaches to metabolism. Some of them give a greater place to the structure of the metabolic network symbolized by its stoichiometry matrix. The constraints in a metabolic network are very strong so that a lot of information can be derived from the stoichiometry matrix without any knowledge of the kinetic parameters of its steps, such that the elementary flux modes and even in some cases, the values of fluxes by the flux balance analysis method. However, a deeper understanding of metabolic network necessitates the values of the kinetic parameters. It is particularly

Advances in Botanical Research, Volume 67
ISSN 0065-2296
http://dx.doi.org/10.1016/B978-0-12-397922-3.00013-7

the case for a fine determination of the fluxes taking into account the metabolic regulations and of the determination of the flux and metabolites control coefficients. A short presentation of softwares and tools that can be used in the study of metabolic networks is presented at the end of this chapter.

1. WHICH RATE EQUATIONS FOR THE REACTIONS OF A METABOLIC NETWORK?

Modelling a metabolic network forces to write how the metabolites are transformed through the different reactions constituting the network. It means writing a rate equation expressing the rate of metabolites transformations as a function of the substrates and products concentrations and some external factors such as temperature, pH, regulators, etc. (Cornish-Bowden, 2012).

1.1. The Michaelis–Menten (Henri) equation

The first acceptable rate equation was given by Henri (Henri, 1902) and under its general from by Briggs and Haldane (Briggs & Haldane, 1925). It is, in my knowledge, one of the first models in biology. It is now known as the Michaelis–Menten equation (Michaelis & Menten, 1913).

$$V = \frac{V_M[S]}{K_M + [S]} \quad \text{with} \quad K_M = \frac{k_{-1} + k_2}{k_1} \quad \text{and} \quad V_M = k_2 \cdot [E]_{\text{total}} \tag{13.1}$$

It corresponds to the simple irreversible reaction of a single substrate according to the scheme of Fig. 13.1 with the representation of the Michaelis and Menten equation.

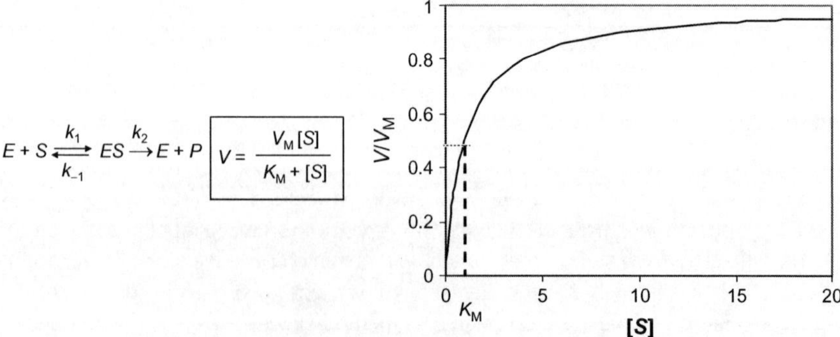

Figure 13.1 Représentation of the Michaelis–Menten equation with $K_M = 1$ (same units as [S]) and $V_M = 1$ (arbitraty units).

This equation and Fig. 13.1 shows the characteristics of an enzymatic reaction: (i) the first-order reaction at low substrate concentration, (ii) the zero order at high substrate concentration, that is, the existence of a saturation depending upon the enzyme concentration, (iii) the fact that $K_M \neq K_{1,eq}$, the equilibrium constant of the first reaction, that is, the binding constant of the substrate *S*.

1.2. The extended Michaelis–Menten reaction

However, it is clear that this equation is too simple and cannot always express the complexity of an enzyme reaction. It is an irreversible reaction involving only one substrate and no product while most of the reactions of the metabolism are reversible bi-substrate bi-product reactions. Nevertheless, its shape is very often experimentally encountered and it is then possible to define a V_M and an apparent K_M, which can vary according to the presence of other metabolites (substrates, products and regulators).

It is thus possible to extend the equation to the case of an irreversible bimolecular reaction under the form (Fig. 13.2):

$$V = \frac{V_f \times \frac{[A]}{K_A} \times \frac{[B]}{K_B}}{\left[1 + \frac{[A]}{K_A}\right] \times \left[1 + \frac{[B]}{K_B}\right]}$$

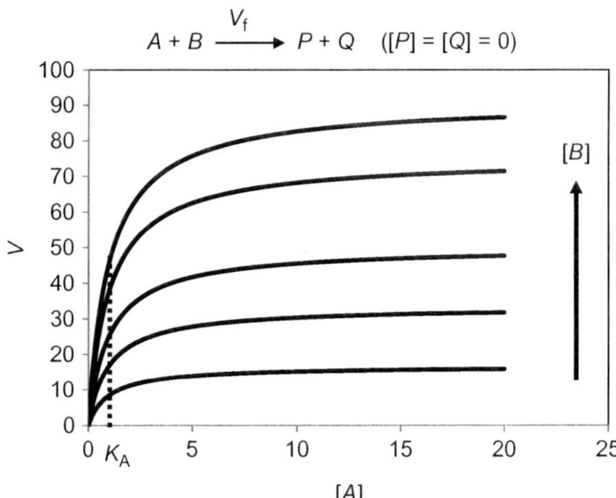

Figure 13.2 Extended irreversible Michaelis–Menten reaction for a bimolecular reaction. Variation of *V* as a function of [*A*] at different concentrations of [*B*]. $K_A = K_B = 1$ (same units as [*A*] and [*B*]) and $V_M = 100$ (arbitrary units).

The extension can be continued to the case where the reaction is reversible:

$$A + B \leftrightarrow P + Q$$

One has to consider a reverse term of the same type for the products with a competitive inhibition for their corresponding substrates:

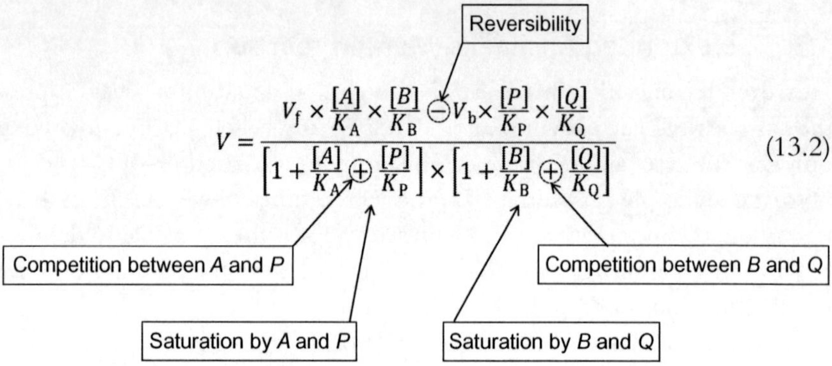

$$V = \frac{V_f \times \frac{[A]}{K_A} \times \frac{[B]}{K_B} \ominus V_b \times \frac{[P]}{K_P} \times \frac{[Q]}{K_Q}}{\left[1 + \frac{[A]}{K_A} \oplus \frac{[P]}{K_P}\right] \times \left[1 + \frac{[B]}{K_B} \oplus \frac{[Q]}{K_Q}\right]} \tag{13.2}$$

Reversibility

Competition between *A* and *P*

Competition between *B* and *Q*

Saturation by *A* and *P*

Saturation by *B* and *Q*

V_f is the maximal forward reaction and V_b is the maximal backward reaction. K_A, K_B, K_P, K_Q are the apparent Michaelis constants for metabolites *A*, *B*, *P* and *Q*.

When the concentration of the products (resp. substrates) is zero, the kinetics is michaelian for one substrate (resp. one product) at fixed concentration of the other. We (and others) (Cornish-Bowden & Hofmeyr, 2005; Raïs et al., 2001) have shown that in many cases this type of equation is sufficient to have a satisfactory representation of the kinetics over a broad range of substrate and product concentrations.

1.3. The Haldane relationship

However, the kinetic parameters are not independent. At equilibrium, $V = 0$ so that:

$$\begin{gathered} V_f \times \frac{[A_{eq}]}{K_A} \times \frac{[B_{eq}]}{K_B} = V_b \times \frac{[P_{eq}]}{K_P} \times \frac{[Q_{eq}]}{K_Q} \text{ and thus :} \\ \frac{[P_{eq}]}{[A_{eq}]} \times \frac{[Q_{eq}]}{[B_{eq}]} = K_{eq} = \frac{V_f}{V_b} \times \frac{K_P}{K_A} \times \frac{K_Q}{K_B} \end{gathered} \tag{13.3}$$

$K_{eq} = \frac{V_f}{V_b} \times \frac{K_P}{K_A} \times \frac{K_Q}{K_B}$ is the Haldane relationship linking the six kinetics parameters to the equilibrium constant K_{eq} which is independent of any catalytic enzyme. In order to not forget this relationship, V_b can be derived from the Haldane relationship and introduced in the rate equation which gives:

$$V = \frac{\left[1 - \frac{1}{K_{eq}} \frac{[P] \cdot [Q]}{[A] \cdot [B]}\right] \left[V_f \times \frac{[A]}{K_A} \times \frac{[B]}{K_B}\right]}{\left[1 + \frac{[A]}{K_A} + \frac{[P]}{K_P}\right] \times \left[1 + \frac{[B]}{K_B} + \frac{[Q]}{K_Q}\right]} \quad (13.2')$$

The drawback of this representation is that the symmetry between substrates and products is broken. The advantage is that the equilibrium constant K_{eq} is taken into account. It is very important to incorporate the equilibrium constants of all reactions in the rate equation, because it participates with the metabolite concentrations to the definition of the directions of the reactions. Depending on the physiological states of the cell, and on the metabolite concentrations, some reactions can change direction (it is also the case in metabolic diseases in human). It is important to know the exact point at which these changes occur. K_{eq} are strong constraints that must be taken into account in metabolic network modelling.

However, in some instances the mechanism is known and the corresponding 'exact' equations can be taken in the place of the generalized Michaelis–Menten equation above. This is the case of the ping-pong mechanism involving an enzyme intermediate and giving rise to simple equations with peculiar behaviours of the double reciprocal plots.

1.4. The Mass-action law

When one wants to just get an idea of the variations of the fluxes as a function of metabolite concentrations, without taking into account the saturation processes, one can simplify Eq. (13.2) in Eq. (13.4):

$$V = V_f \times \frac{[A]}{K_A} \times \frac{[B]}{K_B} - V_b \times \frac{[P]}{K_P} \times \frac{[Q]}{K_Q}$$

which can be simply written:

$$V = V_f \times [A] \times [B] - V_b \times [P] \times [Q] \quad (13.4)$$

K_{eq} can be introduced in Eq. (13.4) like in Eq. (13.2′):

$$V = V_f \left([A] \times [B] - 1/K_{eq} \times [P] \times [Q]\right) \quad (13.4')$$

The advantage of such equations is that they simply express the variations as a function of substrates (increasing function) or as a function of products (decreasing function) and can be used in big networks as a preliminary approach to see how fluxes change and tend to a steady state. The drawback is that these equations are only valid on a limited range of metabolite concentrations because they are non-saturable functions increasing (resp. decreasing) indefinitely as a function of substrate or product concentrations.

1.5. Metabolic regulations

The enzyme activities can be regulated (usually inhibition) by the metabolites of the network including the substrates and products of a reaction. Two types of inhibition have been described:

Competitive inhibition (competitive with a substrate). In this case, the inhibitor competes with the substrate for its site. It is typically the case of the product P (resp. Q) with the substrate A (resp. B) in the reaction of Fig. 13.2. It is usually illustrated in the scheme depicted in Fig. 13.3, where I is the inhibitor competing with the substrate S.

In this case, the effect of the inhibitor is on the K_M, which becomes

$$K'_M = K_M\left(1+\frac{[I]}{K_I}\right) \tag{13.5}$$

where K_I is the inhibition constant (can be the binding constant of I to the substrate site). V_M remains the same because high concentrations of substrate can eliminate the inhibitor.

Non-competitive inhibition (with the substrate). It is depicted in the scheme depicted in Fig. 13.4 in which the inhibitor binds another site than the substrate.

In this case the K_M is unchanged but the V_M becomes:

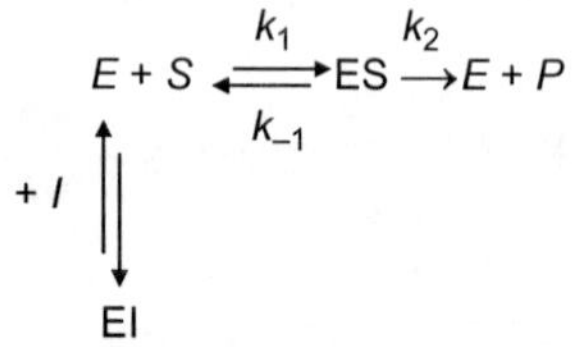

$$\begin{array}{ccc} E+S & \underset{k_{-1}}{\overset{k_1}{\rightleftharpoons}} & ES \overset{k_2}{\longrightarrow} E+P \\ {\scriptstyle +I}\updownarrow & & \\ EI & & \end{array}$$

Figure 13.3 Mechanism of competitive inhibition with substrate.

$$\begin{array}{ccc} E+S & \underset{k_{-1}}{\overset{k_1}{\rightleftharpoons}} & ES \overset{k_2}{\longrightarrow} E+P \\ {\scriptstyle +I}\updownarrow & & {\scriptstyle +I}\updownarrow \\ EI & \rightleftharpoons & ESI \end{array}$$

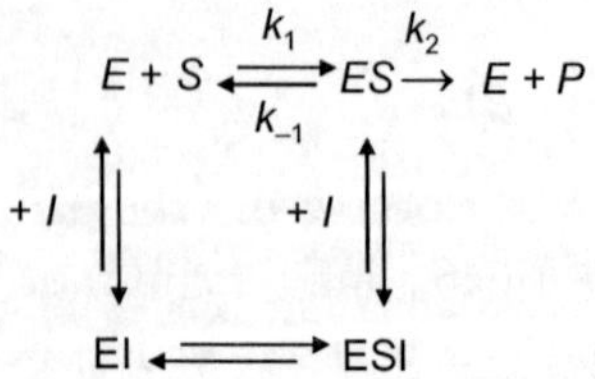

Figure 13.4 Mechanism of non-competitive inhibition with substrate.

$$V'_M = V_M \times \frac{1}{\left(1+\frac{[I]}{K_I}\right)} \qquad (13.6)$$

One of these terms (affecting K_M or V_M) or both can be incorporated in Eq. (13.2) or Eq. (13.4) to account for the known inhibitions in a metabolic network.

1.6. The generalized Hill equation

If most of the kinetic curves exhibit a hyperbolic shape with an asymptote (V_{Max}), some however show a sigmoid behaviour, with a low rate at low substrate concentration and an inflection point around the $V_{0.5}$ value. Equation (13.2) cannot fit such behaviour. However, in 1910, Hill (1910) proposed the following equation for hemoglobin where Hb is free hemoglobin and HbO_2 is hemoglobin complexed with O_2:

$$Y = \frac{HbO_2}{Hb_{tot}} = \frac{[O_2]^n}{K_d^n + [O_2]^n}$$

n is known as Hill number (sometimes written n_h).

Figure 13.5 shows the shape of this curve for $n=2$ and $K_d=1$

This equation can be easily adapted to enzyme kinetics presenting a sigmoidal behaviour for a substrate S under the form (Hofmeyr & Cornish-Bowden, 1997):

$$V = V_M \frac{[S]^n}{K_S^n + [S]^n} = V_M \frac{\left(\frac{S}{K_S}\right)^n}{1+\left(\frac{S}{K_S}\right)^n} \qquad (13.7)$$

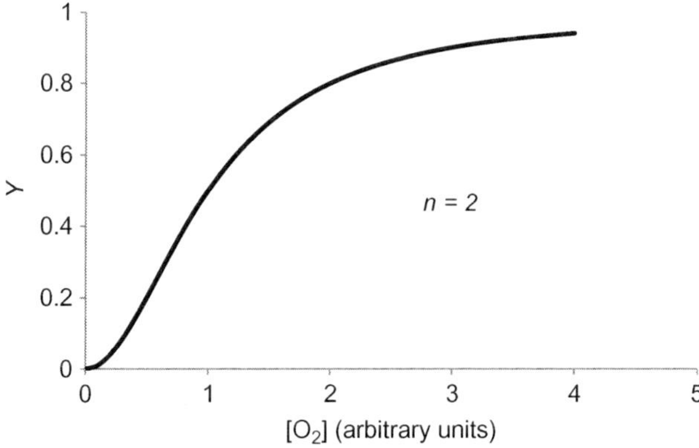

Figure 13.5 Binding of oxygen on haemoglobin ($n \sim 2$).

For each substrate/product of Eq. (13.2) presenting a sigmoidal shape, a factor n_A, n_B, n_P or n_Q can be found to fit the kinetic curve according to the generalized reversible Hill equation:

$$V = V_M \frac{V_f \times \left(\frac{A}{K_A}\right)^{n_A} \times \left(\frac{B}{K_B}\right)^{n_B} - V_b \times \left(\frac{P}{K_P}\right)^{n_P} \times \left(\frac{Q}{K_Q}\right)^{n_Q}}{\left(1 + \left(\frac{A}{K_A}\right)^{n_A} + \left(\frac{P}{K_P}\right)^{n_P}\right) \times \left(1 + \left(\frac{B}{K_B}\right)^{n_B} + \left(\frac{Q}{K_Q}\right)^{n_Q}\right)} \tag{13.8}$$

Of course, the values $n_A = 1$ (resp. $n_B = 1$, etc.), etc., gives back the generalized reversible Michaelis–Menten equation with the hyperbolic shape for the substrate A (resp. B, etc.).

1.7. Sigmoid inhibition

Similar to substrates and products, inhibitors can exhibit sigmoid behaviour. In this case, one can replace the term $(1 + [I]/K_I)$ by $(1 + (I/K_I)^n)$ and incorporate it to K_M or V_M.

Remarque: most of these equations are already prewritten in softwares dedicated to metabolism modelling (Copasi for instance).

2. The Structure of Metabolic Network: The Stoichiometry Matrix

It is useful for a theoretical analysis of the structure of metabolic networks, to represent metabolic networks by a mathematical object that can be theoretically handled. This leads to the concept of stoichiometry matrix (Reder, 1988).

A metabolic network involves 'r' reactions $R_1, R_2, \ldots, R_r$ labelled by their enzymes $E_1, E_2, \ldots, E_r$ and their rates $V_1, V_2, \ldots, V_r$. These reactions in turn involve m metabolites $X_1, X_2, \ldots, X_m$ with stoichiometries $n_{i,j}$, that is, $n_{i,j}$ molecules of the metabolite X_i react in the reaction V_j so that one can write:

$$\frac{dX_i}{dt} = \sum_{j=1}^{r} n_{i,j} V_j$$

The stoichiometries are counted positively for the products (synthesis of the metabolite) and negatively for the substrates (consumption of the metabolite). It means that a direction is defined for each reaction, particularly for the reversible ones; these directions are arbitrary.

All the stoichiometry coefficients can be collected in a table with m rows (the substrates) and r columns (the reactions). Let us give the example of the metabolic network below:

It can be decomposed in the following reactions:

$$
\begin{aligned}
&R_1 :\rightarrow X_1,\\
&R_2 : X_1 :\rightarrow X_2\\
&R_3 : X_2 :\rightarrow\\
&R_4 : X_1 :\rightarrow X_3\\
&R_5 : X_3 :\rightarrow\\
&R_6 : X_3 :\rightarrow X_2
\end{aligned}
$$

The reactions without input or output represent the entry or exit of the metabolic network, defined as an open system and delimited by the dashed frame. It means that we do not care of what is outside the dashed frame. It can be metabolites in large amount (nearly constant, V_1), or products produced at constant rate (V_3 and V_5). The metabolic network is inside the dashed frame and the only variables are the metabolite concentrations X_1, X_2 and X_3.

The stoichiometry matrix is represented below:

$$
\begin{array}{cc}
 & \begin{array}{cccccc} R_1 & R_2 & R_3 & R_4 & R_5 & R_6 \\ \downarrow & \downarrow & \downarrow & \downarrow & \downarrow & \downarrow \end{array} \\
N = \begin{array}{c} X_1 \\ X_2 \\ X_3 \end{array} & \begin{bmatrix} +1 & -1 & 0 & -1 & 0 & 0 \\ 0 & +1 & -1 & 0 & 0 & +1 \\ 0 & +1 & -1 & +1 & -1 & -1 \end{bmatrix}
\end{array}
$$

For instance, in reaction 2, R_2, one molecule of X_1 is consumed (-1) and one molecule of X_2 is produced ($+1$), etc. The stoichiometry coefficients are usually integer but can be fractionals, depending on the way the reaction is represented.

The Table 13.1 gives some examples of stoichiometry matrix:

To understand the interest of the stoichiometry matrix let us take again the example of Fig. 13.6.

The time course of metabolite concentrations is given by the three equations:

$$
\begin{aligned}
\frac{dX_1}{dt} &= V_1 - V_2 - V_4\\
\frac{dX_2}{dt} &= V_2 + V_6 - V_3\\
\frac{dX_3}{dt} &= V_4 - V_5 - V_6
\end{aligned}
\tag{13.9}
$$

We can formally write this ensemble of ordinary differential equations (dynamical system) in the simple form:

Table 13.1 Examples of metabolic network with their associated stoichiometry matrix

	Network	Stoichiometric matrix
Example 1	$* \xrightarrow{R_1} X_1 \xrightarrow{R_2} *$	$\begin{bmatrix} -1 & +1 \end{bmatrix}$
Example 2	$* \xrightarrow{R_1} X_1 \xrightarrow{R_2} X_2 + X_3 \xrightarrow{R_3} *$	$\begin{bmatrix} +1 & -1 & 0 \\ 0 & +1 & -1 \\ 0 & +1 & -1 \end{bmatrix}$
Example 3	$* \xrightarrow{R_1} X_1 \xrightarrow{R_2} 2X_2 \xrightarrow{R_3} *$	$\begin{bmatrix} +1 & 0 & 0 \\ 0 & +2 & -2 \end{bmatrix}$
Example 4	$* \xrightarrow{R_1} X_1$; $X_1 \xrightarrow{R_2} *$; $X_1 \xrightarrow{R_3} *$	$\begin{bmatrix} +1 & -1 & -1 \end{bmatrix}$
Example 5	$X_1 + * \xrightarrow{R_1} X_2$; $X_2 \xrightarrow{R_2} X_3 + *$; $X_3 \xrightarrow{R_3} X_1$	$\begin{bmatrix} -1 & 0 & +1 \\ +1 & -1 & 0 \\ 0 & +1 & -1 \end{bmatrix}$
Example 6	$\xrightarrow{R_1} X_1$; $X_1 \xrightarrow{R_2} X_2$; $X_2 \xrightarrow{R_3}$; $X_1 \xrightarrow{R_4} X_3$; $X_3 \xrightarrow{R_5}$; $X_3 \xrightarrow{R_6} X_2$	$\begin{bmatrix} +1 & -1 & 0 & -1 & 0 & 0 \\ 0 & +1 & -1 & 0 & 0 & +1 \\ 0 & 0 & 0 & +1 & -1 & -1 \end{bmatrix}$

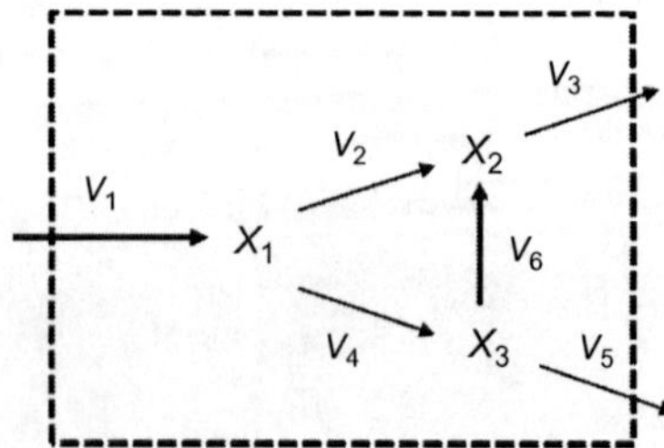

Figure 13.6 Example of a metabolic network (Example 6 of table 13.1). The dashed frame indicates the inner part of the metabolic network considered as an open system with input and output (V_1, V_3 and V_5). The orientation of the arrows is arbitrary.

$$\frac{dX}{dt} = N \cdot V \tag{13.10}$$

where $dX/dt = \begin{bmatrix} dX_1/dt \\ dX_2/dt \\ dX_3/dt \end{bmatrix}$ and $V = \begin{bmatrix} V_1 \\ V_2 \\ V_3 \\ V_4 \\ V_5 \\ V_6 \end{bmatrix}$.

In doing so we replace a dynamical system of several equations (can be several hundreds in large metabolic network) with only one 'matricial' equation. Rules exist to treat such equations.

Four remarks concerning the concept of stoichiometry matrix:

1. The arrow representing an enzyme reaction does not imply that the reaction is irreversible. It simply indicates an arbitrary orientation of the reaction, which allows the definition of a metabolite as a substrate or a product and gives a sign to the rate function V.
2. There is a one-to-one correspondence between a metabolic network and its stoichiometry matrix. We have shown that the determination of the stoichiometry matrix was completely determined by the metabolic network. It is easy to verify that a unique metabolic network can be built from a given stoichiometry matrix.
3. It must be finally emphasized that the stoichiometry coefficients are not necessarily $+1$, -1 or 0. Integer coefficients ($+2$, -3, etc.) can be encountered depending on the reactions. The stoichiometry coefficients can be even fractional, depending on the way the reaction is represented.
4. In the stoichiometry matrix of example 5, the sum of the rows: $[-1 \ \ 0 \ \ +1]$, $[+1 \ \ -1 \ \ 0]$ and $[0 \ \ +1 \ \ -1]$ term by term gives zero; it corresponds to the sum of the differential equations:

$$dX_1/dt + dX_2/dt + dX_3/dt = 0$$

which means $X_1 + X_2 + X_3 = \text{constant}$, because X_1 is transformed in X_2 which is transformed in X_3 which eventually gives X_1 but with their sum remaining constant. It means also that the three variables X_1, X_2 and X_3 are not independent. For instance, X_3 can be determined from X_1 and X_2. One of the variables can be calculated from the other two. The number of independent variable is 2. 2 is also the **rank** of the stoichiometry matrix, that is, the number of independent rows (or columns), that is, the number of rows or columns necessary to calculate the others. It is an important concept that will be dealt with again in the following (see Klipp et al., 2009). The rank is the number of independent variables of the metabolic network; it is the total number of variables minus the number of relationships (mainly conservation relationships) between these variables. The software 'CONTROL' (Letellier, Reder, & Mazat, 1991) calculates the rank of the stoichiometry matrix (but for matrix limited to 20 rows and 20 columns). The software 'Metatool' (http://pinguin.biologie.uni-jena.de/bioinformatik/networks/) calculates the stoichiometry matrix associated to the list of reactions of a metabolic network.

5. Equation (13.10), $dX/dt = N \cdot V$ is an important reaction, which we will take into account later. It shows that it is possible to consider separately the structure of the metabolic network, contained in the stoichiometric matrix N (which tell us if there are loops, branches, linear parts, etc.; see below) and the dynamic of the metabolic network embedded in the rate vector V. In different conditions, in different tissues, in different pathologies, the N matrix will remain the same (the metabolic network does not change) but the rate equations, collected in the rate column vector V, will differ (different isoenzymes, different regulations, etc.), which gives rise to different behaviours (different fluxes).

We will see in the following, some other great interest in introducing the concept of stoichiometry matrix.

3. THE STEADY STATE

3.1. Introduction

In order to calculate the fluxes in a metabolic network, one used to consider the intermediate metabolite concentrations as constant, that is, their rate of consumption balances their production. This hypothesis is particularly well adequate in plants where the variations are rather slow. It does not mean that things are fixed, but just that their changes are slow as compared to the metabolism balance. All is as if we have to analyze a succession of slowly changing quasi steady states.

In the following, we will accept the steady-state principle, knowing that it is a minimal hypothesis not always satisfied.

3.2. The Kernel of the stoichiometry matrix

The fundamental matrix Eq. (13.10) allows us to write the condition of steady state in a simple expression:

$$\frac{d[X]}{dt} = N \cdot V = 0 \tag{13.11}$$

In linear algebra, this equation means that the rate vector V (which is the column of all the rates of the network) belongs to what is called the Kernel of N (Ker(N)), the definition of which is just equation (13.11), that is, the set of vectors V, such that $N \cdot V = 0$.

The concept of Kernel of matrix N depicts in a condensed way, the simple biochemical property that, at steady state, for each internal metabolite the rate of production equals the rate of consumption. The benefit of this representation is that we can take advantage of the general properties (not detailed here) of the Kernel of linear applications represented by matrices. We will use here one property, which is that the Kernel is a vectorial space.

Let us see what it means in the case of example 1, the simplest of all metabolic networks.

In this network, according to the rule of matrix product (elements of the row times the corresponding elements of the column), we obtain:

$$N \cdot V = [-1 \quad +1] * \begin{bmatrix} V_1 \\ V_2 \end{bmatrix} = V_1 - V_2 = 0 \text{ i.e. } V_1 = V_2.$$

In the rate plane (V_1, V_2) (called phase plan), it means that the stationary solutions are on the first bisecting line (Fig. 13.7) (the consumption of X_1 equals its synthesis).

Furthermore, the rate equation V_1 and V_2 are both functions of X_1 so that for each X_1 values, both V_1 and V_2 are determined: at each X_1 value corresponds a pair ($V_1(X_1)$; $V_2(X_1)$), that is, a point in the phase plan (V_1; V_2) of Fig. 13.7. Thus, when X_1 varies in a reasonable range (between 0 and 10 mM for instance), the point ($V_1(X_1)$; $V_2(X_1)$) will describe a curve $V(p)$ in the phase plan (p symbolizes all the parameters involved in the rates equations; enzyme concentration, K_m, temperature, pH, etc.). The steady states, if they exit are situated at the intercept of the Kernel of N (the first bisecting line $V_1 = V_2$) and of $V(p)$.

Mathematically, $V_1 = V_2$, that is, $V_1(X_1, p) = V_2(X_1, p)$ is an equation with the variable X_1, the solution(s) of which, if they exist are the concentrations $X_{1,0}$ at steady state.

When we replace this $X_{1,0}$ value in $V_1(X_{1,0}, p) = F_1$ or $V_2(X_{1,0}, p) = F_2$, one gets the stationary flux already graphically obtained in Fig. 13.7 (with in this case $F_1 = F_2$). Note that steady states do not necessarily exist (if, for instance, the rate of X_1 production is always higher than its consumption, i.e. curve $V(p)$ is on one

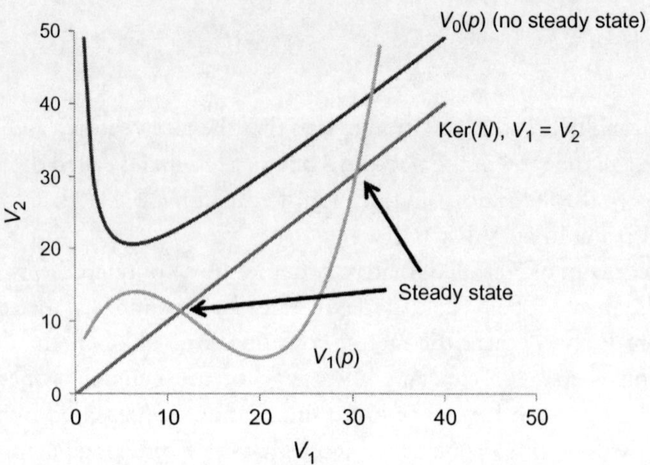

Figure 13.7 The steady states are at the intercepts of the Kernel of the stoichiometry matrix (first bisecting line) and the ensemble of allowed pair of rates (for the same value of X_1, the common metabolite). In the case of $V_0(p)$, there is no intercept and thus no steady state ($V_1 > V_2$ and X_1 increases continuously). In the case of $V_1(p)$, there are two possible steady states. p symbolizes the parameters on which V_1 and V_2 are dependent. (For color version of this figure, the reader is referred to the online version of this chapter.)

side of the bisecting line). When a steady state exists, it is not necessarily unique (the curve $V(p)$ intercepts several times the bisecting line).

In order to understand better the interest of the matrix approach for finding all the steady states of metabolic networks, we will take once more the example 6, a little more complex. In this case, the stoichiometry matrix is the 3×6 matrix (3 metabolites and 6 rates):

$$N = \begin{bmatrix} 1 & -1 & 0 & -1 & 0 & 0 \\ 0 & 1 & -1 & 0 & 0 & 1 \\ 0 & 0 & 0 & 1 & -1 & -1 \end{bmatrix}$$

We will work, as above, in the rate space, the dimension of which is now 6 (6 rate equations $V_1, \ldots, V_6$). N is a linear application from this space in the space of metabolites (or in the space of their derivatives), the dimension of which is less or equal to 3 (the dimension of metabolites can be less than 3 if the metabolites are not independent, which is not the case here); thus the dimension of the image of the rate space (by N) is of a dimension less or equal to 3. It can be easily seen that in our case its dimension is exactly 3 (in N matrix, the column vectors 1, 2 and 4, for instance, are independent).

$$K_1 = \begin{bmatrix} 1 \\ 1 \\ 1 \\ 0 \\ 0 \\ 0 \end{bmatrix} \qquad K_2 = \begin{bmatrix} 1 \\ 0 \\ 0 \\ 1 \\ 1 \\ 0 \end{bmatrix} \qquad K_3 = \begin{bmatrix} 0 \\ -1 \\ 0 \\ 1 \\ 0 \\ 1 \end{bmatrix}$$

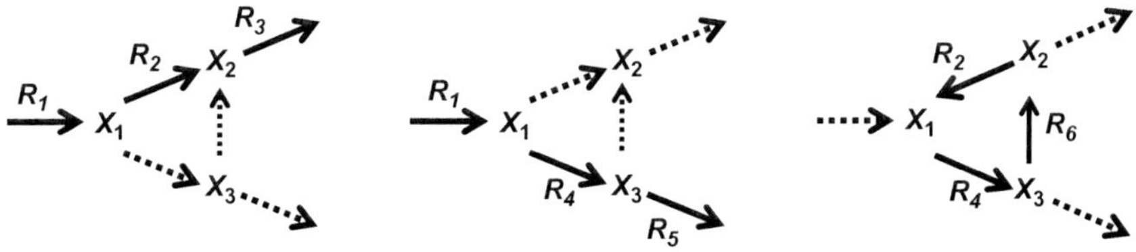

Figure 13.8 Three independent metabolic pathways, represented by the rate vectors above, which can be used to express any flux at steady-state. They constitute a basis of the vectorial space of all flux solution at steady-state (see an example in Fig. 13.9).

It can be demonstrated that:

Dimension of Image space by N (3 in our case, the metabolites) + Dimension of the Kernel of N = Dimension of the rate space (6 in our case, the rates). So that the dimension of Kernel (N) is 3. It means that all steady-state solutions can be expressed as a linear combination of three independent rate vectors. It is rather easy to find three independent vectors of the Kernel (solution of $N \cdot V = 0$) (Fig. 13.8).

Each of these vectors (a basis of the vectorial space) represents a particular metabolic pathway that is represented below each basis vector. It is easy to understand in this figure that any set of fluxes at steady state through this metabolic network (example 6) can be expressed as a linear combination of these particular metabolic pathways, that is, all solutions can be written: $V = \lambda_1 K_1 + \lambda_2 K_2 + \lambda_3 K_3$ where $\lambda_1, \lambda_2, \lambda_3$ are numbers. For instance, the vector:

$$V = \begin{bmatrix} V_1 = 3 \\ V_2 = 2 \\ V_3 = 1 \\ V_4 = 1 \\ V_5 = 2 \\ V_6 = -1 \end{bmatrix} = 1K_1 + 2K_2 - 1K_3$$

is a solution, which can be checked on the right-hand part of Fig. 13.9, on which different rate values are indicated. Note that due to the arbitrary orientation, the reaction 6 is negative because it occurs in the other direction.

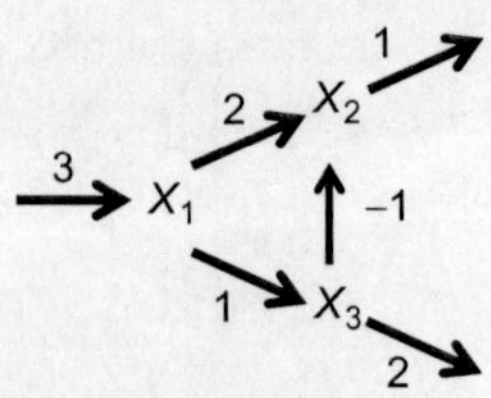

Figure 13.9 Example of possible fluxes in the metabolic network of example 6. The rates are given by $V = 1.\ K_1 + 2.\ K_2 - 1.\ K_3$.

Apparently, the problem of the description of the stationary fluxes in a metabolic network can be solved using a basis of the Kernel of the stoichiometry matrix N: all the stationary fluxes belong to this Kernel and can thus be expressed as a linear combination of basis vectors of the Kernel. Algorithms can be used to find such a basis and to express the fluxes (Letellier et al., 1991).

I said 'apparently' because two problems remain.

The first problem is that biochemists would like to know all the possible pathways in a metabolic network instead of their theoretical description.

The second problem is that it is clear that some theoretical pathways will actually be impossible because some reactions are irreversible or quasi-irreversible. For instance, if reaction V_6 is irreversible in the direction indicated by the arrow, the solution above: $V = 1K_1 + 2K_2 - 1K_3$ is impossible because V_6 will be operating in the wrong direction.

Both problems lead to the concept of elementary flux modes (efms).

4. THE STRUCTURE OF METABOLIC NETWORK: THE ELEMENTARY FLUX MODES (efms)

4.1. Definition of elementary flux modes

The notion of elementary mode is a new, very deep concept in metabolism developed by Stefan Schuster (Schuster & Hilgetag, 1994). It is the right response to an apparent simple question: what is a metabolic pathway? The answer seems to be rather simple and well illustrated by all the biochemistry textbook. However, if we take the example of glycolysis, we immediately realize that there are some problems in defining this well-known metabolic pathway. If the beginning of the pathway is well defined with glucose entry, the end fluctuates between pyruvate and lactate. Even the beginning might well be Glucose-1-phosphate. In the plant, the entry can be defined as sucrose. If we look at all the intermediates of the pathway, we see that they have other possible fates than to continue in glycolysis; the glucose-6-phosphate can feed the pentose phosphate pathway, the breakdown of fructose-1,6-biphosphate can lead to glycerol instead of pyruvate, etc. In addition, the representation of the pathway in the textbooks does not usually take into account the cofactors such as ATP/ADP and NAD/NADH, which participate in many other reactions. The glycolysis of plant biochemists with its metabolism of sugar-phosphate is not exactly the same as the one of animal biochemists. The TCA cycle is not exactly the same for microbiologists and animal biochemists (glyoxylate shunt). The answer to all these questions and a rigorous definition of a metabolic pathway, which can be used in mathematical modelling, was given by the concept of elementary mode.

An efm is a minimal set of reactions that can operate at steady state using irreversible reactions in the right direction (Schuster, Fell, & Dandekar, 2000; Schuster & Hilgetag, 1994; Stelling, Klamt, Bettenbrock, Schuster, & Gilles, 2002). 'Minimal' means that the suppression of any reaction of the set will prevent the reaching of a steady state. In other words, an efm cannot be further reduced nor be included in another bigger efm.

The notion of efm is illustrated in Fig. 13.10 (example 6).

If we consider that all reactions are irreversible in the direction of the arrows, the K_1 and K_2 vectors are elementary modes (with all reactions in the right direction); they are minimal because if one reaction is suppressed, the steady-state property is lost (the pathway is interrupted). On the contrary, K_3 (Fig. 13.8) is not an elementary mode, because

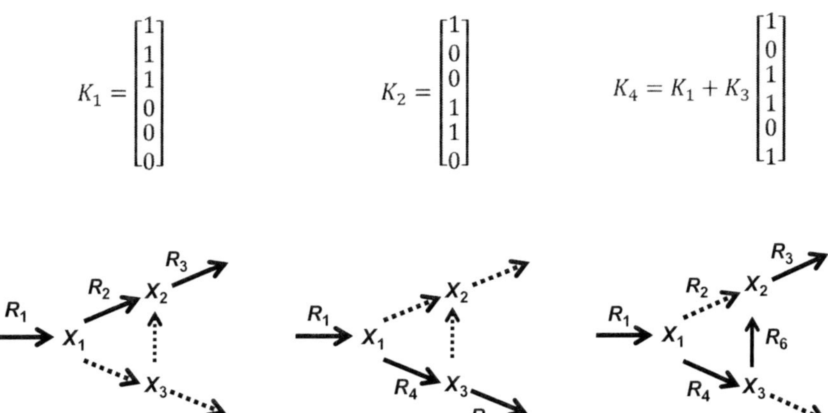

Figure 13.10 Description of all elementary flux modes of example 6 assuming all reactions irreversible in the direction of the arrows.

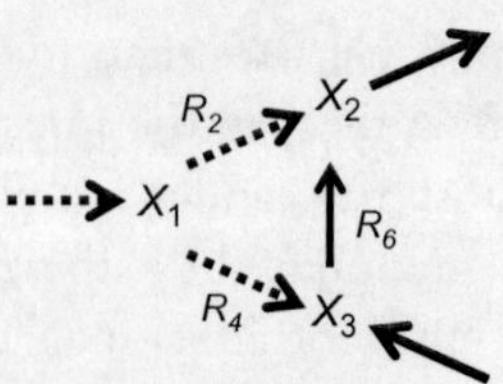

Figure 13.11 Additional efm when the reaction R_5 is considered as reversible.

reaction 2 is operating in the wrong direction. The only other elementary mode is $K_4(= K_1 + K_3)$. It can be shown that K_1, K_2 and K_4 are the only elementary modes of this metabolic network. If it is accepted that some of the reactions are reversible, new elementary modes are found (Fig. 13.11 with reaction R_5 reversible).

For a metabolic network, the set of elementary modes corresponds to all possible (minimal) pathways leading from one (or several) entry metabolite(s) to one (or several) output metabolite(s) or to cycles.

It appears that for some pairs of entry and output metabolites, several pathways (efm) are possible. This is probably one of the reasons for the well-known robustness of metabolic networks and of the observed results that gene inactivation often has little effect on cell survival.

4.2. Usefulness of the analysis of a metabolic network in elementary modes

Elementary modes are the only possible pathways taken in a metabolic network. For this reason, it is useful to know them. The question as to whether a given efm is actually used will depend on the rate functions. For instance, in the case of example 6, if V_4 or V_5 are null or quasi-null, the K_2 efm (pathway) will not be operating.

It has been established that tissue differences, are, at least partly associated with metabolic differences, which correspond to a preferential use of different efms (or group of efms).

The robustness of metabolic networks is due to the presence of several pathways, several efms, connecting two metabolites. For this reason, the knowledge of elementary modes can be useful to predict those mutations that can be tolerated and those that will be lethal. In contrast, in biotechnology, one can be led to favour some particular elementary mode, the yield of which (for a given production) is high (Carlsson, Fell, & Srienc, 2002; Poolman, Assmus, & Fell, 2004; Poolman, Venkatesh, Pidcock, & Fell, 2004).

From a general point of view, all discussion on metabolism functioning has to be done in terms of elementary modes. Examples of such use can be found in Poolman, Assmus, and Fell (2004), Poolman, Olcer, Lloyd, Raines, and Fell (2001), Poolman, Venkatesh, Pidcock, and Fell (2004) and Fell (1998).

An interesting analysis of the TCA cycle in plants is given in Sweetlove, Beard, Nunes-Nesi, Fernie, & Ratcliffe (2010). The authors show that 'the organization of carboxylic acid metabolism in plants is highly dependent on the metabolic and physiological demands of the cell'. They particularly show that non-cyclic TCA flux modes occur in leaves in the light, in some developing oilseeds and in anoxia. Such an analysis in elementary modes is necessary to understand the results of labelling studies or of genes modulation.

Finally, it should be noted that a quasi-equivalent analysis called 'extreme pathways' has been proposed by Palsson and colleagues (Covert et al., 2001; Papin, Reed, & Palsson, 2004; Papin et al., 2004; Schilling, Letscher, & Palsson, 2000).

4.3. The problem of efms analysis

The problem of this approach is that the number of elementary modes increases rapidly with the number of reactions. For example, the metabolic map of mitochondrial energy metabolism involves, according to the tissue or organism, about 40 reactions and about 30 metabolites. The number of efm in this case varies around 8000 (Pérès et al., 2011). Even if one can think that some of them are not possible due to kinetic reasons, a great number of efm can, however, remain. Because it is not possible to look at all the 8000 efm, their classification appears as a pertinent answer to this problem. Several kinds of classification can be envisaged, based on reactions, balance equations, etc. One can also consider efm as a sequence of reactions and treat them like protein or nucleic acid sequences using sequence comparison software. The traditional methods of classification do not suit very well in this case for essentially two reasons. The first reason is that some efm (i.e. some sequences of reactions) are atypical and disturb the classification (but may be interesting to analyze). The second reason is that most of the classification methods are exclusive and do not allow a given efm to belong to several classes. However, we would like indeed to put an efm-involving glycolysis and TCA cycle in a 'glycolysis' class and in a 'TCA-cycle' class. This led Pérès et al. (2011) to develop a new classification algorithm, ACoM (http://savannah.nongnu.org/projects/acom-c), which groups together the efm through common motifs and permitted an efm to belong to several classes. Another advantage of the method is that the result does not depend upon the starting point (starting efm), used to perform the classification. Applied to mitochondrial energetic metabolism, this software allowed the grouping of about 8000 efm into 15–35 classes to which a physiological meaning could be associated most of the time. Generally, ACoM, which can be viewed as a bi-clustering method, helps to give biological meaning to the different efm and to the relatedness between reactions.

5. METABOLIC CONTROL ANALYSIS

In the previous paragraphs, we studied the existence and properties of steady states (supposed stables) of fluxes and metabolites concentrations in metabolic networks. The question is now: how these fluxes change when the conditions are changed (initial conditions, environmental conditions, parameters, etc.), in other words what is the sensitivity of the steady states towards various parameters, that is, towards the variation of the rate function of any step in the metabolism.

Let us count a story. Once upon a time, a geneticist, Henrik Kacser, after having suppressed a step in the arginine biosynthetic pathway in *Neurospora crassa*, looked for revertants from this mutation, that is, strains growing again on a medium without arginine. He found some and checked the return of the activity, which was suppressed in the parent mutant. At his first assay, he found mutants with no return of activity. Surprised, he did the test again more carefully and found a very low activity, 5–10% of the normal one he had missed the first time. He concluded that only a low amount of this activity was enough to sustain the growth. He then decided to study more carefully the phenomenon in inhibiting progressively the activities of each of the four enzymes of the pathway. He obtained curves with the shape of the one in Fig. 13.12, that is, a slow decrease of arginine biosynthesis when the activity of a given step is decreased until a threshold that can be rather high (corresponding to a 70–95% inhibition) after which a strong inhibition is obtained (Flint, Porteous, & Kacser, 1980; Flint et al., 1981). At the beginning, the decrease is quasi linear and permits to easily define a constant slope C_i linking the decrease of the flux of arginine biosynthesis to the decrease in the rate v_i of step I:

$\Delta F = C_i.\ \Delta v_i$. Note that $C_i \ll 1$, which insures that at the beginning the decrease is slightly sensitive to the changes in v_i. Later on, these coefficients defined as $C_i = \Delta F/\Delta v_i$ have been called misunderstandly 'control coefficients' (instead of 'sensitivity coefficient', for instance).

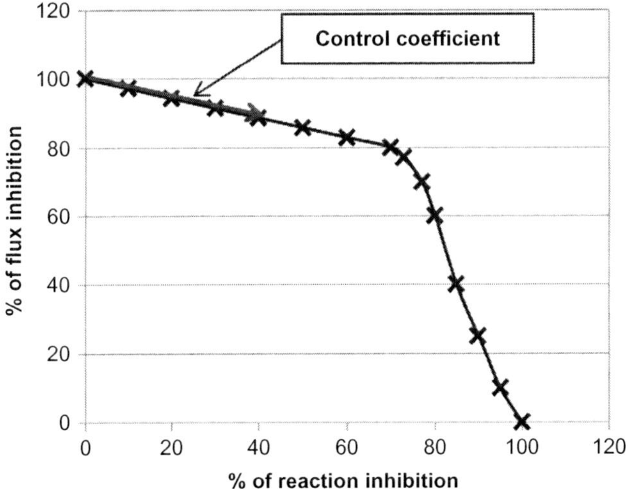

Figure 13.12 Flux inhibition curve by the inhibition of a reaction of the network. The initial slope is the control coefficient of the reaction on the global flux at steady state. (For interpretation of the references to colour in this figure legend, the reader is referred to the online version of this chapter.)

A flux control coefficient can be defined for each step on each flux of the network (r^2 control coefficient if there is r reaction in the network). A flux F_i is defined as the rate v_i at steady state. A change on a rate is made by changing a parameter on which the rate function is dependent (the quantity of enzyme E_i, for instance). The consequences on the rate itself and on the flux (usually another rate at steady state) are not the same, even when one considers a linear pathway, thus leading to $C_i \neq 1$. Kacser and Burns (1973), later on show that the sum of the control coefficient on a flux by all the steps of the network sums up to 1 (summation theorem).

The same concept of control coefficients (called 'control strength') and the same summation relationship was developed independently at the same time by Heinrich and Rapoport (1974). This theorem had immediately important consequences. It shows that the control of a flux can be distributed among the different steps of a metabolic network, each of the reaction changes affecting the global flux proportionally to its control coefficient:

$$\Delta F = C_{v1} \cdot \Delta v_1 + C_{v2} \cdot \Delta v_2 + \cdots + C_{vr} \cdot \Delta v_r.$$

It goes against the widespread belief (particularly in some textbooks and many labs) that in a metabolic pathway, the whole flux through the pathway is controlled by the first step often an irreversible step on which metabolic regulation (feedback inhibition, for instance) operates. Since then, several experimental works showed that indeed the control of flux is distributed among several steps (Chassagnole, Raïs, Quentin, Fell, & Mazat, 2001; Moreno-Sánchez, Saavedra, Rodrıguez-Enrıquez, & Olın-Sandoval, 2008). From then, an actual theory was developed by several authors, Fell (1992), Cascante, Franco, and Canela (1989a, 1989b), Giersch (1988a, 1988b) and under its more general form by Reder (1988). It is not the purpose of this paragraph to write a comprehensive survey of the theory. Let us simply emphasize some key points:

1. As we noticed above, in a metabolic network, there is not necessarily a unique limiting step, but rather the control of the fluxes is distributed among the different steps of the metabolic network. In general, there is not a unique limiting step in a metabolic network, although it is theoretically possible and actually occurs in metabolic diseases when an enzyme activity decreases to the point to be limiting.
2. The experimental measurement of control coefficients in metabolic networks shows that their values are usually low (Mazat, Reder, & Letellier, 1996). It means that even rather big changes of an enzyme activity will have low effects on the flux, that is, the fluxes will be rather insensitive to the changes in enzyme activity. It is a property that may have been selected through evolution. There is a more simple additional reason for this insensitivity, the fact that in a metabolic network, the intermediate metabolites act as buffer of the changes in an enzyme activity. For instance, if the amount of an enzyme is decreased or if the enzyme is inhibited, the concentration of its substrates will tend to increase. If the substrate concentration is not saturating, its increase will increase the enzyme activity, thus partly compensating the decrease in enzyme level or the inhibition.
3. However, theoretically, control coefficients can take any value, particularly negative values. It is the case in a branched network (example 4 of table 13.1) in which C_{R2}^{F3} and C_{R3}^{F2} are negative. C_{R2}^{F3} is the control coefficient of the reaction 2 (v_2) on the flux in reaction 3, that is, $F3 = v_3$ at steady state and vice versa. It means that an increase in v_2, the rate of reaction R_2, will lead to a new steady state in which the flux through R_3, (the rate v_3 at the new steady state) will be decreased and vice versa.
4. The distribution of control coefficients will change when the steady state is changed. It means that a controlling step in one condition may be no longer be controlling in another condition.

The use of metabolic control theory with the measurement of control coefficients is an important tool to understand the behaviour of metabolic networks. It permitted to decipher important biological riddles, such as the recessivity of most of the null mutations (Kacser & Burns, 1981) (in the heterozygote, the activity is roughly half of the normal activity, but because of low control coefficients, it does not affect the fluxes), the control of the mitochondrial respiratory chain (Groen, Wanders, Westerhoff, Van der Meer, & Tager, 1982; Bohnensack, Küster, & Letko, 1982; Letellier, Malgat, & Mazat, 1993; Letellier, Heinrich, Malgat, & Mazat, 1994; Korzeniewski & Mazat, 1996) with later on, the application to mitochondrial diseases and the interest of threshold curves (Mazat et al., 1997).

In plant metabolism, the pioneer experimental work of Mark Stitt must be cited (Neuhaus & Stitt, 1990; Stitt, 1990) followed by many others (Geigenberger, Stitt, & Fernie, 2004; Hill & ap Rees, 1994; Sweetlove & Hill, 2000; Sweetlove, Kossmann, Riesmeier, Trethewey, & Hill, 1998). A recent review (Araújo, Nunes-Nesi, Nikolovski, Sweetlove, & Fernie, 2012) summarizes the work done on the control of the TCA cycle in plant tissues.

6. SOFTWARES AND TOOLS FOR KINETIC MODELLING

Any biologist, without any particular mathematical skill, can now utilize user-friendly bioinformatics tools to build a model and to run simulations of a biological system. Several softwares and work stations are available. The choice must be adapted to the specific problem addressed with metabolic modelling (i.e. structural vs. kinetic modelling, steady-state analysis vs. time-course simulation, parameter scans vs. parameter optimization) and to the personal preference of the user (i.e. graphical user interface (GUI) or command line-based tools). In addition, to facilitate exchange of computer-based models, an XML-based machine-readable language (i.e. the Systems Biology Markup Language, SBML, see http://sbml.org) was actively developed. In theory, a model created in one software can then be exported as SBML and imported in another software. However, in practice, careful modeller should be aware that a significant loss of functionality can be observed during the conversion of a model.

Generically, any system of ODEs can be solved with standard numerical tools such as in Mathematica (http://www.wolfram.com) or Matlab (www.mathworks.com/products/matlab). Specific toolboxes are available that allow the analysis of SBML models in these programs and perform specific functions (e.g. parameter optimization). Moreover, other open-source alternatives such as Octave (http://www.gnu.org/software/octave), SciPy (http://www.scipy.org; Poolman, 2006), PySceS (http://pysces.sourcefoge.net; Olivier, Rohwer, & Hofmeyr, 2005) and Jarnac, part of the Systems Biology Workbench (Sauro et al., 2003), are available. Let us mention the simple software Berkeley Madonna, a general purpose differential equation solver, rapid and very easy to use (http://www.berkeleymadonna.com/). A simple free version exists. Among the most popular software applications of the GUI category is COPASI Hoops et al. (2006). Owing to its enzyme kinetics library, it is comparatively easy to implement biochemical pathways and to parameterize and simulate a model without any programming effort. Moreover, many algorithms of global optimization of multiple parameters have been already implemented in COPASI and recently evaluated (Baker, Schallau, and Junker, 2010).

Finally, the overall functionality of all the tools listed above is equivalent. Indeed, most of them perform, directly or through add-on modules, a large variety of functions, for example, calculation of steady state, time-course simulation, metabolic control analysis, elementary modes analysis, parameter scans and optimization and data visualization (for review, see Alves, Antunes, & Salvador, 2006).

7. CONCLUSION: PROTOCOL FOR THE STUDY OF METABOLIC NETWORKS AT STEADY STATE

The different methods shortly described in the previous paragraphs define a protocol to study metabolic networks at steady state. It will be our conclusion (see also Gombert & Nielsen, 2000).

The first task is of course to write the reactions of the network, which is not always so easy: do we know all the reactions? do we take them all? do we take into account the compartments (mitochondrion, vacuole, plates), etc.? To write these equations in the SBML language should help then to use the softwares to explore the metabolic network.

Once the reactions are written, it is necessary to derive the stoichiometry matrix. For that purpose, the software 'Metatool' or other software developed to calculate the elementary modes of a network is useful because they calculate the stoichiometry matrix as a first step to the determination of the elementary modes. Furthermore, they point on incoherencies, if any in the writing of the equations (metabolites written on different names, etc.), which permit to arrive at a correct metabolic network.

Then it is necessary to derive the Kernel of the stoichiometry matrix, which gives the number of the independent variables (metabolites concentrations) and if necessary to calculate a new stoichiometry matrix (of maximal rank) with these independent variables. The explicit calculation of the Kernel is done in 'Metatool'.

With all that, it is then possible to calculate all the flux elementary modes (fem) (Metatool). If their number is high, it could be possible to class them with any biclustering software. AcOM is particularly adapted to this task because it particularly points to fem, which may be metabolically interesting but tempered the classification.

The FBA approach (Orth, Thiele, & Palsson, 2010) must be included in the protocol when one has a good idea of an objective function and of the constraints on the rate functions in the network. By all means, use of FBA with several objective functions (optimization of the growth rate, of the biomass, of the yield in ATP production, etc.) may be useful to have an idea of how the 'metabolic network' 'reacts' to different promptings.

The ultimate task is to find the steady state(s) (if they exist which is generally the case). For that it is necessary to know the rate functions. Rate functions of Michaelis & Menten type are usually adequate. The K_m values are usually known from the literature. If not, a good estimation is usually to take the metabolite concentration *in vivo* (R. Heinrich, personal communication). On the contrary, the $V_{\max}$, which greatly depend upon the enzyme level, are usually not known and in addition depend upon the tissue, the organism. Ideally, it is necessary to perform the measurement of all the activities in the different conditions on the studied tissue to design a representative model of its metabolism. However, this is not always possible and approximation can be made or other knowledge can be used (over or lower expression in some conditions). If the metabolic network is not too great, monte-carlo method can be used, choosing randomly the $V_{\max}$ and exhibiting great classes of metabolism types (Steuer, 2011), which will correspond more or less to (a combination of) the efms. The same method can be applied to the simplest mass action law.

Finally, the control coefficient of each given steady state can be calculated (the control coefficients depend upon the steady state). It gives an idea of the sensitivity of the metabolic networks towards some steps in the network and on the way the network will evolve when these steps are over/under expressed. If in the same time, some of the control coefficient can be experimentally measured (Araújo, 2012), the comparison of the theoretical values with the experimental ones can give insight in the regulation of the network.

REFERENCES

Alves, R., Antunes, F., & Salvador, A. (2006). Tools for kinetic modeling of biochemical networks. *Nature Biotechnology*, *24*, 667–672.

Araújo, W. L., Nunes-Nesi, A., Nikolovski, Z., Sweetlove, L., & Fernie, A. (2012). Metabolic control and regulation of the tricarboxylic acid cycle in photosynthetic and heterotrophic plant tissues. *Plant Cell and Environment*, *35*, 1–21.

Baker, S. M., Schallau, K., & Junker, B. H. (2010). Comparison of different algorithms for simultaneous estimation of multiple parameters in kinetic metabolic models. *Journal of Integrative Bioinformatics*, 7, 133–141.

Bohnensack, R., Küster, U., & Letko, G. (1982). Rate-controlling steps of oxidative phosphorylation in rat liver mitochondria. A synoptic approach of model and experiment. *Biochimical Biophysical Acta*, *680*, 271–280.

Briggs, G. E., & Haldane, J. B. S. (1925). A note on the kinetics of enzyme action. *Biochemical Journal*, *19*, 338–339.

Carlsson, R., Fell, D., & Srienc, F. (2002). Metabolic pathway analysis of a recombinant yeast for a rational strain development. *Biotechnology and Bioengineering*, *79*, 121–134.

Cascante, M., Franco, R., & Canela, E. I. (1989a). Use of implicit methods from general sensitivity theory to develop a systematic approach to metabolic control. I. Unbranched pathways. *Mathematical Biosciences*, *94*, 271–288.

Cascante, M., Franco, R., & Canela, E. I. (1989b). Use of implicit methods from general sensitivity theory to develop a systematic approach to metabolic control. II. Complex systems. *Mathematical Biosciences*, *94*, 289–309.

Chassagnole, C., Raïs, B., Quentin, E., Fell, D. A., & Mazat, J.-P. (2001). An integrated study of threonine pathway enzyme kinetics in Escherichia coli. *Biochemical Journal*, *356*, 415–423.

Cornish-Bowden, A. (2012). *Fundamentals of enzyme kinetics* (4th ed.). Weinheim: Wiley-Blackwell.

Cornish-Bowden, A., & Hofmeyr, J.-H. S. (2005). Enzymes in context. Kinetic characterization of enzymes for systems biology. *The Biochemist*, *27*, 11–14.

Covert, M. W., Schilling, C. H., Famili, I., Edwards, J. S., Goryanin, I. I., Selkov, E., et al. (2001). Metabolic modelling of microbial strains in silico. *Trends in Biochemical Sciences*, *26*, 179–186.

Fell, D. A. (1992). Metabolic control analysis: A survey of its theoretical and experimental development. *Biochemical Journal*, *286*, 313–330.

Fell, D. A. (1998). Increasing the flux in metabolic pathways: A metabolic control analysis perspective. *Biotechnology and Bioengineering*, *58*, 121–124.

Flint, H. J., Porteous, D. J., & Kacser, H. (1980). Control of the flux in the arginine pathway *of Neurospora crassa*. The flux from citrulline to arginine. *Biochemical Journal*, *190*, 1–15.

Flint, H. J., Tateson, R. W., Barthelmess, I. B., Porteous, D. J., Donachie, D. J., & Kacser, H. (1981). Control of the flux in the arginine pathway *of Neurospora crassa*. Modulations of enzyme activity and concentration. *Biochemical Journal*, *200*, 231–246.

Geigenberger, P., Stitt, M., & Fernie, A. R. (2004). Metabolic control analysis and regulation of the conversion of sucrose to starch in growing potato tubers. *Plant Cell and Environment*, *27*, 655–673.

Giersch, C. (1988a). Control analysis of metabolic networks: 1. Homogeneous functions and the summation theorems for control coefficients. *European Journal of Biochemistry, 174*, 509–513.
Giersch, C. (1988b). Control analysis of metabolic networks: 2. Total differentials and general formulation of the connectivity relations. *European Journal of Biochemistry, 174*, 515–519.
Gombert, A. K., & Nielsen, J. (2000). Mathematical modeling of metabolism. *Current Opinion in Biotechnology, 11*, 180–186.
Groen, A. K., Wanders, R. J. A., Westerhoff, H. V., Van der Meer, R., & Tager, J. M. (1982). Quantification of the contribution of various steps to the control of mitochondrial respiration. *Journal of Biological Chemistry, 257*, 2754–2757.
Heinrich, R., & Rapoport, T. A. (1974). A linear steady-state treatment of enzymatic chains. General properties, control and effector strength. *European Journal of Biochemistry, 42*, 89–95.
Henri, V. (1902). Théorie générale de l'action de quelques diastases. *Comptes Rendus de l'Académie des Sciences, 135*, 916–919.
Hill, A. V. (1910). The possible effects of the aggregation of the molecules of haemoglobin on its dissociation curves. *Journal of Physiology, 40*, iv–vii.
Hill, S. A., & ap Rees, T. (1994). Metabolic control analysis of plant metabolism. *Plant Cell and Environment, 17*, 587–599.
Hofmeyr, J.-H. S., & Cornish-Bowden, A. (1997). The reversible Hill equation: How to incorporate cooperative enzymes into metabolic models? *Computer Applications in the Biosciences, 13*, 377–385.
Hoops, S., Sahle, S., Gauges, R., Lee, C., Pahle, J., Simus, N., et al. (2006). COPASI—A COmplex PAthway SImulator. *Bioinformatics, 22*, 3067–3074.
Kacser, H., & Burns, J. A. (1973). The control of flux. *Symposia of the Society for Experimental Biology, 32*, 65–104.
Kacser, H., & Burns, J. A. (1981). The molecular basis of dominance. *Genetics, 97*, 639–666.
Klipp, E., Leibermeister, W., Wierling, C., Kowald, A., Lehrach, H., & Herwig, R. (2009). *Systems biology* (1st ed.). Weinheim: Wiley-VCH.
Korzeniewski, B., & Mazat, J.-P. (1996). Theoretical studies on the control of oxidative phosphorylation in muscle mitochondria—Application to mitochondrial deficiencies. *Biochemical Journal, 319*, 143–148.
Letellier, T., Heinrich, R., Malgat, M., & Mazat, J.-P. (1994). The kinetic basis of the threshold effects observed in mitochondrial diseases: A systemic approach. *Biochemical Journal, 302*, 171–174.
Letellier, T., Malgat, M., & Mazat, J.-P. (1993). Control of oxidative phosphorylation in rat muscle mitochondrial. Implication to mitochondrial myopathies. *Biochimical Biophysical Acta, 1141*, 58–64.
Letellier, T., Reder, C., & Mazat, J.-P. (1991). CONTROL: A software for the analysis of the control of metabolic networks. *Computer Application in the Biosciences, 7*, 383–390.
Mazat, J.-P., Letellier, T., Bedes, F., Malgat, M., Korzeniewski, B., Jouaville, S. L., et al. (1997). Metabolic control analysis and threshold effect in oxidative phosphorylation. Implications for mitochondrial pathologies. *Molecular and Cellular Biochemistry, 174*, 143–148.
Mazat, J.-P., Reder, C., & Letellier, T. (1996). Why are most flux control coefficients so small? *Journal of Theoretical Biology, 182*, 253–258.
Michaelis, L., & Menten, M. L. (1913). Die kinetik der invertinwirkung. *Biochemische Zeitung, 49*, 333–369.
Moreno-Sánchez, R., Saavedra, E., Rodrıguez-Enrıquez, S., & Olın-Sandoval, V. (2008). Metabolic control analysis: A tool for designing strategies to manipulate metabolic pathways. *Journal of Biomedicine and Biotechnology, 2008*, 597913.
Neuhaus, H. E., & Stitt, M. (1990). Control analysis of photosynthate partitioning—Impact of reduced activity of ADP-Glucose pyrophosphorylase or plastid phosphoglucomutase

on the fluxes to starch and sucrose in *Arabidopsis thaliana* (L.) Heynh. *Planta, 182*, 445–454.

Olivier, B. G., Rohwer, J. M., & Hofmeyr, J. H. S. (2005). Modeling cellular systems with PySCes. *Bioinformatics, 21*, 560–561.

Orth, J. D., Thiele, I., & Palsson, B. O. (2010). What is flux balance analysis? *Nature Biotechnology, 28*, 245–248.

Papin, J. A., Reed, J. L., & Palsson, B. Ø. (2004). Hierarchical thinking in network biology: The unbiased modularisation of biochemical networks. *Trends in Biochemical Sciences, 29*, 641–647.

Papin, J. A., Stelling, J., Price, N. D., Klamt, S., Schuster, S., & Palsson, B. Ø. (2004). Comparison of network-based pathways analysis methods. *Trends in Biotechnology, 22*, 400–405.

Pérès, S., Vallée, F., Beurton-Aimar, M., & Mazat, J.-P. (2011). ACoM: A classification method for elementary flux modes based on motif finding. *Biosystems, 103*, 410–419.

Poolman, M. G. (2006). ScrumPy: Metabolic modelling with Python. *IEE Proceedings in Systems Biology, 153*, 375–378.

Poolman, M. G., Assmus, H. E., & Fell, D. A. (2004). Applications of metabolic modelling to plant metabolism. *Journal of Experimental Botany, 55*, 1177–1186.

Poolman, M. G., Olcer, H., Lloyd, J. C., Raines, C. A., & Fell, D. A. (2001). Computer modelling and experimental evidence for two steady-states in the photosynthetic Calvin cycle. *European Journal of Biochemistry, 268*, 2810–2816.

Poolman, M. G., Venkatesh, K. V., Pidcock, M. K., & Fell, D. A. (2004). A method for the determination of flux in elementary modes, and its application to Lactobacillus rhamnosus. *Biotechnology and Bioengineering, 88*, 601–612.

Raïs, B., Chassagnole, C., Dentellerie, T., Fell, D. A., & Mazat, J.-P. (2001). Threonine synthesis from aspartate in Escherichia coli cell-free extracts: pathway dynamics. *Biochemical Journal, 356*, 425–432.

Reder, C. (1988). Metabolic control theory: A structural approach. *Journal of Theoretical Biology, 135*, 175–201.

Sauro, H. M., Hucka, M., Finney, A., Wellock, C., Bolouri, H., Doyle, J., et al. (2003). Next generation simulation tools: The systems Biology Workbench and BioSPICE integration. *OMICS*, 7, 355–372.

Schilling, C. H., Letscher, D., & Palsson, B. Ø. (2000). Theory for the systemic definition of metabolic pathways and their use in interpreting metabolic function from a pathway-oriented perspective. *Journal of Theoretical Biology, 203*, 229–248.

Schuster, S., Fell, D. A., & Dandekar, T. (2000). A general definition of metabolic pathways useful for systematic organization and analysis of complex metabolic networks. *Nature Biotechnology, 18*, 326–332.

Schuster, S., & Hilgetag, C. (1994). On elementary flux modes in biochemical reaction systems at steady-state. *Journal of Biological Systems, 2*, 165–182.

Stelling, J., Klamt, S., Bettenbrock, K., Schuster, S., & Gilles, E. D. (2002). Metabolic network structure determines key aspects of functionality and regulation. *Nature, 420*, 190–193.

Steuer, R. (2011). Exploring the Dynamics of Large-Scale Biochemical Networks: A Computational Perspective. *The Open Bioinformatics Journal, 5*, 4–15.

Stitt, M. (1990). Application of control analysis to photosynthetic sucrose synthesis. In A. Cornish-Bowden & M. L. Cardenas (Eds.), *Control of metabolic processes* (pp. 363–376). New York: Plenum Publishing Corp.

Sweetlove, L. J., Beard, K. F. M., Nunes-Nesi, A., Fernie, A. R., & Ratcliffe, R. G. (2010). Not just a circle: Flux modes in the plant TCA cycle. *Trends in Plant Sciences, 15*, 462–470.

Sweetlove, L. J., & Hill, S. A. (2000). Source metabolism dominates the control of source to sink carbon flux in tuberizing potato plants throughout the diurnal cycle and under a range of environmental conditions. *Plant Cell and Environment*, *23*, 523–529.

Sweetlove, L. J., Kossmann, J., Riesmeier, J. W., Trethewey, R. N., & Hill, S. A. (1998). The control of source to sink carbon flux during tuber development in potato. *The Plant Journal*, *15*, 697–706.

AUTHOR INDEX

Note: Page numbers followed by "*f*" indicate figures, and "*t*" indicate tables.

D

E

F

G

K

L

N

O

P

S

T

U

V

W

SUBJECT INDEX

Note: Page numbers followed by "*f*" indicate figures, and "*t*" indicate tables.

G

H

M

N

Q

T

U

V

W

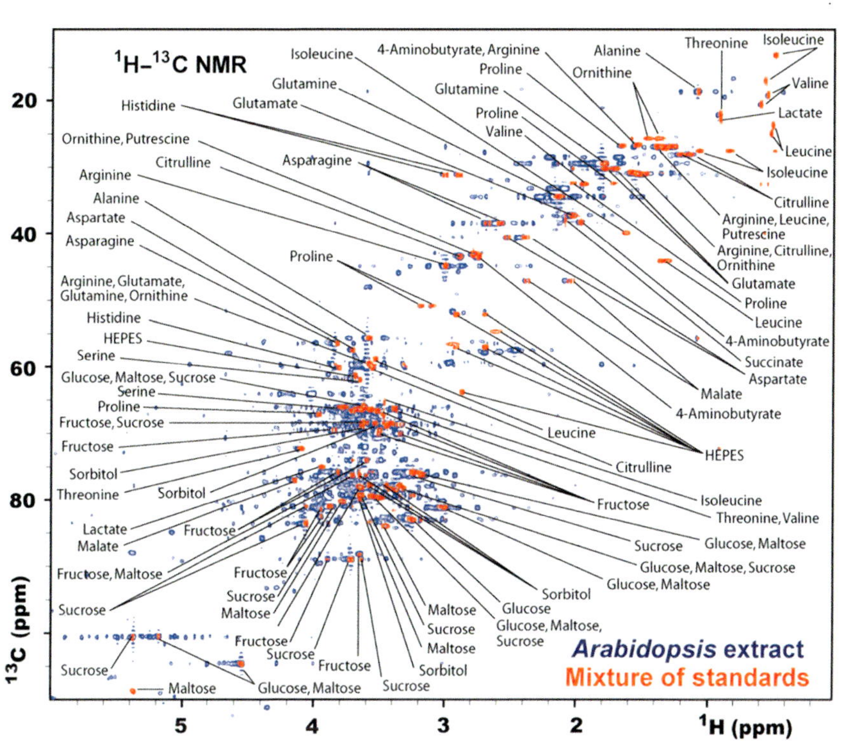

Patrick Giraudeau and Serge Akoka, Figure 3.15 2D $^1H–^{13}C$ HSQC NMR spectrum of a synthetic mixture (equimolar mixture of 26 metabolites, red) overlaid onto a spectrum of aqueous whole-plant extract from *A. thaliana* (blue) (Lewis et al., 2010).

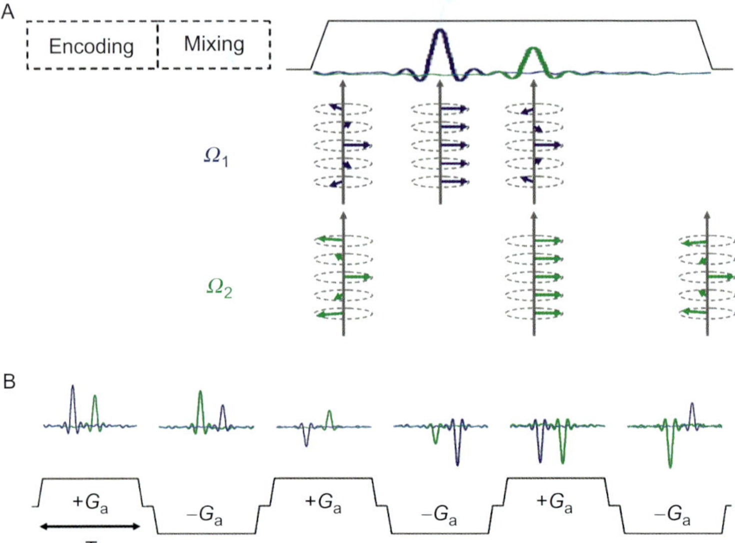

Patrick Giraudeau and Serge Akoka, Figure 3.18 Signal detection in ultrafast 2D NMR. (A) Acquisition of the ultrafast dimension. A magnetic field gradient G_a is applied while the receiver is open, which refocuses the dephasing induced in the course of spatial encoding. It leads to the formation of echo peaks whose positions are proportional to resonance frequencies. (B) Acquisition of the conventional dimension. A series of sub-spectra are detected during a train of bipolar gradient pulses, while the system evolves under the influence of conventional parameters (relaxation, resonance frequency, couplings, etc.). Data rearrangement and Fourier transform along the conventional dimension are necessary to obtain the final 2D spectrum.

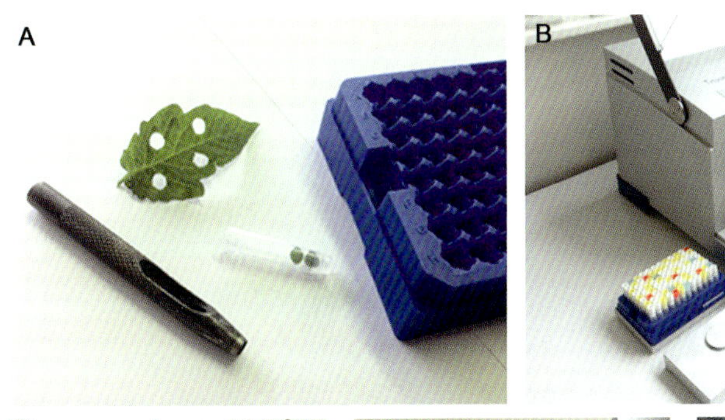

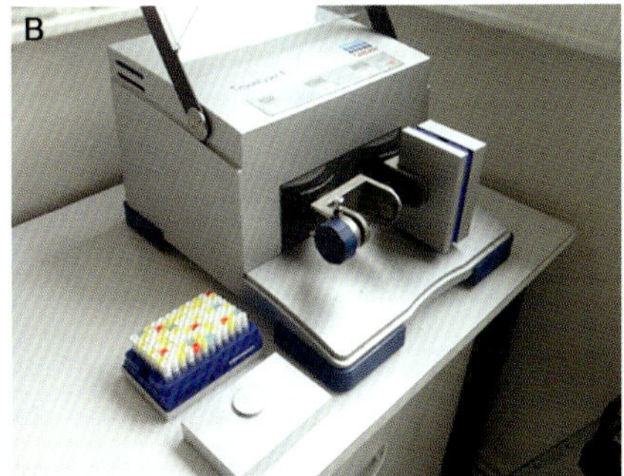

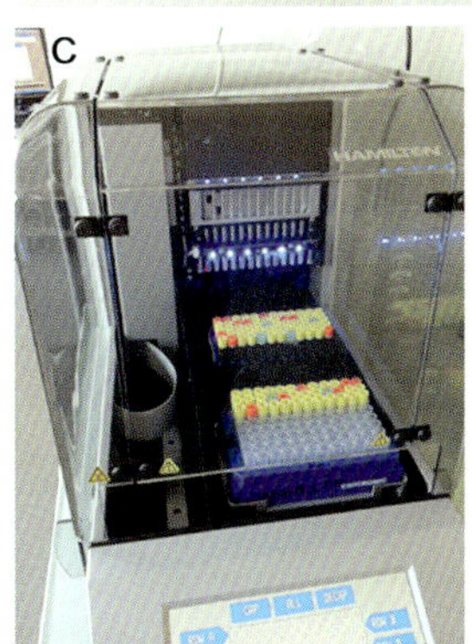

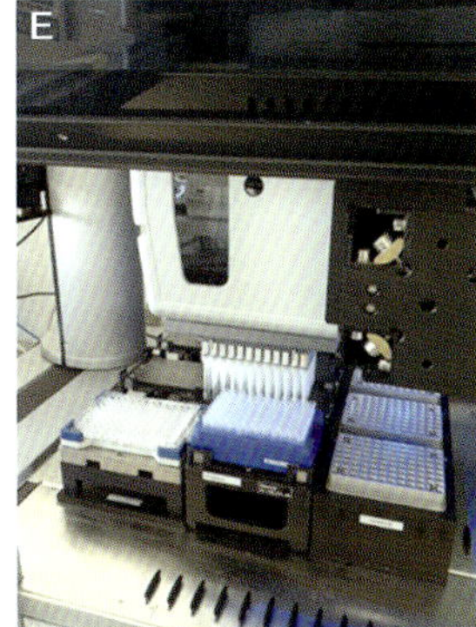

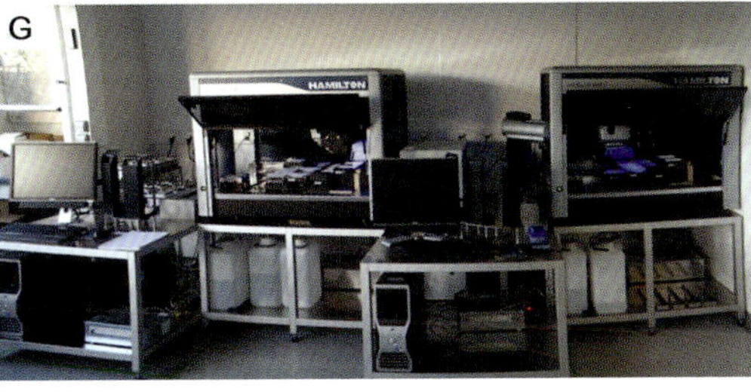

G. Ménard *et al.*, Figure 9.1 Overview of platform equipment. (A) Leaf discs were harvested and snap frozen in liquid nitrogen using micronic® tubes with screw caps. (B) Samples were milled in batch of 96 micronic® using a TissuLyser II (Qiagen®). (C) Automated decapper (Hamilton®) were used to cap/decap the tubes between each step of process. (D) An auxiliary arm iSwap® microplate carrier was used to transport the plate between the liquid handling workstation and the incubators (Hamilton®). (E) Each liquid workstation held 96 tips head Starlet® (Hamilton®). (F) The platform provided eight microplate readers MP96® and one microplate luminometer/fluorimeter reader Xenius® (SAFAS®). (G) Overview of the microplate workstations Starlet® (Hamilton®).

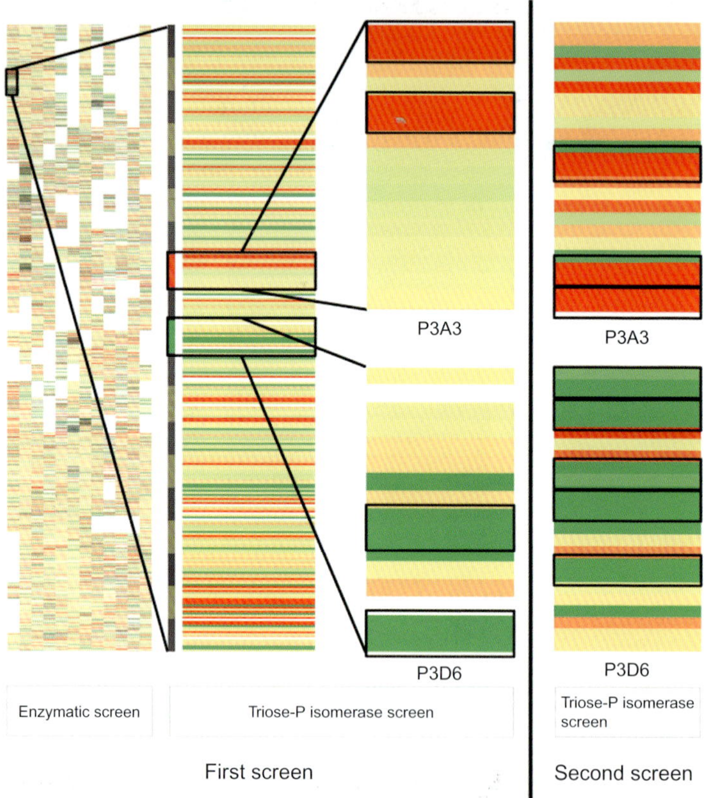

G. Ménard *et al.*, Figure 9.3 EMS mutant collection screen schema. An EMS mutant collection was screened to uncover potential enzymatic mutants. 150 M2 families (12 plants per family) from the collection were harvested and screened. The first screen allowed identifying two families with potential increased (red) or decreased (green) in triose phosphate isomerase activity. A second culture was conducted and screened in the process. This second screen confirmed the alterations initially uncovered.

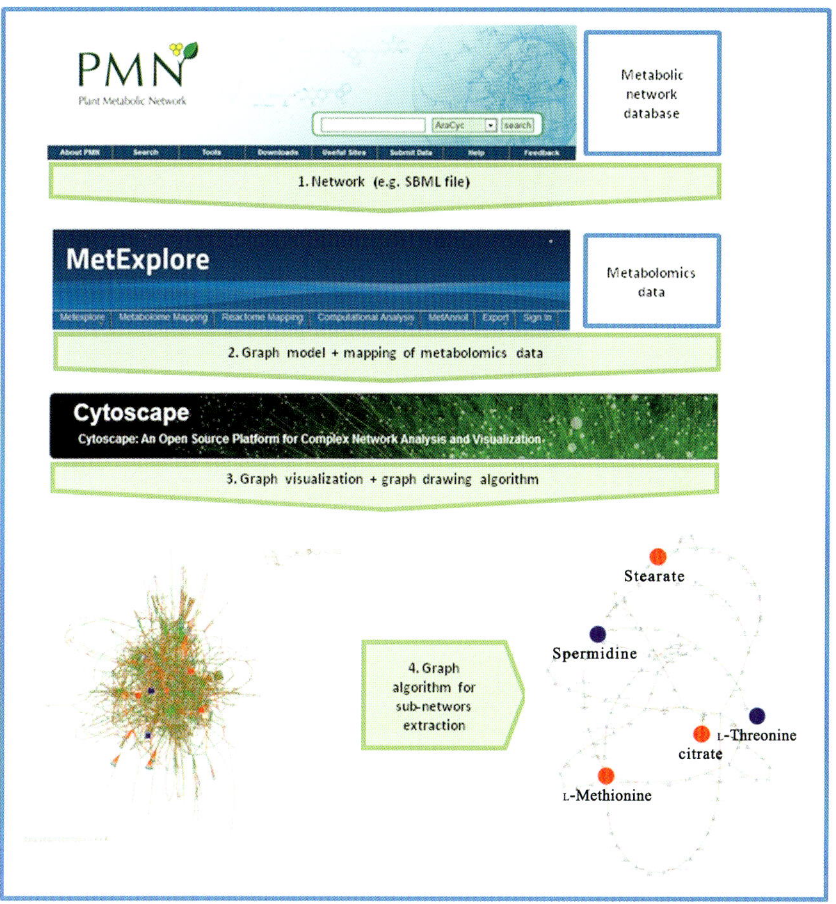

Fabien Jourdan, Figure 12.5 Bioinformatics pipeline used to interpret metabolomics data in the context of metabolic networks. (1) A database containing a metabolic network is chosen (AraCyc). The network is exported in a software readable format (SBML). (2) This file is imported in software in order to generate a graph model of the metabolic network. In addition, metabolomics data are mapped to graph nodes (using MetExplore mapping function). (3) Graph files are imported in biological network visualization software (Cytoscape). Using a graph drawing algorithm (force-directed), the network is visualized. Larger nodes are metabolites identified in metabolomics experiments. (4) The lightest path between each pair of identified metabolites is computed (using the lightest path algorithm). All these paths are merged in a sub-network (112 nodes and 126 edges) showing metabolic processes connecting identified metabolites.